Geometric Computing with Clifford Algebras

Springer
Berlin
Heidelberg
New York
Barcelona
Hong Kong
London
Milan
Paris
Singapore
Tokyo

Gerald Sommer (Ed.)

Geometric Computing with Clifford Algebras

Theoretical Foundations and Applications in Computer Vision and Robotics

With 89 Figures and 16 Tables

Springer

Editor

Gerald Sommer
Institut für Informatik
Lehrstuhl für Kognitive Systeme
Christian-Albrechts-Universität zu Kiel
Preusserstr. 1-9
24105 Kiel
E-mail: gs@ks.informatik.uni-kiel.de

Library of Congress Cataloging-in-Publication data applied for

Die Deutsche Bibliothek - CIP-Einheitsaufnahme

Geometric computing with Clifford algebras: theoretical foundations and applications in computer vision and robotics; with 16 tables/ Gerald Sommer (ed.). - Berlin; Heidelberg; New York; Barcelona; Hong Kong; London; Milan; Paris; Singapore; Tokyo: Springer, 2001
ISBN 3-540-41198-4

ACM Subject Classification (1998): I.4, I.2.9-10, I.3, G.1.3, I.1

ISBN 3-540-41198-4 Springer-Verlag Berlin Heidelberg New York

Springer-Verlag Berlin Heidelberg New York
is a member of BertelsmannSpringer Science+Business Media GmbH

http://www.springer.de

Typesetting: Camera-ready by the editor
Cover Design: d&p, design & production, Heidelberg
Printed on acid-free paper SPIN 10786200 - 06/3142SR - 5 4 3 2 1 0

Preface

This book presents a collection of contributions concerning the task of solving geometry related problems with suitable algebraic embeddings. It is not only directed at scientists who already discovered the power of Clifford algebras for their field, but also at those scientists who are interested in Clifford algebras and want to see how these can be applied to problems in computer science, signal theory, neural computation, computer vision and robotics. It was therefore tried to keep this book accessible to newcomers to applications of Clifford algebra while still presenting up to date research and new developments.

The aim of the book is twofold. It should contribute to shift the fundamental importance of adequate geometric concepts into the focus of attention, but also show the algebraic aspects of formulating a particular problem in terms of Clifford algebra. Using such an universal, general and powerful algebraic frame as Clifford algebra, results in multiple gains, such as completeness, linearity and low symbolic complexity of representations. Even problems which may not usually be classified as geometric, might be better understood by the human mind when expressed in a geometric language.

As a misleading tendency, mathematical education with respect to geometric concepts disappears more and more from curricula of technical subjects. To a certain degree this is caused by the mathematicians themselves. What mathematicians today understand as geometry or algebraic geometry is far from beeing accessible to engineers or computer scientists. This is the more regrettable as the Erlangen program of Felix Klein [FK95] on the strong relations between algebra and geometry is of great potential also for the applied sciences.

This book is a first attempt to overcome this situation. As computer scientists and engineers know in principle of the importance of algebra to gain new qualities of modelling, they will profit from geometric interpretations of their models. This was also the experience the authors of this book had made. However, it is not necessarily trivial to translate a geometry related problem into the language of Clifford algebra. Once translated, it also needs some experience to manipulate Clifford algebra expressions. The many applied problems presented in this book should give engineers, computer scientists and physicists a rich set of examples of how to work with Clifford algebra.

The term 'geometric problem' will at times be understood very loosely. For instance, what relation exists between a Fourier transform or its computation as FFT (fast Fourier transform algorithm) and geometry? It will become clear that the Fourier transform is strongly related to symmetry as geometric entity and that its multidimensional extension necessitates the use of an adequate algebraic frame if all multidimensional symmetries are to be kept accessible.

William K. Clifford (1845–1879) [47] introduced what he called "geometric algebra[1]". It is a generalisation of Hermann G. Grassmann's (1809–1877) exterior algebra and also contains William R. Hamilton's (1805–1865) quaternions. Geometric or Clifford algebra has therefore a strong unifying aspect, since it allows us to view different geometry related algebraic systems as specializations of one "mother algebra". Clifford algebras may therefore find a use in quite different fields of science, while still sharing the same fundamental properties.

David Hestenes was one of the first who revived GA in the mid 1960's and introduced it in different fields of physics with the goal to make it a unified mathematical language that encompasses the system of complex numbers, the quaternions, Grassmann's exterior algebra, matrix algebra, vector, tensor and spinor algebras and the algebra of differential forms. In order to achieve this, he fashioned his GA as a specialization of the general CA which is particularly well suited for the use in physics and, as it turned out, engineering and computer science. It is his merit that GA got widely accepted in diverse fields of physics [112, 109] and engineering. The algebra most similar to Hestenes's GA is probably Grassmann's exterior algebra which is also used by a large part of the physics community [FR97]. Those readers who are interested in the evolution of the relations between algebra and geometry can find a short overview in [YAG88].

I first became interested in Clifford or geometric algebra by reading a short paper in Physics World, written by Anthony Garrett [AG92]. At that time I was searching for a way to overcome some serious problems of complete representations of local, intrinsically multidimensional structures in image processing. I got immediately convinced that the algebraic language presented in this paper would open the door to formulate a real multidimensional and linear signal theory. Since then we not only learned how to proceed in that way, but we also discovered the expressive power of GA for quite different aspects of multidimensional signal structure [34, FSR2000].

In the Cognitive Systems research group in Kiel, Germany, we are working on all aspects concerning the design of seeing robot systems [SO99]. This includes pattern recognition with neural networks, computer vision, multidimensional signal theory and robot kinematics. We found that in all these

[1] Today the terms "geometric algebra" (GA) and "Clifford algebra" (CA) are being used interchangeably.

fields GA is a particularly useful algebraic frame. In the process of this research we made valuable experiences of how to model problems with GA. Several contributions to this book present this work[2].

Designing a seeing robot system is a task where quite a number of different competences have to be modelled mathematically. However, in the end the whole system should appear as one. Furthermore, all competences have to be organized or have to organize themselves in a cycle, which has perception and action as two poles. Therefore, it is important to have a common mathematical language to bring the diverse mathematical disciplines, contributing to the diverse aspects of the perception-action cycle, closer together and eventually to fuse them to a general conception of behaviour based system design. In 1997 we brought to life the international workshop on algebraic frames for the perception-action cycle (AFPAC) [217], with the intention to further this fusion of disciplines under the umbrella of a unified algebraic frame. This workshop brought together researchers from all over the world, many of whom became authors in this book. In this respect this book may be considered a collection of research results inspired by the AFPAC'97. Hopefully, the AFPAC 2000 workshop will be of comparable success.

Another mid-range goal is the design of GA processors for real-time computations in robot vision. Today we have to accept a great gap between the low symbolic complexity on the one hand and the high numeric complexity of coding in GA on the other hand. Because available computers cannot even process complex numbers directly, we have to pay a high computational cost at times, when using GA libraries. In some cases this is already compensated by the gain achieved through a concise problem formulation with GA. Nevertheless, full profit in real-time applications is only possible with adequate processors.

The book is divided into three main sections.

Part I (*A Unified Algebraic Approach for Classical Geometries*) introduces Euclidean, spherical and hyperbolic geometry in the frame of GA. Also the geometric modelling capabilities of GA from a general point of view are outlined. In this first part it will become clear that the language of GA is developing permanently and that by shaping this language, it can be adapted to the problems at hand.

David Hestenes, Hongbo Li and Alyn Rockwood summarize in chapter 1 the basic methods, ideas and rules of GA. This survey will be helpful for the reader as a general reference for all other chapters. Of chapters 2, 3, and 4, written by Hongbo Li et al., I especially want to emphasize two aspects. Firstly, the use of the so-called conformal split as introduced by Hestenes [110] in geometric modelling. Secondly, the proposed unification of classical geometries will become important in modelling catadioptic camera systems (see [GD00]), possessing both reflective and refractive components, for robot vision in a general framework. In chapter 6 Leo Dorst gives an introduction

[2] This research was funded since 1997 by the Deutsche Forschungsgemeinschaft.

to GA which will help to increase the influence of GA on many fields in computer science. Particularly interesting is his discussion of a very general filter scheme. Ambjörn Naeve and Lars Svensson present their own way of constructing GA in chapter 5. Working in the field of computer vision, they choose to demonstrate their framework in applications to geometrical optics.

Part II (*Algebraic Embedding of Signal Theory and Neural Computation*) is devoted to the development of a linear theory of intrinsically multidimensional signals and to make Clifford groups accessible in neural computations with the aim of developing neural networks as experts of basic geometric transformations and thus of the shape of objects.

This part is opened by a contribution of Valeri Labunets and his daughter Ekaterina Rundblad-Labunets, both representing the Russian School of algebraists. In chapter 7 they emphasize two important aspects of image processing in the CA framework. These are the modelling of vector-valued multidimensional signal data, including colour signals, and the formulation of invariants with that respect. Their framework is presented in a very general setting and, hopefully, will be picked up by other researchers to study its application.

The other six chapters of part II are written by the Kiel Cognitive Systems Group. In chapters 8 to 11 Thomas Bülow, Michael Felsberg and Gerald Sommer, partially in cooperation with Vladimir Chernov, Samara (Russia) for the first time are extensively presenting the way to represent intrinsically multidimensional scalar-valued signals in a linear manner by using a GA embedding. Several aspects are considered, as non-commutative and commutative hypercomplex Fourier transforms (chapters 8,9), fast algorithms for their computation (chapter 10), and local, hypercomplex signal representations in chapter 11. As a field of application of the proposed quaternion-valued Gabor transform in the two dimensional case, the problem of texture analysis is considered. In that chapter the old problems of signal theory as missing phase concepts of intrinsically two dimensional signals, embedded in 2D space, and the missing completeness of local symmetry representation (both problems have the same roots) could be overcome. Thus, the way to develop a linear signal theory of intrinsically multidimensional signals is prepared for future research.

Quite a different topic is handled by Sven Buchholz and Gerald Sommer in chapters 12 and 13. This is the design of neurons and neural nets (MLPs) which perform computations in CA. The new quality with respect to modelling neural computation results from the fact that the use of the geometric product in vector spaces induces a structural bias into the neurons. Looking onto the data through the glasses of "CA-neurons" gives valuable contraints while learning the intrinsic (geometric) structure of the data, which results in an excellent generalization ability. As a nearly equally important aspect the complexity of computations is drastically reduced because of the linearization effects of the algebraic embedding. These nets indeed constitute experts for geometric transformations.

Part III (*Geometric Algebra for Computer Vision and Robotics*) is concerned with actual topics of projective geometry in modelling computer vision tasks (chapters 14 to 17) and with the linear modelling of kinematic chains of points and lines in space (chapters 18 to 21).

In chapter 14, Christian Perwass and Joan Lasenby demonstrate a geometrically intuitive way of using the incidence algebra of projective geometry in GA to describe multiple view geometry. Especially the use of reciprocal frames should be emphasized. Many relations which have been derived in matrix algebra and Grassmann-Cayley algebra in the last years can be found here again. An application with respect to 3D reconstruction using vanishing points is laid out in chapter 15. Another application is demonstrated in chapter 16 by Eduardo Bayro-Corrochano and Bodo Rosenhahn with respect to the computation of the intrinsic parameters of a camera. Using the idea of the absolute conic in the context of Pascal's theorem, they develop a method which is comparable to the use of Kruppa equations.

Hongbo Li and Gerald Sommer present in chapter 17 an alternative way to chapter 14 of formulating multiple view geometry. In their approach they use a coordinate-free representation whereby image points are given as bivectors. Using this approach, they discovered new constraints on the trifocal tensor.

Chapters 18-21 are concerned with kinematics. In chapter 18, Eduardo Bayro-Corrochano is developing the framework of screw geometry in the language of motor algebra, a degenerate algebra isomorphic to that of dual quaternions. In contrast to dual quaternions, motors relate translation and rotation as spinors and, thus, result in some cases in simpler expressions. This is the case especially in considering kinematic chains, as is done by Eduardo Bayro-Corrochano and Detlev Kähler in chapter 19. They are modelling the forward and the inverse kinematics of robot arms in that framework. The use of dual quaternions with respect to motion alignment is studied as a tutorial paper by Kostas Daniilidis in chapter 20. His experience with this framework is based on a very successful application with respect to the hand-eye calibration problem in robot vision.

Finally, in chapter 21, Yiwen Zhang, Gerald Sommer and Eduardo Bayro-Corrochano are designing an extended Kalman filter for the tracking of lines. Because the motion of lines is intrinsical to the motor algebra, the authors can demonstrate the performance based on direct observations of such higher order entities. The presented approach can be considered as 3D-3D pose estimation based on lines. A more extensive study of 2D-3D pose estimation based on geometric constraints can be found in [SRZ00].

In summary, this book can serve as a reference of the actual state of applying Clifford algebra as a frame for geometric computing. Furthermore, it shows that the matter is alive and will hopefully grow and mature fast.

Thus, this book is also to be seen as a snapshot of current research and hence as a "workbench" for further developments in geometric computing.

To complete a project like this book requires the cooperation of the contributing authors. My thanks go to all of them. In particular I would like to thank Michael Felsberg for his substantial help with the coordination of this book project. He also prepared the final layout of this book with the help of the student Thomas Jäger. Many thanks to him, as well.

Kiel, December 2000 Gerald Sommer

Table of Contents

Part II. Algebraic Embedding of Signal Theory and Neural Computation

List of Contributors

E. Bayro-Corrochano
Centro de Investigacion en
Matematicas, A.C.
Apartado Postal 402
36000-Guanajuato
Gto. Mexico
edb@fractal.cimat.mx

S. Buchholz
Institute of Computer Science
and Applied Mathematics
Christian-Albrechts-University of Kiel
Preußerstr. 1-9, 24105 Kiel
Germany
sbh@ks.informatik.uni-kiel.de

T. Bülow
Institute of Computer Science
and Applied Mathematics
Christian-Albrechts-University of Kiel
Preußerstr. 1-9, 24105 Kiel
Germany
tbl@ks.informatik.uni-kiel.de

V. M. Chernov
Image Processing System Institute
Russian Academy of Sciences
443001 Samara
Russia
vche@smr.ru

K. Daniilidis
GRASP Laboratory
University of Pennsylvania
3401 Walnut Street, Suite 336C
Philadelphia, PA 19104-6228
USA
kostas@grip.cis.upenn.edu

L. Dorst
Dept. of Computer Science
University of Amsterdam
Kruislaan 403
1012 VE Amsterdam
The Netherlands
leo@wins.uva.nl

M. Felsberg
Institute of Computer Science
and Applied Mathematics
Christian-Albrechts-University of Kiel
Preußerstr. 1-9, 24105 Kiel
Germany
mfe@ks.informatik.uni-kiel.de

D. Hestenes
Dept. of Physics and Astronomy
Arizona State University
Tempe, AZ 85287-1504
USA
hestenes@asu.edu

D. Kähler
Institute of Computer Science
and Applied Mathematics
Christian-Albrechts-University of Kiel
Preußerstr. 1-9, 24105 Kiel
Germany
dek@ks.informatik.uni-kiel.de

V. Labunets
Signal Processing Laboratory
Tampere University of Technology
Tampere
Finland
lab@cs.tut.fi

J. Lasenby
Engineering Department
Cambridge University
Trumpington Street, Cambridge CB2 1PZ
UK
jl@eng.cam.ac.uk

H. Li
Institute of Systems Science
Academia Sinica
Beijing 100080
P. R. China
hli@mmrc.iss.ac.cn

A. Naeve
Dept. of Numerical Analysis and
Computing Science
Royal Institute of Technology
100 44 Stockholm
Sweden
amb@nada.kth.se

C. Perwass
Institute of Computer Science
and Applied Mathematics
Christian-Albrechts-University of Kiel
Preußerstr. 1-9, 24105 Kiel
Germany
chp@ks.informatik.uni-kiel.de

A. Rockwood
Power Take Off Software, Inc.
18375 Highland Estates Dr.
Colorado Springs, CO 80908
USA

B. Rosenhahn
Institute of Computer Science
and Applied Mathematics
Christian-Albrechts-University of Kiel
Preußerstr. 1-9, 24105 Kiel
Germany
bro@ks.informatik.uni-kiel.de

E. Rundblad-Labunets
Signal Processing Laboratory
Tampere University of Technology
Tampere
Finland

G. Sommer
Institute of Computer Science
and Applied Mathematics
Christian-Albrechts-University of Kiel
Preußerstr. 1-9, 24105 Kiel
Germany
gs@ks.informatik.uni-kiel.de

L. Svensson
Dept. of Mathematics
Royal Institute of Technology
100 44 Stockholm
Sweden
larss@math.kth.se

Y. Zhang
Institute of Computer Science
and Applied Mathematics
Christian-Albrechts-University of Kiel
Preußerstr. 1-9, 24105 Kiel
Germany
yz@ks.informatik.uni-kiel.de

Part I

A Unified Algebraic Approach for Classical Geometries

1. New Algebraic Tools for Classical Geometry*

David Hestenes[1], Hongbo Li[1], and Alyn Rockwood[2]

[1] Department of Physics and Astronomy
Arizona State University, Tempe
[2] Power Take Off Software, Inc., Colorado Springs

1.1 Introduction

Classical geometry has emerged from efforts to codify perception of space and motion. With roots in ancient times, the great flowering of classical geometry was in the 19th century, when Euclidean, non-Euclidean and projective geometries were given precise mathematical formulations and the rich properties of geometric objects were explored. Though fundamental ideas of classical geometry are permanently imbedded and broadly applied in mathematics and physics, the subject itself has practically disappeared from the modern mathematics curriculum. Many of its results have been consigned to the seldom-visited museum of mathematics history, in part, because they are expressed in splintered and arcane language. To make them readily accessible and useful, they need to be reexamined and integrated into a coherent mathematical system.

Classical geometry has been making a comeback recently because it is useful in such fields as Computer-Aided Geometric Design (CAGD), CAD/CAM, computer graphics, computer vision and robotics. In all these fields there is a premium on computational efficiency in designing and manipulating geometric objects. Our purpose here is to introduce powerful new mathematical

* This work has been partially supported by NSF Grant RED-9200442.

tools for meeting that objective and developing new insights within a unified algebraic framework. In this and subsequent chapters we show how classical geometry fits neatly into the broader mathematical system of *Geometric Algebra* (GA) and its extension to a complete *Geometric Calculus* (GC) that includes differential forms and much more.

Over the last four decades GC has been developed as a universal geometric language for mathematics and physics. This can be regarded as the culmination of an R & D program innaugurated by Hermann Grassmann in 1844 [94, 111]. Literature on the evolution of GC with extensive applications to math and physics can be accessed from the *GC web site*

<http://ModelingNTS.la.asu.edu/GC_R&D.html>.

Here, we draw on this rich source of concepts, tools and methods to enrich classical geometry by integrating it more fully into the whole system.

This chapter provides a synopsis of basic tools in Geometric Algebra to set the stage for further elaboration and applications in subsequent chapters. To make the synopsis compact, proofs are omitted. Geometric interpretation is emphasized, as it is essential for practical applications.

In classical geometry the primitive elements are points, and *geometric objects* are point sets with properties. The properties are of two main types: structural and transformational. Objects are characterized by structural relations and compared by transformations. In his Erlanger program, Felix Klein [131] classified geometries by the transformations used to compare objects (for example, similarities, projectivities, affinities, etc). Geometric Algebra provides a unified algebraic framework for both kinds of properties and any kind of geometry.

1.2 Geometric Algebra of a Vector Space

The terms "vector space" and "linear space" are ordinarily regarded as synonymous. While retaining the usual concept of linear space, we enrich the concept of vector space by defining a special product among vectors that characterizes their relative directions and magnitudes. The resulting geometric algebra suffices for all the purposes of linear and multilinear algebra. We refer the reader to the extensive treatment of geometric algebra in [113] and to the GC Web site, so that we can give a more concise treatment here.

Basic Definitions

As a rule, we use lower case letters to denote vectors, lower case Greek letters to denote scalars and calligraphic capital letters to denote sets.

First, we define geometric algebra. Let $\mathcal{V}^n$ be an n-dimensional vector space over real numbers $\mathbb{R}$. The *geometric algebra* $\mathcal{G}_n = \mathcal{G}(\mathcal{V}^n)$ is generated from $\mathcal{V}^n$ by defining the *geometric product* as a multilinear, associative product satisfying the *contraction rule*:

$$a^2 = \epsilon_a |a|^2, \text{ for } a \in \mathcal{V}^n, \tag{1.1}$$

where ϵ_a is $1, 0$ or -1, $|a| \geq 0$, and $|a| = 0$ if $a = 0$. The integer ϵ_a is called the *signature* of a; the scalar $|a|$ is its *magnitude*. When $a \neq 0$ but $|a| = 0$, a is said to be a *null vector*.

In the above definition, "multi-linearity" means

$$a_1 \cdots (b_1 + \cdots + b_r) \cdots a_s = (a_1 \cdots b_1 \cdots a_s) + \cdots + (a_1 \cdots b_r \cdots a_s), \tag{1.2}$$

for any vectors $a_1, \cdots, a_s, b_1, \cdots, b_r$ and any position of $b_1 + \cdots + b_r$ in the geometric product. Associativity means

$$a(bc) = (ab)c, \text{ for } a, b, c \in \mathcal{V}^n. \tag{1.3}$$

An element M in $\mathcal{G}_n$ is *invertible* if there exists an element N in $\mathcal{G}_n$ such that $MN = NM = 1$. The element N, if it exists, is unique. It is called the *inverse* of M, and is denoted by M^{-1}. For example, null vectors in $\mathcal{V}^n$ are not invertible, but any non-null vector a is invertible, with $a^{-1} = 1/a = a/a^2$. This capability of GA for division by vectors greatly facilitates computations.

From the geometric product, two new kinds of product can be defined. For $a, b \in \mathcal{V}^n$, the scalar-valued quantity

$$a \cdot b = \tfrac{1}{2}(ab + ba) = b \cdot a \tag{1.4}$$

is called the *inner product*; the (nonscalar) quantity

$$a \wedge b = \tfrac{1}{2}(ab - ba) = -b \wedge a \tag{1.5}$$

is called the *outer product*. Therefore, the geometric product

$$ab = a \cdot b + a \wedge b \tag{1.6}$$

decomposes into symmetric and anti-symmetric parts.

The outer product of r vectors can be defined as the anti-symmetric part of the geometric product of the r vectors. It is called an *r-blade*, or a blade of *grade* r. A linear combination of r-blades is called an *r-vector*. The set of r-vectors is an $\binom{n}{r}$-dimensional subspace of $\mathcal{G}_n$, denoted by $\mathcal{G}_n^r$. The whole of $\mathcal{G}_n$ is given by the subspace sum

$$\mathcal{G}_n = \sum_{i=0}^{n} \mathcal{G}_n^i. \tag{1.7}$$

A generic element in $\mathcal{G}_n$ is called a *multivector*. According to (1.7), every multivector M can be written in the expanded form

$$M = \sum_{i=0}^{n} \langle M \rangle_i, \tag{1.8}$$

where $\langle M\rangle_i$ denotes the i-vector part. The dimension of $\mathcal{G}_n$ is $\sum_{i=0}^{n}\binom{n}{i}=2^n$.

By associativity and multi-linearity, the outer product extended to any finite number of multivectors and to scalars, with the special proviso

$$\lambda\wedge M = M\wedge\lambda = \lambda M,\ \text{ for } \lambda\in\mathbb{R}, M\in\mathcal{G}_n. \tag{1.9}$$

The inner product of an r-blade $a_1\wedge\cdots\wedge a_r$ with an s-blade $b_1\wedge\cdots\wedge b_s$ can be defined recursively by

$$\begin{aligned}&(a_1\wedge\cdots\wedge a_r)\cdot(b_1\wedge\cdots\wedge b_s)\\ &\quad=\begin{cases}((a_1\wedge\cdots\wedge a_r)\cdot b_1)\cdot(b_2\wedge\cdots\wedge b_s) & \text{if } r\geq s\\ (a_1\wedge\cdots\wedge a_{r-1})\cdot(a_r\cdot(b_1\wedge\cdots\wedge b_s)) & \text{if } r<s\end{cases},\end{aligned} \tag{1.10a}$$

and

$$\begin{aligned}&(a_1\wedge\cdots\wedge a_r)\cdot b_1\\ &\quad=\sum_{i=1}^{r}(-1)^{r-i}a_1\wedge\cdots\wedge a_{i-1}\wedge(a_i\cdot b_1)\wedge a_{i+1}\wedge\cdots\wedge a_r,\\ &a_r\cdot(b_1\wedge\cdots\wedge b_s)\\ &\quad=\sum_{i=1}^{s}(-1)^{i-1}b_1\wedge\cdots\wedge b_{i-1}\wedge(a_r\cdot b_i)\wedge b_{i+1}\wedge\cdots\wedge b_s.\end{aligned} \tag{1.10b}$$

By bilinearity, the inner product is extended to any two multivectors, if

$$\lambda\cdot M = M\cdot\lambda = 0,\ \text{ for } \lambda\in\mathbb{R},\ M\in\mathcal{G}_n. \tag{1.11}$$

For any blades A and B with nonzero grades r and s we have

$$A\cdot B = \langle AB\rangle_{|\,r-s\,|}\,, \tag{1.12}$$

$$A\wedge B = \langle AB\rangle_{r+s}\,. \tag{1.13}$$

These relations can be adopted as alternative definitions of inner and outer products or derived as theorems.

An *automorphism* f of $\mathcal{G}_n$ is an invertible linear mapping that preserves the geometric product:

$$f(M_1M_2) = f(M_1)f(M_2),\ \text{ for } M_1, M_2\in\mathcal{G}_n. \tag{1.14}$$

More generally, this defines an *isomorphism* f from one geometric algebra to another.

An *anti-automorphism* g is a linear mapping that reverses the order of geometric products:

$$g(M_1M_2) = g(M_2)g(M_1),\ \text{ for } M_1, M_2\in\mathcal{G}_n. \tag{1.15}$$

The *main anti-automorphism* of $\mathcal{G}_n$, also called *reversion*, is denoted by "$\dagger$", and defined by

$$\langle M^\dagger \rangle_i = (-1)^{\frac{i(i-1)}{2}} \langle M \rangle_i, \text{ for } M \in \mathcal{G}_n, \ 0 \le i \le n. \tag{1.16}$$

An *involution* h is an invertible linear mapping whose composition with itself is the identity map:

$$h(h(M)) = M, \text{ for } M \in \mathcal{G}_n. \tag{1.17}$$

The *main involution* of $\mathcal{G}_n$, also called *grade involution* or *parity conjugation*, is denoted by "$*$", and defined by

$$\langle M^* \rangle_i = (-1)^i \langle M \rangle_i, \text{ for } M \in \mathcal{G}_n, \ 0 \le i \le n. \tag{1.18}$$

For example, for vectors $a_1, \cdots, a_r$, we have

$$(a_1 \cdots a_r)^\dagger = a_r \cdots a_1, \ \ (a_1 \cdots a_r)^* = (-1)^r a_1 \cdots a_r. \tag{1.19}$$

A multivector M is said to be *even*, or have *even parity*, if $M^* = M$; it is *odd*, or has *odd parity*, if $M^* = -M$.

The concept of magnitude is extended from vectors to any multivector M by

$$|M| = \sqrt{\sum_{i=0}^{n} |\langle M \rangle_i|^2}, \tag{1.20a}$$

where

$$|\langle M \rangle_i| = \sqrt{|\langle M \rangle_i \cdot \langle M \rangle_i|}. \tag{1.20b}$$

A natural *scalar product* on the whole of $\mathcal{G}_n$ is defined by

$$\langle M N^\dagger \rangle = \langle N M^\dagger \rangle = \sum_{i=0}^{n} \langle \langle M \rangle_i \langle N \rangle_i^\dagger \rangle, \tag{1.21}$$

where $\langle \cdots \rangle = \langle \cdots \rangle_0$ denotes the "scalar part." When scalar parts are used frequently it is convenient to drop the subscript zero.

In $\mathcal{G}_n$, the maximal grade of a blade is n, and any blade of grade n is called a *pseudoscalar*. The space $\mathcal{V}^n$ is said to be *non-degenerate* if it has a pseudoscalar with nonzero magnitude. In that case the notion of duality can be defined algebraically. Let I_n be a pseudoscalar with magnitude 1, designated as the *unit pseudoscalar*. The *dual* of a multivector M in $\mathcal{G}_n$ is then defined by

$$\widetilde{M} = M^\sim = M I_n^{-1}, \tag{1.22}$$

where I_n^{-1} differs from I_n by at most a sign. The dual of an r-blade is an $(n-r)$-blade; in particular, the dual of an n-blade is a scalar, which is why an n-blade is called a pseudoscalar.

Inner and outer products are *dual* to one another, as expressed by the following identities that hold for any vector a and multivector M:

$$(a \cdot M)I_n = a \wedge (MI_n)\,, \tag{1.23a}$$
$$(a \wedge M)I_n = a \cdot (MI_n)\,. \tag{1.23b}$$

Duality enables us to define the *meet* $M \vee N$ of multivectors M, N with (grade M) + (grade M) $\geq n$ by

$$M \vee N = \widetilde{M} \cdot N\,. \tag{1.24}$$

The meet satisfies the "deMorgan rule"

$$(M \vee N)^{\sim} = \widetilde{M} \wedge \widetilde{N}. \tag{1.25}$$

As shown below, the meet can be interpreted as an algebraic representation for the intersection of vector subspaces. More generally, it can be used to describe "incidence relations" in geometry [114].

Many other products can be defined in terms of the geometric product. The *commutator product* $A \times B$ is defined for any A and B by

$$A \times B \equiv \tfrac{1}{2}(AB - BA) = -B \times A\,. \tag{1.26}$$

Mathematicians classify this product as a "derivation" with respect to the geometric product, because it has the "distributive property"

$$A \times (BC) = (A \times B)C + B(A \times C)\,. \tag{1.27}$$

This implies the *Jacobi identity*

$$A \times (B \times C) = (A \times B) \times C + B \times (A \times C)\,, \tag{1.28}$$

which is a derivation on the commutator product. The relation of the commutator product to the inner and outer products is grade dependent; thus, for a vector a,

$$a \times \langle M \rangle_k = a \wedge \langle M \rangle_k \qquad \text{if } k \text{ is odd}\,, \tag{1.29a}$$
$$a \times \langle M \rangle_k = a \cdot \langle M \rangle_k \qquad \text{if } k \text{ is even}\,. \tag{1.29b}$$

The commutator product is especially useful in computations with bivectors. With any bivector A this product is grade preserving:

$$A \times \langle M \rangle_k = \langle A \times M \rangle_k\,. \tag{1.30}$$

In particular, this implies that the space of bivectors is closed under the commutator product. Consequently, it forms a Lie algebra. The geometric product of bivector A with M has the expanded form

$$AM = A \cdot M + A \times M + A \wedge M \qquad \text{for grade } M \geq 2\,. \tag{1.31}$$

Compare this with the corresponding expansion (1.6) for the product of vectors.

Blades and Subspaces

The elements of $\mathcal{G}_n$ can be assigned a variety of different geometric interpretations appropriate for different applications. For one, geometric algebra characterizes geometric objects composed of points that are represented by vectors in $\mathcal{V}^n$.

To every r-dimensional subspace in $\mathcal{V}^n$, there is an r-blade A_r such that the subspace is the solution set of the equation

$$x \wedge A_r = 0, \quad \text{for } x \in \mathcal{V}^n. \tag{1.32}$$

According to this equation, A_r is unique to within a nonzero scalar factor. This subspace generates a geometric algebra $\mathcal{G}(A_r)$. Conversely, the subspace itself is uniquely determined by A_r. Therefore, as discussed in [113], the blades in $\mathcal{V}^n$ determine an algebra of subspaces for $\mathcal{V}^n$. The blade A_r can be regarded as a directed measure (or r-volume) on the subspace, with magnitude $|A_r|$ and an orientation. Thus, since A_r determines a unique subspace, the two blades A_r and $-A_r$ determine two subspaces of the same vectors but opposite orientation. The blades of grade r form a manifold $G(r,n)$ called a *Grassmannian*. The algebra of blades is therefore an algebraic structure for $G(r,n)$, and the rich literature on Grassmannians [100] can be incorporated into GC.

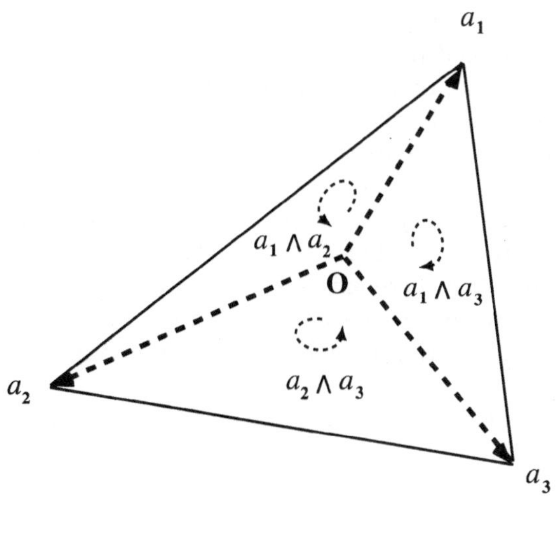

Fig. 1.1. Blades in the space of $a_1 \wedge a_2 \wedge a_3$, where the a_i are vectors

Vectors $a_1, \cdots, a_r$, are linearly dependent if and only if $a_1 \wedge \cdots \wedge a_r = 0$. The r-blade $a_1 \wedge \cdots \wedge a_r$ represents the r-dimensional subspace of $\mathcal{V}^n$ spanned by them, if they are linearly independent. The case $r = 3$ is illustrated in Fig 1.1.

The square of a blade is always a scalar, and a blade is said to be *null* if its square is zero. Null blades represent degenerate subspaces. For a non-degenerate r-subspace, we often use a unit blade, say I_r, to denote it. For any two pseudoscalars A_r, B_r in $\mathcal{G}(I_r)$, since $A_r B_r^{-1} = B_r^{-1} A_r$ is a scalar, we can write it A_r/B_r.

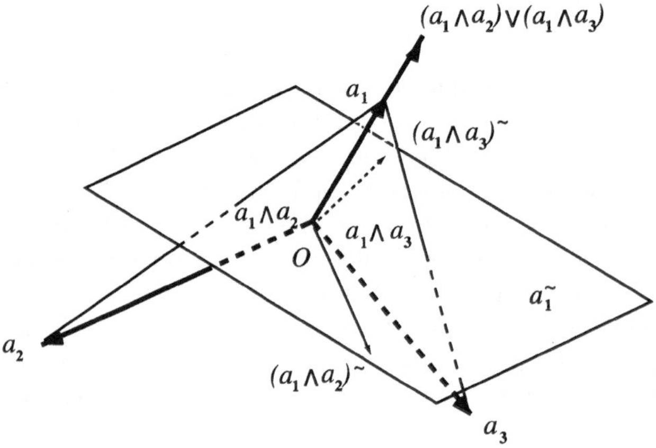

Fig. 1.2. Dual and meet in the space of $a_1 \wedge a_2 \wedge a_3$

Figure 1.2 illustrates duality and meet in $\mathcal{G}_3$ generated by vectors a_1, a_2, a_3. Notice the collinearity of vectors a_1 and $(a_1 \wedge a_3) \vee (a_1 \wedge a_2)$.

In a non-degenerate space, the dual of a blade represents the orthogonal complement of the subspace represented by the blade. The meet of two blades, if nonzero, represents the intersection of the two subspaces represented by the two blades respectively.

Two multivectors are said to be *orthogonal* or *perpendicular* to one another, if their inner product is zero. An r-blade A_r and s-blade B_s are orthogonal if and only if (1) when $r \geq s$ there exists a vector in the space of B_s, that is orthogonal to every vector in the space of A_r, and (2) when $r < s$ there exists a vector in the space of A_r that is orthogonal to every vector in the space of B_s.

Let A_r be a non-degenerate r-blade in $\mathcal{G}_n$. Then any vector $x \in \mathcal{V}^n$ has a *projection* onto the space of A_r defined by

$$P_{A_r}(x) = (x \cdot A_r)\, A_r^{-1}. \tag{1.33}$$

Its *rejection* from the space of A_r is defined by

$$P_{A_r}^{\perp}(x) = (x \wedge A_r)\, A_r^{-1}. \tag{1.34}$$

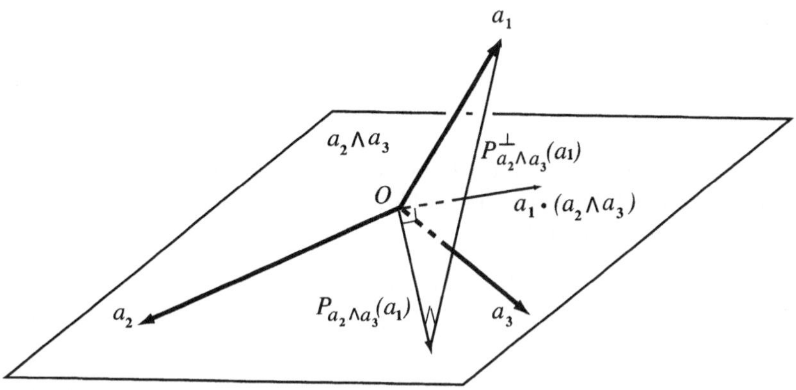

Fig. 1.3. Projection, rejection and inner product in the space of $a_1 \wedge a_2 \wedge a_3$

Therefore,

$$x = P_{A_r}(x) + P^{\perp}_{A_r}(x) \tag{1.35}$$

is the *orthogonal decomposition* of $\mathcal{V}^n$ with respect to the A_r.

Figure 1.3 shows the projection and rejection of vector a_1 with respect to the space $a_2 \wedge a_3$, together with the inner product $a_1 \cdot (a_2 \wedge a_3)$. Note that the vector $a_1 \cdot (a_2 \wedge a_3)$ is perpendicular to the vector $P_{a_2 \wedge a_3}(a_1)$.

Frames

If a blade A_r admits the decomposition

$$A_r = a_1 \wedge a_2 \wedge \ldots \wedge a_r\,, \tag{1.36}$$

the set of vectors $\{a_i\,;\, i = 1, \ldots, r\}$ is said to be a *frame* (or *basis*) for A_r and the vector space it determines. Also, A_r is said to be the *pseudoscalar* for the frame. A *dual frame* $\{a^i\}$ is defined by the equations

$$a^i \cdot a_j = \delta^i_j. \tag{1.37}$$

If A_r is invertible, these equations can be solved for the a^i, with the result

$$a^i = (-1)^{i+1}(a_1 \wedge \ldots \wedge \check{a}_i \wedge \ldots \wedge a_r)A_r^{-1}, \tag{1.38}$$

where $\check{a}_i$ indicates that a_i is omitted from the product. Moreover,

$$A_r^{-1} = a^1 \wedge a^2 \wedge \ldots \wedge a^r. \tag{1.39}$$

In this way geometric algebra greatly facilitates manipulations with frames for vector spaces.

Differentiation

Let $a = \sum_i \alpha^i a_i$ be a *vector variable* with coordinates $\alpha^i = a \cdot a_i$ defined on the space of A_r as specified by (1.36). The *vector derivative* with respect to

a on A_r can be defined by

$$\partial_a = \sum_i a^i \frac{\partial}{\partial \alpha^i}. \tag{1.40}$$

It can be defined without introducing coordinates, but this approach relates it to standard partial derivatives. The following table of derivatives is easily derived. For any vector b in $\mathcal{V}^n$:

$$\partial_a(a \cdot b) = b \cdot \partial_a a = \textstyle\sum_i a^i(a_i \cdot b) = P_{A_r}(b), \tag{1.41a}$$
$$\partial_a a^2 = 2a, \tag{1.41b}$$
$$\partial_a a = \partial_a \cdot a = r, \tag{1.41c}$$
$$\partial_a \wedge a = 0. \tag{1.41d}$$

Of course A_r could be the pseudoscalar for $\mathcal{V}^n$.

[113] generalizes differentiation to any multivector variable defined on $\mathcal{G}_n$ or any of its subspaces. The derivative ∂_X with respect to a multivector variable X is characterized by the basic identity

$$\partial_X \langle XA \rangle = \langle A \rangle_X = \langle A \partial_X \rangle X, \tag{1.42}$$

where A is independent of X, $\langle \ldots \rangle$ means scalar part, and $\langle A \rangle_X$ means "select from A only those grades which are contained in X." If A has the same grade as X then $\langle A \rangle_X = A$. It follows that

$$\partial_X \langle \widetilde{X} A \rangle = \langle \widetilde{A} \rangle_X. \tag{1.43}$$

The operator $\langle A \partial_X \rangle$ is a kind of generalized directional derivative defined by

$$\langle A \partial_X \rangle F(X) \equiv \frac{d}{d\epsilon} F(X + \epsilon A)\big|_{\epsilon=0}, \tag{1.44}$$

where ϵ is a scalar parameter and F is any differentiable function of X. Applied to $F(X) = X$, this yields the right equality in (1.42). If A has the same grade as X, then the left equality in (1.42) gives

$$\partial_X = \partial_A \langle A \partial_X \rangle, \tag{1.45}$$

so the general multivector derivative $\partial_X F(X)$ can be obtained from the "directional derivative" (1.44). From (1.44) one derives the sum, product and chain rules for differential operators. Of course, the vector derivative with its properties (1.40) to (1.41d) is a special case, as is the usual scalar derivative.

Signature

So far, a particular signature has not been attributed to $\mathcal{V}^n$ to emphasize the fact that, for many purposes, signature is inconsequential. To account for signature, we introduce the alternative notation $\mathbb{R}^{p,q,r}$ to denote a real vector space of dimension $n = p + q + r$, where p, q and r are, respectively, the dimensions of subspaces spanned by vectors with positive, negative and null signatures. Let $\mathbb{R}_{p,q,r} = \mathcal{G}(\mathbb{R}^{p,q,r})$ denote the geometric algebra of $\mathbb{R}^{p,q,r}$, and let $\mathbb{R}^k_{p,q,r}$ denote the $\binom{n}{r}$-dimensional subspace of k-vectors, so that

$$\mathbb{R}_{p,q,r} = \sum_{k=0}^{n} \mathbb{R}^k_{p,q,r} \,. \tag{1.46}$$

A pseudoscalar I_n for $\mathbb{R}^{p,q,r}$ factors into

$$I_n = A_p B_q C_r \,, \tag{1.47}$$

where the factors are pseudoscalars for the three different kinds of subspaces.

The algebra is said to be *non-degenerate* if I_n is invertible. That is possible only if $r = 0$, so $n = p + q$ and

$$I_n = A_p B_q \,. \tag{1.48}$$

In that case we write $\mathbb{R}^{p,q} = \mathbb{R}^{p,q,0}$ and $\mathbb{R}_{p,q} = \mathbb{R}_{p,q,0}$. The algebra is said to be *Euclidean* (or *anti-Euclidean*) if $n = p$ (or $n = q$). Then it is convenient to use the notations $\mathbb{R}^n = \mathbb{R}^{n,0}$, $\mathbb{R}^{-n} = \mathbb{R}^{0,n}$, etcetera.

Any degenerate algebra can be embedded in a non-degenerate algebra of larger dimension, and it is almost always a good idea to do so. Otherwise, there will be subspaces without a complete basis of dual vectors, which will complicate algebraic manipulations. The n-dimensional vector spaces of every possible signature are subspaces of $\mathbb{R}^{n,n}$. For that reason, $\mathbb{R}_{n,n}$ is called the *mother algebra* of n-space. As explained in [63], it is the proper arena for the most general approach to linear algebra.

1.3 Linear Transformations

The terms "linear function," "linear mapping" and "linear transformation" are usually regarded as synonymous. To emphasize an important distinction in GA, let us restrict the meaning of the last term to "linear vector-valued functions of a vector variable." Of course, every linear function is isomorphic to a linear transformation. The special importance of the latter derives from the fact that the tools of geometric algebra are available to characterize its structure and facilitate applications. Geometric algebra enables coordinate-free analysis and computations. It also facilitates the use of matrices when they are desired or needed.

To relate a linear transformation $\underline{f}$ on $\mathcal{V}^n$ to its matrix representation $f_i^{\,j}$, we introduce a basis $\{e_i\}$ and its dual $\{e^j\}$, so that

$$\underline{f} e_i = \underline{f}(e_i) = \sum_j e_j f_i^j , \tag{1.49a}$$

and

$$f_i^j = e^j \cdot \underline{f} e_i = (\bar{f} e^j) \cdot e_i . \tag{1.49b}$$

The last equality defines the *adjoint* $\bar{f}$ of $\underline{f}$, so that

$$\bar{f} e^j = \sum_i f_i^j e^i . \tag{1.50}$$

Without reference to a basis the adjoint is defined by

$$b \cdot \underline{f} a = a \cdot \bar{f}\, b , \tag{1.51a}$$

whence,

$$\bar{f} b = \partial_a (b \cdot \underline{f} a) = \sum_i e^i (b \cdot \underline{f}\, e_i) . \tag{1.51b}$$

Within geometric algebra, it seldom helps to introduce matrices unless they have been used in the first place to define the linear transformations of interest, as, for example, in a computer graphics display where coordinates are needed. Some tools to handle linear transformations without matrices are described below.

Outermorphism

Every linear transformation $\underline{f}$ on $\mathcal{V}^n$ extends naturally to a linear function $\underline{f}$ on $\mathcal{G}_n$ with

$$\underline{f}(A \wedge B) = (\underline{f} A) \wedge (\underline{f} B) . \tag{1.52}$$

This extension is called an *outermorphism* because it preserves the outer product. Any ambiguity in using the same symbol $\underline{f}$ for the transformation and its extension can be removed by displaying an argument for the function. For any blade with the form (1.36) we have

$$\underline{f} A_r = (\underline{f} a_1) \wedge (\underline{f} a_2) \wedge \cdots \wedge (\underline{f} a_r) . \tag{1.53}$$

This shows explicitly how the transformation of vectors induces a transformation of blades. By linearity it extends to any multivector.

The outermorphism of the adjoint $\bar{f}$ is easily specified by generalizing (1.51a) and (1.51b); thus, for any multivectors A and B,

$$\langle B \underline{f} A \rangle = \langle A \bar{f} B \rangle . \tag{1.54}$$

By multivector differentiation,

$$\bar{f}B = \bar{f}(B) = \partial_A \langle A\bar{f}(B)\rangle \,. \tag{1.55}$$

We are now equipped to formulate the fundamental theorem:

$$A\cdot(\underline{f}B) = \underline{f}[(\bar{f}A)\cdot B] \qquad \text{or} \qquad (\underline{f}B)\cdot A = \underline{f}[B\cdot\bar{f}A] \,. \tag{1.56}$$

for (grade A) $\leq$ (grade B).

This theorem, first proved in [113], is awkward to formulate without geometric algebra, so it seldom appears (at least in full generality) in the literature on linear algebra. It is important because it is the most general *transformation law for inner products.*

Outermorphisms generalize and simplify the theory of determinants. Let I be a pseudoscalar for $\mathcal{V}^n$. The determinant of $\underline{f}$ is the eigenvalue of $\underline{f}$ on I, as expressed by

$$\bar{f}I = (\det \underline{f})I \,. \tag{1.57}$$

If I is invertible, then

$$I^{-1}\bar{f}I = I^{-1}\cdot(\underline{f}I) = \det \underline{f} = \det \bar{f} = \det f_i^j \,. \tag{1.58}$$

To prove the last equality, we can write $I = e_1 \wedge e_2 \wedge \ldots \wedge e_n$ so that

$$\det \underline{f} = (e^n \wedge \ldots \wedge e^1)\cdot[(\underline{f}e_1)\wedge \ldots \wedge \underline{f}(e_n)] \,. \tag{1.59}$$

Using the identities (1.10a) and (1.10b), the right side of (1.14) can be expanded to get the standard expression for determinants in terms of matrix elements f_i^j. This exemplifies the fact that the Laplace expansion and all other properties of determinants are easy and nearly automatic consequences of their formulation within geometric algebra.

The law (1.56) has the important special case

$$A\bar{f}I = \bar{f}[(\underline{f}A)I] \,. \tag{1.60}$$

For $\det \underline{f} \neq 0$, this gives an explicit expression for the inverse outermorphism

$$\underline{f}^{-1}A = \frac{\bar{f}(AI)I^{-1}}{\det \underline{f}} \,. \tag{1.61}$$

Applying this to the basis $\{e^i\}$ and using (1.38) we obtain

$$\underline{f}^{-1}e^i = \frac{(-1)^{i+1}\bar{f}(e_1\wedge\cdots\wedge\check{a}_i\wedge\cdots\wedge e_n)\cdot(e_1\wedge\cdots\wedge e_n)}{\det \underline{f}} \,. \tag{1.62}$$

Again, expansion of the right side with the help of (1.10a) and (1.10b) gives the standard expression for a matrix inverse in terms of matrix elements.

The composition of linear transformations $\underline{g}$ and $\underline{f}$ can be expressed as an operator product:

$$\underline{h} = \underline{g}\,\underline{f}\,. \tag{1.63}$$

This relation extends to their outermorphism as well. Applied to (1.57), it immediately gives the classical result

$$\det \underline{h} = (\det \underline{g}) \det \underline{f}\,, \tag{1.64}$$

from which many more properties of determinants follow easily.

Orthogonal Transformations

A linear transformation $\underline{U}$ is said to be *orthogonal* if it preserves the inner product of vectors, as specified by

$$(\underline{U}a) \cdot (\underline{U}b) = a \cdot (\overline{U}\underline{U}b) = a \cdot b\,. \tag{1.65}$$

Clearly, this is equivalent to the condition $\underline{U}^{-1} = \overline{U}$. For handling orthogonal transformations geometric algebra is decisively superior to matrix algebra, because it is computationally more efficient and simpler to interpret geometrically. To explain how, some new terminology is helpful.

A *versor* is any multivector that can be expressed as the geometric product of invertible vectors. Thus, any versor U can be expressed in the factored form

$$U = u_k \cdots u_2 u_1\,, \tag{1.66}$$

where the choice of vectors u_i is not unique, but there is a minimal number $k \leq n$. The parity of U is even (odd) for even (odd) k.

Every versor U determines an orthogonal transformation $\underline{U}$ given by

$$\underline{U}(x) = UxU^{*-1} = U^* x U^{-1}, \qquad \text{for} \qquad x \in \mathcal{V}^n\,. \tag{1.67}$$

Conversely, every orthogonal transformation $\underline{U}$ can be expressed in the *canonical form* (1.67). This has at least two great advantages. First, any orthogonal transformation is representable (and therefore, visualizable) as a set of vectors. Second, the composition of orthogonal transformations is reduced to multiplication of vectors.

The outermorphism of (1.67) is

$$\underline{U}(M) = UMU^{-1} \qquad \text{for} \qquad U^* = U\,, \tag{1.68a}$$

or

$$\underline{U}(M) = UM^*U^{-1} \qquad \text{for} \qquad U^* = -U\,, \tag{1.68b}$$

where M is a generic multivector.

An even versor $R = R^*$ is called a *spinor* or *rotor* if

$$RR^\dagger = |\,R\,|^2\,, \tag{1.69a}$$

so that

$$R^{-1} = \frac{1}{R} = \frac{R^\dagger}{RR^\dagger} = \frac{R^\dagger}{|R|^2}. \tag{1.69b}$$

Alternative, but closely related, definitions of "spinor" and "rotor" are common. Often the term rotor presumes the normalization $|R|^2 = 1$. In that case, (1.67) takes the simpler form

$$\underline{R}x = RxR^\dagger \tag{1.70}$$

and $\underline{R}$ is called a *rotation*. Actually, the form with R^{-1} is preferable to the one with $R^\dagger$, because $\underline{R}$ is independent of $|R|$, and normalizing may be inconvenient.

Note that for $U = u_2u_1$, the requirement $|U|^2 = u_2^2u_1^2$ for a rotation implies that the vectors u_1 and u_2 have the same signature. Therefore, when they have opposite signature U is the prime example of an even versor which is not a spinor, and the corresponding linear operator $\underline{U}$ in (1.67) is not a rotation.

In the simplest case where the versor is a single vector $u_1 = -u_1^*$, we can analyze (1.67) with the help of (1.35) as follows:

$$\begin{aligned}\underline{u}_1(x) &= u_1^* x u_1^{-1} = -u_1(P_{u_1}(x) + P_{u_1}^\perp(x))u_1^{-1} \\ &= -P_{u_1}(x) + P_{u_1}^\perp(x).\end{aligned} \tag{1.71}$$

The net effect of the transformation is to re-*verse* direction of the component of x along u_1, whence the name *versor* (which dates back to Hamilton). Since every invertible vector is normal to a hyperplane in $\mathcal{V}^n$, the transformation (1.71) can be described as *reflection in a hyperplane.* In view of (1.66), *every orthogonal transformation can be expressed as the composite of at most n reflections in hyperplanes.* This is widely known as the Cartan-Dieudonné Theorem.

The reflection (1.71) is illustrated in Fig. 1.4 along with the next simplest example, a rotation $\underline{u_3u_2}$ induced by two successive reflections. The figure presumes that the rotation is elliptic (or Euclidean), though it could be hyperbolic (known as a Lorentz transformation in physics), depending on the signature of the plane.

Given that a rotation takes a given vector a into $b = \underline{R}a$, it is often of interest to find the simplest spinor R that generates it. It is readily verified that the solution is

$$R = (a + b)a = b(a + b). \tag{1.72}$$

Without loss of generality, we can assume that a and b are normalized to $a^2 = b^2 = \pm 1$, so that

$$|R|^2 = a^2(a + b)^2 = 2|a \cdot b \pm 1|. \tag{1.73}$$

This is a case where normalization is inconvenient. Besides destroying the simplicity of the unnormalized form (1.72), it would require taking the square root of (1.73), an unnecessary computation because it does not affect $\underline{R}$. Note that $|R| = 0$ and R^{-1} is undefined when a and b are oppositely directed. In that case a and b do not define a unique plane of rotation.

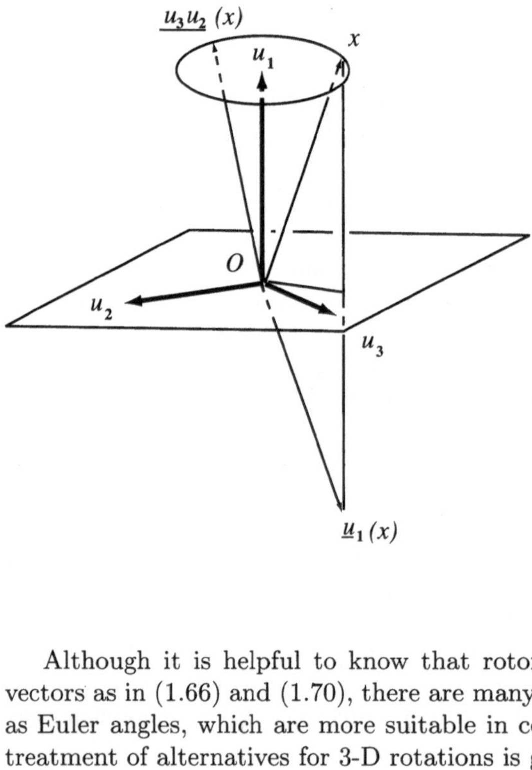

Fig. 1.4. Versor (vector and rotor) actions. Here u_1 is orthogonal to both u_2, u_3

Although it is helpful to know that rotors can be "parameterized" by vectors as in (1.66) and (1.70), there are many other parameterizations, such as Euler angles, which are more suitable in certain applications. A detailed treatment of alternatives for 3-D rotations is given in [112].

The versors in $\mathcal{G}_n$ form a group under the geometric product, called the *versor group*. The versors of unit magnitude form a subgroup, called the *pin group*. Since $\underline{U}$ in (1.67) is independent of the sign and magnitude of U, the two groups are equivalent *double coverings* of the *orthogonal group* $\mathrm{O}(p,q)$, where the signature of $\mathcal{V}^n = \mathbb{R}^{p,q}$ is displayed to enable crucial distinctions. At first sight the pin group seems simpler than the versor group, but we have already noted that it is sometimes more efficient to work with unnormalized versors.

For any $\underline{U}$

$$\det \underline{U} = \pm 1 , \tag{1.74}$$

where the sign corresponds to the parity of the versor U which generates it. Those with positive determinant compose the *special orthogonal group* $\mathrm{SO}(p,r)$. It is doubly covered by the subgroup of even versors. The subgroup of elements in $SO(p,r)$ which are continuously connected to the identity is called the *rotation group* $SO^+(p,r)$. The versor group covering $SO(p,r)$ is called the *spin group* $\mathrm{Spin}(p,r)$. Let us write $\mathrm{Spin}^+(p,r)$ for the group of rotors covering $SO^+(p,r)$. This group is distingiushed by the condition (1.68a) on its elements, and that ensures that the rotations are continuously connected to the identity. The distinction between SO and SO^+ or between Spin and Spin^+ is not always recognized in the literature, but that seldom creates problems. For Euclidean or anti-Euclidean signature there is no distinction.

The spin groups are more general than anyone suspected for a long time. It has been proved in [63] that *every Lie group* can be represented as a spin group in some Geometric Algebra of suitable dimension and signature. The corresponding *Lie algebra* is then represented as an algebra of bivectors under the commutator product (1.26). All this has great theoretical and practical advantages, as it is computationally more efficient than matrix representations. Engineers who compute 3-D rotations for a living are well aware that quaternions (the rotors in $\mathbb{R}_3$) are computationally superior to matrices.

1.4 Vectors as Geometrical Points

The elements of $\mathcal{G}_n$ can be assigned a variety of geometric interpretations appropriate for different applications. The most common practice is to identify vectors with geometric points, so that geometric objects composed of points are represented by point sets in a vector space. In this section we show how geometric algebra can be used to characterize some of the most basic geometric objects. This leads to abstract representations of objects by their properties without reference to the points that compose them. Thus we arrive at a kind of "algebra of geometric properties" which greatly facilitates geometric analysis and applications. We have already taken a large step in this direction by constructing an "algebra of vector subspaces" in Section 1.2. Here we take two more steps. First, we displace the subspaces from the origin to give us as algebra of k-planes. Second, we break the k-planes into pieces to get a "simplicial algebra." In many applications, such as finite element analysis, simplexes are basic building blocks for complex structures. To that end, our objective here is to sharpen the tools for manipulating simplexes. Applications to Geometric Calculus are described in [214].

In physics and engineering the vector space $\mathbb{R}^3$ is widely used to represent Euclidean 3-space $\mathbb{E}^3$ as a model of "physical space." More advanced applications use $\mathbb{R}^{1,3}$ as a model for spacetime with the *"spacetime algebra"* $\mathbb{R}_{1,3}$ to characterize its properties. Both of these important cases are included in the present treatment, which applies to spaces of any dimension and signature.

r-planes

An r-dimensional plane parallel to the subspace of A_r and through the point a is the solution set of

$$(x-a)\wedge A_r = 0, \text{ for } x \in \mathcal{V}^n . \tag{1.75}$$

It is often called an *r-plane*, or *r-flat*. It is called a *hyperplane* when $r = n-1$.

As detailed for $\mathbb{R}^3$ in [112], A_r is the *tangent* of the r-plane, and $M_{r+1} = a\wedge A_r$ is the *moment*. When A_r is invertible, $d = M_{r+1}A_r^{-1}$ is the *directance*. Let n be a unit vector collinear with d, then $d\cdot n$ is the *signed distance* of the r-plane from the origin in the direction of n, or briefly, the *n-distance* of the r-plane.

An r-plane can be represented by

$$A_r + M_{r+1} = (1+d)A_r , \tag{1.76}$$

where A_r is the tangent, M_{r+1} is the moment, and d is the directance. A point x is on the r-plane if and only if $x\wedge M_{r+1} = 0$ and $x\wedge A_r = M_{r+1}$. This representation, illustrated in Fig 1.5, has applications in rigid body mechanics [112]. The representation (1.76) is equivalent to the *Plücker coordinates* for an r-plane [187].

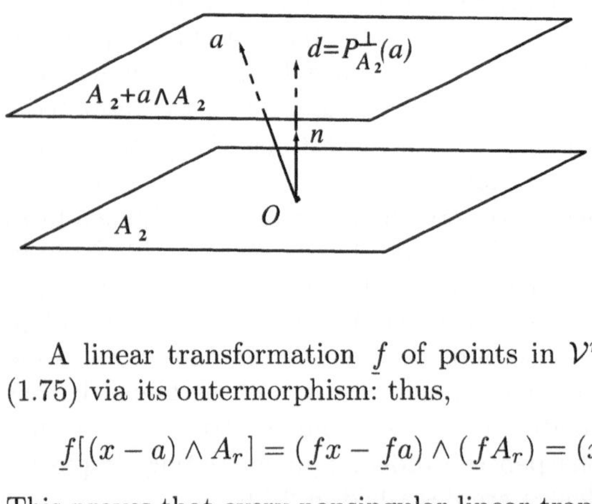

Fig. 1.5. A 2-plane in the space of $a\wedge A_2$

A linear transformation $\underline{f}$ of points in $\mathcal{V}^n$ induces a transformation of (1.75) via its outermorphism: thus,

$$\underline{f}[(x-a)\wedge A_r] = (\underline{f}x - \underline{f}a)\wedge(\underline{f}A_r) = (x'-a')\wedge A_r' = 0 . \tag{1.77}$$

This proves that every nonsingular linear transformation maps straight lines into straight lines and, more generally, k-planes into k-planes. This generalizes trivially to affine transformations.

Simplexes

An r-dimensional *simplex* (*r-simplex*) in $\mathcal{V}^n$ is the convex hull of $r+1$ points, of which at least r are linearly independent. A set $\{a_0, a_1, a_2, \ldots a_r\}$ of defining points is said to be a *frame* for the simplex. One of the points, say a_0, is distinguished and called the *base point* or *place* of the simplex. It will be convenient to introduce the notations

$$A_r \equiv a_0 \wedge a_1 \wedge a_2 \wedge \cdots \wedge a_r = a_0 \wedge \overline{A}_r\,, \tag{1.78a}$$

$$\overline{A}_r \equiv (a_1 - a_0) \wedge (a_2 - a_0) \wedge \cdots \wedge (a_r - a_0)$$
$$= \overline{a}_1 \wedge \overline{a}_2 \wedge \cdots \wedge \overline{a}_r\,, \tag{1.78b}$$

$$\overline{a}_i \equiv a_i - a_0 \qquad \text{for} \qquad i = 1, \ldots, r\,. \tag{1.78c}$$

$\overline{A}_r$ is called the *tangent* of the simplex, because it is tangent for the r-plane in which the simplex lies (See Fig 1.6.). It must be nonzero for the simplex to have a convex hull. We also assume that it is non-degenerate, so we don't deal with complications of null vectors. The tangent assigns a natural *directed measure* $\overline{A}_r/r!$ to the simplex. As shown by Sobczyk [214], this is the appropriate measure for defining integration over chains of simplexes and producing a generalized Stokes Theorem. The scalar *content* (or volume) of the simplex is given by $(r!)^{-1}|\,\overline{A}_r\,|$. For the simplex in Fig 1.6 this is the area of the triangle (3-hedron) with sides $\overline{a}_1$ and $\overline{a}_2$. In general, it is the volume of an $(r+1)$-hedron. The tangent $\overline{A}_r$ assigns a definite *orientation* to the simplex, and interchanging any two of the vectors $\overline{a}_i$ in (1.78b) reverses the orientation. Also, $\overline{A}_r$ is independent of the choice of origin, though A_r is not.

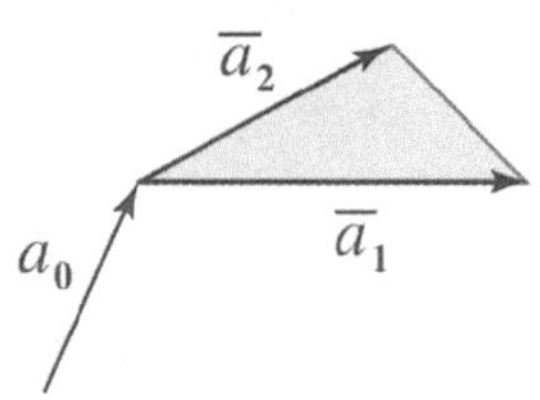

Fig. 1.6. Simplex at a_0 with tangent $\overline{A}_2 = \overline{a}_1 \wedge \overline{a}_2$

In accordance with (1.75), the equation for the plane of the simplex is

$$x \wedge \overline{A}_r = a_0 \wedge \overline{A}_r = A_r\,. \tag{1.79}$$

Thus A_r is the *moment* of the simplex. It will be convenient to use A_r as a name for the simplex, since the expanded form (1.78a) displays all the defining points of the simplex. We also assume $A_r \neq 0$, since it greatly facilitates analysis of simplex properties. However, when the simplex plane (1.79) passes through the origin, its moment $a_0 \wedge \overline{A}_r$ vanishes. There are two ways to deal with this problem. One is to treat it as a degenerate special case. A better way is to remove the origin from $\mathcal{V}^n$ by embedding $\mathcal{V}^n$ as an n-plane in a space of higher dimension. Then all points will be treated on equal footing and the moment $a_0 \wedge \overline{A}_r$ never vanishes. This is tantamount to introducing homogeneous coordinates, an approach which is developed in Chapter 2. Note that the simplex A_r is oriented, since interchanging any pair of vectors in (1.78a) will change its sign.

Since (1.78a) expresses A_r as the pseudoscalar for the simplex frame $\{a_i\}$ it determines a dual frame $\{a^i\}$ given by (1.36). The *face opposite* a_i in simplex A_r is represented by its moment

$$\mathcal{F}_i^r A_r \equiv A_r^i \equiv a^i \cdot A_r = (-1)^{i+1} a_0 \wedge \cdots \wedge \check{a}_i \wedge \cdots \wedge a_r \,. \tag{1.80}$$

This defines a *face operator* $\mathcal{F}_i$ (as illustrated in Fig 1.7). Of course, the face A_r^i is an $(r-1)$-dimensional simplex. Also

$$A_r = a_i \wedge A_r^i \,, \qquad \text{for any} \qquad 0 \le i \le r \,. \tag{1.81}$$

The *boundary* of simplex A_r can be represented formally by the multivector sum

$$\not{\partial} A_r = \sum_{i=0}^{r} A_r^i = \sum_{i=0}^{r} \mathcal{F}_i A_r \,. \tag{1.82}$$

This defines a boundary operator $\not{\partial}$. Obviously,

$$A_r = a_i \wedge \not{\partial} A_r \,, \qquad \text{for any} \qquad 0 \le i \le r \,. \tag{1.83}$$

Comparing the identity

$$(a_1 - a_0) \wedge (a_2 - a_0) \wedge \cdots \wedge (a_r - a_0) = \sum_{i=0}^{r} (-1)^i a_0 \wedge \cdots \wedge \check{a}_i \wedge \cdots \wedge a_r \tag{1.84}$$

with (1.78b) and (1.80) we see that

$$\bar{A}_r = \not{\partial} A_r \,. \tag{1.85}$$

Also, note the symmetry of (1.84) with respect to the choice of base point.

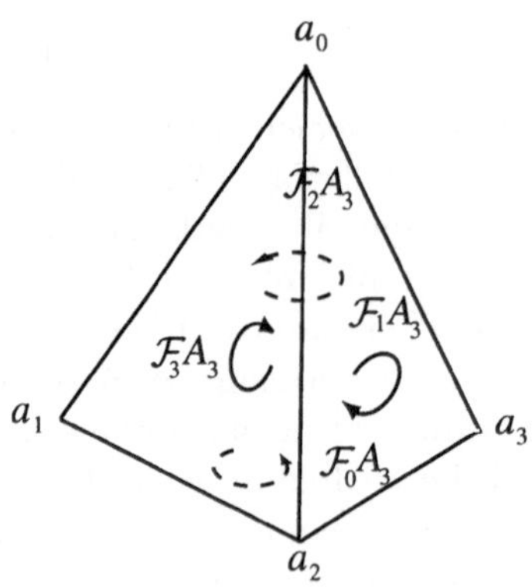

Fig. 1.7. Simplex $a_1 \wedge a_2 \wedge a_3 \wedge a_4$

For the two operators $\mathcal{F}_i$ and $\partial\!\!\!/$ we have

$$\mathcal{F}_i\mathcal{F}_i = 0\,, \tag{1.86a}$$

$$\mathcal{F}_i\mathcal{F}_j = -\mathcal{F}_j\mathcal{F}_i\,, \quad \text{for} \quad i \neq j\,, \tag{1.86b}$$

$$\mathcal{F}_i\partial\!\!\!/ = -\partial\!\!\!/\mathcal{F}_i\,, \tag{1.86c}$$

$$\partial\!\!\!/\partial\!\!\!/ = 0\,. \tag{1.86d}$$

These operator relations are strongly analogous to relations in algebraic topology.

If a point x lies in the r-plane of the simplex, we can write

$$x = \sum_{i=0}^{r} \alpha^i a_i\,, \tag{1.87}$$

where $\alpha^i = x \cdot a^i$. The point lies within the simplex if and only if

$$\sum_{i=0}^{r} \alpha^i = 1 \qquad \text{and} \qquad 0 \leq \alpha^i \leq 1\,. \tag{1.88}$$

Subject to these conditions the α^i are known as *barycentric coordinates* of the point.

From (1.80) it follows that

$$a^i \cdot A_r^i = (a^i \wedge a^i) \cdot A_r = 0 \tag{1.89a}$$

and

$$a^i A_r^i = a^i \wedge A_r^i = A_r\,. \tag{1.89b}$$

Thus, a^i is *normal* to the (moment of) the *i th face* A_r^i and is contained in A_r. In view of (1.88), a^i can be regarded as an *outward normal* to the face.

Precisely analogous relations hold for the tangent $\overline{A}_r$ and its faces. By virtue of (1.78b), the frame $\{\overline{a}_i\}$ has a dual frame $\{\overline{a}^i\}$ of *outward normals* to the faces of $\overline{A}_r$. These normals play a crucial role in an invariant formulation of Stokes Theorem that has the same form for all dimensions ([113, 214]).

1.5 Linearizing the Euclidean Group

The Euclidean group is the group of rigid displacements on $\mathbb{E}^n$. With $\mathbb{E}^n$ represented as $\mathbb{R}^n$ in $\mathbb{R}_n$, any rigid displacement can be expressed in the canonical form

$$\widehat{D}_c : x \to x' = \widehat{D}_c x = \widehat{T}_c \underline{R} x = R x R^{-1} + c\,, \tag{1.90}$$

where, in accordance with (1.67), R is a rotor determining a rotation $\underline{R}$ about the origin and $\widehat{T}_c$ is a translation by vector c. The composition of displacements is complicated by the fact that rotations are multiplicative but translations are additive. We alleviate this difficulty with a device that makes translations multiplicative as well.

We augment $\mathbb{R}^n$ with a null vector e orthogonal to $\mathbb{R}^n$. In other words, we embed $\mathbb{R}^n$ in a vector space $\mathcal{V}^{n+1} = \mathbb{R}^{n,0,1}$, so the degenerate geometric algebra $\mathbb{R}_{n,0,1}$ is at our disposal. In this algebra we can represent the translation $\widehat{T}_c$ as a spinor

$$T_c = e^{\frac{1}{2}ec} = 1 + \tfrac{1}{2}ec\,, \tag{1.91a}$$

with inverse

$$T_c^{-1} = 1 - \tfrac{1}{2}ec = T_c^*\,, \tag{1.91b}$$

where "$*$" is the main involution in $\mathbb{R}_n$, so $c^* = -c$ but $e^* = e$. If we represent the point x by

$$X = 1 + ex\,, \tag{1.92}$$

the translation $\widehat{T}_c$ can be replaced by the equivalent linear operator

$$\underline{T}_c : X \to X' = T_c X T_c^{*-1} = T_c(1 + ex)T_c = 1 + e(x + c)\,. \tag{1.93}$$

Consequently, any rigid displacement $\widehat{D}_c$ has the spinor representation

$$D_c = T_c R\,, \tag{1.94a}$$

and the linear representation

$$\underline{D}_c(X) = D_c X D_c^{*-1} \tag{1.94b}$$

in conformity with (1.67).

We call the representation of points by (1.92) the *affine model for Euclidean space*, because it supports *affine transformations*, which are composites of linear transformations (Section 1.3) with translations. It has the advantage of linearizing the Euclidean group through (1.94b). More important, it gives us the spinor representation (1.94a) for the group elements. This has the great advantage of reducing the group composition to the geometric product. For example, let us construct the spinor R_c for rotation about an arbitrary point c. We can achieve such a rotation by translating c to the origin, rotating about the origin, and then translating back. Thus

$$R_c = T_c R T_c^{-1} = R + ec \times R\,, \tag{1.95}$$

where $\times$ denotes the commutator product.

In $\mathbb{R}^3$ any rigid displacement can be expressed as a *screw displacement*, which consists of a rotation about some point composed with a translation along the rotation axis. This is known as *Chasles Theorem.* It is useful in robotics and other applications of mechanics. The theorem is easily proved with (1.94a), which shows us how to find the screw axis at the same time. Beginning with the displacement (1.94a), we note that the vector direction n for the rotation axis R satisfies

$$RnR^{-1} = n \qquad \text{or} \qquad Rn = nR\,. \tag{1.96}$$

This suggests the decomposition $c = c_{\|} + c_{\perp}$ where $c_{\|} = (c\cdot n)n$. The theorem will be proved if we can find a vector b so that

$$D_c = T_{c_{\|}} T_{c_{\perp}} R = T_{c_{\|}} R_b\,, \tag{1.97}$$

where R_b is given by (1.95). From the null component of $T_{c_{\perp}} R = R_b$ we obtain the condition

$$\tfrac{1}{2} c_{\perp} R = b \times R = \tfrac{1}{2} b(R - R^{\dagger})\,.$$

With R normalized to unity, this has the solution

$$b = c_{\perp}(1 - R^{-2})^{-1} = \tfrac{1}{2} c_{\perp} \frac{1 - R^2}{1 - \langle R^2 \rangle}\,. \tag{1.98}$$

This tells us how to find a point b on the screw axis.

Everything we have done in this section applies without change for reflections as well as rotations. Thus, for any invertible vector n, (1.94a) gives us the versor

$$n_c = n + e c \cdot n\,, \tag{1.99}$$

which represents a reflection in the hyperplane through c with normal n. Note that we have a symmetric product $c \cdot n = n \cdot c$ in (1.99) instead of the skew-symmetric product $c \times R = -R \times c$ in (1.95); this is an automatic consequence of the fact that e anticommutes with vectors n and c but commutes with R.

We have shown how the degenerate model for Euclidean space simplifies *affine geometry.* In Chapter 2 we shall see how it can be generalized to a more powerful model for Euclidean geometry.

Dual Quaternions

Clifford originated the idea of extending the real number system to include an element ϵ with the nilpotence property $\epsilon^2 = 0$. Then any number can be written in the form $\alpha + \epsilon\beta$, where α and β are ordinary numbers. He called them *dual numbers.* Clearly, this has no relation to our use of the term "dual" in geometric algebra, so we employ it only in this brief subsection to explain its connection to the present work.

A *dual quaternion* has the form $Q_1 + \epsilon Q_2$, where, Q_1 and Q_2 are ordinary quaternions. There has been renewed interest in dual quaternions recently, especially for applications to rigid motions in robotics. A direct relation to the present approach stems from the fact that the quaternions can be identified with the spinors in $\mathbb{R}_3$. To demonstrate, it will suffice to consider (1.94a). The so-called "vector part" of a quaternion actually corresponds to a bivector in $\mathbb{R}_3$. The dual of every vector c in $\mathbb{R}_3$ is a bivector $C = cI^\dagger = -cI$, where I is the unit pseudoscalar. Therefore, writing $\epsilon = eI = -Ie$, we can put (1.94a) in the equivalent "dual quaternion form"

$$R_c = R + \epsilon C \times R\,, \tag{1.100}$$

where $\epsilon^2 = 0$, and ϵ commutes with both C and R. In precisely the same way, for $\mathbb{E}^3$ the representation of points by (1.92) can be reexpressed as a dual quaternion, so we have a *dual quaternion model* of $\mathbb{E}^3$.

The drawback of quaternions is that they are limited to 3-D applications, and even there they fail to make the important distinction between vectors and bivectors. This can be remedied somewhat by resorting to complex quaternions, because they are isomorphic to the whole algebra $\mathbb{R}_3$. However, the quaternion nomenclature (dual complex or otherwise) is not well-tailored to geometric applications, so we advise against it. It should be clear that geometric algebra retains all of the advantages and none of the drawbacks of quaternions, while extending the range of applications enormously.

2. Generalized Homogeneous Coordinates for Computational Geometry*

Hongbo Li[1], David Hestenes[1], and Alyn Rockwood[2]

[1] Department of Physics and Astronomy
Arizona State University, Tempe
[2] Power Take Off Software, Inc., Colorado Springs

2.1 Introduction

The standard algebraic model for Euclidean space $\mathbb{E}^n$ is an n-dimensional real vector space $\mathbb{R}^n$ or, equivalently, a set of real coordinates. One trouble with this model is that, algebraically, the origin is a distinguished element, whereas all the points of $\mathbb{E}^n$ are identical. This deficiency in the *vector space model* was corrected early in the 19th century by removing the origin from the plane and placing it one dimension higher. Formally, that was done by introducing *homogeneous coordinates* [110]. The vector space model also lacks adequate representation for Euclidean points or lines at infinity. We solve both problems here with a new model for $\mathbb{E}^n$ employing the tools of geometric algebra. We call it the *homogeneous model* of $\mathbb{E}^n$.

Our "new model" has its origins in the work of F. A. Wachter (1792–1817), a student of Gauss. He showed that a certain type of surface in hyperbolic geometry known as a *horosphere* is metrically equivalent to Euclidean space, so it constitutes a non-Euclidean model of Euclidean geometry. Without knowledge of this model, the technique of *comformal and projective splits*

* This work has been partially supported by NSF Grant RED-9200442.

needed to incorporate it into geometric algebra were developed by Hestenes in [110]. The *conformal split* was developed to linearize the conformal group and simplify the connection to its spin representation. The *projective split* was developed to incorporate all the advantages of homogeneous coordinates in a "coordinate-free" representation of geometrical points by vectors.

Andraes Dress and Timothy Havel [65] recognized the relation of the conformal split to Wachter's model as well as to classical work on *distance geometry* by Menger [168], Blumenthal [23, 24] and Seidel [205, 204]. They also stressed connections to classical *invaraint theory*, for which the basics have been incorporated into geometric algebra in [114] and [113]. The present work synthesizes all these developments and integrates conformal and projective splits into a powerful algebraic formalism for representing and manipulating geometric concepts. We demonstrate this power in an explicit construction of the new *homogeneous model* of $\mathbb{E}^n$, the characterization of geometric objects therein, and in the proofs of geometric theorems.

The truly new thing about our model is the algebraic formalism in which it is embedded. *This integrates the representational simplicity of synthetic geometry with the computational capabilities of analytic geometry.* As in synthetic geometry we designate points by letters a, b, $\ldots$, but we also give them algebraic properties. Thus, the outer product $a \wedge b$ represents the line determined by a and b. This notion was invented by Hermann Grassmann [94] and applied to projective geometry, but it was incorporated into geometric algebra only recently [114]. To this day, however, it has not been used in Euclidean geometry, owing to a subtle defect that is corrected by our homogeneous model. We show that in our model $a \wedge b \wedge c$ represents the circle through the three points. If one of these points is a null vector e representing the point at infinity, then $a \wedge b \wedge e$ represents the straight line through a and b as a circle through infinity. This representation was not available to Grassmann, because he did not have the concept of null vector.

Our model also solves another problem that perplexed Grassmann throughout his life. He was finally forced to conclude that it is impossible to define a geometrically meaningful inner product between points. The solution eluded him because it requires the concept of indefinite metric that accompanies the concept of null vector. Our model supplies an inner product $a \cdot b$ that directly represents the Euclidean distance between the points. This is a boon to distance geometry, because it greatly facilitates computation of distances among many points. Havel [106] has used this in applications of geometric algebra to the theory of molecular conformations. The present work provides a framework for significantly advancing such applications.

We believe that our homogeneous model provides the first ideal framework for computational Euclidean geometry. The concepts and theorems of synthetic geometry can be translated into algebraic form without the unnecessary complexities of coordinates or matrices. Constructions and proofs can be done by direct computations, as needed for practical applications in computer vision and similar fields. The spin representation of conformal transformations greatly facilitates their composition and application. We aim to develop the basics and examples in sufficient detail to make applications in Euclidean geometry fairly straightforward. As a starting point, we presume familiarity with the notations and results of chapter 1

We have confined our analysis to Euclidean geometry, because it has the widest applicability. However, the algebraic and conceptual framework applies to geometrics of any signature. In particular, it applies to modeling spacetime geometry, but that is a matter for another time.

2.2 Minkowski Space with Conformal and Projective Splits

The real vector space $\mathbb{R}^{n,1}$ (or $\mathbb{R}^{1,n}$) is called a *Minkowski space*, after the man who introduced $\mathbb{R}^{3,1}$ as a model of spacetime. Its signature $(n, 1)$ $(1, n)$ is called the *Minkowski signature.* The orthogonal group of Minkowski space is called the *Lorentz group*, the standard name in relativity theory. Its elements are called *Lorentz transformations.* The special orthogonal group of Minkowski space is called the *proper Lorentz group*, though the adjective "proper" is often dropped, especially when reflections are not of interest. A good way to remove the ambiguity is to refer to rotations in Minkowski space as *proper Lorentz rotations* composing the *proper Lorentz rotation group.*

As demonstrated in many applications to relativity physics (beginning with [109]) the *"Minkowski algebra"* $\mathbb{R}_{n,1} = \mathcal{G}(\mathbb{R}^{n,1})$ is the ideal instrument for characterizing geometry of Minkowski space. In this paper we study its surprising utility for Euclidean geometry. For that purpose, the simplest Minkowski algebra $\mathbb{R}_{1,1}$ plays a special role.

The *Minkowski plane* $\mathbb{R}^{1,1}$ has an orthonormal basis $\{e_+, e_-\}$ defined by the properties

$$e_\pm^2 = \pm 1\,, \qquad e_+ \cdot e_- = 0\,. \tag{2.1}$$

A *null basis* $\{e_0, e\}$ can be introduced by

$$e_0 = \tfrac{1}{2}(e_- - e_+)\,, \tag{2.2a}$$

$$e = e_- + e_+\,. \tag{2.2b}$$

Alternatively, the null basis can be defined directly in terms of its properties

$$e_0^2 = e^2 = 0\,, \qquad e \cdot e_0 = -1\,. \tag{2.3}$$

A unit pseudoscalar E for $\mathbb{R}_{1,1}$ is defined by

$$E = e \wedge e_0 = e_+ \wedge e_- = e_+ e_-\,. \tag{2.4}$$

We note the properties

$$E^2 = 1\,, \qquad E^\dagger = -E\,, \tag{2.5a}$$
$$Ee_\pm = -e_\mp\,, \tag{2.5b}$$
$$Ee = -eE = -e\,, \qquad Ee_0 = -e_0E = e_0\,, \tag{2.5c}$$
$$1 - E = -ee_0\,, \qquad 1 + E = -e_0e\,. \tag{2.5d}$$

The basis vectors and null lines in $\mathbb{R}^{1,1}$ are illustrated in Fig. 2.1. It will be seen later that the asymmetry in our labels for the null vectors corresponds to an asymmetry in their geometric interpretation.

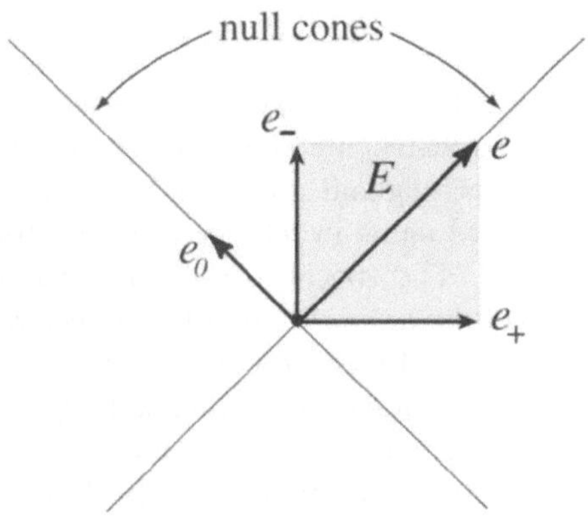

Fig. 2.1. Basis vectors null lines in the Minkowski plane. The shaded area represents the unit pseudoscalar E

The Lorentz rotation group for the Minkowski plane is represented by the rotor

$$U_\varphi = e^{\frac{1}{2}\varphi E}\,, \tag{2.6}$$

where φ is a scalar parameter defined on the entire real line, and use of the symbol e to denote the exponential function will not be confused with the null vector e. Accordingly, the Lorentz rotation $\underline{U}$ of the basis vectors is given by

$$\begin{aligned} \underline{U}_\varphi e_\pm &= U_\varphi e_\pm U_\varphi^{-1} = U_\varphi^2 e_\pm \\ &= e_\pm \cosh\varphi - e_\mp \sinh\varphi \equiv e'_\pm\,, \end{aligned} \tag{2.7}$$
$$\underline{U}_\varphi e = e^{\varphi E}e = e\,e^{-\varphi E} \equiv e'\,, \tag{2.8}$$
$$\underline{U}_\varphi e_0 = e^{\varphi E}e_0 \equiv e'_0\,. \tag{2.9}$$

The rotation is illustrated in Fig 2.2. Note that the null lines are invariant, but the null vectors are rescaled.

The complete spin group in $\mathbb{R}_{1,1}$ is

$$\mathrm{Spin}(1,1) = \{e^{\lambda E},\ E\}\,. \tag{2.10}$$

Note that E cannot be put in exponential form, so it is not continuously connected to the identity within the group. On any vector $a \in \mathbb{R}^{1,1}$ it generates the orthogonal transformation

$$\underline{E}(a) = EaE = -a = a^*\,. \tag{2.11}$$

Hence $\underline{E}$ is a discrete operator interchanging opposite branches of the null cone.

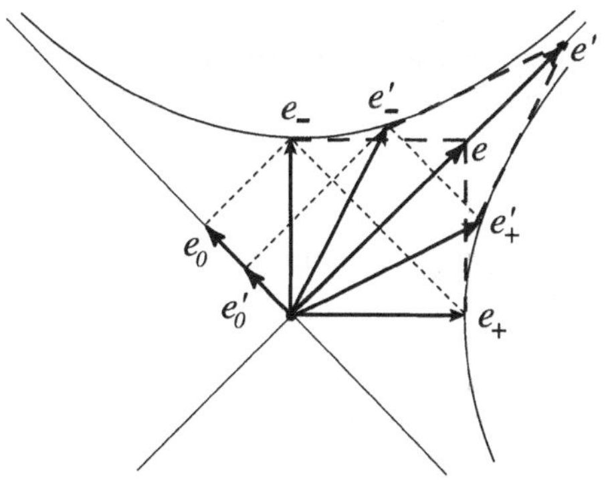

Fig. 2.2. Fig 2.2. Lorentz rotations slide unit vectors along hyperbolas in the Minkowski plane, and they rescale null vectors

It is of interest to know that the Minkowski algebra $\mathbb{R}_{1,1}$ is isomorphic to the algebra $L_2(\mathbb{R})$ of real 2×2 matrices. The general linear and special linear groups have the following isomorphisms to multiplicative subgroups in $\mathbb{R}_{1,1}$

$$\{G \in \mathbb{R}_{1,1} \mid G^*G^\dagger \neq 0\} \simeq GL_2(\mathbb{R})\,, \tag{2.12}$$

$$\{G \in \mathbb{R}_{1,1} \mid G^*G^\dagger = 1\} \simeq SL_2(\mathbb{R})\,. \tag{2.13}$$

The matrix representations are worked out in [110], but they have little practical value when geometric algebra is available. The group (2.13) is a 3-parameter group whose structure is revealed by the following canonical decomposition:

$$G = K_\alpha T_\beta U_\varphi\,, \tag{2.14}$$

where $U_\varphi = U_\varphi^*$ is defined by (2.6), and

$$K_\alpha \equiv 1 + \alpha e_0 = K_\alpha^\dagger\,, \tag{2.15a}$$

$$T_\beta \equiv 1 + \beta e = T_\beta^\dagger\,. \tag{2.15b}$$

The form (2.14) holds for all values of the scalar parameters α, β, φ in the interval $[-\infty, \infty]$. Our interest in (2.14) stems from its relation to the conformal group described later.

Throughout the rest of this paper we will be working with $\mathbb{R}^{n+1,1}$, often decomposed into the direct sum

$$\mathbb{R}^{n+1,1} = \mathbb{R}^n \oplus \mathbb{R}^{1,1} . \tag{2.16}$$

This decomposition was dubbed a *conformal split* in [110], because it relates to the conformal group on $\mathbb{R}^n$ in an essential way. It will be convenient to represent vectors or vector components in $\mathbb{R}^n$ by boldface letters and employ the null basis $\{e_0, e\}$ for $\mathbb{R}^{1,1}$. Accordingly, any vector $a \in \mathbb{R}^{n+1,1}$ admits the split

$$a = \mathbf{a} + \alpha e_0 + \beta e . \tag{2.17}$$

The conformal split is uniquely determined by the pseudoscalar E for $\mathbb{R}^{1,1}$. Let I denote the pseudoscalar for $\mathbb{R}^{n+1,1}$, then

$$\widetilde{E} = EI^{-1} = -EI^{\dagger} \tag{2.18}$$

is a unit pseudoscalar for $\mathbb{R}^n$, and we can express the split as

$$a = P_E(a) + P_E^{\perp}(a) , \tag{2.19}$$

where the projection operators P_E and $P_E^{\perp}$ are given by

$$P_E(a) = (a \cdot E)E = \alpha e_0 + \beta e \in \mathbb{R}^{1,1} , \tag{2.20a}$$

$$P_E^{\perp}(a) = (a \cdot \widetilde{E})\widetilde{E}^{\dagger} = (a \wedge E)E = \mathbf{a} \in \mathbb{R}^n . \tag{2.20b}$$

The Minkowski plane for $\mathbb{R}^{1,1}$ is referred to as the *E-plane*, since, as (2.20b) shows, it is uniquely determined by E. The projection $P_E^{\perp}$ can be regarded as a *rejection* from the E-plane.

It is worth noting that the conformal split was defined somewhat differently in [110]. There the points $\mathbf{a}$ in $\mathbb{R}^n$ were identified with trivectors $(a \wedge E)$ in (2.20b). Each of these two alternatives has its own advantages, but their representations of $\mathbb{R}^n$ are isomorphic, so the choice between them is a minor matter of convention.

The idea underlying *homogeneous coordinates* for "points" in $\mathbb{R}^n$ is to remove the troublesome origin by embedding $\mathbb{R}^n$ in a space of higher dimension. An efficient technique for doing this with geometric algebra is the *projective split* introduced in [110]. We use it here as well. Let e be a vector in the E-plane. Then for any vector $a \in \mathbb{R}^{n+1,1}$ with $a \cdot e \neq 0$, the projective split with respect to e is defined by

$$ae = a \cdot e + a \wedge e = a \cdot e\Big(1 + \frac{a \wedge e}{a \cdot e}\Big) . \tag{2.21}$$

This represents vector a with the bivector $a \wedge e/a \cdot e$. The representation is independent of scale, so it is convenient to fix the scale by the condition

$a \cdot e = e_0 \cdot e = -1$. This condition does not affect the components of a in $\mathbb{R}^n$. Accordingly, we refer to $e \wedge a = -a \wedge e$ as a *projective representation* for a. The classical approach to homogeneous coordinates corresponds to a projective split with respect to a non-null vector. We shall see that there are great advantages to a split with respect to a null vector. The result is a kind of "generalized" homogeneous coordinates.

A *hyperplane* $\mathbb{P}^{n+1}(n, a)$ with normal n and containing point a is the solution set of the equation

$$n \cdot (x - a) = 0, \qquad x \in \mathbb{R}^{n+1,1}. \tag{2.22}$$

As explained in chapter 1, this can be alternatively described by

$$\widetilde{n} \wedge (x - a) = 0, \qquad x \in \mathbb{R}^{n+1,1}. \tag{2.23}$$

where $\widetilde{n} = nI^{-1}$ is the $(n+1)$-vector dual to n.

The "normalization condition" $x \cdot e = e \cdot e_0 = -1$ for a projective split with respect to the null vector e is equivalent to the equation $e \cdot (x - e_0) = 0$; thus x lies on the hyperplane

$$\mathbb{P}^{n+1}(e, e_0) = \{x \in \mathbb{R}^{n+1,1} \mid e \cdot (x - e_0) = 0\}. \tag{2.24}$$

This fulfills the primary objective of homogeneous coordinates by displacing the origin of $\mathbb{R}^n$ by e_0. One more condition is needed to fix x as representation for a unique $\mathbf{x}$ in $\mathbb{R}^n$.

2.3 Homogeneous Model of Euclidean Space

The set $\mathbb{N}^{n+1}$ of all null vectors in $\mathbb{R}^{n+1,1}$ is called the *null cone*. We complete our definition of *generalized homogeneous coordinates* for points in $\mathbb{R}^n$ by requiring them to be null vectors, and lie in the intersection of $\mathbb{N}^{n+1}$ with the hyperplane $\mathbb{P}^{n+1}(e, e_0)$ defined by (2.24). The resulting surface

$$\mathbb{N}^n_e = \mathbb{N}^{n+1} \cap \mathbb{P}^{n+1}(e, e_0) = \{x \in \mathbb{R}^{n+1,1} \mid x^2 = 0, \quad x \cdot e = -1\} \tag{2.25}$$

is a parabola in $\mathbb{R}^{2,1}$, and its generalization to higher dimensions is called a *horosphere* in the literature on hyperbolic geometry. Applying the conditions $x^2 = 0$ and $x \cdot e = -1$ to determine the parameters in (2.17), we get

$$x = \mathbf{x} + \tfrac{1}{2}\mathbf{x}^2 e + e_0. \tag{2.26}$$

This defines a bijective mapping of $\mathbf{x} \in \mathbb{R}^n$ to $x \in \mathbb{N}^n_e$. Its inverse is the rejection (2.20b). Its projection onto the E-plane (2.20a) is shown in Fig. 2.3.

Since $\mathbb{R}^n$ is isomorphic to $\mathbb{E}^n$, so is $\mathbb{N}^n_e$, and we have proved

Theorem 2.3.1.

$$\mathbb{E}^n \simeq \mathbb{N}_e^n \simeq \mathbb{R}^n \,. \tag{2.27}$$

We call $\mathbb{N}_e^n$ the *homogeneous model* of $\mathbb{E}^n$ (or $\mathbb{R}^n$), since its elements are (generalized) *homogeneous coordinates* for points in $\mathbb{E}^n$ (or $\mathbb{R}^n$). In view of their isomorphism, it will be convenient to identify $\mathbb{N}_e^n$ with $\mathbb{E}^n$ and refer to the elements of $\mathbb{N}_e^n$ simply as (*homogeneous*) *points.* The adjective homogeneous will be employed when it is necessary to distinguish these points from points in $\mathbb{R}^n$, which we refer to as *inhomogeneous points.* Our notations x and $\mathbf{x}$ in (2.26) are intended to maintain this distinction.

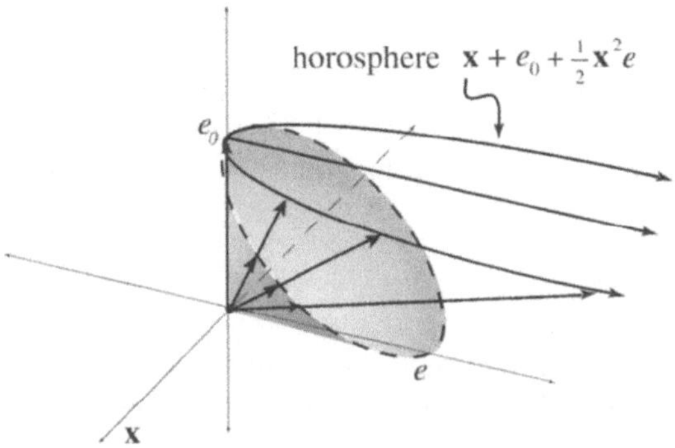

Fig. 2.3. The horosphere $\mathbb{N}_e^n$ and its projection onto the E-plane

We have framed our discussion in terms of "homogeneous coordinates" because that is a standard concept. However, geometric algebra enables us to characterize a point as a single vector without ever decomposing a vector into a set of coordinates for representational or computational purposes. It is preferable, therefore, to speak of "homogeneous points" rather than "homogeneous coordinates."

By setting $\mathbf{x} = 0$ in (2.26) we see that e_0 is the homogeneous point corresponding to the origin of $\mathbb{R}^n$. From

$$\frac{x}{-x \cdot e_0} = e + 2\Big(\frac{\mathbf{x} + e_0}{\mathbf{x}^2}\Big) \xrightarrow[\mathbf{x}^2 \to \infty]{} e\,, \tag{2.28}$$

we see that e represents the point at infinity.

As introduced in (2.21), the projective representation for the point (2.26) is

$$e \wedge x = \frac{e \wedge x}{-e \cdot x} = e\mathbf{x} + e \wedge e_0 \,. \tag{2.29}$$

Note that $e \wedge \mathbf{x} = e\mathbf{x} = -\mathbf{x}e$ since $e \cdot \mathbf{x} = 0$. By virtue of (2.5a) and (2.5c),

$$(e \wedge x)\, E = 1 + e\mathbf{x}\,. \tag{2.30}$$

This is identical to the representation for a point in the *affine model* of $\mathbb{E}^n$ introduced in chapter 1. Indeed, the homogeneous model maintains and generalizes all the good features of the affine model.

Lines, planes and simplexes

Before launching into a general treatment of geometric objects, we consider how the homogeneous model characterizes the simplest objects and relations in Euclidean geometry. Using (2.26) we expand the geometric product of two points a and b as

$$ab = \mathbf{ab} + (\mathbf{a} - \mathbf{b})\, e_0 - \tfrac{1}{2}\left[(\mathbf{a}^2 + \mathbf{b}^2) + (\mathbf{ba}^2 - \mathbf{ab}^2)\, e + (\mathbf{b}^2 - \mathbf{a}^2)E\right]\,. \tag{2.31}$$

From the bivector part we get

$$e \wedge a \wedge b = e \wedge (\mathbf{a} + e_0) \wedge (\mathbf{b} + e_0) = e\mathbf{a} \wedge \mathbf{b} + (\mathbf{b} - \mathbf{a})E\,. \tag{2.32}$$

From chapter 1, we recognize $\mathbf{a} \wedge \mathbf{b} = \mathbf{a} \wedge (\mathbf{b} - \mathbf{a})$ as the moment for a line through point $\mathbf{a}$ with tangent $\mathbf{a} - \mathbf{b}$, so $e \wedge a \wedge b$ characterizes the line completely. Accordingly, we interpret $e \wedge a \wedge b$ as a *line passing through points* a and b, or, more specifically, as a *1-simplex with endpoints* a and b.

The scalar part of (2.31) gives us

$$a \cdot b = -\tfrac{1}{2}(\mathbf{a} - \mathbf{b})^2\,. \tag{2.33}$$

Thus, the inner product of two homogeneous points gives directly the squared Euclidean distance between them. Since $a^2 = b^2 = 0$, we have

$$(a - b)^2 = -2\, a \cdot b = (\mathbf{a} - \mathbf{b})^2\,. \tag{2.34}$$

Incidentally, this shows that the embedding (2.26) of $\mathbb{R}^n$ in $\mathbb{N}_e^n$ is isometric. The *squared content* of the line segment (2.32) is given by

$$\begin{aligned}(e \wedge a \wedge b)^2 &= -(b \wedge a \wedge e) \cdot (e \wedge a \wedge b)\\ &= [(b \wedge a) \cdot e] \cdot [e \cdot (a \wedge b)]\\ &= [a - b] \cdot [a - b] = (a - b)^2\,,\end{aligned} \tag{2.35}$$

which equals the squared Euclidean length of the segment, as it should. In evaluating (2.35) we used identities from chapter 1 as well as the special properties $e^2 = 0$ and $e \cdot a = e \cdot b = -1$. Alternatively, one could use (2.32) to evaluate $(e \wedge a \wedge b)^2$ in terms of inhomogeneous points.

Again using (2.26) we find from (2.32)

$$e \wedge a \wedge b \wedge c = e\mathbf{a} \wedge \mathbf{b} \wedge \mathbf{c} + E(\mathbf{b} - \mathbf{a}) \wedge (\mathbf{c} - \mathbf{a})\,. \tag{2.36}$$

We recognize $\mathbf{a}\wedge\mathbf{b}\wedge\mathbf{c}$ as the moment of a plane with tangent $(\mathbf{b}-\mathbf{a})\wedge(\mathbf{c}-\mathbf{a})$. Therefore $e \wedge a \wedge b \wedge c$ represents a plane through points a, b, c, or, more specifically, the triangle (2-simplex) with these points as vertices. The squared content of the triangle is obtained directly from

$$(e \wedge a \wedge b \wedge c)^2 = [(\mathbf{b} - \mathbf{a}) \wedge (\mathbf{c} - \mathbf{a})]^2 , \tag{2.37}$$

the negative square of twice the area of the triangle, as anticipated.

Spheres

The equation for a sphere of radius ρ centered at point $\mathbf{p}$ in $\mathbb{R}^n$ can be written

$$(\mathbf{x} - \mathbf{p})^2 = \rho^2 . \tag{2.38}$$

Using (2.33), we can express this as an equivalent equation in terms of homogeneous points:

$$x \cdot p = -\tfrac{1}{2}\rho^2 . \tag{2.39}$$

Using $x \cdot e = -1$, we can simplify this equation to

$$x \cdot s = 0 , \tag{2.40}$$

where

$$s = p - \tfrac{1}{2}\rho^2 e = \mathbf{p} + e_0 + \frac{\mathbf{p}^2 - \rho^2}{2} e . \tag{2.41}$$

The vector s has the properties

$$s^2 = \rho^2 > 0 , \tag{2.42a}$$
$$e \cdot s = -1 . \tag{2.42b}$$

From these properties the form (2.41) and center p can be recovered. Therefore, every sphere in $\mathbb{R}^n$ is completely characterized by a unique vector s in $\mathbb{R}^{n+1,1}$. According to (2.42b), s lies in the hyperplane $\mathbb{P}^{n+1,1}(e, e_0)$, but (2.42a) says that s has positive signature, so it lies outside the null cone. Our analysis shows that every such vector determines a sphere.

Alternatively, a sphere can be described by the $(n+1)$-vector $\widetilde{s} = sI^{-1}$ dual to s. Since

$$I^\dagger = (-1)^\epsilon I = -I^{-1} , \tag{2.43}$$

where $\epsilon = \frac{1}{2}(n+2)(n+1)$, we can express the constraints (2.42a) and (2.42b) in the form

$$s^2 = -\widetilde{s}^\dagger \widetilde{s} = \rho^2 , \tag{2.44a}$$
$$s \cdot e = e \cdot (\widetilde{s} I) = (e \wedge \widetilde{s}) I = -1 . \tag{2.44b}$$

The equation (2.40) for the sphere has the dual form

$$x \wedge \widetilde{s} = 0 . \tag{2.45}$$

As seen later, the advantage of $\widetilde{s}$ is that it can be calculated directly from points on the sphere. Then s can be obtained by duality to find the center of the sphere. This duality of reprentations for a sphere is very powerful both computationally and conceptually. We do not know if it has been recognized before. In any case, we doubt that it has ever been expressed so simply.

Euclidean Plane Geometry

The advantages of the homogeneous model for $\mathbb{E}_2$ are best seen in an example:

Simson's Theorem. *Let ABC be a triangle and D be a point in the plane. Draw lines from D perpendicular to the three sides of the triangle and intersecting at points A_1, B_1, C_1. The points A_1, B_1, C_1 lie on a straight line if and only if D lies on the circle circumscribing triangle ABC.*

Analysis and proof of the theorem is facilitated by constructing *Simson's triangle* A_1, B_1, C_1 as shown in Fig. 4. Then the collinearity of points is linked to vanishing area of Simson's triangle.

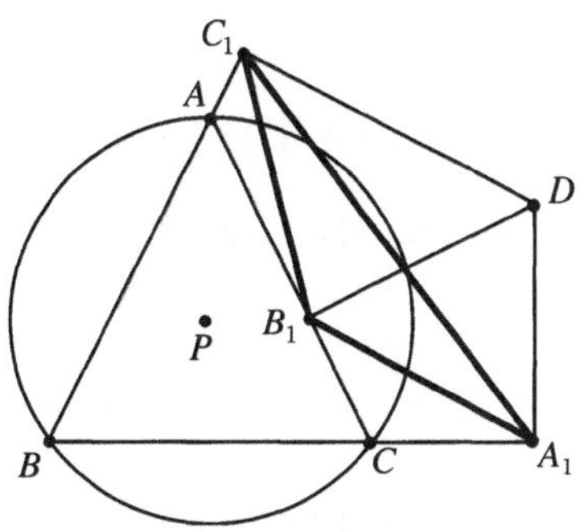

Fig. 2.4. Construction of Simson's Triangle

Suspending for the moment our convention of representing vectors by lower case letters, we interpret the labels in Fig. 2.4 as homogeneous points in $\mathbb{E}^2$. We have geometric algebra to express relations and facilitate analysis. We can speak of *triangle* $e \wedge A \wedge B \wedge C$ and its *side* $e \wedge A \wedge B$. This fuses the expressive advantages of synthetic geometry with the computational power of geometric algebra, as we now show.

Before proving Simson's theorem, we establish some basic results of general utility in Euclidean geometry. First, the relation between a triangle $e \wedge A \wedge B \wedge C$ and its circumcircle is

$$\widetilde{s} = A \wedge B \wedge C\,. \tag{2.46}$$

A general proof that this does indeed represent a circle (=sphere in $\mathbb{E}^2$) through the three points is given in the next section, so we take it for granted here. However, (2.46) is an unnormalized representation, so to calculate the circle radius ρ we modify (2.44a) and (2.44b) to

$$\rho^2 = \frac{s^2}{(s \cdot e)^2} = \frac{\widetilde{s}^\dagger \widetilde{s}}{(e \wedge \widetilde{s})^2} = \frac{(C \wedge B \wedge A) \cdot (A \wedge B \wedge C)}{(e \wedge A \wedge B \wedge C) \cdot (e \wedge A \wedge B \wedge C)}\,. \tag{2.47}$$

The right side of (2.47) is the ratio of two determinants, which, when expanded, express ρ^2 in terms of the distances between points, in other words, the lengths of the sides of the triangle. Recalling (2.34), the numerator gives

$$\begin{aligned}(A \wedge B \wedge C)^2 &= -\begin{vmatrix} 0 & A \cdot B & A \cdot C \\ B \cdot A & 0 & B \cdot C \\ C \cdot A & C \cdot B & 0 \end{vmatrix} = -2A \cdot B\, B \cdot C\, C \cdot A \\ &= \tfrac{1}{4}(A - B)^2 (B - C)^2 (C - A)^2 \\ &= \tfrac{1}{4}(\mathbf{A} - \mathbf{B})^2 (\mathbf{B} - \mathbf{C})^2 (\mathbf{C} - \mathbf{A})^2\,. \end{aligned} \tag{2.48}$$

The denominator is obtained from (2.37), which relates it to the area of the triangle and expands to

$$\begin{aligned}(e \wedge A \wedge B \wedge C)^2 &= -4(\text{area})^2 \\ &= [(\mathbf{B} - \mathbf{A}) \cdot (\mathbf{C} - \mathbf{A})]^2 - (\mathbf{B} - \mathbf{A})^2 (\mathbf{C} - \mathbf{A})^2 \\ &= [(B - A) \cdot (C - A)]^2 - 4(A \cdot B)^2 (A \cdot C)^2\,. \end{aligned} \tag{2.49}$$

By normalizing $A \wedge B \wedge C$ and taking its dual, we find the center P of the circle from (2.41); thus

$$\frac{-(A \wedge B \wedge C)^{\sim}}{(e \wedge A \wedge B \wedge C)^{\sim}} = P - \tfrac{1}{2}\rho^2\, e\,. \tag{2.50}$$

This completes our characterization of the intrinsic properties of a triangle.

To relate circle $A \wedge B \wedge C$ to a point D, we use

$$(A \wedge B \wedge C) \vee D = (A \wedge B \wedge C)^{\sim} \cdot D = -(A \wedge B \wedge C \wedge D)^{\sim}$$

with (2.50) to get

$$A \wedge B \wedge C \wedge D = \frac{\rho^2 - \delta^2}{2}\, e \wedge A \wedge B \wedge C\,, \tag{2.51}$$

where

$$\delta^2 = -2P \cdot D \tag{2.52}$$

is the squared distance between D and P. According to (2.45), the left side of (2.51) vanishes when D is on the circle, in conformity with $\delta^2 = \rho^2$ on the right side of (2.51).

To construct the Simson triangle algebraically, we need to solve the problem of finding the "perpendicular intersection" B_1 of point D on line $e \wedge A \wedge C$ (Fig. 2.4). Using inhomogeneous points we can write the condition for perpendicularity as

$$(\mathbf{B}_1 - \mathbf{D}) \cdot (\mathbf{C} - \mathbf{A}) = 0 . \tag{2.53}$$

Therefore

$$(\mathbf{B}_1 - \mathbf{D})(\mathbf{C} - \mathbf{A}) = (\mathbf{B}_1 - \mathbf{D}) \wedge (\mathbf{C} - \mathbf{A}) = (\mathbf{A} - \mathbf{D}) \wedge (\mathbf{C} - \mathbf{A}) .$$

Dividing by $(\mathbf{C} - \mathbf{A})$,

$$\begin{aligned} \mathbf{B}_1 - \mathbf{D} &= [(\mathbf{A} - \mathbf{D}) \wedge (\mathbf{C} - \mathbf{A})] \cdot (\mathbf{C} - \mathbf{A})^{-1} \\ &= \mathbf{A} - \mathbf{D} - (\mathbf{A} - \mathbf{D}) \cdot (\mathbf{C} - \mathbf{A})^{-1}(\mathbf{C} - \mathbf{A}) . \end{aligned} \tag{2.54}$$

Therefore

$$\mathbf{B}_1 = \mathbf{A} + \frac{(\mathbf{D} - \mathbf{A}) \cdot (\mathbf{C} - \mathbf{A})}{(\mathbf{C} - \mathbf{A})^2} (\mathbf{C} - \mathbf{A}) . \tag{2.55}$$

We can easily convert this to a relation among homogeneous points. However, we are only interested here in Simson's triangle $e \wedge A_1 \wedge B_1 \wedge C_1$, which by (2.36) can be represented in the form

$$\begin{aligned} e \wedge A_1 \wedge B_1 \wedge C_1 &= E(\mathbf{B}_1 - \mathbf{A}_1) \wedge (\mathbf{C}_1 - \mathbf{A}_1) \\ &= E(\mathbf{A}_1 \wedge \mathbf{B}_1 + \mathbf{B}_1 \wedge \mathbf{C}_1 + \mathbf{C}_1 \wedge \mathbf{A}_1) . \end{aligned} \tag{2.56}$$

Calculations are simplified considerably by identifying $\mathbf{D}$ with the origin in $\mathbb{R}^n$, which we can do without loss of generality. Then equation (2.52) becomes $\delta^2 = -2P \cdot D = \mathbf{p}^2$. Setting $\mathbf{D} = 0$ in (2.55) and determining the analogous expressions for $\mathbf{A}_1$ and $\mathbf{C}_1$, we insert the three points into (2.56) and find, after some calculation,

$$e \wedge A_1 \wedge B_1 \wedge C_1 = \left(\frac{\rho^2 - \delta^2}{4\rho^2} \right) e \wedge A \wedge B \wedge C . \tag{2.57}$$

The only tricky part of the calculation is getting the coefficient on the right side of (2.57) in the form shown. To do that the expanded form for ρ^2 in (2.47) to (2.49) can be used.

Finally, combining (2.57) with (2.51) we obtain the identity

$$e \wedge A_1 \wedge B_1 \wedge C_1 = \frac{A \wedge B \wedge C \wedge D}{2\rho^2} . \tag{2.58}$$

This proves Simson's theorem, for the right side vanishes if and only if D is on the circle, while the left side vanishes if and only if the three points lie on the same line.

2.4 Euclidean Spheres and Hyperspheres

A hyperplane through the origin is called a *hyperspace.* A hyperspace $\mathbb{P}^{n+1}(s)$ in $\mathbb{R}^{n+1,1}(s)$ with Minkowski signature is called a *Minkowski hyperspace.* Its normal s must have positive signature.

Theorem 2.4.1. *The intersection of any Minkowski hyperspace $\mathbb{P}^{n+1}(s)$ with the horosphere $\mathbb{N}_e^{n+1}(s) \simeq \mathbb{E}^n$ is a sphere or hyperplane*

$$\mathcal{S}(s) = \mathbb{P}^{n+1}(s) \cap \mathbb{N}_e^{n+1} \tag{2.59}$$

in $\mathbb{E}^n$ (or $\mathbb{R}^n$), and every Euclidean sphere or hyperplane can be obtained in this way. $\mathcal{S}(s)$ is a sphere if $e \cdot s < 0$ or a hyperplane if $e \cdot s = 0$.

Corollary. *Every Euclidean sphere or hyperplane can be represented by a vector s (unique up to scale) with $s^2 > 0$ and $s \cdot e \leq 0$.*

From our previous discussion we know that the sphere $\mathcal{S}(s)$ has radius ρ given by

$$\rho^2 = \frac{s^2}{(s \cdot e)^2}, \tag{2.60}$$

and it is centered at point

$$p = \frac{s}{-s \cdot e} + \tfrac{1}{2}\rho^2 e. \tag{2.61}$$

Therefore, with the normalization $s \cdot e = -1$, each sphere is represented by a unique vector. With this normalization, the set $\{\mathbf{x} = P_E^{\perp}(x) \in \mathbb{R}^n | x \cdot s > 0\}$ represents the interior of the sphere, and we refer to (2.61) as the *standard form* for the representation of a sphere by vector s.

To prove Theorem 2, it suffices to analyze the two special cases. These cases are distinguished by the identity

$$(s \cdot e)^2 = (s \wedge e)^2 \geq 0, \tag{2.62}$$

which follows from $e^2 = 0$. We have already established that $(e \cdot s)^2 > 0$ characterizes a sphere. For the case $e \cdot s = 0$, we observe that the component of s in $\mathbb{R}^n$ is given by

$$\mathbf{s} = P_E^{\perp}(s) = (s \wedge E)E = s + (s \cdot e_0)e. \tag{2.63}$$

Therefore

$$s = |\, s\,|(\mathbf{n} + e\delta), \tag{2.64}$$

where $\mathbf{n}^2 = 1$ and $\delta = -s \cdot e_0/|\,s\,|$. Set $|\,s\,| = 1$. The equation for a point x on the surface $\mathcal{S}(s)$ is then

$$x \cdot s = \mathbf{n} \cdot \mathbf{x} - \delta = 0\,. \tag{2.65}$$

This is the equation for a hyperplane in $\mathbb{R}^n$ with unit normal $\mathbf{n}$ and signed distance δ from the origin. Since $x \cdot e = 0$, the "point at infinity" e lies on $\mathcal{S}(s)$. Therefore, a *hyperplane* $\mathbb{E}^n$ *can be regarded as a sphere that "passes through" the point at infinity.*

With $|\,s\,| = 1$, we refer to (2.64) as the *standard form* for representation of a hyperplane by vector s.

Theorem 2.4.2. *Given homogeneous points $a_0, a_1, a_2, \ldots, a_n$ "in" $\mathbb{E}^n$ such that*

$$\widetilde{s} = a_0 \wedge a_1 \wedge a_2 \wedge \cdots \wedge a_n \neq 0\,, \tag{2.66}$$

then the $(n+1)$-blade $\widetilde{s}$ represents a Euclidean sphere if

$$(e \wedge \widetilde{s})^2 \neq 0\,. \tag{2.67}$$

or a hyperplane if

$$(e \wedge \widetilde{s})^2 = 0\,. \tag{2.68}$$

A point x is on the sphere/hyperplane $\mathcal{S}(s)$ if and only if

$$x \wedge \widetilde{s} = 0\,. \tag{2.69}$$

Since (2.66) is a condition for linear independence, we have the converse theorem that every $\mathcal{S}(s)$ is uniquely determined by $n+1$ linearly independent points.

By duality, Theorem 3 is an obvious consequence of Theorem 2 where $\widetilde{s}$ is dual to the normal s of the hyperspace $\mathbb{P}^{n+1}(s)$, so it is a tangent for the hyperspace.

For a hyperplane, we can always employ the point at infinity so the condition (2.66) becomes

$$\widetilde{s} = e \wedge a_1 \wedge a_2 \wedge \cdots \wedge a_n \neq 0\,. \tag{2.70}$$

Therefore only n linearly independent finite points are needed to define a hyperplane in $\mathbb{E}^n$.

2.5 Multi-dimensional Spheres, Planes, and Simplexes

We have seen that $(n+1)$-blades of Minkowski signature in $\mathbb{R}^{n+1,1}$ represent spheres and hyperplanes in $\mathbb{R}^n$, so the following generalization is fairly obvious

Theorem 2.5.1. *For $2 \le r \le n+1$, every r-blade A_r of Minkowski signature in $\mathbb{R}^{n+1,1}$ represents an $(r-2)$-dimensional sphere in $\mathbb{R}^n$ (or $\mathbb{E}^n$).*

There are three cases to consider:

Case 1. $e \wedge A_r = e_0 \wedge A_r = 0$, A_r represents an $(r-2)$-plane through the origin in $\mathbb{R}^n$ with *standard form*

$$A_r = E\mathbf{I}_{r-2}\,, \tag{2.71}$$

where $\mathbf{I}_{r-2}$ is unit tangent for the plane.

Case 2. A_r represents an $(r-2)$-plane when $e \wedge A_r = 0$ and

$$A_{r+1} = e_0 \wedge A_r \neq 0\,. \tag{2.72}$$

We can express A_r as the dual of a vector s with respect to A_{r+1}:

$$A_r = sA_{r+1} = (-1)^{\epsilon}\,\widetilde{s} \vee A_{r+1}\,. \tag{2.73}$$

In this case $e \cdot s = 0$ but $s \cdot e_0 \neq 0$, so we can write s in the standard form $s = \mathbf{n} + \delta e$ for the hyperplane $\widetilde{s}$ with unit normal $\mathbf{n}$ in $\mathbb{R}^n$ and $\mathbf{n}$-distance δ from the origin. Normalizing A_{r+1} to unity, we can put A_r into the *standard form*

$$A_r = (\mathbf{n} + e\delta)E\mathbf{I}_{r-1} = E\mathbf{n}\mathbf{I}_{r-1} + e\delta\mathbf{I}_{r-1}\,. \tag{2.74}$$

This represents an $(r-2)$-plane with unit tangent $\mathbf{n}\mathbf{I}_{r-1} = \mathbf{n} \cdot \mathbf{I}_{r-1}$ and moment $\delta\mathbf{I}_{r-1}$. Its directance from the origin is the vector $\delta\mathbf{n}$.

As a corollary to (2.74), the r-plane passing through point $\mathbf{a}$ in $\mathbb{R}^n$ with unit r-blade $\mathbf{I}_r$ as tangent has the *standard form*

$$A_{r+1} = e \wedge a \wedge \mathbf{I}_r\,, \tag{2.75}$$

where $\mathbf{a} = P_E^{\perp}(a)$ is the inhomogeneous point.

Case 3. A_r represents an $(r-2)$-dimensional sphere if

$$A_{r+1} \equiv e \wedge A_r \neq 0\,. \tag{2.76}$$

The vector

$$s = A_r A_{r+1}^{-1} \tag{2.77}$$

has positive square and $s \cdot e \neq 0$, so its dual $\widetilde{s} = sI^{-1}$ represents an $(n-1)$-dimensional sphere

$$A_r = sA_{r+1} = (\widetilde{s}I) \cdot A_{r+1} = (-1)^{\epsilon}\,\widetilde{s} \vee A_{r+1}\,, \tag{2.78}$$

where the (inessential) sign is determined by (2.43). As shown below, condition (2.76) implies that A_{r+1} represents an $(r-1)$-plane in $\mathbb{R}^n$. Therefore the meet product $\widetilde{s} \vee A_{r+1}$ in (2.78) expresses the $(r-2)$-sphere A_r as the intersection of the $(n-1)$-sphere $\widetilde{s}$ with the $(r-1)$-plane A_{r+1}.

With suitable normalization, we can write $s = c - \frac{1}{2}\rho^2 e$ where c is the center and ρ is the radius of sphere $\widetilde{s}$. Since $s \wedge A_{r+1} = e \wedge A_{r+1} = 0$, the sphere A_r is also centered at point c and has radius ρ.

Using (2.74) for the standard form of A_{r+1}, we can represent an $(r-2)$-sphere on a plane in the *standard form*

$$A_r = (c - \tfrac{1}{2}\rho^2 e) \wedge (\mathbf{n} + e\delta)E\mathbf{I}_r\,, \tag{2.79}$$

where $|\,\mathbf{I}_r\,| = 1$, $\mathbf{c} \wedge \mathbf{I}_r = \mathbf{n} \wedge \mathbf{I}_r = 0$ and $\mathbf{c} \cdot \mathbf{n} = \delta$.

In particular, we can represent an $(r-2)$-sphere in a space in the *standard form*

$$A_r = (c - \tfrac{1}{2}\rho^2 e)E\mathbf{I}_{r-1}\,, \tag{2.80}$$

where $E = e \wedge e_0$ and $\mathbf{I}_{r-1}$ is a unit $(r-1)$-blade in $\mathbb{R}_n$. In (2.80) the factor $E\mathbf{I}_{r-1}$ has been normalized to unit magnitude. Both (2.78) and (2.80) express A_r as the dual of vector s with respect to A_{r+1}. Indeed, for $r = n+1$, $\mathbf{I}_n$ is a unit pseudoscalar for $\mathbb{R}_n$, so (2.78) and (2.80) give the dual form $\widetilde{s}$ that we found for spheres in the preceding section.

This completes our classification of standard representations for spheres and planes in $\mathbb{E}^n$.

Simplexes and spheres

Now we examine geometric objects determined by linearly independent homogeneous points $a_0, a_1, \ldots, a_r$, with $r \le n$ so that $a_0 \wedge a_1 \wedge \cdots \wedge a_r \ne 0$. Introducing inhomogeneous points by (2.26), a simple computation gives the *expanded form*

$$a_0 \wedge a_1 \wedge \cdots \wedge a_r = \mathbf{A}_r + e_0\mathbf{A}_r^+ + \tfrac{1}{2}e\mathbf{A}_r^- - \tfrac{1}{2}E\mathbf{A}_r^\pm\,, \tag{2.81}$$

where, for want of a better notation,

$$\begin{aligned}
\mathbf{A}_r &= \mathbf{a}_0 \wedge \mathbf{a}_1 \wedge \cdots \wedge \mathbf{a}_r,\\
\mathbf{A}_r^+ &= \sum_{i=0}^{r}(-1)^i \mathbf{a}_0 \wedge \cdots \wedge \check{\mathbf{a}}_i \wedge \cdots \wedge \mathbf{a}_r = (\mathbf{a}_1 - \mathbf{a}_0) \wedge \cdots \wedge (\mathbf{a}_r - \mathbf{a}_0),\\
\mathbf{A}_r^- &= \sum_{i=0}^{r}(-1)^i \mathbf{a}_i^2 \mathbf{a}_0 \wedge \cdots \wedge \check{\mathbf{a}}_i \wedge \cdots \wedge \mathbf{a}_r,\\
\mathbf{A}_r^\pm &= \sum_{i=0}^{r}\sum_{j=i+1}^{r}(-1)^{i+j}(\mathbf{a}_i^2 - \mathbf{a}_j^2)\mathbf{a}_0 \wedge \cdots \wedge \check{\mathbf{a}}_i \wedge \cdots \wedge \check{\mathbf{a}}_j \wedge \cdots \wedge \mathbf{a}_r.
\end{aligned} \tag{2.82}$$

Theorem 2.5.2. *The expanded form* (2.81)

(1) *determines an r-simplex if* $\mathbf{A}_r \neq 0$,

(2) *represents an* $(r-1)$*-simplex in a plane through the origin if* $\mathbf{A}_r^+ = \mathbf{A}_r^- = 0$,

(3) *represents an* $(r-1)$*-sphere if and only if* $\mathbf{A}_r^+ \neq 0$.

We establish and analyze each of these three cases in turn.

From our study of simplexes in Chapter 1, we recognize $\mathbf{A}_r$ as the *moment* of a simplex with *boundary* (or *tangent*) $\mathbf{A}_r^+$. Therefore,

$$e \wedge a_0 \wedge a_1 \wedge \cdots \wedge a_r = e\mathbf{A}_r + E\mathbf{A}_r^+ \tag{2.83}$$

represents an r-simplex. The *volume* (or *content*) of the simplex is $k!|\,\mathbf{A}_r^+\,|$, where

$$\begin{aligned} |\,\mathbf{A}_r^+\,|^2 &= (\mathbf{A}_r^+)^\dagger \mathbf{A}_r^+ = -(a_r \wedge \cdots \wedge a_0 \wedge e)\cdot(e \wedge a_0 \wedge \cdots \wedge a_r) \\ &= -(-\tfrac{1}{2})^r \begin{vmatrix} 0 & 1 \cdots 1 \\ 1 & \\ \vdots & d_{ij}^2 \\ 1 & \end{vmatrix} \end{aligned} \tag{2.84}$$

and $d_{ij} = |\,\mathbf{a}_i - \mathbf{a}_j\,|$ is the pairwise interpoint distance. The determinant on the right side of (2.84) is called the *Cayley-Menger determinant*, because Cayley found it as an expression for volume in 1841, and nearly a century later Menger [168] used it to reformulate Euclidean geometry with the notion of interpoint distance as a primitive.

Comparison of (2.83) with (2.74) gives the directed distance from the origin in $\mathbb{R}^n$ to the plane of the simplex in terms of the points:

$$\delta\mathbf{n} = \mathbf{A}_r(\mathbf{A}_r^+)^{-1}\,. \tag{2.85}$$

Therefore, the squared distance is given by the ratio of determinants:

$$\delta^2 = \frac{|\,\mathbf{A}_r\,|^2}{|\,\mathbf{A}_r^+\,|^2} = \frac{(\mathbf{a}_r \wedge \cdots \wedge \mathbf{a}_0)\cdot(\mathbf{a}_0 \wedge \cdots \wedge \mathbf{a}_r)}{(\bar{\mathbf{a}}_r \wedge \cdots \wedge \bar{\mathbf{a}}_1)\cdot(\bar{\mathbf{a}}_1 \wedge \cdots \wedge \bar{\mathbf{a}}_r)}\,, \tag{2.86}$$

where $\bar{\mathbf{a}}_i = \mathbf{a}_i - \mathbf{a}_0$ for $i = 1, \ldots, r$, and the denominator is an alternative to (2.84).

When $\mathbf{A}_r^+ = \mathbf{A}_r^- = 0$, (2.81) reduces to

$$a_0 \wedge \cdots \wedge a_r = -\tfrac{1}{2}E\mathbf{A}_r^\pm\,. \tag{2.87}$$

Comparing with (2.83) we see that this degenerate case represents an $(r-1)$-simplex with volume $\frac{1}{2}k!|\,A_r^\pm\,|$ in an $(r-1)$-plane through the origin. To get an arbitrary $(r-1)$-simplex from $a_0 \wedge \cdots \wedge a_r$ we must place one of the points, say a_0, at ∞. Then we have $e \wedge a_1 \wedge a_2 \wedge \cdots \wedge a_r$, which has the same form as (2.83).

We get more insight into the expanded form (2.81) by comparing it with the standard forms (2.79), (2.80) for a sphere. When $\mathbf{A}_r = 0$, then $\mathbf{A}_r^+ \neq 0$ for $a_0 \wedge \cdots \wedge a_r$ to represent a sphere. Since

$$a_0 \wedge \cdots \wedge a_r = -[e_0 - \tfrac{1}{2} e \mathbf{A}_r^- (\mathbf{A}_r^+)^{-1} + \tfrac{1}{2} \mathbf{A}_r^\pm (\mathbf{A}_r^+)^{-1}] E \mathbf{A}_r^+ \,, \tag{2.88}$$

we find that the sphere is in the space represented by $E\mathbf{A}_r^+$, with center and squared radius

$$\mathbf{c} = \tfrac{1}{2} \mathbf{A}_r^\pm (\mathbf{A}_r^+)^{-1} \,, \tag{2.89a}$$

$$\rho^2 = \mathbf{c}^2 + \mathbf{A}_r^- (\mathbf{A}_r^+)^{-1} \,. \tag{2.89b}$$

When $\mathbf{A}_r \neq 0$, then $\mathbf{A}_r^+ \neq 0$ because of (2.95b) below. Since

$$a_0 \wedge \cdots \wedge a_r = \frac{(\mathbf{A}_r + e_0 \mathbf{A}_r^+ + \frac{1}{2} e \mathbf{A}_r^- - \frac{1}{2} E \mathbf{A}_r^\pm)(e\mathbf{A}_r + E\mathbf{A}_r^+)^\dagger}{(e\mathbf{A}_r + E\mathbf{A}_r^+)^\dagger (e\mathbf{A}_r + E\mathbf{A}_r^+)} (e\mathbf{A}_r + E\mathbf{A}_r^+) \,, \tag{2.90}$$

and the numerator equals

$$\mathbf{A}_r^+ (\mathbf{A}_r^+)^\dagger \left[e_0 + \frac{2\mathbf{A}_r^+ (\mathbf{A}_r)^\dagger + \mathbf{A}_r^\pm (\mathbf{A}_r^+)^\dagger}{2\mathbf{A}_r^+ (\mathbf{A}_r^+)^\dagger} + \frac{2\mathbf{A}_r (\mathbf{A}_r)^\dagger - \mathbf{A}_r^- (\mathbf{A}_r^+)^\dagger}{2\mathbf{A}_r^+ (\mathbf{A}_r^+)^\dagger} e \right] \,, \tag{2.91}$$

we find that the sphere is on the plane represented by $e\mathbf{A}_r + E\mathbf{A}_r^+$, with center and squared radius

$$\mathbf{c} = \frac{2\mathbf{A}_r^+ (\mathbf{A}_r)^\dagger + \mathbf{A}_r^\pm (\mathbf{A}_r^+)^\dagger}{2\mathbf{A}_r^+ (\mathbf{A}_r^+)^\dagger} \,, \tag{2.92a}$$

$$\rho^2 = \mathbf{c}^2 + \frac{\mathbf{A}_r^- (\mathbf{A}_r^+)^\dagger - 2\mathbf{A}_r (\mathbf{A}_r)^\dagger}{\mathbf{A}_r^+ (\mathbf{A}_r^+)^\dagger} \,. \tag{2.92b}$$

We see that (2.92a), (2.92b) congrue with (2.89a), (2.89b) when $\mathbf{A}_r = 0$.

Having shown how the expanded form (2.81) represents spheres or planes of any dimension, let us analyze relation among the $\mathbf{A}$'s. In (2.82) $\mathbf{A}_r^+$ is already represented as a blade; when $\mathbf{a}_i \neq 0$ for all i, the analogous representation for $\mathbf{A}_r^-$ is

$$\mathbf{A}_r^- = \Pi_r (\mathbf{a}_1^{-1} - \mathbf{a}_0^{-1}) \wedge (\mathbf{a}_2^{-1} - \mathbf{a}_0^{-1}) \wedge \cdots \wedge (\mathbf{a}_r^{-1} - \mathbf{a}_0^{-1}) \,, \tag{2.93}$$

where

$$\Pi_r = \mathbf{a}_0^2 \mathbf{a}_1^2 \cdots \mathbf{a}_r^2 . \tag{2.94}$$

From this we see that $\mathbf{A}_r^+$ and $\mathbf{A}_r^-$ are interchanged by inversions $\mathbf{a}_i \to \mathbf{a}_i^{-1}$, of all inhomogeneous points.

Using the notation for the boundary of a simplex from chapter 1, we have

$$\mathbf{A}_r^+ = \partial\!\!\!/ \mathbf{A}_r , \qquad \mathbf{A}_r^- / \Pi_r = \partial\!\!\!/ (\mathbf{A}_r / \Pi_r) , \tag{2.95a}$$

$$\mathbf{A}_r^\pm = -\partial\!\!\!/ \mathbf{A}_r^- , \qquad \mathbf{A}_r^\pm / \Pi_r = \partial\!\!\!/ (\mathbf{A}_r^+ / \Pi_r) . \tag{2.95b}$$

An immediate corollary is that all $\mathbf{A}$'s are blades, and if $\mathbf{A}_r^\pm = 0$ then all other $\mathbf{A}$'s are zero.

If $\mathbf{A}_r \neq 0$, then we have the following relation among the four $\mathbf{A}$'s:

$$\mathbf{A}_r^+ \vee \mathbf{A}_r^- = -\widetilde{\mathbf{A}}_r \mathbf{A}_r^\pm , \tag{2.96}$$

where the meet and dual are defined in $\mathcal{G}(\mathbf{A}_r)$. Hence when $\mathbf{A}_r \neq 0$, the vector spaces defined by $\mathbf{A}_r^+$ and $\mathbf{A}_r^-$ intersect and the intersection is the vector space defined by $\mathbf{A}_r^\pm$.

Squaring (2.81) we get

$$\begin{aligned} |\, a_0 \wedge \cdots \wedge a_r \,|^2 &= \det(a_i \cdot a_j) = (-\tfrac{1}{2})^{r+1} \det(|\, \mathbf{a}_i - \mathbf{a}_j \,|^2) \\ &= |\, \mathbf{A}_r \,|^2 - (\mathbf{A}_r^+)^\dagger \cdot \mathbf{A}_r^- - \tfrac{1}{4} |\, \mathbf{A}_r^\pm \,|^2 . \end{aligned} \tag{2.97}$$

For $r = n+1$, $\mathbf{A}_r$ vanishes and we obtain

Ptolemy's Theorem: *Let* $\mathbf{a}_0, \mathbf{a}_1, \ldots, \mathbf{a}_{n+1}$ *be points in* $\mathbb{R}^n$*, then they are on a sphere or a hypersphere if and only if* $\det(|\, \mathbf{a}_i - \mathbf{a}_j \,|^2)_{(n+2)\times(n+2)} = 0$.

2.6 Relation among Spheres and Hyperplanes

In Section 2.3 we learned that every sphere or hyperplane in $\mathbb{E}^n$ is *uniquely* represented by some vector s with $s^2 > 0$ or by its dual $\widetilde{s}$. It will be convenient, therefore, to use s or $\widetilde{s}$ as *names* for the surface they represent. We also learned that spheres and hyperplanes are distinguished, respectively, by the conditions $s \cdot e > 0$ and $s \cdot e = 0$, and the latter tells us that a hyperplane can be regarded as a sphere through the point at infinity. This intimate relation between spheres and hyperplanes makes it easy to analyze their common properties.

A main advantage of the representation by s and $\widetilde{s}$ is that it can be used directly for algebraic characterization of both qualitative and quantitative properties of surfaces without reference to generic points on the surfaces. In this section we present important examples of qualitative relations among spheres and hyperplanes that can readily be made quantitative. The simplicity of these relations and their classifications should be of genuine value in computational geometry, especially in problems of constraint satisfaction.

Intersection of spheres and hyperplanes

Let $\widetilde{s}_1$ and $\widetilde{s}_2$ be two different spheres or hyperplanes of $\mathbb{R}^n$ (or $\mathbb{E}^n$). Both $\widetilde{s}_1$ and $\widetilde{s}_2$ are tangent $(n+1)$-dimensional Minkowski subspaces of $\mathbb{R}^{n+1,1}$. These subspaces intersect in an n-dimensional subspace with n-blade tangent given algebraically by the *meet* product $\widetilde{s}_1 \vee \widetilde{s}_2$ defined in chapter 1. This illustrates how the homogeneous model of $\mathbb{E}^n$ reduces the computations of intersections of spheres and planes of any dimension to intersections of linear subspaces in $\mathbb{R}^{n+1,1}$, which are computed with the meet product.

To classify topological relations between two spheres or hyperplanes, it will be convenient to work with the dual of the meet:

$$(\widetilde{s}_1 \vee \widetilde{s}_2)^{\sim} = s_1 \wedge s_2\,. \tag{2.98}$$

There are three cases corresponding to the possible signatures of $s_1 \wedge s_2$:

Theorem 2.6.1. *Two spheres or hyperplanes $\widetilde{s}_1$, $\widetilde{s}_2$ intersect, are tangent or parallel, or do not intersect if and only if $(s_1 \wedge s_2)^2$ <, =, > 0, respectively.*

Let us examine the various cases in more detail.
When $\widetilde{s}_1$ and $\widetilde{s}_2$ are both spheres, then

- *if they intersect*, the intersection $(s_1 \wedge s_2)^{\sim}$ is a sphere, as $e \wedge (s_1 \wedge s_2)^{\sim} \neq 0$. The center and radius of the intersection are the same with those of the sphere $(P_{s_1 \wedge s_2}(e))^{\sim}$. The intersection lies on the hyperplane $(e \cdot (s_1 \wedge s_2))^{\sim}$.
- *if they are tangent*, the tangent point is proportional to the null vector $P^{\perp}_{s_1}(s_2) = (s_2 \wedge s_1)s_1^{-1}$.
- *if they do not intersect*, there are two points $\mathbf{a}$, $\mathbf{b} \in \mathbb{R}^n$, called *Poncelet points* [197], which are inversive to each other with respect to both spheres $\widetilde{s}_1$ and $\widetilde{s}_2$. The reason is, since $s_1 \wedge s_2$ is Minkowski, it contains two noncollinear null vectors $|s_1 \wedge s_2|s_1 \pm |s_1|s_2|P^{\perp}_{s_1}(s_2)$, which correspond to $\mathbf{a}, \mathbf{b} \in \mathbb{R}^n$ respectively. Let $s_i = \lambda_i a + \mu_i b$, where λ_i, μ_i are scalars. Then the inversion of a homogeneous point a with respect to the sphere $\mathbf{s}_i$ gives the point $\underline{s_i}\, a = (-\mu_i/\lambda_i)b$, as shown in the section on conformal transformations.

When $\widetilde{s}_1$ is a hyperplane and $\widetilde{s}_2$ is a sphere, then

- *if they intersect*, the intersection $(s_1 \wedge s_2)^{\sim}$ is a sphere, since $e \wedge (s_1 \wedge s_2)^{\sim} \neq 0$. The center and radius of the intersection are the same with those of the sphere $(P^{\perp}_{s_1}(s_2))^{\sim}$.
- *if they are tangent*, the tangent point corresponds to the null vector $P^{\perp}_{s_1}(s_2)$. When a sphere $\widetilde{s}$ and a point a on it is given, the tangent hyperplane of the sphere at a is $(s + s \cdot ea)^{\sim}$.
- *if they do not intersect*, there are two points $\mathbf{a}$, $\mathbf{b} \in \mathbb{R}^n$ as before, called *Poncelet points* [197], which are symmetric with respect to the hyperplane $\widetilde{s}_1$ and also inversive to each other with respect to the sphere $\widetilde{s}_2$.

When $\widetilde{s}_1$ and $\widetilde{s}_2$ are both hyperplanes, they always intersect or are parallel, as $(s_1 \wedge s_2)^\sim$ always contains e, and therefore cannot be Euclidean. For the two hyperplanes,

- *if they intersect*, the intersection $(s_1 \wedge s_2)^\sim$ is an $(n-2)$-plane. When both $\widetilde{s}_1$ and $\widetilde{s}_2$ are hyperspaces, the intersection corresponds to the $(n-2)$-space $(s_1 \wedge s_2)\mathbf{I}_n$ in $\mathbb{R}^n$, where $\mathbf{I}_r$ is a unit pseudoscalar of $\mathbb{R}^n$; otherwise the intersection is in the hyperspace $(e_0 \cdot (s_1 \wedge s_2))^\sim$ and has the same normal and distance from the origin as the hyperplane $(P_{s_1 \wedge s_2}(e_0))^\sim$.
- *if they are parallel*, the distance between them is $|e_0 \cdot P^{\perp}_{s_2}(s_1)|/|s_1|$.

Now let us examine the geometric significance of the inner product $s_1 \cdot s_2$. For spheres and hyperspaces $\widetilde{s}_1, \widetilde{s}_2$, the scalar $s_1 \cdot s_2/|s_1||s_2|$ is called the *inversive product* [120] and denoted by $s_1 * s_2$. Obviously, it is invariant under orthogonal transformations in $\mathbb{R}^{n+1,1}$, and

$$(s_1 * s_2)^2 = 1 + \frac{(s_1 \wedge s_2)^2}{s_1^2 s_2^2}. \tag{2.99}$$

Let us assume that $\widetilde{s}_1$ and $\widetilde{s}_2$ are normalized to standard form. Following [120], p. 40, 8.7, when $\widetilde{s}_1$ and $\widetilde{s}_2$ intersect, let $\mathbf{a}$ be a point of intersection, and let $\mathbf{m}_i$, $i = 1, 2$, be the respective outward unit normal vector of $\widetilde{s}_i$ at $\mathbf{a}$ if it is a sphere, or the negative of the unit normal vector in the standard form of $\widetilde{s}_i$ if it is a hyperplane; then

$$s_1 * s_2 = \mathbf{m}_1 \cdot \mathbf{m}_2. \tag{2.100}$$

The above conclusion is proved as follows: For $i = 1, 2$, when $\widetilde{s}_i$ represents a sphere with standard form $s_i = c_i - \frac{1}{2}\rho_i^2 e$ where c_i is its center, then

$$s_1 * s_2 = \frac{\rho_1^2 + \rho_2^2 - |\mathbf{c}_1 - \mathbf{c}_2|^2}{2\rho_1\rho_2}, \tag{2.101}$$

$$\mathbf{m}_1 \cdot \mathbf{m}_2 = \frac{(\mathbf{a} - \mathbf{c}_1)}{|\mathbf{a} - \mathbf{c}_1|} \cdot \frac{(\mathbf{a} - \mathbf{c}_2)}{|\mathbf{a} - \mathbf{c}_2|} = \frac{\rho_1^2 + \rho_2^2 - |\mathbf{c}_1 - \mathbf{c}_2|^2}{2\rho_1\rho_2}. \tag{2.102}$$

When s_2 is replaced by the standard form $\mathbf{n}_2 + \delta_2 e$ for a hyperplane, then

$$s_1 * s_2 = \frac{\mathbf{c}_1 \cdot \mathbf{n}_2 - \delta_2}{\rho_1}, \tag{2.103}$$

$$\mathbf{m}_1 \cdot \mathbf{m}_2 = \frac{(\mathbf{a} - \mathbf{c}_1)}{|\mathbf{a} - \mathbf{c}_1|} \cdot (-\mathbf{n}_2) = \frac{\mathbf{c}_1 \cdot \mathbf{n}_2 - \delta_2}{\rho_1}; \tag{2.104}$$

For two hyperspheres $s_i = \mathbf{n}_i + \delta_i f$; then

$$s_1 * s_2 = \mathbf{n}_1 \cdot \mathbf{n}_2, \tag{2.105}$$

$$\mathbf{m}_1 \cdot \mathbf{m}_2 = \mathbf{n}_1 \cdot \mathbf{n}_2. \tag{2.106}$$

An immediate consequence of this result is that orthogonal transformations in $\mathbb{R}^{n+1,1}$ induce angle-preserving transformations in $\mathbb{R}^n$. These are the conformal transformations discussed in the next section.

Relations among Three Points, Spheres or Hyperplanes

Let s_1, s_2, s_3 be three distinct nonzero vectors of $\mathbb{R}^{n+1,1}$ with non-negative square. Then the sign of

$$\Delta = s_1 \cdot s_2 \; s_2 \cdot s_3 \; s_3 \cdot s_1 \tag{2.107}$$

is invariant under the rescaling $s_1, s_2, s_3 \to \lambda_1 s_1, \lambda_2 s_2, \lambda_3 s_3$, where the λ's are nonzero scalars. Geometrically, when $s_i^2 > 0$, then $\widetilde{s}_i$ represents either a sphere or a hyperplane; when $s_i^2 = 0$, then s_i represents either a finite point or the point at infinity e. So the sign of Δ describes some geometric relationship among points, spheres or hyperplanes. Here we give a detailed analysis of the case when $\Delta < 0$.

When the s's are all null vectors, then $\Delta < 0$ is always true, as long as no two of them are linearly dependent.

When $s_1 = e$, s_2 is null, and $s_3^2 > 0$, then $\Delta < 0$ implies $\widetilde{s}_3$ to represent a sphere. Our previous analysis shows that $\Delta < 0$ if and only if the point s_2 is outside the sphere $\widetilde{s}_3$.

When s_1, s_2 are finite points and $s_3^2 > 0$, a simple analysis shows that $\Delta < 0$ if and only if the two points by s_1, s_2 are on the same side of the sphere or hyperplane $\widetilde{s}_3$.

When $s_1 = e$, $s_2^2, s_3^2 > 0$, then $\Delta < 0$ implies $\widetilde{s}_2$, $\widetilde{s}_3$ to represent two spheres. For two spheres with centers $\mathbf{c}_1, \mathbf{c}_2$ and radii ρ_1, ρ_2 respectively, we say they are (1) *near* if $|\mathbf{c}_1 - \mathbf{c}_2|^2 < \rho_1^2 + \rho_2^2$, (2) *far* if $|\mathbf{c}_1 - \mathbf{c}_2|^2 > \rho_1^2 + \rho_2^2$, and (3) *orthogonal* if $|\mathbf{c}_1 - \mathbf{c}_2|^2 = \rho_1^2 + \rho_2^2$. According to the first equation of (2.6), $\Delta < 0$ if and only if the two spheres $\widetilde{s}_2$ and $\widetilde{s}_3$ are far.

When s_1 is a finite point and $s_2^2, s_3^2 > 0$, then

- if $\widetilde{s}_2$ and $\widetilde{s}_3$ are hyperplanes, then $\Delta < 0$ implies that they are neither orthogonal nor identical. When the two hyperplanes are parallel, then $\Delta < 0$ if and only if the point s_1 is between the two hyperplanes. When the hyperplanes intersect, then $\Delta < 0$ if and only if s_1 is in the wedge domain of the acute angle in $\mathbb{R}^n$ formed by the two intersecting hyperplanes.
- if $\widetilde{s}_2$ is a hyperplane and $\widetilde{s}_3$ is a sphere, then $\Delta < 0$ implies that they are non-orthogonal, i.e., the center of the sphere does not lie on the hyperplane. If the center of a sphere is on one side of a hyperplane, we also say that the sphere is on that side of the hyperplane. If the point s_1 is outside the sphere $\widetilde{s}_3$, then $\Delta < 0$ if and only if s_1 and the sphere $\widetilde{s}_3$ are on the same side of the hyperplane $\widetilde{s}_2$; if the point is inside the sphere $\widetilde{s}_3$, then $\Delta < 0$ if and only if the point and the sphere are on opposite sides of the hyperplane.
- if $\widetilde{s}_2, \widetilde{s}_3$ are spheres, then $\Delta < 0$ implies that they are non-orthogonal. If they are far, then $\Delta < 0$ if and only if the point s_1 is either inside both of

them or outside both of them. If they are near, then $\Delta < 0$ if and only if s_1 is inside one sphere and outside the other.

When s_1, s_2, s_3 are all of positive square, then $\Delta < 0$ implies that no two of them are orthogonal or identical.

- If they are all hyperplanes, with normals $\mathbf{n}_1$, $\mathbf{n}_2$, $\mathbf{n}_3$ respectively, then $\Delta < 0$ implies that no two of them are parallel, as the sign of Δ equals that of $\mathbf{n}_1 \cdot \mathbf{n}_2\ \mathbf{n}_2 \cdot \mathbf{n}_3\ \mathbf{n}_3 \cdot \mathbf{n}_1$. $\Delta < 0$ if and only if a normal vector of $\widetilde{s}_1$ with its base point at the intersection of the two hyperplanes $\widetilde{s}_2$ and $\widetilde{s}_3$, has its end point in the wedge domain of the acute angle in $\mathbb{R}^n$ formed by the two intersecting hyperplanes.
- If $\widetilde{s}_1$, $\widetilde{s}_2$ are hyperplanes and $\widetilde{s}_3$ is a sphere, then when the hyperplanes are parallel, $\Delta < 0$ if and only if the sphere's center is between the two hyperplanes. When the hyperplanes intersect, $\Delta < 0$ if and only if the sphere's center is in the wedge domain of the acute angle in $\mathbb{R}^n$ formed by the two intersecting hyperplanes.
- If $\widetilde{s}_1$ is a hyperplane and $\widetilde{s}_2, \widetilde{s}_3$ are spheres, then when the two spheres are far, $\Delta < 0$ if and only if the spheres are on the same side of the hyperplane. When the spheres are near, $\Delta < 0$ if and only if they are on opposite sides of the hyperplane.
- If all are spheres, then either they are all far from each other, or two spheres are far and the third is near to both of them.

Bunches of Spheres and Hyperplanes

In previous sections, we proved that Minkowski subspaces of $\mathbb{R}^{n+1,1}$ represent spheres and planes of various dimensions in $\mathbb{R}^n$. In this subsection we consider subspaces of $\mathbb{R}^{n+1,1}$ containing only their normals, which are vectors of positive square. Such subspaces are dual to Minkowski hyperspaces that represent spheres or hyperplanes. Therefore the tangent blade for a subspace A_r of $\mathbb{R}^{n+1,1}$ can be used to represent a set of spheres and hyperplanes, where each of them is represented by a vector of positive square. Or dually, the dual of A_r represents the intersection of a set of spheres and hyperplanes.

The simplest example is a pencil of spheres and hyperplanes. Let $\widetilde{s}_1, \widetilde{s}_2$ be two different spheres or hyperplanes, then the *pencil* of spheres/hyperplanes determined by them is the set of spheres/hyperplanes $(\lambda_1 s_1 + \lambda_2 s_2)^\sim$, where λ_1, λ_2 are scalars satisfying

$$(\lambda_1 s_1 + \lambda_2 s_2)^2 > 0. \tag{2.108}$$

The entire pencil is represented by the blade $A_2 = s_1 \wedge s_2$ or its dual $(s_1 \wedge s_2)^\sim$. There are three kinds of pencils corresponding to the three possible signatures of the blade $s_1 \wedge s_2$:

1. *Euclidean*, $(s_1 \wedge s_2)^2 < 0$. The space $(s_1 \wedge s_2)^\sim$, which is a subspace of any of the spaces $(\lambda_1 s_1 + \lambda_2 s_2)^\sim$, is Minkowski, and represents an

$(n-2)$-dimensional sphere or plane in $\mathbb{R}^n$. If the point at infinity e is in the space, then the pencil $(s_1 \wedge s_2)^\sim$ is composed of hyperplanes passing through an $(n-2)$-dimensional plane. We call it a *concurrent pencil.* If e is not in the space $(s_1 \wedge s_2)^\sim$, there is an $(n-2)$-dimensional sphere that is contained in every sphere or hyperplane in the pencil $(s_1 \wedge s_2)^\sim$. We call it an *intersecting pencil.*

2. *Degenerate,* $(s_1 \wedge s_2)^2 = 0$. The space $(s_1 \wedge s_2)^\sim$ contains a one-dimensional null subspace, spanned by $P_{s_1}^\perp(s_2)$. If e is in the space, then the pencil is composed of hyperplanes parallel to each other. We call it a *parallel pencil.* If e is not in the space $(s_1 \wedge s_2)^\sim$, the pencil is composed of spheres tangent to each other at the point in $\mathbb{R}^n$ represented by the null vector $P_{s_1}^\perp(s_2)$. We call it a *tangent pencil.*
3. *Minkowski,* $(s_1 \wedge s_2)^2 > 0$. The Minkowski plane $s_1 \wedge s_2$ contains two non-collinear null vectors $|s_1 \wedge s_2| s_1 \pm |s_1||s_2| P_{s_1}^\perp(s_2)$. The two one-dimensional null spaces spanned by them are conjugate with respect to any of the vectors $\lambda_1 s_1 + \lambda_2 s_2$, which means that the two points represented by the two null vectors are inversive with respect to any sphere or hyperplane in the pencil $(s_1 \wedge s_2)^\sim$. If e is in the space $s_1 \wedge s_2$, then the pencil is composed of spheres centered at the point represented by the other null vector in the space. We call it a *concentric pencil.* If e is outside the space $s_1 \wedge s_2$, the two points represented by the two null vectors in the space are called *Poncelet points.* The pencil now is composed of spheres and hyperplanes with respect to which the two points are inversive. We call it a *Poncelet pencil.*

This finishes our classification of pencils. From the above analysis we also obtain the following corollary:

- The concurrent (or intersecting) pencil passing through an $(n-2)$-dimensional plane (or sphere) represented by Minkowski subspace A_n is $\widetilde{A}_n$.
- The parallel pencil containing a hyperplane $\widetilde{s}$ is $(e \wedge s)^\sim$. In particular, the parallel pencil normal to vector $\mathbf{n} \in \mathbb{R}^n$ is $(e \wedge \mathbf{n})^\sim$.
- The tangent pencil containing a sphere or hyperplane $\widetilde{s}$ and having tangent point $\mathbf{a} = P_E^\perp(a) \in \mathbb{R}^n$ is $(a \wedge s)^\sim$. In particular, the tangent pencil containing a hyperplane normal to $\mathbf{n} \in \mathbb{R}^n$ and having tangent point $\mathbf{a}$ is $(a \wedge (\mathbf{n} + \mathbf{a} \cdot \mathbf{n} e))^\sim$.
- The concentric pencil centered at $\mathbf{a} = P_E^\perp(a) \in \mathbb{R}^n$ is $(e \wedge a)^\sim$.
- The Poncelet pencil with Poncelet points $\mathbf{a}, \mathbf{b} \in \mathbb{R}^n$ is $(a \wedge b)^\sim$.

The generalization of a pencil is a bunch. A *bunch* of spheres and hyperplanes determined by r spheres and hyperplanes $\widetilde{s}_1, \ldots, \widetilde{s_r}$ is the set of spheres and hyperplanes $(\lambda_1 s_1 + \cdots + \lambda_r s_r)^\sim$, where the λ's are scalars and satisfy

$$(\lambda_1 s_1 + \cdots + \lambda_r s_r)^2 > 0. \tag{2.109}$$

When $s_1 \wedge \cdots \wedge s_r \neq 0$, the integer $r-1$ is called the *dimension* of the bunch, and the bunch is represented by $(s_1 \wedge \cdots \wedge s_r)^{\sim}$. A pencil is a one-dimensional bunch. The dimension of a bunch ranges from 1 to $n-1$.

The classification of bunches is similar to that of pencils. Let $(s_1 \wedge \cdots \wedge s_r)^{\sim}$, $2 \leq r \leq n$, be a bunch. Then the signature of the space $(s_1 \wedge \cdots \wedge s_r)^{\sim}$ has three possibilities:

1. *Minkowski.* The space $(s_1 \wedge \cdots \wedge s_r)^{\sim}$ corresponds to an $(n-r)$-dimensional sphere or plane of $\mathbb{R}^n$, and is contained in any of the spheres and hyperplanes $(\lambda_1 s_1 + \cdots + \lambda_r s_r)^{\sim}$. If e is in the space, then the bunch is composed of hyperplanes passing through an $(n-r)$-dimensional plane. We call it a *concurrent bunch.* If e is not in the space, there is an $(n-r)$-dimensional sphere that are on any sphere or hyperplane in the bunch. We call it an *intersecting bunch.*
2. *Degenerate.* The space $(s_1 \wedge \cdots \wedge s_r)^{\sim}$ contains a one-dimensional null subspace, spanned by the vector $(s_1 \wedge \cdots \wedge s_r) \cdot (s_1 \wedge \cdots \wedge \check{s}_i \wedge \cdots \wedge s_r)$, where the omitted vector s_i is chosen so that $(s_1 \wedge \cdots \wedge \check{s}_i \wedge \cdots \wedge s_r)^2 \neq 0$. If e is in the space $(s_1 \wedge \cdots \wedge s_r)^{\sim}$, then the bunch is composed of hyperplanes normal to an $(r-1)$-space of $\mathbb{R}^n$ represented by the blade $e_0 \cdot (s_1 \wedge \cdots \wedge s_r)$. We call it a *parallel bunch.* If e is not in the space, the bunch is composed of spheres and hyperplanes passing through a point $\mathbf{a}_i \in \mathbb{R}^n$ represented by the null vector of the space, at the same time orthogonal to the $(r-1)$-plane of $\mathbb{R}^n$ represented by $e \wedge a \wedge (e \cdot (s_1 \wedge \cdots \wedge s_r))$. We call it a *tangent bunch.*
3. *Euclidean.* The Minkowski space $s_1 \wedge \cdots \wedge s_r$ corresponds to an $(r-2)$-dimensional sphere or plane. It is orthogonal to all of the spheres and hyperplanes $(\lambda_1 s_1 + \cdots + \lambda_r s_r)^{\sim}$. If e is in the space $s_1 \wedge \cdots \wedge s_r$, then the pencil is composed of hyperplanes perpendicular to the $(r-2)$-plane represented by $s_1 \wedge \cdots \wedge s_r$, together with spheres whose centers are in the $(r-2)$-plane. We call it a *concentric bunch.* If e is outside the space, the $(r-2)$-sphere represented by $s_1 \wedge \cdots \wedge s_r$ is called a *Poncelet sphere.* Now the pencil is composed of spheres and hyperplanes orthogonal to the Poncelet sphere, called a *Poncelet bunch.*

Finally we discuss duality between two bunches. Let A_r, $2 \leq r \leq n$, be a blade. Then it represents an $(n-r+1)$-dimensional bunch. Its dual, $\widetilde{A}_r$, represents an $(r-1)$-dimensional bunch. Any bunch and its dual bunch are orthogonal, i.e., any sphere or hyperplane in a bunch A_r is orthogonal to a sphere or hyperplane in the bunch $\widetilde{A}_r$. Table 2.1 provides details of the duality.

2.7 Conformal Transformations

A transformation of geometric figures is said to be *conformal* if it *preserves shape*; more specifically, it preserves angles and hence the shape of straight

Table 2.1. Bunch dualities

Geometric conditions	Bunch A_r	Bunch $\widetilde{A}_r$
$A_r \cdot A_r^\dagger < 0\,,$ $e \wedge A_r = 0$	Concurrent bunch, concurring at the $(r-2)$-plane A_r	Concentric bunch, centered at the $(r-2)$-plane A_r
$A_r \cdot A_r^\dagger < 0\,,$ $e \wedge A_r \neq 0$	Intersecting bunch, at the $(r-2)$-sphere A_r	Poncelet bunch, with Poncelet sphere A_r
$A_r \cdot A_r^\dagger = 0\,,$ $e \wedge A_r = 0$	Parallel bunch, normal to the $(n-r+1)$-space $(e_0 \cdot A_r)^\sim$	Parallel bunch, normal to the $(r-1)$-space $e \wedge e_0 \wedge (e_0 \cdot A_r)$
$A_r \cdot A_r^\dagger = 0$, $e \wedge A_r \neq 0$, assuming a is a null vector in the space A_r	Tangent bunch, at point a and orthogonal to the $(n-r+1)$-plane $(e \cdot A_r)^\sim$	Tangent bunch, at point $\mathbf{a}$ and orthogonal to the $(r-1)$-plane $e \wedge a \wedge (e \cdot A_r)$
$A_r \cdot A_r^\dagger > 0\,,$ $e \wedge A_r = 0$	Concentric bunch, centering at the $(n-r)$-plane $\widetilde{A}_r$	Concurrent bunch, concurring at the $(n-r)$-plane $\widetilde{A}_r$
$A_r \cdot A_r^\dagger > 0\,,$ $e \wedge A_r \neq 0$	Poncelet bunch, with Poncelet sphere $\widetilde{A}_r$	Intersecting bunch, at the $(n-r)$-sphere $\widetilde{A}_r$

lines and circles. As first proved by Liouville [151] for $\mathbb{R}^3$, any conformal transformation on the whole of $\mathbb{R}^n$ can be expressed as a composite of *inversions* in spheres and *reflections* in hyperplanes. Here we show how the homogeneous model of $\mathbb{E}^n$ simplifies the formulation of this fact and thereby facilitates computations with conformal transformations. Simplification stems from the fact that the conformal group on $\mathbb{R}^n$ is isomorphic to the Lorentz group on $\mathbb{R}^{n+1}$. Hence nonlinear conformal transformations on $\mathbb{R}^n$ can be linearized by representing them as Lorentz transformation and thereby further simplified as versor representations. The present treatment follows, with some improvements, [110], where more details can be found.

From chapter 1, we know that any Lorentz transformation $\underline{G}$ of a generic point $x \in \mathbb{R}^{n+1}$ can be expressed in the form

$$\underline{G}(x) = Gx(G^*)^{-1} = \sigma x'\,, \tag{2.110}$$

where G is a versor and σ is a scalar. We are only interested in the action of $\underline{G}$ on homogeneous points of $\mathbb{E}^n$. Since the null cone is invariant under $\underline{G}$, we have $(x')^2 = x^2 = 0$. However, for fixed e, $x \cdot e$ is not Lorentz invariant, so a scale factor σ has been introduced to ensure that $x' \cdot e = x \cdot e = -1$ and

Table 2.2. Conformal transformations and their versor representations (see text for explanation)

Type	$g(\mathbf{x})$ on $\mathbb{R}^n$	Versor in $\mathbb{R}_{n+1,1}$	$\sigma(\mathbf{x})$
Reflection	$-\mathbf{n}\mathbf{x}\mathbf{n} + 2\mathbf{n}\delta$	$s = \mathbf{n} + e\delta$	1
Inversion	$\dfrac{\rho^2}{\mathbf{x}-\mathbf{c}} + \mathbf{c}$	$s = c - \frac{1}{2}\rho^2 e$	$\left(\dfrac{\mathbf{x}-\mathbf{c}}{\rho}\right)^2$
Rotation	$R(\mathbf{x}-\mathbf{c})R^{-1} + \mathbf{c}$	$R_{\mathbf{c}} = R + e(\mathbf{c}\times R)$	1
Translation	$\mathbf{x} - \mathbf{a}$	$T_{\mathbf{a}} = 1 + \frac{1}{2}\mathbf{a}e$	1
Transversion	$\dfrac{\mathbf{x} - \mathbf{x}^2\mathbf{a}}{\sigma(\mathbf{x})}$	$K_{\mathbf{a}} = 1 + \mathbf{a}e_0$	$1 - 2\mathbf{a}\cdot\mathbf{x} + \mathbf{x}^2\mathbf{a}^2$
Dilation	$\lambda\mathbf{x}$	$D_\lambda = e^{-\frac{1}{2}E\ln\lambda}$	λ^{-1}
Involution	$\mathbf{x}^* = -\mathbf{x}$	$E = e \wedge e_0$	-1

x' remains a point in $\mathbb{E}^n$. Expressing the right equality in (2.110) in terms of homogeneous points we have the expanded form

$$G[\mathbf{x} + \tfrac{1}{2}\mathbf{x}^2 e + e_0](G^*)^{-1} = \sigma[\mathbf{x}' + \tfrac{1}{2}(\mathbf{x}')^2 e + e_0]\,, \tag{2.111}$$

where

$$\mathbf{x}' = g(\mathbf{x}) \tag{2.112}$$

is a conformal transformation on $\mathbb{R}^n$ and

$$\sigma = -e\cdot(\underline{G}x) = -\langle\, eG^*xG^{-1}\,\rangle\,. \tag{2.113}$$

We study the simplest cases first.

For reflection by a vector $s = -s^*$ (2.110) becomes

$$\underline{s}(x) = -sxs^{-1} = x - 2(s\cdot x)s^{-1} = \sigma x'\,, \tag{2.114}$$

where $sx + xs = 2s\cdot x$ has been used. Both inversions and reflections have this form as we now see by detailed examination.

Inversions. We have seen that a circle of radius ρ centered at point $c = \mathbf{c} + \frac{1}{2}\mathbf{c}^2 e + e_0$ is represented by the vector

$$s = c - \tfrac{1}{2}\rho^2 e\,. \tag{2.115}$$

We first examine the important special case of the unit sphere centered at the origin in $\mathbb{R}^n$. Then s reduces to $e_0 - \frac{1}{2}e$, so $-2s \cdot x = \mathbf{x}^2 - 1$ and (2.114) gives

$$\sigma x' = (\mathbf{x} + \tfrac{1}{2}\mathbf{x}^2 e + e_0) + (\mathbf{x}^2 - 1)(e_0 - \tfrac{1}{2}e) = \mathbf{x}^2[\mathbf{x}^{-1} + \tfrac{1}{2}\mathbf{x}^{-2}e + e_0]\,. \tag{2.116}$$

Whence the inversion

$$g(\mathbf{x}) = \mathbf{x}^{-1} = \frac{1}{\mathbf{x}} = \frac{\mathbf{x}}{|\mathbf{x}|^2}\,. \tag{2.117}$$

Note how the coefficient of e_0 has been factored out on the right side of (2.116) to get $\sigma = \mathbf{x}^2$. This is usually the best way to get the rescaling factor, rather than separate calculation from (2.113). Actually, we seldom care about σ, but it must be factored out to get the proper normalization of $g(\mathbf{x})$.

Turning now to inversion with respect to an arbitrary circle, from (2.115) we get

$$s \cdot x = c \cdot x - \tfrac{1}{2}\rho^2 e \cdot x = -\tfrac{1}{2}[(\mathbf{x} - \mathbf{c})^2 - \rho^2]\,. \tag{2.118}$$

Insertion into (2.114) and a little algebra yields

$$\sigma x' = \left(\frac{\mathbf{x} - \mathbf{c}}{\rho}\right)^2 [g(\mathbf{x}) + \tfrac{1}{2}[g(\mathbf{x})]^2 e + e_0]\,, \tag{2.119}$$

where

$$g(\mathbf{x}) = \frac{\rho^2}{\mathbf{x} - \mathbf{c}} + \mathbf{c} = \frac{\rho^2}{(\mathbf{x} - \mathbf{c})^2}(\mathbf{x} - \mathbf{c}) + \mathbf{c} \tag{2.120}$$

is the inversion in $\mathbb{R}^n$.

Reflections. We have seen that a hyperplane with unit normal $\mathbf{n}$ and signed distance δ from the origin in $\mathbb{R}^n$ is represented by the vector

$$s = \mathbf{n} + e\delta\,. \tag{2.121}$$

Inserting $s \cdot x = \mathbf{n} \cdot \mathbf{x} - \delta$ into (2.114) we easily find

$$g(\mathbf{x}) = \mathbf{nxn}^* + 2\mathbf{n}\delta = \mathbf{n}(\mathbf{x} - \mathbf{n}\delta)\mathbf{n}^* + \mathbf{n}\delta\,. \tag{2.122}$$

We recognize this as equivalent to a reflection $\mathbf{nxn}^*$ at the origin translated by δ along the direction of $\mathbf{n}$. A point $\mathbf{c}$ is on the hyperplane when $\delta = \mathbf{n} \cdot \mathbf{c}$, in which case (2.121) can be written

$$s = \mathbf{n} + e\mathbf{n} \cdot \mathbf{c}\,. \tag{2.123}$$

Via (2.122), this vector represents reflection in a hyperplane through point $\mathbf{c}$.

With a minor exception to be explained, all the basic conformal transformations in Table 2.2 can be generated from inversions and reflections. Let us see how.

Translations. We have already seen in chapter 1 that versor $T_{\mathbf{a}}$ in Table 2.2 represents a translation. Now notice

$$(\mathbf{n} + e\delta)\mathbf{n} = 1 + \tfrac{1}{2}\mathbf{a}e = T_{\mathbf{a}} \tag{2.124}$$

where $\mathbf{a} = 2\delta\mathbf{n}$. This tells us that the composite of reflections in two parallel hyperplanes is a translation through twice the distance between them.

Transversions. The transversion $K_{\mathbf{a}}$ in Table 2.2 can be generated from two inversions and a translation; thus, using $e_0 e e_0 = -2e_0$ from (2.5c) and (2.5d), we find

$$e_+ T_{\mathbf{a}} e_+ = (\tfrac{1}{2}e - e_0)(1 + \tfrac{1}{2}\mathbf{a}e)(\tfrac{1}{2}e - e_0) = 1 + \mathbf{a}e_0 = K_{\mathbf{a}}\,. \tag{2.125}$$

The transversion generated by $K_{\mathbf{a}}$ can be put in the various forms

$$g(\mathbf{x}) = \frac{\mathbf{x} - \mathbf{x}^2\mathbf{a}}{1 - 2\mathbf{a}\cdot\mathbf{x} + \mathbf{x}^2\mathbf{a}^2} = \mathbf{x}(1 - \mathbf{a}\mathbf{x})^{-1} = (\mathbf{x}^{-1} - \mathbf{a})^{-1}\,. \tag{2.126}$$

The last form can be written down directly as an inversion followed by a translation and another inversion as asserted by (2.125). That avoids a fairly messy computation from (2.111).

Rotations. Using (2.123), the composition of reflections in two hyperplanes through a common point $\mathbf{c}$ is given by

$$(\mathbf{a} + e\mathbf{a}\cdot\mathbf{c})(\mathbf{b} + e\mathbf{b}\cdot\mathbf{c}) = \mathbf{a}\mathbf{b} + e\mathbf{c}\cdot(\mathbf{a}\wedge\mathbf{b})\,, \tag{2.127}$$

where $\mathbf{a}$ and $\mathbf{b}$ are unit normals. Writing $R = \mathbf{a}\mathbf{b}$ and noting that $\mathbf{c}\cdot(\mathbf{a}\wedge\mathbf{b}) = \mathbf{c}\times R$, we see that (2.127) is equivalent to the form for the rotation versor in Table 2.2 that we found in chapter 1. Thus we have established that the product of two reflections at any point is equivalent to a rotation about that point.

Dilations. Now we prove that the composite of two inversions centered at the origin is a dilation (or dilatation). Using (2.5d) we have

$$(e_0 - \tfrac{1}{2}e)(e_0 - \tfrac{1}{2}\rho^2 e) = \tfrac{1}{2}(1 - E) + \tfrac{1}{2}(1 + E)\rho^2\,. \tag{2.128}$$

Normalizing to unity and comparing to (2.6) with $\rho = e^{\varphi}$, we get

$$D_\rho = \tfrac{1}{2}(1 + E)\rho + \tfrac{1}{2}(1 - E)\rho^{-1} = e^{E\varphi}\,, \tag{2.129}$$

where D_ρ is the square of the versor form for a dilation in table 2.2. To verify that D_ρ does indeed generate a dilation, we note from (2.8) that

$$\underline{D}_\rho(e) = D_\rho e D_\rho^{-1} = D_\rho^2 e = \rho^{-2} e\,.$$

Similarly

$$\underline{D}_\rho(e_0) = \rho^2 e_0 .$$

Therefore,

$$D_\rho(\mathbf{x} + \tfrac{1}{2}\mathbf{x}^2 e + e_0)D_\rho^{-1} = \rho^2[\rho^{-2}\mathbf{x} + \tfrac{1}{2}(\rho^{-2}\mathbf{x})^2 e + e_0] . \tag{2.130}$$

Thus $g(\mathbf{x}) = \rho^{-2}\mathbf{x}$ is a dilation as advertised.

We have seen that every vector with positive signature in $\mathbb{R}^{n+1,1}$ represents a sphere or hyperplane as well as an inversion or reflection in same. They compose a multiplicative group which we identify as the versor representation of the full *conformal group* $\mathrm{C}(n)$ of $\mathbb{E}^n$. Subject to a minor proviso explained below, our construction shows that this conformal group is equivalent to the *Lorentz group* of $\mathbb{R}^{n+1,1}$. Products of an even number of these vectors constitute a subgroup, the spin group $\mathrm{Spin}^+(n+1,1)$. It is known as the spin representation of the *proper Lorentz group*, the orthogonal group $\mathrm{O}^+(n+1,1)$. This, in turn, is equivalent to the *special orthogonal group* $\mathrm{SC}^+(n+1,1)$.

Our constructions above show that translations, transversions, rotations, dilations belong to $\mathrm{SC}^+(n)$. Moreover, every element of $\mathrm{SC}^+(n)$ can be generated from these types. This is easily proved by examining our construction of their spin representations $T_\mathbf{a}$, $K_\mathbf{b}$, $R_\mathbf{c}$, D_λ from products of vectors. One only needs to show that every other product of two vectors is equivalent to some product of these. Not hard! Comparing the structure of $T_\mathbf{a}$, $K_\mathbf{b}$, $R_\mathbf{c}$, D_λ exhibited in Table 2.2 with equations (2.6) through (2.15b), we see how it reflects the structure of the Minkowski plane $\mathbb{R}_{1,1}$ and groups derived therefrom.

Our construction of $\mathrm{Spin}^+(n+1,n)$ from products of vectors with positive signature excludes the bivector $E = e_+e_-$ because $e_-^2 = -1$. Extending $\mathrm{Spin}^+(n+1,n)$ by including E we get the full spin group $\mathrm{Spin}(n+1,n)$. Unlike the elements of $\mathrm{Spin}^+(n+1,n)$, E is not parametrically connected to the identity, so its inclusion gives us a double covering of $\mathrm{Spin}^+(n+1,n)$. Since E is a versor, we can ascertain its geometric significance from (2.111); thus, using (2.5c) and (2.5a), we easily find

$$E(\mathbf{x} + \tfrac{1}{2}\mathbf{x}^2 e + e_0)E = -[\,-\mathbf{x} + \tfrac{1}{2}\mathbf{x}^2 e + e_0] . \tag{2.131}$$

This tells us that E represents the *main involution* $\mathbf{x}^* = -\mathbf{x}$ of $\mathbb{R}_n$, as shown in table 2.2. The conformal group can be extended to include involution, though this is not often done. However, in even dimensions involution can be achieved by a rotation so the extension is redundant.

Including E in the versor group brings all vectors of negative signature along with it. For $e_- = Ee_+$ gives us one such vector, and $D_\lambda e_-$ gives us (up to scale factor) all the rest in the E-plane. Therefore, extension of the versor group corresponds only to extension of $\mathrm{C}(n)$ to include involution.

Since every versor G in $\mathbb{R}_{n+1,1}$ can be generated as a product of vectors, expression of each vector in the expanded form (2.17) generates the expanded form

$$G = e(-e_0 A + B) - e_0(C + eD) \tag{2.132}$$

where A, B, C, D are versors in $\mathbb{R}_n$ and a minus sign is associated with e_0 for algebraic convenience in reference to (2.5d) and (2.2a). To enforce the versor property of G, the following conditions must be satisfied

$$AB^{\dagger},\ BD^{\dagger},\ CD^{\dagger},\ AC^{\dagger} \in \mathbb{R}^n\,, \tag{2.133}$$

$$GG^{\dagger} = AD^{\dagger} - BC^{\dagger} = \pm|\,G\,|^2 \neq 0\,. \tag{2.134}$$

Since G must have a definite parity, we can see from (2.132) that A and D must have the same parity which must be opposite to the parity of C and D. This implies that the products in (2.133) must have odd parity. The stronger condition that these products must be vector-valued can be derived by generating G from versor products or from the fact that the conformal transformation generated by G must be vector-valued. For $G \in \text{Spin}^+(n+1, n)$ the sign of (2.134) is always positive, but for $G \in \text{Spin}(n+1, n)$ a negative sign may derive from a vector of negative signature.

Adopting the normalization $|\,G\,| = 1$, we find

$$G^{*\dagger} = \pm(G^*)^{-1} = -(A^{*\dagger}e_0 + B^{*\dagger})e + (C^{*\dagger} - D^{*\dagger}e)e_0\,, \tag{2.135}$$

and inserting the expanded form for G into (2.111), we obtain

$$g(\mathbf{x}) = (A\mathbf{x} + B)(C\mathbf{x} + D)^{-1} \tag{2.136}$$

with the rescaling factor

$$\sigma = \sigma_g(\mathbf{x}) = (C\mathbf{x} + D)(C^*\mathbf{x} + D^*)^{\dagger}\,. \tag{2.137}$$

In evaluating (2.134) and (2.113) to get (2.137) it is most helpful to use the property $\langle\, MN\,\rangle = \langle\, NM\,\rangle$ for the scalar part of any geometric product.

The general *homeographic form* (2.136) for a conformal transformation on $\mathbb{R}^n$ is called a *Möbius transformation* by Ahlfors [1]. Because of its nonlinear form it is awkward for composing transformations. However, composition can be carried out multiplicatively with the versor form (2.132) with the result inserted into the homeographic form. As shown in [110], the versor (2.136) has a 2×2 matrix representation

$$[\,G\,] = \begin{bmatrix} A & B \\ C & D \end{bmatrix}, \tag{2.138}$$

so composition can be carried out by matrix multiplication. Ahlfors [2] has traced this matrix representation back to Vahlen [232].

The apparatus developed in this section is sufficient for efficient formulation and solution of any problem or theorem in conformal geometry. As an example, consider the problem of deriving a conformal transformation on the whole of $\mathbb{R}^n$ from a given transformation of a finite set of points.

Let $\mathbf{a}_1, \cdots, \mathbf{a}_{n+2}$ be distinct points in $\mathbb{R}^n$ spanning the whole space. Let $\mathbf{b}_1, \cdots, \mathbf{b}_{n+2}$ be another set of such points. If there is a Möbius transformation g changing $\mathbf{a}_i$ into $\mathbf{b}_i$ for $1 \leq i \leq n+2$, then g must be induced by a Lorentz transformation $\underline{G}$ of $\mathbb{R}^{n+1,1}$, so the corresponding homogeneous points are related by

$$\underline{G}(a_i) = \lambda_i b_i, \qquad \text{for} \qquad 1 \leq i \leq n+2. \tag{2.139}$$

Therefore $a_i \cdot a_j = (\lambda_i b_i) \cdot (\lambda_j b_j)$ and g exists if and only if the λ's satisfy

$$(\mathbf{a}_i - \mathbf{a}_j)^2 = \lambda_i \lambda_j (\mathbf{b}_i - \mathbf{b}_j)^2 \,.\ \text{for } 1 \leq i \neq j \leq n+2, \tag{2.140}$$

This sets $(n+2)(n-1)/2$ constraints on the $\mathbf{b}$'s from which the λ's can be computed if they are satisfied.

Now assuming that g exists, we can employ (2.139) to compute $g(\mathbf{x})$ for a generic point $\mathbf{x} \in \mathbb{R}^n$. Using the a_i as a basis, we write

$$x = \sum_{i=1}^{n+1} x^i \mathbf{a}_i, \tag{2.141}$$

so $\underline{G}(x) = \sum_{i=1}^{n+1} x^i \lambda_i b_i$, and

$$g(\mathbf{x}) = \frac{\sum_{i=1}^{n+1} x^i \lambda_i \mathbf{b}_i}{\sum_{i=1}^{n+1} x^i \lambda_i}. \tag{2.142}$$

The x's can be computed by employing the basis dual to $\{a_i\}$ as explained in chapter 1.

If we are given, instead of $n+2$ pairs of corresponding points, two sets of points, spheres and hyperplanes, say $t_1, \cdots, t_{n+2}$, and $u_1, \cdots, u_{n+2}$, where $t_i^2 \geq 0$ for $1 \leq i \leq n+2$ and where both sets are linearly independent vectors in $\mathbb{R}^{n+1,1}$, then we can simply replace the $\mathbf{a}$'s with the t's and the $\mathbf{b}$'s with the u's to compute g.

3. Spherical Conformal Geometry with Geometric Algebra *

Hongbo Li[1], David Hestenes[1], and Alyn Rockwood[2]

[1] Department of Physics and Astronomy
Arizona State University, Tempe
[2] Power Take Off Software, Inc., Colorado Springs

3.1 Introduction

The recorded study of spheres dates back to the first century in the book *Sphaerica* of Menelaus. Spherical trigonometry was thoroughly developed in modern form by Euler in his 1782 paper [75]. Spherical geometry in n-dimensions was first studied by Schläfli in his 1852 treatise, which was published posthumously in [202]. The most important transformation in spherical geometry, the Möbius transformation, was considered by Möbius in his 1855 paper [171].

Hamilton was the first to apply vectors to spherical trigonometry. In 1987 Hestenes [112] formulated a spherical trigonometry in terms of Geometric Algebra, and that remains a useful supplement to the present treatment.

This chapter is a continuation of the preceding chapter. Here we consider the homogeneous model of spherical space, which is similar to that of Euclidean space. We establish conformal geometry of spherical space in this model, and discuss several typical conformal transformations.

Although it is well known that the conformal groups of n-dimensional Euclidean and spherical spaces are isometric to each other, and are all iso-

* This work has been partially supported by NSF Grant RED-9200442.

metric to the group of isometries of hyperbolic $(n+1)$-space [129, 130] spherical conformal geometry has its unique conformal transformations, and it can provide good understanding for hyperbolic conformal geometry. It is an indispensible part of the unification of all conformal geometries in the homogeneous model, which is addressed in the next chapter.

3.2 Homogeneous Model of Spherical Space

In the previous chapter, we saw that, given a null vector $e \in \mathbb{R}^{n+1,1}$, the intersection of the null cone $\mathbb{N}^n$ of $\mathbb{R}^{n+1,1}$ with the hyperplane $\{x \in \mathbb{R}^{n+1,1} \mid x \cdot e = -1\}$ represents points in $\mathbb{R}^n$. This representation is established through the projective split of the null cone with respect to null vector e.

What if we replace the null vector e with any nonzero vector in $\mathbb{R}^{n+1,1}$? This section shows that when e is replaced by a unit vector p_0 of negative signature, then the set

$$\mathbb{N}^n_{p_0} = \{x \in \mathbb{N}^n | x \cdot p_0 = -1\} \tag{3.1}$$

represents points in the n-dimensional spherical space

$$\mathcal{S}^n = \{x \in \mathbb{R}^{n+1} | x^2 = 1\}. \tag{3.2}$$

The space dual to p_0 corresponds to $\mathbb{R}^{n+1} = \tilde{p}_0$, an $(n+1)$-dimensional Euclidean space whose unit sphere is $\mathcal{S}^n$.

Applying the orthogonal decomposition

$$x = P_{p_0}(x) + P_{\widetilde{p_0}}(x) \tag{3.3}$$

to vector $x \in \mathbb{N}^n_{p_0}$, we get

$$x = p_0 + \mathbf{x} \tag{3.4}$$

where $\mathbf{x} \in \mathcal{S}^n$. This defines a bijective map $i_{p_0} : \mathbf{x} \in \mathcal{S}^n \longrightarrow x \in \mathbb{N}^n_{p_0}$. Its inverse map is $P^{\perp}_{p_0} = P_{\tilde{p}_0}$.

Theorem 3.2.1.

$$\mathbb{N}^n_{p_0} \simeq \mathcal{S}^n. \tag{3.5}$$

We call $\mathbb{N}^n_{p_0}$ the *homogeneous model* of $\mathcal{S}^n$. Its elements are called *homogeneous points.*

Distances

For two points $\mathbf{a}, \mathbf{b} \in \mathcal{S}^n$, their *spherical distance* $d(\mathbf{a}, \mathbf{b})$ is defined as

$$d(\mathbf{a}, \mathbf{b}) = \cos^{-1}(\mathbf{a} \cdot \mathbf{b}). \tag{3.6}$$

We can define other equivalent distances. Distances d_1, d_2 are said to be equivalent if for any two pairs of points $\mathbf{a}_1, \mathbf{b}_1$ and $\mathbf{a}_2, \mathbf{b}_2$, then $d_1(\mathbf{a}_1, \mathbf{b}_1) = d_1(\mathbf{a}_2, \mathbf{b}_2)$ if and only if $d_2(\mathbf{a}_1, \mathbf{b}_1) = d_2(\mathbf{a}_2, \mathbf{b}_2)$. The *chord distance* measures the length of the chord between $\mathbf{a}, \mathbf{b}$:

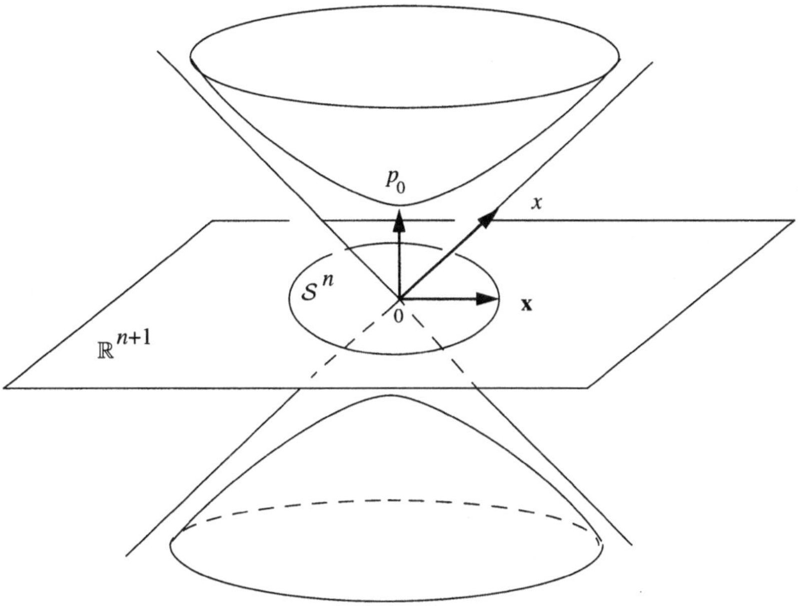

Fig. 3.1. The homogeneous model of $\mathcal{S}^n$

$$d_c(\mathbf{a}, \mathbf{b}) = |\mathbf{a} - \mathbf{b}|. \tag{3.7}$$

The *normal distance* is

$$d_n(\mathbf{a}, \mathbf{b}) = 1 - \mathbf{a} \cdot \mathbf{b}. \tag{3.8}$$

It equals the distance between points $\mathbf{a}, \mathbf{b}'$, where $\mathbf{b}'$ is the projection of $\mathbf{b}$ onto $\mathbf{a}$. The *stereographic distance* measures the distance between the origin 0 and $\mathbf{a}'$, the intersection of the line connecting $-\mathbf{a}$ and $\mathbf{b}$ with the hyperspace of $\mathbb{R}^{n+1}$ parallel with the tangent hyperplane of $\mathcal{S}^n$ at $\mathbf{a}$:

$$d_s(\mathbf{a}, \mathbf{b}) = \frac{|\mathbf{a} \wedge \mathbf{b}|}{1 + \mathbf{a} \cdot \mathbf{b}}. \tag{3.9}$$

Some relations among these distances are:

$$\begin{aligned}
d_c(\mathbf{a}, \mathbf{b}) &= 2 \sin \frac{d(\mathbf{a}, \mathbf{b})}{2}, \\
d_n(\mathbf{a}, \mathbf{b}) &= 1 - \cos d(\mathbf{a}, \mathbf{b}), \\
d_s(\mathbf{a}, \mathbf{b}) &= \tan \frac{d(\mathbf{a}, \mathbf{b})}{2}, \\
d_s^2(\mathbf{a}, \mathbf{b}) &= \frac{d_n(\mathbf{a}, \mathbf{b})}{2 - d_n(\mathbf{a}, \mathbf{b})}.
\end{aligned} \tag{3.10}$$

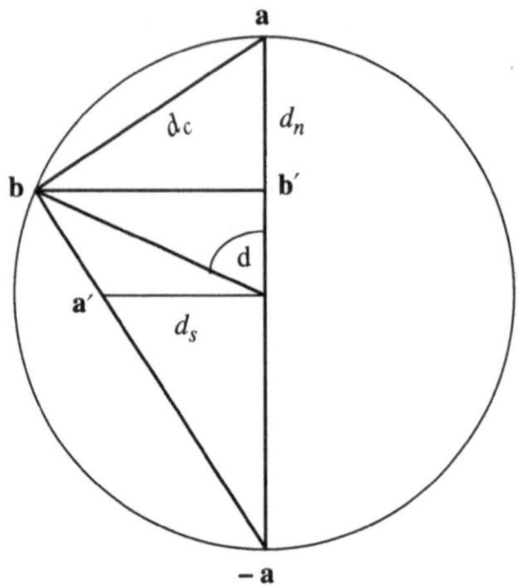

Fig. 3.2. Distances in $\mathcal{S}^n$

For two points $\mathbf{a}, \mathbf{b}$ in $\mathcal{S}^n$, we have

$$a \cdot b = \mathbf{a} \cdot \mathbf{b} - 1 = -d_n(\mathbf{a}, \mathbf{b}). \tag{3.11}$$

Therefore the inner product of two homogeneous points a, b characterizes the normal distance between the two points.

Spheres and hyperplanes

A sphere is a set of points having equal distances with a fixed point in $\mathcal{S}^n$. A sphere is said to be *great*, or *unit*, if it has normal radius 1. In this chapter, we call a great sphere a *hyperplane*, or an $(n-1)$-*plane*, of $\mathcal{S}^n$, and a non-great one a *sphere*, or an $(n-1)$-*sphere*.

The intersection of a sphere with a hyperplane is an $(n-2)$-dimensional sphere, called $(n-2)$-*sphere*; the intersection of two hyperplanes is an $(n-2)$-dimensional plane, called $(n-2)$-*plane*. In general, for $1 \leq r \leq n-1$, the intersection of a hyperplane with an r-sphere is called an $(r-1)$-*sphere*; the intersection of a hyperplane with an r-plane is called an $(r-1)$-*plane*. A 0-plane is a pair of antipodal points, and a 0-sphere is a pair of non-antipodal ones.

We require that the normal radius of a sphere be less than 1. In this way a sphere has only one center. For a sphere with center $\mathbf{c}$ and normal radius ρ, its *interior* is

$$\{\mathbf{x} \in \mathcal{S}^n | d_n(\mathbf{x}, \mathbf{c}) < \rho\}; \tag{3.12}$$

its *exterior* is

$$\{\mathbf{x} \in \mathcal{S}^n | \rho < d_n(\mathbf{x}, \mathbf{c}) \leq 2\}. \tag{3.13}$$

A sphere with center $\mathbf{c}$ and normal radius ρ is characterized by the vector

$$s = c - \rho p_0 \tag{3.14}$$

of positive signature. A point $\mathbf{x}$ is on the sphere if and only if $x \cdot c = -\rho$, or equivalently,

$$x \wedge \tilde{s} = 0. \tag{3.15}$$

Form (3.14) is called the *standard form* of a sphere.

A hyperplane is characterized by its *normal vector* $\mathbf{n}$; a point $\mathbf{x}$ is on the hyperplane if and only if $\mathbf{x} \cdot \mathbf{n} = 0$, or equivalently,

$$x \wedge \tilde{\mathbf{n}} = 0. \tag{3.16}$$

Theorem 3.2.2. *The intersection of any Minkowski hyperspace $\tilde{s}$ with $\mathbb{N}^n_{p_0}$ is a sphere or hyperplane in $\mathcal{S}^n$, and every sphere or hyperplane of $\mathcal{S}^n$ can be obtained in this way. Vector s has the standard form*

$$s = \mathbf{c} + \lambda p_0, \tag{3.17}$$

where $0 \leq \lambda < 1$. It represents a hyperplane if and only if $\lambda = 0$.

The dual theorem is:

Theorem 3.2.3. *Given homogeneous points $a_0, \ldots, a_n$ such that*

$$\tilde{s} = a_0 \wedge \cdots \wedge a_n, \tag{3.18}$$

then the $(n+1)$-blade $\tilde{s}$ represents a sphere in $\mathcal{S}^n$ if

$$p_0 \wedge \tilde{s} \neq 0, \tag{3.19}$$

or a hyperplane if

$$p_0 \wedge \tilde{s} = 0. \tag{3.20}$$

The above two theorems also provide an approach to compute the center and radius of a sphere in $\mathcal{S}^n$. Let $\tilde{s} = a_0 \wedge \cdots \wedge a_n \neq 0$, then it represents the sphere or hyperplane passing through points $\mathbf{a}_0, \ldots, \mathbf{a}_n$. When it represents a sphere, let $(-1)^\epsilon$ be the sign of $s \cdot p_0$. Then the center of the sphere is

$$(-1)^{\epsilon+1} \frac{P^{\perp}_{p_0}(s)}{|P^{\perp}_{p_0}(s)|}, \tag{3.21}$$

and the normal radius is

$$1 - \frac{|s \cdot p_0|}{|s \wedge p_0|}. \tag{3.22}$$

3.3 Relation between Two Spheres or Hyperplanes

Let $\tilde{s}_1, \tilde{s}_2$ be two distinct spheres or hyperplanes in $\mathcal{S}^n$. The signature of the blade

$$s_1 \wedge s_2 = (\tilde{s}_1 \vee \tilde{s}_2)^{\sim} \tag{3.23}$$

characterizes the relation between the two spheres or hyperplanes:

Theorem 3.3.1. *Two spheres or hyperplanes $\tilde{s}_1, \tilde{s}_2$ intersect, are tangent, or do not intersect if and only if $(s_1 \wedge s_2)^2$ is less than, equal to or greater than 0, respectively.*

There are three cases:

Case 1. For two hyperplanes represented by $\tilde{\mathbf{n}}_1, \tilde{\mathbf{n}}_2$, since $\mathbf{n}_1 \wedge \mathbf{n}_2$ has Euclidean signature, the two hyperplanes always intersect. The intersection is an $(n-2)$-plane, and is normal to both $P^{\perp}_{\mathbf{n}_1}(\mathbf{n}_2)$ and $P^{\perp}_{\mathbf{n}_2}(\mathbf{n}_1)$.

Case 2. For a hyperplane $\tilde{\mathbf{n}}$ and a sphere $(\mathbf{c} + \lambda p_0)^{\sim}$, since

$$(\mathbf{n} \wedge (\mathbf{c} + \lambda p_0))^2 = (\lambda + |\mathbf{c} \wedge \mathbf{n}|)(\lambda - |\mathbf{c} \wedge \mathbf{n}|), \tag{3.24}$$

then:

- If $\lambda < |\mathbf{c} \wedge \mathbf{n}|$, they intersect. The intersection is an $(n-2)$-sphere with center $\dfrac{P^{\perp}_{\mathbf{n}}(\mathbf{c})}{|P^{\perp}_{\mathbf{n}}(\mathbf{c})|}$ and normal radius $1 - \dfrac{\lambda}{|\mathbf{c} \wedge \mathbf{n}|}$.
- If $\lambda = |\mathbf{c} \wedge \mathbf{n}|$, they are tangent at the point $\dfrac{P^{\perp}_{\mathbf{n}}(\mathbf{c})}{|P^{\perp}_{\mathbf{n}}(\mathbf{c})|}$.
- If $\lambda > |\mathbf{c} \wedge \mathbf{n}|$, they do not intersect. There is a pair of points in $\mathcal{S}^n$ which are inversive with respect to the sphere, while at the same time symmetric with respect to the hyperplane. They are $\dfrac{P^{\perp}_{\mathbf{n}}(\mathbf{c}) \pm \mu \mathbf{n}}{\lambda}$, where $\mu = \sqrt{\lambda^2 + (\mathbf{c} \wedge \mathbf{n})^2}$.

Case 3. For two spheres $(\mathbf{c}_i + \lambda_i p_0)^{\sim}$, $i = 1, 2$, since

$$((\mathbf{c}_1 + \lambda_1 p_0) \wedge (\mathbf{c}_2 + \lambda_2 p_0))^2 = (\mathbf{c}_1 \wedge \mathbf{c}_2)^2 + (\lambda_2 \mathbf{c}_1 - \lambda_1 \mathbf{c}_2)^2, \tag{3.25}$$

then:

- If $|\mathbf{c}_1 \wedge \mathbf{c}_2| > |\lambda_2 \mathbf{c}_1 - \lambda_1 \mathbf{c}_2|$, they intersect. The intersection is an $(n-2)$-sphere on the hyperplane of $\mathcal{S}^n$ represented by

$$(\lambda_2 \mathbf{c}_1 - \lambda_1 \mathbf{c}_2)^{\sim}. \tag{3.26}$$

The intersection has center

$$\frac{\lambda_1 P^{\perp}_{\mathbf{c}_2}(\mathbf{c}_1)+\lambda_2 P^{\perp}_{\mathbf{c}_1}(\mathbf{c}_2)}{|\lambda_1\mathbf{c}_2-\lambda_2\mathbf{c}_1||\mathbf{c}_1\wedge\mathbf{c}_2|} \tag{3.27}$$

and normal radius

$$1-\frac{|\lambda_2\mathbf{c}_1-\lambda_1\mathbf{c}_2|}{|\mathbf{c}_1\wedge\mathbf{c}_2|}. \tag{3.28}$$

– If $|\mathbf{c}_1\wedge\mathbf{c}_2|=|\lambda_2\mathbf{c}_1-\lambda_1\mathbf{c}_2|$, they are tangent at the point

$$\frac{\lambda_1 P^{\perp}_{\mathbf{c}_2}(\mathbf{c}_1)+\lambda_2 P^{\perp}_{\mathbf{c}_1}(\mathbf{c}_2)}{|\mathbf{c}_1\wedge\mathbf{c}_2|^2}. \tag{3.29}$$

– If $|\mathbf{c}_1\wedge\mathbf{c}_2|<|\lambda_2\mathbf{c}_1-\lambda_1\mathbf{c}_2|$, they do not intersect. There is a pair of points in $\mathcal{S}^n$ which are inversive with respect to both spheres. They are

$$\frac{\lambda_1 P^{\perp}_{\mathbf{c}_2}(\mathbf{c}_1)+\lambda_2 P^{\perp}_{\mathbf{c}_1}(\mathbf{c}_2)\pm\mu(\lambda_2\mathbf{c}_1-\lambda_1\mathbf{c}_2)}{(\lambda_2\mathbf{c}_1-\lambda_1\mathbf{c}_2)^2}, \tag{3.30}$$

where $\mu=\sqrt{(\mathbf{c}_1\wedge\mathbf{c}_2)^2+(\lambda_2\mathbf{c}_1-\lambda_1\mathbf{c}_2)^2}$. The two points are called the *Poncelet points* of the spheres.

The scalar

$$s_1 * s_2=\frac{s_1\cdot s_2}{|s_1||s_2|} \tag{3.31}$$

is called the *inversive product* of vectors s_1 and s_2. Obviously, it is invariant under orthogonal transformations in $\mathbb{R}^{n+1,1}$. We have the following conclusion for the geometric interpretation of the inversive product:

Theorem 3.3.2. *Let* $\mathbf{a}$ *be a point of intersection of two spheres or hyperplanes* $\tilde{s}_1$ *and* $\tilde{s}_2$*, let* m_i*,* $i=1,2$*, be the respective outward unit normal vector at* $\mathbf{a}$ *of* $\tilde{s}_i$ *if it is a sphere, or* $s_i/|s_i|$ *if it is a hyperplane, then*

$$s_1 * s_2 = m_1\cdot m_2. \tag{3.32}$$

Proof. Given that s_i has the standard form $\mathbf{c}_i+\lambda_i p_0$. When $\tilde{s}_i$ is a sphere, its outward unit normal vector at point $\mathbf{a}$ is

$$\mathbf{m}_i=\frac{\mathbf{a}(\mathbf{a}\wedge\mathbf{c}_i)}{|\mathbf{a}\wedge\mathbf{c}_i|}, \tag{3.33}$$

which equals $\mathbf{c}_i$ when $\tilde{s}_i$ is a hyperplane. Point $\mathbf{a}$ is on both $\tilde{s}_1$ and $\tilde{s}_2$ and yields

$$\mathbf{a}\cdot\mathbf{c}_i=\lambda_i,\ \text{for } i=1,2, \tag{3.34}$$

so

$$\mathbf{m}_1 \cdot \mathbf{m}_2 = \frac{\mathbf{c}_1 - \mathbf{a}\cdot\mathbf{c}_1\mathbf{a}}{\sqrt{1-(\mathbf{a}\cdot\mathbf{c}_1)^2}} \cdot \frac{\mathbf{c}_2 - \mathbf{a}\cdot\mathbf{c}_2\mathbf{a}}{\sqrt{1-(\mathbf{a}\cdot\mathbf{c}_2)^2}} = \frac{\mathbf{c}_1\cdot\mathbf{c}_2 - \lambda_1\lambda_2}{\sqrt{(1-\lambda_1^2)(1-\lambda_2^2)}}. \quad (3.35)$$

On the other hand,

$$s_1 * s_2 = \frac{(\mathbf{c}_1 + \lambda_1 p_0)\cdot(\mathbf{c}_2 + \lambda_2 p_0)}{|\mathbf{c}_1 + \lambda_1 p_0||\mathbf{c}_2 + \lambda_2 p_0|} = \frac{\mathbf{c}_1\cdot\mathbf{c}_2 - \lambda_1\lambda_2}{\sqrt{(1-\lambda_1^2)(1-\lambda_2^2)}}. \quad (3.36)$$

An immediate corollary is that any orthogonal transformation in $\mathbb{R}^{n+1,1}$ induces an angle-preserving transformation in $\mathcal{S}^n$. This conformal transformation will be discussed in the last section.

3.4 Spheres and Planes of Dimension r

We have the following conclusion similar to that in Euclidean geometry:

Theorem 3.4.1. *For $2 \le r \le n+1$, every r-blade A_r of Minkowski signature in $\mathbb{R}^{n+1,1}$ represents an $(r-2)$-dimensional sphere or plane in $\mathcal{S}^n$.*

Corollary *The $(r-2)$-dimensional sphere passing through r points $\mathbf{a}_1, \ldots, \mathbf{a}_r$ in $\mathcal{S}^n$ is represented by $a_1 \wedge \cdots \wedge a_r$; the $(r-2)$- plane passing through $r-1$ points $\mathbf{a}_1, \ldots, \mathbf{a}_{r-1}$ in $\mathcal{S}^n$ is represented by $p_0 \wedge a_1 \wedge \cdots \wedge a_{r-1}$.*

There are two possibilities:

Case 1. When $p_0 \wedge A_r = 0$, A_r represents an $(r-2)$-plane in $\mathcal{S}^n$. After normalization, the *standard form* of the $(r-2)$-plane is

$$p_0 \wedge \mathbf{I}_{r-1}, \quad (3.37)$$

where $\mathbf{I}_{r-1}$ is a unit $(r-1)$-blade of $\mathcal{G}(\mathbb{R}^{n+1})$ representing the minimal space in $\mathbb{R}^{n+1}$ supporting the $(r-2)$-plane of $\mathcal{S}^n$.

Case 2. A_r represents an $(r-2)$-dimensional sphere if

$$A_{r+1} = p_0 \wedge A_r \neq 0. \quad (3.38)$$

The vector

$$s = A_r A_{r+1}^{-1} \quad (3.39)$$

has positive square and $p_0 \cdot s \neq 0$, so its dual $\tilde{s}$ represents an $(n-1)$-dimensional sphere. According to Case 1, A_{r+1} represents an $(r-1)$-dimensional plane in $\mathcal{S}^n$, therefore

$$A_r = sA_{r+1} = (-1)^{\epsilon}\tilde{s} \vee A_{r+1}, \quad (3.40)$$

where $\epsilon = \frac{(n+2)(n+1)}{2} + 1$ represents the intersection of $(n-1)$-sphere $\tilde{s}$ with $(r-1)$-plane A_{r+1}.

With suitable normalization, we can write $s = c - \rho p_0$. Since $s \wedge A_{r+1} = p_0 \wedge A_{r+1} = 0$, the sphere A_r is also centered at c and has normal radius ρ. Accordingly we represent an $(r-2)$-dimensional sphere in the *standard form*

$$(c - \rho p_0)\,(p_0 \wedge \mathbf{I}_r), \tag{3.41}$$

where $\mathbf{I}_r$ is a unit r-blade of $\mathcal{G}(\mathbb{R}^{n+1})$ representing the minimal space in $\mathbb{R}^{n+1}$ supporting the $(r-2)$-sphere of $\mathcal{S}^n$.

This completes our classification of standard representations for spheres and planes in $\mathcal{S}^n$.

Expanded form

For $r+1$ homogeneous points $a_0, \ldots, a_r$ "in" $\mathcal{S}^n$, where $0 \le r \le n+1$, we have

$$A_{r+1} = a_0 \wedge \cdots \wedge a_r = \mathbf{A}_{r+1} + p_0 \wedge \mathbf{A}_r, \tag{3.42}$$

where

$$\begin{aligned} \mathbf{A}_{r+1} &= \mathbf{a}_0 \wedge \cdots \wedge \mathbf{a}_r, \\ \mathbf{A}_r &= \not{\partial}\mathbf{A}_{r+1}. \end{aligned} \tag{3.43}$$

When $\mathbf{A}_{r+1} = 0$, A_{r+1} represents an $(r-1)$-plane, otherwise A_{r+1} represents an $(r-1)$-sphere. In the latter case, $p_0 \wedge \mathbf{A}_{r+1} = p_0 \wedge A_{r+1}$ represents the support plane of the $(r-1)$-sphere in $\mathcal{S}^n$, and $p_0 \wedge \mathbf{A}_r$ represents the $(r-1)$-plane normal to the center of the $(r-1)$-sphere in the support plane. The center of the $(r-1)$-sphere is

$$\frac{\mathbf{A}_r \mathbf{A}_{r+1}^{\dagger}}{|\mathbf{A}_r||\mathbf{A}_{r+1}|}, \tag{3.44}$$

and the normal radius is

$$1 - \frac{|\mathbf{A}_{r+1}|}{|\mathbf{A}_r|}. \tag{3.45}$$

Since

$$\begin{aligned} A_{r+1}^{\dagger} \cdot A_{r+1} &= \det(a_i \cdot a_j)_{(r+1)\times(r+1)} \\ &= (-\tfrac{1}{2})^{r+1} \det(|\mathbf{a}_i - \mathbf{a}_j|^2)_{(r+1)\times(r+1)}; \end{aligned} \tag{3.46}$$

thus, when $r = n+1$, we obtain Ptolemy's Theorem for spherical geometry:

Theorem 3.4.2 (Ptolemy's Theorem). *Let* $\mathbf{a}_0, \cdots, \mathbf{a}_{n+1}$ *be points in* $\mathcal{S}^n$*, then they are on a sphere or hyperplane of* $\mathcal{S}^n$ *if and only if* $\det(|\mathbf{a}_i - \mathbf{a}_j|^2)_{(n+2)\times(n+2)} = 0$.

3.5 Stereographic Projection

In the homogeneous model of $\mathcal{S}^n$, let $\mathbf{a}_0$ be a fixed point on $\mathcal{S}^n$. The space $\mathbb{R}^n = (\mathbf{a}_0 \wedge p_0)^\sim$, which is parallel to the tangent spaces of $\mathcal{S}^n$ at points $\pm\mathbf{a}_0$, is Euclidean. By the stereographic projection of $\mathcal{S}^n$ from point $\mathbf{a}_0$ to the space $\mathbb{R}^n$, the ray from $\mathbf{a}_0$ through $\mathbf{a} \in \mathcal{S}^n$ intersects the space at the point

$$j_{\mathcal{S}\mathbb{R}}(\mathbf{a}) = \frac{\mathbf{a}_0(\mathbf{a}_0 \wedge \mathbf{a})}{1 - \mathbf{a}_0 \cdot \mathbf{a}} = 2(\mathbf{a} - \mathbf{a}_0)^{-1} + \mathbf{a}_0. \tag{3.47}$$

Many advantages of Geometric Algebra in handling stereographic projections are demonstrated in [113].

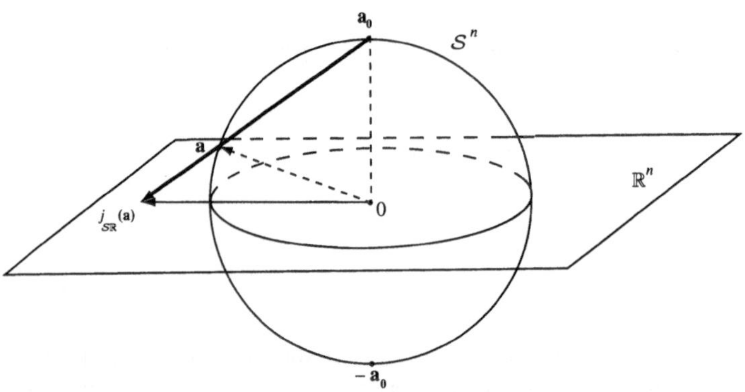

Fig. 3.3. Stereographic projection of $\mathcal{S}^n$ from $\mathbf{a}_0$ to the space normal to $\mathbf{a}_0$

We note the following facts about the stereographic projection:

1. A hyperplane passing through $\mathbf{a}_0$ and normal to $\mathbf{n}$ is mapped to the hyperspace in $\mathbb{R}^n$ normal to $\mathbf{n}$.
2. A hyperplane normal to $\mathbf{n}$ but not passing through $\mathbf{a}_0$ is mapped to the sphere in $\mathbb{R}^n$ with center $c = \mathbf{n} - \dfrac{\mathbf{a}_0}{\mathbf{n} \cdot \mathbf{a}_0}$ and radius $\rho = \dfrac{1}{\sqrt{|\mathbf{n} \cdot \mathbf{a}_0|}}$. Such a sphere has the feature that
$$\rho^2 = 1 + c^2. \tag{3.48}$$
Its intersection with the unit sphere of $\mathbb{R}^n$ is a unit $(n-2)$-dimensional sphere. Conversely, given a point c in $\mathbb{R}^n$, we can find a unique hyperplane in $\mathcal{S}^n$ whose stereographic projection is the sphere in $\mathbb{R}^n$ with center c and radius $\sqrt{1+c^2}$. It is the hyperplane normal to $a_0 - c$.
3. A sphere passing through $\mathbf{a}_0$, with center $\mathbf{c}$ and normal radius ρ, is mapped to the hyperplane in $\mathbb{R}^n$ normal to $P^{\perp}_{\mathbf{a}_0}(\mathbf{c})$ and with $\dfrac{1-\rho}{\sqrt{1-(1-\rho)^2}}$ as the signed distance from the origin.

4. A sphere not passing through $\mathbf{a}_0$, with center $\mathbf{c}$ and normal radius ρ, is mapped to the sphere in $\mathbb{R}^n$ with center $\dfrac{(1-\rho)\mathbf{p}_0 + P^{\perp}_{\mathbf{a}_0}(\mathbf{c})}{d_n(\mathbf{c},\mathbf{a}_0)-\rho}$ and radius $\dfrac{\sqrt{1-(1-\rho)^2}}{|d_n(\mathbf{c},\mathbf{a}_0)-\rho|}$.

It is a classical result that the map $j_{\mathcal{S}\mathbb{R}}$ is a conformal map from $\mathcal{S}^n$ to $\mathbb{R}^n$. The conformal coefficient λ is defined by

$$|j_{\mathcal{S}\mathbb{R}}(\mathbf{a}) - j_{\mathcal{S}\mathbb{R}}(\mathbf{b})| = \lambda(\mathbf{a},\mathbf{b})|\mathbf{a}-\mathbf{b}|, \quad \text{for } \mathbf{a},\mathbf{b} \in \mathcal{S}^n. \tag{3.49}$$

We have

$$\lambda(\mathbf{a},\mathbf{b}) = \frac{1}{\sqrt{(1-\mathbf{a}_0\cdot\mathbf{a})(1-\mathbf{a}_0\cdot\mathbf{b})}}. \tag{3.50}$$

Using the null cone of $\mathbb{R}^{n+1,1}$ we can construct the conformal map $j_{\mathcal{S}\mathbb{R}}$ trivially: it is nothing but a rescaling of null vectors.

Let

$$e = \mathbf{a}_0 + p_0, \quad e_0 = \frac{-\mathbf{a}_0 + p_0}{2}, \quad E = e \wedge e_0. \tag{3.51}$$

For $\mathbb{R}^n = (e\wedge e_0)^{\sim} = (\mathbf{a}_0 \wedge p_0)^{\sim}$, the map $i_E : x \in \mathbb{R}^n \mapsto e_0 + x + \frac{x^2}{2}e \in \mathrm{N}^n_e$ defines a homogeneous model for the Euclidean space.

Any null vector h in $\mathcal{S}^n$ represents a point in the homogeneous model of $\mathcal{S}^n$, while in the homogeneous model of $\mathbb{R}^n$ it represents a point or point at infinity of $\mathbb{R}^n$. The rescaling transformation $k_{\mathbb{R}} : \mathrm{N}^n \longrightarrow \mathrm{N}^n_e$ defined by

$$k_{\mathbb{R}}(h) = -\frac{h}{h\cdot e}, \quad \text{for } h \in \mathrm{N}^n, \tag{3.52}$$

induces the conformal map $j_{\mathcal{S}\mathbb{R}}$ through the following commutative diagram:

$$\begin{array}{ccc} \mathbf{a}+p_0 \in \mathrm{N}^n_{p_0} & \overset{k_{\mathbb{R}}}{- - - \longrightarrow} & \dfrac{\mathbf{a}+p_0}{1-\mathbf{a}\cdot\mathbf{a}_0} \in \mathrm{N}^n_e \\ i_{p_0} \uparrow & & \downarrow P^{\perp}_E \\ \mathbf{a} \in \mathcal{S}^n & \overset{j_{\mathcal{S}\mathbb{R}}}{- - - \longrightarrow} & \dfrac{\mathbf{a}_0(\mathbf{a}_0\wedge\mathbf{a})}{1-\mathbf{a}\cdot\mathbf{a}_0} \in \mathbb{R}^n \end{array} \tag{3.53}$$

i.e., $j_{\mathcal{S}\mathbb{R}} = P^{\perp}_E \circ k_{\mathbb{R}} \circ i_{p_0}$. The conformal coefficient λ is derived from the following identity: for any vector x and null vectors h_1, h_2,

$$\left| -\frac{h_1}{h_1\cdot x} + \frac{h_2}{h_2\cdot x} \right| = \frac{|h_1 - h_2|}{\sqrt{|(h_1\cdot x)(h_2\cdot x)|}}. \tag{3.54}$$

The inverse of the map $j_{\mathcal{S}\mathbb{R}}$, denoted by $j_{\mathbb{R}\mathcal{S}}$, is

$$j_{\mathbb{R}\mathcal{S}}(u) = \frac{(u^2-1)\mathbf{a}_0 + 2u}{u^2+1} = 2(u-\mathbf{a}_0)^{-1} + \mathbf{a}_0, \quad \text{for } u \in \mathbb{R}^n. \tag{3.55}$$

According to [113], (3.55) can also be written as

$$j_{\mathbb{R}\mathcal{S}}(u) = -(u-\mathbf{a}_0)^{-1}\mathbf{a}_0(u-\mathbf{a}_0). \tag{3.56}$$

From the above algebraic construction of the stereographic projection, we see that the null vectors in $\mathbb{R}^{n+1,1}$ have geometrical interpretations in both $\mathcal{S}^n$ and $\mathbb{R}^n$, as do the Minkowski blades of $\mathbb{R}^{n+1,1}$. Every vector in $\mathbb{R}^{n+1,1}$ of positive signature can be interpreted as a sphere or hyperplane in both spaces. We will discuss this further in the next chapter.

3.6 Conformal Transformations

In this section we present some results on the conformal transformations in $\mathcal{S}^n$. We know that the conformal group of $\mathcal{S}^n$ is isomorphic with the Lorentz group of $\mathbb{R}^{n+1,1}$. Moreover, a Lorentz transformation in $\mathbb{R}^{n+1,1}$ is the product of at most $n+2$ reflections with respect to vectors of positive signature. We first analyze the induced conformal transformation in $\mathcal{S}^n$ of such a reflection in $\mathbb{R}^{n+1,1}$.

3.6.1 Inversions and Reflections

After normalization, any vector in $\mathbb{R}^{n+1,1}$ of positive signature can be written as $s = \mathbf{c} + \lambda p_0$, where $0 \leq \lambda < 1$. For any point $\mathbf{a}$ in $\mathcal{S}^n$, the reflection of a with respect to s is

$$\frac{1+\lambda^2 - 2\lambda\mathbf{c}\cdot\mathbf{a}}{1-\lambda^2}\mathbf{b}, \tag{3.57}$$

where

$$\mathbf{b} = \frac{(1-\lambda^2)\mathbf{a} + 2(\lambda - \mathbf{c}\cdot\mathbf{a})\mathbf{c}}{1+\lambda^2 - 2\lambda\mathbf{c}\cdot\mathbf{a}}. \tag{3.58}$$

If $\lambda = 0$, i.e., if $\tilde{s}$ represents a hyperplane of $\mathcal{S}^n$, then (3.58) gives

$$\mathbf{b} = \mathbf{a} - 2\mathbf{c}\cdot\mathbf{a}\,\mathbf{c}, \tag{3.59}$$

i.e., $\mathbf{b}$ is the reflection of $\mathbf{a}$ with respect to the hyperplane $\tilde{\mathbf{c}}$ of $\mathcal{S}^n$.

If $\lambda \neq 0$, let $\lambda = 1-\rho$, then from (3.58) we get

$$\left(\frac{\mathbf{c}\wedge\mathbf{a}}{1+\mathbf{c}\cdot\mathbf{a}}\right)^{\dagger}\left(\frac{\mathbf{c}\wedge\mathbf{b}}{1+\mathbf{c}\cdot\mathbf{b}}\right) = \frac{\rho}{2-\rho}. \tag{3.60}$$

Since the right-hand side of (3.60) is positive, $\mathbf{c}, \mathbf{a}, \mathbf{b}$ and $-\mathbf{c}$ are on a half great circle of $\mathcal{S}^n$. Using (3.9) and (3.10) we can write (3.60) as

$$d_s(\mathbf{a}, \mathbf{c})d_s(\mathbf{b}, \mathbf{c}) = \rho_s^2, \tag{3.61}$$

where ρ_s is the stereographic distance corresponding to the normal distance ρ. We say that $\mathbf{a}, \mathbf{b}$ are *inversive* with respect to the sphere with center $\mathbf{c}$ and stereographic radius ρ_s. This conformal transformation is called an *inversion* in $\mathcal{S}^n$.

An inversion can be easily described in the language of Geometric Algebra. The two inversive homogeneous points a and b correspond to the null directions in the 2-dimensional space $a \wedge (c - \rho p_0)$, which is degenerate when a is on the sphere represented by $(c - \rho p_0)^\sim$, and Minkowski otherwise.

Any conformal transformation in $\mathcal{S}^n$ is generated by inversions with respect to spheres, or reflections with respect to hyperplanes.

3.6.2 Other Typical Conformal Transformations

Antipodal transformation

By an antipodal transformation a point $\mathbf{a}$ of $\mathcal{S}^n$ is mapped to point $-\mathbf{a}$. This transformation is induced by the versor p_0.

Rotations

A rotation in $\mathcal{S}^n$ is just a rotation in $\mathbb{R}^{n+1}$. Any rotation in $\mathcal{S}^n$ can be induced by a spinor in $\mathcal{G}(\mathbb{R}^{n+1})$.

Given a unit 2-blade $\mathbf{I}_2$ in $\mathcal{G}(\mathbb{R}^{n+1})$ and $0 < \theta < 2\pi$, the spinor $e^{\mathbf{I}_2\theta/2}$ induces a rotation in $\mathcal{S}^n$. Using the orthogonal decomposition

$$\mathbf{x} = P_{\mathbf{I}_2}(\mathbf{x}) + P_{\mathbf{I}_2}^{\perp}(\mathbf{x}), \text{ for } \mathbf{x} \in \mathcal{S}^n, \tag{3.62}$$

we get

$$e^{-\mathbf{I}_2\theta/2}\mathbf{x}e^{\mathbf{I}_2\theta/2} = P_{\mathbf{I}_2}(\mathbf{x})e^{\mathbf{I}_2\theta} + P_{\tilde{\mathbf{I}}_2}(\mathbf{x}). \tag{3.63}$$

Therefore when $n > 1$, the set of fixed points under this rotation is the $(n-2)$-plane in $\mathcal{S}^n$ represented by $\tilde{\mathbf{I}}_2$. It is called the *axis* of the rotation, where θ is the angle of rotation for the points on the line of $\mathcal{S}^n$ represented by $p_0 \wedge \mathbf{I}_2$. This line is called the *line of rotation.*

For example, the spinor $\mathbf{a}(\mathbf{a}+\mathbf{b})$ induces a rotation from point $\mathbf{a}$ to point $\mathbf{b}$, with $p_0 \wedge \mathbf{a} \wedge \mathbf{b}$ as the line of rotation. The spinor $(\mathbf{c} \wedge \mathbf{a})(\mathbf{c} \wedge (\mathbf{a} + \mathbf{b}))$, where $\mathbf{a}$ and $\mathbf{b}$ have equal distances from $\mathbf{c}$, induces a rotation from point $\mathbf{a}$ to point $\mathbf{b}$ with $p_0 \wedge P_{\mathbf{c}}^{\perp}(\mathbf{a}) \wedge P_{\mathbf{c}}^{\perp}(\mathbf{b})$ as the line of rotation.

Rotations belong to the orthogonal group $O(\mathcal{S}^n)$. A versor in $\mathcal{G}(\mathbb{R}^{n+1,1})$ induces an orthogonal transformation in $\mathcal{S}^n$ if and only if it leaves $\{\pm p_0\}$ invariant.

Tidal transformations

A tidal transformation of coefficient $\lambda \neq \pm 1$ with respect to a point $\mathbf{c}$ in $\mathcal{S}^n$ is a conformal transformation induced by the spinor $1 + \lambda p_0 \wedge \mathbf{c}$. It changes a point $\mathbf{a}$ to point

$$\mathbf{b} = \frac{(1-\lambda^2)\mathbf{a} + 2\lambda(\lambda \mathbf{a}\cdot\mathbf{c} - 1)\mathbf{c}}{1+\lambda^2 - 2\lambda \mathbf{a}\cdot\mathbf{c}}. \tag{3.64}$$

Points $\mathbf{a}, \mathbf{b}$ and $\mathbf{c}$ are always on the same line. Conversely, from $\mathbf{a}, \mathbf{b}$ and $\mathbf{c}$ we obtain

$$\lambda = \frac{d_n^2(\mathbf{b},\mathbf{a})}{d_n^2(\mathbf{b}-\mathbf{c}) - d_n^2(\mathbf{a}-\mathbf{c})}. \tag{3.65}$$

By this transformation, any line passing through point $\mathbf{c}$ is invariant, and any sphere with center $\mathbf{c}$ is transformed into a sphere with center $\mathbf{c}$ or $-\mathbf{c}$, or the hyperplane normal to $\mathbf{c}$. The name "tidal transformation" arises from interpreting points $\pm\mathbf{c}$ as the source and influx of the tide.

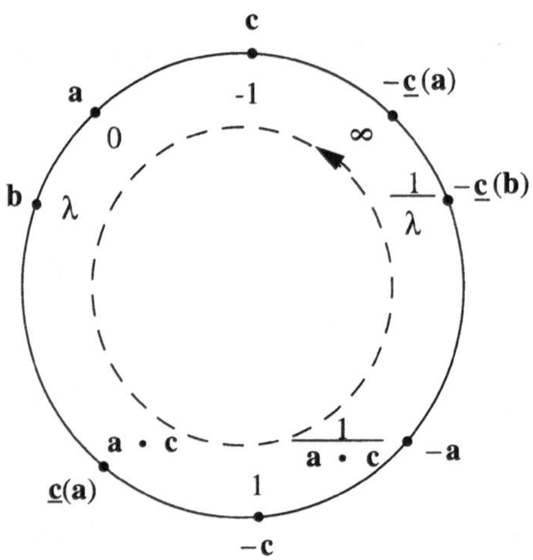

Fig. 3.4. $\lambda = \lambda(\mathbf{b})$ of a tidal transformation

Given points $\mathbf{a}, \mathbf{c}$ in $\mathcal{S}^n$, which are neither identical nor antipodal, let point $\mathbf{b}$ move on line $\mathbf{ac}$ of $\mathcal{S}^n$, then $\lambda = \lambda(\mathbf{b})$ is determined by (3.65). This function has the following properties:

1. $\lambda \neq \pm 1$, i.e., $\mathbf{b} \neq \pm\mathbf{c}$. This is because if $\lambda = \pm 1$, then $1 + \lambda p_0 \wedge \mathbf{c}$ is no longer a spinor.
2. Let $\underline{\mathbf{c}}(\mathbf{a})$ be the reflection of $\mathbf{a}$ with respect to $\mathbf{c}$, then

$$\underline{\mathbf{c}}(\mathbf{a}) = \mathbf{a} - 2\mathbf{a}\cdot\mathbf{c}\mathbf{c}^{-1}, \tag{3.66}$$

and

$$\lambda(-\underline{\mathbf{c}}(\mathbf{a})) = \infty, \quad \lambda(\underline{\mathbf{c}}(\mathbf{a})) = \mathbf{a}\cdot\mathbf{c}. \tag{3.67}$$

3. When $\mathbf{b}$ moves from $-\underline{\mathbf{c}}(\mathbf{a})$ through $\mathbf{c}$, $\mathbf{a}$, $-\mathbf{c}$ back to $-\underline{\mathbf{c}}(\mathbf{a})$, λ increases strictly from $-\infty$ to ∞.
4. $$\lambda(-\underline{\mathbf{c}}(\mathbf{b})) = \frac{1}{\lambda(\mathbf{b})}. \tag{3.68}$$
5. When $\mathbf{c} \cdot \mathbf{a} = 0$ and $0 < \lambda < 1$, then $\mathbf{b}$ is between $\mathbf{a}$ and $-\mathbf{c}$, and
$$\lambda = d_s(\mathbf{a}, \mathbf{b}). \tag{3.69}$$
When $0 > \lambda > -1$, then $\mathbf{b}$ is between $\mathbf{a}$ and $\mathbf{c}$, and
$$\lambda = -d_s(\mathbf{a}, \mathbf{b}). \tag{3.70}$$

When $|\lambda| > 1$, a tidal transformation is the composition of an inversion with the antipodal transformation, because

$$1 + \lambda p_0 \wedge \mathbf{c} = -p_0(p_0 - \lambda \mathbf{c}). \tag{3.71}$$

4. A Universal Model for Conformal Geometries of Euclidean, Spherical and Double-Hyperbolic Spaces*

Hongbo Li[1], David Hestenes[1], and Alyn Rockwood[2]

[1] Department of Physics and Astronomy
Arizona State University, Tempe
[2] Power Take Off Software, Inc., Colorado Springs

4.1 Introduction

The study of relations among Euclidean, spherical and hyperbolic geometries dates back to the beginning of last century. The attempt to prove Euclid's fifth postulate led C. F. Gauss to discover hyperbolic geometry in the 1820's. Only a few years passed before this geometry was rediscovered independently by N. Lobachevski (1829) and J. Bolyai (1832). The strongest evidence given by the founders for its consistency is the duality between hyperbolic and spherical trigonometries. This duality was first demonstrated by Lambert in his 1770 memoir [139] Some theorems, for example the law of sines, can be stated in a form that is valid in spherical, Euclidean, and hyperbolic geometries [25].

To prove the consistency of hyperbolic geometry, people built various analytic models of hyperbolic geometry on the Euclidean plane. E. Beltrami [21] constructed a Euclidean model of the hyperbolic plane, and using differential geometry, showed that his model satisfies all the axioms of hyperbolic plane geometry. In 1871, F. Klein gave an interpretation of Beltrami's model in terms of projective geometry. Because of Klein's interpretation, Beltrami's

* This work has been partially supported by NSF Grant RED-9200442.

model is later called Klein's disc model of the hyperbolic plane. The generalization of this model to n-dimensional hyperbolic space is now called the Klein ball model [37].

In the same paper Beltrami constructed two other Euclidean models of the hyperbolic plane, one on a disc and the other on a Euclidean half-plane. Both models are later generalized to n-dimensions by H. Poincaré [188], and are now associated with his name.

All three of the above models are built in Euclidean space, and the latter two are conformal in the sense that the metric is a point-to-point scaling of the Euclidean metric. In his 1878 paper [127], Killing described a hyperboloid model of hyperbolic geometry by constructing the stereographic projection of Beltrami's disc model onto the hyperbolic space. This hyperboloid model was generalized to n-dimensions by Poincaré.

There is another model of hyperbolic geometry built in spherical space, called hemisphere model, which is also conformal. Altogether there are five well-known models for the n-dimensional hyperbolic geometry:

- the half-space model,
- the conformal ball model,
- the Klein ball model,
- the hemisphere model,
- the hyperboloid model.

The theory of hyperbolic geometry can be built in a unified way within any of the models. With several models one can, so to speak, turn the object around and scrutinize it from different viewpoints. The connections among these models are largely established through stereographic projections. Because stereographic projections are conformal maps, the conformal groups of n-dimensional Euclidean, spherical, and hyperbolic spaces are isometric to each other, and are all isometric to the group of isometries of hyperbolic $(n+1)$-space, according to observations of Klein [129, 130].

It seems that everything is worked out for unified treatment of the three spaces. In this chapter we go further. We unify the three geometries, together with the stereographic projections, various models of hyperbolic geometry, in such a way that we need only one Minkowski space, where null vectors represent points or points at infinity in any of the three geometries and any of the models of hyperbolic space, where Minkowski subspaces represent spheres and hyperplanes in any of the three geometries, and where stereographic projections are simply rescaling of null vectors. We call this construction the *homogeneous model*. It serves as a sixth analytic model for hyperbolic geometry.

We constructed homogeneous models for Euclidean and spherical geometries in previous chapters. There the models are constructed in Minkowski space by projective splits with respect to a fixed vector of null or negative signature. Here we show that a projective split with respect to a fixed vector of positive signature produces the homogeneous model of hyperbolic geometry.

Because the three geometries are obtained by interpreting null vectors of the same Minkowski space differently, natural correspondences exist among geometric entities and constraints of these geometries. In particular, there are correspondences among theorems on conformal properties of the three geometries. Every algebraic identity can be interpreted in three ways and therefore represents three theorems. In the last section we illustrate this feature with an example.

The homogeneous model has the significant advantage of simplifying geometric computations, because it employs the powerful language of *Geometric Algebra.* Geometric Algebra was applied to hyperbolic geometry by H. Li in [148], stimulated by Iversen's book [120] on the algebraic treatment of hyperbolic geometry and by the paper of Hestenes and Zielger [114] on projective geometry with Geometric Algebra.

4.2 The Hyperboloid Model

In this section we introduce some fundamentals of the hyperboloid model in the language of Geometric Algebra. More details can be found in [148].

In the Minkowski space $\mathbb{R}^{n,1}$, the set

$$\mathcal{D}^n = \{x \in \mathbb{R}^{n,1} | x^2 = -1\} \tag{4.1}$$

is called an n-dimensional *double-hyperbolic space*, any element in it is called a *point.* It has two connected branches, which are symmetric to the origin of $\mathbb{R}^{n+1,1}$. We denote one branch by $\mathcal{H}^n$ and the other by $-\mathcal{H}^n$. The branch $\mathcal{H}^n$ is called the *hyperboloid model* of n-dimensional hyperbolic space.

4.2.1 Generalized Points

Distances between two points

Let $\mathbf{p}, \mathbf{q}$ be two distinct points in $\mathcal{D}^n$, then $\mathbf{p}^2 = \mathbf{q}^2 = -1$. The blade $\mathbf{p} \wedge \mathbf{q}$ has Minkowski signature, therefore

$$0 < (\mathbf{p} \wedge \mathbf{q})^2 = (\mathbf{p} \cdot \mathbf{q})^2 - \mathbf{p}^2\mathbf{q}^2 = (\mathbf{p} \cdot \mathbf{q})^2 - 1. \tag{4.2}$$

From this we get

$$|\mathbf{p} \cdot \mathbf{q}| > 1. \tag{4.3}$$

Since $\mathbf{p}^2 = -1$, we can prove

Theorem 4.2.1. *For any two points* $\mathbf{p}, \mathbf{q}$ *in* $\mathcal{H}^n$ *(or* $-\mathcal{H}^n$*),*

$$\mathbf{p} \cdot \mathbf{q} < -1. \tag{4.4}$$

As a corollary, there exists a unique nonnegative number $d(\mathbf{p},\mathbf{q})$ such that

$$\mathbf{p}\cdot\mathbf{q} = -\cosh d(\mathbf{p},\mathbf{q}). \tag{4.5}$$

$d(\mathbf{p},\mathbf{q})$ is called the *hyperbolic distance* between $\mathbf{p},\mathbf{q}$.

Below we define several other equivalent distances. Let $\mathbf{p},\mathbf{q}$ be two distinct points in $\mathcal{H}^n$ (or $-\mathcal{H}^n$). The positive number

$$d_n(\mathbf{p},\mathbf{q}) = -(1+\mathbf{p}\cdot\mathbf{q}) \tag{4.6}$$

is called the *normal distance* between $\mathbf{p},\mathbf{q}$. The positive number

$$d_t(\mathbf{p},\mathbf{q}) = |\mathbf{p}\wedge\mathbf{q}| \tag{4.7}$$

is called the *tangential distance* between $\mathbf{p},\mathbf{q}$. The positive number

$$d_h(\mathbf{p},\mathbf{q}) = |\mathbf{p}-\mathbf{q}| \tag{4.8}$$

is called the *horo-distance* between $\mathbf{p},\mathbf{q}$. We have

$$\begin{aligned} d_n(\mathbf{p},\mathbf{q}) &= \cosh d(\mathbf{p},\mathbf{q}) - 1, \\ d_t(\mathbf{p},\mathbf{q}) &= \sinh d(\mathbf{p},\mathbf{q}), \\ d_h(\mathbf{p},\mathbf{q}) &= 2\sinh\frac{d(\mathbf{p},\mathbf{q})}{2}. \end{aligned} \tag{4.9}$$

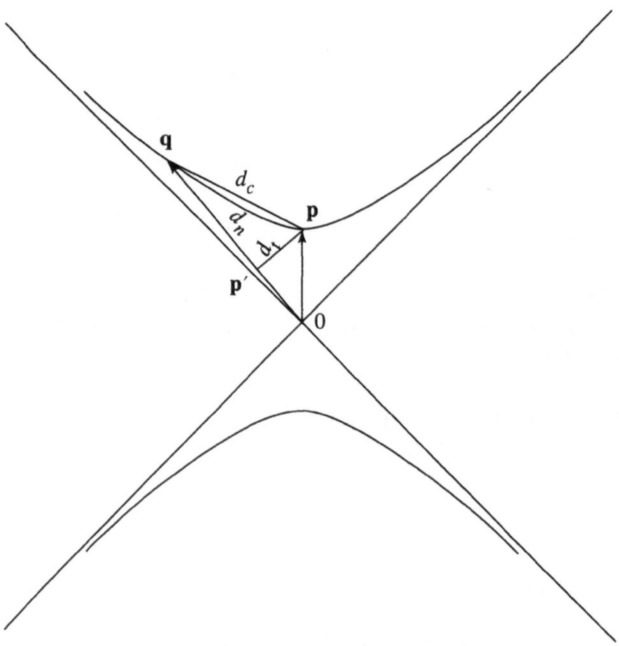

Fig. 4.1. Distances in hyperbolic geometry

Points at infinity

A *point at infinity* of $\mathcal{D}^n$ is a one-dimensional null space. It can be represented by a single null vector uniquely up to a nonzero scale factor.

The set of points at infinity in $\mathcal{D}^n$ is topologically an $(n-1)$-dimensional sphere, called the *sphere at infinity* of $\mathcal{D}^n$. The null cone

$$\mathcal{N}^{n-1} = \{x \in \mathbb{R}^{n,1} | x^2 = 0, x \neq 0\} \tag{4.10}$$

of $\mathbb{R}^{n,1}$ has two branches. Two null vectors h_1, h_2 are on the same connected component if and only if $h_1 \cdot h_2 < 0$. One branch $\mathcal{N}_+^{n-1}$ has the property: for any null vector h in $\mathcal{N}_+^{n-1}$, any point $\mathbf{p}$ in $\mathcal{H}^n$, $h \cdot \mathbf{p} < 0$. The other branch of the null cone is denoted by $\mathcal{N}_-^{n-1}$.

For a null vector h, the *relative distance* between h and point $\mathbf{p} \in \mathcal{D}^n$ is defined as

$$d_r(h, \mathbf{p}) = |h \cdot \mathbf{p}|. \tag{4.11}$$

Imaginary points

An *imaginary point* of $\mathcal{D}^n$ is a one-dimensional Euclidean space. It can be represented by a vector of unit square in $\mathbb{R}^{n+1,1}$.

The dual of an imaginary point is a hyperplane. An *r-plane* in $\mathcal{D}^n$ is the intersection of an $(r+1)$-dimensional Minkowski space of $\mathbb{R}^{n,1}$ with $\mathcal{D}^n$. A *hyperplane* is an $(n-1)$-plane.

Let a be an imaginary point, $\mathbf{p}$ be a point. There exists a unique line, a 1-plane in $\mathcal{D}^n$, which passes through $\mathbf{p}$ and is perpendicular to the hyperplane $\tilde{a}$ dual to a. This line intersects the hyperplane at a pair of antipodal points $\pm\mathbf{q}$. The *hyperbolic, normal and tangent distances* between $a, \mathbf{p}$ are defined as the respective distances between $\mathbf{p}, \mathbf{q}$. We have

$$\begin{aligned} \cosh d(a, \mathbf{p}) &= |a \wedge \mathbf{p}|, \\ d_n(a, \mathbf{p}) &= |a \wedge \mathbf{p}| - 1, \\ d_t(a, \mathbf{p}) &= |a \cdot \mathbf{p}|. \end{aligned}$$

A *generalized point* of $\mathcal{D}^n$ refers to a point, or a point at infinity, or an imaginary point.

Oriented generalized points and signed distances

The above definitions of generalized points are from [148], where the topic was $\mathcal{H}^n$ instead of $\mathcal{D}^n$, and where $\mathcal{H}^n$ was taken as $\mathcal{D}^n$ with antipodal points identified, instead of just a connected component of $\mathcal{D}^n$. When studying double-hyperbolic space, it is useful to distinguish between null vectors h and $-h$ representing the same point at infinity, and vectors a and $-a$ representing the same imaginary point. Actually it is indispensible when we study generalized spheres in $\mathcal{D}^n$. For this purpose we define oriented generalized points.

Any null vector in $\mathbb{R}^{n,1}$ represents an *oriented point at infinity* of $\mathcal{D}^n$. Two null vectors in $\mathcal{D}^n$ are said to represent the same oriented point at infinity if

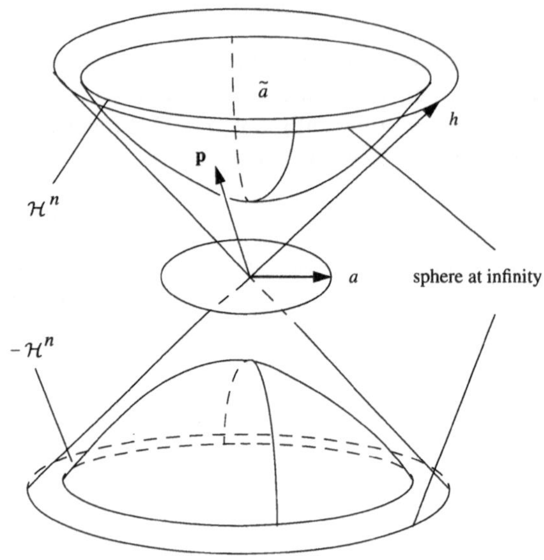

Fig. 4.2. Generalized points in $\mathcal{D}^n$: $\mathbf{p}$ is a point, h is a point at infinity, and a is an imaginary point

and only they differ by a positive scale factor; in other words, null vectors f and $-f$ represent two antipodal oriented points at infinity.

Any unit vector in $\mathbb{R}^{n,1}$ of positive signature represents an *oriented imaginary point* of $\mathcal{D}^n$. Two unit vectors a and $-a$ of positive signature represent two antipodal oriented imaginary points. The dual of an oriented imaginary point is an oriented hyperplane of $\mathcal{D}^n$.

A point in $\mathcal{D}^n$ is already oriented.

We can define various *signed distances* between two oriented generalized points, for example,

– the signed normal distance between two points $\mathbf{p}, \mathbf{q}$ is defined as

$$-\mathbf{p} \cdot \mathbf{q} - 1, \tag{4.12}$$

which is nonnegative when $\mathbf{p}, \mathbf{q}$ are on the same branch of $\mathcal{D}^n$ and ≤ -2 otherwise;

– the signed relative distance between point $\mathbf{p}$ and oriented point at infinity h is defined as

$$-h \cdot \mathbf{p}, \tag{4.13}$$

which is positive for $\mathbf{p}$ on one branch of $\mathcal{D}^n$ and negative otherwise;

– the signed tangent distance between point $\mathbf{p}$ and oriented imaginary point a is defined as

$$-a \cdot \mathbf{p}, \tag{4.14}$$

which is zero when $\mathbf{p}$ is on the hyperplane $\tilde{a}$, positive when $\mathbf{p}$ is on one side of the hyperplane and negative otherwise.

4.2.2 Total Spheres

A *total sphere* of $\mathcal{D}^n$ refers to a hyperplane, or the sphere at infinity, or a generalized sphere. An *r-dimensional total sphere* of $\mathcal{D}^n$ refers to the intersection of a total sphere with an $(r+1)$-plane.

A *generalized sphere* in $\mathcal{H}^n$ (or $-\mathcal{H}^n$, or $\mathcal{D}^n$) refers to a sphere, or a horosphere, or a hypersphere in $\mathcal{H}^n$ (or $-\mathcal{H}^n$, or $\mathcal{D}^n$). It is defined by a pair (c, ρ), where c is a vector representing an oriented generalized point, and ρ is a positive scalar.

1. When $c^2 = -1$, i.e., c is a point, then if c is in $\mathcal{H}^n$, the set
$$\{\mathbf{p} \in \mathcal{H}^n | d_n(\mathbf{p}, c) = \rho\} \tag{4.15}$$
is the *sphere* in $\mathcal{H}^n$ with center c and normal radius ρ; if c is in $-\mathcal{H}^n$, the set
$$\{\mathbf{p} \in -\mathcal{H}^n | d_n(\mathbf{p}, c) = \rho\} \tag{4.16}$$
is a *sphere* in $-\mathcal{H}^n$.
2. When $c^2 = 0$, i.e., c is an oriented point at infinity, then if $c \in \mathcal{N}_+^{n-1}$, the set
$$\{\mathbf{p} \in \mathcal{H}^n | d_r(\mathbf{p}, c) = \rho\} \tag{4.17}$$
is the *horosphere* in $\mathcal{H}^n$ with center c and relative radius ρ; otherwise the set
$$\{\mathbf{p} \in -\mathcal{H}^n | d_r(\mathbf{p}, c) = \rho\} \tag{4.18}$$
is a *horosphere* in $-\mathcal{H}^n$.
3. When $c^2 = 1$, i.e., c is an oriented imaginary point, the set
$$\{\mathbf{p} \in \mathcal{D}^n | \mathbf{p} \cdot c = -\rho\} \tag{4.19}$$
is the *hypersphere* in $\mathcal{D}^n$ with center c and tangent radius ρ; its intersection with $\mathcal{H}^n$ (or $-\mathcal{H}^n$) is a *hypersphere* in $\mathcal{H}^n$ (or $-\mathcal{H}^n$). The hyperplane $\tilde{c}$ is called the *axis* of the hypersphere.
A hyperplane can also be regarded as a hypersphere with zero radius.

4.3 The Homogeneous Model

In this section we establish the homogeneous model of the hyperbolic space. Strictly speaking, the model is for the double-hyperbolic space, as we must take into account both branches.

Fixing a vector a_0 of positive signature in $\mathbb{R}^{n+1,1}$, assuming $a_0^2 = 1$, we get
$$\mathcal{N}_{a_0}^n = \{x \in \mathbb{R}^{n+1,1} | x^2 = 0, x \cdot a_0 = -1\}. \tag{4.20}$$
Applying the orthogonal decomposition

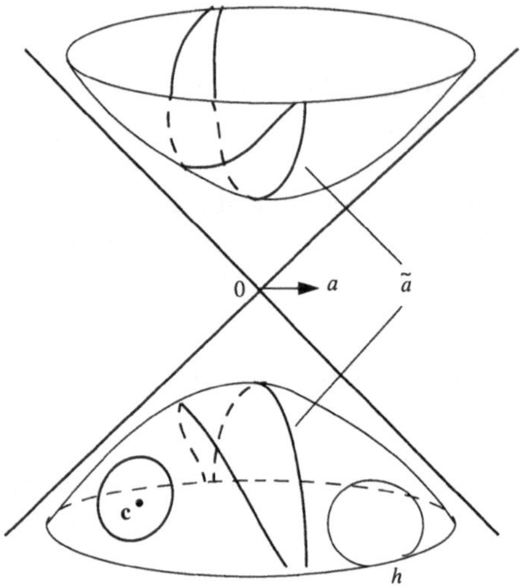

Fig. 4.3. Generalized spheres in $\mathcal{D}^n$: $\mathbf{p}$ the center of a sphere, h the center of a horosphere, a the center of a hypersphere

$$x = P_{a_0}(x) + P_{\tilde{a}_0}(x) \tag{4.21}$$

to vector $x \in \mathbb{N}^n_{a_0}$, we get

$$x = -a_0 + \mathbf{x} \tag{4.22}$$

where $\mathbf{x} \in \mathcal{D}^n$, the negative unit sphere of the Minkowski space represented by $\tilde{a}_0$. The map $i_{a_0} : \mathbf{x} \in \mathcal{D}^n \mapsto x \in \mathbb{N}^n_{a_0}$ is bijective. Its inverse map is $P^{\perp}_{a_0}$.

Theorem 4.3.1.

$$\mathbb{N}^n_{a_0} \simeq \mathcal{D}^n. \tag{4.23}$$

We call $\mathbb{N}^n_{a_0}$ the *homogeneous model* of $\mathcal{D}^n$. Its elements are called *homogeneous points.*

We use $\mathcal{H}^n$ to denote the intersection of $\mathcal{D}^n$ with $\mathcal{H}^{n+1}$, and $-\mathcal{H}^n$ to denote the intersection of $\mathcal{D}^n$ with $-\mathcal{H}^{n+1}$. Here $\pm\mathcal{H}^{n+1}$ are the two branches of $\mathcal{D}^{n+1}$, the negative unit sphere of $\mathbb{R}^{n+1,1}$.

4.3.1 Generalized Points

Let $\mathbf{p}, \mathbf{q}$ be two points in $\mathcal{D}^n$. Then for homogeneous points $\mathbf{p}, \mathbf{q}$

$$p \cdot q = (-a_0 + \mathbf{p}) \cdot (-a_0 + \mathbf{q}) = 1 + \mathbf{p} \cdot \mathbf{q}. \tag{4.24}$$

Thus the inner product of two homogeneous points "in" $\mathcal{D}^n$ equals the negative of the signed normal distance between them.

An oriented point at infinity of $\mathcal{D}^n$ is represented by a null vector h of $\mathbb{R}^{n+1,1}$ satisfying

$$h \cdot a_0 = 0. \tag{4.25}$$

For a point $\mathbf{p}$ of $\mathcal{D}^n$, we have

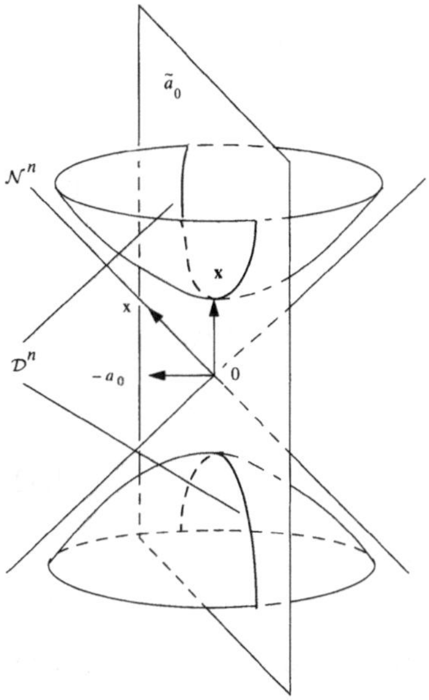

Fig. 4.4. The homogeneous model of $\mathcal{D}^n$

$$h \cdot p = h \cdot (-a_0 + \mathbf{p}) = h \cdot \mathbf{p}. \tag{4.26}$$

Thus the inner product of an oriented point at infinity with a homogeneous point equals the negative of the signed relative distance between them.

An oriented imaginary point of $\mathcal{D}^n$ is represented by a vector a of unit square in $\mathbb{R}^{n+1,1}$ satisfying

$$a \cdot a_0 = 0. \tag{4.27}$$

For a point $\mathbf{p}$ of $\mathcal{D}^n$, we have

$$a \cdot p = a \cdot (-a_0 + \mathbf{p}) = a \cdot \mathbf{p}. \tag{4.28}$$

Thus the inner product of a homogeneous point and an oriented imaginary point equals the negative of the signed tangent distance between them.

4.3.2 Total Spheres

Below we establish the conclusion that any $(n+1)$-blade of Minkowski signature in $\mathbb{R}^{n+1,1}$ corresponds to a total sphere in $\mathcal{D}^n$.

Let s be a vector of positive signature in $\mathbb{R}^{n+1,1}$.

1. If $s \wedge a_0 = 0$, then s equals a_0 up to a nonzero scalar factor. The blade $\tilde{s}$ represents the sphere at infinity of $\mathcal{D}^n$.
2. If $s \wedge a_0$ has Minkowski signature, then $s \cdot a_0 \neq 0$. Let $(-1)^\epsilon$ be the sign of $s \cdot a_0$. Let

$$\mathbf{c} = (-1)^{1+\epsilon} \frac{P_{a_0}^{\perp}(s)}{|P_{a_0}^{\perp}(s)|}, \tag{4.29}$$

then $\mathbf{c} \in \mathcal{D}^n$. Let

$$s' = (-1)^{1+\epsilon} \frac{s}{|a_0 \wedge s|}, \tag{4.30}$$

then

$$s' = (-1)^{1+\epsilon} \frac{P_{a_0}^{\perp}(s)}{|a_0 \wedge s|} + (-1)^{1+\epsilon} \frac{P_{a_0}(s)}{|a_0 \wedge s|} = \mathbf{c} - (1+\rho)a_0, \tag{4.31}$$

where

$$\rho = \frac{|a_0 \cdot s|}{|a_0 \wedge s|} - 1 > 0 \tag{4.32}$$

because $|a_0 \wedge s|^2 = (a_0 \cdot s)^2 - s^2 < (a_0 \cdot s)^2$.
For any point $\mathbf{p} \in \mathcal{D}^n$,

$$s' \cdot p = (\mathbf{c} - (1+\rho)a_0) \cdot (\mathbf{p} - a_0) = \mathbf{c} \cdot \mathbf{p} + 1 + \rho. \tag{4.33}$$

So $\tilde{s}$ represents the sphere in $\mathcal{D}^n$ with center $\mathbf{c}$ and normal radius ρ; a point $\mathbf{p}$ is on the sphere if and only if $p \cdot s = 0$.
The standard form of a sphere in $\mathcal{D}^n$ is

$$c - \rho a_0. \tag{4.34}$$

3. If $s \wedge a_0$ is degenerate, then $(s \wedge a_0)^2 = (s \cdot a_0)^2 - s^2 = 0$, so $|s \cdot a_0| = |s| \neq 0$. As before, $(-1)^\epsilon$ is the sign of $s \cdot a_0$. Let

$$c = (-1)^{1+\epsilon} P_{a_0}^{\perp}(s), \tag{4.35}$$

then $c^2 = 0$ and $c \cdot a_0 = 0$, so c represents an oriented point at infinity of $\mathcal{D}^n$. Let

$$s' = (-1)^{1+\epsilon} s. \tag{4.36}$$

Then

$$s' = (-1)^{1+\epsilon}(P_{a_0}^{\perp}(s) + P_{a_0}(s)) = c - \rho a_0, \tag{4.37}$$

where

$$\rho = |a_0 \cdot s| = |s| > 0. \tag{4.38}$$

For any point $\mathbf{p} \in \mathcal{D}^n$,

$$s' \cdot p = (c - \rho a_0) \cdot (\mathbf{p} - a_0) = c \cdot \mathbf{p} + \rho, \tag{4.39}$$

so $\tilde{s}$ represents the horosphere in $\mathcal{D}^n$ with center c and relative radius ρ; a point $\mathbf{p}$ is on the sphere if and only if $p \cdot s = 0$.
The standard form of a horosphere in $\mathcal{D}^n$ is

$$c - \rho a_0. \tag{4.40}$$

4. The term $s \wedge a_0$ is Euclidean, but $s \cdot a_0 \neq 0$. Let

$$c = (-1)^{1+\epsilon} \frac{P^{\perp}_{a_0}(s)}{|P^{\perp}_{a_0}(s)|}, \tag{4.41}$$

then $c^2 = 1$ and $c \cdot a_0 = 0$, i.e., c represents an oriented imaginary point of $\mathcal{D}^n$. Let

$$s' = (-1)^{1+\epsilon} \frac{s}{|a_0 \wedge s|}, \tag{4.42}$$

then

$$s' = (-1)^{1+\epsilon} \frac{P^{\perp}_{a_0}(s)}{|a_0 \wedge s|} + (-1)^{1+\epsilon} \frac{P_{a_0}(s)}{|a_0 \wedge s|} = c - \rho a_0, \tag{4.43}$$

where

$$\rho = \frac{|a_0 \cdot s|}{|a_0 \wedge s|} > 0. \tag{4.44}$$

For any point $\mathbf{p} \in \mathcal{D}^n$,

$$s' \cdot p = (c - \rho a_0) \cdot (\mathbf{p} - a_0) = c \cdot \mathbf{p} + \rho, \tag{4.45}$$

so $\tilde{s}$ represents the hypersphere in $\mathcal{D}^n$ with center c and tangent radius ρ; a point $\mathbf{p}$ is on the hypersphere if and only if $p \cdot s = 0$.
The standard form of a hypersphere in $\mathcal{D}^n$ is

$$c - \rho a_0. \tag{4.46}$$

5. If $s \cdot a_0 = 0$, then $s \wedge a_0$ is Euclidean, because $(s \wedge a_0)^2 = -s^2 < 0$. For any point $\mathbf{p} \in \mathcal{D}^n$, since

$$s \cdot p = s \cdot \mathbf{p}, \tag{4.47}$$

$\tilde{s}$ represents the hyperplane of $\mathcal{D}^n$ normal to vector s; a point $\mathbf{p}$ is on the hyperplane if and only if $p \cdot s = 0$.

From the above analysis we come to the following conclusion:

Theorem 4.3.2. *The intersection of any Minkowski hyperspace of $\mathbb{R}^{n+1,1}$ represented by $\tilde{s}$ with $\mathcal{N}^n_{a_0}$ is a total sphere in $\mathcal{D}^n$, and every total sphere can be obtained in this way. A point $\mathbf{p}$ in $\mathcal{D}^n$ is on the total sphere if and only if $p \cdot s = 0$.*

The dual of the above theorem is:

Theorem 4.3.3. *Given $n+1$ homogeneous points or points at infinity of $\mathcal{D}^n$: $a_0, \ldots, a_n$ such that*

$$\tilde{s} = a_0 \wedge \cdots \wedge a_n. \tag{4.48}$$

This $(n+1)$-blade $\tilde{s}$ represents a total sphere passing through these points or points at infinity. It is a hyperplane if

$$a_0 \wedge \tilde{s} = 0, \tag{4.49}$$

the sphere at infinity if

$$a_0 \cdot \tilde{s} = 0, \tag{4.50}$$

a sphere if

$$(a_0 \cdot \tilde{s})^\dagger (a_0 \cdot \tilde{s}) > 0, \tag{4.51}$$

a horosphere if

$$a_0 \cdot \tilde{s} \neq 0, \text{ and } (a_0 \cdot \tilde{s})^\dagger (a_0 \cdot \tilde{s}) = 0, \tag{4.52}$$

or a hypersphere if

$$(a_0 \cdot \tilde{s})^\dagger (a_0 \cdot \tilde{s}) < 0. \tag{4.53}$$

The scalar

$$s_1 * s_2 = \frac{s_1 \cdot s_2}{|s_1||s_2|} \tag{4.54}$$

is called the *inversive product* of vectors s_1 and s_2. Obviously, it is invariant under orthogonal transformations in $\mathbb{R}^{n+1,1}$. We have the following conclusion for the inversive product of two vectors of positive signature:

Theorem 4.3.4. *When total spheres $\tilde{s}_1$ and $\tilde{s}_2$ intersect, let p be a point or point at infinity of the intersection. Let m_i, $i = 1, 2$, be the respective outward unit normal vector of $\tilde{s}_i$ at p if it is a generalized sphere and p is a point, or let m_i be $s_i/|s_i|$ otherwise, then*

$$s_1 * s_2 = m_1 \cdot m_2. \tag{4.55}$$

Proof. The case when p is a point at infinity is trivial, so we only consider the case when p is a point, denoted by $\mathbf{p}$. The total sphere $\tilde{s}_i$ has the standard form $(c_i - \lambda_i a_0)^\sim$, where $c_i \cdot a_0 = 0$, $\lambda_i \geq 0$ and $(c_i - \lambda_i a_0)^2 = c_i^2 + \lambda_i^2 > 0$. Hence

$$s_1 * s_2 = \frac{c_1 \cdot c_2 + \lambda_1 \lambda_2}{|c_1 - \lambda_1 a_0||c_2 - \lambda_2 a_0|} = \frac{c_1 \cdot c_2 + \lambda_1 \lambda_2}{\sqrt{(c_1^2 + \lambda_1^2)(c_2^2 + \lambda_2^2)}}. \tag{4.56}$$

On the other hand, at point $\mathbf{p}$ the outward unit normal vector of generalized sphere $\tilde{s}_i$ is

$$m_i = \frac{\mathbf{p}(\mathbf{p} \wedge c_i)}{|\mathbf{p} \wedge c_i|}, \tag{4.57}$$

which equals $c_i = s_i/|s_i|$ when $\tilde{s}_i$ is a hyperplane. Since point $\mathbf{p}$ is on both total spheres, $\mathbf{p} \cdot c_i = -\lambda_i$, so

$$m_1 \cdot m_2 = \frac{(c_1 - \lambda_1 a_0) \cdot (c_2 - \lambda_2 a_0)}{|\mathbf{p} \wedge c_1||\mathbf{p} \wedge c_2|} = \frac{c_1 \cdot c_2 + \lambda_1 \lambda_2}{\sqrt{(c_1^2 + \lambda_1^2)(c_2^2 + \lambda_2^2)}}. \tag{4.58}$$

An immediate corollary is that any orthogonal transformation in $\mathbb{R}^{n+1,1}$ induces an angle-preserving transformation in $\mathcal{D}^n$.

4.3.3 Total Spheres of Dimensional r

Theorem 4.3.5. *For $2 \leq r \leq n+1$, every r-blade A_r of Minkowski signature in $\mathbb{R}^{n+1,1}$ represents an $(r-2)$-dimensional total sphere in $\mathcal{D}^n$.*

Proof. There are three possibilities:
Case 1. When $a_0 \wedge A_r = 0$, A_r represents an $(r-2)$-plane in $\mathcal{D}^n$. After normalization, the *standard form* of an $(r-2)$-plane is

$$a_0 \wedge \mathbf{I}_{r-2,1}, \tag{4.59}$$

where $\mathbf{I}_{r-2,1}$ is a unit Minkowski $(r-1)$-blade of $\mathcal{G}(\mathbb{R}^{n,1})$, and where $\mathbb{R}^{n,1}$ is represented by $\tilde{a}_0$.
Case 2. When $a_0 \cdot A_r = 0$, A_r represents an $(r-2)$-dimensional sphere at infinity of $\mathcal{D}^n$. It lies on the $(r-1)$-plane $a_0 \wedge A_r$. After normalization, the *standard form* of the $(r-2)$-dimensional sphere at infinity is

$$\mathbf{I}_{r-1,1}, \tag{4.60}$$

where $\mathbf{I}_{r-1,1}$ is a unit Minkowski r-blade of $\mathcal{G}(\mathbb{R}^{n,1})$.
Case 3. When both $a_0 \wedge A_r \neq 0$ and $a_0 \cdot A_r \neq 0$, A_r represents an $(r-2)$-dimensional generalized sphere. This is because

$$A_{r+1} = a_0 \wedge A_r \neq 0, \tag{4.61}$$

and the vector

$$s = A_r A_{r+1}^{-1} \tag{4.62}$$

has positive square with both $a_0 \cdot s \neq 0$ and $a_0 \wedge s \neq 0$, so $\tilde{s}$ represents an $(n-1)$-dimensional generalized sphere. According to Case 1, A_{r+1} represents an $(r-1)$-dimensional plane in $\mathcal{D}^n$. Therefore, with $\epsilon = \frac{(n+2)(n+1)}{2} + 1$,

$$A_r = sA_{r+1} = (-1)^{\epsilon}\tilde{s} \vee A_{r+1} \tag{4.63}$$

represents the intersection of $(n-1)$-dimensional generalized sphere $\tilde{s}$ with $(r-1)$-plane A_{r+1}, which is an $(r-2)$-dimensional generalized sphere.

With suitable normalization, we can write

$$s = c - \rho a_0. \tag{4.64}$$

Since $s \wedge A_{r+1} = p_0 \wedge A_{r+1} = 0$, the generalized sphere A_r is also centered at c and has normal radius ρ, and it is of the same type as the generalized sphere represented by $\tilde{s}$. Now we can represent an $(r-2)$-dimensional generalized sphere in the *standard form*

$$(c - \lambda a_0)\ (a_0 \wedge \mathbf{I}_{r-1,1}), \tag{4.65}$$

where $\mathbf{I}_{r-1,1}$ is a unit Minkowski r-blade of $\mathcal{G}(\mathbb{R}^{n,1})$.

Corollary: *The $(r-2)$-dimensional total sphere passing through r homogeneous points or points at infinity $p_1, \ldots, p_r$ in $\mathcal{D}^n$ is represented by $A_r = p_1 \wedge \cdots \wedge p_r$; the $(r-2)$-plane passing through $r-1$ homogeneous points or points at infinity $p_1, \ldots, p_{r-1}$ in $\mathcal{D}^n$ is represented by $a_0 \wedge p_1 \wedge \cdots \wedge p_{r-1}$.*

When the p's are all homogeneous points, we can expand the inner product $A_r^{\dagger} \cdot A_r$ as

$$A_r^{\dagger} \cdot A_r = \det(p_i \cdot p_j)_{r\times r} = (-\frac{1}{2})^r \det((\mathbf{p}_i - \mathbf{p}_j)^2)_{r\times r}. \tag{4.66}$$

When $r = n+2$, we obtain Ptolemy's Theorem for double-hyperbolic space:

Theorem 4.3.6 (Ptolemy's Theorem). *Let $\mathbf{p}_1, \cdots, \mathbf{p}_{n+2}$ be points in $\mathcal{D}^n$, then they are on a generalized sphere or hyperplane of $\mathcal{D}^n$ if and only if* $\det((\mathbf{p}_i - \mathbf{p}_j)^2)_{(n+2)\times(n+2)} = 0$.

4.4 Stereographic Projection

In $\mathbb{R}^{n,1}$, let $\mathbf{p}_0$ be a fixed point in $\mathcal{H}^n$. The space $\mathbb{R}^n = (a_0 \wedge \mathbf{p}_0)^{\sim}$, which is parallel to the tangent hyperplanes of $\mathcal{D}^n$ at points $\pm\mathbf{p}_0$, is Euclidean. By the *stereographic projection* of $\mathcal{D}^n$ from point $-\mathbf{p}_0$ to the space $\mathbb{R}^n$, every affine line of $\mathbb{R}^{n,1}$ passing through points $-\mathbf{p}_0$ and $\mathbf{p}$ intersects $\mathbb{R}^n$ at point

$$j_{\mathcal{D}\mathbb{R}}(\mathbf{p}) = \frac{\mathbf{p}_0(\mathbf{p}_0 \wedge \mathbf{p})}{\mathbf{p}_0 \cdot \mathbf{p} - 1} = -2(\mathbf{p} + \mathbf{p}_0)^{-1} - \mathbf{p}_0. \tag{4.67}$$

Any point at infinity of $\mathcal{D}^n$ can be written in the form $\mathbf{p}_0 + a$, where a is a unit vector in $\mathbb{R}^n$ represented by $(a_0 \wedge \mathbf{p}_0)^{\sim}$. Every affine line passing through point $-p_0$ and point at infinity $\mathbf{p}_0 + a$ intersects $\mathbb{R}^n$ at point a. It is a classical result that the map $j_{\mathcal{D}\mathbb{R}}$ is a conformal map from $\mathcal{D}^n$ to $\mathbb{R}^n$.

We show that in the homogeneous model we can construct the conformal map j_{SR} trivially; it is nothing but a rescaling of null vectors.

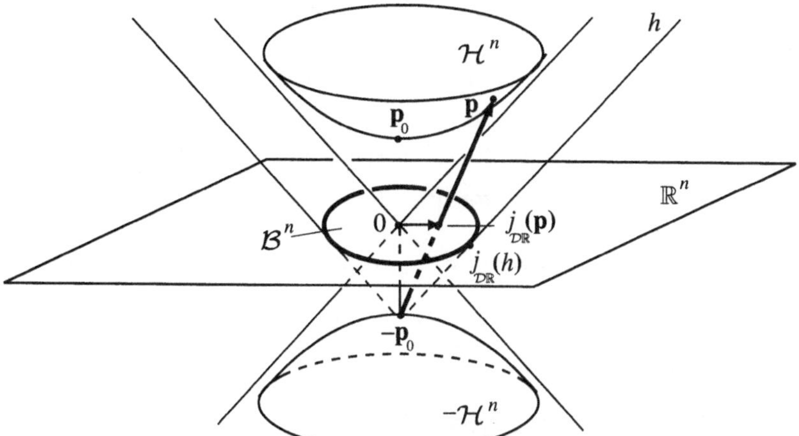

Fig. 4.5. Stereographic projection of $\mathcal{D}^n$ from $-\mathbf{p}_0$ to $\mathbb{R}^n$

Let

$$e = a_0 + \mathbf{p}_0, \quad e_0 = \frac{-a_0 + \mathbf{p}_0}{2}, \quad E = e \wedge e_0. \tag{4.68}$$

Then for $\mathbb{R}^n = (e \wedge e_0)^\sim = (a_0 \wedge \mathbf{p}_0)^\sim$, the map

$$i_E : x \in \mathbb{R}^n \mapsto e_0 + x + \frac{x^2}{2} e \in \mathcal{N}_e^n \tag{4.69}$$

defines a homogeneous model for Euclidean space.

Any null vector h in $\mathbb{R}^{n+1,1}$ represents a point or point at infinity in both homogeneous models of $\mathcal{D}^n$ and $\mathbb{R}^n$. The rescaling transformation $k_\mathbb{R} : \mathcal{N}^n \longrightarrow \mathcal{N}_e^n$ defined by

$$k_\mathbb{R}(h) = -\frac{h}{h \cdot e}, \quad \text{for } h \in \mathcal{N}^n, \tag{4.70}$$

where $\mathcal{N}^n$ represents the null cone of $\mathbb{R}^{n+1,1}$, induces the stereographic projection $j_{\mathcal{D}\mathbb{R}}$ through the following commutative diagram:

$$\begin{array}{ccc}
\mathbf{p} - a_0 \in \mathcal{N}_{a_0}^n & \overset{k_\mathbb{R}}{- - - - \longrightarrow} & \dfrac{\mathbf{p} - a_0}{1 - \mathbf{p} \cdot \mathbf{p}_0} \in \mathcal{N}_e^n \\
\uparrow i_{a_0} & & \downarrow P_E^\perp \\
\mathbf{p} \in \mathcal{D}^n & \overset{j_{\mathcal{D}\mathbb{R}}}{- - - - \longrightarrow} & \dfrac{\mathbf{p}_0(\mathbf{p}_0 \wedge \mathbf{p})}{\mathbf{p} \cdot \mathbf{p}_0 - 1} \in \mathbb{R}^n
\end{array}$$

i.e., $j_{\mathcal{D}\mathbb{R}} = P_E^\perp \circ k_\mathbb{R} \circ i_{a_0}$. Since a point at infinity $\mathbf{p}_0 + a$ of $\mathcal{D}^n$ belongs to $\mathcal{N}_e^n$, we have

$$j_{\mathcal{D}\mathbb{R}}(\mathbf{p}_0 + a) = P_E^{\perp}(\mathbf{p}_0 + a) = a. \tag{4.71}$$

The inverse of the map $j_{\mathcal{D}\mathbb{R}}$, denoted by $j_{\mathbb{R}\mathcal{D}}$, is

$$j_{\mathbb{R}\mathcal{D}}(u) = \begin{cases} \dfrac{(1+u^2)\mathbf{p}_0 + 2u}{1-u^2}, & \text{for } u^2 \neq 1, u \in \mathbb{R}^n, \\ \mathbf{p}_0 + u, & \text{for } u^2 = 1, u \in \mathbb{R}^n, \end{cases} \tag{4.72}$$

When u is not on the unit sphere of $\mathbb{R}^n$, $j_{\mathbb{R}\mathcal{D}}(u)$ can also be written as

$$j_{\mathbb{R}\mathcal{D}}(u) = -2(u + \mathbf{p}_0)^{-1} - \mathbf{p}_0 = (u + \mathbf{p}_0)^{-1}\mathbf{p}_0(u + \mathbf{p}_0). \tag{4.73}$$

4.5 The Conformal Ball Model

The standard definition of the conformal ball model [120] is the unit ball $\mathcal{B}^n$ of $\mathbb{R}^n$ equipped with the following metric: for any $u, v \in \mathcal{B}^n$,

$$\cosh d(u,v) = 1 + \frac{2(u-v)^2}{(1-u^2)(1-v^2)}. \tag{4.74}$$

This model can be derived through the stereographic projection from $\mathcal{H}^n$ to $\mathbb{R}^n$. Recall that the sphere at infinity of $\mathcal{H}^n$ is mapped to the unit sphere of $\mathbb{R}^n$, and $\mathcal{H}^n$ is projected onto the unit ball $\mathcal{B}^n$ of $\mathbb{R}^n$. Using the formula (4.72) we get that for any two points u, v in the unit ball,

$$|j_{\mathbb{R}\mathcal{D}}(u) - j_{\mathbb{R}\mathcal{D}}(v)| = \frac{2|u-v|}{\sqrt{(1-u^2)(1-v^2)}}, \tag{4.75}$$

which is equivalent to (4.74) since

$$\cosh d(u,v) - 1 = \frac{|j_{\mathbb{R}\mathcal{D}}(u) - j_{\mathbb{R}\mathcal{D}}(v)|^2}{2}. \tag{4.76}$$

The following correspondences exist between the hyperboloid model and the conformal ball model:

1. A hyperplane normal to a and passing through $-\mathbf{p}_0$ in $\mathcal{D}^n$ corresponds to the hyperspace normal to a in $\mathbb{R}^n$.
2. A hyperplane normal to a but not passing through $-\mathbf{p}_0$ in $\mathcal{D}^n$ corresponds to the sphere orthogonal to the unit sphere $\mathcal{S}^{n-1}$ in $\mathbb{R}^n$; it has center $-\mathbf{p}_0 - \dfrac{a}{a \cdot \mathbf{p}_0}$ and radius $\dfrac{1}{|a \cdot \mathbf{p}_0|}$.
3. A sphere with center c and normal radius ρ in $\mathcal{D}^n$ and passing through $-\mathbf{p}_0$ corresponds to the hyperplane in $\mathbb{R}^n$ normal to $P_{\mathbf{p}_0}^{\perp}(c)$ with signed distance from the origin $-\dfrac{1+\rho}{\sqrt{(1+\rho)^2 - 1}} < -1$.

4. A sphere not passing through $-\mathbf{p}_0$ in $\mathcal{D}^n$ corresponds to a sphere disjoint with $\mathcal{S}^{n-1}$.
5. A horosphere with center c and relative radius ρ in $\mathcal{D}^n$ passing through $-\mathbf{p}_0$ corresponds to the hyperplane in $\mathbb{R}^n$ normal to $P^{\perp}_{\mathbf{p}_0}(c)$ and with signed distance -1 from the origin.
6. A horosphere not passing through $-\mathbf{p}_0$ in $\mathcal{D}^n$ corresponds to a sphere tangent with $\mathcal{S}^{n-1}$.
7. A hypersphere with center c and tangent radius ρ in $\mathcal{D}^n$ passing through $-\mathbf{p}_0$ corresponds to the hyperplane in $\mathbb{R}^n$ normal to $P^{\perp}_{\mathbf{p}_0}(c)$ and with signed distance from the origin $-\dfrac{\rho}{\sqrt{1+\rho^2}} > -1$.
8. A hypersphere not passing through $-\mathbf{p}_0$ in $\mathcal{D}^n$ corresponds to a sphere intersecting but not perpendicular with $\mathcal{S}^{n-1}$.

The homogeneous model differs from the hyperboloid model only by a rescaling of null vectors.

4.6 The Hemisphere Model

Let a_0 be a point in $\mathcal{S}^n$. The hemisphere model [37] is the hemisphere $\mathcal{S}^n_+$ centered at $-a_0$ of $\mathcal{S}^n$, equipped with the following metric: for two points a, b,

$$\cosh d(a,b) = 1 + \frac{1 - a \cdot b}{(a \cdot a_0)(b \cdot a_0)}. \tag{4.77}$$

This model is traditionally obtained as the stereographic projection $j_{\mathcal{S}\mathbb{R}}$ of $\mathcal{S}^n$ from a_0 to $\mathbb{R}^n$, which maps the hemisphere $\mathcal{S}^n_+$ onto the unit ball of $\mathbb{R}^n$. Since the stereographic projection $j_{\mathcal{D}\mathbb{R}}$ of $\mathcal{D}^n$ from $-\mathbf{p}_0$ to $\mathbb{R}^n$ also maps $\mathcal{H}^n$ onto the unit ball of $\mathbb{R}^n$, the composition

$$j_{\mathcal{D}\mathcal{S}} = j^{-1}_{\mathcal{S}\mathbb{R}} \circ j_{\mathcal{D}\mathbb{R}} : \mathcal{D}^n \longrightarrow \mathcal{S}^n \tag{4.78}$$

maps $\mathcal{H}^n$ to $\mathcal{S}^n_+$, and maps the sphere at infinity of $\mathcal{H}^n$ to $\mathcal{S}^{n-1}$, the boundary of $\mathcal{S}^n_+$, which is the hyperplane of $\mathcal{S}^n$ normal to a_0. This map is conformal and bijective. It produces the hemisphere model of the hyperbolic space.

The following correspondences exist between the hyperboloid model and the hemisphere model:

1. A hyperplane normal to a and passing through $-\mathbf{p}_0$ in $\mathcal{D}^n$ corresponds to the hyperplane normal to a in $\mathcal{S}^n$.
2. A hyperplane normal to a but not passing through $-\mathbf{p}_0$ in $\mathcal{D}^n$ corresponds to a sphere with center on $\mathcal{S}^{n-1}$.
3. A sphere with center $\mathbf{p}_0$ (or $-\mathbf{p}_0$) in $\mathcal{D}^n$ corresponds to a sphere in $\mathcal{S}^n$ with center $-a_0$ (or a_0).
4. A sphere in $\mathcal{D}^n$ corresponds to a sphere disjoint with $\mathcal{S}^{n-1}$.

5. A horosphere corresponds to a sphere tangent with $\mathcal{S}^{n-1}$.
6. A hypersphere with center c, relative radius ρ in $\mathcal{D}^n$ and axis passing through $-\mathbf{p}_0$ corresponds to the hyperplane in $\mathcal{S}^n$ normal to $c - \rho a_0$.
7. A hypersphere whose axis does not pass through $-\mathbf{p}_0$ in $\mathcal{D}^n$ corresponds to a sphere intersecting with $\mathcal{S}^{n-1}$.

The hemisphere model can also be obtained from the homogeneous model by rescaling null vectors. The map $k_{\mathcal{S}} : \mathcal{N}^n \longrightarrow \mathcal{N}^n_{p_0}$ defined by

$$k_{\mathcal{S}}(h) = -\frac{h}{h \cdot p_0}, \text{ for } h \in \mathcal{N}^n \tag{4.79}$$

induces a conformal map $j_{\mathcal{DS}}$ through the following commutative diagram:

$$\begin{array}{ccc}
\mathbf{p} - a_0 \in \mathcal{N}^n_{a_0} & \overset{k_{\mathcal{S}}}{- - - - \longrightarrow} & -\dfrac{\mathbf{p} - a_0}{\mathbf{p} \cdot \mathbf{p}_0} \in \mathcal{N}^n_{\mathbf{p}_0} \\
\uparrow i_{a_0} & & \downarrow P^{\perp}_{\mathbf{p}_0} \\
\mathbf{p} \in \mathcal{D}^n & \overset{j_{\mathcal{DS}}}{- - - - \longrightarrow} & \dfrac{a_0 + \mathbf{p}_0(\mathbf{p}_0 \wedge \mathbf{p})}{\mathbf{p} \cdot \mathbf{p}_0} \in \mathcal{S}^n
\end{array}$$

i.e., $j_{\mathcal{DS}} = P^{\perp}_{\mathbf{p}_0} \circ k_{\mathcal{S}} \circ i_{a_0}$. For a point $\mathbf{p}$ in $\mathcal{D}^n$,

$$j_{\mathcal{DS}}(\mathbf{p}) = \frac{a_0 + \mathbf{p}_0(\mathbf{p}_0 \wedge \mathbf{p})}{\mathbf{p}_0 \cdot \mathbf{p}} = -\mathbf{p}_0 - \frac{\mathbf{p} - a_0}{\mathbf{p} \cdot \mathbf{p}_0}. \tag{4.80}$$

For a point at infinity $\mathbf{p}_0 + a$, we have

$$j_{\mathcal{DS}}(\mathbf{p}_0 + a) = P^{\perp}_{\mathbf{p}_0}(\mathbf{p}_0 + a) = a. \tag{4.81}$$

We see that $\pm\mathbf{p}_0$ corresponds to $\mp a_0$. Let $\mathbf{p}$ correspond to a in $\mathcal{S}^n$. Then

$$\mathbf{p} \cdot \mathbf{p}_0 = -\frac{1}{a \cdot a_0}. \tag{4.82}$$

The inverse of the map $j_{\mathcal{DS}}$, denoted by $j_{\mathcal{SD}}$, is

$$j_{\mathcal{SD}}(a) = \begin{cases} a_0 - \dfrac{\mathbf{p}_0 + a}{a_0 \cdot a}, & \text{for } a \in \mathcal{S}^n, a \cdot a_0 \neq 0, \\ \mathbf{p}_0 + a, & \text{for } a \in \mathcal{S}^n, a \cdot a_0 = 0. \end{cases} \tag{4.83}$$

4.7 The Half-Space Model

The standard definition of the half-space model [120] is the half space $\mathbb{R}^n_+$ of $\mathbb{R}^n$ bounded by $\mathbb{R}^{n-1}$, which is the hyperspace normal to a unit vector a_0, contains point $-a_0$, and is equipped with the following metric: for any $u, v \in \mathbb{R}^n_+$,

$$\cosh d(u,v) = 1 + \frac{(u-v)^2}{2(u \cdot a_0)(v \cdot a_0)}. \tag{4.84}$$

This model is traditionally obtained from the hyperboloid model as follows: The stereographic projection $j_{\mathcal{S}\mathbb{R}}$ of $\mathcal{S}^n$ is from a_0 to $\mathbb{R}^{n+1,1}$. As an alternative "north pole" select a point b_0, which is orthogonal to a_0. This pole determines a stereographic projection j_{b_0} with projection plane is $\mathbb{R}^n = (b_0 \wedge \mathbf{p}_0)^{\sim}$. The map $j_{\mathcal{DS}} : \mathcal{D}^n \longrightarrow \mathcal{S}^n$ maps $\mathcal{H}^n$ to the hemisphere $\mathcal{S}^n_+$ centered at $-a_0$. The map j_{b_0} maps $\mathcal{S}^n_+$ to $\mathbb{R}^n_+$. As a consequence, the map

$$j_{\mathcal{H}\mathbb{R}} = j_{b_0} \circ j_{\mathcal{DS}} : \mathcal{D}^n \longrightarrow \mathbb{R}^n \tag{4.85}$$

maps $\mathcal{H}^n$ to $\mathbb{R}^n_+$, and maps the sphere at infinity of $\mathcal{D}^n$ to $\mathbb{R}^{n-1}$.

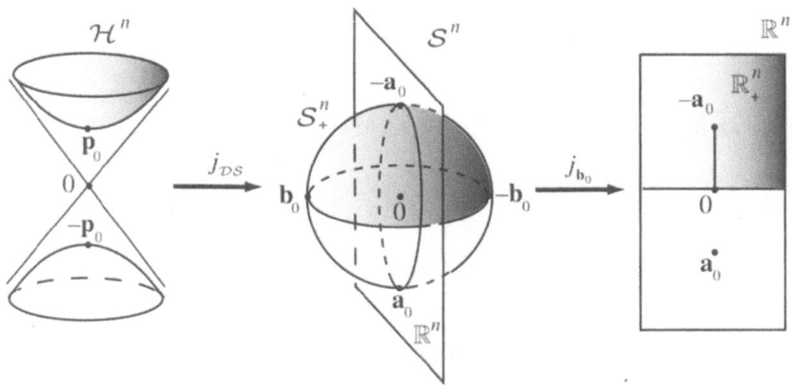

Fig. 4.6. The hemisphere model and the half-space model

The half-space model can also be derived from the homogeneous model by rescaling null vectors. Let $\mathbf{p}_0$ be a point in $\mathcal{H}^n$ and h be a point at infinity of $\mathcal{H}^n$, then $h \wedge \mathbf{p}_0$ is a line in $\mathcal{H}^n$, which is also a line in $\mathcal{H}^{n+1}$, the $(n+1)$-dimensional hyperbolic space in $\mathbb{R}^{n+1,1}$. The Euclidean space $\mathbb{R}^n = (h \wedge \mathbf{p}_0)^{\sim}$ is in the tangent hyperplane of $\mathcal{H}^{n+1}$ at $\mathbf{p}_0$ and normal to the tangent vector $P^{\perp}_{\mathbf{p}_0}(h)$ of line $h \wedge \mathbf{p}_0$. Let

$$e = -\frac{h}{h \cdot \mathbf{p}_0}, \quad e_0 = \mathbf{p}_0 - \frac{e}{2}. \tag{4.86}$$

Then $e^2 = e_0^2 = 0$, $e \cdot e_0 = e \cdot \mathbf{p}_0 = -1$, and $e \wedge \mathbf{p}_0 = e \wedge e_0$. The unit vector

$$b_0 = e - \mathbf{p}_0 \tag{4.87}$$

is orthogonal to both $\mathbf{p}_0$ and a_0, and can be identified with the pole b_0 of the stereographic projection j_{b_0}. Let $E = e \wedge e_0$. The rescaling map $k_R : \mathcal{N}^n \longrightarrow \mathcal{N}^n_e$ induces the map $j_{\mathcal{H}\mathbb{R}}$ through the following commutative diagram:

$$\begin{array}{ccc}
\mathbf{p} - a_0 \in \mathcal{N}^n_{a_0} & \overset{k_{\mathbb{R}}}{\dashrightarrow} & -\dfrac{\mathbf{p} - a_0}{\mathbf{p} \cdot e} \in \mathcal{N}^n_e \\
\Big\uparrow i_{a_0} & & \Big\downarrow P_E^{\perp} \\
\mathbf{p} \in \mathcal{D}^n & \overset{j_{\mathcal{H}\mathbb{R}}}{\dashrightarrow} & \dfrac{a_0 - P_E^{\perp}(\mathbf{p})}{\mathbf{p} \cdot e} \in \mathbb{R}^n
\end{array}$$

i.e., $j_{\mathcal{H}\mathbb{R}} = P_E^{\perp} \circ k_R \circ i_{a_0}$. For a point $\mathbf{p}$ in $\mathcal{D}^n$, we have

$$j_{\mathcal{H}\mathbb{R}}(\mathbf{p}) = \frac{a_0 - P^{\perp}_{e \wedge \mathbf{p}_0}(\mathbf{p})}{\mathbf{p} \cdot e}. \tag{4.88}$$

For a point at infinity $\mathbf{p}_0 + a$ in $\mathcal{D}^n$, we have

$$j_{\mathcal{H}\mathbb{R}}(\mathbf{p}_0 + a) = \frac{a + e \cdot a(\mathbf{p}_0 - e)}{1 - e \cdot a}. \tag{4.89}$$

The inverse of the map $j_{\mathcal{H}\mathbb{R}}$ is denoted by $j_{\mathbb{R}\mathcal{H}}$:

$$j_{\mathbb{R}\mathcal{H}}(u) = \begin{cases} a_0 - \dfrac{e_0 + u + \frac{u^2}{2} e}{a_0 \cdot u}, & \text{for } u \in \mathbb{R}^n, u \cdot a_0 \neq 0, \\[2ex] e_0 + u + \dfrac{u^2}{2} e, & \text{for } u \in \mathbb{R}^n, u \cdot a_0 = 0. \end{cases} \tag{4.90}$$

The following correspondences exist between the hyperboloid model and the half-space model:

1. A hyperplane normal to a and passing through e in $\mathcal{D}^n$ corresponds to the hyperplane in $\mathbb{R}^n$ normal to $a + a \cdot \mathbf{p}_0 e$ with signed distance $-a \cdot \mathbf{p}_0$ from the origin.
2. A hyperplane not passing through e in $\mathcal{D}^n$ corresponds to a sphere with center on $\mathbb{R}^{n-1}$.
3. A sphere in $\mathcal{D}^n$ corresponds to a sphere disjoint with $\mathbb{R}^{n-1}$.
4. A horosphere with center e (or $-e$) and relative radius ρ corresponds to the hyperplane in $\mathbb{R}^n$ normal to a_0 with signed distance $-1/\rho$ (or $1/\rho$) from the origin.
5. A horosphere with center other than $\pm e$ corresponds to a sphere tangent with $\mathbb{R}^{n-1}$.

6. A hypersphere with center c, tangent radius ρ in $\mathcal{D}^n$ and axis passing through e corresponds to the hyperplane in $\mathbb{R}^n$ normal to $c - \rho a_0 + c \cdot \mathbf{p}_0 e$ with signed distance $-\dfrac{c \cdot \mathbf{p}_0}{\sqrt{1+\rho^2}}$ from the origin.
7. A hypersphere whose axis does not pass through e in $\mathcal{D}^n$ corresponds to a sphere intersecting with $\mathbb{R}^{n-1}$.

4.8 The Klein Ball Model

The standard definition of the Klein ball model [120] is the unit ball $\mathcal{B}^n$ of $\mathbb{R}^n$ equipped with the following metric: for any $u, v \in \mathcal{B}^n$,

$$\cosh d(u,v) = \frac{1 - u \cdot v}{\sqrt{(1-u^2)(1-v^2)}}. \tag{4.91}$$

This model is not conformal, contrary to all the previous models, and is valid only for $\mathcal{H}^n$, not for $\mathcal{D}^n$.

The standard derivation of this model is through the central projection of $\mathcal{H}^n$ to $\mathbb{R}^n$. Recall that when we construct the conformal ball model, we use the stereographic projection of $\mathcal{D}^n$ from $-\mathbf{p}_0$ to the space $\mathbb{R}^n = (a_0 \wedge \mathbf{p}_0)^\sim$. If we replace $-\mathbf{p}_0$ with the origin, replace the space $(a_0 \wedge \mathbf{p}_0)^\sim$ with the tangent hyperplane of $\mathcal{H}^n$ at point $\mathbf{p}_0$, and replace $\mathcal{D}^n$ with its branch $\mathcal{H}^n$, then every affine line passing through the origin and point $\mathbf{p}$ of $\mathcal{H}^n$ intersects the tangent hyperplane at point

$$j_K(\mathbf{p}) = \frac{\mathbf{p}_0(\mathbf{p}_0 \wedge \mathbf{p})}{\mathbf{p}_0 \cdot \mathbf{p}}. \tag{4.92}$$

Every affine line passing through the origin and a point at infinity $\mathbf{p}_0 + a$ of $\mathcal{H}^n$ intersects the tangent hyperplane at point a.

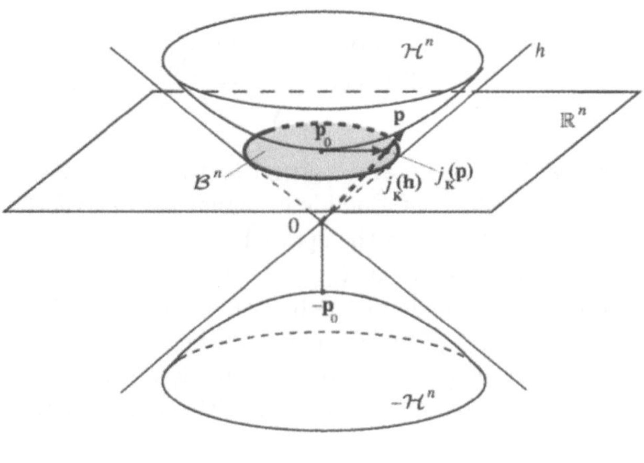

Fig. 4.7. The Klein ball model

The projection $j_{\mathcal{HB}}$ maps $\mathcal{H}^n$ to $\mathcal{B}^n$, and maps the sphere at infinity of $\mathcal{H}^n$ to the unit sphere of $\mathbb{R}^n$. This map is one-to-one and onto. Since it is central projection, every r-plane of $\mathcal{H}^n$ is mapped to an r-plane of $\mathbb{R}^n$ inside $\mathcal{B}^n$.

Although not conformal, the Klein ball model can still be constructed in the homogeneous model. We know that $j_{\mathcal{DS}}$ maps $\mathcal{H}^n$ to $\mathcal{S}^n_+$, the hemisphere of $\mathcal{S}^n$ centered at $-a_0$. A stereographic projection of $\mathcal{S}^n$ from a_0 to $\mathbb{R}^n$, yields a model of $\mathcal{D}^n$ in the whole of $\mathbb{R}^n$. Now instead of a stereographic projection, use a parallel projection $P_{\tilde{a}_0} = P^{\perp}_{a_0}$ from $\mathcal{S}^n_+$ to $\mathbb{R}^n = (a_0 \wedge \mathbf{p}_0)^{\sim}$ along a_0. The map

$$j_K = P^{\perp}_{a_0} \circ j_{\mathcal{DS}} : \mathcal{H}^n \longrightarrow \mathcal{B}^n \tag{4.93}$$

is the central projection and produces the Klein ball model.

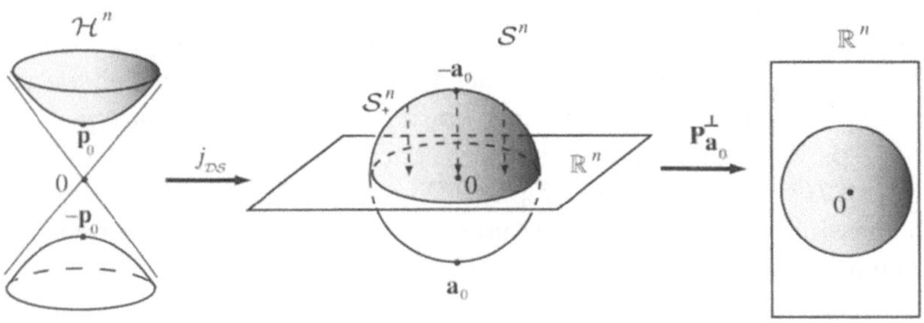

Fig. 4.8. The Klein ball model derived from parallel projection of $\mathcal{S}^n$ to $\mathbb{R}^n$

The following are some properties of the map j_K. There is no correspondence between spheres in $\mathcal{H}^n$ and $\mathcal{B}^n$ because the map is not conformal.

1. A hyperplane of $\mathcal{H}^n$ normal to a is mapped to the hyperplane of $\mathcal{B}^n$ normal to $P^{\perp}_{\mathbf{p}_0}(a)$ and with signed distance $-\dfrac{a \cdot \mathbf{p}_0}{\sqrt{1 + (a \cdot \mathbf{p}_0)^2}}$ from the origin.
2. An r-plane of $\mathcal{H}^n$ passing through $\mathbf{p}_0$ and normal to the space of $\mathbf{I}_{n-r}$, where $\mathbf{I}_{n-r}$ is a unit $(n-r)$-blade of Euclidean signature in $\mathcal{G}(\mathbb{R}^n)$, is mapped to the r-space of $\mathcal{B}^n$ normal to the space $\mathbf{I}_{n-r}$.
3. An r-plane of $\mathcal{H}^n$ normal to the space of $\mathbf{I}_{n-r}$ but not passing through $\mathbf{p}_0$, where $\mathbf{I}_{n-r}$ is a unit $(n-r)$-blade of Euclidean signature in $\mathcal{G}(\mathbb{R}^n)$, is mapped to an r-plane $\mathcal{L}$ of $\mathcal{B}^n$. The plane $\mathcal{L}$ is in the $(r+1)$-space, which is normal to the space of $\mathbf{p}_0 \cdot \mathbf{I}_{n-r}$ of $\mathbb{R}^n$, and is normal to the vector $\mathbf{p}_0 + (P_{\mathbf{I}_{n-r}}(\mathbf{p}_0))^{-1}$ in the $(r+1)$-space, with signed distance

$-\dfrac{1}{\sqrt{1+(P_{\mathbf{I}_{n-r}}(\mathbf{p}_0))^{-2}}}$ from the origin.

The inverse of the map $j_{\mathcal{HB}}$ is

$$j_K^{-1}(u) = \begin{cases} \dfrac{u+\mathbf{p}_0}{|u+\mathbf{p}_0|}, & \text{for } u \in \mathbb{R}^n,\ u^2 < 1, \\ u+\mathbf{p}_0, & \text{for } u \in \mathbb{R}^n,\ u^2 = 1. \end{cases} \tag{4.94}$$

The following are some properties of this map:

1. A hyperplane of $\mathcal{B}^n$ normal to n with signed distance δ from the origin is mapped to the hyperplane of $\mathcal{H}^n$ normal to $n - \delta\mathbf{p}_0$.
2. An r-space $\mathbf{I}_r$ of $\mathcal{B}^n$, where $\mathbf{I}_r$ is a unit r-blade in $\mathcal{G}(\mathbb{R}^n)$, is mapped to the r-plane $a_0 \wedge \mathbf{p}_0 \wedge \mathbf{I}_r$ of $\mathcal{H}^n$.
3. An r-plane in the $(r+1)$-space $\mathbf{I}_{r+1}$ of $\mathcal{B}^n$, normal to vector n in the $(r+1)$-space with signed distance δ from the origin, where $\mathbf{I}_{r+1}$ is a unit $(r+1)$-blade in $\mathcal{G}(\mathbb{R}^n)$, is mapped to the r-plane $(n-\delta\mathbf{p}_0)\,(a_0 \wedge \mathbf{p}_0 \wedge \mathbf{I}_{r+1})$ of $\mathcal{H}^n$.

4.9 A Universal Model for Euclidean, Spherical, and Hyperbolic Spaces

We have seen that spherical and Euclidean spaces and the five well-known analytic models of the hyperbolic space, all derive from the null cone of a Minkowski space, and are all included in the homogeneous model. Except for the Klein ball model, these geometric spaces are conformal to each other. No matter how viewpoints are chosen for projective splits, the correspondences among spaces projectively determined by common null vectors and Minkowski blades are always conformal. This is because for any nonzero vectors c, c' and any null vectors $h_1, h_2 \in \mathcal{N}^n_{c'}$, where

$$\mathcal{N}^n_{c'} = \{x \in \mathcal{N}^n | x \cdot c' = -1\}, \tag{4.95}$$

we have

$$\left| -\frac{h_1}{h_1 \cdot c} + \frac{h_2}{h_2 \cdot c} \right| = \frac{|h_1 - h_2|}{\sqrt{|(h_1 \cdot c)(h_2 \cdot c)|}}, \tag{4.96}$$

i.e., the rescaling is conformal with conformal coefficient $1/\sqrt{|(h_1 \cdot c)(h_2 \cdot c)|}$.

Recall that in previous constructions of the geometric spaces and models in the homogeneous model, we selected special viewpoints: $\mathbf{p}_0$, a_0, b_0, $e = \mathbf{p}_0 + a_0$, $e_0 = \frac{\mathbf{p}_0 - a_0}{2}$, etc. We can select any other nonzero vector c in $\mathbb{R}^{n+1,1}$ as the viewpoint for projective split, thereby obtaining a different realization for one of these spaces and models. For the Euclidean case, we can select any null vector in $\mathcal{N}^n_e$ as the origin e_0. This freedom in choosing viewpoints for projective and conformal splits establishes an equivalence among geometric theorems in conformal geometries of these spaces and models. From a single theorem, many "new" theorems can be generated in this way. We illustrate this with a simple example.

The original Simson's Theorem in plane geometry is as follows:

Theorem 4.9.1 (Simson's Theorem). *Let ABC be a triangle, D be a point on the circumscribed circle of the triangle. Draw perpendicular lines from D to the three sides AB, BC, CA of triangle ABC. Let C_1, A_1, B_1 be the three feet respectively. Then A_1, B_1, C_1 are collinear.*

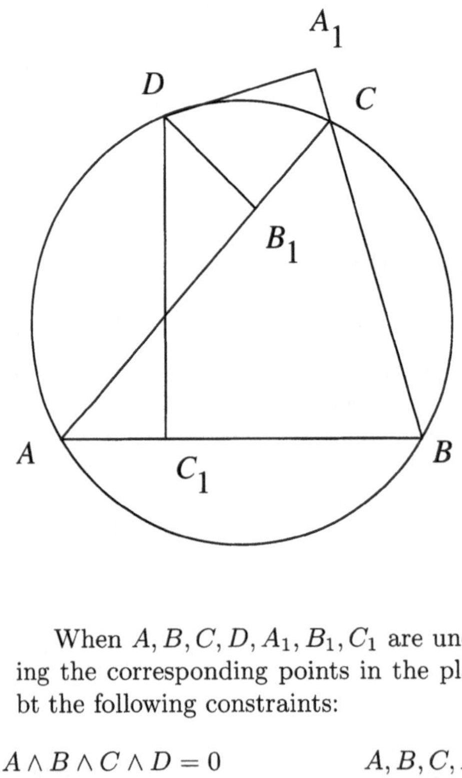

Fig. 4.9. Original Simson's Theorem

When $A, B, C, D, A_1, B_1, C_1$ are understood to be null vectors representing the corresponding points in the plane, the hypothesis can be expressed bt the following constraints:

$$
\begin{array}{ll}
A \wedge B \wedge C \wedge D = 0 & A, B, C, D \text{ are on the same circle} \\
e \wedge A \wedge B \wedge C \neq 0 & ABC \text{ is a triangle} \\
e \wedge A_1 \wedge B \wedge C = 0 & A_1 \text{ is on line } BC \\
(e \wedge D \wedge A_1) \cdot (e \wedge B \wedge C) = 0 & \text{Lines } DA_1 \text{ and } BC \text{ are perpendicular} \\
e \wedge A \wedge B_1 \wedge C = 0 & B_1 \text{ is on line } CA \\
(e \wedge D \wedge B_1) \cdot (e \wedge C \wedge A) = 0 & \text{Lines } DB_1 \text{ and } CA \text{ are perpendicular} \\
e \wedge A \wedge B \wedge C_1 = 0 & C_1 \text{ is on line } AB \\
(e \wedge D \wedge C_1) \cdot (e \wedge A \wedge B) = 0 & \text{Lines } DC_1 \text{ and } AB \text{ are perpendicular}
\end{array}
\tag{4.97}
$$

The conclusion can be expressed as

$$e \wedge A_1 \wedge B_1 \wedge C_1 = 0. \tag{4.98}$$

Both the hypothesis and the conclusion are invariant under rescaling of null vectors, so this theorem is valid for all three geometric spaces, and is free of the requirement that $A, B, C, D, A_1, B_1, C_1$ represent points and e represents the point at infinity of $\mathbb{R}^n$. Various "new" theorems can be produced by interpreting the algebraic equalities and inequalities in the hypothesis and conclusion of Simson's theorem differently.

First let us exchange the roles that D, e play in Euclidean geometry. The hypothesis becomes

$$\begin{cases} e \wedge A \wedge B \wedge C = 0 \\ A \wedge B \wedge C \wedge D \neq 0 \\ A_1 \wedge B \wedge C \wedge D = 0 \\ (e \wedge D \wedge A_1) \cdot (D \wedge B \wedge C) = 0 \\ A \wedge B_1 \wedge C \wedge D = 0 \\ (e \wedge D \wedge B_1) \cdot (D \wedge C \wedge A) = 0 \\ A \wedge B \wedge C_1 \wedge D = 0 \\ (e \wedge D \wedge C_1) \cdot (D \wedge A \wedge B) = 0 \end{cases} \tag{4.99}$$

and the conclusion becomes

$$A_1 \wedge B_1 \wedge C_1 \wedge D = 0. \tag{4.100}$$

This "new" theorem can be stated as follows:

Theorem 4.9.2. *Let DAB be a triangle, C be a point on line AB. Let A_1, B_1, C_1 be the symmetric points of D with respect to the centers of circles DBC, DCA, DAB respectively. Then D, A_1, B_1, C_1 are on the same circle.*

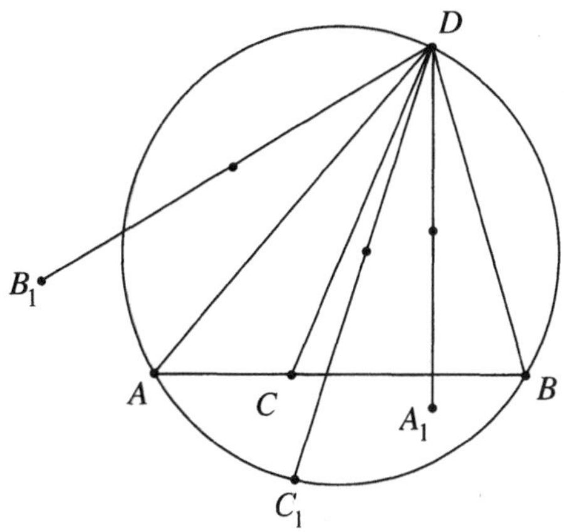

Fig. 4.10. Theorem 4.9.2

We can get another theorem by interchanging the roles of A, e. The constraints become

$$\begin{array}{l} e \wedge B \wedge C \wedge D = 0 \\ e \wedge A \wedge B \wedge C \neq 0 \\ A \wedge A_1 \wedge B \wedge C = 0 \\ (A \wedge D \wedge A_1) \cdot (A \wedge B \wedge C) = 0 \\ e \wedge A \wedge B_1 \wedge C = 0 \\ (A \wedge D \wedge B_1) \cdot (e \wedge C \wedge A) = 0 \\ e \wedge A \wedge B \wedge C_1 = 0 \\ (A \wedge D \wedge C_1) \cdot (e \wedge A \wedge B) = 0\,, \end{array} \tag{4.101}$$

and the conclusion becomes

$$A \wedge A_1 \wedge B_1 \wedge C_1 = 0. \tag{4.102}$$

This "new" theorem can be stated as follows:

Theorem 4.9.3. *Let ABC be a triangle, D be a point on line AB. Let EF be the perpendicular bisector of line segment AD, which intersects AB, AC at E, F respectively. Let C_1, B_1 be the symmetric points of A with respect to points E, F respectively. Let AG be the tangent line of circle ABC at A, which intersects EF at G. Let A_1 be the intersection, other than A, of circle ABC with the circle centered at G and passing through A. Then A, A_1, B_1, C_1 are on the same circle.*

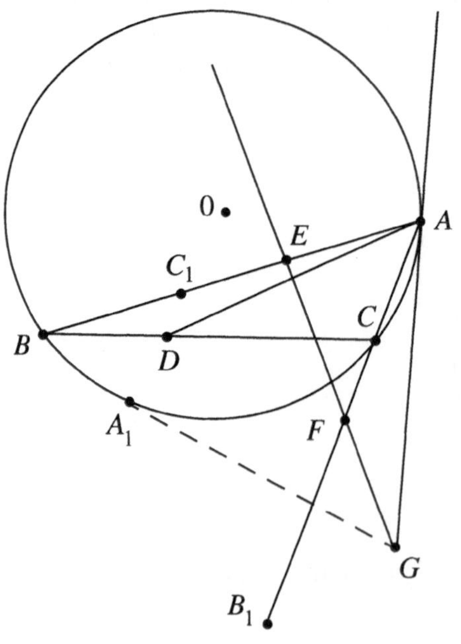

Fig. 4.11. Theorem 4.9.3

There are equivalent theorems in spherical geometry. We consider only one case. Let $e = -D$. A "new" theorem as follows:

Theorem 4.9.4. *Within the sphere there are four points A, B, C, D on the same circle. Let A_1, B_1, C_1 be the symmetric points of $-D$ with respect to the centers of circles $(-D)BC$, $(-D)CA$, $(-D)AB$ respectively. Then $-D, A_1, B_1, C_1$ are on the same circle.*

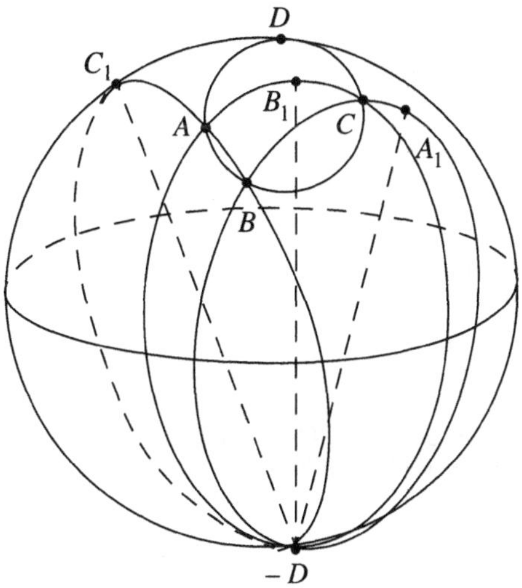

Fig. 4.12. Theorem 4.9.4

There are various theorems in hyperbolic geometry that are also equivalent to Simson's theorem because of the versatility of geometric entities. We present one case here. Let A, B, C, D be points on the same branch of $\mathcal{D}^2$, $e = -D$.

Theorem 4.9.5. *Let A, B, C, D be points in the Lobachevski plane $\mathcal{H}^2$ and be on the same generalized circle. Let L_A, L_B, L_C be the axes of hypercycles (1-dimensional hyperspheres) $(-D)BC$, $(-D)CA$, $(-D)AB$ respectively. Let A_1, B_1, C_1 be the symmetric points of D with respect to L_A, L_B, L_C respectively. Then $-D, A_1, B_1, C_1$ are on the same hypercycle.*

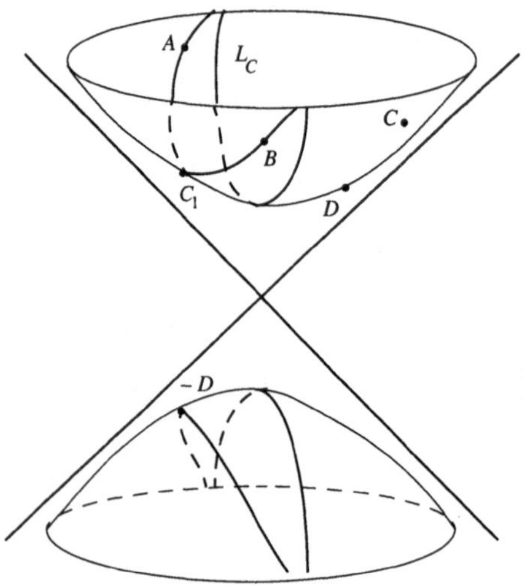

Fig. 4.13. Construction of C_1 from A, B, D in Theorem 4.9.5

5. Geo-MAP Unification

Ambjörn Naeve and Lars Svensson

Dept. of Numerical Analysis and Computing Science / Dept. of Mathematics
Royal Institute of Technology, Stockholm

5.1 Introduction

The aim of this chapter is to contribute to David Hestenes' vision - formulated on his web-site[1] - of desiging a universal geometric calculus based on geometric algebra. To this purpose we introduce a framework for geometric computations which we call geo-MAP (geo-Metric-Affine-Projective) unification. It makes use of geometric algebra to embed the representation of Euclidean, affine and projective geometry in a way that enables coherent shifts between these different perspectives. To illustrate the versatility and usefulness of this framework, it is applied to a classical problem of plane geometrical optics, namely how to compute the envelope of the rays emanating from a point source of light after they have been reflected in a smoothly curved mirror.

Moreover, in the appendix, we present a simple proof of the fact that the 'natural basis candidate' of a geometric algebra - the set of finite subsets of its formal variables - does in fact form a vector space basis for the algebra. This theorem opens the possibility of a deductive presentation of geometric algebra to a wider audience.

---

[1] <http://phy.asu.edu/directory/Fac_Pages/Hestenes.html>

5.2 Historical Background

Applying polynomial algebra to geometry is called algebraic geometry if the polynomials commute with each other, and geometric algebra if they don't.[2] Let us take a brief look at the historical process that has developed the design of present day relationships between geometry and algebra.[3]

With his work La Geometrie[4], published in 1637, Descartes wielded together the two subjects of algebra and geometry, which had each been limping along on its own one leg. In this process he created the two-legged subject of analytic geometry, which turned out to be capable of moving forward in big strides and even giant leaps - such as e.g. the one manifested by the differential and integral calculus of Newton and Leibnitz[5] towards the end of the 17:th century - building on work by Fermat, Wallis and Barrow[6].

But these tremendous advancements came at a considerable price. As Hestenes points out in [112], the careful separation between number and one-dimensional un-oriented magnitude that had been so meticulously upheld by Euclid, was thrown away by Descartes, an identification that has had fundamental consequences for the design of 'vectorial' mathematics. By giving up the difference between order and magnitude, Descartes in effect created a 1-dimensional concept of direction, where only 1-dimensional geometric objects - like lines or line segments - could be oriented in a coordinate free way.

As we know, two centuries later, Grassmann discovered how to represent oriented multi-dimensional magnitudes and introduced a calculus for their manipulation[7]. Unfortunately, the fundamental nature of his contributions were not acknowledged by his contemporaries, and in the great battle over how to express multi-dimensional direction that was eventually fought between the 'ausdehnungen' of Grassmann [93] on the one hand and the 'quaternions' of Hamilton [98] on the other, the victory was carried off by the vector calculus of Gibbs [89] and Heaviside [107] - under heavy influence from the electro-magnetic theory of Maxwell [163]. This fact has fundamentally shaped the way we think about vectors today, and it has lead to concepts such as the familiar cross-product that we teach our mathematics students at the universities[8].

Another example of a dominating design-pattern within present day geometrical mathematics is provided by the field of algebraic geometry. Within this field of mathematics, one describes non-commutative geometrical config-

---

[2] Hence, in these mathematical field descriptions, the words 'algebra' and 'geometry' do not themselves commute.

[3] Hestenes [112] gives an excellent description of this process.

[4] Descartes [61]

[5] See Newton [178, 177] and Leibnitz [145, 146].

[6] See [27] for more detail.

[7] See Grassmann [94]

[8] An interesting account of this development is given by Crowe in [52].

urations in terms of commutative polynomial rings - using so called "varieties of zero-sets" for different sets of polynomials[9]. The urge to factor these polynomials creates an urge to algebraically close the field of coefficients, which leads to complex-valued coefficients, where one has factorization of single-variable polynomials into linear factors, and where Hilbert's so called 'nullstellensatz'[10] provides the fundamental linkage between the maximal ideals of the polynomial ring and the points of the underlying geometric space.

By adhering to this design one gains a lot of analytical power, but one also achieves two important side effects that have major consequences for the application of this kind of mathematics to geometric problems: First, one loses contact with real space, which means that algebraic geometry does not have much to say about real geometry at all, since there is no nullstellensatz to build on here. Second, since the interesting structures (= the varieties) are zero-sets of polynomials, they are mostly impossible to compute exactly, and hence must be approached by some form of iterational scheme - often using the pseudo-inverse (singular values) decomposition which is the flag-ship of computational linear algebra.

We could argue other cases in a similar way. The point is not to criticise the corresponding mathematical structures per se, but rather to underline the importance of discussing the concept of mathematical design in general - especially within the scientific communities that make use of mathematical models to represent the phenomena which they study. Bringing a powerful mathematical theory into a scientific field of study most often leads to interesting applications of that theory, but it always carries with it the risk of getting caught up in the difficulties introduced by the theory itself - rather than using it as a tool to handle the difficulties of the initial problem domain. Of course there is never a sharp distinction between these two cases, but rather a trade-off between the beneficial and the obstructional aspects of any mathematical tool.

In short, the historical process described above has resulted in a large variety of algebraic systems dealing with vectors - systems such as matrix algebra, spinor algebra, differential forms, tensor algebra, etc. etc. For many years David Hestenes has pointed out that such a multitude of computational systems for closely related conceptual entities indicate underlying weaknesses of the corresponding mathematical designs. As members of the scientific community we all share a dept of gratitude to people like David Hestenes and Garret Sobczyk, who have devoted a large part of their research to exploring various alternative designs - designs that aim to develop the legacy of Grassmann [94, 93] and Clifford [46, 45, 47] into its full algebraic and geometric potential[11]. In fact, it was the works of Hestenes [112] and Hestenes

[9] See e.g. Hartshorne: "What is algebraic geometry" [105], pp.55-59.

[10] See e.g. Atiyah-Macdonald [10], p. 85.

[11] In this battle Hestenes and Sobczyk have been joined by a number of people. They include Rota and his co-workers Barabei and Brini [14], who were instru-

& Sobczyk [113] that brought the present authors into the field, and made us interested enough to take up active research in geometric algebra.

5.3 Geometric Background

5.3.1 Affine Space

As a background to the following discussion we begin by introducing the abstract concept of an affine space followed by a concrete model of such a space to which we can apply geometric algebra. Our presentation of affine spaces follows essentially that of Snapper & Troyer [213], which starts out by discussing affine spaces in abstract and axiomatic mathematical terms:

Definition 5.3.1. *The n-dimensional affine space over a field K consists of a non-empty set X, an n-dimensional vector space V over K, and an 'action' of the additive group of V on X. The elements of X are called points, the elements of V are called vectors and the elements of K are called scalars.*

Definition 5.3.2. *To say that the additive group of the vector space V acts on the set X means that, for every vector $v \in V$ and every point $x \in X$ there is defined a point $vx \in X$ such that*

1. *If $v, w \in V$ and $x \in X$, then $(v + w)x = v(wx)$.*
2. *If 0 denotes the zero vector of V, then $0x = x$ for all $x \in X$.*
3. *For every ordered pair (x, y) of points of X, there is one and only one vector $v \in V$ such that $vx = y$.*

The unique vector v with $vx = y$ is denoted by $\boldsymbol{xy}$ and we write

$$v \equiv \boldsymbol{xy} \equiv y - x \tag{5.1}$$

Also, it is convenient to introduce the following

Notation: The affine space defined by X, V, K and the action of the additive group of V on X is denoted by (X, V, K).

From now on, we will restrict K to be the field of real numbers $\mathbb{R}$. The corresponding affine space $(X, V, \mathbb{R})$ is called *real affine space.* We now introduce a model for real affine space - a model which is in fact often taken as a definition of such a space.

Let V be an n-dimensional vector space over $\mathbb{R}$. For the set X, we choose the vectors of V, that is, $X = V$, where V is considered only as a set. The action of the additive group of V on the set V is defined as follows:

$$\text{If } v \in V \text{ and } w \in V, \text{ then } v \circ w = v + w \tag{5.2}$$

It is an easy exercise to verify that

mental in reviving Grassmann's original ideas, and Sommer [217, 215], who plays an important role in bringing these ideas in contact with the engineering community, thus contributing to the growing number of their applications.

Proposition 5.3.1. *The space $(V, V, \mathbb{R})$ as defined above, is a model for n-dimensional real affine space, in other words, the three conditions of Def. (5.3.2) are satisfied.*

In this case one says that the vectors of V act on the points of V by *translation* - thereby giving rise to the affine space $(V, V, \mathbb{R})$. In linear algebra one becomes accustomed to regarding the *vector* v of the vector space V as an arrow, starting at the point of origin. When V is regarded as an affine space, that is, $X = V$, the *point* v should be regarded as the end of that arrow.

To make the distinction between a vector space and its corresponding affine space more visible, it is customary to talk of *direction vectors* when referring to elements of the vector space V and *position vectors* when referring to elements of the corresponding affine space $(V, V, \mathbb{R})$.

5.3.2 Projective Space

Definition 5.3.3. *Let V be an $(n+1)$-dimensional vector space. The n-dimensional projective space $P(V)$ is the set of all non-zero subspaces of V.*

To each non-zero k-blade $B = b_1 \wedge b_2 \wedge \ldots \wedge b_k$ we can associate the linear span $\bar{B} = Linspan[b_1, b_2, \ldots, b_k]$. Hence we have the mapping from the set $\mathbf{B}$ of non-zero blades to $P(V)$ given by

$$\mathbf{B} \ni B \to \bar{B} \in P(V), \tag{5.3}$$

which takes non-zero k-blades to k-dimensional subspaces of V.

As is well known, $P(V)$ carries a natural lattice structure. Let S and T be two subspaces of V. We denote by $S \wedge T$ the subspace $S \cap T$, and by $S \vee T$ the subspace $S + T$. Moreover, let us recall the geometric algebra dual $\vee$ of the outer product $\wedge$, defined by

$$x \vee y = (\tilde{x} \wedge \tilde{y})I = (xI^{-1}) \wedge (yI^{-1})I. \tag{5.4}$$

We can now state the following important result:

Proposition 5.3.2. *Let A and B be non-zero blades in the geometric algebra G. Then*

$$\begin{aligned} \overline{A \wedge B} &= \overline{A} \vee \overline{B}, \ \textit{if}\ \overline{A} \wedge \overline{B} = 0, \\ \overline{A \vee B} &= \overline{A} \wedge \overline{B}, \ \textit{if}\ \overline{A} \vee \overline{B} = V. \end{aligned} \tag{5.5}$$

Proof. See e.g. Hestenes & Sobczyk [113] or Svensson [225].

In the so called *double algebra* - also known as the *Grassmann-Cayley algebra* - the lattice structure of $P(V)$ is exploited in order to express the *join* (= sum) and *meet* (= intersection) of its subspaces.[12] In order to obtain the same

[12] See e.g. Barabei & Brini & Rota [14] or Svensson [225].

computational capability within the geometric algebra, we can introduce an alternating multilinear map called the *bracket* (or the *determinant*), given by

$$\begin{aligned} &V \times \ldots \times V \to \mathbb{R} \\ &(v_1, \ldots, v_n) \mapsto (v_1 \wedge \ldots \wedge v_n)I^{-1} = |v_1, \ldots, v_n|. \end{aligned} \tag{5.6}$$

As an example, which we will make use of below, we have the following

Proposition 5.3.3. *If* $A, B, C, D \in G_3^1$, *then*

$$(A \wedge B) \vee (C \wedge D) = |A, B, C|D - |A, B, D|C. \tag{5.7}$$

Proof. Since both sides of (5.7) are multilinear in A, B, C and D, it is enough to verify the equality for $\{A, B, C, D\} \subset \{e_1, e_2, e_3\}$. This is left as a routine exercise.

5.4 The Unified Geo-MAP Computational Framework

As is well-known, the present day vector concept is surrounded by a great deal of confusion, and we argued above that this is an indicator of its weakness of design. Symptoms range from the inability of students to discriminate between direction vectors and position vectors to heated discussions among experts as to which type of algebra that is best suited to represent vectors. Since any representational perspective has its own inherent strengths and weaknesses, it is important to be able to move between such perspectives in a consistent way, which means to remain within the same computational framework in spite of the change of representation.

In this section we demonstrate how geometric algebra provides a common background for such movement. We will explain how this background can be used to handle the interplay between Euclidean (direction) vectors, affine (position) vectors and homogeneous (sub-space) vectors - such as the ones used in projective geometry.

The technique for doing this we have termed *geo-Metric-Affine-Projective unification.* It is important because it allows passing from the Euclidean vector space algebra into the Grassmann-Cayley algebra and then back again without changing the frame of reference. Later we will make use of the geo-MAP unification technique to compute intersections of affine sets in cartesian coordinates.

5.4.1 Geo-MAP Unification

Let V be a n-dimensional Euclidean real vector space with the vectors $\{e_1, \ldots, e_n\}$ as an orthonormal basis. Denote by $G(I)$ the corresponding geometric algebra, where $I = e_1 e_2 \ldots e_n$. Moreover, let O denote an arbitrary (fixed) point of the affine space $(V, V, \mathbb{R})$. We can represent O by introducing a unit vector e orthogonal to $\{e_1, \dots, e_n\}$ and consider the geometric algebra $G(J)$, with unit pseudo-scalar $J = Ie$. Then it follows directly from Def. (5.3.2) that for each affine point $p \in (V, V, \mathbb{R})$ there is a unique vector such that

$$p = e + x. \tag{5.8}$$

Moreover, the additive action of the vectors $x, y \in V$ is given by

$$(e + x) + y = e + (x + y). \tag{5.9}$$

Now, by construction, we have $V = G^1(I) = \overline{I}$. Let us introduce the vector space W with the corresponding relation to J:

$$W = G^1(J) = \overline{J}. \tag{5.10}$$

Then it is clear that

$$A = e + V = \{e + v : v \in V\} \tag{5.11}$$

is an affine set in W.

We now introduce the two mappings

$$V \ni x \mapsto {}^*x = e + x \in A, \tag{5.12}$$

and

$$W \backslash V \ni y \mapsto {}_*y = (y \cdot e)^{-1} y - e \in V. \tag{5.13}$$

Note that $y \notin V$ ensures that $y \cdot e \neq 0$. Since the right hand side of (5.13) is invariant under scaling of y, it follows that this mapping can be extended to $P_1(W)$ (= the 1-dimensional subspaces of W), excluding $P_1(V)$ (=the 1-dimensional subspaces of V). Hence (5.13) induces a mapping from the affine part of projective space:

$$P_1(W) \backslash P_1(V) \ni \bar{y} \mapsto {}_*y \in V. \tag{5.14}$$

We can now make the following important observation:

$${}_*(\overline{{}^*x}) = {}_*(\overline{x + e}) = ((x + e) \cdot e)^{-1}(x + e) - e = x. \tag{5.15}$$

The relation (5.15) embodies the essence of the unified geo-MAP computational framework. It shows how to pass from a point x of Euclidean space - via affine space - into projective space, and then how to get back again to the original starting point x. In this way the upper and lower left star operators bridge the gap betweeen the metric, affine and projective perspectives on geometry and unifies them within the same computational framework. The configuration is illustrated in Figure (5.1).

Using the geo-MAP unification technique, we can start with a Euclidean direction vector problem and translate it into projective space, where we can apply e.g. the techniques of the double algebra in order to find intersections of affine sets. These intersections can then be transported back to the Euclidean representation and deliver e.g. cartesian coordinates of the intersection elements. In the next paragraph will apply the geo-MAP unification technique in this way.

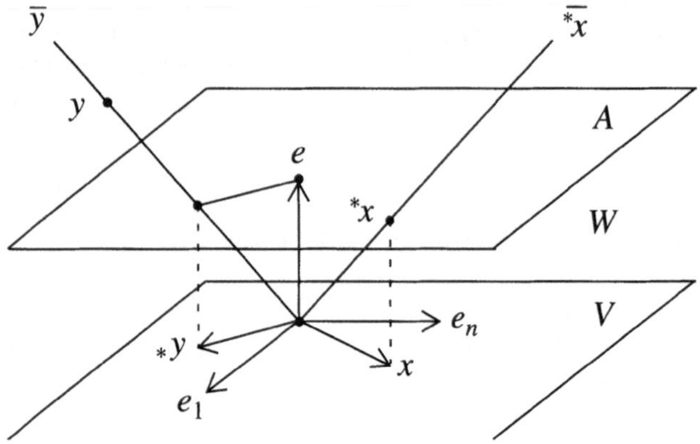

Fig. 5.1. The unified geo-MAP computational framework

5.4.2 A Simple Example

To illustrate how it works, will now apply the geo-MAP computational technique to the simple problem of finding the intersection of two lines in the affine plane.

Let $v_0, v_1, w_0, w_1 \in G^1(e_1, e_2)$ and let the two lines be determined respectively by the two point-pairs $V_i = {}^*v_i$ and $W_i = {}^*w_i, i = 0, 1$.

Making use of (5.7), we can express the point p_{vw} of intersection between these two lines as:

$$p_{vw} = {}_*((V_0 \wedge V_1) \vee (W_0 \wedge W_1)) = {}_*(|V_0, V_1, W_0| W_1 - |V_0, V_1, W_1| W_0) \tag{5.16}$$

and the two brackets that appear in (5.16) can be written as:

$$\begin{aligned} |V_0, V_1, W_0| &= |v_0 + e, v_1 + e, w_0 + e| \\ &= |v_0 - w_0, v_1 - w_0, w_0 + e| \\ &= |v_0 - w_0, v_1 - w_0| \\ &= ((v_0 - w_0) \wedge (v_1 - w_0))e_2 e_1 \end{aligned} \tag{5.17}$$

and analogously

$$|V_0, V_1, W_1| = ((v_0 - w_1) \wedge (v_1 - w_1))e_2 e_1. \tag{5.18}$$

Writing $\alpha = |V_0, V_1, W_0|$ and $\beta = |V_0, V_1, W_1|$, (5.16) takes the form:

$$\begin{aligned} p_{vw} &= {}_*(\alpha W_1 - \beta W_0) \\ &= ((\alpha W_1 - \beta W_0) \cdot e)^{-1}(\alpha W_1 - \beta W_0) - e \\ &= (\alpha - \beta)^{-1}(\alpha w_1 + \alpha e - \beta w_0 - \beta e) - e \\ &= \frac{\alpha w_1 - \beta w_0}{\alpha - \beta}. \end{aligned} \tag{5.19}$$

Here we have a good example of the geo-MAP unification technique at work. Taking the upper star of the Euclidean direction vectors w_0 and w_1, they are brought into the affine part of projective space. Here they can be subjected to the double algebraic lattice operations of $\wedge$ and $\vee$, in this particular case in the combination expressed by (5.7).

5.4.3 Expressing Euclidean Operations in the Surrounding Geometric Algebra

We will end this section by showing how to embed some important Euclidean direction vector operations within the surrounding geometric algebra G_4 - namely the difference and the cross-product operations. In this way the ordinary Euclidean algebra of (cartesian) direction vectors can be emulated in geometric algebra.

Consider the Euclidean direction vector $x \in G_3^1 \subset G_4^1$ with its cartesian coordinate expansion given by $x = x_1 e_1 + x_2 e_2 + x_3 e_3$. Recall that $I = e_1 e_2 e_3$ and $J = Ie$. By (5.14), the corresponding affine position vector $X \in (V, V, \mathbb{R}) \subset G_4^1$ is expressible as $X = x + e$.

With these definitions of x and X, and the corresponding definitions for y and Y, we will now deduce two formulas that connect the Euclidean direction vector algebra of Gibbs with the surrounding geometric algebra G_4:

Proposition 5.4.1. *The Euclidean cross product vector can be expressed in G_4 as:*

$$x \times y = (e \wedge X \wedge Y)J. \tag{5.20}$$

Proof. From the definition of the cross product, it follows directly that

$$x \times y = (x \wedge y)I^{-1}. \tag{5.21}$$

Plugging $X = x + e$ and $Y = y + e$ into the right-hand-side of (5.20) now gives

$$\begin{aligned}(e \wedge X \wedge Y)J &= (e \wedge (x+e) \wedge (y+e))J = (e \wedge x \wedge y)J \\ &= e(x \wedge y)J = (x \wedge y)eIe = (x \wedge y)I^{-1} \\ &= x \times y. \end{aligned} \tag{5.22}$$

Proposition 5.4.2. *The Euclidean difference vector can be expressed in G_4 as:*

$$y - x = e \cdot (X \wedge Y). \tag{5.23}$$

Proof. Expanding again the right-hand-side of (5.23), we can write

$$\begin{aligned} e \cdot (X \wedge Y) &= e \cdot ((x+e) \wedge (y+e)) = e \cdot (x \wedge y + xe + ey) \\ &= e \cdot (xe + ey) = e \cdot (ey - ex) = \{ey \text{ and } ex \in G_2\} \\ &= \langle eey \rangle_1 - \langle eex \rangle_1 = y - x. \end{aligned}$$

Formulas such as (5.20) and (5.23) are useful for translating a geometric problem from one representation into another. Moreover, since for a 1-vector v and a blade B we have $v \cdot B + v \wedge B = vB$, it follows that (5.20) and (5.23) can be combined into:

$$e(X \wedge Y) = y - x + (x \times y)J^{-1} \tag{5.24}$$

or

$$X \wedge Y = e(y - x) + (x \times y)I. \tag{5.25}$$

The expression (5.25) indicates interesting relationships between ordinary direction vector algebra and various forms of generalized complex numbers. However, to pursue this topic further is outside the scope of the current text.

5.5 Applying the Geo-MAP Technique to Geometrical Optics

In order to illustrate the workings of the unified geo-MAP computational technique, we will now apply it to a classical problem of geometrical optics. It was first treated by Tschirnhausen and is known as the problem of *Tschirnhausen's caustics.*

5.5.1 Some Geometric-Optical Background

Since light is considered to emanate from each point of a visible object, it is natural to study optics in terms of collections of point sources. In geometrical optics, a point source is considered as a set of (light) rays - i.e. a set of directed half-lines - through a point. But a point source does not (in general) retain its 'pointness' as it travels through an optical system of mirrors and lenses. When a point source of in-coming light is reflected by a mirror or refracted by a lens, the out-going rays will in general not pass through a point. Instead, they will be tangent to two different surface patches, together called the *focal surface* of the rays[13].

The importance of focal surfaces in geometrical optics is tied up with a famous theorem due to Malus and Dupin[14]. In order to understand what this theorem says, we introduce a geometric property that is sometimes possessed by a set of lines:

Definition 5.5.1. *A two-parameter family of curves K is said to form a normal congruence, if there exists a one-parameter family of smooth surfaces Γ such that each surface of the family Γ is orthogonal to each curve of the family K.*

[13] Fo a survey ov the theory of focal surfaces, we refer the reader to Naeve [175].
[14] see e.g. Lie [150]

A surface in Γ is called an *orthogonal trajectory* to the curves of the family K. A point field K_P is of course an example of a *normal congruence of lines*, the orthogonal trajectories being the one-parameter family of concentric spheres Γ_P with centre P. Furthermore, the rays of K_P carry a direction, which varies continuously when we pass from one ray to its neighbours. Such a family of lines is called a *directed normal congruence.*

Now, the theorem of Malus and Dupin can be formulated as follows:

Proposition 5.5.1. *A directed normal congruence of lines remains directed and normal when passing through an arbitrary optical system.*

In optics, the orthogonal trajectories of the light rays are called *wave fronts*, and the Malus-Dupin theorem can be expressed by stating that any *optical system preserves the existence of wave fronts.*

5.5.2 Determining the Second Order Law of Reflection for Planar Light Rays

In what follows below we will restrict to the plane and consider one-parameter families of rays that emanate from a planar point source and then are reflected by a curved mirror in the same plane. We will deduce an expression that connects the curvatures of the in-coming and out-going wave fronts with the curvature of the mirror at the point of impact. Since curvature is a second-order phenomenon, it is natural to call this expression the second order law of reflection - as opposed to the first order law, that expresses only the direction of an outgoing ray as a function of the direction of the in-coming ray and the direction of the mirror normal.

Let us begin by recalling some classical concepts from the differential geometry of plane curves.

Definition 5.5.2. *Consider a one-parameter family of smooth curves $F(c)$ in the same plane (with c as the parameter). If there is a curve Γ which has the property of being tangent to every curve of the family $F(c)$ in such a way that each point $\Gamma(t_0)$ is a point of tangency of exactly one curve $F(c_0)$, then the curve Γ is called the envelope of the family of curves $F(c)$.*

Definition 5.5.3. *Consider a smooth plane curve M. At each point $m(s)$ of this curve there is an osculating circle with centre point $r(s)$, called the centre of curvature of the point $m(s)$. When we vary s (i.e. when we move along the curve M) the point $r(s)$ will describe a curve E called the evolute of the original curve M. Reciprocally, the curve M is called the evolvent (or the involute) of the curve E.*

In 2-dimensional geometrical optics, a point source of light corresponds to a *pencil of rays.* After having been reflected or refracted by various mirrors and lenses, these rays will in general be tangent to a curve, called a *caustic curve* in optics[15]. This is the kind of bright, shining curve that we can observe when sunlight is reflected e.g. in a cup of tea.

Definition 5.5.4. *A certain angular sector of a plane pencil of light-rays is made to be incident on a smoothly curved mirror M in the same plane. After being reflected by M, the light-rays of this sector are all tangent to the caustic curve forming their envelope. Such a 1-parameter family of light-rays will be referred to as a tangential sector of rays.*

Note: In view of the discussion above, we can conclude that the caustic curve of a tangential sector of rays is at the same time the *envelope of the rays* and the *evolute of their orthogonal trajectories.*

Let us consider a tangential sector of rays $L_{12} = \{l(s) : s_1 < s < s_2\}$ with caustic curve C_{in} whose rays are incident on a smoothly curved mirror M between the points $m(s_1)$ and $m(s_2)$ as depicted in Figure (5.2).

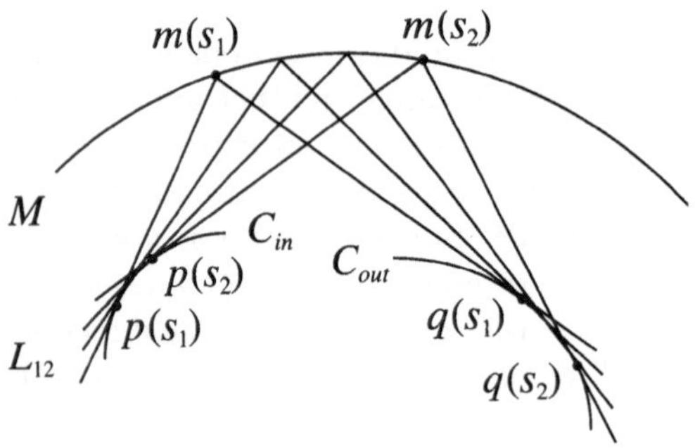

Fig. 5.2. The tangential sector L_{12} is reflected by the mirror M

Two closely related rays $l(s)$ and $l(s + ds)$ will intersect in a point that is close to the caustic curve C_{in} and when $l(s + ds)$ is brought to coincide with $l(s)$ by letting $ds \to 0$ their point of intersection will in the limit fall on the caustic point $p(s)$. Hence we can regard the caustic curve passing through p as the locus of intersection of neighbouring rays, where the term 'neighbouring' refers to the limiting process just described[16].

[15] See e.g. Cornbleet [48] or Hecht & Zajac [108].

[16] This is an example of an ituitive (and very useful) way to think about geometrical entities that has been used by geometers ever since the days of Archimedes. Unfortunately it has no formal foundation in classical analysis. Since infinitesimal entities (like dx and dy) do not exist in the world of real numbers, a line in a continuous, (real), one-parameter family can habe no neighbour. However, the concept of neighbouring geometrical objects can be made rigorous by the use of so called non-standard analysis.

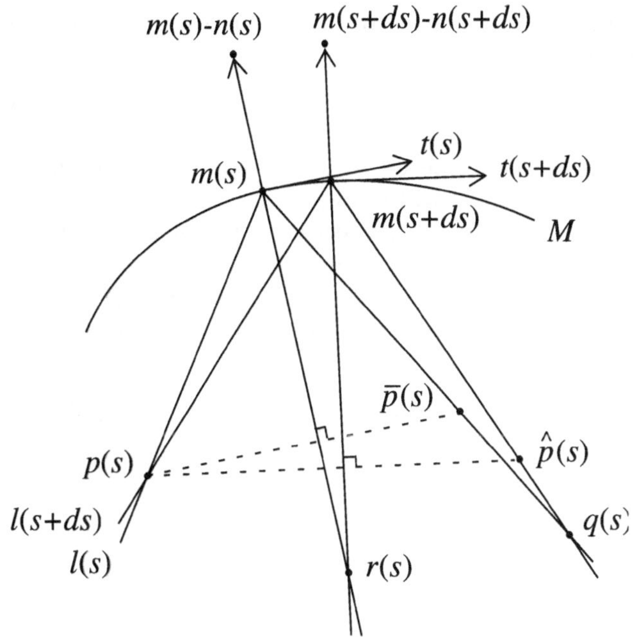

Fig. 5.3. Two neighboring rays $l(s)$ and $l(s+ds)$ intersecting at $p(s)$ and their respective reflections intersecting at $q(s)$

In Figure (5.3) the symbols t (=tangent) and n (=normal) denote Euclidean direction vectors, and m, p, q, r denote affine points (= position vectors). The symbol s denotes the parameter that connects corresponding points and lines. The two rays $l(s)$ and $l(s+ds)$ can be thought of as forming an infinitesimal sector with its vertex p on the caustic curve of l. Within this sector, the corresponding (infinitesimal) parts of the wave-fronts are circular and concentric around p. The point p can therefore be regarded as the focal point of this infinitesimal sector, i.e. the *local focal point* of the wave fronts in the direction given by $l(s)$.

Having thus established some terminology and notation, we now turn to Tschirnhausen's problem, which is concerned with determining the point $q(s)$ on the reflected caustic C_{out} that corresponds to the point $p(s)$ on the in-caustic C_{in}. It can be solved by making use of the theory of envelopes[17], but here we will give a more intuitive and straight-forward solution that makes use of the unified geo-MAP technique - in combination with ordinary Taylor expansion - to compute a point of the reflected caustic C_{out} as the intersection of two neighbouring reflected rays.

From Sec. (5.4.2) we recall the expression (5.19) for the point of intersection of the two lines determined respectively by the two pairs of points $V_i = {}^*v_i$ and $W_i = {}^*w_i, i = 0, 1$, where the determinants α and β appearing in (5.19) are given by (5.17) and (5.18).

[17] see Kowalewski [132], pp. 50-54.

For the sake of convenience (with respect to the computations that follow) let us fix the coordinate system so that

$$m(s) = O, t(s) = e_1, n(s) = -e_2. \tag{5.26}$$

Note: Since the point m is to be considered as the point O of origin for the space of direction vectors, we will write p and q in order to denote the direction vectors $p - m$ and $q - m$ respectively. This will shorten the presentation of the computations considerably. When we have finished, we will restore the correct direction vector expressions and present the desired result in a way that distinguishes clearly between position vectors and direction vectors.

Following these preliminaries, we will now make use of (5.19) to compute the point of intersection $q(s)$ of two neighbouring reflected rays.

From classical differential geometry[18] we recall the so called Frenét equations for a curve M:

$$\begin{aligned} \dot{t}(s) &= \frac{1}{\rho} n(s) \\ \dot{n}(s) &= -\frac{1}{\rho} t(s). \end{aligned} \tag{5.27}$$

Here $t(s)$ and $n(s)$ are the unit tangent respectively the unit normal to M at the point $m(s)$, and $\rho = \rho(s) = |r(s) - m(s)|$ is the radius of curvature of M at this point.

Moreover, since s denotes arc-length on M, we have $\dot{m}(s) = t(s) = e_1(s)$, and (5.27) can be written in the form:

$$\begin{aligned} \dot{e}_1(s) &= -\frac{1}{\rho} e_2(s) \\ \dot{e}_2(s) &= \frac{1}{\rho} e_1(s). \end{aligned} \tag{5.28}$$

The reflected rays corresponding to the parameter values s and $s + ds$ are determined by the two point-pairs $\{m(s), \bar{p}(s)\}$ respectively $\{m(s+ds), \hat{p}(s)\}$, where the points $\bar{p}(s)$ and $\hat{p}(s)$ are constructed by reflecting the point $p(s)$ in the tangent $t(s)$ at the point $m(s)$ respectively the tangent $t(s + ds)$ at the point $m(s + ds)$. Using the reflection formula for vectors[19], we can write

$$\begin{aligned} \bar{p}(s) - m(s) &= -e_1(s)(p(s) - m(s))e_1(s)^{-1} \\ \hat{p}(s) - m(s + ds) &= -e_1(d + ds)(p(s) - m(s + ds))e_1(s + ds)^{-1}. \end{aligned} \tag{5.29}$$

Suppressing the dependence of s and recalling from (5.26) that $m(s) = 0$, (5.29) takes the form

[18] See e.g. Eisenhart [68] or Struik [221]

[19] See Hestenes [112], p. 278]

$$\bar{p} = -e_1 p e_1$$
$$\hat{p} - \dot{m} ds = -(e_1 + \dot{e}_1 ds)(p - \dot{m} ds)(e_1 + \dot{e}_1 ds) + O(ds^2) \tag{5.30}$$

where $O(ds^2)$ denotes the well-known 'big-oh' ordo-class of functions f, that is

$$f \in O(ds^2) \Leftrightarrow |f(s + ds) - f(s)| \leq K|ds|^2 \tag{5.31}$$

for some constant $K = K(s)$. Expanding the right hand side of the second equation of (5.30) gives

$$\hat{p} - \dot{m} ds = -(e_1 p e_1 + (e_1 p \dot{e}_1 + \dot{e}_1 p e_1 - e_1 \dot{m} e_1) ds) + O(ds^2). \tag{5.32}$$

Now $\dot{e}_1 p e_1 = e_1 p \dot{e}_1$, and since $\dot{m} = e_1$, we have $e_1 \dot{m} e_1 = e_1$. Therefore we can write

$$\begin{aligned} \hat{p} &= -e_1 p e_1 + (e_1 + e_1 - e_1 p \dot{e}_1 - \dot{e}_1 p e_1) ds + O(ds^2) \\ &= -e_1 p e_1 + 2(e_1 - e_1 p \dot{e}_1) ds + O(ds^2). \end{aligned} \tag{5.33}$$

In order to make use of the intersection formula (5.19), we first compute

$$\begin{aligned} \alpha &= (v_0 - w_0) \wedge (v_1 - w_0) e_2 e_1 = -(\dot{m} ds) \wedge (\bar{p} - \dot{m} ds) e_2 e_1 + O(ds^2) \\ &= -((\dot{m} ds) \wedge \bar{p}) e_2 e_1 + O(ds^2) = (\bar{p} \wedge \dot{m}) e_2 e_1 ds + O(ds^2) \end{aligned} \tag{5.34}$$

and

$$\begin{aligned} \beta &= (v_0 - w_1) \wedge (v_1 - w_1) e_2 e_1 = (0 - \hat{p}) \wedge (\bar{p} - \hat{p}) e_2 e_1 \\ &= -(\hat{p} \wedge \bar{p}) e_2 e_1 = (\bar{p} \wedge \hat{p}) e_2 e_1. \end{aligned} \tag{5.35}$$

Moreover, if we split the vector p into the components $p = \psi_1 e_1 + \psi_2 e_2$ and make use of the fact that $\dot{m} = e_1$, we get from (5.34):

$$\alpha = -\psi_2 ds + O(ds^2), \tag{5.36}$$

and a rather lengthy but straightforward calculation gives

$$\bar{p} \wedge \hat{p} = -2(|\dot{e}_1| p^2 + \psi_2) e_1 e_2 ds + O(ds^2). \tag{5.37}$$

Plugging (5.37) into (5.35) gives

$$\beta = -2(|\dot{e}_1| p^2 + \psi_2) ds + O(ds^2). \tag{5.38}$$

Finally, by making use of the intersection formula (5.19), we arrive at the following expression for the local focal point $q = q(s)$ of the reflected wave front corresponding to the local focal point $p = p(s)$ of the incident wave front:

$$q = \frac{1}{\alpha - \beta}(\alpha w_1 - \beta w_0) = \frac{1}{\alpha - \beta}(\alpha \hat{p} - \beta m(s + ds))$$
$$= \frac{(\psi_1 e_1 - \psi_2 e_2)}{1 + 2|\dot{e}_1|\frac{p^2}{\psi_2}} + O(ds). \tag{5.39}$$

In order to restore this result - as we promised above - to a logically consistent and coordinate-free form, we must now substitute $p - m$ for p and $q - m$ for q in (5.39). Performing this substitution, we get

$$q - m = \frac{(\psi_1 e_1 - \psi_2 e_2)}{1 + 2|\dot{e}_1|\frac{(p-m)^2}{\psi_2}}. \tag{5.40}$$

Observe that if $|\dot{e}_1| \to 0$, i.e. if the mirror becomes plane, (5.40) reduces to the familiar law of planar reflection:

$$q_t - m = \psi_1 e_1 - \psi_2 e_2, \tag{5.41}$$

where the point q_t is the reflection of p in the straight line mirror t that is tangent to the mirror M at the point m. Hence, recalling that $\psi_i = (p-m) \cdot e_i$ and that $|\dot{e}_1| = |\ddot{m}| = 1/\rho$, we can express the relationship between the corresponding points $p \in C_{in}$ and $q \in C_{out}$ in the following way:

$$q - m = \frac{((p-m) \cdot e_1)e_1 - ((p-m) \cdot e_2)e_2}{1 + 2|\ddot{m}|\frac{(p-m)^2}{(p-m) \cdot e_2}}$$
$$= \frac{((p-m) \cdot t)t - ((p-m) \cdot n)n}{1 - 2|\ddot{m}|\frac{(p-m)^2}{(p-m) \cdot n}}. \tag{5.42}$$

This relation expresses the second order law of reflection for plane geometrical optics. We summarize Tschirnhausens result in the following

Proposition 5.5.2. *Let C_{in} be the caustic curve of a plane tangential sector of rays that is incident on a plane-curve mirror M (located in the same plane) in such a way that the ray which touches C_{in} at the point p is intercepted by M in the point m, where the unit-tangent, unit-normal Frenét frame for the curve M is given by the vectors t and n (according to an arbitrarily chosen incremental parametric direction of M).*

Under these conditions, the point q which corresponds to p, that is the point q where the reflected ray from m touches the caustic curve C_{out}, is given by the expression

$$q - m = \frac{((p-m) \cdot t)t - ((p-m) \cdot n)n}{1 - 2|\ddot{m}|\frac{(p-m)^2}{(p-m) \cdot n}}. \tag{5.43}$$

5.5.3 Interpreting the Second Order Law of Reflection Geometrically

In order to illustrate the geometric significance of the second order reflection law given by (5.43), we will interpret it in projective geometric terms. In Figure (5.4), $p, \bar{p}, m$ and q have the same meaning as before, and $\bar{r}$ denotes the result of projecting the point r orthogonally onto the reflected ray through the point m with direction $\boldsymbol{mq} = q - m$.

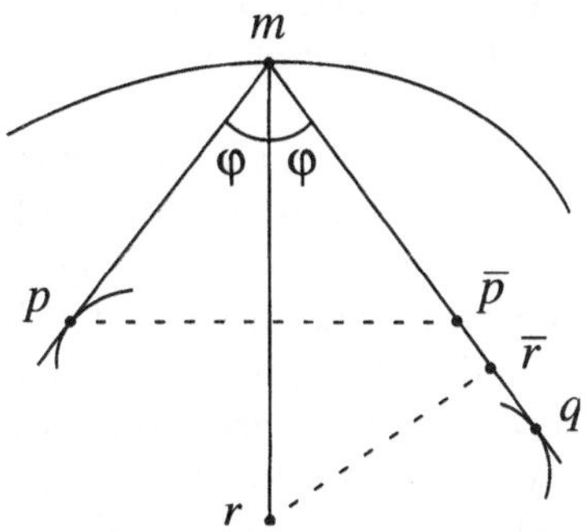

Fig. 5.4. Overview of the Tschirnhausen configuration

Introducing the angle φ between the incident ray of direction $\boldsymbol{pm} = m - p$ and the corresponding mirror normal e_2, we note the following relations between the participating magnitudes:

$$\begin{aligned} q_t - m &= m - \bar{p} \\ (p - m) \cdot e_2 &= -|p - m| \cos\varphi \\ |\bar{r} - m| &= |r - m| \cos\varphi = \rho \cos\varphi \\ |\bar{p} - m| &= |p - m|. \end{aligned} \tag{5.44}$$

Taking (5.44) into accout, the reflection law (5.43) can be expressed as

$$q - m = \frac{\bar{p} - m}{\frac{2|p-m|}{\rho \cos\varphi} - 1} \tag{5.45}$$

and taking the modulus of both sides of (5.45), we can write

$$\frac{1}{|\bar{p} - m|} \pm \frac{1}{|q - m|} = \frac{2}{\rho \cos\varphi} = \frac{2}{|\bar{r} - m|}. \tag{5.46}$$

The sign in the left hand side of (5.46) corresponds to the sign of the denominator in the right hand side of (5.45).

Since, by (5.26), our coordinate system $\{e_1, e_2\}$ has its point of origin at m, the formula (5.46) expresses the fact that the points $\bar{p}$ and q separate the points m and $\bar{r}$ harmonically, that is, these two pairs of points constitute a harmonic 4-tuple. This is the form in which Tschirnhausen presented his reflection law[20].

[20] See Kowalewski [132]. p.51.

5.6 Summary and Conclusions

5.6.1 The Geo-MAP Unification Technique

In Sec. (5.4.1) we introduced the unified geo-MAP computational framework - inspired by classical projective line geometry.[21] We then demonstrated how the geo-MAP framework provides a way to represent the metric (= Euclidean), affine and projective aspects of geometry within the same geometric algebra, and how this representation creates a computational background for performing coherent shifts between these three different geometrical systems.

In (5.12) we showed how to pass from a Euclidean point (= direction vector), to the corresponding affine part of projective space, and in (5.13) we showed how to get back again from the finite part of projective space to the original Euclidean point that we started with. The proof that this works was provided by (5.15). Formulas (5.12) and (5.13) are key formulas underlying many of our later computations. Because of their great practical utility in combining the powers of the ordinary Euclidean direction vector algebra with those of the Grassmann-Cayley algebra, we feel that they should be of particular interest to the engineering community.

In Sec. (5.4.3) we showed how to embed the basic (Euclidean) direction vector algebra into the surrounding geometric algebra. The formulas (5.20) and (5.23) - combined in (5.24) or (5.25) - illustrate the interplay between the Euclidean operations of vector addition and Gibbs' cross-product on the one hand - and the operations of geometric, outer and inner product on the other. Such formulas as these we have not seen anywhere else.

As an illustrative application of the unified geo-MAP computational technique, we applied it in Sec. (5.5) to a classical problem of plane geometrical optics called 'Tschirnhausens problem', which is concerned with determining the envelope of the rays from a point source of light after their reflection in a smoothly curved mirror.

Using the geo-MAP technique in combination with ordinary Taylor expansion, we computed the desired envelope as the locus of intersection of 'neighboring' rays, i.e. rays that differ infinitesimally from one another. In this way we deduced the expression (5.43), which could be termed the "second order law of reflection", since it expresses the curvature relations between the in-coming and out-going wave fronts and the curved mirror.

Although, in the planar case, the same result can be achieved using envelopes, the geo-MAP framework has the advantage of being applicable in higher dimensions. For example, in 3 dimensions, it can be used in order to perform the corresponding computations - relating the points on the respec-

[21] see e.g. Sauer [201], Naeve [174] or Naeve & Eklundh [176].

tive focal surfaces of an in-coming and out-going normal congruence of rays to the corresponding points on the focal surface of the normals to the mirror. However, the complexity of such computations have made it necessary to exclude them here.

5.6.2 Algebraic and Combinatorial Construction of a Geometric Algebra

As a didactic comment on how to teach geometric algebra, we present - in the appendix - a constructional proof of the fact that the 'expected basis elements' of a geometric algebra G -i.e. the set of finite subsets of its formal variables - actually do form a basis for G. This is done in Prop. (5.8.1) and Prop. (5.8.2), leading up to Def. (5.8.3), where we define a geometric algebra by constructing it.

Our construction enables the possibility of a logically self-contained description of geometric algebra which does not require such high levels of abstraction as in the traditional tensor algebra approach, and which should therefore be accessible to a wider audience. In our opinion, the main reason for the lack of such a presentation in the present literature is the difficulties encountered in establishing a vector space basis for a geometric algebra.

Using this approach to presenting geometric algebra, we do not have to worry about the question of whether there exists any algebraic structure that is capable of implementing the desired specifications. We are therefore free to take the 'basis approach', both to defining different operations on the algebra as well as to proving their structural properties. In our opinion this greatly simplifies a deductive presentation of geometric algebra to students.

5.7 Acknowledgements

We want to express our gratitude to professors Jan-Olof Eklundh and Harold Shapiro at the Royal Institute of Technology in Stockholm/Sweden, for providing the supportive environment in which we have been able to perform our academic research over the years. We also would like to thank professor Gerald Sommer at the Christian-Albrechts-University in Kiel/Germany for inviting us to contribute to this book.

5.8 Appendix: Construction of a Geometric Algebra

Let R be a commutative ring, and let $\{E, <\}$ be a totally ordered set. The non-commutative ring of polynomials over R in the formal variables E is denoted by $R\{E\}$, and the set of monomials and the set of terms in the ring $R\{E\}$ is denoted by M respectively by T.

Moreover, let $S = \{-1, 0, 1\} \subset R$, and let $\text{sgn} : T \to R$ be any given mapping.

Definition 5.8.1. *We say that the pair $(e, e') \in E^2$ is an involution in the term t, if there exist terms t', t'', t''' with $t = t'et''e't'''$ such that either $e' < e$ or $e' = e$, $sgn(e) = -1$.*

Notation: The number of inversions in the term t is denoted by $\text{inv}(t)$.

We now define a mapping $\mu : T \to R$ in the following way:

- $\mu(rm) = r\mu(m)$, where $r \in R$, $m \in M$.
- $\mu(m) = 0$, if m contains at least two occurrences of some $e \in E$ with $\text{sgn}(e) = 0$.
- $\mu(m) = (-1)^{\text{inv}(m)}$ otherwise.

We also introduce a reduction rule $\to$, i.e. a binary relation on T by making the following

Definition 5.8.2. *$t \to t'$, where $t, t' \in T$, if there exist terms $t_1, t_2 \in T$ and $e_1, e_2 \in E$ such that $t = t_1 e_2 e_1 t_2$, $e_1 \leq e_2$, and where*
$t' = -t_1 e_1 e_2 t_2$, if $e_1 < e_2$, or
$t' = sgn(e_1) t_1 t_2$, if $e_1 = e_2$.

Notation: If no t' exists in T such that $t \to t'$, we write $t|$, and if $t \to t_1 \to \ldots \to t_k$ we write $t \stackrel{*}{\to} t_k$.

By inspection, we observe that

$$\text{inv}(m_1 e e m_2) = \text{inv}(m_1 m_2) + \text{inv}(ee) + 2N, \text{ for some } N \in \mathbb{N}, \tag{5.47}$$

and that

$$\text{inv}(m_1 e_2 e_1 m_2) = \text{inv}(m_1 e_1 e_2 m_2) + 1, \text{ if } e_1 < e_2. \tag{5.48}$$

From (5.47) and (5.48) we can conclude that if $t \to t'$, we have

$$\mu(t) = \mu(t'). \tag{5.49}$$

We can now state the following

Proposition 5.8.1. *For each t in T there exists a unique $t' = red(t)$ in T, such that $t \to t'|$.*

Proof. We start by proving uniqueness. If $\mu(t) = 0$ then obviously $\mu(t') = 0$. Let $t = rm \stackrel{*}{\to} r'm'|$, where $\mu(m) \neq 0$. Then, by inspection, $m' = e_1 e_2 \ldots e_n$, where $e_1 < e_2 < \ldots < e_n$, and $\{e_1, \ldots, e_n\}$ is the set of $e : s$ in E occurring an *odd* number of times in m. If this set is empty, we put $m' = 1$. Hence, m' is unique. Moreover, $\mu(t) = r\mu(m) = r'\mu(m')$ which shows that r' is unique. This finishes the uniqueness part of the proof.

For the proof of the existence part, we observe that if $t_1 \to t_2$, then we have

$$\deg(t_1) + \mathrm{inv}(t1) > \deg(t_2) + \mathrm{inv}(t_2). \tag{5.50}$$

Hence every reduction chain $t \xrightarrow{*} t_k$ is finite, which proves the existence of t'. □

From Prop. (5.8.1) we can directly conclude:

Proposition 5.8.2. *Let B denote the set of monomials m in $R\{E\}$ such that $m|$. Then B is in one-to-one correspondence with the set of finite subsets of E.*

Notation: Let B_n denote the set of monomials in B of degree n. The R-modules generated by B and B_n are denoted by G and G_n respectively.

We now turn G into a ring by introducing an R-bilinear mapping (multiplication)

$$\begin{aligned} &G \times G \to G \\ &(x, y) \to x \circ y \end{aligned} \tag{5.51}$$

in the following way:

By R-bilinearity, it is enough to define $m_1 \circ m_2$ for $m_1, m_2 \in B$. We do so by defining

$$m_1 \circ m_2 = \mathrm{red}(m_1 m_2). \tag{5.52}$$

We then have

$$m_1 m_2 m_3 \xrightarrow{*} m_1(m_2 \circ m_3) \xrightarrow{*} m_1 \circ (m_2 \circ m_3) = \mathrm{red}(m_1 m_2 m_3)|, \tag{5.53}$$

and

$$m_1 m_2 m_3 \xrightarrow{*} (m_1 \circ m_2) m_3) \xrightarrow{*} (m_1 \circ m_2) \circ m_3) = \mathrm{red}(m_1 m_2 m_3)|. \tag{5.54}$$

Since $\mathrm{red}(m_1 m_2 m_3)$ is unique, it follows that the product $\circ$ is associative.

Definition 5.8.3. *The ring $(G, \circ)$ is called a geometric algebra (or a Clifford algebra).*

Notation: The product $\circ$ is called the *geometric product*, and it is usually written as a concatenation. Following this convention, we will from now on write xy for the product $x \circ y$, i.e.

$$xy \equiv x \circ y. \tag{5.55}$$

We can reformulate Prop. (5.8.2) as

Proposition 5.8.3. *Let G be a geometric algebra over R with formal variables E. Then G has an R-module basis consisting of the set of all finite subsets of E.*

Moreover, it can be shown that the following holds:

Proposition 5.8.4. *Let E' be another set, totally ordered by $<'$, and let the mapping $sgn' : E' \to R$ satisfy the condition $card(E^s) = card(E'^s)$, where $E^s = \{e \in E : sgn(e) = s\}$ and $E'^s = \{e \in E' : sgn'(e') = s\}$. Then G and G' are isomorphic as geometric algebras.*

One way to establish this isomorphism is to show that if J is the ideal generated by $\{e^2 - \text{sgn}(e), ee' + e'e : e, e' \in E, e \neq e'\}$, then G is isomorphic to $R\{E\}/J$.

6. Honing Geometric Algebra for Its Use in the Computer Sciences

Leo Dorst

Dept. of Computer Science, University of Amsterdam

6.1 Introduction

A computer scientist first pointed to geometric algebra as a promising way to 'do geometry' is likely to find a rather confusing collection of material, of which very little is experienced as immediately relevant to the kind of geometrical problems occurring in practice. Literature ranges from highly theoretical mathematics to highly theoretical physics, with relatively little in between apart from some papers on the projective geometry of vision [143]. After perusing some of these, the computer scientist may well wonder what all the fuss is about, and decide to stick with the old way of doing things, i.e. in every application a bit of linear algebra, a bit of differential geometry, a bit of vector calculus, each sensibly used, but *ad hoc* in their connections. That this approach tends to split up actual issues in the application into modules that match this traditional way of doing geometry (rather than into natural divisions matching the nature of the problem) is seen as 'the way things are'.

However, if one spends at least a year in absorbing this material, a different picture emerges. One obtains increased clarity and prowess in handling geometry. This is due to being able to do computations without using coordinates; and by having elements of computation which are higher-dimensional than vectors, and thus collate geometrical coherence. The operators that can be applied are at the same time more limited in number, and more powerful in purity and general validity. Through this, one obtains the confidence to tackle higher-dimensional parameter spaces with the intuition obtained from 3-dimensional geometry. Programs written are magically insensitive to the dimensionality of the embedding space, or of the objects they act on. The concept of a 'split' endows the limited set of operators with a varied semantics, which begins to suggests an applicability to all geometries one is likely to encounter.

The hardest part in achieving such a re-appraisal is actually letting go of the usual geometrical concepts, and embracing new ones. It is not hard to rewrite, say, linear algebra into geometric algebra; but it is a different matter altogether to use the full power of geometric algebra to solve problems for which one would otherwise employ linear algebra. This overhaul of the mind takes time; and would be greatly aided by material aimed towards computer scientists. This will doubtlessly appear, for an evangelical zeal appears to be common to all who have been touched by geometric algebra, but at the moment it is scarce.

So geometric algebra can (and will) change computer science; but vice versa, the need for a clear syntax and semantics for the geometric objects and operators in a specification language requires a rigor beyond the needs of its current applications in physics, and this is where computer science may affect geometric algebra. Imposing this necessary formalization – always with the applications in mind – reveals some ambiguities in the structure of geometric algebra which need to be repaired. This paper reports on some issues encountered when preparing the wealth of geometric algebra for its application in the computer sciences. They involve simply making the internal structure explicit (section 6.2); redesigning the operators (even the rather basic inner product can be improved, in section 6.3); the development of new techniques to enable the user to adapt the structure to his or her needs (section 6.4); and making mathematical isomorphisms explicit in applicable operators (section 6.5).

When this is done, many individual 'tricks' occurring in different branches of classical geometry become unified (this is shown for the 'meet' in section 6.6), and therefore implementable in a generic toolbox structuring the thinking and increasing the capabilities of the geometrical computer scientist. This is an ongoing effort; as a consequence, this paper is still directed more towards the *developers* of geometric algebra than towards its *users*. Yet it should help potential users to assess these exciting new developments in unified geometrical computation.

6.2 The Internal Structure of Geometric Algebra

The monolithic term 'geometric algebra' hides an internal structure that consists of various levels, each of which are relevant to the computer scientist desiring to use it in an application. It is important to distinguish them, for various branches of literature deal with different levels – so you may not find

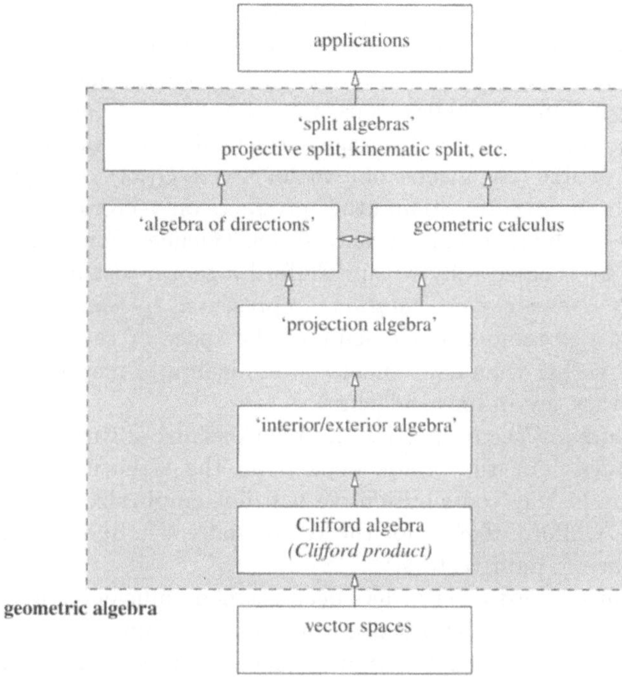

Fig. 6.1. Levels in geometric algebra with their operators (non-standard terms in quotes).

what you need in any one book or paper. I have found the levels sketched in table 6.1 useful in expositions on the subject, since they explicitly indicate the scope of each part of the formalism. They are depicted in a 'bottom-up' manner from the mathematics of Clifford algebras (at the basis) to various applications (at the top). (Some levels and their names are my own suggestions, for the purpose of this chapter; they are denoted in quotes throughout.)

- **Clifford algebra**
  At the basis of all geometric algebra is Clifford algebra. This introduces a *(Clifford) product* in a vector space V^n over a field of scalars K, thereby extending it to a 2^n-dimensional linear space of multivectors.[1] This product is commonly introduced using a bilinear form $\langle\, , \rangle : V^n \times V^n \to K$ or a quadratic form $Q : V^n \to K$, to satisfy the axioms:
  1. *scalars commute with everything*: $\alpha\, u = u\, \alpha$, for $\alpha \in K$, $u \in \mathcal{C}\ell_n$.
  2. *vectors* $\mathbf{x} \in V^n$ *obey:* $\mathbf{x}\mathbf{x} = Q(\mathbf{x})$ (which is a scalar!).

[1] Several levels higher, the geometric semantics of this product suggests itself so strongly that it has become custom in geometric algebra to denote the Clifford product as a 'geometric product'; but at this basic level that is not obvious yet, and leads to confusion.

3. *algebraic properties:* geometric product is linear in both factors, associative, and distributive over +. Do *not* demand commutativity!

Repeated application of the geometric product then produces the basic elements for the whole Clifford algebra, consisting of scalars, vectors, bivectors, etcetera. A big mathematical advantage of the Clifford product is that it is in principle invertible (the inverse of a vector $\mathbf{x}$ is $\mathbf{x}/Q(\mathbf{x})$). This gives a much richer algebraic structure than other products on vectors (such as the inner product) – with far-reaching practical consequences. For instance, a subject that can be studied fully within Clifford algebra, just using the Clifford product, is *n-dimensional rotations*, represented by *spinors.* Rotations are directly represented as elements of the space of the algebra, just as vectors are, rather than as elements of an algebra of mappings on a vector space (as they are in linear algebra).

When one starts studying the properties and relationships of various Clifford algebras, it turns out that these depend on the signature of the quadratic form; but in this contribution we will not emphasize this, using $\mathcal{C}\ell_n$ to denote a Clifford algebra for the vector space V^n, and (slightly casually) for its space of multivectors.[2]

The mathematics of Clifford algebra has been studied sufficiently for all immediate purposes in computer science, and good accounts exist (try [192], chapter 1). The style of explanation in such accounts is often 'permutation of indices' rather than 'geometrically motivated construction', a consequence of its close (and, to mathematicians, interesting) relationship to *tensor algebra.* Although this is somewhat off-putting at first, it does give a clear indication to the computer scientist of how the basic operations can be implemented efficiently, and how their syntax is defined independent of any geometric semantics we might choose to impose later.

- **'interior /exterior algebra'**

In derivations in Clifford algebra, one often uses commutativity or anti-commutativity of Clifford products. This occurs so often that it makes sense to decompose the Clifford product of vectors into a symmetric and anti-symmetric part under commutation, and use those as higher level 'macros' to develop higher level insights. There is an unambiguous and agreed-upon choice for the anti-symmetric *outer product* $\wedge$ which is defined by:

$$\begin{array}{ll} 1. & \mathbf{x} \wedge u = \frac{1}{2}(\mathbf{x}\, u + u^* \,\mathbf{x}) \quad \text{for } \mathbf{x} \in V^n, u \in \mathcal{C}\ell_n \\ 2. & \wedge \text{ is linear in both arguments, and associative.} \end{array} \tag{6.1}$$

[2] It is a dilemma, when learning Clifford algebra, whether you should do algebras of purely Euclidean spaces first (most intuitive!), or learn it in its full generality from the start (most general!). In any case, a practitioner will have to learn non-Euclidean Clifford algebras eventually, because the projective split (section 6.5) and the recent development of the homogeneous model of Euclidean space (Chapter 2) show that non-Euclidean Clifford algebras are a very convenient representation for computations on the geometry of purely Euclidean spaces!

(Here $\cdot^*$ denotes the main involution of $\mathcal{C}\ell_n$.) For the symmetric counterpart there are two choices, the *inner product* '$\cdot$' or the *contraction* '$\rfloor$', both agreeing on vectors:

$$\mathbf{x} \cdot u = \mathbf{x} \rfloor u = \tfrac{1}{2}(\mathbf{x}\, u - u^*\, \mathbf{x}) \quad \text{for} \quad \mathbf{x} \in V^n, u \in \mathcal{C}\ell_n,$$

but differing in action on general multivectors (details later). In geometric algebra as developed for physics [113], the inner product '$\cdot$' is used. We will argue below why the contraction '$\rfloor$' is preferable for computer science since it gives a cleaner algebraic computational structure, without exceptions or conditions to geometrically meaningful results.

– **'projection algebra'**
The fresh contribution of Clifford algebras to the way we compute in geometry is the treatment of *composite geometrical objects (lines, planes, spheres) as basic elements of computation.* This leads to new geometric and computational insights, and new methods, even for such basic constructions as, say, the intersection of two lines (section 6.6). The main consequence is that the use of geometric algebra makes our algorithms coordinate-free, valid in or extendible to n-dimensional spaces, and fully specific on direction parity (which is useful for consistent treatment of inside/outside, a notorious issue in computer graphics).
A k-dimensional subspace of a vector space V^n is characterized by a *blade* i, which is an outer product of k independent vectors in that subspace:

$$i = \mathbf{a}_1 \wedge \mathbf{a}_2 \wedge \cdots \wedge \mathbf{a}_k, \quad \text{with} \quad \mathbf{a}_i \in V^n. \tag{6.2}$$

Such a multivector is called *simple*; its magnitude is the directed volume spanned by the $\mathbf{a}_i$. The subspace spanned by i is denoted $\mathcal{G}(i)$. Eq.(6.2) explains the relevance of the outer product: it codifies 'linear (in)dependence' in an operational manner. The interaction of the *non-invertible* outer product with the *invertible* Clifford product produces compact notation and computation for algorithms involving *orthogonality.* For instance, the determination of a vector of $\mathcal{C}\ell^n$ perpendicular to the subspace $\mathcal{G}(i)$ is:

$$\mathrm{P}_i^{\perp}(\mathbf{x}) \equiv (\mathbf{x} \wedge i)\, i^{-1}.$$

This leads to a compact and computable formulation of such algorithms as 'Gram-Schmidt orthogonalization'. Also, we can construct the *dual* $\widetilde{a} \equiv a I^{-1}$ of a simple multivector a within a subspace I, and interpret it as the blade of the subspace *perpendicular* to a in $\mathcal{G}(I)$.
These constructions have very intuitive geometrical interpretations (see figure **??**); it is at this level that it becomes natural to call the Clifford product a *geometric product.*

– **'algebra of directions'**
Closely related to the above, but often used more qualitatively, is the idea of union and intersection of subspaces to produce higher or lower dimensional subspaces. The operations that do this are known as the *join* and *meet* operations. They are a precise extension of set union and set intersection

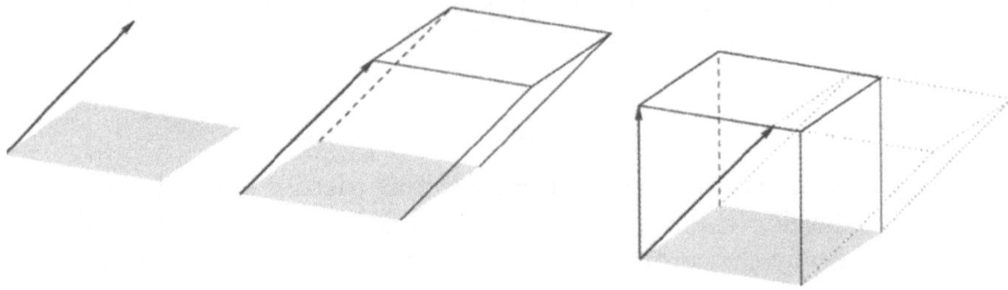

Fig. 6.2. The perpendicular component $\mathbf{x}_\perp$ of a vector $\mathbf{x}$ to a subspace characterized by a blade $\mathbf{A}$: make the volume $\mathbf{x} \wedge \mathbf{A}$, straighten it in your mind (to view it as geometric product), then factor out $\mathbf{A}$ by division – but beware that division is not commutative, so compute it as $\mathbf{x}_\perp = (\mathbf{x} \wedge \mathbf{A})\mathbf{A}^{-1}$.

for directed subspaces; usually, they are treated modulo positive scalar factors since blades signify a directed subspace modulo such a factor. We will argue in section 6.6 that there is a quantitative structure to these operations which is very useful in computations, since it determines how well-conditioned the operations are (similar to the use of the condition number of a matrix equation in numerical linear algebra) on the basis of 'distance measures' between the subspaces.
Join $\dot{\wedge}$ and meet $\vee$ of spaces are definable in terms of outer product and the contraction (or the inner product):

$$a \,\dot{\wedge}\, b = b \wedge a \quad \text{and} \quad a \vee_i b = (ai^{-1}) \rfloor b,$$

but their geometrical use involves some care, as we will see in section 6.6. Hestenes [113] pg. 19 calls this use of geometric algebra an 'algebra of directions', since the relationships between the blades implement the lattice of k-dimensional directed subspaces of a vector space V^n.

- **geometric calculus**
  Differentiation operators in geometric algebra are associated with multivectors; as a consequence, they have both properties of calculus and of geometry [113, 64]. The geometrical properties need to obey the various product rules sketched before for multivectors; so differentiation with respect to a (multi)vector has commutation rules, decomposition rules, and orthogonality properties that fit the above scheme. This leads to a powerful calculus, which can usefully redo and extend the constructions of differential geometry. The popular differential forms, for instance, can be viewed advantageously within the more general framework of geometric algebra.
- **'split algebras'**
  The above gives the framework of basic techniques in geometric algebra. This needs to be augmented by specific techniques for *mapping the geometric structure of an application* to a properly identified algebra. This is the *modeling* step, which is part of the application domain more than

of the algebraic mathematics. It is of course highly important to applied computer science.
There is an important construction technique which brings some unification in these various required embeddings: Hestenes' *split* [110]. This is a technique in which the geometric algebra of an $(n+k)$-dimensional space is used to model the geometry of an n-dimensional space V^n. The split explicitly relates multivectors in the two spaces. The advantage is that the 'orthogonality algebra' and 'directed intersection algebra' of $\mathcal{C}\ell_{n+k}$ (which were developed for *homogeneous*, because simple, multivectors), now can describe the *non*-homogeneous quantities of projective and affine geometry in V^n (using a *projective split*) and of kinematics in V^n (using a *conformal split*) [110]. Mathematically, the split makes an n-dimensional vector *isomorphic* to, say, an $n+1$-dimensional bivector. This is often denoted as '=' in literature. We will see in section 6.5 that to actually use the split in an implementation, it is more proper to be explicit about the mapping relating the elements of the algebras $\mathcal{C}\ell_{n+k}$ and $\mathcal{C}\ell_n$.

With these refinements of the monolithic term 'geometric algebra' into various levels of meaning and associated operators, we can better state its relevance to computer scientists needing to 'do' geometry.

> *Geometric algebra is a collection of computation rules and techniques relevant to doing computations in models of the geometric aspects of applications. Its structure contains several distinct but exactly related levels, each with its own syntax of operators, and an accompanying interpretation. A specific application will probably need them all; fortunately they are generic in their construction. It is thus advantageous to connect to this framework, both for unified theoretical developments and for the actual software performing the calculations.*

If our hopes come true, geometric algebra does away with the *internal interface problem* between geometric computational modules (typically arising when solving part of one's application by techniques of linear algebra using matrices, and then having to translate them to differential forms to treat other aspects, all proceeded by the projective geometry of processing and interpreting actual visual observations). It will replace all this with a common language in which all these specialized modules can communicate. and in which algorithms can be specified and developed. The *modeling problem* ('which geometric model for which application') remains, but the choices are limited (one of the Clifford algebras), and can all be implemented in advance, in a standard manner, with generic data structures. We can then focus on *what* we need to compute in our applications, rather than on *how* to compute it.

6.3 The Contraction: An Alternative Inner Product

The Clifford product is the unambiguous basis of all geometric algebra, and from it are constructed derived products which are useful for 'orthogonality algebra'. Using such products, we would expect to prove lemmas which are universally valid 'total identities', into which we can plug any element of the geometric algebra. The currently used inner product of [113], however, is riddled with lemmas containing conditions, mostly on the relative grade of its operands. These problems were recognized (see [113] pg.20), but not resolved until recently, when Lounesto [153] called attention to a different way of introducing an inner-product-like operation into geometric algebra. He calls this the *contraction* and denotes by '$\rfloor$'; his suggestion does not seem to have been followed in the applied literature. Yet the contraction may be a great improvement to geometric algebra, since it simplifies the algebraic structure without sacrificing any of the geometric meaning – as will now be shown.

Here is the definition. Assume that you have already defined the Clifford product based on a bilinear form $\langle , \rangle$ on vectors, and have based on that the outer product, as in eq.(6.1): by means of what it does on vectors, and demanding bilinearity and associativity. Now extend the bilinear form $\langle\, , \,\rangle$ to arbitrary multivectors, as follows.

1. For scalars:
$$\langle \alpha, \beta \rangle = \alpha\beta \quad \text{for } \alpha, \beta \in K. \tag{6.3}$$
2. For two multivectors of the form $a = \mathbf{a}_1 \wedge \mathbf{a}_2 \wedge \cdots \wedge \mathbf{a}_k$ and $b = \mathbf{b}_1 \wedge \mathbf{b}_2 \wedge \cdots \wedge \mathbf{b}_\ell$:
$$\langle a, b \rangle = \begin{cases} \det\left(\langle \mathbf{a}_i, \mathbf{b}_j \rangle\right) & \text{if } \ k = \ell \\ 0 & \text{if } \ k \neq \ell \end{cases} \tag{6.4}$$
Here $(\langle \mathbf{a}_i, \mathbf{b}_j \rangle)$ denotes the matrix of which the (i,j)-th element equals $\langle \mathbf{a}_i, \mathbf{b}_j \rangle$; its determinant is just used as a convenient shorthand for the anti-symmetric construction of the bilinear form.
3. The bilinear form is to be linear in both arguments.

Note that this is symmetrical, i.e. $\langle a, b \rangle = \langle b, a \rangle$. As a consequence of the imposed orthogonality of this extended bilinear form, a set of equalities $\langle x, a \rangle = \langle b, a \rangle$ for all a in (a basis of) $\mathcal{Cl}_n$ implies $x = b$.

With this bilinear form, define the contraction as *adjoint* to the outer product:

$$\langle u \rfloor v, w \rangle \equiv \langle v, u^\dagger \wedge w \rangle \quad \text{for all} \quad u, v, w \in \mathcal{Cl}_n \tag{6.5}$$

(where the reversion $u^\dagger$ of u is used to absorb some inconvenient signs). Now one can prove the following properties (see also [153]):

$$(a)\ \alpha\rfloor\beta = \alpha\beta, \quad \alpha\rfloor\mathbf{x} = \alpha\mathbf{x} \ \text{ and } \ \mathbf{x}\rfloor\alpha = 0, \qquad \text{for } \mathbf{x} \in V^n, \alpha, \beta \in K$$
$$(b)\ \mathbf{x}\rfloor\mathbf{y} = \langle \mathbf{x}, \mathbf{y} \rangle \quad \text{for } \mathbf{x}, \mathbf{y} \in V^n$$
$$(c)\ \mathbf{x}\rfloor(u \wedge v) = (\mathbf{x}\rfloor u) \wedge v + u^* \wedge (\mathbf{x}\rfloor v) \quad \text{for } \mathbf{x} \in V^n, \ \ u, v \in \mathcal{C}\!\ell_n$$
$$(d)\ (u \wedge v)\rfloor w = u\rfloor(v\rfloor w), \quad u, v, w \in \mathcal{C}\!\ell_n$$

Property (a) shows that a contraction involving scalars is not symmetric, as opposed to their inner product for which [113] explicitly demands $\alpha \cdot \mathbf{x} = \mathbf{x} \cdot \alpha = 0$. Property (b) shows that for vectors the contraction corresponds with the inner product. Property (c) shows that it is like a derivation, and the common inner product satisfies it as well. Property (d) is a duality between outer product and contraction, valid for *all* multivectors; the corresponding statement for the inner product has much more limited validity (more about this below).

These properties, together with *linearity in both arguments*, are sufficient to compute the contraction of *any* two multivectors, more conveniently than by the formal definition eq.(6.5). An important difference with the inner product '$\cdot$' is that the contraction '$\rfloor$' is *not* symmetric in its arguments (property (a) is one example). This means that many of the useful constructions and proofs of [113] which use the inner product need to be redone. When we do so, we find that the asymmetrical *conditions* under which the proof worked for the inner product (for instance on the relative grades of arguments) are now elegantly absorbed in the contraction operator (outside the range of the conditions on $\cdot$, the expression with $\rfloor$ then automatically produces 0). Thus the useful results from [113], ch.1 and its sequels are not only 'rescued', but also expressed more concisely. And many results obtain an expanded range of validity, due to the nice algebraic properties of the new inner product. We give some examples.

Examples:

1. *Duality statements.* In [113](1-1.25b) we find for *homogeneous* multivectors a_r, b_s, c_t of grades r, s, t, respectively, the property:

$$a_r \cdot (b_s \cdot c_t) = (a_r \wedge b_s) \cdot c_t \quad \text{for } \ r + s \le t \text{ and } r, s \ge 0 \tag{6.6}$$

With the contraction rather than the inner product, we can prove the much stronger:

$$u\rfloor(v\rfloor i) = (u \wedge v)\rfloor i, \quad \text{for } \ u \in \mathcal{G}(i), \ \ v \text{ arbitrary.} \tag{6.7}$$

Here u and v are general (*not just homogeneous!*) multivectors, i is a blade (and therefore homogeneous), and the only condition is that u is in the geometric algebra of the subspace with pseudoscalar i. Note that it is permitted to have v in a space exceeding $\mathcal{G}(i)$; if it is, both sides are 0 and hence still equal. Thus this structural property in 'interior/exterior algebra', and the algebras built on it, has a *much enlarged scope of validity.*

Other duality statements from [113] generalize similarly, we will prove an example in section 6.4.2. The most extreme is property (*d*) above (which is similar to [113](1-2.17b), but now valid for *all* multivectors u, v, w).

2. *Expansion of geometric products.* In [113](1-1.63) we find for the expansion of a geometric product of a bivector **B** with a multivector u:

$$\mathbf{B}\,u = \mathbf{B}\cdot u + \tfrac{1}{2}(\mathbf{B}\,u - u\,\mathbf{B}) + \mathbf{B}\wedge u \quad \text{if} \quad \mathbf{B} = \langle \mathbf{B}\rangle_2 \text{ and } \langle u\rangle_1 = 0$$

Note the demand $\langle u\rangle_1 = 0$: this formula does *not* work for vectors. So we have an identity of which the validity depends on the grade of an operand. In (subtle) contrast, using the contraction, we can prove:

$$\mathbf{B}\,u = \mathbf{B}\rfloor u + \tfrac{1}{2}(\mathbf{B}\,u - u\,\mathbf{B}) + \mathbf{B}\wedge u \quad \text{if} \quad \mathbf{B} = \langle \mathbf{B}\rangle_2,$$

a formula that is now valid for *all* u, since $\mathbf{B}\rfloor u = 0$ for the scalar and vector parts of u. As before, the contraction operator automatically takes care of the conditions. This formula is part of a series of expansion formulas, for a scalar α, vector $\mathbf{x}$ and bivector **B** we get:

$$\alpha\,u = \alpha\rfloor u = \alpha\wedge u, \qquad \mathbf{x}\,u = \mathbf{x}\rfloor u + \mathbf{x}\wedge u$$

$$\mathbf{B}\,u = \mathbf{B}\rfloor u + \tfrac{1}{2}(\mathbf{B}\,u - u\,\mathbf{B}) + \mathbf{B}\wedge u.$$

Each higher order obtains one more term. In [113], the statement for bivectors takes the exception stated above, while that for scalars reads $\alpha\,u = \alpha\cdot u + \alpha\wedge u = \alpha\wedge u$ since $\alpha\cdot u = 0$ by definition.

3. Continuing in In this manner, it is indeed possible to reproduce *all* geometric constructions from [113], chapter 1, using the contraction to replace the inner product. This demonstrates that the contraction can also be used as a basis for a full geometric algebra.

In summary, an equally or more powerful structure is created by using $\rfloor$ rather than $\cdot$, *in which known results are simultaneously generalized and more simply expressible*, without conditional exceptions. This cleaner algebraic structure will lead to simpler geometric software, since no exception handling is required.[3]

6.4 The Design of Theorems and 'Filters'

Since the contraction operator reduces conditions in expressions, it becomes possible to develop a technique for designing 'geometric filters', i.e. expressions in geometric algebra that perform certain desired tasks. Let us call

[3] Moreover, the extended bilinear form permits us to repair some inelegancies in the common definitions of basic concepts, such as the use of grade operators to define elementary concepts like the norm of a multivector u by $\langle u^\dagger\, u\rangle_0$. Having defined the extended bilinear form we can simply define: $|u|^2 = \langle u, u\rangle$; the same in value, but arguably more elegant, requiring fewer operators.

this technique the 'index set method', since it designs the filters based on which independent orthogonal basis vectors (characterized by indices) occur in input and output of the filter. Such indices may get passed, they may be cancelled, or they may lead to a zero result. For instance, in $\mathbf{e}_1\mathbf{e}_2$, both indices 1 and 2 occur in the result; in $\mathbf{e}_1\mathbf{e}_2\mathbf{e}_1 = -\mathbf{e}_1^2\mathbf{e}_2 = \alpha\,\mathbf{e}_2$, index 1 has been absorbed in a scalar α; and in $\mathbf{e}_1 \rfloor \mathbf{e}_2$, the combination of index 1 for the first argument and index 2 for the second results in 0 (remember that the $\mathbf{e}_i$ are orthogonal). We denote the index set of a simple multivector a by $\mathcal{I}(a)$. Despite the index-based nature of this procedure, linearity guarantees that the final results are coordinate free, independent of the basis on which they were derived.

Figure 6.3 and 6.4 present the different filters of two and three terms, using only geometric product, outer product and contraction between terms (some reduction of the full range was made using the symmetry of geometric product and outer product on index sets).

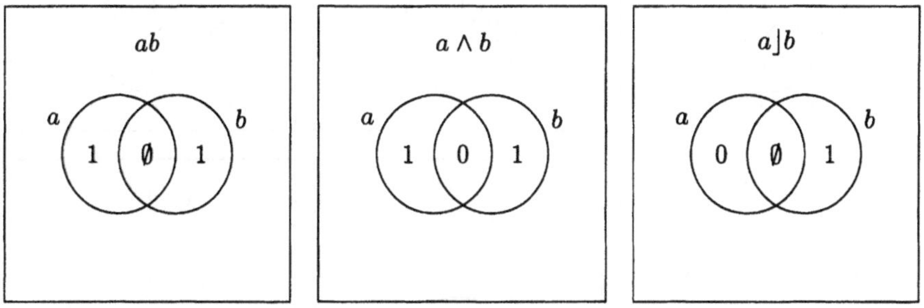

Fig. 6.3. The index sets of the basic products. Notation: '1' denotes that indices in this subset appear in the result, '0' denotes that indices appearing in this set make the whole result 0, and '∅' denotes that indices in this subset do not appear in the result (but neither do they make the result zero).

6.4.1 Proving Theorems

We can prove identities and their conditions by the following method:

1. First assume that the arguments are simple multivectors.
2. Draw up the outcome diagrams of both sides (using figures 6.3 and 6.4 to compose them quickly).
3. Make a composite diagram retaining only those subsets in which no conflict exists between outcomes.
4. In the areas with outcome 0, the identity obviously (but rather trivially) holds. Construct general simple multivectors for each of the arguments, taking a representative from each non-zero area, taking care to satisfy

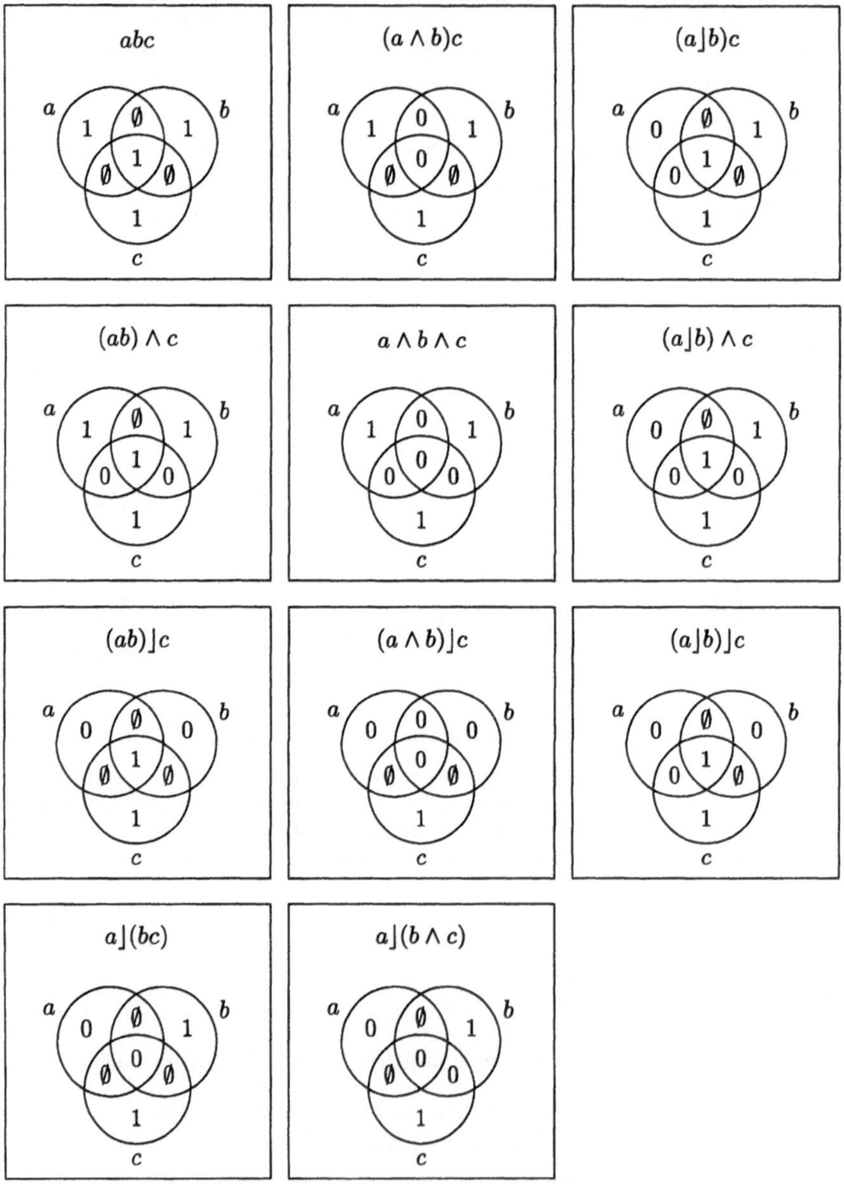

Fig. 6.4. The complete catalog of different three-term products. Notation as in figure 6.3.

the containment relationships of the diagram's construction in step 3. (Details below.)

5. Evaluate both sides of the identity for these sample multivectors. If it holds the identity has been proven for all simple multivectors; if it does not, this computation shows which scalar factor needs to be introduced.
6. For those arguments in which the identity is linear, extend it to general multivectors, within the derived preconditions of step 3.

The method most clearly saves work in step 5, since the exceptional cases messing up the computations have already been taken out in steps3 and 4. This is best illustrated by an example.

6.4.2 Example: Proof of a Duality

Let us investigate the validity of the identity $u \wedge (vw) = (u\rfloor v)w$, a form of duality between $\wedge$ and $\rfloor$.

1. We focus first on the identity for simple multivectors a, b, c, so on the identity $a \wedge (bc) = (a\rfloor b)c$.
2. The diagrams for both sides of the possible identity can be gleaned from the earlier figures as:

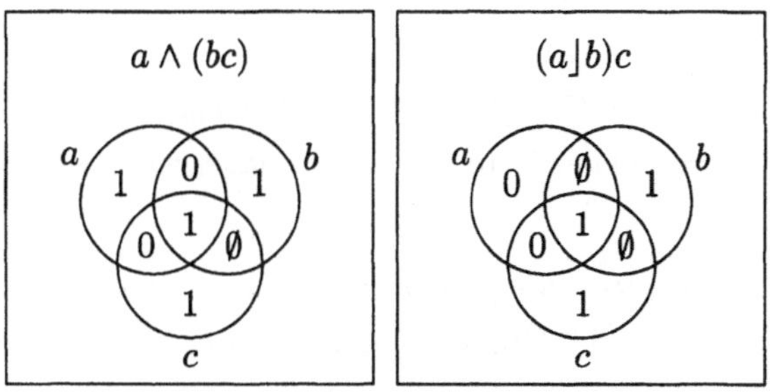

3. The composite diagram must contain the consistent parts of both. We observe that the parts where $\mathcal{I}(a) \not\subseteq \mathcal{I}(c)$ are not consistent, and therefore redraw the diagram to exclude this, noting the condition $\mathcal{I}(a) \subseteq \mathcal{I}(c)$:

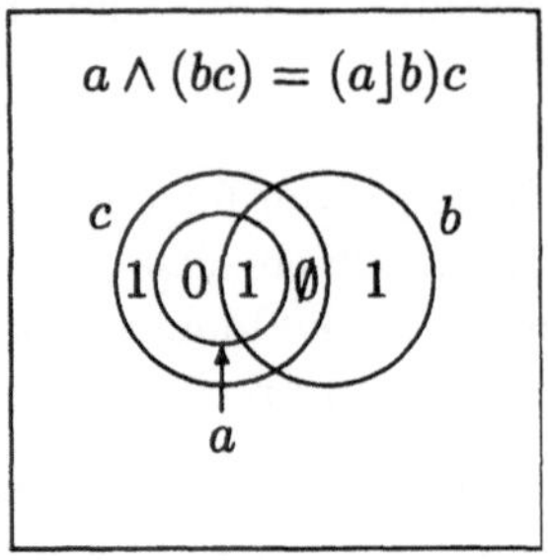

4. To establish the full identity, we now have to inspect the scalar factors. To do so, take a representative simple multi-vector in each of the subsets of the diagram that do *not* lead to a zero result, to compose a typical example. In our diagram this implies, for instance: $a = a_k$, $b = a^\dagger{}_k b_\ell d_n$, $c = a_k d^\dagger{}_n c_m$, with $a_k = \mathbf{e}_1 \cdots \mathbf{e}_k$, $b_\ell = \mathbf{e}_{k+1} \cdots \mathbf{e}_{k+\ell}$, $c_m = \mathbf{e}_{k+\ell+1} \cdots \mathbf{e}_{k+\ell+m}$, $d_n = \mathbf{e}_{k+\ell+m+1} \cdots \mathbf{e}_{k+\ell+m+n}$; the index of a_k, b_ℓ, c_m, d_n indicates the grade, and the components of each are orthogonal basic vectors. The reversions in the expressions for b and c were merely put in for convenience in the computations below; since they only involve a scalar factor ± 1 on both sides of the identity, this is permitted.
5. With these sample multivectors, we obtain:

$$\begin{aligned} a \wedge (bc) &= a_k \wedge (a^\dagger{}_k b_\ell d_n a_k d^\dagger{}_n c_m) \\ &= a_k \wedge ((-1)^{kn} (d_n d^\dagger{}_n) a^\dagger{}_k b_\ell a_k c_m) \\ &= (-1)^{k(n+\ell)} (d_n d^\dagger{}_n)(a^\dagger{}_k a_k) a_k b_\ell c_m, \end{aligned}$$

and

$$\begin{aligned} (a \rfloor b)c &= (a_k \rfloor (a^\dagger{}_k b_\ell d_n)) a_k d^\dagger{}_n c_m = (a_k a^\dagger{}_k) b_\ell d_n a_k d^\dagger{}_n c_m \\ &= (-1)^{k(n+\ell)} (a_k a^\dagger{}_k)(d_n d^\dagger{}_n) a_k b_\ell c_m. \end{aligned}$$

This establishes that the two results are indeed identical under the condition found in step 3:

$$a \wedge (bc) = (a \rfloor b)c \quad \text{if} \quad \mathcal{I}(a) \subseteq \mathcal{I}(c). \tag{6.8}$$

6. The two sides in eq.(6.8) are linear in all arguments. The precondition assumes that all indices in $\mathcal{I}(a)$ are in $\mathcal{I}(c)$. The *simplest* linear extension is obtained by keeping c simple, making it in effect the pseudoscalar i of the space in which a and its linear extensions u reside. Then the precondition $\mathcal{I}(a) \subseteq \mathcal{I}(c)$ extends to $u \in \mathcal{G}(i)$. Thus we have proved an identity for two general multivectors u and v and a pseudoscalar i of the u-space:

$$u \wedge (v\, i) = (u \rfloor v) i \quad \text{if} \quad u \in \mathcal{G}(i) \tag{6.9}$$

By carefully keeping track of indices, further extensions of the result for simple multivectors may be possible, but they are hard to phrase and are less useful because of that.

It should be clear that we can use the method also to come up with new theorems – this now becomes a routine exercise for any practitioner of geometric algebra (as it should be).

6.4.3 Filter Design to Specification

Since 'filters' are merely 'useful expressions', their method of design is very similar:

1. Focus first on simple multivectors.
2. Specify the desired outcome set in terms of a diagram.
3. Identify this diagram in an exhaustive table of outcomes (such as figure 6.3 or 6.4). It may be a *sub-diagram* of an entry.
4. Identify the conditions on the arguments that select this (sub-)diagram.
5. These conditions, applied to the equation defining the diagram, give the desired 'filter' expression.

6.4.4 Example: The Design of the Meet Operation

We illustrate the design of the important *directed intersection operator* in the 'algebra of directions'.

1. The desired outcome of the intersection on two index sets $\mathcal{I}(a)$ and $\mathcal{I}(b)$ is that indices 'pass' when they are in the intersection of the index sets, are indifferent when in either of them, and zero outside. This is thus:

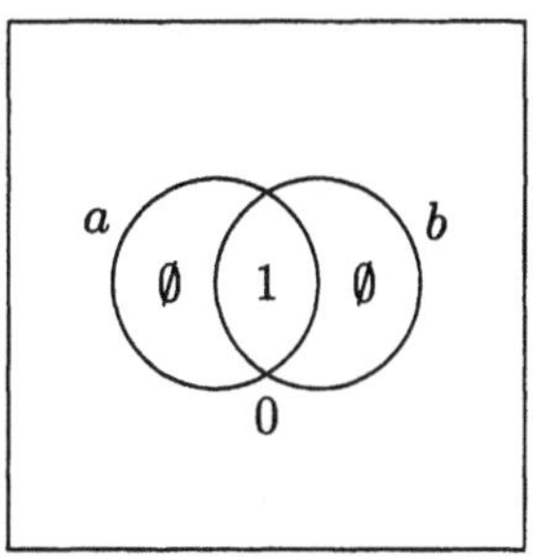

2. Such a filter cannot be made by simply combining the two index sets; all possibilities of that were indicated in figure 6.3, and it is not among

them. Thus we look in the three-argument filters of figure 6.4. We find the desired possibility as a subset of the diagram of $(ab)\rfloor c$:

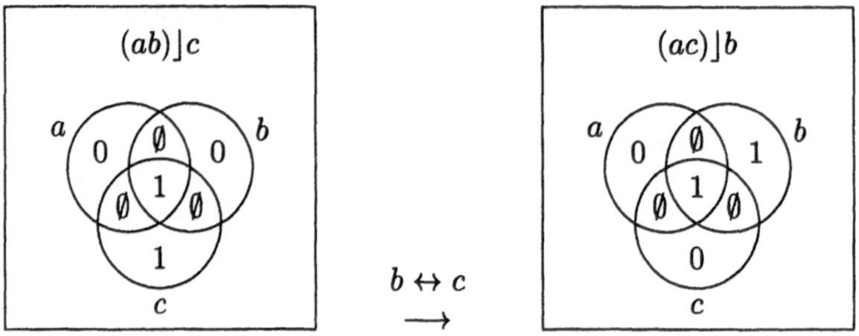

(We interchanged the dummy filter parameters b and c, to make the parameters and diagram correspond with our choice under step 1.)

3. From this diagram, we produce the desired result by demanding: $\mathcal{I}(a) \subseteq \mathcal{I}(c)$ and $\mathcal{I}(b) \subseteq \mathcal{I}(c)$. So c must contain both a and b in this sense; the simplest is if c is a pseudoscalar for the space containing both a and c. The new diagram is:

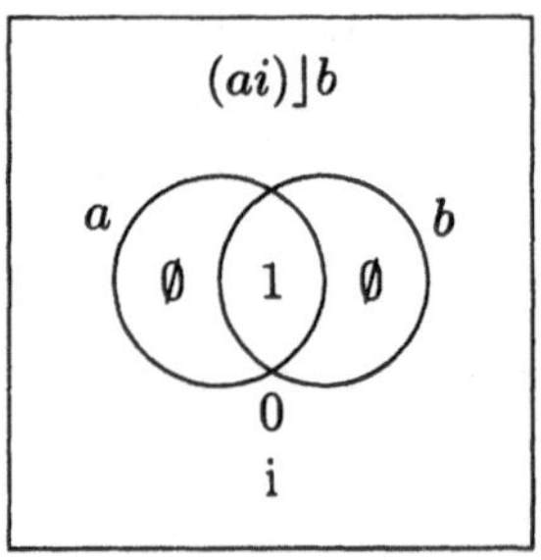

4. This shows that a *non-trivial result* (i.e. non-zero) is only achieved when i is a pseudoscalar of the *smallest* space containing both a and b. Then $(ai)\rfloor b$ is a pseudoscalar for the subspace common to $\mathcal{G}^1(a)$ and $\mathcal{G}^1(b)$, since it contains only indices from $\mathcal{I}(a) \cap \mathcal{I}(b)$.

The operation we have constructed in this example is proportional to the *meet* of subspaces, conventionally defined by $a \vee_i b = (ai^{-1})\rfloor b$ (which differs by an admissible scalar sign from our filter); more about this, its geometrical interpretations and the importance of scalar factors in section 6.6.

6.5 Splitting Algebras Explicitly

As we stated in section 6.2, 'splitting' is a generic operation that helps in translating geometrical structures in an application to an appropriate Clifford algebra. A split is based on the following fact: the space of k-vectors in a Clifford algebra $\mathcal{C}\ell_n$ contains $\binom{n}{k}$ elements. The identity

$$\binom{n+1}{k} = \binom{n}{k-1} + \binom{n}{k} \tag{6.10}$$

suggests that k-vectors of the Clifford algebra $\mathcal{C}\ell_{n+1}$ could be mapped onto $(k-1)$-vectors and k-vectors in the algebra $\mathcal{C}\ell_n$. We can indeed make this explicit. An important example is the split of a k-vector a of $\mathcal{C}\ell_{n+1}$ relative to a fixed vector e_0 in $\mathcal{C}\ell_{n+1}$, which can be considered as a decomposition according to the identity:

$$a = e_0^{-1} \wedge (e_0 \rfloor a) + e_0^{-1} \rfloor (e_0 \wedge a). \tag{6.11}$$

In this equation, $e_0 \rfloor a$ is a $(k-1)$ vector in $\mathcal{C}\ell_{n+1}$, which is moreover constrained to the n-dimensional subspace $\mathcal{G}(\widetilde{e_0})$, perpendicular to e_0 (Proof: $e_0 \rfloor (e_0 \rfloor a) = (e_0 \wedge e_0) \rfloor a = 0 \rfloor a = 0$). The second term $e_0^{-1} \rfloor (e_0 \wedge a)$ is a k-vector in this same subspace (Proof: $e_0 \rfloor (e_0^{-1} \rfloor (e_0 \wedge a)) = (e_0 \wedge e_0^{-1}) \rfloor (e_0 \wedge a) = 0$). Thus if we identify this subspace with the vector space generating the lower-dimensional Clifford algebra $\mathcal{C}\ell_n$, then we have explicitly constructed a mapping from $\mathcal{C}\ell_{n+1}$ onto $\mathcal{C}\ell_n$. We follow custom in denoting the elements of $\mathcal{C}\ell_n$ or its isomorphic subspace in **bold** font, the other elements of $\mathcal{C}\ell_{n+1}$ in normal *math* font.

Doing the split for $k = 1$, we see that a vector x of $\mathcal{C}\ell_{n+1}$ corresponds to a vector $\mathbf{x}$ in $\mathcal{C}\ell_n$ of the form $e_0^{-1} \rfloor (e_0 \wedge x)/\alpha$, with α a scalar or scalar-valued function.[4] This gives, conversely:

$$x = (e_0^{-1} \rfloor x) e_0 + \alpha \mathbf{x} = x_0 e_0 + \alpha \mathbf{x} \tag{6.12}$$

(defining $x_0 \equiv e_0^{-1} \rfloor x$). By choosing different α, we can implement different splits. A particularly useful way is the *projective split* obtained by setting $\alpha = x_0 = e_0^{-1} \rfloor x$. This gives:

$$x = x_0(e_0 + \mathbf{x}) \quad \text{and} \quad \mathbf{x} = e_0^{-1} \rfloor \left(\frac{e_0 \wedge x}{e_0^{-1} \rfloor x}\right) = e_0 \rfloor \left(\frac{e_0 \wedge x}{e_0 \rfloor x}\right). \tag{6.13}$$

---

[4] There is a second way to embed vectors, using $k = 2$: out of x, construct a bivector $e_0^{-1} \wedge x$, then map that according to the first term of eq.(6.11) as $e_0 \rfloor (e_0^{-1} \wedge x)$. The result is equal to $e_0^{-1} \rfloor (e_0 \wedge x)$, and thus the same; this second way may to be more indirect, but it is actually sounder algebraically, see [110].

In the projective split, any x representing a vector $\mathbf{x}$ can thus be written as a member of a 1-scalar-parameter family of representatives, namely as some scalar multiple of the vector $e_0 + \mathbf{x}$. It is called 'projective' since it is useful in doing projective geometry (see [110]), but it has other uses too. For some of those, we must perform the projective split embedding in a *canonical* way, representing $\mathbf{x}$ by $x = e_0 + \mathbf{x}$, taking the arbitrary constant x_0 in eq.(6.13) equal to 1; since $x_0 = e_0{}^{-1} \rfloor x$, we can view this as an embedding to the hyperplane $e_0{}^{-1} \rfloor x = 1$ in $\mathcal{C}\ell_{n+1}$. For some other embeddings, we require normalization of higher-order blades.

The basic usefulness of the projective split is that the embedding of $\mathcal{C}\ell_n$ into $\mathcal{C}\ell_{n+1}$ simplifies computations on non-homogeneous affine subspaces. Take, for instance, the equation for the set of points on a directed line from $\mathbf{p}$ to $\mathbf{q}$ in V^n of $\mathcal{C}\ell_n$. This set is

$$\{\mathbf{x} \in V^n \mid (\mathbf{x} - \mathbf{p}) \wedge (\mathbf{q} - \mathbf{p}) = 0\}. \tag{6.14}$$

(for this is the set of all points $\mathbf{x}$ such that $\mathbf{x} - \mathbf{p}$ and $\mathbf{q} - \mathbf{p}$ have the same direction). Under the projective split, this set is represented by the bivector $\ell = p \wedge q$ (with $p \equiv p_0(e_0 + \mathbf{p})$ and $q \equiv q_0(e_0 + \mathbf{q})$ the representations of $\mathbf{p}$ and $\mathbf{q}$, respectively), in the sense that $\mathbf{x}$ is on the line iff its representative x is in the space of this pseudoscalar ℓ:

$$(\mathbf{x} - \mathbf{p}) \wedge (\mathbf{q} - \mathbf{p}) = 0 \quad \Longleftrightarrow \quad x \wedge (p \wedge q) = 0, \tag{6.15}$$

as is easily verified. This result extends to higher order affine linear subspaces: they are *simple* multivectors of $\mathcal{C}\ell_{n+1}$, to be interpreted by the projective split. It generalizes the 'trick' of *homogeneous coordinates* (well-known in applied linear algebra) from mere vectors to all of the higher-dimensional subspaces. A further extension has been achieved recently, see the Chapter 2 on generalized homogeneous coordinates.

In the literature on projective splits (such as [110, 180]), the actual embedding is hidden in isomorphisms stating the *equivalence* of certain bivectors in $\mathcal{C}\ell_{n+1}$ and vectors in $\mathcal{C}\ell_n$. As a consequence, the grades of the various elements get confusing (a bivector can be 'equal' to a vector!), and the operation $\wedge$ acquires a meaning which subtly changes with the operands (if you write $\mathbf{x} = e_0 \wedge x$ and $\mathbf{y} = e_0 \wedge y$, then formally $\mathbf{x} \wedge \mathbf{y} = 0$ for any $\mathbf{x}$, $\mathbf{y}$!). The above shows that there is no need for this: we can make the mappings totally explicit. The resulting cleaner algebraic structure leads to cleaner software.[5]

[5] This corresponds in spirit to Stolfi's careful treatment of the implementation of oriented projective geometry [220]. He also recommends the introduction of a 0 for every grade k (denoted $\mathbf{0}^k$), to make all theorems on grades universally valid.

6.6 The Rich Semantics of the Meet Operation

We are now ready to discuss the meet operation from the 'algebra of directions' in more detail, and to apply it to the intersection of directed affine linear subspaces by combining it with the projective split interpretation.

In literature, the meet is often treated as a 'qualitative' operation. The reason is probably that its most useful application is when a and b in $a \vee_i b$ are blades, and that these in turn have their most useful application when they are considered the representatives of affine subspaces in the projective split. Since the projective split contains an arbitrary scalar for the embedding (such as x_0 in eq.(6.13)), this then leads one to neglect all scalar factors (or, when done more carefully, all *positive* scalar factors) [180]. This qualitative approach, however, is also necessarily *binary*: subspaces either intersect or they don't, and there is no measure of the *relevance* of the intersection. This is a problem in applications where geometrical data has an associated *uncertainty*. For instance, when intersecting two observed planes that are almost co-directional, the location of the intersecting line is ill-determined and this should be expressed in the error margin; it may even require the observed planes to be considered as two observations of the same plane, making the intersection line physically meaningless. We thus need a way to express 'intersection strength' as well as the intersection result.

Traditionally, the 'meet' operation is just taken as providing the intersection subspace, and not the intersection strength. We now show that it can give both, with the *magnitude of the meet* giving such diverse measures of intersection strength as the *distance measure between subspaces* (known from numerical linear algebra), and even (in the explicit projective split interpretation of subspaces of $\mathcal{C}\ell_{n+1}$) the *Euclidean distance between non-intersecting affine subspaces* in $\mathcal{C}\ell_n$!

6.6.1 Meeting Blades

We first need to understand the meet in more detail, especially being more careful about scalar factors (including signs) than is common in literature.

For blades a and b, the meet $a \vee_i b$ is a blade of their intersection, with a sign and magnitude that depends on those of a, b and i. For the standard definition $a \vee_i b \equiv (ai^{-1}) \rfloor b$, involving the *inverse* of the pseudoscalar, this is as follows.

Let a and b be simple multivectors with a common factor c. Then defining i through:

$$i = (b\,c^{-1}) \wedge c \wedge (c^{-1}a) \tag{6.16}$$

we have:

$$a \vee_i b = c \tag{6.17}$$

(The proof is straightforward using the methods of section 6.4.1; the use of i^{-1} in the meet causes the somewhat unfortunate reversion of the arguments in the definition of i.)[6] Note that there is no such thing as '*the*' meet of a and b; replacing c by $-c$ gives an opposite sign. It is therefore necessary *always* to denote the pseudoscalar relative to which the meet is taken, and any suggestion that it can be omitted or defined objectively from a and b (such as found in [180]) is wrong.

There is less confusion about the *join* $\dot{\wedge}$ of two spaces, an operation that is used to give a blade for the common subspace spanned by two blades a and b. If a and b have *no common factors* (so the corresponding subspaces have only the element 0 in common), then the join is given by:

$$a \dot{\wedge} b = b \wedge a \tag{6.18}$$

(the reversal of the arguments is done to prevent stray signs when using this in combination with the meet, and is again due to the use of i^{-1} in the definition of the meet.) The join is then a *directed union* of the subspaces.

If a and b *do* have a common subspace, then an objective definition of their join can *not* be given ([114],pg.34): there is an ambiguity of sign which can *not* be resolved explicitly, as in the case of the meet (see [220] for a clear explanation of this counterintuitive issue). Thus a directed union can then not exist, and eq.(6.18) correctly yields 0 (the only blade representing a non-directed subspace).

For readers familiar with the wonderfully illustrated book on oriented projective geometry by Stolfi [220], note that his meet (denoted there by $\wedge_i$, presumably following [14]) differs from the above standard definition in geometric algebra. He defines it [220] pg. 50 (modulo a positive scalar) through equations which in our notation would effectively read:

$$a \wedge_i b = c \iff i = (ac^{-1}) \wedge c \wedge (c^{-1}b) \tag{6.19}$$

Thus Stolfi's meet $a \wedge_i b$ is identical to our meet $b \vee_i a$ (same i!) – and similarly, his join $a \vee b$ is identical to our join $b \dot{\wedge} a$. His delightful graphic constructions are therefore applicable to the 'algebra of directions' with a simple interchange of the operands.

6.6.2 Meets of Affine Subspaces

Affinely translated subspaces of $\mathcal{C}\ell_n$ are represented by blades of $\mathcal{C}\ell_{n+1}$ under the projective split; the meet of these blades can then be interpreted in terms of quantities of $\mathcal{C}\ell_n$ as signifying the directed intersection of affine subspaces.

When we compute the meet of two non-homogeneous linear subspaces of a space $\mathcal{G}(\mathbf{i})$, represented as $a = (e_0 + \mathbf{a}) \wedge \mathbf{A}$ and $b = (e_0 + \mathbf{b}) \wedge \mathbf{B}$ of $\mathcal{G}(e_0\mathbf{i})$

[6] The above can be used to correct an error in Pappas [180], who uses a pseudoscalar and decomposition that should have made his meet equal to $(-1)^{\mathrm{grade}(A')(\mathrm{grade}(C)+\mathrm{grade}(B'))}C^{-1}$ rather than C, in his notation.

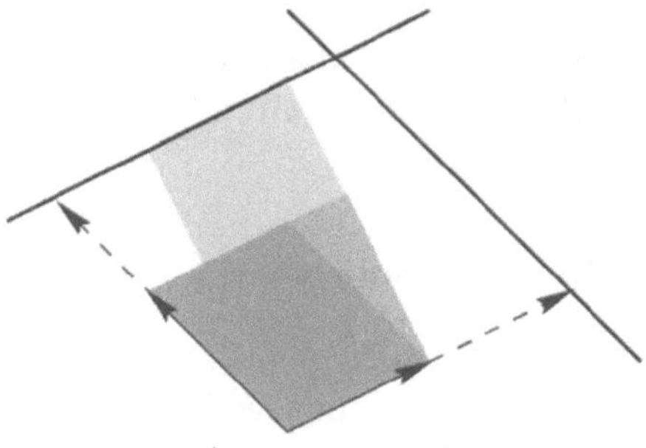

Fig. 6.5. The directed intersection of two lines in $\mathbb{R}^2$

in the homogeneous projective split representation (so $\mathbf{A}$ and $\mathbf{B}$ are blades indicating the tangents, and $\mathbf{a}$ and $\mathbf{b}$ are translational offsets to create the affine subspaces parallel to these), we obtain:

$$\begin{aligned}
&a \vee_{e_0\mathbf{i}} b \\
&= \big(((e_0 + \mathbf{a}) \wedge \mathbf{A})\mathbf{i}^{-1}{e_0}^{-1}\big) \rfloor ((e_0 + \mathbf{b}) \wedge \mathbf{B}) \\
&= (e_0\mathbf{A}\mathbf{i}^{-1}{e_0}^{-1})\rfloor(e_0\mathbf{B} + \mathbf{b} \wedge \mathbf{B}) + \big((\mathbf{a} \wedge \mathbf{A})\mathbf{i}^{-1}{e_0}^{-1}\big)\rfloor(e_0\mathbf{B} + \mathbf{b} \wedge \mathbf{B}) \\
&= (\mathbf{A}^*{\mathbf{i}^*}^{-1})\rfloor(e_0\mathbf{B} + \mathbf{b} \wedge \mathbf{B}) + \big((\mathbf{a} \wedge \mathbf{A})\mathbf{i}^{-1}\big)\rfloor\mathbf{B} \\
&= -e_0(\mathbf{A}^* \vee_{\mathbf{i}^*} \mathbf{B}) + (\mathbf{A}^* \vee_{\mathbf{i}^*} (\mathbf{b} \wedge \mathbf{B}) + (\mathbf{a} \wedge \mathbf{A}) \vee_{\mathbf{i}} \mathbf{B}) \\
&= \Big(e_0 + (\mathbf{A}^* \vee_{\mathbf{i}^*} (\mathbf{b} \wedge \mathbf{B}) + (\mathbf{a} \wedge \mathbf{A}) \vee_{\mathbf{i}} \mathbf{B})\,(-\mathbf{A}^* \vee_{\mathbf{i}^*} \mathbf{B})^{-1}\Big) \\
&\quad \wedge(-\mathbf{A}^* \vee_{\mathbf{i}^*} \mathbf{B}),
\end{aligned}$$

where the last step assumes that $(\mathbf{A}^* \vee_{\mathbf{i}^*} \mathbf{B})$ is invertible; which is the case if $\mathbf{i}$ is at most a pseudoscalar for the smallest common subspace of $\mathbf{A}$ and $\mathbf{B}$. Under this condition, the projective split interpretation of the result of the meet is thus an affine subspace with tangent $(-\mathbf{A}^* \vee_{\mathbf{i}^*} \mathbf{B})$, translated over the position vector $(\mathbf{A}^* \vee_{\mathbf{i}^*} (\mathbf{b} \wedge \mathbf{B}) + (\mathbf{a} \wedge \mathbf{A}) \vee_{\mathbf{i}} \mathbf{B})\,(-\mathbf{A}^* \vee_{\mathbf{i}^*} \mathbf{B})^{-1}$.

Example: The directed intersection of two lines in $\mathbb{R}^2$: $\ell = e_0 \wedge \mathbf{a} + \mathbf{A}$ and $m = e_0 \wedge \mathbf{b} + \mathbf{B}$ in the projective split representation, see figure 6.5. We compute, with $\mathbf{i}$ taken as a pseudoscalar for $\mathbb{R}^2$:

$$\begin{aligned}
p = \ell \vee_{e_0\mathbf{i}} m &= -e_0(\widetilde{\mathbf{a}^*}\rfloor\mathbf{b}) + \widetilde{\mathbf{a}^*}\rfloor\mathbf{B} + \widetilde{\mathbf{A}}\rfloor\mathbf{b} \\
&= e_0(\mathbf{b}\rfloor\widetilde{\mathbf{a}}) - (\widetilde{\mathbf{a}} \wedge \widetilde{\mathbf{B}})\mathbf{i} + \widetilde{\mathbf{A}}\mathbf{b} \\
&= e_0(\mathbf{b} \wedge \mathbf{a})^{\sim} - \widetilde{\mathbf{B}}\mathbf{a} + \widetilde{\mathbf{A}}\mathbf{b}.
\end{aligned}$$

Since $(\mathbf{b} \wedge \mathbf{a})^{\sim}$ is scalar, this corresponds to the intersection *point*:

$$\mathbf{p} = \frac{\widetilde{\mathbf{B}}}{(\mathbf{a} \wedge \mathbf{b})^{\sim}}\,\mathbf{a} + \frac{\widetilde{\mathbf{A}}}{(\mathbf{b} \wedge \mathbf{a})^{\sim}}\,\mathbf{b}$$

if $(\mathbf{b} \wedge \mathbf{a})^{\sim}$ is non-zero. Figure 6.5 graphically demonstrates the correctness of this result: $\mathbf{a}$ and $\mathbf{b}$ are weighted by ratios of areas.

The geometric algebra framework validates the intersection results in any dimension, and in a computational representation that does not require exceptional data structures: points, lines, planes, etcetera are all admissible outcomes.

6.6.3 Scalar Meets Yield Distances between Subspaces

We have seen in eq.(6.17) that the meet $a \vee_i b$ normally gives a *blade* as its result, and that this is interpretable as the space of intersection of a and b. There is also an interpretation when the result of the meet is a *scalar.* The subspaces then intersect in the origin only; i.e. they are complementary in the smallest common space (with pseudoscalar i), though not necessarily orthogonal. When the meet is a scalar, we can rewrite it as:

$$\begin{aligned} a \vee_i b &= \widetilde{a} \rfloor b = \langle \widetilde{a} \rfloor b, 1 \rangle = \langle b, \widetilde{a}^* \rangle = \langle \widetilde{a}^*, b \rangle = \langle \widetilde{a}, b^* \rangle = \langle b \rfloor \widetilde{a}, 1 \rangle = b \rfloor \widetilde{a} \\ &= (b \wedge a)^{\sim}, \end{aligned}$$

Thus in such a case, the meet equals the *volume* of the commonly spanned space, relative to the standard pseudoscalar i. This is a useful measure if we take all blades involved (a, b and i) to be *unit* blades. Then the meet varies continuously from 1 to -1, and is zero when the two subspaces have some subspace in common (they do *not* necessarily coincide: *any* common factor in a and b makes $b \wedge a$ equal to zero). We can interpret the values ± 1 as: the subspaces are *orthogonal* in the embedding space $\mathcal{G}^1(i)$. *The magnitude of a scalar meet of unit blades is thus a measure for the 'parallelism' of the spaces they represent.*

> **Example:** Consider two vectors $\mathbf{x}$ and $\mathbf{y}$ in $\mathbb{R}^2$ with pseudoscalar i. Then $\mathbf{x} \vee_i \mathbf{y} = (\mathbf{y} \wedge \mathbf{x})^{\sim} = (-|\mathbf{y}|\,|\mathbf{x}|\, i \sin\phi)\, i^{-1} = |\mathbf{x}|\,|\mathbf{y}| \sin\phi$, with ϕ the angle from $\mathbf{x}$ to $\mathbf{y}$ in i. If both are unit vectors this yields $\sin\phi$. Thus the meet has the largest absolute value, 1, when $\mathbf{x}$ and $\mathbf{y}$ are *orthogonal* (with $+1$ when $\mathbf{y}$ is in the positive direction from $\mathbf{x}$, so $\mathbf{y} = \mathbf{x}i$, and -1 for the opposite direction), and goes *continuously to zero* when $\mathbf{x}$ and $\mathbf{y}$ become more and more parallel.

In linear algebra, a commonly used distance measure between subspaces is the sine of the angle between them, see e.g. [92]. It can be shown by generalization from the 1-dimensional example above that this is indeed what $(b \wedge a)^{\sim}$ is, for unit a, b and i. Thus the meet contains this common practice in numerical linear algebra, casting an interesting light on its essential nature.

6.6.4 The Euclidean Distance between Affine Subspaces

If we are in a projective split representation, a and b in $\mathcal{C}\ell_{n+1}$ represent affinely translated linear subspaces of $\mathcal{C}\ell_n$. If these subspaces are complementary in $\mathcal{C}\ell_{n+1}$ as in the previous section, then their meet is scalar; this complemen-

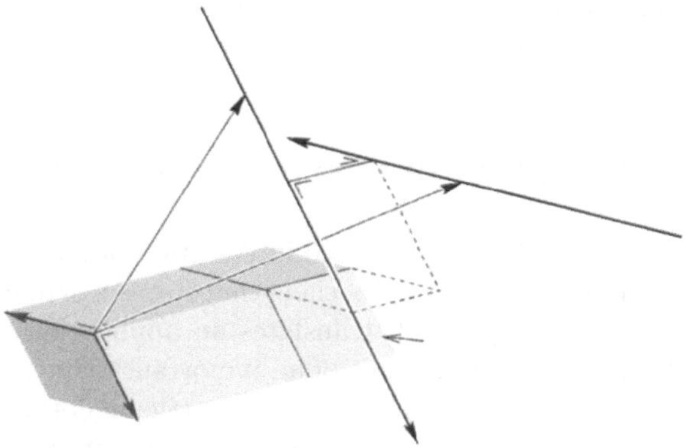

Fig. 6.6. The Euclidean distance between two affine subspaces in $\mathbb{R}^n$ whose ranks add to $n+1$.

tarity means their ranks in $\mathcal{C}\ell_n$ add up to $n+1$. For two such spaces (a point and a line in 2-D, two lines in 3-D) we can define their (directed) distance in $\mathcal{C}\ell_n$ as the length of the (directed) mutual perpendicular connecting them. This turns out to be proportional to *the meet of their projective split representatives*!

Let us write the affine subspace represented by the blade a as the translation by a vector $\mathbf{a}$ of a subspace with unit blade $\mathbf{A}$, i.e. the set $\{(\mathbf{x}-\mathbf{a}) \wedge \mathbf{A} = 0\}$. Its canonical projective split representation is $a = (e_0 + \mathbf{a}) \wedge \mathbf{A}$, and similarly for b we have $b = (e_0 + \mathbf{b}) \wedge \mathbf{B}$. Let $\mathbf{i}$ be the unit pseudoscalar of $\mathcal{C}\ell_n$, and $I = e_0\mathbf{i}$ the pseudoscalar for $\mathcal{C}\ell_{n+1}$ (note the order!). Then with the assumed complementarity of a and b in I, their meet relative to I is:

$$\begin{aligned} a \vee_I b = (b \wedge a)I^{-1} &= ((e_0 + \mathbf{b}) \wedge \mathbf{B} \wedge (e_0 + \mathbf{a}) \wedge \mathbf{A})I^{-1} \\ &= (e_0 \wedge \mathbf{B} \wedge \mathbf{a} \wedge \mathbf{A} + \mathbf{b} \wedge \mathbf{B} \wedge e_0 \wedge \mathbf{A})I^{-1} \\ &= e_0(\mathbf{B} \wedge \mathbf{a} \wedge \mathbf{A} - \mathbf{B} \wedge \mathbf{b} \wedge \mathbf{A})I^{-1} \\ &= (\mathbf{B} \wedge \mathbf{a} \wedge \mathbf{A} - \mathbf{B} \wedge \mathbf{b} \wedge \mathbf{A})\mathbf{i}^{-1} \end{aligned} \tag{6.20}$$

This is a quantity that is entirely computable in $\mathcal{C}\ell_n$. It is proportional to the *orthogonal directed Euclidean distance* between the two subspaces represented by a and b, by a proportionality factor of $(\mathbf{B} \wedge \mathbf{A})\mathbf{i}^{-1}$ (which is the 'distance' between the directional elements in the sense of section 6.6.3). This is depicted in figure 6.6: the expression in brackets is a difference of two volumes in $\mathbf{i}$-space, which can be viewed as being spanned by $\mathbf{B}$, $\mathbf{A}$ and a vector in the direction of their perpendicular connection; the difference relative to $\mathbf{i}$ is the directed length of this vector.

We thus find yet another classical distance measure embedded in the intersection operation of geometric algebra. Note that eq.(6.20) is valid in any finite-dimensional space, and coordinate free. It is thus well-suited for implementation in a generic geometric software package!

6.7 The Use and Interpretation of Geometric Algebra

The meet was designed as a straightforward 'directed intersection operation' for geometric algebra in section 6.6. The examples show that it has a semantics that depends on the modeling step which translates an application to appropriate geometric algebra. This is an instance of an important principle: there is no *unique* interpretation of Clifford algebra or geometric algebra.[7]

This is not a weakness of geometric algebra, but rather a sign of its strength: a limited number of generic constructions in the mathematically consistent theory of geometric algebra suffices to implement what used to be seen as disparate geometrical tricks in different applications. If you view geometric algebra as giving an exhaustive library of advanced computational techniques, then once you have made the mapping between your application and geometric algebra, the meaning of these techniques is automatic, and gives a complete set of internally consistent operators in the application. This seeming restriction will prevent you from going astray (not just anything is permitted) and can help to inspire you (since it gives advanced and consistent constructive techniques). Also, since these can be defined in generic terms of Clifford algebra, they need only be implemented once – the only responsibility of the practicing computer scientist is then the explicit implementation of the mapping between the application and this generic body. After that, computations are automatic.

Having said that, we need to show that the library of techniques in geometric algebra is indeed sufficient for such purposes, and extend it where possible. The challenge is not necessarily to do new things using geometric algebra (though that is always nice!), but rather to show that a single framework encompasses all previously known results, and does so compactly. To mathematicians, this may not be a very exciting task; to computer scientists and physicists, its completion would be immensely gratifying. It would give us a box of integrated geometrical power tools which we could use to perceive, describe and direct objects in the world without being hampered by interface problems between incompatible sets of mathematical instruments (as we now so often are).

[7] Hestenes frequently points out that the use of a bilinear form in the Clifford algebra does not automatically imply that it *only* applies to metric spaces: it all depends on how you use it.

6.8 Geometrical Models of Multivectors

We have seen how we could understand some of the formulas coming out of our 'meet' computations by drawing a picture of the situation, and representing the multivectors involved by directed lines, directed areas, directed volumes, etc. Once you have done this for a while, you will find that this tends to reverse: the pictures soon become a natural construction tool for the design of formulas and algorithms. Unfortunately, few authors using geometric algebra appear to find a need for such pictorial explanations and constructions (an exception is [112]). Why? Any explanation of a powerful framework for 'doing geometry' that does not contain pictures must be less than convincing to the intended audience! In my experience, pictorial constructions such as figure **??** immediately instill a desire to learn more about geometric algebra in an audience of novices, and they are therefore immensely helpful.

It may indeed be possible to give a proper grammar for the construction of these pictures, which would turn this into a sound design procedure, and one that could be taught to the graphically inclined. There is some work to be done, though, to find a proper pictorial model: are vectors better viewed as emanating from the origin (i.e. as positions), or should we treat them as 'free vectors' (i.e. as directions)? Should we represent a bivector as a reshapable homogeneous plane element of a certain magnitude (as in [154]) or as a stack of planes with a stacking density (as in [170])? Do the answers depend on the 'model' of the geometry we are using (e.g. the 'free vector' image for affine directions, the 'fixed vectors' for their projective split representation)?

Whatever the answers, they are worth some research: the use of pictorial representations by proponents of differential forms (e.g. [170]) has helped them in 'spreading the faith', since the pictures effectively convey the intuition behind the computations and instill confidence in their consistency. Geometric algebra and geometric calculus could and should use a similar route to speedy introduction to a wider audience.

6.9 Conclusions

Clifford algebra is not useful by itself; it is just a consistent mathematical structure. Its surprising power comes from the discovery that this structure can be used to represent very many geometrical phenomena; indeed, maybe even *all* of geometry. It does so in *geometric algebra* which reorganizes the structure of Clifford algebra at various levels, guided by geometrical significance (see figure 6.1). This provides a framework that is immediately computational, rather than an arcane abstraction (not to be confused with algebraic geometry!). This has clear advantages: it *unites* geometry, and this is very important to the computer sciences, for a unified framework minimizes conversion between modules. It also gives a richer conceptual structure to design geometric algorithms, mainly since we do not need to express everything in terms of vectors (or, worse, coordinates) before we can make it computable. This makes advanced geometrical techniques more accessible to non-geometers.

In this contribution I argued that the user-oriented development of geometric algebra requires some new approaches, or changes in emphasis:

- it is insightful to the novice to convey explicitly the ordering of geometric ideas involved in turning Clifford algebra into geometric algebra (section 6.2);
- the algebraic structure of geometric algebra should be cleaned up to make operators operand-independent; we demonstrated this principle in the substitution of the contraction for the inner product (section 6.3), and in the totally explicit formulation of the mapping implementing the projective split isomorphism (section 6.5)
- we need a convenient design strategy to construct geometric 'filters' tuned to specific purposes, empowering the users to develop their own 'theory' as needed (section 6.4);
- we need to map traditionally useful concepts to their counterparts in geometrical algebra; and conversely, we should interpret the basic operators in geometric algebra in classical terms (section 6.6)
- it would be helpful to have a standardized pictorial representation of the basic concepts (section 6.8)

A lot of the work that has been done in geometric algebra is immediately relevant to these goals; notably the work of Hestenes and his followers, who have focussed on spreading the faith among physicists and mathematicians. Similar work now needs to be undertaken to promote its application to the geometrical issues in such computer sciences as vision, graphics and robotics.

Part II

Algebraic Embedding of Signal Theory and Neural Computation

7. Spatial–Color Clifford Algebras for Invariant Image Recognition*

Ekaterina Rundblad-Labunets and Valeri Labunets

Signal Processing Laboratory, Tampere University of Technology

7.1 Introduction

One of the main and interesting problems of information science is the clarification of how human eyes and brain recognize objects of the real world. Practice shows that they successfully cope with the problem of recognizing objects at different locations, of different views and illumination, and in different orders of blurring. But how is this done by the brain? How do we see? How do we recognize constantly moving and changing objects of the surrounding world?

The phenomenon of moving objects recognition is as obvious as incomprehensible because moving objects are fixed in the retina in the form of a sequence of images each of which in its own right does not permit to conclude on the true shape of an object. But it is beyond question that this sequential set of images appearing in the retina must contain something constant, thanks to which we see and realize the object as something constant. This "something" constant is called *invariant* [3], [4]. An old problem in pattern recognition is how to achieve various kinds of invariances. In order that an artifical pattern recognition system performs in the same way as any biological sensory systems does, the recognition result, to a limited extent at least,

* This work was supperted by INTAS, grant no. INTAS–94–708, and RFFR–98/99–01–0002.

should be invariant with respect to various *transformation groups* of the patterns such as translation, rotation, size variation, and change in illumination.

This chapter describes new methods of image recognition based on algebraic–geometrical theory of invariants. Changes in the surrounding world which cause object shape transformations (e.g. translation, rotation, reflection, scaling, etc.) can be treated as the action of some Clifford numbers in image space, appearing in the eye's retina. The sequential development of this idea may emphasize the number–theoretical approach to recognition. But it can also be considered as a purely geometrical approach.

In 1872 F. Klein delivered his famous lecture "Comparative Review of the Modern Geometrical Researches," known as the "Erlangen Program" now. In this lecture, geometry of any specific form is connected with a group and vice versa. From this point of view, any number–theoretical research on recognition methods can be considered as devoted to the study of the visual space geometry. This visual space differs from the usual Euclidean–Newtonean physical space by its properties.

We adopt the philosophy that geometry is the basis for computer vision, and agree with F. Klein on his Erlangen program that geometry is the study of those properties of objects that remain invariant under particular groups of transformations. Klein's idea proved to be fruitful not only in mathematics but also in modern physics. For example, space–time physics may be regarded as the Minkowski geometry corresponding to the group of Lorentz transformations which made Einstein consider at one time the name "Invariantentheorie" for his special theory of relativity.

In the "geometrical" direction [90], [91] of pattern recognition theory, ideas of applying different non–Euclidean geometries for simulation and explanation of structural properties of a hypothetical visual space have long been stated (see review on this problem in [126]). But nothing was done to check the applicability of pseudo–Euclidean geometry and the problem remained open till the works of R. Luneburg [155], [156] published in 1947–1950. In these papers the idea is stated for the first time that the visual space of a human being is characterized by Lobatchevsky geometry. In essence, the author states a "perceptual theory of relativity" unifying the perceived space and time similarly to physical space–time as a whole analogously to the physical space–time in Einstein's special theory of relativity.

This chapter is devoted to the elaboration of new methods of image invariant recognition in Euclidean and non–Euclidean 2-D, 3-D and n–dimensional spaces, based on the theory of Clifford hypercomplex numbers that allow the calculation of efficient algorithms of computing moments and invariants. We will give special emphasis to the representation problem of color images.

7.2 Groups of Transformations and Invariants

Variants and invariants are intrinsic in any change of the surrounding world. Variant is something that changes under any transformation while invariant is something that stays invariable in such case. Invariants and variants exist so far as the transformations of the surrounding world exist. Such changes in the surrounding world as object shape transformations (e.g. by translation, rotation, reflection, dilation, etc.) can be treated as actions of some transformation group in image space. The theory of continuous transformation groups developed by the great Norwegian mathematician Sofus Lie is the adequate mathematical method for the description of such changes. Let

$$\begin{cases} x_1' = g_1(x_1, x_2, \dots, x_n; a_1, a_2, \dots, a_r), \\ x_2' = g_2(x_1, x_2, \dots, x_n; a_1, a_2, \dots, a_r), \\ \quad \vdots \\ x_n' = g_n(x_1, x_2, \dots, x_n; a_1, a_2, \dots, a_r), \end{cases} \tag{7.1}$$

or more briefly $\mathbf{x}' = \mathbf{g}(\mathbf{x}, \mathbf{a}) := \mathbf{g}(\mathbf{a}) \circ \mathbf{x}$, some continuous one–to–one transformation of the n–dimensional vector space $\mathbb{R}^n$, where $\mathbf{x} := (x_1, \dots, x_n) \in \mathbb{R}^n$ and $\mathbf{a} := (a_1, \dots, a_r) \in \mathbb{R}^r$ is a vector of r parameters.

Definition 7.2.1. *The family (7.1) of transformations* $\mathbf{G}_r^n = \{\mathbf{g}(\mathbf{a}) \mid \mathbf{a} \in \mathbb{R}^r\}$ *forms an r–parameter group of transformations, if the following conditions are fulfilled:*

1) *Together with every transformation* $\mathbf{g}(\mathbf{a})$ *the inverse transformation* $\mathbf{g}^{-1}(\mathbf{a})$ *belongs to the family* $\mathbf{G}_r^n$.
2) *The identity transformation belongs to the family* $\mathbf{G}_r^n$.
3) *The sequential action of two transformations of the form (7.1) is some transformation of the same family:* $\mathbf{x}' = \mathbf{g}(\mathbf{x}, \mathbf{a})$, $\mathbf{x}'' = \mathbf{g}(\mathbf{x}', \mathbf{b}) = \mathbf{g}(\mathbf{g}(\mathbf{x}, \mathbf{a})\mathbf{b}) = \mathbf{g}(\mathbf{x}, \mathbf{c})$; $\mathbf{c} = \varphi(\mathbf{a}, \mathbf{b})$. *Using other notation:*

$$\mathbf{x}' = \mathbf{g}(\mathbf{a}) \circ \mathbf{x}, \quad \mathbf{x}'' = \mathbf{g}(\mathbf{b}) \circ (\mathbf{g}(\mathbf{a}) \circ \mathbf{x}) = \mathbf{g}(\mathbf{c}) \circ \mathbf{x}; \quad \mathbf{c} = \varphi(\mathbf{a}, \mathbf{b}).$$

Here, $\varphi(\mathbf{a}, \mathbf{b}) = \varphi(a_1, \dots, a_r;\ b_1, \dots, b_r)$ *are r fixed functions specifying a rule of the multiplication in every group.*

Multiple examples of transformation groups were also known before Sofus Lie, but he was the first one to give a general definition of an abstract group.

7.3 Pattern Recognition

Let $\mathbb{R}^n$ be an n–dimensional vector space over $\mathbb{R}$ and $f(\mathbf{x})$ an arbitrary n–D image. Let some group $\mathbf{G}_r^n$ of transformations $\mathbf{g}(\mathbf{a}) : \mathbb{R}^n \longrightarrow \mathbb{R}^n$ act on $\mathbb{R}^n$. Under the action of the transformation $\mathbf{g}(\mathbf{a})$ the image $f(\mathbf{x})$ will be mapped onto the new image $f(\mathbf{g}(\mathbf{a}) \circ \mathbf{x})$.

The notion of an invariant is one of the most general and important in mathematics together with the notions of numbers, sets, functions, transformations, etc. The term *invariant* stands for everything that stays unchanged for some transformations of the considered mathematical objects, being connected with them in definite way.

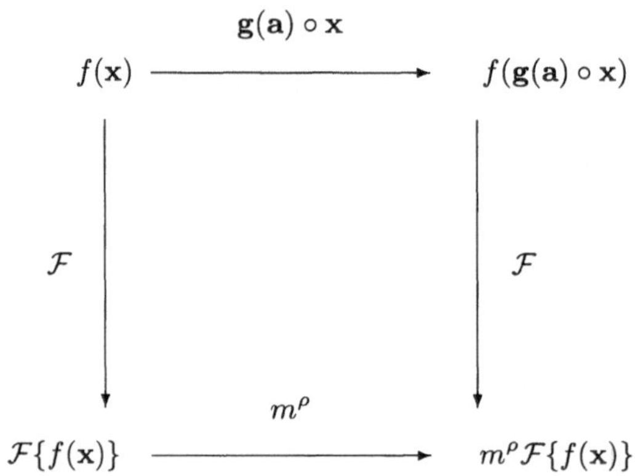

Fig. 7.1. Transformation of relative invariants with respect to the group $\mathbf{G}_r^n$

Definition 7.3.1. *The scalar–valued functional $\mathcal{J} = F\{f(\mathbf{x})\}$ is called the classical relative invariant with the weight ρ of the image $f(\mathbf{x})$ with respect to the group $\mathbf{G}_r^n$ if the following is true*

$$\mathcal{J} = F\{f(\mathbf{g}(\mathbf{a}) \circ \mathbf{x})\} = m^\rho F\{f(\mathbf{x})\}$$

for every transformation $\mathbf{g}(\mathbf{a}) \in \mathbf{G}_r^n$, where $m = \mathrm{Jcob}(\mathbf{g}(\mathbf{a}))$ is the Jacobian of the transformation $\mathbf{g}(\mathbf{a})$ (see fig. 7.1). If $m = 0$, then $\mathcal{J}$ is called the absolute invariant and is denoted as $\mathcal{I}$. The factor m^ρ is called the multiplicator.

The problem of invariant pattern recognition is formulated as follows.

Definition 7.3.2. *We will say that two images $f(\mathbf{x})$ and $f'(\mathbf{x})$ have the "$\mathbf{G}_r^n$-equivalent shape" if there exists such an element $\mathbf{g} \in \mathbf{G}_r^n$ that $f'(\mathbf{x}) = f(\mathbf{g} \circ \mathbf{x})$.*

The relation "$\mathbf{G}_r^n$*–equivalent shape*" defines an equivalence relation on the space of all images.

Definition 7.3.3. *The set all images having the "$\mathbf{G}_r^n$-equivalent shape" is called an equivalence class.*

Let the set of T images $\{f_i(\mathbf{x}) \mid i = 1, 2, \ldots, T\}$ be given, which will be called templates, and let the group $\mathbf{G}$ of their transformations be known.

Definition 7.3.4. *For an image $f_i(\mathbf{x})$ we define the set (equivalence class) $\mathcal{O}_i$, containing all "$\mathbf{G}_r^n$-equivalent shape" images, which can be obtained from the template $f_i(\mathbf{x})$ under the action of the transformations $\mathbf{g} \in \mathbf{G}_r^n$:*

$$\mathcal{O}_i = \{f_i(\mathbf{g} \circ \mathbf{x}) \mid \forall \mathbf{g} \in \mathbf{G}_r^n\}, \ i = 1, 2, \dots, T.$$

The set $\mathcal{O}_i$ is called the orbit of the template image $f_i(\mathbf{x})$.

A basic pattern recognition problem is to find an index $i_0 \in \{1, 2, \dots, T\}$ so that the current image $f^c(\mathbf{x}) \in \mathcal{O}_{i_0}$.

Definition 7.3.5. *The set of functionals $I_k[f(\mathbf{x})]$ is called the complete system of invariants, if every functional is constant on the orbit:*

$$I_k(f_i(\mathbf{g} \circ \mathbf{x}) = I_k(f_i(\mathbf{x})), \ \forall \mathbf{g} \in \mathbf{G}, \quad \forall i = 1, \dots, T, \ \forall k = 1, \dots, \infty$$

and has different values on at least two different orbits $\forall k \ , i, j \ (I_k(\mathcal{O}_i) \neq I_k(\mathcal{O}_j))$.

The property completeness of an invariant system means that all functionals together can "distinguish" between all orbits.

Let $\mathbf{V} := \{\mathbf{I} := (I_1, I_2, \dots, I_K, I_{K+1}, \dots) \mid I_k := I_k(f(\mathbf{x}))\}$ be the set of functional independent invariants and let $\mathbf{V}_K$ be the subset of $\mathbf{V}$ consisting of K-tuples $\mathbf{I}_K := (I_1, I_2, \dots, I_K) : \mathbf{V}_K := \{\mathbf{I}_K := (I_1, I_2, \dots, I_K) \mid I_k = I_k(f(\mathbf{x}))\}$. If on the $\mathbf{V}$ we introduce the metric $d[\mathbf{I}, \mathbf{I}'] = \sum_{k=1}^{\infty} \mid I_k - I_k' \mid$, then $\mathbf{V}$ and $\mathbf{V}_K$ are transformed into metric spaces. Let $\delta(K) := \sum_{k=K+1}^{\infty} |I_k|$.

Definition 7.3.6. *A set $\mathbf{S}(\mathbf{I}_0, \delta(K)) := \{\mathbf{I} \mid d[\mathbf{I}_0, \mathbf{I}] \leq \delta(K)\}$ is called the sphere of radius $\delta(K)$ with the center $\mathbf{I}_0$.*

Technical devices may evaluate a finite number of K invariants. These define a radius $\delta(K)$ of its representing sphere $S_i(\delta(K)) \subset \mathbf{V}$. The problem of pattern recognition in this case can be solved only if spheres do not intersect pair-wise, i.e. $S_i(\delta(K)) \cap S_j(\delta(K)) = \emptyset, \ i, j = 1, 2, \dots, T$.

In the classical methods the problem of recognition is solved the following way. At the preliminary step the centers $\mathbf{I}_K^i = (I_1^i, I_2^i, \dots, I_K^i), \ i = 1, 2, \dots, T$, of the spheres $\mathbf{S}(\mathbf{I}_i, \delta(K)$ in the space of invariants the $\mathbf{V}_K$ are evaluated for T template images $f_i(\mathbf{x}), \ i = 1, 2, \dots, T$. Coordinates $I_k^i, \ k = 1, 2, \dots, K$, of these centers are stored in the computer's memory.

When the current image $f^c(\mathbf{x})$ is brought to the system, the set of K of its invariants $I_k^c = I_k\{f^c(\mathbf{x})\}, \ k = 1, 2, \dots, K$, is estimated, that is, the point $\mathbf{I}_K^c$ is determined in $\mathbf{V}_K$. Then T distances are computed

$$d_i = d\left[\mathbf{I}_K^i, \mathbf{I}_K^c\right] = \sum_{k=1}^{K} \mid I_k^i - I_k^c \mid, \quad i = 1, 2, \dots, T. \tag{7.2}$$

The smallest distance $d_{i_0} = \min_{i=1,T} \ (d_i)$ is found among them, which indicates that the current image $f^c(\mathbf{x})$ lies nearest to the i_0-th orbit $\mathcal{O}_{i_0}$. Thus, the current image is considered as a distorted version of the i_0-th template image $f_{i_0}(\mathbf{x})$.

7.4 Clifford Algebras as Unified Language for Pattern Recognition

We suppose that a brain calculates hypercomplex–valued invariants of an image when recognizing it. Of course, the algebraic nature of hypercomplex numbers must correspond to the spaces with respect to the geometrically perceivable properties. For recognition of 2–D, 3–D and n–D images we turn the spaces $\mathbb{R}^2$, $\mathbb{R}^3$, $\mathbb{R}^n$ into corresponding algebras of hypercomplex numbers. Here, we present a brief introduction to the conventions of geometrical algebra that are used in this paper. A more comprehensive introduction can be found in Hestenes and Sobczyk [113].

7.4.1 Clifford Algebras as Models of Geometrical and Perceptual Spaces

Algebra and Geometry of 2–D Spaces. We start with the space $\mathbb{R}^2$ and provide it with the algebraic frame of the algebra of generalized complex numbers:

$$\mathbb{R}^2 \longrightarrow \mathcal{A}_2(\mathbb{R}) := \mathbb{R} + \mathbb{R}I = \{\mathbf{z} = x_1 + Ix_2 \mid x_1, x_2 \in \mathbb{R}\},$$

where I is a generalized imaginary unit.

If $I^2 = -1$, i.e. $I = i$, then $\mathcal{A}_2(\mathbb{R})$ is *the field of complex numbers*

$$\mathbf{COM} := \{x_1 + ix_2 \mid x_1, x_2 \in \mathbb{R};\ i^2 = -1\}.$$

If $I^2 = +1$, i.e $I = e$, then $\mathcal{A}_2(\mathbb{R})$ is *the ring of double numbers*

$$\mathbf{DOU} := \{x_1 + ex_2 \mid x_1, x_2 \in \mathbb{R};\ e^2 = 1\}.$$

If $I^2 = 0$, i.e $I = \varepsilon$ then $\mathcal{A}_2(\mathbb{R})$ is *the ring of dual numbers*

$$\mathbf{DUA} := \{x_1 + \varepsilon x_2 \mid x_1, x_2 \in \mathbb{R};\ \varepsilon^2 = 0\}.$$

In $\mathcal{A}_2(\mathbb{R})$ we introduce a conjugation operation which maps every element $\mathbf{z} = x_1 + Ix_2$ to the element $\overline{\mathbf{z}} = x_1 - Ix_2$. Now, the generalized complex plane is turned into a pseudometric space: $\mathcal{A}_2(\mathbb{R}) \longrightarrow \mathcal{GC}_2^{p,q,r}$ if one defines the pseudodistance

$$\rho(\mathbf{z}_1, \mathbf{z}_2) = \sqrt{(\mathbf{z}_2 - \mathbf{z}_1)\overline{(\mathbf{z}_2 - \mathbf{z}_1)}} = \begin{cases} \sqrt{(x_2 - x_1)^2 + (y_2 - y_1)^2}, & \mathbf{z} \in \mathbf{COM}, \\ \sqrt{(x_2 - x_1)^2 - (y_2 - y_1)^2}, & \mathbf{z} \in \mathbf{DOU}, \\ |x_2 - x_1|, & \mathbf{z} \in \mathbf{DUA}, \end{cases}$$

where $\mathbf{z}_1 := (x_1, x_2) = x_1 + Ix_2$, $\mathbf{z}_2 := (y_1, y_2) = y_1 + Iy_2$ and the three superscripts p, q, r are denoting the signature of the space. So, the plane of classical complex numbers is a 2–dimensional Euclidean space $\mathcal{GC}_2^{2,0,0}$, the double numbers plane is a 2–dimensional Minkowskian space $\mathcal{GC}_2^{1,1,0}$ and the dual numbers plane is a 2- dimensional Galilean space $\mathcal{GC}_2^{1,0,1}$.

When one speaks about all three algebras (or geometries) simultaneously, then the corresponding algebra (or geometry) is that of *generalized complex numbers*, that is denoted as $\mathcal{A}_2^{p,q,r}$ (or $\mathcal{GC}_2^{p,q,r}$).

Algebra and Geometry of 3–D Spaces. Quaternions, as constructed by Hamilton, form 4–D algebra

$$\mathcal{A}_4(\mathbb{R}|1,i,j,k) = \mathcal{A}_4(\mathbb{R}) = \mathcal{A}_4 := \mathbb{R} + \mathbf{R}i + \mathbb{R}j + \mathbb{R}k$$

spanned on four hyperimaginary units $1, i, j, k$. The following identities are valid for these units: $i^2 = j^2 = k^2 = -1$, $ij = -ji = k$. It can be set $i^2 = j^2 = k^2 = \delta \in \{-1, 0, 1\}$. Here, the two latter values (0 and 1) result in non–classical quaternions that were proposed by Clifford [47]. Introducing notations I, J, K for the three new hyperimaginary units we get nine spatial algebras of generalized quaternions

$$\mathcal{A}_4(\mathbb{R}|1,I,J,K) := \mathcal{A}_4 = \mathbb{R} + \mathbb{R}I + \mathbb{R}J + \mathbb{R}K \tag{7.3}$$

depending on which of the nine possibilities resulting from $I^2 \in \{1, 0, -1\}$, $J^2 \in \{1, 0, -1\}$ is valid for two independent hyperimaginary units.

Every generalized quaternion $\mathbf{q}$ has a unique representation in the form $\mathbf{q} = q_0 + q_1 I + q_2 J + q_3 K = \mathrm{Sc}(\mathbf{q}) + \mathrm{Vec}(\mathbf{q})$, where q_0, q_1, q_2, q_3 are real numbers, $\mathrm{Sc}(\mathbf{q}) := q_0$, $\mathrm{Vec}(\mathbf{q}) := q_1 I + q_2 J + q_3 K$ are scalar and vector parts of the quaternion $\mathbf{q}$, respectively.

In $\mathcal{A}_4(\mathbb{R})$ we introduce a conjugation operation which maps every quaternion $\mathbf{q} = q_0 + Iq_1 + Jq_2 + Kq_3$ to the element $\overline{\mathbf{q}} = q_0 - Iq_1 - Jq_2 - Kq_3$. If the pseudodistance $\rho(\mathbf{p}, \mathbf{q})$ between two generalized quaternions $\mathbf{p}$ and $\mathbf{q}$ is defined as modulus of their difference $\mathbf{u} = \mathbf{p} - \mathbf{q} = t + xI + yJ + zK$:

$$\rho(\mathbf{p},\mathbf{q}) = \begin{cases} \sqrt{(t^2+x^2+y^2+z^2)}, & \text{if } I^2 = -1, J^2 = -1, \\ \sqrt{(t^2+y^2)}, & \text{if } I^2 = 0, J^2 = -1, \\ \sqrt{(t^2-x^2+y^2-z^2)}, & \text{if } I^2 = 1, J^2 = -1, \\ \sqrt{(t^2+x^2)}, & \text{if } I^2 = -1, J^2 = 0, \\ \sqrt{(t^2)} = |t|, & \text{if } I^2 = 0, J^2 = 0, \\ \sqrt{(t^2-x^2)}, & \text{if } I^2 = 1, J^2 = 0, \\ \sqrt{(t^2+x^2-y^2-z^2)}, & \text{if } I^2 = -1, J^2 = 1, \\ \sqrt{(t^2-y^2)}, & \text{if } I^2 = 0, J^2 = 1, \\ \sqrt{(t^2-x^2-y^2+z^2)}, & \text{if } I^2 = 1, J^2 = 1, \end{cases}$$

then the nine spatial algebras $\mathcal{A}_4(\mathbb{R})$ are transformed into nine 4–D pseudometric spaces designed as $\mathcal{GH}_4^{p,q,r}$, where p, q and r stand for the number of basis vectors which square to 1, -1 and 0, respectively and fulfill $p+q+r=n$. Thus, the pseudodistance can take positive, negative and pure imaginary values. There are only 5 different geometries $\mathcal{GH}_4^{p,q,r}$: $\mathcal{GH}_4^{4,0,0}$, $\mathcal{GH}_4^{2,2,0}$, $\mathcal{GH}_4^{2,0,2}$, $\mathcal{GH}_4^{1,3,0}$, $\mathcal{GH}_4^{1,2,1}$.

The subspaces of vector generalized quaternions $xI+yJ+zK$ are 3–D spaces $\mathcal{GR}_3^{p,q,r} := \mathbf{Vec}\{\mathcal{GH}_4^{p,q,r}\}$. The pseudometrics introduced in $\mathcal{GH}_4^{p,q,r}$ induce only three different pseudometrics in $\mathcal{GR}_3^{p,q,r}$:

$\rho(\mathbf{Vec}\{\mathbf{p}\}, \mathbf{Vec}\{\mathbf{q}\}) = |\mathbf{Vec}\{\mathbf{p}-\mathbf{q}\}| = |\mathbf{Vec}\{\mathbf{u}\}| =$

$$= \sqrt{\|xI+yJ+zK\|_{\mathcal{GH}}} = \begin{cases} \sqrt{(x^2+y^2+z^2)}, \\ \sqrt{(x^2-y^2-z^2)}, \\ \sqrt{x^2} = |x|. \end{cases}$$

The corresponding 3–D metrical spaces will be denoted as $\mathcal{GR}_3^{3,0,0}$, $\mathcal{GR}_3^{1,0,2}$, and $\mathcal{GR}_3^{1,2,0}$. They form Euclidean, Minkowskian, and Galilean 3–D pseudometric spaces, respectively.

Algebra and Geometry of n–D Spaces. Now, let us consider an n–D space $\mathbb{R}^n$ spanned on the orthonormal basis of n hyperimaginary units I_i, $i=1,2,\ldots,n$. We suppose $I_i^2=+1$ for $i=1,2,\ldots,p$, $I_i^2=-1$ for $i=p+1,2,\ldots,p+q$, $I_i^2=0$ for $i=p+q+1,2,\ldots,p+q+r=n$ and $I_iI_j=-I_jI_i$. Now, we construct the "big" 2^n–D space $\mathbb{R}^{2^n}$ as a direct sum of subspaces of dimensions $C_n^0, C_n^1, \ldots, C_n^p, \ldots C_n^n$:

$$\mathbb{R}^{2^n} = \mathbb{R}^{C_n^0} \oplus \mathbb{R}^{C_n^1} \oplus \ldots \oplus \mathbb{R}^{C_n^p} \oplus \ldots \oplus \mathbb{R}^{C_n^n},$$

where the subspaces $\mathbb{R}^{C_n^p}$ $(p=0,1,\ldots,n)$ are spanned on the p–products of units $I_{i_1}I_{i_2}\ldots I_{i_p}$ $(i_1<i_2<\ldots<i_p)$. By definition, we suppose that $I_{i_0}:=1$ is the classical real unit 1. So

$$\mathbb{R}^{C_n^0} := \{xI_{i_0} \mid x \in \mathbb{R}\},$$

$$\mathbb{R}^{C_n^1} := \{x_1I_1 + x_2I_2 + \ldots + x_nI_n \mid x_i \in \mathbb{R}, \forall i = 1,\ldots,n\},$$

$$\mathbb{R}^{C_n^2} := \{x_{12}I_1I_2 + x_{13}I_1I_3 + \ldots + x_{n-1,n}I_{n-1}I_n \mid$$
$$\mid x_{i_1i_2} \in \mathbb{R},\ \forall i_1, i_2 = 1,\ldots,n\},$$

$$\mathbb{R}^{C_n^3} := \{x_{123}I_1I_2I_3 + x_{124}I_1I_2I_4 + \ldots + x_{n-2,n-1,n}I_{n-2}I_{n-1}I_n \mid$$
$$\mid x_{i_1i_2i_3} \in \mathbb{R},\ \forall i_1, i_2, i_3 = 1,\ldots,n\},$$

$$\ldots\ldots\ldots,$$

$$\mathbb{R}^{C_n^n} := \{x_{12\ldots n}I_1I_2\cdots I_n \mid x_{12\ldots n} \in \mathbb{R}\}.$$

Example 7.4.1. $\mathbb{R}^8 = \mathbb{R}^1 \oplus \mathbb{R}^3 \oplus \mathbb{R}^3 \oplus \mathbb{R}^1 = \mathbb{R}I_0 \oplus \{\mathbb{R}I_1 \oplus \mathbb{R}I_2 \oplus \mathbb{R}I_3\} \oplus$

$$\oplus \{\mathbb{R}I_1 I_2 \oplus \mathbb{R}I_1 I_3 \oplus \mathbb{R}I_2 I_3\} \oplus \mathbb{R}I_1 I_2 I_3.$$

Every element $\mathcal{C} \in \mathbb{R}^{2^n}$ has the following representation

$$\mathcal{C} = \sum_{p=0}^{n} \left(\sum_{i_1, i_2, \ldots i_p = 1}^{n} x_{i_1 i_1 \ldots i_p} I_{i_1} I_{i_2} \ldots I_{i_p} \right) =$$

$$= \mathrm{Vec}^0(\mathcal{C}) + \mathrm{Vec}^1(\mathcal{C}) + \mathrm{Vec}^2(\mathcal{C}) + \ldots + \mathrm{Vec}^n(\mathcal{C}),$$

where $\mathrm{Vec}^0(\mathcal{C}) \equiv \mathrm{Sc}(\mathcal{C}) \in \mathbb{R}^{C_n^0}$ is the scalar part of the Clifford number $\mathcal{C}$, $\mathrm{Vec}^1(\mathcal{C}) \in \mathbb{R}^{C_n^1}$ is its vector part, $\mathrm{Vec}^2(\mathcal{C}) \in \mathbb{R}^{C_n^2}$ is its bivector part,... and $\mathrm{Vec}^n(\mathcal{C}) \in \mathbb{R}^{C_n^n}$ is its n-vector part.

If $\mathcal{C}_1, \mathcal{C}_2 \in \mathbb{R}^{2^n}$, then we can define their product as

$$\mathcal{C}_1 \mathcal{C}_2 := \left\{ \sum_{p=0}^{n} \sum_{i_1, i_2, \ldots i_p = 1}^{n} x_{i_1 i_1 \ldots i_p} I_{i_1} I_{i_2} \ldots I_{i_p} \right\} \circ$$

$$\circ \left\{ \sum_{q=0}^{n} \sum_{j_j, j_2, \ldots j_q = 1}^{n} y_{i_1 i_1 \ldots i_q} I_{j_1} I_{j_2} \ldots I_{j_q} \right\} =$$

$$= \sum_{r=0}^{n} \sum_{k_1, k_2, \ldots k_r = 1}^{n} z_{k_1 k_1 \ldots k_r} I_{k_1} I_{k_2} \ldots I_{k_r},$$

where $I_{k_1} I_{k_2} \ldots I_{k_r}$ are obtained from products $I_{i_1} I_{i_2} \ldots I_{i_p} I_{j_1} I_{j_2} \ldots I_{j_q}$ with the conditions $I_i^2 = \delta$ and $I_i I_j = -I_j I_i$. There are 3^n possibilities for $I_i^2 = +1, 0, -1$, $\forall i = 1, 2, \ldots, n$. Every possibility generates one algebra. Consequently, the space $\mathbb{R}^{2^n}$ with 3^n rules of multiplication forms 3^n different 2^n–D algebras $\mathcal{A}_{2^n}^{p,q,r}(\mathbb{R})$, which are called *Clifford algebras*.

In $\mathcal{A}_{2^n}^{p,q,r}(\mathbb{R})$ we introduce the conjugation operation which maps every Clifford number $\mathcal{C}$ to the number $\overline{\mathcal{C}}$. If the pseudodistance between two Clifford numbers $\mathcal{A}$ and $\mathcal{B}$ is defined as modulus of their difference

$$\rho(\mathcal{A}, \mathcal{B}) = |\mathcal{A} - \mathbf{B}| = \sqrt{(\mathcal{A} - \mathcal{B})\overline{(\mathcal{A} - \mathcal{B})}},$$

then the algebras $\mathcal{A}_{2^n}^{p,q,r}(\mathbb{R})$ are transformed into 2^n–D pseudometric spaces designed as $\mathcal{CL}_{2^n}^{p,q,r}$.

The subspaces of pure vector Clifford numbers $x_1 I_1 + \ldots + x_n I_n \in \mathrm{Vec}^1(\mathcal{A}_{2^n}^{p,q,r}(\mathbb{R}))$ are n–D spaces $\mathbb{R}^n := \mathcal{GR}_n^{p,q,r}$. The pseudometrics constructed in $\mathcal{CL}_{2^n}^{p,q,r}$ induces in $\mathcal{GR}_n^{p,q,r}$ the corresponding pseudometrics.

Every algebra $\mathcal{A}_{2^n}^{p,q,r}(\mathbb{R})$ has an even subalgebra

$$^{ev}\mathcal{A}_{2^n}^{p,q,r}(\mathbb{R}) = \left\{ \mathcal{C} \mid \mathcal{C} := \sum_{p=even} \sum_{i_1,i_2,\ldots i_p=1}^{n} x_{i_1 i_1 \ldots i_p} I_{i_1} I_{i_2} \ldots I_{i_p} \right\}.$$

All Clifford numbers $\mathcal{E} \in {}^{ev}\mathcal{A}_{2^n}^{p,q,r}$ of unit modulus represent the rotation group of the corresponding space $\mathcal{GR}_n^{p,q,r}$ which is called the *spinor group* and is denoted as $\mathrm{Spin}(\mathcal{A}_{2^n}^{p,q,r})$.

7.4.2 Clifford Algebra of Motion and Affine Groups of Metric Spaces

Generalized complex numbers and quaternions of unit modulus have the form

$$\mathbf{z} = e^{I\varphi} = \cos\varphi + I \sin\varphi,$$

$$\mathbf{q} = e^{\mathbf{u}_0\varphi} = \cos\varphi + \mathbf{u}_0 \sin\varphi,$$

where $\cos\varphi$ and $\sin\varphi$ are trigonometric functions in the corresponding n–D $\mathcal{GC}_n^{p,q,r}$–geometries, φ is a rotation angle around the vector–valued quaternion $\mathbf{u}_0$ of unit modulus ($|\mathbf{u}_0| = 1$, $\mathbf{u}_0 = -\overline{\mathbf{u}_0}$).

Clifford numbers $\mathcal{E} \in \mathrm{Spin}(\mathcal{A}_{2^n}^{p,q,r})$ with unit modulus have analogous form

$$\mathcal{E} = e^{\mathbf{u}_0\varphi} = \cos\varphi + \mathbf{u}_0 \sin\varphi \in \mathrm{Spin}(\mathcal{A}_{2^n}^{p,q,r}).$$

Theorem 7.4.1. *[47]. The transformations*

$$\mathbf{z}' = e^{I\varphi/2}\mathbf{z}e^{I\varphi/2} + \mathbf{y} = e^{I\varphi}\mathbf{z} + \mathbf{y},$$

$$\mathbf{q}' = e^{u\varphi/2}\mathbf{q}e^{-u\varphi/2} + \mathbf{p}, \quad \mathcal{Q}' = \mathcal{E}^{1/2}\mathcal{Q}\mathcal{E}^{-1/2} + \mathcal{P}$$

form the groups of motions $\mathbf{Mov}(\mathcal{GC}_2^{p,q,r})$, $\mathbf{Mov}(\mathcal{GH}_4^{p,q,r})$, *and* $\mathbf{Mov}(\mathcal{CL}_{2^n}^{p,q,r})$ *of the spaces* $\mathcal{GC}_2^{p,q,r}$, $\mathcal{GH}_4^{p,q,r}$ *and* $\mathcal{CL}_{2^n}^{p,q,r}$ *respectively.*

Theorem 7.4.2. *[47]. All motions of 2-D, 3-D and n–D spaces* $\mathcal{GR}_2^{p,q,r}$, $\mathcal{GR}_3^{p,q,r}$, $\mathcal{GR}_n^{p,q,r}$ *are represented in the form*

$$\mathbf{z}' = \mathbf{e}\mathbf{z} + \mathbf{y}, \quad \mathbf{z}, \mathbf{y} \in \mathcal{GR}_2^{p,q,r},$$

$$\mathbf{x}' = \mathbf{q}^{1/2}\mathbf{x}\mathbf{q}^{-1/2} + \mathbf{y}, \quad \mathbf{x}, \mathbf{y} \in \mathcal{GR}_3^{p,q,r},$$

$$\mathbf{x}' = \mathcal{E}^{1/2}\mathbf{x}\mathcal{E}^{-1/2} + \mathbf{y}, \quad \mathbf{x}, \mathbf{y} \in \mathcal{GR}_n^{p,q,r},$$

where $\mathbf{e} := e^{I\varphi}$, $|\mathbf{q}| = 1$, $|\mathcal{E}| = 1$.

If $|\mathbf{e}|, |\mathbf{q}|, |\mathcal{E}| \neq \mathbf{1}$ then latter transformations form the affine groups $\mathbf{Aff}(\mathcal{GR}_2^{p,q,r})$, $\mathbf{Aff}(\mathcal{GR}_3^{p,q,r})$ and $\mathbf{Aff}(\mathcal{GR}_n^{p,q,r})$.

7.4.3 Algebraic Models of Perceptual Color Spaces

The aim of this subsection is to present, as an example, algebraic models of the subjective perceptual color space. Note, that the perceived color is the result of the human mind, not a physical property of an object.

The color representation we are using is based on Young's theory (1802), asserting that any color can be visually reproduced by a property combination of three colors, referred to as primary colors. Later studies have been developed on the basis of the physiological discovery that the human eye retina mainly contains three different color receptors (cones), tuned to three overlapping intervals of the visible light spectrum:

$$f_R(x,y) = \int_\lambda s(\lambda) H_R(\lambda) d\lambda,$$

$$f_G(x,y) = \int_\lambda s(\lambda) H_G(\lambda) d\lambda,$$

$$f_B(x,y) = \int_\lambda s(\lambda) H_B(\lambda) d\lambda,$$

where $s(\lambda)$ is the color spectrum received from the object, $H_R(\lambda)$, $H_B(\lambda)$, $H_R(\lambda)$ are three photoreceptor (cone or sensor) sensitivity functions, and λ is the wavelength.

Usually, the three primary colors are chosen as Red, Green, Blue. Once the primary colors have been chosen, every color is associated to a point of a 3-D color space. The color of this point represents the weight sum of the primary colors.

A color image can be considered as a 2–dimensional vector–valued (R, G, B) function

$$\mathbf{f}_{col}(x,y) = f_R(x,y)\mathbf{i} + f_G(x,y)\mathbf{j} + f_B(x,y)\mathbf{k}.$$

We use a color transformation to separate the color image into two terms: 1–D luminance (intensity) term and 2–D chromaticity term (color information). This color transformation is simply a linear projection of the color vector-valued image on a plane of the color space, which is orthogonal to the diagonal vector $(1,1,1)$. Then we get

$$\mathbf{f}_{col}(x,y) = f_{lu}(x,y)\mathbf{e} + \mathbf{f}_{ch}(x,y),$$

where $\mathbf{e} := (\mathbf{i}+\mathbf{j}+\mathbf{k})/\sqrt{3}$, $\langle \mathbf{e} \mid \mathbf{f}_{ch}(x,y) \rangle = 0$.

The same result is obtained if we consider a color image as hypercomplex–valued (triplet–valued) function:

$$\mathbf{f}_{col}(x,y) = f_R(x,y) 1_{col} + f_G(x,y)\varepsilon_{col} + f_B(x,y)\varepsilon_{col}^2,$$

where $1_{col}, \varepsilon_{col}, \varepsilon_{col}^2$ are hyperimaginary units, $\varepsilon_{col}^3 = 1$. Note, that the numbers of the form $a1_{col} + b\varepsilon_{col} + c\varepsilon_{col}^2$ are called *triplets* or *3–cycle numbers*. They form a *color triplet (or 3–cycle) algebra*:

$$\mathcal{A}_3^{col} = \mathcal{A}_3^{col}(\mathbb{R} \mid 1, \varepsilon_{col}, \varepsilon_{col}^2) := \mathbb{R}1_{col} + \mathbb{R}\varepsilon_{col} + \mathbb{R}\varepsilon_{col}^2, \quad \varepsilon_{col}^3 = 1.$$

One can show that the color triplet algebra is the direct sum of the real numbers field $\mathbb{R}$ and of the complex numbers field $\mathbf{C}$: $\mathcal{A}_3^{col} = \mathbb{R}\mathbf{e}_{lu} \oplus \mathbf{C}\mathbf{E}_{ch}$, where $\mathbf{e}_{lu} := (1_{col} + \varepsilon_{col} + \varepsilon_{col}^2)/\sqrt{3}$, $\mathbf{E}_{ch} := (1_{col} + \omega\varepsilon_{col} + \omega^2\varepsilon_{col}^2)/\sqrt{3}$ are new "luminance" and "chromaticity" hyperimaginary units, respectively, $\omega := e^{\frac{2\pi}{3}}$, $\mathbf{e}_{lu}^2 = \mathbf{e}_{lu}$, $\mathbf{E}_{ch}^2 = \mathbf{E}_{ch}$, $\mathbf{e}_{lu}\mathbf{E}_{ch} = \mathbf{E}_{ch}\mathbf{e}_{lu} = 0$.

So, every triplet $\mathcal{C} \in \mathcal{A}_3^{col}$ is a linear combination of the "scalar" part $a\mathbf{e}_{lu}$ and the "complex" part $\mathbf{A}\mathbf{E}_{ch}$: $\mathcal{C} = a\mathbf{e}_{lu} + \mathbf{A}\mathbf{E}_{ch}$. The real numbers $a \in \mathbb{R}$ we call *intensity numbers* and complex numbers $\mathbf{A} \in \mathbf{C}$ we call *chromaticity numbers.* We introduce a generalized color triplet algebra ${}^{p,q,r}\mathcal{A}_3^{col} := \mathbb{R}\mathbf{e}_{lu} \oplus {}^{p,q,r}\mathcal{A}_2\mathbf{E}_{ch}$. For briefly, this algebra will be denoted as $\mathcal{A}_3^{col}$.

Definition 7.4.1. *A color image of the form*

$$\begin{aligned}\mathbf{f}_{col}(x,y) = f_R(x,y)1_{col} + f_G(x,y)\varepsilon_{col} + f_B(x,y)\varepsilon_{col}^2 \\ = f_{lu}(x,y)\mathbf{e}_{lu} + \mathbf{f}_{ch}(x,y)\mathbf{E}_{ch}\end{aligned}$$

is called a triplet–valued color image.

The intensity (luminance) term is $f_{lu}(x,y)$ and the chromaticity term is $\mathbf{f}_{ch}(x,y)$. Such a model of color images will be called *3–cycle model* (or *model of odd grade*).

Definition 7.4.2. *The color triplet (3–cycle) algebra $\mathcal{A}_3^{col}$ is called perceptive color space.*

However, the representation of one perceived color by three numerical values is not complete, due to the fact that only a subset of colors is generated by a combination of three assigned "primary" sources. A more complete model of the visual perception is based on the lateral geniculate nucleus (LGN), at which the optical nerve terminates. According to de Valois et al. [233], there are many types of color sensitive cells in the LGN. Spectrally opponent cells respond to wide uniform fields by increasing their firing rate within the same wavelength region, and by descreasing the firing rate for other wavelengths. Depending on the firing threshold, these cells are called: "–Red+Blue", "+Red–Blue", "–Green+Red", "+Green–Red", "–Green+Blue", "+Green–Blue", where "+" means excitation and "–" means inhibition. These opponent cells map R,G,B components on the 4–D unit sphere

$$f_{Bl}^2 + f_{RG}^2 + f_{RB}^2 + f_{GB}^2 = 1,$$

where $f_{Bl}, f_{RG}, f_{RB}, f_{GB}$ are black, red–green, red–blue and green–blue components, respectively.

Now, let us consider the 3-D color space $\mathbb{R}^3_{col} := \mathbb{R}I_R + \mathbb{R}I_G + \mathbb{R}I_B$ spanned on basis I_R, I_G, I_B. We suppose $I_R^2 = +1, 0, -1$, $I_G^2 = +1, 0, -1$, $I_B^2 = +1, 0, -1$. Now, we construct a new color Clifford algebra

$$\mathcal{A}_8^{col} := \mathbb{R}I_{Bl} + (\mathbb{R}I_R + \mathbb{R}I_G + \mathbb{R}I_B) +$$

$$+(\mathbb{R}I_{RG} + \mathbb{R}I_{RB} + \mathbb{R}I_{GB}) + \mathbb{R}I_{Wh},$$

where $I_{Bl} = 1$, $I_{Wh} = I_R I_G I_B$ are black and white units, $I_{RG} := I_R I_G$, $I_{RB} := I_R I_B$, $I_{GB} := I_G I_B$. Therefore, resulting from the capacity of this algebraic model of color images, we can formulate the spin–valued function

$$\mathbf{f}_{col} := f_{Bl} I_{Bl} + f_{RG} I_{RG} + f_{RB} I_{RB} + f_{GB} I_{GB},$$

which has values in the spin part $\text{Spin}(\mathcal{A}_8^{col})$ of the color Clifford algebra.

Definition 7.4.3. *Functions of the form*

$$\mathbf{f}_{col} : \mathbb{R}^2 \longrightarrow Spin(\mathcal{A}_8^{col}), \quad \mathbf{f}_{col} : \mathbb{R}^n \longrightarrow Spin(\mathcal{A}_8^{col})$$

will be called spin–valued color 2–D and n–D images, respectively.

This is an even grade model of color images.

The color vision model of primates is based on the existence of three different types of photoreceptors in the eye. However, Cronin et al. [51] showed recently that mantis shrimps have at least ten spectral types of photoreceptors in their eyes giving them the capability to recognize fine spectral details. This example illustrates the need of more general approaches for color analysis and representation than is given by three broadband filter representations.

Note that for a long time multispectral measurements have been used for region classification in satellite images. More recently, Swain and Ballard [226] proposed a method called color indexing to demonstrate that color distributions without geometric information could be used to recognize objects efficiently from a large database of models.

The multicomponent color image is measured as k–component vector

$$\mathbf{f}_{mcol}(\mathbf{x}) := \begin{bmatrix} f_1(x,y) \\ f_2(x,y) \\ \dots \\ f_k(x,y) \end{bmatrix} = \begin{bmatrix} \int_\lambda s(\lambda) H_1(\lambda) d\lambda \\ \int_\lambda s(\lambda) H_2(\lambda) d\lambda \\ \dots \\ \int_\lambda s(\lambda) H_k(\lambda) d\lambda \end{bmatrix},$$

where $f_1(x,y), f_2(x,y) \dots, f_k(x,y)$ are sensor sensitivity functions. For instance, using 5nm resolution between 400nm and 700nm, k equals 60, and for 20nm resolution, k equals 16.

We will interprete such images as hypercomplex–valued signals

$$\mathbf{f}_{mcol} = f_0 1 + f_1 \varepsilon^1_{col} + \dots + f_{k-1} \varepsilon^{k-1}_{col}$$

which takes values in the k–cycle algebra $\mathcal{A}(\mathbf{R}|1, \varepsilon_{col}^1, \ldots, \varepsilon_{col}^{k-1})$. This is an odd grade model of multicolor images.

To form the even grade model we consider the k–D color space $\mathbb{R}_{col}^k := \mathbb{R}I_1 + \mathbb{R}I_2 + \ldots + \mathbb{R}I_k$ spanned on the basis $I_i,\ i = 1, 2, \ldots, k$. We suppose $I_i^2 = +1, 0, -1,\ i = 1, 2, \ldots, k$. This color space generates the Clifford algebra $\mathcal{A}_{2^k}^{col} := {}^{p,q,r}\mathcal{A}_{2^k}^{col}(I_1, I_2, \ldots, I_k|\mathbb{R})$.

Definition 7.4.4. *Functions which take values in the spin part of the color Clifford algebra*

$$\mathbf{f}_{col} : \mathbb{R}^2 \longrightarrow Spin(\mathcal{A}_{2^k}^{col}), \quad \mathbf{f}_{col} : \mathbb{R}^n \longrightarrow Spin(\mathcal{A}_{2^k}^{col})$$

will be called spin–valued color 2–D and n–D images, respectively.

Further, we interpret an image as an embedding of a manifold in a higher dimensional spatial–color Clifford algebra. The embedding manifold is a hybrid space that includes spatial coordinates as well as color coordinates.

A grey level image is considered in this framework as a 2–D surface (i.e. graph of $f(x, y)$) embedded in the 3–D spatial–color space

$$\mathbb{R}_3^{SpCol}(I_1^{Sp}, I_2^{Sp}; I_f^{gl}) = (\mathbb{R}I_1^{Sp} + \mathbb{R}I_2^{Sp}) + \mathbb{R}I_f^{gl}$$

whose coordinates are (x, y, f), where $x \in \mathbb{R}I_1^{Sp}$, $y \in \mathbb{R}I_2^{Sp}$, $f \in \mathbb{R}I_f^{gl}$. A color image is accordingly considered as a 3–D manifold in the 5–D spatial–color space

$$\mathbb{R}_5^{SpCol}(I_1^{Sp}, I_2^{Sp}; I_R^{Col}, I_G^{Col}, I_B^{Col}) =$$
$$= (\mathbb{R}I_1^{Sp} + \mathbb{R}I_2^{Sp}) + (\mathbb{R}I_R^{Col} + \mathbb{R}I_G^{Col} + \mathbb{R}I_B^{Col}) = \mathbb{R}_2^{Sp} + \mathbb{R}_3^{col},$$

whose coordinates are (x, y, f_R, f_G, f_B), where $x \in \mathbb{R}I_1^{Sp}$, $y \in \mathbb{R}I_2^{Sp}$ are spatial coordinates and $f_R \in \mathbb{R}I_R^{Col}$, $f_G \in \mathbb{R}I_R^{Col}$, $f_B \in \mathbb{R}I_R^{Col}$ are color coordinates.

For n–D k–color images we have an $n + k$ spatial–color space

$$\mathbb{R}_{n+k}^{SpCol}(I_1^{Sp}, \ldots, I_n^{Sp}; I_1^{Col}, I_2^{Col}, \ldots, I_k^{Col}) =$$
$$= (\mathbb{R}I_1^{Sp} + \ldots + \mathbb{R}I_n^{Sp}) + (\mathbb{R}I_1^{Col} + \mathbb{R}I_2^{Col} + \ldots + \mathbb{R}I_k^{Col}) = \mathbb{R}_n^{Sp} + \mathbb{R}_k^{Col},$$

whose coordinates are $(x_1, \ldots, x_n; f_1, f_2, \ldots, f_k)$, where $x_i \in \mathbb{R}I_I^{Sp}$, $i = 1, \ldots, n$ are spatial coordinates and $f_j \in \mathbb{R}I_j^{Col}$, $j = 1, 2, \ldots, k$.

It is clear that the geometrical, color and spatial–color spaces $\mathbb{R}_n^{Sp}$, $\mathbb{R}_k^{Col}$, $\mathbb{R}_{n+k}^{SpCol}$ generate *spatial, color and spatial–color Clifford algebras*

$$\mathcal{A}_{2^n}^{Sp}, \quad \mathcal{A}_{2^k}^{Col}, \quad \mathcal{A}_{2^{n+k}}^{SpCol} = \mathcal{A}_{2^n}^{Sp} \otimes \mathcal{A}_{2^k}^{Col}, \tag{7.4}$$

respectively, where $\otimes$ is the symbol of the tensor product.

7.5 Hypercomplex–Valued Moments and Invariants

The recognition of an object independent of its position, size and orientation is an important problem in pattern recognition. In the last two decades a number of techniques have been developed to extract image features which are invariant under translation, scale change and rotation caused by the image formation process. In particular, moment invariants are used as image description for object recognition, image classification and scene matching.

7.5.1 Classical ℝ–Valued Moments and Invariants

The classical geometrical moments $m_{\mathbf{p}}$ of the grey level image $f(\mathbf{x}) = f(x_1, x_2, \ldots, x_n)$ are integrals

$$m_{\mathbf{p}} = \int_{\mathbb{R}^n} \mathbf{x}^{\mathbf{p}} f(\mathbf{x}) \mathbf{dx},$$

where $\mathbf{dx} := dx_1 dx_2 \cdots dx_n$, $\mathbf{x}^{\mathbf{p}} := x_1^{p_1} x_2^{p_2} \cdots x_n^{p_n}$.

The invariants of an image $f(\mathbf{x})$ are usually constructed in two steps [118] – [227]. At the second step the invariants are computed by inserting the moments $m_{\mathbf{p}}$ into quite definite polynomial expressions. The form of these expressions depends on the view of the group $\mathbf{G}_n^r$.

Definition 7.5.1. *If* $\mathbf{c}$ *is the centroid of the image* $f(\mathbf{x})$, *then functionals*

$$\dot{m}_{\mathbf{p}} = \int_{\mathbb{R}^n} f(\mathbf{x})(\mathbf{x} - \mathbf{c})^{\mathbf{p}} \mathbf{dx}$$

are called central moments.

Theorem 7.5.1. *[118]. Central moments are absolute invariants with respect to the group of translation* $\mathbf{Tr}(\mathbb{R}^n)$ *(see fig. 7.2):*

$$\dot{m}_{\mathbf{p}}\{f(\mathbf{x} + \mathbf{a})\} = \dot{m}_{\mathbf{p}}\{f(\mathbf{x})\}, \ \forall \mathbf{a} \in \mathbf{Tr}(\mathbb{R}^n).$$

Therefore, there is an infinite set of invariants with respect to translation: $I_{\mathbf{p}}\{\mathbf{Tr}(\mathbb{R}^n) \mid f(\mathbf{x} + \mathbf{a})\} = I_{\mathbf{p}}\{\mathbf{Tr}(\mathbb{R}^n) \mid f(\mathbf{x})\} := \dot{m}_{\mathbf{p}}$.

Now, let us investigate how the central moments change with respect to scaling $\mathbf{x}' = \lambda \mathbf{x}$. Let $f_\lambda(\mathbf{x}) := f(\lambda \mathbf{x})$. Then we have

$$\dot{m}_{\mathbf{p}}\{f_\lambda\} = \int\limits_{\mathbf{x} \in \mathbb{R}^n} \mathbf{x}^{\mathbf{p}} f(\lambda \mathbf{x}) \ \mathbf{dx} = \left(\frac{1}{\lambda}\right)^{p_1 + p_2 + \ldots p_n + n} \dot{m}_{\mathbf{p}}\{f\}. \tag{7.5}$$

Theorem 7.5.2. *[118]. The central moments of an n–D image* $f(\mathbf{x})$ *are the relative invariants with respect to the scaling transformation* $\mathbf{MR}^n$ *with multiplicators* $(1/\lambda)^{p_1 + p_2 + \cdots + p_n + n}$ *(see fig 7.3).*

Definition 7.5.2. *[118]. Expressions* $\dot{m}_{\mathbf{p}} / [\dot{m}_{00\ldots0}]^{(p_1 + p_2 + \cdots + p_n + n)/n}$ *are called normalized central moments.*

The normalized moments being invariants will be denoted as $I_{\mathbf{p}}\{\mathbf{MR}^n | f\}$.

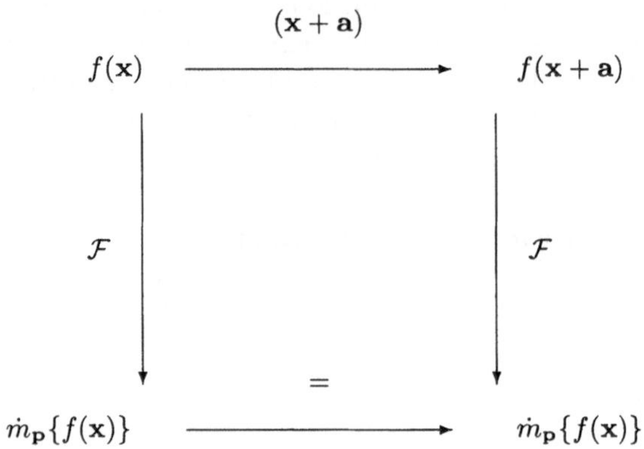

Fig. 7.2. Transformations of central moments with respect to the group $\mathbf{Tr}(\mathbf{R}^n)$

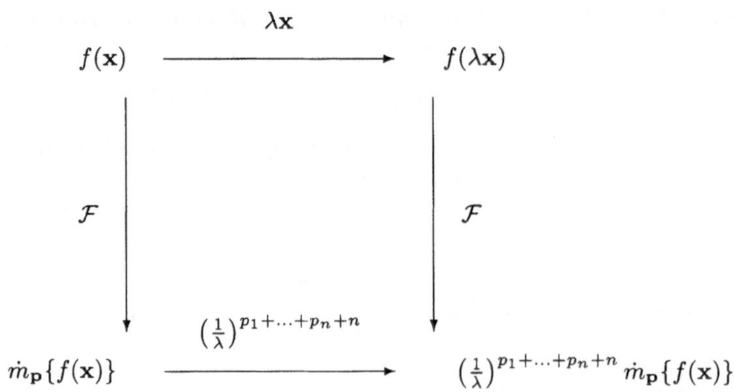

Fig. 7.3. Transformation of central moments with respect to the group $\mathbf{MR}^n$

Theorem 7.5.3. *Normalized central moments $I_{\mathbf{p}}\{\mathbf{MR}^n|f\}$ are absolute invariants with respect to the group of similarities.*

The invariants with respect to rotations are more difficult to calculate. Here we consider only the case $n = 2$. Let now act the group of rotations $\mathbf{SO}_2$ in the plane $\mathbb{R}^2$. If $f_\varphi(x,y) := f(x\cos\varphi - y\sin\varphi, x\sin\varphi + y\cos\varphi)$ is a copy of the initial image, rotated by the angle φ, then we have

$$\dot{m}_{pq}\{f_\varphi\} =$$

$$= \sum_{r=0}^{p}\sum_{s=0}^{q}\binom{p}{r}\binom{q}{s}(-1)^r \cos^{p-r+s}(\varphi)\ \sin^{q+r-s}(\varphi)\ \dot{m}_{p+q-r-s,r+s}\{f\}. \tag{7.6}$$

We see that the set of moments $\dot{m}_{pq}$ of the given order $k = p+q$ is transformed via the set of moments of the same order

$$\dot{m}_{k-q,q}\{f_\varphi\} = \sum_{n=0}^{k-q}\sum_{s=0}^{q}\binom{k-q}{n-s}\binom{q}{s}(-1)^{n-s}\cos^{k-q-n+2s}(\varphi)\times$$

$$\times \sin^{q+n-2s}(\varphi)\ \dot{m}_{k-n,n}\{f\}, \tag{7.7}$$

which clarifies the rule of moment transformations under rotation.

Let us give a matrix interpretation of this result. For this purpose consider an infinite sequence of vectors of increasing dimension: $\dot{\mathbf{m}}^0 := (\dot{m}_{00}) \in \mathbb{R}^1$, $\dot{\mathbf{m}}^1 := (\dot{m}_{01}, \dot{m}_{10}) \in \mathbb{R}^2$, $\dot{\mathbf{m}}^2 := (\dot{m}_{20}, \dot{m}_{11}, \dot{m}_{02}) \in \mathbb{R}^3$, ..., $\dot{\mathbf{m}}^k := (\dot{m}_{k,0}, \dot{m}_{k-1,1}, ..., \dot{m}_{1,k-1}, \dot{m}_{0,k}) \in \mathbf{R}^{k+1}$. According to (7.7), every vector

$$\mathbf{m}^k = (m_{0,k}, m_{1,k-1}, \dots, m_{k,0}) =$$

$$= \int_{\mathbb{R}^2} (x^0y^k, x^1y^{k-1}\cdots, x^ky^0) f(x,y)dxdy \tag{7.8}$$

is transformed into itself:

$$\dot{\mathbf{m}}^k\{f_\varphi\} = \mathbf{M}^k(\varphi)\dot{\mathbf{m}}^k\{f\},$$

where the matrix elements of the $(k+1)\times(k+1)$–matrix $\mathbf{M}^k(\varphi)$ are defined by the inner sum in expression (7.7).

Theorem 7.5.4. *[118]. The vector–valued moments $\dot{\mathbf{m}}^k$, $k = 0, 1, 2, \dots$ are the relative vector–valued $\mathbf{SO}_2$–invariants*

$$\mathbf{J}\{f_\varphi\} := \dot{\mathbf{m}}^k\{f_\varphi\} = \mathbf{M}^k(\varphi)\dot{\mathbf{m}}^k\{f\},$$

with matrix multiplicators $\mathbf{M}^k(\varphi)$ (see fig. 7.4).

Definition 7.5.3. *Let $A : \mathbb{R}^2\{x_1, x_2\} \longrightarrow \mathbb{R}^2\{x_1, x_2\}$ be a transformation. Then it induces another transformation $A^{[k]}$ acting into the $(k+1)$–D space spanned by the basis vectors $\{x^py^q | p+q = k\}$. This transformation $A^{[k]}$ is called the k–symmetric tensor power of the matrix A.*

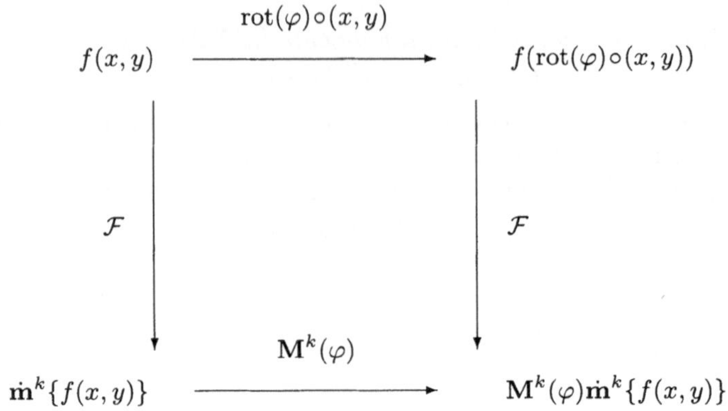

Fig. 7.4. Transformation of vector–valued moments with respect to the group $\mathbf{SO}_2$

In particular, when $A = \mathrm{rot}(\varphi) := \mathbf{M}^{[1]}(\varphi)$, we have

$$[M^{[k]}_{mn}(\varphi)] = \sum_{s=0}^{m} \binom{k-m}{n-s}\binom{m}{s} (-1)^{n-s} \cos^{k-m-n+2s}(\varphi)\ \sin^{m+n-2s}(\varphi).$$

Comparing (7.9) with (7.7) shows that $[M^{[k]}_{mn}(\varphi)]$ actually are matrix elements of the matrix $\mathbf{M}^{[k]}$. For example,

$$\mathbf{M}^2(\varphi) = \begin{bmatrix} \cos\varphi \ \sin\varphi \\ -\sin\varphi \ \cos\varphi \end{bmatrix}^{[2]}$$
$$= \begin{bmatrix} \cos^2(\varphi) & 2\cos(\varphi)\sin(\varphi) & \sin^2(\varphi) \\ -\cos(\varphi)\sin(\varphi) & \cos^2(\varphi)-\sin^2(\varphi) & \cos(\varphi)\sin(\varphi) \\ \sin^2(\varphi) & -2\cos(\varphi)\sin(\varphi) & \cos^2(\varphi) \end{bmatrix}. \quad (7.9)$$

Theorem 7.5.5. *[20]. The following equations are valid:* $(AB)^{[k]} = A^{[k]}B^{[k]}$ *and if* $AB = I$, *then* $A^{[k]}B^{[k]} = I^{[k]}$.

Now, we will use these expressions to simplify (7.9). Note then that

$$\mathbf{M}(\varphi)] = \begin{bmatrix} \cos(\varphi) & \sin(\varphi) \\ -\sin(\varphi) & \cos(\varphi) \end{bmatrix} = \Delta(i)H_2^{[1]}\Delta(\varphi)H_2^{[1]}\Delta(-i), \quad (7.10)$$

where $\Delta(i) := \begin{bmatrix} 1 & \\ & i \end{bmatrix}$, $H_2^{[1]} = \begin{bmatrix} 1 & 1 \\ 1 & -1 \end{bmatrix}$, $\Delta(\varphi) := \begin{bmatrix} e^{i\varphi} & \\ & e^{-i\varphi} \end{bmatrix}$.
Raising (7.10) into the k–symmetric tensor power we get

$$\mathbf{M}^{[k]}(\varphi) = \Delta^{[k]}(i)H_2^{[k]}\Delta^{[k]}(\varphi)H_2^{[k]}\Delta^{[k]}(-i), \tag{7.11}$$

where $\Delta^{[k]}(i) = [\delta_{mn}i^m]_{m,n=0}^k$, $\Delta^{[k]}(\varphi) = [\delta_{mn}e^{i\varphi(k-2m)}]_{m,n=0}^k$,

$$H_2^{[k]} = [K_{mn}]_{m,n=0}^k = \left[\sum_{s=0}^{m}(-1)^s \binom{k-m}{n-s}\binom{m}{s}\right]_{m,n=0}^k .$$

The transformation $H_2^{[k]}$ is called *Kravchuk transform* and is widely used in theory of codes [161]. For example,

$$H_2^{[1]} = \begin{bmatrix} 1 & 1 \\ 1 & -1 \end{bmatrix}, \ H_2^{[2]} = \begin{bmatrix} 1 & 2 & 1 \\ 1 & 0 & -1 \\ 1 & -2 & 1 \end{bmatrix}.$$

From (7.7), taking in consideration (7.11), we get

$$\dot{\mathbf{m}}^k\{f_\varphi\} = \mathbf{M}^k(\varphi)\dot{\mathbf{m}}^k\{f\} =$$

$$= \Delta^{[k]}(i)H_2^{[k]}\Delta^{[k]}(\varphi)H_2^{[k]}\Delta^{[k]*}(i)\dot{\mathbf{m}}^k\{f\},$$

where

$$\dot{\mathbf{m}}^k\{f_\varphi\} := (\dot{m}_{k,0}\{f_\varphi\},\ \dot{m}_{k-1,1}\{f_\varphi\}, \ldots,\ \dot{m}_{0,k}\{f_\varphi\}),$$

$$\dot{\mathbf{m}}^k\{f\} := (\dot{m}_{k,0}\{f\},\ \dot{m}_{k-1,1}\{f\}, \ldots, \dot{m}_{0,k}\{f\}).$$

Hence, we get the final expression

$$H_2^{[k]}\Delta^{[k]}(-i)\dot{\mathbf{m}}^k\{f_\varphi\} = \Delta^{[k]}(\varphi)[H_2^{[k]}\Delta^{[k]}(-i)\dot{\mathbf{m}}^k\{f\}]. \tag{7.12}$$

Theorem 7.5.6. *[135]. The spectral coefficients $\mathbf{J}^k\{\mathbf{SO}_2|f\}$ of the Kravchuk transform $H_2^{[k]}\Delta^{[k]}(-i)\dot{\mathbf{m}}^k\{f\}$ of the vector–valued moment $\dot{\mathbf{m}}^k\{f\}$ are the relative $\mathbf{SO}_2$–invariants of the image f : $\mathbf{J}^k\{f_\varphi\} = \Delta^{[k]}(\varphi)\mathbf{J}^k\{f\}$, i.e.*

$$\left(J_{k,o}^k, J_{k-1,1}^k, \ldots, J_{o,k}^k,\right)\{f_\varphi\} =$$

$$= diag(e^{-i(k-2m)\varphi})\left(J_{k,o}^k, J_{k-1,1}^k, \ldots, J_{o,k}^k,\right)\{f\} \tag{7.13}$$

(see fig. 7.5).

We find now the direct expressions for the relative invariants. From (7.13) it follows

$$\mathbf{J}_{k-m,m}^k\{f)\} = \sum_{n=0}^{k}\left[\sum_{s=0}^{m}(-1)^s(-i)^n\binom{k-m}{n-s}\binom{m}{s}\right]\dot{m}_{k-n,n}^k\{f\},$$

which is the desired expression. For example,

$$\begin{bmatrix} \mathbf{J}_{20}^2 \\ \mathbf{J}_{11}^2 \\ \mathbf{J}_{02}^2 \end{bmatrix} = \begin{bmatrix} 1 & 2 & 1 \\ 1 & 0 & -1 \\ 1 & -2 & 1 \end{bmatrix}\begin{bmatrix} \dot{m}_{02} \\ -i\dot{m}_{11} \\ -\dot{m}_{20} \end{bmatrix} = \begin{bmatrix} (\dot{m}_{20} - \dot{m}_{02}) - 2i\dot{m}_{11} \\ (\dot{m}_{20} + \dot{m}_{02}) \\ (\dot{m}_{20} - \dot{m}_{02}) + 2i\dot{m}_{11} \end{bmatrix}.$$

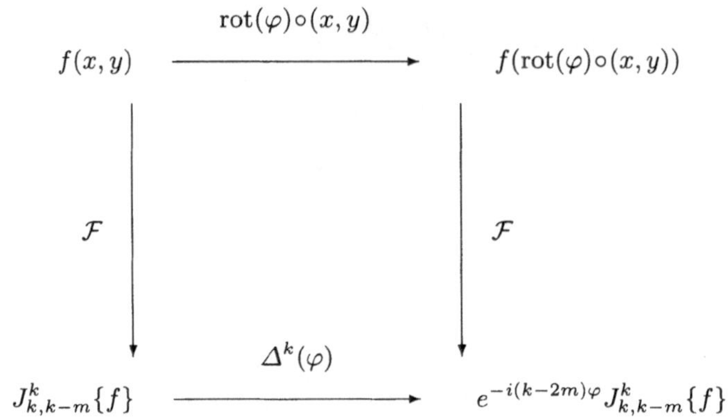

Fig. 7.5. Transformation of relative invariants with respect to the group $\mathbf{SO}_2$

As we have stated, when an image is rotated counter–clockwise by the angle φ, the new set of the relative invariants is connected with the old one according to (7.13) by the relation

$$\mathbf{J}^k_{k-m,m}\{f_\varphi\} = e^{-i(k-2m)\varphi}\mathbf{J}^k_{k-m,m}\{f\},\ 0 \leq m \leq k, \tag{7.14}$$

which we repeat here for convenience. As can be seen from this, rotations influence only the phase of relative invariants and do not change their absolute value. Hence, the absolute values of relative invariants are indeed the absolute invariants of rotations:

$$\mathbf{HMI}^k_{k-m,m} := \mid \mathbf{J}^k_{k-m,m}\{f_\varphi\} \mid \ = \ \mid \mathbf{J}^k_{k-m,m}\{f)\} \mid \tag{7.15}$$

(where **HMI** is the abridged notation of **H**u **m**oment **i**nvariants).

As the relative invariant $\mathbf{J}^k_{k-m,m}$ is the complex conjugate to $\mathbf{J}^k_{m,k-m}$, then only $(k/2)+1$ independent binary absolute invariants can be obtained from $k+1$ relative invariants of the k–th order. That is explained by the information lost by choosing as invariant a real number and not a complex one.

Complex–valued functionals of images can also be used as invariants because they do not change under the transformation. Hence, if we compose the product of the powers of two relative invariants, the phases of which are changing in the opposite directions by the same amount, then such a product will give a complex–valued invariant. For example, the product

$$\mathbf{HMI}^{s_1;s_2}_{k_1-m_1,m_1;k_2-m_2,m_2} = [\mathbf{J}^{k_1}_{k_1-m_1,m_1}]^{s_1}[\mathbf{J}^{k_2}_{k_2-m_2,m_2}]^{s_2}$$

is such an invariant if $s_1(k_1-2m_1)+s_2(k_2-2m_2)=0$. This expression can be easily generalized to n-ary absolute invariants of the $(s_1+s_2+...+s_n)$–th power:

$$\mathbf{HMI}^{s_1;s_2;\ldots;s_n}_{k_1-m_1,m_1;k_2-m_2,m_2;\ldots;k_n-m_n,m_n} =$$
$$= [\mathbf{J}^{k_1}_{k_1-m_1,m_1}]^{s_1}[\mathbf{J}^{k_2}_{k_2-m_2,m_2}]^{s_2}\ldots[\mathbf{J}^{k_n}_{k_n-m_n,m_n}]^{s_n},$$

where $s_1(k_1-2m_1)+s_2(k_2-2m_2)+\ldots+s_n(k_n-2m_n)=0$.

7.5.2 Generalized Complex Moments and Invariants

Let $\mathcal{GC}_2^{Sp} := \{\mathbf{z} = x_1 + Ix_2 \mid x_1, x_2 \in \mathbb{R};\ I^2 = -1, 0, +1\}$ be a generalized spatial complex plane. Then the grey–level image $f(x,y)$ can be considered as a function of a generalized complex variable, i.e. $f(\mathbf{z})$, $\mathbf{z} \in \mathcal{GC}_2^{Sp}$.

Definition 7.5.4. *If* $\mathbf{c}$ *is the centroid of the image* $f(\mathbf{z})$, *then the functionals*

$$\dot{\mathbf{m}}_p\{f\} := \int_{\mathbf{z}\in\mathcal{GC}_2^{Sp}} (\mathbf{z}-\mathbf{c})^p f(\mathbf{z})\mathbf{dz}$$

are called one–index central $\mathcal{A}_2^{Sp}$*–valued moments of the image* $f(\mathbf{z})$, *where* $\mathbf{dz} := dxdy$, $p \in \mathbf{Q}$ *is an arbitrary rational number.*

Let us now clarify the rules of moment transformations under geometrical distortions of the initial images. We will consider translation, rotation, and scaling transformations. If $f(\mathbf{z})$ is the initial image, then $f_{\mathbf{v},\mathbf{a}}(\mathbf{z})$ denotes its geometrical distorted copy $f_{\mathbf{v},\mathbf{a}}(\mathbf{z}) = f(\mathbf{v}(\mathbf{z}+\mathbf{a})) = f(\mathbf{z}^*)$. Here $\mathbf{v}$ and $\mathbf{a}$ are arbitrary fixed generalized numbers, $\mathbf{z}^* = \mathbf{v}(\mathbf{z}+\mathbf{a})$. Summing $\mathbf{z}$ with $\mathbf{a}$ brings us to image translation, and multiplication of $\mathbf{z}+\mathbf{a}$ by $\mathbf{v}$ is equivalent to the rotation (in Euclidean plane — classical rotation, in Minkowskian plane — hyperbolic (Lorentz) rotation, in Galilean plane — Galilei rotation) by an angle equal to $\arg(\mathbf{v})$ and with a dilatation given by $|\mathbf{v}|$.

Theorem 7.5.7. *Central moments of the image* $f(\mathbf{z})$ *are relative complex–valued invariants*

$$\mathcal{J}_p\{f_{\mathbf{v},\mathbf{a}}\} := \dot{\mathbf{m}}_p\{f_{\mathbf{v},\mathbf{a}}\} = \mathbf{v}^p|\mathbf{v}|^2\dot{\mathbf{m}}_p\{f\} \qquad (7.16)$$

with respect to the affine group $\mathbf{Aff}(\mathcal{GC}_2^{Sp})$ *of the generalized complex plane* $\mathcal{GC}_2^{Sp}$ *with* $\mathcal{A}_2^{Sp}$*–valued multiplicators* $\mathbf{v}^p|\mathbf{v}|^2$ *(see fig. 7.6).*

To get absolute invariants of the affine group based on relative ones, it is necessary to "neutralize" multiplicators by any way.

Definition 7.5.5. *The products* $\dot{\mathbf{m}}^{k_1}_{p_1}\dot{\mathbf{m}}^{k_2}_{p_2}\cdots\dot{\mathbf{m}}^{k_s}_{p_s}$ *are called* s*–ary* $\mathcal{A}_2^{Sp}$*–valued central moments, where* $k_i \in \mathbf{Q}$.

Theorem 7.5.8. *The* s*–ary central moments of the image* $f(\mathbf{z})$ *are relative* $\mathcal{A}_2^{Sp}$*–valued invariants*

$$\mathcal{J}^{k_1,\ \ldots,\ k_s}_{p_1,\ \ldots,\ p_s}\{f_{\mathbf{v},\mathbf{w}}\} = \dot{\mathbf{m}}^{k_1}_{p_1}\dot{\mathbf{m}}^{k_2}_{p_2}\cdots\dot{\mathbf{m}}^{k_s}_{p_s}\{f_{\mathbf{v},\mathbf{w}}\} =$$
$$= \mathbf{v}^{(p_1k_1+\ldots+p_sk_s)}|\mathbf{v}|^{2(k_1+\ldots+k_s)}\dot{\mathbf{m}}^{k_1}_{p_1}\dot{\mathbf{m}}^{k_2}_{p_2}\cdots\dot{\mathbf{m}}^{k_s}_{p_s}\{f\}$$

with respect to the group $\mathbf{Aff}(\mathcal{GC}_2^{Sp})$ *with multiplicators* $\mathbf{v}^{(p_1k_1+\ldots+p_sk_s)}\ |\mathbf{v}|^{2(k_1+\ldots+k_s)}$, *which have* s *free parameters* $k_1,\ldots,k_s$.

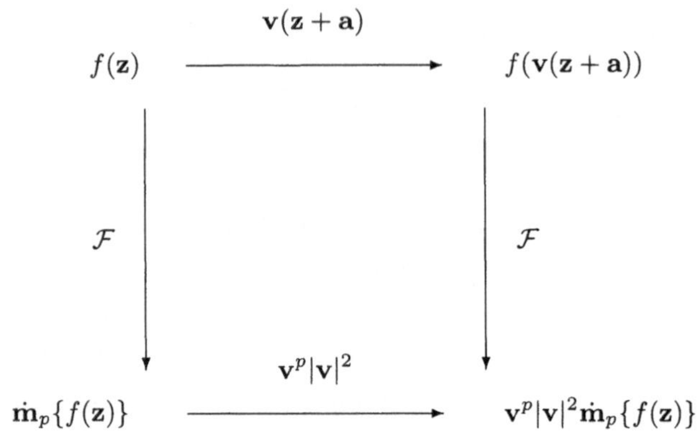

Fig. 7.6. Transformation of $\mathcal{A}_2^{Sp}$–valued moments with respect to the group $\mathbf{Aff}(\mathcal{GC}_2^{Sp})$

The appropriate choice of these parameters can make the multiplicators equal to 1.

Definition 7.5.6. *S–ary moments of the form* $\dot{\mathbf{m}}_{p_1}^{k_1}\dot{\mathbf{m}}_{p_2}^{k_2}\cdots\dot{\mathbf{m}}_{p_s}^{k_s}$, *where* $k_1p_1+\ldots+k_sp_s=0$ *and* $k_1+\ldots+k_s=0$, *are called normalized central one–index moments.*

Normalized central one–index moments are by definition absolute $\mathcal{A}_2^{Sp}$-valued invariants of the affine group $\mathbf{Aff}(\mathcal{GC}_2^{Sp})$. Being invariants, they will be denoted as $\mathcal{I}_{p_1,\ \ldots,\ p_s}^{k_1,\ \ldots,\ k_s}\{f\}$.

The system of equations $k_1p_1+\ldots+k_sp_s=0, \quad k_1+\ldots+k_s=0$ has no solution for $s=1,2$ in the ring $\mathbf{Z}$ and in the field $\mathbf{Q}$.

Theorem 7.5.9. *2–D images have no absolute* $\mathcal{A}_2^{Sp}$*–valued invariants among unary and binary normalized central moments and have such only among s–ary* ($s\geq 3$) *moments with respect to the affine group* $\mathbf{Aff}(\mathcal{A}_2^{Sp})$.

Example 7.5.1. For $s=3$ we have

$$\mathcal{I}_{p_1,\ \ p_2,\ \ p_3}^{\frac{p_3-p_2}{p_2-p_1},\frac{p_3-p_1}{p_2-p_1},1}\{f\}:=\left(\dot{\mathbf{m}}_{p_1}\right)^{\frac{p_3-p_2}{p_2-p_1}}\left(\dot{\mathbf{m}}_{p_2}\right)^{\frac{p_3-p_1}{p_2-p_1}}\left(\dot{\mathbf{m}}_{p_3}\right)^{1}\{f\},$$

where $k_1=\frac{p_3-p_2}{p_2-p_1}$, $k_2=\frac{p_3-p_1}{p_2-p_1}$, $k_3=1$.

Now, let us return to binary relative invariants. We will show that among them there are absolute $\mathcal{A}_2^{Sp}$–valued invariants of the most important subgroups of affine group.

1. For the "small affine subgroup" $\mathbf{aff}(\mathcal{GC}_2^{Sp}):=\mathbf{M}*\mathbf{Tr}(\mathcal{GC}_2^{Sp})$ in (7.16) we have $\mathbf{v}=\lambda\in\mathbb{R}$, where $*$ is the symbol of the semidirect product. Hence, the central moments of the image $f(\mathbf{z})$ are relative $\mathcal{A}_2^{Sp}$–valued invariants

$$\mathcal{J}_p\{f_{\lambda,\mathbf{w}}\} := \dot{\mathbf{m}}_p\{f_{\lambda,\mathbf{w}}\} = \lambda^{p+2}\dot{\mathbf{m}}_p\{f\} \tag{7.17}$$

with respect to the group $\mathbf{aff}(\mathcal{GC}_2^{Sp})$ with scalar-valued multiplicators λ^{p+2}.

Theorem 7.5.10. *The binary moments of the form*

$$\mathcal{I}_{p_1,p_2}^{k_1,k_2}\{f\} := \dot{\mathbf{m}}_{p_1}^{k_1}\dot{\mathbf{m}}_{p_2}^{k_2}\{f\} = \dot{\mathbf{m}}_{p_1}^{k_1}\dot{\mathbf{m}}_{p_2}^{-\frac{p_1+2}{p_2+2}}\{f\}$$

are absolute $\mathcal{A}_2^{Sp}$–valued invariants of the group $\mathbf{aff}(\mathcal{GC}_2^{Sp})$*, where $(p_1+2)k_1+(p_2+2)k_2 = 0$. In particular,*

$$\mathcal{I}_{p,0}^{1,(p+2)/2}\{f\} := \dot{\mathbf{m}}_p^1\{f\}/\dot{\mathbf{m}}_0^{(p+2)/2}\{f\}$$

are such invariants, where $p_1 := p$, $k_1 = 1$, $p_2 = 0$.

2. For the subgroup of motion $\mathbf{Mov}(\mathcal{GC}_2^{Sp})$ in (7.16) we have $\mathbf{v} = e^{I\varphi}$ and $|\mathbf{v}| = 1$. Hence, the central moments of the image $f(\mathbf{z})$ are relative $\mathcal{A}_2^{Sp}$–valued invariants

$$\mathcal{J}_p\{f_{\varphi,\mathbf{w}}\} := \dot{\mathbf{m}}_p\{f_{\varphi,\mathbf{w}}\} = e^{Ip\varphi}\dot{\mathbf{m}}_p\{f\} \tag{7.18}$$

of this subgroup with $\mathcal{A}_2^{Sp}$–valued multiplicators $e^{Ip\varphi}$.

Theorem 7.5.11. *The binary moments of the form*

$$\mathcal{I}_{p_1,p_2}^{k_1,-p_1k_1/p_2}\{f\} := \dot{\mathbf{m}}_{p_1}^{k_1}\{f\}/\dot{\mathbf{m}}_{p_2}^{p_1k_1/p_2}\{f\}$$

are absolute $\mathcal{A}_2^{Sp}$–valued invariants with respect to $\mathbf{Mov}(\mathcal{GC}_2^{Sp})$*. In particular, absolute values of central moments $\mathcal{I}_{p,p}^{1,-1} := |\dot{\mathbf{m}}_p\{f_{v,w}\}| = |\dot{\mathbf{m}}_p\{f\}|$ are real–valued invariants with respect to the group* $\mathbf{Mov}(\mathcal{A}_2^{Sp})$.

7.5.3 Triplet Moments and Invariants of Color Images

Changes in the surrounding world as such of intensity, color or illumination can be treated in the language of the triplet algebra as action of some transformation group in the perceivable color space.

Let $\mathcal{A} = a\mathbf{e} + \mathbf{AE}$ be an arbitrary triplet number. Let us clarify the rules of moment transformations under distortions of chromaticity and geometry of initial images. If $\mathbf{f}_{col}(\mathbf{z})$ is the initial image then

$$_{\mathbf{v},\mathbf{w}}f_{col}^{\mathcal{A}}(z) = \mathbf{f}_{col}^{\mathcal{A}}(\mathbf{v}(\mathbf{z}+\mathbf{w})) = a\mathbf{f}_I(v(z+w)\mathbf{e} + \mathbf{A}\mathbf{f}_{Ch}(\mathbf{v}(\mathbf{z}+\mathbf{w})\mathbf{E}$$

denotes its distorted copy.

Here we need a correct definition of the multiplication of spatial numbers by colors numbers. Let $\mathbf{z} \in \mathcal{A}_2^{Sp}$ be spatial numbers and $\mathcal{A} \in \mathcal{A}_3^{Col}$ be color triplet numbers, then all products $\mathbf{z}\mathcal{A}$ will be called *spatial–color numbers.* They form a *space–color algebra* (see (7.4))

$$\mathcal{A}_6^{SpCol} := \mathcal{A}_2^{Sp} \otimes \mathcal{A}_3^{Col} = \mathcal{A}_2^{Sp} \otimes \left(\mathbb{R}\mathbf{e}_{lu} \oplus \mathcal{A}_2^{ch}\mathbf{E}_{ch}\right) =$$

$$= \left(\mathcal{A}_2^{Sp} \otimes \mathbb{R}\right)\mathbf{e}_{lu} \oplus \left(\mathcal{A}_2^{Sp} \otimes \mathcal{A}_2^{ch}\right)\mathbf{E}_{ch} =$$

$$= \mathcal{A}_2^{Sp}\mathbf{e}_{lu} \oplus \left(\mathcal{A}_2^{Sp} \otimes \mathcal{A}_2^{ch}\right)\mathbf{E}_{ch} = \mathcal{A}_2^{Sp}\mathbf{e}_{lu} \oplus \mathcal{A}_4^{SpCh}\mathbf{E}_{ch},$$

where $\mathcal{A}_4^{SpCh}\mathbf{E}_{ch} := (\mathcal{A}_2^{Sp} \otimes \mathcal{A}_2^{ch})\mathbf{E}_{ch}$ and $\otimes$ is the tensor product.

We suppose that all *spatial hyperimaginary units* commute with all *chromaticity units*. Therefore, $\mathcal{A}_6^{SpCol} = \mathcal{A}_2^{Sp} \otimes \mathcal{A}_3^{Col} = \mathcal{A}_3^{Col} \otimes \mathcal{A}_2^{Sp}$.

Definition 7.5.7. *If* $\mathbf{c}$ *is the centroid of the image* $\mathbf{f}_{col}$*, then the functionals*

$$\dot{\mathcal{M}}_p := \int\limits_{\mathbf{z}\in\mathcal{A}_2^{Sp}} (\mathbf{z}-\mathbf{c})^p \mathbf{f}_{col}(\mathbf{z})d\mathbf{z}$$

$$= \left(\int\limits_{\mathbf{z}\in\mathcal{A}_2^{Sp}} (\mathbf{z}-\mathbf{c})^p \mathbf{f}_{lu}(\mathbf{z})d\mathbf{z} \right)\mathbf{e}_{lu} + \left(\int\limits_{\mathbf{z}\in\mathcal{A}_2^{Sp}} (\mathbf{z}-\mathbf{c})^q \mathbf{f}_{ch}(\mathbf{z})d\mathbf{z} \right)\mathbf{E}_{ch} =$$

$$= \dot{\mathbf{m}}_p\mathbf{e}_{lu} + \dot{\mathbf{M}}_p\mathbf{E}_{ch}$$

are called central $\mathcal{A}_6^{SpCol}$*–valued moments of the color image* $\mathbf{f}_{col}(\mathbf{z})$*, where* $\dot{\mathbf{m}}_p$ *are* $\mathcal{A}_2^{Sp}$*–valued central moments of the intensity term and* $\dot{\mathbf{M}}_p$ *are* $\mathcal{A}_4^{SpCh}$*–valued central moments of the cromaticity term.*

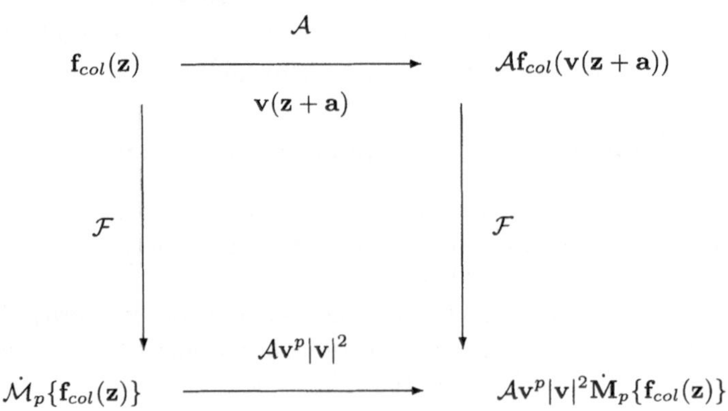

Fig. 7.7. Transformations of relative $\mathcal{A}_6^{SpCol}$–valued moments with respect to the geometrical group $\mathbf{Aff}(\mathcal{GC}_2^{Sp})$ and the color group of similarities $\mathbf{M} \times \mathbf{SO}(\mathcal{A}_3^{Col})$ of the perceptual color space.

Theorem 7.5.12. *The central moments $\dot{\mathcal{M}}_p$ of the color image $\mathbf{f}_{col}(\mathbf{z})$ are relative*
$\mathcal{A}_6^{SpCol}$*–valued invariants*

$$\mathcal{J}_p\{_{\mathbf{v},\mathbf{a}}\mathbf{f}_{col}^{\mathcal{A}}\} := \dot{\mathcal{M}}_p\{_{\mathbf{v},\mathbf{a}}\mathbf{f}_{col}^{\mathcal{A}}\} = \mathcal{A}\mathbf{v}^p|\mathbf{v}|^2\dot{\mathcal{M}}_p\{\mathbf{f}\} \tag{7.19}$$

with respect to both the geometrical group $\mathbf{Aff}(\mathcal{GC}_2^{Sp})$ *and the color group of similarities* $\mathbf{M}\times\mathbf{SO}(\mathcal{A}_3^{Col})$ *of the perceptual color space with* $\mathcal{A}_6^{SpCol}$*–valued multiplicators* $\mathcal{A}\mathbf{v}^p|\mathbf{v}|^2$ *(see fig. 7.7).*

Definition 7.5.8. *The products* $\dot{\mathcal{M}}_{p_1}^{k_1}\dot{\mathcal{M}}_{p_2}^{k_2}\cdots\dot{\mathcal{M}}_{p_s}^{k_s}$ *are called s–ary* $\mathcal{A}_6^{SpCol}$*–valued central moments, where* $k_i\in\mathbf{Q}$.

Theorem 7.5.13. *The s–ary central moments of the image* $f(\mathbf{z})$ *are relative* $\mathcal{A}_6^{SpCol}$*–valued invariants*

$$\mathcal{J}_{p_1,\ldots,p_s}^{k_1,\ldots,k_s}\{\ _{\mathbf{v},\mathbf{a}}\mathbf{f}_{col}^{\mathcal{A}}\} = \dot{\mathcal{M}}_{p_1}^{k_1}\dot{\mathcal{M}}_{p_2}^{k_2}\cdots\dot{\mathcal{M}}_{p_s}^{k_s}\{_{\mathbf{v},\mathbf{a}}\mathbf{f}_{col}^{\mathbf{A}}\} =$$
$$= \mathbf{v}^{(p_1k_1+\ldots+p_sk_s)}(\mathcal{A}|\mathbf{v}|^2)^{(k_1+\ldots+k_s)}\dot{\mathcal{M}}_{p_1}^{k_1}\dot{\mathcal{M}}_{p_2}^{k_2}\cdots\dot{\mathcal{M}}_{p_s}^{k_s}\{\mathbf{f}_{col}\}$$

with respect to the groups $\mathbf{Aff}(\mathcal{GC}_2^{Sp})$ *and* $\mathbf{M}\times\mathbf{SO}(\mathcal{A}_3^{Col})$ *with multiplicators* $\mathbf{v}^{(p_1k_1+\ldots+p_sk_s)}(\mathcal{A}|\mathbf{v}|^2)^{(k_1+\ldots+k_s)}$*, which have s free parameters* $k_1,\ldots,k_s$.

Definition 7.5.9. *The S–ary moments of the form* $\dot{\mathcal{M}}_{p_1}^{k_1}\dot{\mathcal{M}}_{p_2}^{k_2}\cdots\dot{\mathcal{M}}_{p_s}^{k_s}$*, where* $k_1p_1+\ldots+k_sp_s=0$*, and* $k_1+\ldots+k_s=0$*, are called normalized central one–index moments.*

Normalized central one–index moments are by definition absolute complex-valued invariants with respect to the groups $\mathbf{Aff}(\mathcal{GC}_2^{Sp})$ and $\mathbf{M}\times\mathbf{SO}(\mathcal{A}_3^{Col})$. Being invariants, they will be denoted as $\mathcal{I}_{p_1,\ldots,p_s}^{k_1,\ldots,k_s}\{\mathbf{f}_{col}\}$.

The system of equations $k_1p_1+\ldots+k_sp_s=0$, $k_1+\ldots+k_s=0$, has no solution for $s=1,2$ in the ring $\mathbf{Z}$ and in the field $\mathbf{Q}$.

Theorem 7.5.14. *2–D color images have no absolute* $\mathcal{A}_6^{SpCol}$*–valued invariant among unary and binary moments with respect to the groups* $\mathbf{Aff}(\mathcal{GC}_2^{Sp})$, $\mathbf{M}\times\mathbf{SO}(\mathcal{A}_3^{Col})$ *and have such among only s–ary* ($s\geq 3$) *moments.*

Example 7.5.2. For $s=3$ we have

$$\mathcal{I}_{p_1,\ \ p_2,\ \ p_3}^{\frac{p_3-p_2}{p_2-p_1},\frac{p_3-p_1}{p_2-p_1},1}\{\mathbf{f}_{col}\} := \left(\dot{\mathcal{M}}_{p_1}\right)^{\frac{p_3-p_2}{p_2-p_1}}\dot{\mathcal{M}}_{p_2}^{\frac{p_3-p_1}{p_2-p_1}}\left(\dot{\mathcal{M}}_{p_3}\right)^1\{\mathbf{f}_{col}\},$$

where $k_1=\frac{p_3-p_2}{p_2-p_1}$, $k_2=\frac{p_3-p_1}{p_2-p_1}$, $k_3=1$.

Now, we return to binary relative invariants. We will show that among them there are absolute invariants of the most important subgroups of the geometrical and chromaticity affine groups.

1. For the "small geometrical affine subgroup" $\mathbf{aff}(\mathcal{GC}_2^{Sp}) := \mathbf{M}*\mathbf{Tr}(\mathcal{GC}_2^{Sp})$ and for the color group of similarities $\mathbf{M}\times\mathbf{SO}(\mathcal{A}_3^{Col})$ in (7.19) we have

$\mathbf{v} = \lambda \in \mathbb{R}$. Hence, the central moments of the image $\mathbf{f}_{col}(\mathbf{z})$ are relative $\mathcal{A}_6^{SpCol}$–valued invariants

$$\mathcal{J}_p\{_{\lambda,\mathbf{a}}\mathbf{f}_{col}^{\mathcal{A}}\} := \dot{\mathcal{M}}_p\{_{\lambda,\mathbf{a}}\mathbf{f}_{col}^{\mathcal{A}}\} = \mathcal{A}\lambda^{p+2}\dot{\mathcal{M}}_p\{\mathbf{f}_{col}\} \tag{7.20}$$

with respect to these groups with triplet–valued multiplicators $\mathcal{A}\lambda^{p+2}$.

Let by definition $\dot{\mathcal{N}}_p := \dot{\mathcal{M}}_p/\dot{\mathcal{M}}_0$, then from (7.19) we have

$$\mathcal{N}_p\{_{\lambda,\mathbf{w}}\mathbf{f}_{col}^{\mathcal{A}}\} := \mathcal{A}\lambda^{p+2}\dot{\mathcal{M}}_p\{\mathbf{f}_{col}\}/\mathcal{A}\lambda^2\mathcal{M}_0\{\mathbf{f}_{col}\} = \lambda^{p+2}\dot{\mathcal{N}}_p\{\mathbf{f}_{col}\}. \tag{7.21}$$

Theorem 7.5.15. *The binary moments*

$$\mathcal{I}_{p_1,p_2}^{k_1,k_2}\{\mathbf{f}_{col}\} := \dot{\mathcal{N}}_{p_1}^{k_1}\dot{\mathcal{N}}_{p_2}^{k_2}\{\mathbf{f}_{col}\} = \dot{\mathcal{N}}_{p_1}^{k_1}\dot{\mathcal{N}}_{p_2}^{-\frac{p_1}{p_2}}\{\mathbf{f}_{col}\},$$

where $p_1k_1 + p_2k_2 = 0$, are absolute $\mathcal{A}_6^{SpCol}$–valued invariants with respect to the groups $\mathbf{Aff}(\mathcal{GC}_2^{Sp})$ and $\mathbf{M} \times \mathbf{SO}(\mathcal{A}_3^{Col})$.

2. For the geometrical group $\mathbf{Mov}(\mathcal{GC}_2^{Sp})$ and the color group $\mathbf{SO}(\mathcal{A}_3^{Col})$ in (7.19) we have $\mathbf{v} = e^{I\varphi}$, $|\mathbf{v}| = 1$ and $\mathcal{A} = 1\mathbf{e}_{lu} + \mathbf{A}\mathbf{E}_{ch}$, where $|\mathbf{A}| = 1$. Hence, the central moments of the image $f_{col}(\mathbf{z})$ are relative $\mathcal{A}_6^{SpCol}$–valued invariants of the form

$$\mathcal{J}_p\{_{\varphi,\mathbf{a}}\mathbf{f}_{col}\} = \dot{\mathcal{M}}_p\{_{\varphi,\mathbf{a}}\mathbf{f}_{col}\} = \mathcal{A}e^{Ip\varphi}\dot{\mathcal{N}}_p\{\mathbf{f}_{col}\} \tag{7.22}$$

with respect to these groups with $\mathcal{A}_6^{SpCol}$–valued multiplicators $\mathcal{A}e^{Ip\varphi}$, where $\mathcal{A} \in \mathcal{A}_3^{Col}$ and $e^{Ip\varphi} \in \mathcal{A}_2^{Sp}$. From (7.22) we have

$$\dot{\mathcal{N}}_p\{_{\lambda,\mathbf{a}}\mathbf{f}_{col}^{\mathcal{A}}\} = \mathcal{A}e^{Ip\varphi}\dot{\mathcal{M}}_p\{\mathbf{f}_{col}\}/\mathcal{A}\mathcal{M}_0\{\mathbf{f}_{col}\} = e^{Ip\varphi}\dot{\mathcal{N}}_p\{\mathbf{f}_{col}\}. \tag{7.23}$$

Theorem 7.5.16. *The binary moments*

$$\overline{\mathcal{I}}_{p_1,p_2}^{k_1,k_2}\{\mathbf{f}_{col}\} := \dot{\mathcal{N}}_{p_1}^{k_1}\dot{\mathcal{N}}_{p_2}^{k_2}\{\mathbf{f}_{col}\} = \dot{\mathcal{N}}_{p_1}^{k_1}\dot{\mathcal{N}}_{p_2}^{-\frac{p_1}{p_2}}\{\mathbf{f}_{col}\},$$

where $p_1k_1 + p_2k_2 = 0$, are absolute $\mathcal{A}_6^{SpCol}$–valued with respect to the geometrical group $\mathbf{Mov}(\mathcal{GC}_2^{Sp})$ and the color group $\mathbf{SO}(\mathcal{A}_3^{Col})$, respectively. In particular, the absolute values of the central moments $|\mathcal{I}_{p,p}^{1,-1}|\{_{v,a}\mathbf{f}_{col}\} := |\dot{\mathcal{N}}_p|\{_{v,a}\mathbf{f}_{col}\}| = |\dot{\mathcal{N}}_p|\{\mathbf{f}_{col}\}$ are real–valued invariants.

7.5.4 Quaternionic Moments and Invariants of 3–D Images

Let be $\mathbf{f}_{col}(\mathbf{q})$ a color 3-D image depending on the pure vector generalized quaternion $\mathbf{q} \in \mathbf{Vec}\{\mathcal{A}_4^{Sp}\} = \mathcal{GR}_3$.

If $\mathbf{q} \in \mathcal{GR}_3 \subset \mathcal{A}_4^{Sp}$ are generalized quaternions and $\mathcal{A} \in \mathcal{A}^{Col}$ are color numbers then all products of the form $\mathbf{q}\mathcal{A}$ will be called *spatial–color numbers.* They form a *space–color algebra*

$$\mathcal{A}^{SpCol} := \mathcal{A}_4^{Sp} \otimes \mathcal{A}^{Col} = \mathcal{A}^{Col} \otimes \mathcal{A}_4^{Sp},$$

where all spatial hyperimaginary units commute with all color units. Here, $\mathcal{A}^{SpCol} := \mathcal{A}_{12}^{SpCol} = \mathcal{A}_4^{Sp} \otimes \mathcal{A}_3^{Col}$ if a 3–cycle model of the color image is used and $\mathcal{A}^{Sp} := \mathcal{A}_{16}^{SpCol} = \mathcal{A}_4^{Sp} \otimes \mathcal{A}_4^{Col}$ if a spin–valued model of the color image is used.

Definition 7.5.10. *If* $\mathbf{c}$ *is the centroid of the image* $\mathbf{f}_{col}(\mathbf{q})$, *then the functionals of the form*

$$\dot{\mathcal{M}}_p\{\mathbf{f}_{col}\} := \int_{\mathbf{q}\in\mathcal{GR}^3} (\mathbf{q} - \mathbf{c}) f_{col}(\mathbf{q})\mathbf{dq} \tag{7.24}$$

are called one–index central $\mathcal{A}^{SpCol}$*–valued moments of the 3-D color image* $\mathbf{f}_{col}(\mathbf{q})$, *where* $p \in \mathbf{Q}$, $\mathbf{dq} := dxdydz$.

Now, let us find rules of moments changing with respect to geometrical and chromatical distortions of the initial image:

$$\mathbf{f}_{\lambda\mathcal{Q}\mathbf{a}}^{\mathcal{A}}(\mathbf{q}) := \mathcal{A}\mathbf{f}_{col}(\lambda\mathcal{Q}(\mathbf{q}+\mathbf{a})\mathcal{Q}^{-1}), \tag{7.25}$$

where $\mathbf{q}^* := \lambda\mathcal{Q}(\mathbf{q}+\mathbf{a})\mathcal{Q}^{-1}$, $\mathcal{A} \in \mathcal{A}_3^{Col}$ (or $\mathcal{A} \in \mathcal{A}_4^{Col}$), $\mathcal{A} = |\mathcal{A}|e^{u\varphi}$.

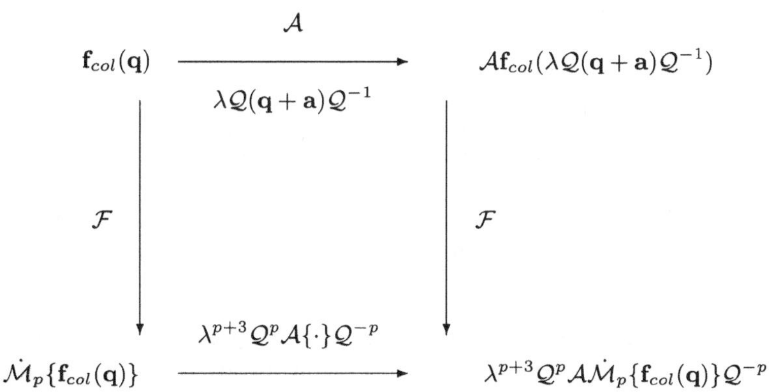

Fig. 7.8. Transformation of $\mathcal{A}^{SpCol}$–valued moments with respect to the groups $\mathbf{Aff}(\mathcal{GR}_3)$ and $\mathbf{M} \times \mathbf{SO}(\mathcal{A}^{Col})$

Theorem 7.5.17. *The central moments* $\dot{\mathcal{M}}_p\{\mathbf{f}_{col}\}$ *are relative* $\mathcal{A}^{SpCol}$*–valued invariants*

$$\mathcal{J}_p\left\{\mathbf{f}_{\lambda\mathcal{Q}\mathbf{a}}^{\mathcal{A}}\right\} := \dot{\mathcal{M}}_p\{\mathbf{f}_{\lambda\mathcal{Q}\mathbf{a}}^{\mathcal{A}}\} = \mathcal{A}\lambda^{p+3}\mathcal{Q}^p\dot{\mathcal{M}}_p\{f\}\mathcal{Q}^{-p}$$

with respect to the groups $\mathbf{Aff}(\mathcal{GR}_3)$ *and* $\mathbf{M} \times \mathbf{SO}(\mathcal{A}^{Col})$ *with left* $\mathcal{A}^{SpCol}$*–valued multiplicators* $\mathcal{A}\lambda^{p+3}\mathcal{Q}^p$ *and with right Spin*$(\mathcal{A}_4^{Sp})$*–valued multiplicators* $\mathcal{Q}^{-p}$, *respectively (see fig. 7.8).*

As relative invariants obtain both left and right multiplicators, common multiplication does not result in absolute invariants from s–ary moments. Hence, 3–D k–color image has no absolute invariants among s–ary $\mathcal{A}^{SpCol}$–valued moments with respect to the groups $\mathbf{Aff}(\mathcal{GR}_3)$ and $\mathbf{M} \times \mathbf{SO}(\mathcal{A}^{Col})$.

Theorem 7.5.18. *The modulus of the central moments $\dot{\mathcal{M}}_p\{f\}$ are the relative $\mathbb{R}$–valued invariants*

$$|\mathcal{J}_p\{\mathbf{f}^{\mathcal{A}}_{\lambda\mathcal{Q}\mathbf{a}}(\mathbf{q})\}| := \dot{\mathcal{M}}_p\{\mathbf{f}^{\mathcal{A}}_{\lambda\mathcal{Q}\mathbf{a}}\} = \lambda^{p+3}|\mathcal{A}||\dot{\mathcal{M}}_p\{f\}| \tag{7.26}$$

with respect to the groups $\mathbf{Aff}(\mathcal{GR}_3)$, $\mathbf{M} \times \mathbf{SO}(\mathcal{A}^{Col})$ with scalar–valued multiplicators $\lambda^{p+3}|\mathcal{A}|$.

Let by definition $\dot{\mathcal{N}}_p := |\dot{\mathcal{M}}_p|/|\dot{\mathcal{M}}_0|$, then from (7.26) we have

$$\mathcal{N}_p\{_{\lambda,\mathbf{a}}\mathbf{f}^{\mathcal{A}}_{col}\} := \mathcal{A}\lambda^{p+3}|\dot{\mathcal{M}}_p|\{\mathbf{f}_{col}\}/\mathcal{A}\lambda^3|\dot{\mathcal{M}}_0|\{\mathbf{f}_{col}\} = \lambda^p\dot{\mathcal{N}}_p\{\mathbf{f}_{col}\}. \tag{7.27}$$

Definition 7.5.11. *The products $\dot{\mathcal{N}}^{k_1}_{p_1}\dot{\mathcal{N}}^{k_2}_{p_2}\cdots\dot{\mathcal{N}}^{k_s}_{p_s}$ are called s–ary $\mathbb{R}$–valued modular central moments, where $k_i \in \mathbf{Q}$.*

Theorem 7.5.19. *The s–ary modular central moments of the image $\mathbf{f}_{col}(\mathbf{z})$ are relative $\mathbb{R}$–valued invariants*

$$|\mathcal{J}^{k_1,\ldots,k_s}_{p_1,\ldots,p_s}|\{f_{\lambda\mathcal{Q}\mathbf{a}}\} = \dot{\mathcal{N}}^{k_1}_{p_1}\cdots\dot{\mathcal{N}}^{k_s}_{p_s}\{\mathbf{f}^{\mathcal{A}}_{\lambda\mathcal{Q}\mathbf{a}}\} = \lambda^{p_1k_1+\ldots+p_sk_s}\dot{\mathcal{N}}^{k_1}_{p_1}\cdots\dot{\mathcal{N}}^{k_s}_{p_s}\{\mathbf{f}_{col}\}$$

with respect to the groups $\mathbf{Aff}(\mathcal{GR}_3)$ and $\mathbf{M} \times \mathbf{SO}(\mathcal{A}^{Col})$ and with scalar–valued multiplicators $\lambda^{p_1k_1+\cdots+p_sk_s}$, which have s free parameters $k_1,\ldots,k_s$.

Definition 7.5.12. *The modulus of s–ary moments of the form $\dot{\mathcal{N}}^{k_1}_{p_1}\cdots\dot{\mathcal{N}}^{k_s}_{p_s}$, where $p_1k_1+\ldots+p_sk_s = 0$, are called normalized modular central moments.*

The normalized modular central moments are by definition absolute $\mathbb{R}$-valued invariants with respect to the groups $\mathbf{Aff}(\mathcal{GR}_3)$ and $\mathbf{M} \times \mathbf{SO}(\mathcal{A}^{Col})$.

The normalized modular moments being invariants will be denoted as $|\mathcal{I}^{k_1,\ldots,k_s}_{p_1,\ldots,p_s}\{\mathbf{f}_{col}\}|$.

The system of equations $k_1p_1 + \ldots + k_sp_s = 0$ has an infinite number of solutions for $s \geq 2$ in the ring $\mathbf{Z}$ and in the field $\mathbf{Q}$, for example, $k_1 = -k_2\frac{p_1+3}{p_2+3}$.

Theorem 7.5.20. *The binary modular moments*

$$\mathcal{I}^{k_1,-k_1\frac{p_1+3}{p_2+3}}_{p_1,p_2}\{\mathbf{f}_{col}(\mathbf{q})\} := |\dot{\mathcal{M}}^{k_1}_{p_1}|/|\dot{\mathcal{M}}^{k_1\frac{p_1+3}{p_2+3}}_{p_2}|$$

are absolute $\mathbb{R}$–valued invariants with respect to the groups $\mathbf{Aff}(\mathcal{GR}_3)$ and $\mathbf{M} \times \mathbf{SO}(\mathcal{A}^{Col})$.

Analogously we can construct absolute invariants for important subgroups of these groups.

7.5.5 Hypercomplex–Valued Invariants of n–D Images

Let us assume that $\mathbf{f}_{mcol}(\mathbf{x}) : \mathbb{R}^n \longrightarrow \mathbb{R}^k$ is an image of some k-color n–D signal as an element of the image space $\mathbf{L}(\mathbb{R}^n, \mathbb{R}^k)$, and $\mathcal{J}$ is a descriptor of some feature of the image as an element of some spatial–multicolor Clifford algebra

$$\mathcal{A}_{2^{k+n}}^{SpCol} = \mathcal{A}_{2^n}^{p,q,r} \otimes {}^{p,q,r}\mathcal{A}_{2^k}^{Col},$$

where all spatial hyperimaginary units commute with all color units. Here, $\mathcal{A}^{SpCol} := \mathcal{A}_{2^n k}^{SpCol} = \mathcal{A}_{2^n}^{Sp} \otimes \mathcal{A}_k^{Col}$ if a k–cycle model of the color image is used and $\mathcal{A}^{SpCol} := \mathcal{A}_{2^{n+k}}^{SpCol} = \mathcal{A}_{2^n}^{Sp} \otimes \mathcal{A}_{2^k}^{Col}$ if a spin–valued model of the color image is used.

Formally we can write $\mathcal{J} = \mathcal{F}\big[f(\mathbf{x})\big]$, where $\mathcal{F} : \mathbf{L}(\mathbb{R}^n, \mathbb{R}^k) \longrightarrow \mathcal{A}^{SpCol}$ represents a feature map. In $\mathbb{R}^n$ let some group of transformations act: $\mathbf{G} : \mathbb{R}^n \longrightarrow \mathbb{R}^n$.

Definition 7.5.13. *The $\mathcal{A}^{SpCol}$–valued functional $\mathcal{J} = \mathcal{F}\big[\mathbf{f}_{mcol}(\mathbf{x})\big]$ of the image $\mathbf{f}_{mcol}(\mathbf{x})$ is called relative $\mathbf{G}$-invariant with respect to some transformation group $\mathbf{G}$ if*

$$\mathcal{J} = \mathcal{F}\big[f(\mathbf{g} \circ \mathbf{x})\big] = \mathcal{C}(\mathbf{g}) \cdot \mathcal{F}\big[f(\mathbf{x})\big] \cdot \mathcal{C}^{-1}(\mathbf{g}), \ \forall \mathbf{g} \in \mathbf{G},$$

where $\mathcal{C}, \mathcal{C}^{-1}$ are left and right $\mathcal{A}^{SpCol}$–valued multiplicators. If $\mathcal{C} = 1$ then $\mathcal{J}$ is called absolute invariant and is denoted as $\mathcal{I}$.

The function $\mathbf{f}_{mcol}(x_1, \dots, x_n)$ can be considered as a function of Clifford variables: $\mathbf{f}_{mcol}(\mathbf{x})$, $\mathbf{x} \in \mathrm{Vec}^1(\mathcal{A}_{2^n}^{Sp}) := \mathcal{GR}_n$.

Definition 7.5.14. *If $\mathbf{c}$ is the centroid of the image $\mathbf{f}_{mcol}(\mathbf{x})$, then the functionals*

$$\dot{\mathcal{M}}_p\{\mathbf{f}_{mcol}\} := \int_{\mathbf{x} \in \mathcal{GR}_n} (\mathbf{x} - \mathbf{c})^p \mathbf{f}_{mcol}(\mathbf{x}) d\mathbf{x}$$

will be called one–index central $\mathcal{A}^{SpCol}$–valued moments of the n–D k-color image $\mathbf{f}_{mcol}(\mathbf{x})$.

Let us clarify the rules of moment transformations under geometrical and multicolor distortions of the initial images:

$$\mathbf{f}_{\lambda \mathcal{Q} \mathbf{a}}^{\mathcal{A}}(\mathbf{x}) := \mathcal{A} \mathbf{f}_{mcol}\left(\lambda \mathcal{Q}(\mathbf{x} + \mathbf{a}) \mathcal{Q}\right),$$

where λ is a scale factor, $\mathbf{x}^* = r\mathcal{Q}(\mathbf{x} + \mathbf{a})\mathcal{Q}$, $\mathbf{x}^*, \mathbf{x}, \mathbf{a} \in \mathcal{GR}_n^{p,q,r}$, $|\mathcal{Q}| = 1$.

Theorem 7.5.21. *(**Main theorem**). The central moments $\dot{\mathcal{M}}_p\{\mathbf{f}_{mcol}\}$ are the relative $\mathcal{A}^{SpCol}$–valued invariants*

$$\mathcal{J}_p\{\mathbf{f}_{\lambda \mathcal{Q} \mathbf{a}}^{\mathcal{A}}\} := \dot{\mathcal{M}}_p\{\mathbf{f}_{\lambda \mathcal{Q} \mathbf{a}}^{\mathcal{A}}\} = \mathcal{A} \lambda^{p+n} \mathcal{Q}^p \dot{\mathcal{M}}_p\{\mathbf{f}_{col}\} \mathcal{Q}^{-p}$$

with respect to the groups $\mathbf{Aff}(\mathcal{GR}_n)$ and $\mathbf{M} \times \mathbf{SO}(\mathcal{A}^{Col})$ with $\mathcal{A}^{SpCol}$–valued left multiplicators $\mathcal{A}\lambda^{p+3}\mathcal{Q}^p$ and $\mathcal{A}_{2^n}^{Sp}$–valued right multiplicators $\mathcal{Q}^{-p}$, respectively (see fig. 7.9).

$$
\begin{array}{ccc}
\mathbf{f}_{mcol}(\mathbf{x}) & \xrightarrow[\lambda\mathcal{Q}(\mathbf{x}+\mathbf{a})\mathcal{Q}^{-1}]{\mathcal{A}} & \mathcal{A}\mathbf{f}_{mcol}(\lambda\mathcal{Q}(\mathbf{x}+\mathbf{a})\mathcal{Q}^{-1}) \\
\Big\downarrow \mathcal{F} & & \Big\downarrow \mathcal{F} \\
\dot{\mathcal{M}}_p\{\mathbf{f}_{mcol}(\mathbf{x})\} & \xrightarrow{\lambda^{p+n}\mathcal{Q}^p\mathcal{A}\{\cdot\}\mathcal{Q}^{-p}} & \lambda^{p+n}\mathcal{Q}^p\mathcal{A}\dot{\mathcal{M}}_p\{\mathbf{f}_{mcol}(\mathbf{x})\}\mathcal{Q}^{-p}
\end{array}
$$

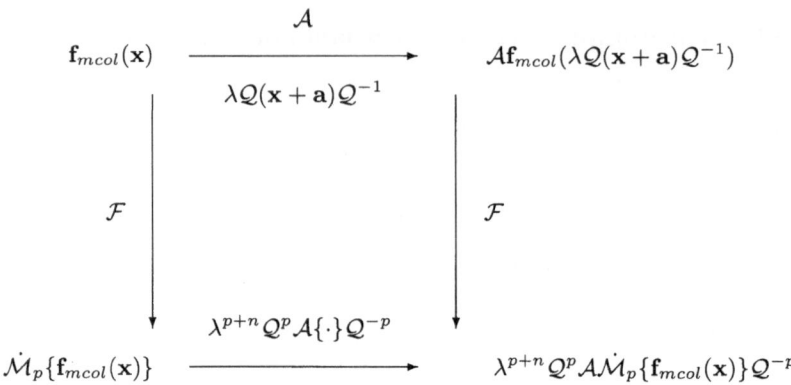

Fig. 7.9. Transformation of $\mathcal{A}^{SpCol}$–valued moments with respect to the groups $\mathbf{Aff}(\mathcal{GR}_n)$ and $\mathbf{M} \times \mathbf{SO}(\mathcal{A}^{Col})$

As relative invariants obtain both left and right multiplicators, common multiplication does not result in absolute invariants from s–ary moments. Hence, a 3–D color image has no absolute invariants among s–ary $\mathcal{A}^{SpCol}$–valued moments with respect to the groups $\mathbf{Aff}(\mathcal{GR}_n)$ and $\mathbf{M} \times \mathbf{SO}(\mathcal{A}^{Col})$.

Theorem 7.5.22. *The modulus of the central moments $\dot{\mathcal{M}}_p\{\mathbf{f}_{mcol}\}$ are the relative $\mathbb{R}$–valued invariants*

$$|\mathcal{J}_p|\{\mathbf{f}^{\mathcal{A}}_{\lambda\mathcal{Q}\mathbf{a}}\} := |\dot{\mathcal{M}}_p|\{\mathbf{f}^{\mathcal{A}}_{\lambda\mathcal{Q}\mathbf{a}}\} = \lambda^{p+3}|\mathcal{A}||\dot{\mathcal{M}}_p|\{\mathbf{f}_{mcol}\} \tag{7.28}$$

with respect to the groups $\mathbf{Aff}(\mathcal{GR}_n)$ and $\mathbf{M} \times \mathbf{SO}(\mathcal{A}^{Col})$ and with scalar–valued multiplicators $\lambda^{p+3}|\mathcal{A}|$.

Let by definition be $\dot{\mathcal{N}}_p := |\dot{\mathcal{M}}_p|/|\dot{\mathcal{M}}_0|$, then from (7.26) we have

$$\mathcal{N}_p\{ {}_{\lambda,\mathbf{a}}\mathbf{f}^{\mathcal{A}}_{mcol}\} :=$$

$$= \mathcal{A}\lambda^{p+3}|\dot{\mathcal{M}}_p|\{\mathbf{f}_{mcol}\}/\mathcal{A}\lambda^3|\mathcal{M}_0|\{\mathbf{f}_{mcol}\} = \lambda^p\dot{\mathcal{N}}_p\{\mathbf{f}_{mcol}\}. \tag{7.29}$$

Definition 7.5.15. *The products $\dot{\mathcal{N}}_{p_1}^{k_1}\dot{\mathcal{N}}_{p_2}^{k_2}\cdots\dot{\mathcal{N}}_{p_s}^{k_s}$ are called s–ary $\mathbb{R}$–valued modular central moments, where $k_i \in \mathbf{Q}$.*

Theorem 7.5.23. *The s–ary modular central moments of the image $\mathbf{f}_{mcol}(\mathbf{x})$ are relative $\mathbb{R}$–valued invariants*

$$|\mathcal{J}^{k_1,\ \dots,\ k_s}_{p_1,\ \dots,\ p_s}|\{f_{\lambda\mathcal{Q}\mathbf{a}}\} =$$

$$= \dot{\mathcal{N}}_{p_1}^{k_1}\cdots\dot{\mathcal{N}}_{p_s}^{k_s}\{\mathbf{f}^{\mathcal{A}}_{\lambda\mathcal{Q}\mathbf{a}}\} = \lambda^{p_1k_1+\ldots+p_sk_s}\dot{\mathcal{N}}_{p_1}^{k_1}\cdots\dot{\mathcal{N}}_{p_s}^{k_s}\{\mathbf{f}_{mcol}\}$$

with respect to the groups $\mathbf{Aff}(\mathcal{GR}_n)$ and $\mathbf{M} \times \mathbf{SO}(\mathcal{A}^{Col})$ and with scalar–valued multiplicators $\lambda^{p_1k_1+\ldots+p_sk_s}$, which have s free parameters $k_1, \ldots, k_s$.

Definition 7.5.16. *The modulus of s–ary moments of the form* $\dot{\mathcal{N}}_{p_1}^{k_1}\cdots\dot{\mathcal{N}}_{p_s}^{k_s}$, *where* $p_1k_1+\ldots+p_sk_s=0$, *are called normalized modular central moments.*

Normalized modular central moments are by definition absolute $\mathbb{R}$-valued invariants with respect to the groups $\mathbf{Aff}(\mathcal{GR}_n)$ and $\mathbf{M}\times\mathbf{SO}(\mathcal{A}^{Col})$.

The normalized modular moments being invariants will be denoted as $|\mathcal{I}_{p_1,\ \ldots,\ p_s}^{k_1,\ \ldots,\ k_s}|\{\mathbf{f}_{mcol}\}$.

The system of equations $k_1p_1+\ldots+k_sp_s=0$ has an infinite number of solutions for $s\geq 2$ in the ring $\mathbf{Z}$ and in the field $\mathbf{Q}$, for example, $k_1=-k_2\frac{p_1+3}{p_2+3}$.

Theorem 7.5.24. *The binary modular moments*

$$\mathcal{I}_{p_1,p_2}^{k_1,-k_1\frac{p_1+3}{p_2+3}}\{\mathbf{f}_{col}(\mathbf{q})\}:=|\dot{\mathcal{M}}_{p_1}^{k_1}|/|\dot{\mathcal{M}}_{p_2}^{k_1\frac{p_1+3}{p_2+3}}|$$

are absolute $\mathbb{R}$*–valued invariants with respect to the groups* $\mathbf{Aff}(\mathcal{GR}_n)$ *and* $\mathbf{M}\times\mathbf{SO}(\mathcal{A}^{Col})$.

Analogously we can construct absolute invariants for important subgroups of these groups.

7.6 Conclusion

We have shown how Clifford algebra can be used in formation and computation of invariants of 2-D, 3-D and n–D color and multicolor objects of different Euclidean and non–Euclidean geometries. The proved theorems show how simple and efficient the methods of calculation of invariants are that use spatial and color Clifford algebra.

Note that digital computers use Boolean algebra. This is a Clifford algebra $\mathcal{A}_{2^n}(\mathbf{GF}(2))$ over the Galois field $\mathbf{GF}(2)$. But are there analog computers working in Clifford algebra $\mathcal{A}_{2^n}^{p,q,r}(\mathbb{R})$? Now we will try to answer this question by asking a question which may be a subtitle of this chapter: "Is the visual cortex a *Clifford Algebra Computer*"? Yes, it can be if this visual cortex belongs to a human being living even in n–D non–Euclidean space $\mathcal{GR}_n^{p,q,r}$. But, how fast can the invariants be calculated? The answer to this question the interested reader can find in [136]–[138], where Fourier–Clifford and Fourier–Hamilton number theoretical transforms are used for this purpose.

8. Non-commutative Hypercomplex Fourier Transforms of Multidimensional Signals*

Thomas Bülow, Michael Felsberg, and Gerald Sommer

Institute of Computer Science and Applied Mathematics,
Christian-Albrechts-University of Kiel

8.1 Introduction

Harmonic transforms, and among those especially the Fourier transform, play an essential role in mathematical analysis, in almost any part of modern physics, as well as in electrical engineering. The analysis of the following four chapters is motivated by the use of the Fourier transform in signal processing. It turns out that some powerful concepts of one-dimensional signal theory can hardly be carried over to the theory of n-dimensional signals by using the complex Fourier transform. We start by introducing and studying the hypercomplex Fourier transforms in the following two chapters. In this chapter representations in non-commutative algebras are investigated, while chapter 9 is concerned with representations in commutative hypercomplex algebras. After these rather theoretical investigations we turn towards practice in chapter 10 where fast algorithms for the transforms are presented and in chapter 11 where local quaternion-valued LSI-filters based on the quaternionic Fourier transform are introduced and applied to image processing tasks.

* This work has been supported by German National Merit Foundation and by DFG Grants So-320-2-1, So-320-2-2, and Graduiertenkolleg No. 357.

In this chapter we will introduce hypercomplex Fourier transforms. The main motivation lies in the following two facts:

1. *The basis functions of the complex Fourier transform of arbitrary dimension n are intrinsically one-dimensional.*
2. *The symmetry selectivity of the 1-D complex Fourier transform is not carried forward completely to n-D by a complex transform.*

The first point refers to the fact that the basis functions of the complex Fourier transform look like plane waves, i.e. they vary along one orientation while being constant within the orthogonal $(n-1)$-dimensional hyperplane. This turns out to be a severe restriction in the analysis of the local structure of multidimensional signals. The implications of introducing transforms with intrinsically multidimensional basis functions will be regarded in chapter 11 for the case $n = 2$. The second point is important since the phase concept depends on the symmetry selectivity of the transform. E.g. the local phase of a signal is defined as the angular phase of the complex number made up of a real filter-response of the locally even signal component and the imaginary filter-response of the locally odd component. Extending the phase concept to higher dimensions we need a representation handling more than two symmetry components separately. This second point is directly related to the first one, since the introduction of a transform with higher symmetry selectivity leads directly to intrinsically multidimensional basis functions.

The structure of this chapter is as follows. In section 8.2 we consider several 1-D harmonic transforms. Among these transforms are those which map real-valued functions to real-valued, to complex-valued, and to vector-valued ones. We compare these transforms and thus motivate the introduction of multidimensional transforms with values in hypercomplex algebras. In 2-D the quaternionic Fourier transform (QFT) is such a transform. The QFT will be introduced and compared to real- and complex-valued transforms in section 8.3. In section 8.4 a hierarchy of 2-D transforms is introduced which is based on the symmetry selectivity of the transforms. Furthermore the main theorems for the QFT like the shift-theorem, Rayleigh's theorem, and some more are proven in this section. The definition of n-D transforms with values in Clifford algebras, i.e. in non-commutative hypercomplex algebras, is given in section 8.5 as an extension of the QFT. Before concluding this chapter we give a short overview of the literature on hypercomplex Fourier transforms in section 8.6 which seems of interest because the field is rather disjointed still.

8.2 1-D Harmonic Transforms

Before delving into the theory of hypercomplex transforms, we present some of the well-known harmonic transforms. In this section we restrict ourselves to 1-D signals and the corresponding 1-D transforms. The signals considered are assumed to be square integrable and real-valued: $f \in L^2(\mathbb{R}, \mathbb{R})$. For these signals the transforms considered in the following are guaranteed to exist.

Definition 8.2.1 (Cosine transform and sine transform).
For $f \in L^2(\mathbb{R}, \mathbb{R})$

$$F_c(u) = 2\int_0^\infty f(x)\cos(2\pi ux)dx$$

is called the cosine transform of f. Analogously

$$F_s(u) = 2\int_0^\infty f(x)\sin(2\pi ux)dx$$

defines the sine transform of f.

The trigonometric transforms as defined above take no account of f to the left of the origin. Thus, since in signal processing we are interested in complete transforms, we modify definition 8.2.1 slightly.

Definition 8.2.2 (C-transform and S-transform). *For $f \in L^2(\mathbb{R}, \mathbb{R})$ we define the two transforms $\mathcal{C} : L^2(\mathbb{R}, \mathbb{R}) \to L^2(\mathbb{R}, \mathbb{R})$ and $\mathcal{S} : L^2(\mathbb{R}, \mathbb{R}) \to L^2(\mathbb{R}, \mathbb{R})$, where*

$$\mathcal{C}\{f\}(u) = C(u) = \int_{\mathbb{R}} f(x)\cos(2\pi ux)dx$$

is called the C-transform of f. Analogously

$$\mathcal{S}\{f\}(u) = S(u) = \int_{\mathbb{R}} f(x)\sin(2\pi ux)dx$$

defines the S-transform of f.

Since each transform takes account either of the even or the odd part of f neither the C- nor the S-transform is invertible. If a complete transform is desired, both transforms have to be combined. We will show three different combinations of the C- and the S-transform, all of which lead to complete and thus invertible transforms.

Definition 8.2.3 (1-D Hartley transform).
Consider $f \in L^2(\mathbb{R}, \mathbb{R})$. Then, $\mathcal{H} : L^2(\mathbb{R}, \mathbb{R}) \to L^2(\mathbb{R}, \mathbb{R})$, with

$$\mathcal{H}\{f\}(u) = H(u) = \mathcal{C}\{f\}(u) + \mathcal{S}\{f\}(u)$$

is called the Hartley transform of f.

Definition 8.2.4 (1-D Fourier transform). *Let $f \in L^2(\mathbb{R}, \mathbb{R})$. Then, $\mathcal{F} : L^2(\mathbb{R}, \mathbb{R}) \to L^2(\mathbb{R}, \mathbb{C})$, with*

$$\mathcal{F}\{f\}(u) = F(u) = \mathcal{C}\{f\}(u) - i\mathcal{S}\{f\}(u)$$

is the Fourier transform of f.

Finally, we introduce a transform which results from combining the C- and the S-transform into a vector.

Definition 8.2.5 (Trigonometric vector transform).
For any square-integrable one-dimensional real signal $f \in L^2(\mathbb{R}, \mathbb{R})$ we define a vector-valued transform $\mathcal{V} : L^2(\mathbb{R}, \mathbb{R}) \to L^2(\mathbb{R}, \mathbb{R}^2)$ by

$$\mathcal{V}\{f\}(u) = V(u) = \begin{pmatrix} \mathcal{C}\{f\}(u) \\ \mathcal{S}\{f\}(u) \end{pmatrix}.$$

Theorem 8.2.1. *The Hartley transform, the Fourier transform, and the trigonometric vector transform are invertible.*

The main difference between the Hartley transform on the one hand and the Fourier and the vector-valued transform on the other hand is that the Hartley transform does not separate even signal components from odd ones while the others do.

A question that often arises when talking about hypercomplex spectral transforms is: *Do we really need this complicated mathematics of hypercomplex algebras? Or can we do the same using real numbers or vectors?* The answer is: We can do the same using real numbers or vectors, but in fact using hypercomplex numbers makes things easier and more natural rather than complicated. This is at least true for the applications we have in mind and which we will demonstrate in the following chapters. We can partly explain this on the example of the complex Fourier transform and the vector-valued transform. Both transforms are complete, both transforms separate even from odd signal components. Thus, insofar the transforms are equivalent to each other. However, there are properties of the transforms which can be expressed very naturally only using the complex transform. For demonstration purpose we merely mention the shift theorem and the Hermite symmetry of a real signal's Fourier transform. The shift theorem of the Fourier transform describes how the transform of a signal varies when the signal is shifted. If the signal f is shifted by d, its Fourier transform is multiplied by a phase factor $\exp(-2\pi i\, du)$. Thus, a shift in spatial domain corresponds to multiplication of the complex transform by a complex number. Expressing this theorem for the vector-valued transform is of course possible. However, the algebraic frame would have to be extended to include not only vectors but also square matrices. The Hermite symmetry of the Fourier transform of a real signal is expressed by $F^*(u) = F(-u)$ which immediately explains the redundancy of the spectrum. There is no special notation for expressing this in vector algebra. These two simple examples already explain why the complex Fourier transform is more convenient than the vector-valued transform. Similar arguments apply for the introduction of hypercomplex numbers for signals of higher dimension.

8.3 2-D Harmonic Transforms

8.3.1 Real and Complex Harmonic Transforms

Again, we start with defining real trigonometric transforms from which we will derive the transforms of interest in this chapter.

Definition 8.3.1.
Let f be a real two-dimensional square-integrable signal $f \in L^2(\mathbb{R}^2, \mathbb{R})$. Then we define the transforms $\mathcal{CC}, \mathcal{SC}, \mathcal{CS}, \mathcal{SS} : L^2(\mathbb{R}^2, \mathbb{R}) \to L^2(\mathbb{R}^2, \mathbb{R})$ by

$$\mathcal{CC}\{f\}(\boldsymbol{u}) = CC(\boldsymbol{u}) = \int_{\mathbb{R}^2} f(\boldsymbol{x}) \cos(2\pi ux) \cos(2\pi vy) d^2\boldsymbol{x} \tag{8.1}$$

$$\mathcal{SC}\{f\}(\boldsymbol{u}) = SC(\boldsymbol{u}) = \int_{\mathbb{R}^2} f(\boldsymbol{x}) \sin(2\pi ux) \cos(2\pi vy) d^2\boldsymbol{x} \tag{8.2}$$

$$\mathcal{CS}\{f\}(\boldsymbol{u}) = CS(\boldsymbol{u}) = \int_{\mathbb{R}^2} f(\boldsymbol{x}) \cos(2\pi ux) \sin(2\pi vy) d^2\boldsymbol{x} \tag{8.3}$$

$$\mathcal{SS}\{f\}(\boldsymbol{u}) = SS(\boldsymbol{u}) = \int_{\mathbb{R}^2} f(\boldsymbol{x}) \sin(2\pi ux) \sin(2\pi vy) d^2\boldsymbol{x}. \tag{8.4}$$

We could have started in def. 8.3.1 with an 2-D C- and S-transform. However, the four transforms allow the construction of more general transforms than the C- and S-transform, which can in fact be constructed from the transforms of def. 8.3.1 by linear combination due to the addition theorem of the sine and cosine function. Actually, the introduction of the four separable transforms is crucial four the following analysis.

As it is possible to construct the 1-D Hartley- and Fourier-transform from the C- and the S-transform, we can combine the separable trigonometric transforms given in def. 8.3.1 in different ways to yield the well-known 2-D spectral transforms.

Definition 8.3.2 (2-D Hartley and Fourier transform). *Let f be a real 2-D signal $f \in L^2(\mathbb{R}^2, \mathbb{R})$. The 2-D Hartley transform of f is then given by*

$$\mathcal{CC}\{f\}(\boldsymbol{u}) + \mathcal{SC}\{f\}(\boldsymbol{u}) + \mathcal{CS}\{f\}(\boldsymbol{u}) + \mathcal{SS}\{f\}(\boldsymbol{u}) = \mathcal{H}\{f\}(\boldsymbol{u}) = H(\boldsymbol{u}).$$

The 2-D Fourier transform of f is

$$\mathcal{CC}\{f\}(\boldsymbol{u}) - \mathcal{SS}\{f\}(\boldsymbol{u}) - i(\mathcal{CS}\{f\}(\boldsymbol{u}) + \mathcal{SC}\{f\}(\boldsymbol{u}) = \mathcal{F}\{f\}(\boldsymbol{u}) = F(\boldsymbol{u}).$$

Definition 8.3.3 (2-D Trigonometric vector transform).
A vector-valued transform of $\mathcal{V} : L^2(\mathbb{R}^2, \mathbb{R}) \to L^2(\mathbb{R}^2, \mathbb{R}^4)$ is given by

$$\mathcal{V}\{f\}(\boldsymbol{u}) = V(\boldsymbol{u}) = \begin{pmatrix} \mathcal{CC}\{f\}(\boldsymbol{u}) \\ \mathcal{SC}\{f\}(\boldsymbol{u}) \\ \mathcal{CS}\{f\}(\boldsymbol{u}) \\ \mathcal{SS}\{f\}(\boldsymbol{u}) \end{pmatrix}.$$

In section 8.2 we saw that it is advantageous to replace the the transform $\mathcal{V}$ with values in $\mathbb{R}^2$ by the Fourier transform with values in $\mathbb{C}$. Actually, $\mathbb{C}$ and $\mathbb{R}^2$ are isomorphic as vector spaces. However, $\mathbb{C}$ has an additional algebraic structure. In the following section we will introduce a 2-D transform which adds an algebraic structure to the values of the 2-D $\mathcal{V}$-transform by replacing $\mathbb{R}^4$ by $\mathbb{H}$.

8.3.2 The Quaternionic Fourier Transform (QFT)

Definition 8.3.4 (Quaternionic Fourier transform). *The quaternionic Fourier transform* $\mathcal{F}_q : L^2(\mathbb{R}^2, \mathbb{R}) \to L^2(\mathbb{R}^2, \mathbb{H})$ *is given by*

$$\begin{aligned}\mathcal{F}_q\{f\}(\boldsymbol{u}) &= F^q(\boldsymbol{u}) \\ &= \mathcal{CC}\{f\}(\boldsymbol{u}) - i\,\mathcal{SC}\{f\}(\boldsymbol{u}) - j\,\mathcal{CS}\{f\}(\boldsymbol{u}) + k\,\mathcal{SS}\{f\}(\boldsymbol{u}).\end{aligned}$$

The three symbols i, j, and k denote the imaginary units of the algebra of quaternions. The choice of the signs in Def. 8.3.4 will become clear below. We shortly review the quaternion algebra in the following.

The quaternions are a special Clifford algebra, namely $\mathbb{R}_{0,2}$. Historically, the algebra of quaternions is one of the predecessors of Clifford's geometric algebra. In 1843 quaternions were first introduced by Hamilton. To his honor the algebra is commonly denoted by the letter $\mathbb{H}$.

Definition 8.3.5. *The set*

$$\mathbb{H} = \{a + bi + cj + dk|\, a, b, c, d \in \mathbb{R}\}$$

together with the multiplication rules

$$ij = -ji = k \quad \text{and} \quad i^2 = j^2 = k^2 = -1,$$

as well as component-wise addition and multiplications by real numbers form an associative $\mathbb{R}$*-algebra, called the quaternions.*

The impulse for introducing quaternions was the quest for an algebra which was able to represent rotations in three-dimensional space. Later, when considering the polar representation of a quaternion, we will exploit this relationship between rotations and quaternions.

For later use we present some definitions and properties concerning $\mathbb{H}$. The *conjugate* of a quaternion

$$q = a + i\,b + j\,c + k\,d$$

is given by

$$\bar{q} = a - i\,b - j\,c - k\,d.$$

The *norm* of q is given by $|q| = \sqrt{q\bar{q}}$. It can be shown that $\mathbb{H}$ is a *normed algebra*, i.e. for $q_1, q_2 \in \mathbb{H}$ we have $|q_1||q_2| = |q_1 q_2|$. $\mathbb{H}$ forms a group under multiplication, i.e. there exist a unit element, namely $e = 1 \in \mathbb{H}$, and to each $q \in \mathbb{H}$ there exists a multiplicative inverse q^{-1} with $qq^{-1} = q^{-1}q = e$. The multiplicative inverse is given by $q^{-1} = \bar{q}/|q|^2$. For the components of a quaternion $qa + i\,b + j\,c + k\,d$ we sometimes write

$$a = \mathcal{R}q, \quad b = \mathcal{I}q, \quad c = \mathcal{J}q, \quad d = \mathcal{K}q.$$

There are three non-trivial involutions defined on $\mathbb{H}$:

$$\begin{aligned} \alpha &: \mathbb{H} \to \mathbb{H}, q \mapsto \alpha(q) = -iqi = a + ib - jc - kd, \\ \beta &: \mathbb{H} \to \mathbb{H}, q \mapsto \beta(q) = -jqj = a - ib + jc - kd, \\ \gamma &: \mathbb{H} \to \mathbb{H}, q \mapsto \gamma(q) = -kqk = a - ib - jc + kd,. \end{aligned} \tag{8.5}$$

These involutions will be used in order to extend the notion of Hermite symmetry from complex to quaternion-valued functions. A function $f : \mathbb{R}^n \to \mathbb{C}$ is called *Hermite symmetric* or *hermitian* if $f(\boldsymbol{x}) = f^*(-\boldsymbol{x})$ for all $\boldsymbol{x} \in \mathbb{R}^n$. The notion of Hermite symmetry of a function is useful in the context of Fourier transforms since the Fourier transform of a real function owes this property.

Definition 8.3.6 (Quaternionic Hermite symmetry). *A function f : $\mathbb{R}^2 \to \mathbb{H}$ is called quaternionic hermitian if:*

$$f(-x, y) = \beta(f(x, y)) \quad \text{and} \quad f(x, -y) = \alpha(f(x, y)) \quad , \tag{8.6}$$

for each $(x, y) \in \mathbb{R}^2$.

One main subject of chapter 11 is the local quaternionic phase of a signal. In order to define the phase we introduce the angular phase of a quaternion as follows.

Theorem 8.3.1. *Each quaternion q can be represented in the form*

$$q = |q|e^{i\phi}e^{k\psi}e^{j\theta} \;\text{with}\; (\phi, \theta, \psi) \in [-\pi, \pi[\times[-\pi/2, \pi/2[\times[-\pi/4, \pi/4]. \tag{8.7}$$

The triple (ϕ, θ, ψ) is called the angular phase of q.

The angular phase of a quaternion can be understood in terms of rotations. Any 3-D rotation about the origin can be expressed in terms of quaternions. The set of unit quaternions is the 3D unit hypersphere

$$S^3 = \{q \in \mathbb{H} |\, |q| = 1\}.$$

Let $q \in S^3$ be given by $q = \cos(\phi) + \boldsymbol{n}\sin(\phi)$, where $\boldsymbol{n}$ is a pure unit quaternion. Further let $\boldsymbol{x}$ be a pure quaternion, representing the three-dimensional vector $(x_1, x_2, x_3)^\top$. A rotation about the axis defined by $\boldsymbol{n}$ through the angle 2ϕ takes $\boldsymbol{x}$ to $\boldsymbol{x}' = q\boldsymbol{x}q^{-1}$. Thus, any unit quaternion q represents a rotation in $\mathbb{R}^3$.

In this interpretation the angles $\phi/2$, $\theta/2$ and $\psi/2$ are the Euler angles of the corresponding rotation[1].

[1] Note that the definition of the Euler angles in not unique. The above representation corresponds to a concatenation of rotations about the y-axis, the z-axis, and the x-axis.

Table 8.1. How to calculate the quaternionic phase-angle representation from a quaternion given in Cartesian representation

<table>
<tr><td colspan="3">$q = a + bi + cj + dk, \quad \lvert q \rvert = 1$</td></tr>
<tr><td colspan="3">$\psi = -\frac{arcsin(2(bc-ad))}{2}$</td></tr>
<tr><td rowspan="2">if $\psi \in]-\frac{\pi}{4}, \frac{\pi}{4}[$

$\phi = \frac{arg_i(q\beta(\bar{q}))}{2}$

$\theta = \frac{arg_j(\alpha(\bar{q})q)}{2}$</td><td colspan="2">if $\psi = \pm\frac{\pi}{4}$
choose</td></tr>
<tr><td>$\phi = 0$

$\theta = \frac{arg_j(\gamma(\bar{q})q)}{2}$</td><td>or

$\theta = 0$

$\phi = \frac{arg_i(q\gamma(\bar{q}))}{2}$</td></tr>
<tr><td>if $e^{i\phi}e^{k\psi}e^{j\theta} = -q$
and $\phi \geq 0$

$\phi \to \phi - \pi$</td><td colspan="2">if $e^{i\phi}e^{k\psi}e^{j\theta} = -q$
and $\phi < 0$

$\phi \to \phi + \pi$</td></tr>
</table>

8.4 Some Properties of the QFT

8.4.1 The Hierarchy of Harmonic Transforms

Before analyzing some properties of the QFT we present what we call the *hierarchy of harmonic transforms.* The hierarchy of transforms is understood in terms of selectivity of the transforms with respect to the specular symmetry of an analyzed signal: Let $L^2_s(\mathbb{R}^n, \mathbb{R})$ be the set of functions in $L^2(\mathbb{R}^n, \mathbb{R})$ with symmetry $s \in S_n = \{(s_1, \ldots, s_n), s_i \in \{e, o\}\}$, where s_i is the symmetry (even or odd) with respect to x_i, $i \in \{1, \ldots, n\}$. Furthermore, let $\mathcal{T}$ be an n-D harmonic transform, e.g. the Fourier transform:

$$\mathcal{T} : L^2(\mathbb{R}^n, \mathbb{R}) \to L^2(\mathbb{R}^n, V) \tag{8.8}$$

$$\mathcal{T} : L^2_s(\mathbb{R}^n, \mathbb{R}) \to L^2_s(\mathbb{R}^n, V_s). \tag{8.9}$$

Since all the transforms considered here are based on trigonometric integral kernels, the transforms preserve the symmetries of signals (see eq. (8.9)). The values of the transformed signal functions $\mathcal{T}\{f\}$ are supposed to lie in the real vector space V. In case of algebra-valued functions, V is the underlying $\mathbb{R}$-vector-space, e.g. $\mathbb{R}^2$ for complex-valued functions. V_s is supposed to be the smallest possible subspaces of V fulfilling (8.9). If the V_r and V_s, $r, s \in S$ intersect only in the zero-vector $V_r \cap V_s = \{(0, \ldots, 0)\}$, $\mathcal{T}$ is said to separate signal components with symmetry s from those of symmetry r. The more symmetry components are separated by a transform, the higher this transform stands in the hierarchy of harmonic transforms.

In the case $n = 1$ we merely have to consider the Hartley transform $\mathcal{H}$, the trigonometric vector transform $\mathcal{V}$, and the Fourier transform $\mathcal{F}$. For the Hartley transform $\mathcal{H}$ we find $V = \mathbb{R}$. The even and odd components of a signal f are mixed in the transform $H(u)$ since $V_e = V_o = \mathbb{R}$. In contrast the 1-D Fourier transform and the trigonometric vector transform separate even from odd components of a real signal: While $V = \mathbb{R}^2$ in these cases, we find $V_e = \{(a,0)|a \in \mathbb{R}\} =: P_1\mathbb{R}^2$ and $V_o = \{(0,b)|b \in \mathbb{R}\} =: P_2\mathbb{R}^2$, thus $V_e \cap V_o = \{(0,0)\}$.

The symmetry selectivity of the Fourier transform, is also expressed by the fact, that the Fourier transform of a real signal is hermitian, i.e. $F(u) = F^*(-u)$. Thus, the real part of F is even, while its imaginary part is odd.

In the case $n = 2$ we consider the four transforms $\mathcal{H}$, $\mathcal{F}$, $\mathcal{F}^q$, and $\mathcal{V}$. For the Hartley and the Fourier transform we get similar results as for $n = 1$: For $\mathcal{H}$, we have $V = \mathbb{R}$ and $V_s = \mathbb{R}$ for all $s \in S_2$. For $\mathcal{F}$ we find $V = \mathbb{R}^2$, while $V_{ee} = V_{oo} = P_1\mathbb{R}^2$ and $V_{oe} = V_{eo} = P_2\mathbb{R}^2$. Thus, the 2-D Fourier transform separates the four symmetry components of a signal into two subspaces. In this case it is more natural to write $V_e = P_1\mathbb{R}^2$ and $V_o = P_2\mathbb{R}^2$. Here the indices e and o mean even and odd with respect to the *argument vector* of an n-D function $f : \mathbb{R}^n \to \mathbb{R}$. I.e. $f_e(\boldsymbol{x}) = f_e(-\boldsymbol{x})$ and $f_o(\boldsymbol{x}) = -f_o(-\boldsymbol{x})$. Finally, for $\mathcal{V}$ and $\mathcal{F}^q$ we get $V = \mathbb{R}^4$ and the four symmetry components are completely separated:

$$V_{ee} = P_1\mathbb{R}^4 \tag{8.10}$$
$$V_{oe} = P_2\mathbb{R}^4 \tag{8.11}$$
$$V_{eo} = P_3\mathbb{R}^4 \tag{8.12}$$
$$V_{oo} = P_4\mathbb{R}^4. \tag{8.13}$$

Thus, we get a three-level hierarchy of 2-D harmonic transforms, on the lowest level of which stands the Hartley transform. On the second level we find the complex Fourier transform while on the highest level the quaternionic Fourier transform and the trigonometric vector transform can be found. This hierarchy is visualized in figure 8.1.

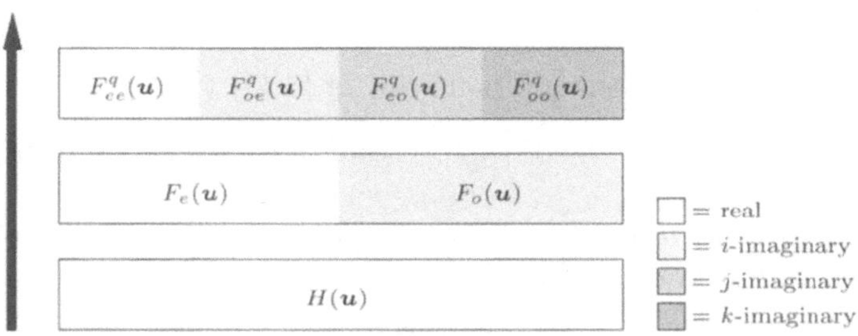

Fig. 8.1. The hierarchy of 2-D harmonic transforms

8.4.2 The Main QFT-Theorems

All harmonic transforms share some important properties. In his famous book on the Fourier transform Bracewell states that for every theorem about the Fourier transform there is a corresponding Hartley transform theorem ([28], p. 391). In order to put the QFT on a theoretically firm basis we derive the most important QFT-analogies to Fourier theorems in the following. First of all we rewrite the definition of the QFT given in def. 8.3.4.

Theorem 8.4.1. *The QFT of a 2-D signal is given by*

$$F^q(\boldsymbol{u}) = \int_{\mathbb{R}^2} e^{-i2\pi u_1 x_1} f(\boldsymbol{x}) e^{-j2\pi u_2 x_2} d^2\boldsymbol{x}. \tag{8.14}$$

Proof. Euler's equation $exp(\boldsymbol{n}\phi) = \cos(\phi) + \boldsymbol{n}\sin(\phi)$ holds for any *pure unit* quaternion $\boldsymbol{n}$. Thus, it applies to the two exponentials in theorem 8.4.1 where we have $\boldsymbol{n} = i$ and $\boldsymbol{n} = j$, respectively. Expanding the product of the two exponentials expressed as sums via Euler's equality gives the expression in def. 8.3.4. □

For clarity we depict the basis functions of the complex Fourier transform and the QFT in figures 8.4.2 and 8.4.2, respectively. The small images show the real part of the basis functions in the spatial domain for fixed frequency $\boldsymbol{u}$. The frequency-parameter varies from image to image. Since only the real component is shown, in case of the complex Fourier transform the imaginary component is missing while in case of the quaternionic Fourier transform three imaginary components exist which are not shown. It can be seen that the basis functions of the complex Fourier transform are intrinsically 1-D. They resemble plane waves. In contrast, the basis functions of the quaternionic Fourier transform are intrinsically 2-D. As the complex Fourier transform the quaternionic Fourier transform is an invertible transform.

Theorem 8.4.2 (Inverse QFT). *The QFT is invertible. The transform $\mathcal{G}$ given by*

$$\mathcal{G}\{F^q\}(\boldsymbol{x}) = \int_{\mathbb{R}^2} e^{i2\pi ux} F^q(\boldsymbol{u}) e^{j2\pi vy} d^2\boldsymbol{u} \tag{8.15}$$

is the inverse of the QFT.

Proof. By inserting (8.14) into the right hand side of (8.15) we get

$$\mathcal{G}\{F^q\}(\boldsymbol{x}) = \int_{\mathbb{R}^2}\int_{\mathbb{R}^2} e^{i2\pi ux} e^{-i2\pi ux'} f(\boldsymbol{x}') e^{-j2\pi vy'} e^{j2\pi vy} d^2\boldsymbol{u}\, d^2\boldsymbol{x}'. \tag{8.16}$$

Integrating with respect to $\boldsymbol{u}$ and taking into account the orthogonality of harmonic exponential functions this simplifies to

$$\begin{aligned} \mathcal{G}\{F^q\}(\boldsymbol{u}) &= \int_{\mathbb{R}^2} \delta(x - x') f(\boldsymbol{x}') \delta(y - y') d^2\boldsymbol{x}' \\ &= f(\boldsymbol{x}), \end{aligned} \tag{8.17}$$

thus $\mathcal{G} = \mathcal{F}_q^{-1}$. □

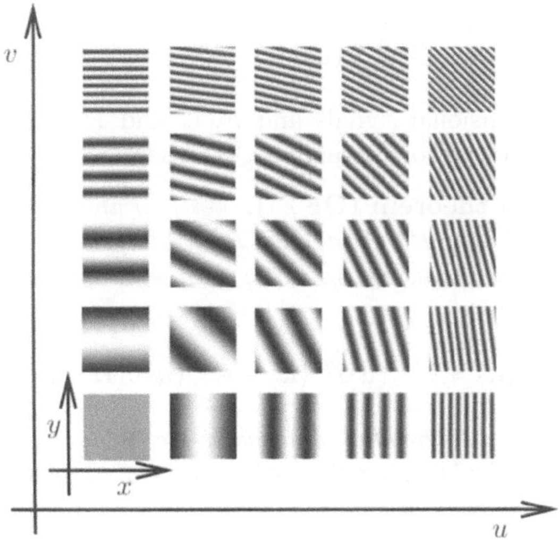

Fig. 8.2. The basis functions of the complex 2-D Fourier transform

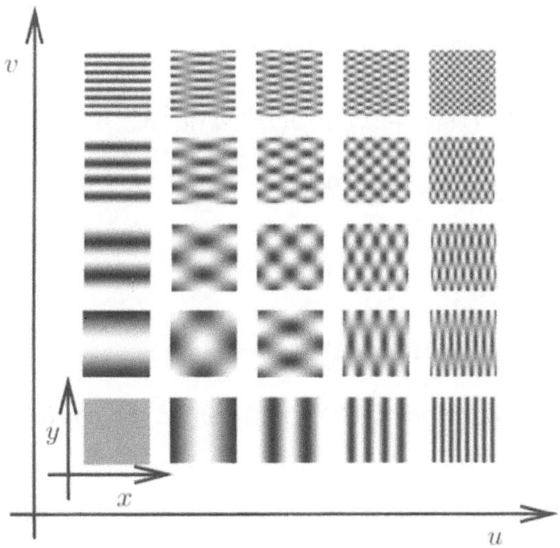

Fig. 8.3. The basis functions of the quaternionic Fourier transform

The convolution theorem of the Fourier transform states that convolution of two signals in the spatial domain corresponds to their pointwise multiplication in the frequency domain, i.e.

$$f(\boldsymbol{x}) = (g * h)(\boldsymbol{x}) \quad \Leftrightarrow \quad F(\boldsymbol{u}) = G(\boldsymbol{u})H(\boldsymbol{u}) \tag{8.18}$$

where f, g and h are two-dimensional signals and F, G and H are their Fourier transforms. We now give the corresponding QFT theorem.

Theorem 8.4.3 (Convolution theorem (QFT)). *Let f, g and h be two-dimensional signals and F^q, G^q and H^q their QFT's. In the following g is assumed to be real-valued, while h and consequently f may be quaternion-valued. Then,*

$$f(\boldsymbol{x}) = (g * h)(\boldsymbol{x}) \iff F^q(\boldsymbol{u}) = G^q_{\cdot e}(\boldsymbol{u})H^q(\boldsymbol{u}) + G^q_{\cdot o}(\boldsymbol{u})\beta(H^q(\boldsymbol{u})).$$

Here β denotes one of the three non-trivial automorphisms of the quaternion algebra as defined in (8.5). $G_{\cdot e}$ and $G_{\cdot o}$ are the components of G which are even or odd with respect to the second argument.

Proof. We prove the convolution theorem by directly calculating the QFT of the convolution integral:

$$\begin{aligned}
F^q(\boldsymbol{u}) &= \int_{\mathbb{R}^2} e^{-2\pi i x u} \left[\int_{\mathbb{R}^2} (g(\boldsymbol{x}')h(\boldsymbol{x} - \boldsymbol{x}'))d^2\boldsymbol{x}' \right] e^{-2\pi j y v} d^2\boldsymbol{x} \\
&= \int_{\mathbb{R}^2} e^{-2\pi i x' u} g(\boldsymbol{x}')H^q(\boldsymbol{u})e^{-2\pi j y' v} d^2\boldsymbol{x}' \\
&= \int_{\mathbb{R}^2} e^{-2\pi i x' u} g(\boldsymbol{x}')\cos(-2\pi y' v)H^q(\boldsymbol{u})d^2\boldsymbol{x}' \\
&\quad + \int_{\mathbb{R}^2} e^{-2\pi i x' u} g(\boldsymbol{x}')j\sin(-2\pi y' v)\beta(H^q(\boldsymbol{u}))d^2\boldsymbol{x}' \\
&= G^q_{\cdot e}(\boldsymbol{u})H^q(\boldsymbol{u}) + G^q_{\cdot o}(\boldsymbol{u})\beta(H^q(\boldsymbol{u})),
\end{aligned} \tag{8.19}$$

which completes the proof. □

Analogously it can be shown, that

$$F^q(\boldsymbol{u}) = H^q(\boldsymbol{u})G^q_{e\cdot}(\boldsymbol{u}) + \alpha(H^q(\boldsymbol{u}))G^q_{o\cdot}(\boldsymbol{u}).$$

If h is a quaternion-valued function which QFT is real-valued the convolution theorem simplifies to

$$f(\boldsymbol{x}) = (g * h)(\boldsymbol{x}) \Rightarrow F^q(\boldsymbol{u}) = G^q(\boldsymbol{u})H^q(\boldsymbol{u}), \tag{8.20}$$

which is of the same form as the convolution theorem of the two-dimensional Fourier transform. This is an important fact, since later we will convolve real-valued signals with quaternionic Gabor filters, which QFT's are real-valued.

According to (8.20) in this case the convolution theorem can be applied as usually.

The energy of a signal is defined as the integral (or sum in the case of discrete signals) over the squared magnitude of the signal. Rayleigh's theorem states that the signal energy is preserved by the Fourier transform:

$$\int_{\mathbb{R}^2} |f(\boldsymbol{x})|^2 d^2\boldsymbol{x} = \int_{\mathbb{R}^2} |F(\boldsymbol{u})|^2 d^2\boldsymbol{u} \quad , \tag{8.21}$$

where $F(\boldsymbol{u})$ is the Fourier transform of $f(\boldsymbol{x})$. Rayleigh's theorem is valid for arbitrary integer dimension of the signal. In mathematical terms Rayleigh's theorem states that the L^2-norm of a signal is invariant under the Fourier transform. We will show that the analogous statement for the QFT is true.

Theorem 8.4.4 (Rayleigh's theorem (QFT)).
The quaternionic Fourier transform preserves the L^2-norm of any real two-dimensional signal $f(\boldsymbol{x})$:

$$\int_{\mathbb{R}^2} |f(\boldsymbol{x})|^2 d^2\boldsymbol{x} = \int_{\mathbb{R}^2} |F^q(\boldsymbol{u})|^2 d^2\boldsymbol{u}, \tag{8.22}$$

where $F^q(\boldsymbol{u})$ is the QFT of $f(\boldsymbol{x})$.

Proof. We make use of Rayleigh's theorem for the two-dimensional Fourier transform. Thus, we only have to prove that

$$\int_{\mathbb{R}^2} |F(\boldsymbol{u})|^2 d^2\boldsymbol{u} - \int_{\mathbb{R}^2} |F^q(\boldsymbol{u})|^2 d^2\boldsymbol{u} = 0 \quad . \tag{8.23}$$

Regarding the integrands and using the addition theorems of the sine and the cosine function we find out that

$$|F(\boldsymbol{u})|^2 = (\mathcal{CC}\{f\}(\boldsymbol{u}) - \mathcal{SS}\{f\}(\boldsymbol{u}))^2 + (\mathcal{SC}\{f\}(\boldsymbol{u}) + \mathcal{CS}\{f\}(\boldsymbol{u}))^2 , \tag{8.24}$$

while

$$|F^q(\boldsymbol{u})|^2 = \mathcal{CC}\{f\}^2(\boldsymbol{u}) + \mathcal{SC}\{f\}^2(\boldsymbol{u}) + \mathcal{CS}\{f\}^2(\boldsymbol{u}) + \mathcal{SS}\{f\}^2(\boldsymbol{u}). \tag{8.25}$$

Thus, the left hand side of (8.23) can be evaluated as follows:

$$\begin{aligned} &\int_{\mathbb{R}^2} |F(\boldsymbol{u})|^2 d^2\boldsymbol{u} - \int_{\mathbb{R}^2} |F^q(\boldsymbol{u})|^2 d^2\boldsymbol{u} \\ &= 2\int_{\mathbb{R}^2} (\mathcal{SC}\{f\}(\boldsymbol{u})\mathcal{CS}\{f\}(\boldsymbol{u}) - \mathcal{CC}\{f\}(\boldsymbol{u})\mathcal{SS}\{f\}(\boldsymbol{u}))\, d^2\boldsymbol{u}. \end{aligned} \tag{8.26}$$

The integrand in (8.26) is odd with respect to both arguments (since $\mathcal{S}$-terms are odd). Thus, the integral is zero which completes the proof. □

The shift theorem of the Fourier transform describes how the transform of a signal varies when the signal is shifted. If the signal f is shifted by $\boldsymbol{d}$, it is known that its Fourier transform is multiplied by a phase factor $\exp(-2\pi\, i\, \boldsymbol{d} \cdot \boldsymbol{x})$. How the QFT of f is affected by the shift is described by the following theorem.

Theorem 8.4.5 (Shift theorem (QFT)). *Let*

$$F^q(\boldsymbol{u}) = \int_{\mathbb{R}^2} e^{-i2\pi ux} f(\boldsymbol{x}) e^{-j2\pi vy} d^2\boldsymbol{x} \tag{8.27}$$

and

$$F^q_s(\boldsymbol{u}) = \int_{\mathbb{R}^2} e^{-i2\pi ux} f(\boldsymbol{x} - \boldsymbol{d}) e^{-j2\pi vy} d^2\boldsymbol{x} \tag{8.28}$$

be the QFT's of a 2-D signal f and a shifted version of f, respectively. Then, $F^q(\mathbf{u})$ and $F^q_s(\boldsymbol{u})$ are related by

$$F^q_s(\boldsymbol{u}) = e^{-i2\pi ud_1} F^q(\boldsymbol{u}) e^{-j2\pi vd_2} \quad . \tag{8.29}$$

If we denote the phase of $F^q(\boldsymbol{u})$ by $(\phi(\boldsymbol{u}), \theta(\boldsymbol{u}), \psi(\boldsymbol{u}))^\top$ then, as a result of the shift, the first and the second component of the phase undergo a phase-shift

$$\begin{pmatrix} \phi(\boldsymbol{u}) \\ \theta(\boldsymbol{u}) \\ \psi(\boldsymbol{u}) \end{pmatrix} \quad \rightarrow \quad \begin{pmatrix} \phi(\boldsymbol{u}) - 2\pi u d_1 \\ \theta(\boldsymbol{u}) - 2\pi v d_2 \\ \psi(\boldsymbol{u}) \end{pmatrix} . \tag{8.30}$$

Proof. Equation (8.29) follows from (8.27) and (8.28) by substituting $(\boldsymbol{x} - \boldsymbol{d})$ with $\boldsymbol{x}'$. If $F^q(\boldsymbol{u})$ has the polar representation

$$F^q(\boldsymbol{u}) = |F^q(\boldsymbol{u})| e^{i\phi(\boldsymbol{u})} e^{k\psi(\boldsymbol{u})} e^{j\theta(\boldsymbol{u})},$$

we find for the polar representation of $F^q_s(\boldsymbol{u})$

$$\begin{aligned} F^q_s(\boldsymbol{u}) &= e^{-i2\pi ud_1} F^q(\boldsymbol{u}) e^{-j2\pi vd_2} \\ &= e^{-i2\pi ud_1} |F^q(\boldsymbol{u})| e^{i\phi(\boldsymbol{u})} e^{k\psi(\boldsymbol{u})} e^{j\theta(\boldsymbol{u})} e^{-j2\pi vd_2} \\ &= |F^q(\boldsymbol{u})| e^{i(\phi(\boldsymbol{u}) - 2\pi ud_1)} e^{k\psi(\boldsymbol{u})} e^{j(\theta(\boldsymbol{u}) - 2\pi vd_2)} . \end{aligned}$$

This proves (8.30). □

In the shift theorem a shift of the signal in the spatial domain is considered. The effect of such a shift are the modulation factors shown in (8.29). In the following theorem we regard the converse situation: the signal is modulated in the spatial domain, and we ask for the effect in the quaternionic frequency domain.

$$\begin{aligned}
&\frac{1}{4}(F^q(\boldsymbol{u}+\boldsymbol{u}_0)+F^q(u-u_0,v+v_0) \qquad (8.35)\\
&+F^q(u+u_0,v-v_0)+F^q(\boldsymbol{u}-\boldsymbol{u}_0))\\
&=\frac{1}{4}(I_{ee}(\boldsymbol{u}_0)+I_{oe}(\boldsymbol{u}_0)+I_{eo}(\boldsymbol{u}_0)+I_{oo}(\boldsymbol{u}_0))\\
&+\frac{1}{4}(I_{ee}(-u_0,v_0)+I_{oe}(-u_0,v_0)+I_{eo}(-u_0,v_0)+I_{oo}(-u_0,v_0))\\
&+\frac{1}{4}(I_{ee}(u_0,-v_0)+I_{oe}(u_0,-v_0)+I_{eo}(u_0,-v_0)+I_{oo}(u_0,-v_0))\\
&+\frac{1}{4}(I_{ee}(-\boldsymbol{u}_0)+I_{oe}(-\boldsymbol{u}_0)+I_{eo}(-\boldsymbol{u}_0)+I_{oo}(-\boldsymbol{u}_0))\\
&=I_{ee}(\boldsymbol{u}_0)=\mathcal{F}_q\{\cos(2\pi u_0 x)f(\boldsymbol{x})\cos(2\pi v_0 y)\}, \qquad (8.36)
\end{aligned}$$

which completes the proof. □

Theorem 8.4.7 (Derivative theorem (QFT)).
Let f be a real two-dimensional signal, F^q its QFT, and $n = p + r, p, r \in \mathbb{N}$. Then

$$\mathcal{F}_q\left\{\frac{\partial^n}{\partial x^p \partial y^r} f\right\}(\boldsymbol{u}) = (2\pi)^n (iu)^p F^q(\boldsymbol{u})(jv)^r.$$

Proof. We prove the theorem for $(p,r) = (1,0)$ and $(p,r) = (0,1)$ starting with the first case. We have

$$f(\boldsymbol{x}) = \int_{\mathbb{R}^2} e^{i2\pi ux} F^q(\boldsymbol{u}) e^{j2\pi vy}. \qquad (8.37)$$

Thus, it follows that

$$\frac{\partial}{\partial x} f(\boldsymbol{x}) = \frac{\partial}{\partial x}\int_{\mathbb{R}^2} e^{i2\pi ux} F^q(\boldsymbol{u}) e^{j2\pi vy} = \int_{\mathbb{R}^2} e^{i2\pi ux}(i2\pi u F^q(\boldsymbol{u}))e^{j2\pi vy}.$$

Therefore, we have

$$\mathcal{F}_q\left\{\frac{\partial}{\partial x} f\right\}(\boldsymbol{u}) = i2\pi u F^q(\boldsymbol{u}). \qquad (8.38)$$

Analogously we derive

$$\frac{\partial}{\partial y} f(\boldsymbol{x}) = \frac{\partial}{\partial y}\int_{\mathbb{R}^2} e^{i2\pi ux} F^q(\boldsymbol{u}) e^{j2\pi vy} = \int_{\mathbb{R}^2} e^{i2\pi ux}(F^q(\boldsymbol{u}) j2\pi v)e^{j2\pi vy},$$

which shows that

$$\mathcal{F}_q\left\{\frac{\partial}{\partial y} f\right\}(\boldsymbol{u}) = 2\pi F^q(\boldsymbol{u})(jv). \qquad (8.39)$$

For general derivatives the theorem follows from successive application of first order derivatives. □

Theorem 8.4.6 (Modulation theorem (QFT)).
Let $f(\boldsymbol{x})$ be a quaternion-valued signal and $F^q(\boldsymbol{u})$ its QFT. Further, let be the following modulated version of $f(\boldsymbol{x})$:

$$f_m(\boldsymbol{x}) = e^{i2\pi u_0 x} f(\boldsymbol{x}) e^{j2\pi v_0 y}.$$

The QFT of $f_m(\boldsymbol{x})$ is then given by

$$\mathcal{F}_q\{f_m\}(\boldsymbol{u}) = F^q(\boldsymbol{u} - \boldsymbol{u}_0).$$

If $f_m(\boldsymbol{x})$ is a real modulated version of $f(\boldsymbol{x})$, i.e.

$$f_m(\boldsymbol{x}) = f(\boldsymbol{x}) \cos(2\pi x u_0) \cos(2\pi y v_0),$$

the QFT of $f_m(\boldsymbol{x})$ is given by

$$\begin{aligned}\mathcal{F}_q\{f_m\}(\boldsymbol{u}) = \frac{1}{4}(F^q(\boldsymbol{u} + \boldsymbol{u}_0) + F^q(u - u_0, v + v_0) \\ + F^q(u + u_0, v - v_0) + F^q(\boldsymbol{u} - \boldsymbol{u}_0)).\end{aligned}$$

Proof. First, we consider the QFT of

$$f_m(\boldsymbol{x}) = e^{i2\pi u_0 x} f(\boldsymbol{x}) e^{j2\pi v_0 y}.$$

By inserting f_m into the definition of the QFT we obtain

$$\begin{aligned}F_m^q(\boldsymbol{u}) &= \int_{\mathbb{R}^2} e^{-i2\pi u x} f_m(\boldsymbol{x}) e^{-j2\pi v y} d^2\boldsymbol{x} \\ &= \int_{\mathbb{R}^2} e^{-i2\pi(u - u_0)x} f(\boldsymbol{x}) e^{-j2\pi(v - v_0)y} d^2\boldsymbol{x} \\ &= F^q(\boldsymbol{u} - \boldsymbol{u}_0).\end{aligned}$$

For the second part of the proof we introduce the abbreviation $e^{-i2\pi u x} f(\boldsymbol{x}) e^{-j2\pi v y}$. Further, we use the notation

$$\begin{aligned}I_{ee}(\boldsymbol{u}_0) &= \int_{\mathbb{R}^2} \cos(2\pi u_0 x) \mathfrak{f}(\boldsymbol{x}) \cos(2\pi v_0 y) d^2\boldsymbol{x} \\ I_{oe}(\boldsymbol{u}_0) &= i \int_{\mathbb{R}^2} \sin(2\pi u_0 x) \mathfrak{f}(\boldsymbol{x}) \cos(2\pi v_0 y) d^2\boldsymbol{x} \\ I_{eo}(\boldsymbol{u}_0) &= j \int_{\mathbb{R}^2} \cos(2\pi u_0 x) \mathfrak{f}(\boldsymbol{x}) \sin(2\pi v_0 y) d^2\boldsymbol{x} \\ I_{oo}(\boldsymbol{u}_0) &= k \int_{\mathbb{R}^2} \sin(2\pi u_0 x) \mathfrak{f}(\boldsymbol{x}) \sin(2\pi v_0 y) d^2\boldsymbol{x},\end{aligned}$$

where $I_{ee}(\boldsymbol{u}_0)$ is even with respect to both u_0 and to v_0, $I_{oe}(\boldsymbol{u}_0)$ is o(respect to u_0 and even with respect to v_0 and so on. We can then wr

Theorem 8.4.8. *The QFT of a real two-dimensional signal f is quaternionic hermitian.*

Proof. We have shown before that the QFT of a real signal has the form

$$F^q(\boldsymbol{u}) = F^q_{ee}(\boldsymbol{u}) + iF^q_{oe}(\boldsymbol{u}) + jF^q_{eo}(\boldsymbol{u}) + kF^q_{oo}(\boldsymbol{u}).$$

Applying the automorphisms α and β yields

$$\begin{aligned}\alpha(F^q(\boldsymbol{u})) &= F^q_{ee}(\boldsymbol{u}) + iF^q_{oe}(\boldsymbol{u}) - jF^q_{eo}(\boldsymbol{u}) - kF^q_{oo}(\boldsymbol{u}) \\ &= F^q_{ee}(u,-v) + iF^q_{oe}(u,-v) + jF^q_{eo}(u,-v) + kF^q_{oo}(u,-v) \\ &= F^q(u,-v) \qquad (8.40) \\ \beta(F^q(\boldsymbol{u})) &= F^q_{ee}(\boldsymbol{u}) - iF^q_{oe}(\boldsymbol{u}) + jF^q_{eo}(\boldsymbol{u}) - kF^q_{oo}(\boldsymbol{u}) \\ &= F^q_{ee}(-u,v) + iF^q_{oe}(-u,v) + jF^q_{eo}(-u,v) + kF^q_{oo}(-u,v) \\ &= F^q(-u,v), \qquad (8.41)\end{aligned}$$

which proves the theorem according to definition 8.3.6. □

It can often happen that a signal undergoes an affine transformation in the spatial domain, which can be written as $f(\boldsymbol{x}) \rightarrow f(A\boldsymbol{x}+\boldsymbol{b})$, where $\boldsymbol{b} \in \mathbb{R}^2$ and $A \in Gl(2,\mathbb{R})$. In these cases it is desirable to know how this transformation affects the frequency representation F^q of f. The effect of the shift by $\boldsymbol{b}$ is already known from the shift theorem. It remains to work out how the frequency representation is transformed under a linear transformation of the spatial domain: $f(\boldsymbol{x}) \rightarrow f(A\boldsymbol{x})$. This is done by the following theorem.

Theorem 8.4.9 (Affine theorem (QFT)).
Let $f(\boldsymbol{x})$ be a real 2D signal and $F^q(\boldsymbol{u}) = \mathcal{F}_q\{f(\boldsymbol{x})\}(\boldsymbol{u})$ its QFT. Further, let A be the real regular 2×2 matrix

$$A = \begin{pmatrix} a & b \\ c & d \end{pmatrix}, \quad \text{with} \quad \det(A) = ad - bc \neq 0.$$

The QFT of $f(A\boldsymbol{x})$ is then given by

$$\begin{aligned}\mathcal{F}_q\{f(A\boldsymbol{x})\}(\boldsymbol{u}) = \frac{1}{2\det(A)} & \left(F^q(\det(B)B^{-1\top}\boldsymbol{u}) + F^q(B^{\top t}\boldsymbol{u}) \right. \qquad (8.42) \\ & \left. + i(F^q(\det(B)B^{-1\top}\boldsymbol{u}) - F^q(B^{\top t}\boldsymbol{u}))j \right).\end{aligned}$$

where we introduced the matrix

$$B = \begin{pmatrix} a' & b' \\ c' & d' \end{pmatrix} =: \frac{A}{\det(A)}.$$

Furthermore, $A^\top$ denotes the transpose of A and A^t the transpose of A according to the minor diagonal:

$$A = \begin{pmatrix} a & b \\ c & d \end{pmatrix} \Rightarrow A^\top = \begin{pmatrix} a & c \\ b & d \end{pmatrix}, \quad A^t = \begin{pmatrix} d & b \\ c & a \end{pmatrix}.$$

Proof. The inverse of A is given by

$$A^{-1} = \frac{1}{\det(A)} \begin{pmatrix} d & -b \\ -c & a \end{pmatrix}.$$

For the transformed coordinates we introduce the notation

$$A\boldsymbol{x} = \boldsymbol{x}' = \begin{pmatrix} x' \\ y' \end{pmatrix} = \begin{pmatrix} ax+by \\ cx+dy \end{pmatrix} \Rightarrow \begin{pmatrix} x \\ y \end{pmatrix} = \frac{1}{\det(A)} \begin{pmatrix} dx'-by' \\ -cx'+ay' \end{pmatrix}.$$

Now we can express $\mathcal{F}_q\{f(A\boldsymbol{x})\}(\boldsymbol{u})$ using the coordinates $\boldsymbol{x}'$ in the following way:

$$\begin{aligned}
\mathcal{F}_q\{f(A\boldsymbol{x})\}(\boldsymbol{u}) &= \int_{\mathbb{R}^2} e^{-i2\pi ux} f(A\boldsymbol{x}) e^{-j2\pi vy} d^2\boldsymbol{x} \\
&= \frac{1}{\det(A)} \int_{\mathbb{R}^2} e^{-i2\pi u(d'x'-b'y')} f(\boldsymbol{x}') e^{-j2\pi v(-c'x'+a'y')} d^2\boldsymbol{x}' \\
&= \frac{1}{\det(A)} \int_{\mathbb{R}^2} e^{-i2\pi u(d'x-b'y)} f(\boldsymbol{x}) e^{-j2\pi v(-c'x+a'y)} d^2\boldsymbol{y}.
\end{aligned}$$

In order to complete the proof we still have to show that

$$\begin{aligned}
e^{-i2\pi u(d'x-b'y)} e^{-j2\pi v(-c'x+a'y)} = \frac{1}{2} \big(& e^{-i2\pi x(d'u+c'v)} e^{-j2\pi y(b'u+a'v)} \\
& + e^{-i2\pi x(d'u-c'v)} e^{-j2\pi y(-b'u+a'v)} \\
& - i e^{i2\pi x(-d'u+c'v)} e^{-j2\pi y(-b'u+a'v)} j \\
& + i e^{i2\pi x(-d'u-c'v)} e^{-j2\pi y(b'u+a'v)} j \big).
\end{aligned}$$

For a more compact form of (8.43) we introduce the abbriviations

$$\alpha = 2\pi vya', \quad \beta = 2\pi uyb', \quad \gamma = 2\pi vxc', \quad \delta = 2\pi uxd'$$

and we get the following expression:

$$\begin{aligned}
e^{-i(\delta-\beta)} e^{-j(-\gamma+\alpha)} = \frac{1}{2} \Big(& e^{i(-\delta-\gamma)} e^{j(-\beta-\alpha)} + e^{i(-\delta+\gamma)} e^{j(\beta-\alpha)} \\
& - k e^{i(\delta-\gamma)} e^{j(\beta-\alpha)} + k e^{i(\delta+\gamma)} e^{j(-\beta-\alpha)} \Big).
\end{aligned} \tag{8.43}$$

We evaluate the right-hand side:

$$\begin{aligned}
& \frac{1}{2} \Big(e^{i(-\delta-\gamma)} e^{j(-\beta-\alpha)} + e^{i(-\delta+\gamma)} e^{j(\beta-\alpha)} \\
& - i e^{-i(\delta-\gamma)} e^{j(\beta-\alpha)} j + i e^{-i(\delta+\gamma)} e^{j(-\beta-\alpha)} j \Big) \\
&= \frac{1}{2} e^{-i\delta} \left(e^{-i\gamma} e^{-j\beta} + e^{i\gamma} e^{j\beta} - i e^{i\gamma} e^{j\beta} j + i e^{-i\gamma} e^{-j\beta} j \right) e^{-j\alpha} \\
&= e^{-i\delta} e^{i\beta} e^{j\gamma} e^{-j\alpha}.
\end{aligned} \tag{8.44}$$

Obviously, this final result equals the left-hand side of 8.43 which completes the proof. □

Example 1. As an example we will demonstrate the effect of a rotation of the original signal. The transformation matrix A is then given by

$$A = \begin{pmatrix} \cos(\phi) & -\sin(\phi) \\ \sin(\phi) & \cos(\phi) \end{pmatrix} \Rightarrow \det(A) = 1, \quad B = A^t = A, \tag{8.45}$$

$$A^\top = A^{-1} = \begin{pmatrix} \cos(\phi) & \sin(\phi) \\ -\sin(\phi) & \cos(\phi) \end{pmatrix}. \tag{8.46}$$

$$\mathcal{F}_q\{f(A\boldsymbol{x})\}(\boldsymbol{u}) = \frac{1}{2} \left(F^q(A\boldsymbol{u}) + F^q(A^{-1}\boldsymbol{u}) \right. \\ \left. + i(F^q(A\boldsymbol{u}) - F^q(A^{-1}\boldsymbol{u}))j \right). \tag{8.47}$$

Example 2. Here we regard a pure dilation of the original signal with different scaling factors for the x-axis and the y-axis. In this case the transformation matrix takes the form:

$$A = \begin{pmatrix} a & 0 \\ 0 & b \end{pmatrix} \Rightarrow \det(A) = ab, \tag{8.48}$$

$$B = B^\top = \begin{pmatrix} 1/b & 0 \\ 0 & 1/a \end{pmatrix}, \quad B^t = \frac{1}{ab}B^{-1} = \begin{pmatrix} 1/a & 0 \\ 0 & 1/b \end{pmatrix}. \tag{8.49}$$

$$\mathcal{F}_q\{f(A\boldsymbol{x})\}(\boldsymbol{u}) = \frac{1}{2ab} \left(F^q\left(\frac{u}{a}, \frac{v}{b}\right) + F^q\left(\frac{u}{a}, \frac{v}{b}\right) \right. \tag{8.50}$$

$$\left. + i\left(F^q\left(\frac{u}{a}, \frac{v}{b}\right) - F^q\left(\frac{u}{a}, \frac{v}{b}\right)\right) j \right) \tag{8.51}$$

$$= \frac{1}{ab} F^q\left(\frac{u}{a}, \frac{v}{b}\right). \tag{8.52}$$

This result has the same form as the analogue result for the 2-D Fourier transform. The affine theorem of the Hartley transform [29] is like the version for the QFT more complicated than the affine theorem of the Fourier transform.

8.5 The Clifford Fourier Transform

Above we developed the QFT which applies to images or other 2-D signals. When one wants to deal with volumetric data, image sequences or any other signals of higher dimensions, the QFT has to be extended. For this reason we introduce the Clifford Fourier transform for signals of arbitrary dimension n. Which Clifford algebra has to be used depends on the signal's dimension n.

We recall the QFT in the form given in theorem 8.4.1:

$$F^q(\boldsymbol{u}) = \int_{\mathbb{R}^2} e^{-i2\pi u_1 x_1} f(\boldsymbol{x}) e^{-j2\pi u_2 x_2}.$$

As mentioned earlier, the position of the signal f between the exponential functions is of no importance as long as f is real-valued.

Definition 8.5.1 (Clifford Fourier transform).
The Clifford Fourier transform $\mathcal{F}_c : L^2(\mathbb{R}^n, \mathbb{R}_{0,n}) \to L^2(\mathbb{R}^n, \mathbb{R}_{0,n})$ *of an n–dimensional signal* $f(\boldsymbol{x})$ *is defined by*

$$F^c(\boldsymbol{u}) = \int_{\mathbb{R}^n} f(\boldsymbol{x}) \prod_{k=1}^{n} \exp(-e_k 2\pi u_k x_k) d^n\boldsymbol{x} \quad . \tag{8.53}$$

where $\boldsymbol{u} = (u_1, u_2, \dots, u_n)$, $\boldsymbol{x} = (x_1, x_2, \dots, x_n)$ *and* $e_1, e_2, \dots, e_n$ *are the basis vectors of the Clifford algebra* $\mathbb{R}_{0,n}$ *as defined in chapter 1. The product is meant to be performed in a fixed order:* $\prod_{j=1}^{n} a_j = a_1 a_2 \cdots a_n$.

For real signals and $n = 2$ the Clifford Fourier transform is identical to the QFT. For $n = 1$ it is the complex Fourier transform.

Theorem 8.5.1 (Inverse Clifford Fourier transform). *The inverse Clifford Fourier transform is obtained by*

$$\mathcal{F}_c^{-1}\{F^c\}(\boldsymbol{x}) = \int_{\mathbb{R}^n} F^c(\boldsymbol{u}) \prod_{k=0}^{n-1} \exp(e_{n-k} 2\pi u_{n-k} x_{n-k}) d^n\boldsymbol{u}. \tag{8.54}$$

Proof. Inserting term (8.53) into the formula (8.54) yields

$$\begin{aligned}
&\int_{\mathbb{R}^{2n}} f(\boldsymbol{x}') \prod_{j=1}^{n} \exp(-e_j 2\pi u_j x'_j) d^n\boldsymbol{x}' \prod_{k=0}^{n-1} \exp(e_{n-k} 2\pi u_{n-k} x_{n-k}) d^n\boldsymbol{u} \\
&= \int_{\mathbb{R}^n} f(\boldsymbol{x}')\delta^n(\boldsymbol{x} - \boldsymbol{x}') d^n\boldsymbol{x}' \\
&= f(\boldsymbol{x}),
\end{aligned}$$

where the orthogonality of the harmonic exponential functions is used. □

In chapter 9 we will introduce a corresponding transform using an n-D commutative hypercomplex algebra.

8.6 Historical Remarks

Although hypercomplex spectral signal representations are of special interest for image processing tasks, the Clifford Fourier transform does not seem to have attracted a lot of attention yet. The reason may lie in the fact that articles on the subject are spread in the literature of many different fields and are not easily accessible. For this reason it is not surprising that the authors of this chapter first thought to have "invented" the QFT and the Clifford Fourier transform in [35]. Since the literature on the QFT is rather disjointed the following review may be of interest to researchers in this field.

The first appearance we could trace the QFT back to is an article by the Nobel laureate R.R. Ernst et al. which appeared in 1976 [72]. The scope of this work is 2-D NMR spectroscopy. In the analysis of molecular systems transfer functions of perturbed systems are recorded, which leads to 2-D spectra. Ernst shows that for the analysis of so called *quadruple phase 2-D Fourier transform spectroscopy* the introduction of a hypercomplex Fourier transform is necessary. The transform introduced in [72] could be the same as the QFT. However, the algebra involved is not completely defined: The elements i and j are given as imaginary units $i^2 = j^1 = -1$, and a new element ji is introduced. There is nothing said about $(ji)^2$ and on whether ji equals ij or not. This work has again been reported in [73] and [59] where the used algebra is specified to a commutative algebra with $ij = ji$.

In mathematical literature the Clifford Fourier transform was introduced by Brackx et al. [30] in the context of Clifford analysis. This branch of mathematics is concerned with the extension of results of the theory of complex functions to Clifford-valued functions.

In 1992 the QFT was reinvented by Ell for the analysis of 2-D partial-differential systems [69, 70]. This work was taken up and adapted to the use in color image processing by Sangwine [198, 199, 200]. Sangwine represents an RGB color image as a pure quaternion-valued function

$$f(\boldsymbol{x}) = i\, r(\boldsymbol{x}) + j\, g(\boldsymbol{x}) + k\, b(\boldsymbol{x})$$

which can be transformed into the frequency domain by the QFT. This allows to transform color images holistically instead of transforming each color component separately using a complex Fourier transform. A more extensive discussion of algebraic embeddings of color images can be found in Chap. 7 of this book.

The discrete QFT or DQFT has been used by Chernov in order to develop fast algorithms for the 2-D discrete complex Fourier transform [40]. Chernov reduces the size of a real image by assigning to each pixel a quaternion made up from four real pixel-values of the original image. This method is called overlapping. The shrunk image is transformed by the DQFT. The result is expanded to the DFT of the input signal using simple automorphisms of the quaternion algebra.

8.7 Conclusion

The quaternionic Fourier transform (QFT) has been introduced as an alternative to the 2-D complex Fourier transform. It has been shown that the main theorems of the complex Fourier transform have their analogues in case of the QFT. An n-D Clifford Fourier transform has been introduced as an alternative to the complex Fourier transform. It has been shown that there is a hierarchy of harmonic transforms. Actually, all lower level transforms can be easily derived from the higher level transforms. Whereas here mainly theoretical considerations were made, we will demonstrate the impact of the quaternionic Fourier transform on image processing in chapter 11.

9. Commutative Hypercomplex Fourier Transforms of Multidimensional Signals*

Michael Felsberg, Thomas Bülow, and Gerald Sommer

Institute of Computer Science and Applied Mathematics,
Christian-Albrechts-University of Kiel

9.1 Introduction

In Chap. 8 the approach of the Clifford Fourier transform (CFT) and of the quaternionic Fourier transform (QFT) have been introduced. We have shown that the CFT yields an extended and more efficient multi-dimensional signal theory compared to the theory based on complex numbers. Though the CFT of a *real* signal does not include new information (the complex Fourier transform is a *complete* transform in the mathematical sense), the Clifford spectrum is a richer representation with respect to the *symmetry concepts* of n-D signals than the complex spectrum. Furthermore, the possibility of designing *Clifford-valued filters* represents a fundamental extension in multi-dimensional signal theory. Our future aim is to develop principles for the design of hypercomplex filters. The first method is introduced in Chap. 11, where the quaternionic Gabor filters are explored.

One main property of Clifford algebras is the non-commutativity of the Clifford product. This property is impractical in some cases of analytic and numerical calculations. Some theorems are very complicated to formulate in

* This work has been supported by German National Merit Foundation and by DFG Grants So-320-2-1, So-320-2-2, and Graduiertenkolleg No. 357.

higher dimensions, e.g. the affine theorem (The. 8.4.9) and the convolution theorem (The. 8.4.3). Similar problems occur in the derivation of fast algorithms (Chap. 10), because for the decimation of space method, the exponential functions need to be separated. Due to non-commutativity, the additional exponential terms cannot be sorted, and hence, no closed formulation of the partial spectra is obtained.

Therefore, we have generalized the approach of Davenport [58], who introduces 'a commutative hypercomplex algebra with associated function theory'. Davenport uses the $\mathbb{C}^2$ algebra (commutative ring with unity) in order to extend the classical complex analysis for treating four-dimensional variables, which are similar to quaternions. Ell [69] applies this approach to the quaternionic Fourier transform in order to simplify the convolution theorem.

We have picked up this idea to develop fast algorithms for the CFT (Chap. 10). For the separation of the CFT-kernel, we need a commutative algebra. Therefore, we have designed a new transform which is based on a different algebra, but yields the same spectrum as the CFT for real signals. For hypercomplex valued signals the spectrum differs from the Clifford spectrum.

Though it seems that the commutative hypercomplex Fourier transform (HFT) is no more than a tool for easier or faster calculation of the CFT, we will show in this chapter, that the HFT has the same right to exist as the CFT, because neither transform can be considered to be *the* correct extension of the complex Fourier transform (both yield the complex FT in the 1-D case). Up to now, there is no fundamental reason that determines which transform to use. Therefore, we study the properties of both transforms.

In this chapter we show several important properties of the algebra that generalizes Davenport's approach. After introducing the algebraic framework, we define the HFT and prove several theorems. We do this to motivate the reader to make his own experiments. This chapter together with Chap. 10 should form a base for further analytic and numerical investigations.

9.2 Hypercomplex Algebras

In this section, we define the algebraic framework for the rest of the chapter. The term *hypercomplex algebra* is explained and a specific four-dimensional algebra is introduced.

9.2.1 Basic Definitions

In general, a hypercomplex algebra is generated by a hypercomplex number system and a multiplication which satisfies the algebra axioms (see [125]).

To start with, we define what is meant by the term *hypercomplex number* (see also Chap. 7):

Definition 9.2.1 (Hypercomplex numbers). *A hypercomplex number of dimension n is an expression of the form*

$$\boldsymbol{a} = a_0 + a_1 i_1 + a_2 i_2 + \ldots + a_{n-1} i_{n-1} \tag{9.1}$$

where $a_j \in \mathbb{R}$ for all $j \in \{0, \ldots, n-1\}$ and i_j ($j \in \{1, \ldots, n-1\}$) are formal symbols (often called imaginary units). Two hypercomplex numbers

$$\boldsymbol{a} = a_0 + a_1 i_1 + \ldots + a_{n-1} i_{n-1} \qquad \text{and}$$
$$\boldsymbol{b} = b_0 + b_1 i_1 + \ldots + b_{n-1} i_{n-1}$$

are equal if and only if $a_j = b_j$ for all $j \in \{0, \ldots, n-1\}$.

Take, for example, $n = 2$. In this case we obtain numbers of the form $a_0 + a_1 i_1$ – this could be the complex numbers, dual numbers or double numbers. If $n = 4$ we obtain numbers of the form $a_0 + a_1 i_1 + a_2 i_2 + a_3 i_3$. This could be the quaternions or the commutative algebra which we will introduce at the end of this section.

Definition 9.2.2 (Addition, subtraction, and multiplication). *The addition of two hypercomplex numbers $\boldsymbol{a}$ and $\boldsymbol{b}$ is defined by*

$$\begin{aligned}\boldsymbol{a} + \boldsymbol{b} &= (a_0 + a_1 i_1 + \ldots + a_{n-1} i_{n-1}) + (b_0 + b_1 i_1 + \ldots + b_{n-1} i_{n-1}) \\ &= (a_0 + b_0) + (a_1 + b_1) i_1 + \ldots + (a_{n-1} + b_{n-1}) i_{n-1}\end{aligned} \tag{9.2}$$

and their subtraction is defined by

$$\begin{aligned}\boldsymbol{a} - \boldsymbol{b} &= (a_0 + a_1 i_1 + \ldots + a_{n-1} i_{n-1}) - (b_0 + b_1 i_1 + \ldots + b_{n-1} i_{n-1}) \\ &= (a_0 - b_0) + (a_1 - b_1) i_1 + \ldots + (a_{n-1} - b_{n-1}) i_{n-1} \,.\end{aligned} \tag{9.3}$$

The multiplication of two hypercomplex numbers is defined by an $(n-1) \times (n-1)$ multiplication table with the entries

$$i_\alpha i_\beta = p_0^{\alpha\beta} + p_1^{\alpha\beta} i_1 + \ldots + p_{n-1}^{\alpha\beta} i_{n-1} \tag{9.4}$$

where $\alpha, \beta \in \{1, \ldots, n-1\}$. The product

$$\boldsymbol{ab} = (a_0 + a_1 i_1 + \ldots + a_{n-1} i_{n-1})(b_0 + b_1 i_1 + \ldots + b_{n-1} i_{n-1}) \tag{9.5}$$

is evaluated by using the distributive law and the multiplication table.

The sum and the difference of two hypercomplex numbers are calculated like in an n-dimensional vectorspace with the base vectors $1, i_1, i_2, \ldots, i_{n-1}$. The product is more general than a vectorspace product: we can embed the commonly used products in this hypercomplex product.

If we, for example, consider the scalar product according to the Euclidean norm, then we have $p_j^{\alpha\beta} = 0$ for $j \neq 0$ and $p_0^{\alpha\beta} = 1$ for $\alpha = \beta$.

Standard algebra products are covered by the hypercomplex product, too. For example, the product of the algebra of complex numbers is obtained for $n = 2$, $p_0^{11} = -1$ and $p_1^{11} = 0$. The quaternion product is obtained by the following table 9.1. According to this table, we have $p_0^{jj} = -1$ ($j = 1, 2, 3$), $p_k^{jj} = 0$ ($j = 1, 2, 3$, $k = 1, 2, 3$), etc..

Table 9.1. Multiplication table of the quaternion algebra

	i_1	i_2	i_3
i_1	-1	i_3	$-i_2$
i_2	$-i_3$	-1	i_1
i_3	i_2	$-i_1$	-1

A *hypercomplex number system* of dimension n consists of all numbers of the form (9.1) of dimension n and the operations which are defined in (9.2),(9.3), and (9.5).

A hypercomplex number system contains even more structure than it seems so far. In the following theorem, we show that a hypercomplex number system forms an associative algebra.

Theorem 9.2.1 (Hypercomplex algebra). *All hypercomplex number systems fulfill the following properties and therefore they are associative algebras:*

1. *the product is bilinear, i.e.*

$$(a\boldsymbol{u})\boldsymbol{v} = a(\boldsymbol{u}\boldsymbol{v}) = \boldsymbol{u}(a\boldsymbol{v}) \tag{9.6a}$$

$$(\boldsymbol{v}+\boldsymbol{w})\boldsymbol{u} = \boldsymbol{v}\boldsymbol{u} + \boldsymbol{w}\boldsymbol{u} \tag{9.6b}$$

$$\boldsymbol{u}(\boldsymbol{v}+\boldsymbol{w}) = \boldsymbol{u}\boldsymbol{v} + \boldsymbol{u}\boldsymbol{w} \quad , \tag{9.6c}$$

2. *and the product is associative, i.e.*

$$\boldsymbol{u}(\boldsymbol{v}\boldsymbol{w}) = (\boldsymbol{u}\boldsymbol{v})\boldsymbol{w} \ , \tag{9.7}$$

where $\boldsymbol{u}, \boldsymbol{v}, \boldsymbol{w}$ *are hypercomplex numbers and* $a \in \mathbb{R}$.

Proof. The theorem is proved by elementary calculations using Def. 9.2.2. □

Therefore, complex numbers and their product form the algebra of complex numbers, the quaternions and their product form the algebra of quaternions, etc..

9.2.2 The Commutative Algebra $\mathcal{H}_2$

In the following we consider a further, specific four-dimensional hypercomplex algebra, which is commutative and somehow similar to the algebra of quaternions. The new algebra denoted by $\mathcal{H}_2$ is formed by the space

$$\mathrm{span}(1, \mathbf{e}_1\wedge\mathbf{e}_3, \mathbf{e}_2\wedge\mathbf{e}_4, \mathbf{e}_1\wedge\mathbf{e}_3\wedge\mathbf{e}_2\wedge\mathbf{e}_4)$$

and the geometric product. Consequently, $\mathcal{H}_2$ is a subalgebra of $\mathbb{R}^+_{4,0}$ and we have the following multiplication table (Tab. 9.2):

The same multiplication table is obtained for the two-fold tensor product of the complex algebra ($\mathbb{C}\otimes\mathbb{C}$). In this case, we have the basis elements $\{1\otimes 1, i\otimes 1, 1\otimes i, i\otimes i\}$. Since $\mathcal{H}_2$ and $\mathbb{C}\otimes\mathbb{C}$ have the same dimension

Table 9.2. Multiplication table of $\mathcal{H}_2$

	$\mathbf{e}_1 \wedge \mathbf{e}_3$	$\mathbf{e}_2 \wedge \mathbf{e}_4$	$\mathbf{e}_1 \wedge \mathbf{e}_3 \wedge \mathbf{e}_2 \wedge \mathbf{e}_4$
$\mathbf{e}_1 \wedge \mathbf{e}_3$	-1	$\mathbf{e}_1 \wedge \mathbf{e}_3 \wedge \mathbf{e}_2 \wedge \mathbf{e}_4$	$-\mathbf{e}_2 \wedge \mathbf{e}_4$
$\mathbf{e}_2 \wedge \mathbf{e}_4$	$\mathbf{e}_1 \wedge \mathbf{e}_3 \wedge \mathbf{e}_2 \wedge \mathbf{e}_4$	-1	$-\mathbf{e}_1 \wedge \mathbf{e}_3$
$\mathbf{e}_1 \wedge \mathbf{e}_3 \wedge \mathbf{e}_2 \wedge \mathbf{e}_4$	$-\mathbf{e}_2 \wedge \mathbf{e}_4$	$-\mathbf{e}_1 \wedge \mathbf{e}_3$	1

and the multiplication tables[1] are the same, $\mathcal{H}_2$ and $\mathbb{C} \otimes \mathbb{C}$ are isomorphic as algebras by the mapping $f^2 : \mathcal{H}_2 \to \mathbb{C} \otimes \mathbb{C}$ and $f(1) = 1 \otimes 1$, $f(\mathbf{e}_1 \wedge \mathbf{e}_3) = i \otimes 1$, $f(\mathbf{e}_2 \wedge \mathbf{e}_4) = 1 \otimes i$, and $f(\mathbf{e}_1 \wedge \mathbf{e}_3 \wedge \mathbf{e}_2 \wedge \mathbf{e}_4) = i \otimes i$.

Since the multiplication table is symmetric with respect to the major diagonal, the algebra $\mathcal{H}_2$ is commutative. Furthermore, Tab. 9.2 is equal to the multiplication table of the quaternion algebra (Tab. 9.1) in the cells $(1,1), (1,2), (1,3), (2,2)$ and $(3,2)$. In particular, we obtain for $(a+ib)(c+jd)$ in the quaternion algebra the same coefficients as for $(a+b\mathbf{e}_1 \wedge \mathbf{e}_3)(c+d\mathbf{e}_2 \wedge \mathbf{e}_4)$ in the algebra $\mathcal{H}_2$:

$$(a + ib)(c + jd) = ac + ibc + jad + kbd \tag{9.8a}$$

$$\begin{aligned}(a + b\mathbf{e}_1 \wedge \mathbf{e}_3)(c + d\mathbf{e}_2 \wedge \mathbf{e}_4) &= ac + bc\mathbf{e}_1 \wedge \mathbf{e}_3 + ad\mathbf{e}_2 \wedge \mathbf{e}_4 \\ &\quad + bd\mathbf{e}_1 \wedge \mathbf{e}_3 \wedge \mathbf{e}_2 \wedge \mathbf{e}_4 \ .\end{aligned} \tag{9.8b}$$

From this fact we will conclude The. 9.3.1 about the commutative hypercomplex Fourier transform (HFT2) of a real signal in the following section.

9.3 The Two-Dimensional Hypercomplex Fourier Analysis

In this section, we firstly define an integral transform which is based on the commutative algebra $\mathcal{H}_2$ and acts on hypercomplex 2-D signals. This transform which is denoted HFT2 yields the same spectrum as the QFT (8.3.4) for real signals. We reformulate the affine theorem, the convolution theorem, the symmetry theorem, and the shift theorem. Additionally, we prove that the algebra $\mathcal{H}_2$ and the two-fold Cartesian product of the complex numbers are isomorphic as algebras (see also [58]).

9.3.1 The Two-Dimensional Hypercomplex Fourier Transform

We introduce the HFT2 according to Ell [69] in the following. Furthermore, we make some fundamental considerations about this transform.

Definition 9.3.1 (Commutative hypercomplex Fourier transform). *The two-dimensional commutative hypercomplex Fourier transform (HFT2) of a two-dimensional signal $f(x, y)$ is defined by*

[1] Note that both algebras are multilinear.

$$F^h(u,v) = \int_{-\infty}^{\infty}\int_{-\infty}^{\infty} f(x,y)e^{-2\pi(xu\mathbf{e}_1\wedge\mathbf{e}_3+yv\mathbf{e}_2\wedge\mathbf{e}_4)}\,dx\,dy\ . \tag{9.9}$$

Note that due to the commutativity of $\mathcal{H}_2$ we have the identity

$$e^{-2\pi(xu\mathbf{e}_1\wedge\mathbf{e}_3+yv\mathbf{e}_2\wedge\mathbf{e}_4)} = e^{-2\pi xu\mathbf{e}_1\wedge\mathbf{e}_3}e^{-2\pi yv\mathbf{e}_2\wedge\mathbf{e}_4}\ .$$

The commutativity implies that all commutator terms in the Campbell-Hausdorff formula (see e.g. [206]) vanish.

As mentioned at the end of the last section, the product of two quaternions and the product of two $\mathcal{H}_2$ elements are equal if the multiplication in the algebra $\mathbb{H}$ is ordered wrt. the index set of the basis. That means that no product of the form $\mathbf{e}_i\mathbf{e}_j$ with $i \geq j$ appears. This is the case for the quaternionic Fourier transform of real signals:

Theorem 9.3.1 (Correspondence of HFT2 and QFT). *The 2-D commutative hypercomplex Fourier transform of a real 2-D signal $f(x,y)$ yields the same coefficients as the QFT of $f(x,y)$.*

Proof. The coefficient of the spectra are the same, because all multiplications have the form (9.8a,b). □

In particular, we can decompose both transforms into four real-valued transforms: a cos-cos-transform, a cos-sin-transform, a sin-cos-transform, and a sin-sin-transform. Then, we can take the real valued transforms as coefficients of the QFT and the HFT2 spectrum (see Def. 8.3.1):

$$F^q = \mathcal{CC}\{f\} - \mathcal{SC}\{f\}i - \mathcal{CS}\{f\}j + \mathcal{SS}\{f\}k \tag{9.10a}$$

$$F^h = \mathcal{CC}\{f\} - \mathcal{SC}\{f\}\mathbf{e}_1\wedge\mathbf{e}_3 - \mathcal{CS}\{f\}\mathbf{e}_2\wedge\mathbf{e}_4 + \mathcal{SS}\{f\}\mathbf{e}_1\wedge\mathbf{e}_3\wedge\mathbf{e}_2\wedge\mathbf{e}_4\ . \tag{9.10b}$$

The HFT2 (9.9) yields a geometric interpretation concerning the spatial and the frequency domain. If we span the spatial domain by $\mathbf{e}_1$ and $\mathbf{e}_2$, i.e. each point is represented by $x\mathbf{e}_1 + y\mathbf{e}_2 = \boldsymbol{x} + \boldsymbol{y}$, and same with the frequency domain ($u\mathbf{e}_3 + v\mathbf{e}_4 = \boldsymbol{u} + \boldsymbol{v}$), we can rewrite (9.9) as

$$F^h(u,v) = \int_{-\infty}^{\infty}\int_{-\infty}^{\infty} f(x,y)e^{2\pi(\boldsymbol{u}\wedge\boldsymbol{x}+\boldsymbol{v}\wedge\boldsymbol{y})}\,dx\,dy\ . \tag{9.11}$$

Both, the spatial and the frequency domain are 2-D vectorspaces, which are orthogonal with respect to each other (see Fig. 9.1).

The hypercomplex spectrum includes a scalar part, two bivector parts ($\mathbf{e}_1\wedge\mathbf{e}_3$ and $\mathbf{e}_2\wedge\mathbf{e}_4$) and a four-vector part. Therefore, the spectral values are denoted in the same algebra as the coordinates! This is an obvious advantage of Def. 9.11 and therefore, we use this definition for proving theorems in the following. Nevertheless, all results can easily be transferred to the Def. 9.9.

One property which is important in signal theory is the uniqueness of a transform and the existence of an inverse transform. Otherwise, the identification and manipulation of the signal in frequency representation would not be possible.

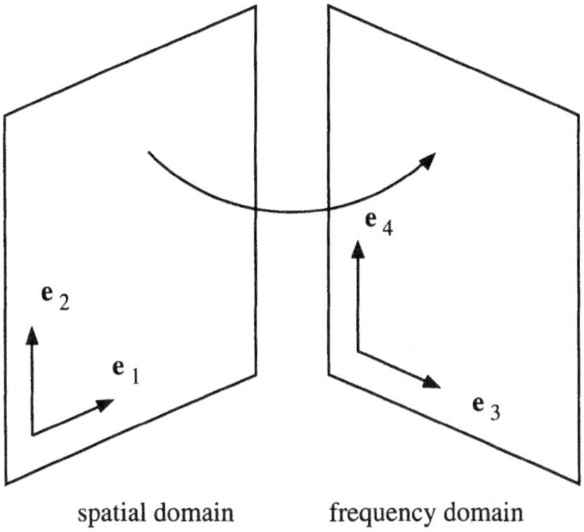

Fig. 9.1. The HFT2 visualized

Theorem 9.3.2 (The HFT2 is unique and invertible). *The HFT2 of a signal f is unique and invertible.*

Proof. In order to show the uniqueness, we prove that the kernel of the transform consists of orthogonal functions. We do so by reducing the exponential function to sine and cosine functions:

$$\begin{aligned} e^{2\pi(\boldsymbol{u}\wedge\boldsymbol{x}+\boldsymbol{v}\wedge\boldsymbol{y})} &= e^{2\pi\boldsymbol{u}\wedge\boldsymbol{x}} e^{2\pi\boldsymbol{v}\wedge\boldsymbol{y}} \\ &= \big(\cos(2\pi ux) - \mathbf{e}_1\wedge\mathbf{e}_3\sin(2\pi ux)\big)\big(\cos(2\pi vy) - \mathbf{e}_2\wedge\mathbf{e}_4\sin(2\pi vy)\big) \end{aligned}$$

Since the sine and cosine functions are orthogonal, the HFT2 is unique and furthermore, the inverse transform reads

$$f(x,y) = \int_{\mathbb{R}^2} F^h(u,v) e^{2\pi(\boldsymbol{x}\wedge\boldsymbol{u}+\boldsymbol{y}\wedge\boldsymbol{v})}\, du\, dv \ ,$$

which can be verified by a straightforward calculation. □

One nice property of the HFT2 is the fact that both, the transform and the inverse transform, are formulated in the same way[2]. The minus sign which we have for the complex Fourier transform and for the QFT can be omitted for the HFT2.

If we recall the isomorphism between $\mathcal{H}_2$ and $\mathbb{C}\otimes\mathbb{C}$, we can rewrite the HFT2-kernel as

$$e^{-2\pi(xu\mathbf{e}_1\wedge\mathbf{e}_3+yv\mathbf{e}_2\wedge\mathbf{e}_4)} \cong e^{-i2\pi xu} \otimes e^{-i2\pi yv} \ . \tag{9.12}$$

Consider now a real-valued, separable signal $f(x,y) = f^x(x)f^y(y)$. Then, due to multilinearity, the HFT2 of $f(x,y)$ itself can be written as

[2] We know such property from the Hartley transform.

$$f(x,y) \circ\!\!-\!\!\bullet\, F^h(u,v) \cong F^x(u) \otimes F^y(v) \tag{9.13}$$

where $F^x(u)$ is the 1-D Fourier transform of the signal $f(x,y)$ wrt. the x-coordinate and $F^y(v)$ accordingly to the y-coordinate.

This notation introduces another interpretation of the HFT2: we obtain the HFT2 of a real, separable signal by the tensor product of the complex 1-D spectra. Note that this is not valid for hypercomplex signals, because in that case we cannot exchange the tensor product and the product between signal and kernel[3]. Nevertheless, since the coefficients of the quaternionic spectrum and the $\mathcal{H}_2$ spectrum of a real signal are the same, the QFT of a separable signal can be interpreted as the tensor product of complex 1-D Fourier transforms wrt. to x and y as well.

9.3.2 Main Theorems of the HFT2

In this section, we consider some theorems for the HFT2. For the QFT, some of the main theorems are more complicated compared to those of the complex FT. We will show that this drawback is less crucial for the HFT2.

In the shift theorem of the QFT (Eq. 8.29) one exponential factor moves to the left (the i term) and one moves to the right (the j term). This is necessary since the algebra of quaternions is not commutative. But what can we do for Clifford transforms of higher dimension? There are only two ways of multiplication: one from the left and one from the right. The great advantage of commutative algebras is the fact that neither the order nor the direction of multiplication is relevant. Hence, the shift theorem for the HFT yields two exponential factors which can be placed arbitrarily or even composed in one exponential factor:

Theorem 9.3.3 (Shift theorem). *Let $F^h(u,v)$ be the HFT2 of a signal $f(x,y)$. Then, the HFT2 of the signal $f'(x,y) = f(x-\xi, y-\eta)$ reads $F^{h'}(u,v) = e^{2\pi(\boldsymbol{u}\wedge\boldsymbol{\xi}+\boldsymbol{v}\wedge\boldsymbol{\eta})}F^h(u,v)$ where $\boldsymbol{\xi} = \xi\mathbf{e}_1$ and $\boldsymbol{\eta} = \eta\mathbf{e}_2$.*

Proof. We prove this theorem by straightforward calculation:

$$\begin{aligned}
&F^{h'}(u,v)\\
&= \int_{\mathbb{R}^2} f(x-\xi, y-\eta)\, e^{2\pi(\boldsymbol{u}\wedge\boldsymbol{x}+\boldsymbol{v}\wedge\boldsymbol{y})}\, dx\, dy\\
&\overset{\substack{\boldsymbol{x}-\boldsymbol{\xi}=\boldsymbol{x}'\\ \boldsymbol{y}-\boldsymbol{\eta}=\boldsymbol{y}'}}{=} \int_{\mathbb{R}^2} f(x',y')\, e^{2\pi(\boldsymbol{u}\wedge\boldsymbol{x}'+\boldsymbol{v}\wedge\boldsymbol{y}')} e^{2\pi(\boldsymbol{u}\wedge\boldsymbol{\xi}+\boldsymbol{v}\wedge\boldsymbol{\eta})}\, dx\, dy\\
&= e^{2\pi(\boldsymbol{u}\wedge\boldsymbol{\xi}+\boldsymbol{v}\wedge\boldsymbol{\eta})} F^h(u,v)
\end{aligned}$$

Hence, the theorem is proved. □

[3] Note that, since we use the field $\mathbb{R}$, the multilinearity is only valid for real factors.

The shift theorem of the complex FT is closely related to the modulation theorem. The relation is even more general: we have a so-called symmetry theorem, which yields the Fourier transform of a signal simply by the inverse Fourier transform of the signal. We can formulate this theorem for the HFT2 as well:

Theorem 9.3.4 (Symmetry of the HFT2). *Let $f(x,y)$ be a $\mathcal{H}_2$-valued signal and $F^h(u,v)$ its HFT2. Then, the HFT2 of ${F^h}^\dagger(x,y)$ reads $f^\dagger(u,v)$ (where $\cdot^\dagger$ indicates the reversion of the underlying geometric algebra).*

Proof. In this proof we notate the exponents in the form $ux\mathbf{e}_3 \wedge \mathbf{e}_1$ instead of $\boldsymbol{u} \wedge \boldsymbol{x}$. The HFT2 of ${F^h}^\dagger(x,y)$ reads

$$\int_{\mathbb{R}^2} {F^h}^\dagger(x,y)\, e^{2\pi(ux\mathbf{e}_3\wedge\mathbf{e}_1+vy\mathbf{e}_4\wedge\mathbf{e}_2)}\,dx\,dy$$

$$= \int_{\mathbb{R}^4} (f(u',v')\, e^{2\pi(u'x\mathbf{e}_3\wedge\mathbf{e}_1+v'y\mathbf{e}_4\wedge\mathbf{e}_2)})^\dagger\,du'dv'\, e^{2\pi(ux\mathbf{e}_3\wedge\mathbf{e}_1+vy\mathbf{e}_4\wedge\mathbf{e}_2)}\,dx\,dy$$

$$= \int_{\mathbb{R}^2} f^\dagger(u',v') \int_{\mathbb{R}^2} e^{2\pi(u'x\mathbf{e}_1\wedge\mathbf{e}_3+v'y\mathbf{e}_2\wedge\mathbf{e}_4)}\, e^{2\pi(ux\mathbf{e}_3\wedge\mathbf{e}_1+vy\mathbf{e}_4\wedge\mathbf{e}_2)}\,dx\,dy\,du'dv'$$

$$= \int_{\mathbb{R}^2} f^\dagger(u',v')\delta(u-u')\delta(v-v')\,du'dv' = f^\dagger(u,v)$$

Note that in the commutative algebra $\mathcal{H}_2$ the order of the factors is not inverted by the reversion $\cdot^\dagger$ (so the reversion is an automorphism in $\mathcal{H}_2$). □

The shift theorem together with the symmetry theorem yield the modulation theorem of the HFT2 which we do not formulate explicitly.

Up to now there is no significant improvement in the formulation of the theorems, although some formulation might be more elegant. However, the next theorem shows that in the commutative algebra a closed formulation of the convolution theorem is possible. The 2-D convolution is defined as follows.

Definition 9.3.2 (2-D convolution). *Let $f(x,y), g(x,y)$ be two 2-D signals. The 2-D convolution $f * g$ is then defined by*

$$(f * g)(x,y) = \int_{\mathbb{R}^2} f(\xi,\eta)g(x-\xi,y-\eta)\,d\xi\,d\eta\ . \tag{9.14}$$

In contrast to the convolution theorem of the QFT, the convolution theorem of the HFT2 can be formulated similarly to the convolution theorem of the complex Fourier transform.

Theorem 9.3.5 (Convolution theorem of the FHT2). *Let $f(x,y)$ and $g(x,y)$ be two 2-D signals and let $F^h(u,v)$ and $G^h(u,v)$ be their HFT2s, respectively. Then, the HFT2 of $f * g$ is equivalent to the pointwise product of F^h and G^h, i.e.*

$$\int_{\mathbb{R}^2} (f * g)(x,y)e^{2\pi(\boldsymbol{u}\wedge\boldsymbol{x}+\boldsymbol{v}\wedge\boldsymbol{y})}\,dx\,dy = F^h(u,v)G^h(u,v)\ . \tag{9.15}$$

Proof. We obtain by straightforward calculation

$$\begin{aligned}
&\int_{\mathbb{R}^2} (f * g)(x,y) e^{2\pi(\boldsymbol{u}\wedge\boldsymbol{x}+\boldsymbol{v}\wedge\boldsymbol{y})}\,dx\,dy \\
&= \int_{\mathbb{R}^2}\int_{\mathbb{R}^2} f(\xi,\eta) g(x-\xi, y-\eta)\,d\xi\,d\eta\, e^{2\pi(\boldsymbol{u}\wedge\boldsymbol{x}+\boldsymbol{v}\wedge\boldsymbol{y})}\,dx\,dy \\
&\overset{\substack{x-\xi=x'\\ y-\eta=y'}}{=} \int_{\mathbb{R}^2}\int_{\mathbb{R}^2} f(\xi,\eta) g(x',y') e^{2\pi(\boldsymbol{u}\wedge\boldsymbol{\xi}+\boldsymbol{v}\wedge\boldsymbol{\eta})} e^{2\pi(\boldsymbol{u}\wedge\boldsymbol{x}'+\boldsymbol{v}\wedge\boldsymbol{y}')}\,d\xi\,d\eta\,dx'dy' \\
&= F^h(u,v) G^h(u,v)
\end{aligned}$$

and therefore, the theorem is proved. □

Of course, the convolution defined in Def. 9.3.2 can be formulated for discrete signals as well. Note that we obtain the *cyclic* convolution by the pointwise product in the frequency domain and not the *linear* convolution. If the latter is needed, the signal must be filled up by zeroes.

9.3.3 The Affine Theorem of the HFT2

The next theorem states an isomorphism between $\mathcal{H}_2$ and the two-fold (Cartesian) product of the complex algebra ($\mathbb{C}^2$, see also [58]). Though this theorem seems to be a pure mathematical result, it will be important for the subsequent theorems.

Theorem 9.3.6 ($\mathcal{H}_2 \cong \mathbb{C}^2$). *The commutative hypercomplex algebra $\mathcal{H}_2$ is isomorphic to the two-fold (Cartesian) product of the complex algebra $\mathbb{C}^2$. For an arbitrary element $Z = a + b\mathbf{e}_1\wedge\mathbf{e}_3 + c\mathbf{e}_2\wedge\mathbf{e}_4 + d\mathbf{e}_1\wedge\mathbf{e}_3\wedge\mathbf{e}_2\wedge\mathbf{e}_4$ we obtain the representation $(\xi,\eta) = \big((a-d)+i(b+c)\,,\,(a+d)+i(b-c)\big) \in \mathbb{C}^2$.*

Proof. Consider the matrix representations of $Z \in \mathcal{H}_2$ and $z = (x+iy) \in \mathbb{C}$

$$Z \cong a\begin{bmatrix}1&0&0&0\\0&1&0&0\\0&0&1&0\\0&0&0&1\end{bmatrix} + b\begin{bmatrix}0&-1&0&0\\1&0&0&0\\0&0&0&-1\\0&0&1&0\end{bmatrix} + c\begin{bmatrix}0&0&-1&0\\0&0&0&-1\\1&0&0&0\\0&1&0&0\end{bmatrix} + d\begin{bmatrix}0&0&0&1\\0&0&-1&0\\0&-1&0&0\\1&0&0&0\end{bmatrix}$$

$$z \cong x\begin{bmatrix}1&0\\0&1\end{bmatrix} + y\begin{bmatrix}0&-1\\1&0\end{bmatrix}$$

which can both easily be verified to be isomorphic.

The eigenvectors of the matrix representation of Z read

$$e_1 = \begin{bmatrix}1\\-i\\-i\\-1\end{bmatrix} \quad e_2 = \begin{bmatrix}1\\i\\i\\-1\end{bmatrix} \quad e_3 = \begin{bmatrix}1\\-i\\i\\1\end{bmatrix} \quad e_4 = \begin{bmatrix}1\\i\\-i\\1\end{bmatrix} \tag{9.16}$$

and since these eigenvectors are independent of the coefficients a, b, c, d, they yield an eigenvalue transform which turns the matrix representation of any Z into diagonal form. The eigenvalues read

$$\xi = (a-d) + i(b+c) \qquad \xi^* = (a-d) - i(b+c)$$
$$\eta = (a+d) + i(b-c) \qquad \eta^* = (a+d) - i(b-c)$$

and therefore, the matrix multiplication yields a pointwise product on (ξ, η). □

Note 9.3.1. The proof of The. 9.3.6 is only sketched because it is a special case of The. 9.4.1. In the following, we use this theorem in a less formal way, since we replace the i of the complex numbers by $\mathbf{e}_1 \wedge \mathbf{e}_3$ in the following. The reason for this is that we can write $Z = \xi b_1 + \eta b_2$ now, where $b_1 = (1 - \mathbf{e}_1 \wedge \mathbf{e}_3 \wedge \mathbf{e}_2 \wedge \mathbf{e}_4)/2$ and $b_2 = (1 + \mathbf{e}_1 \wedge \mathbf{e}_3 \wedge \mathbf{e}_2 \wedge \mathbf{e}_4)/2$.

Theorem 9.3.6 can be used to state a relation between the complex Fourier transform and the HFT2. Due to the isomorphism we can map the signal and the kernel to $\mathbb{C}^2$ and can perform all calculations in this representation. Afterwards, the HFT2-spectrum can be obtained from the two complex spectra. In this context, it is not surprising that the representation of the HFT2 kernel in $\mathbb{C}^2$ consists of two Fourier kernels.

Theorem 9.3.7 (Relation between FT2 and HFT2). *Let $f(x,y)$ be a two-dimensional $\mathcal{H}_2$-valued signal and let $(f_\xi(x,y), f_\eta(x,y))$ be its representation in $\mathbb{C}^2$. Furthermore, let $F_\xi(u,v)$ and $F_\eta(u,v)$ be the complex Fourier transforms of $f_\xi(x,y)$ and $f_\eta(x,y)$, respectively. Then, the HFT2 of $f(x,y)$ reads*

$$F^h(u,v) = F_\xi(u,v)b_1 + F_\eta(u,-v)b_2 \; . \tag{9.17}$$

Proof. We can rewrite the kernel of the HFT2 by

$$e^{2\pi(xu\mathbf{e}_3\wedge\mathbf{e}_1 + yv\mathbf{e}_4\wedge\mathbf{e}_2)} = e^{2\pi(xu+yv)\mathbf{e}_3\wedge\mathbf{e}_1}b_1 + e^{2\pi(xu-yv)\mathbf{e}_3\wedge\mathbf{e}_1}b_2 \; . \tag{9.18}$$

Therefore, we obtain for the HFT2 of $f(x,y) = f_\xi(x,y)b_1 + f_\eta(x,y)b_2$ by

$$\begin{aligned}
F^h(u,v) &= \int_{\mathbb{R}^2} f(x,y)\, e^{2\pi(xu\mathbf{e}_3\wedge\mathbf{e}_1 + yv\mathbf{e}_4\wedge\mathbf{e}_2)}\, dx\, dy \\
&= \int_{\mathbb{R}^2} (f_\xi(x,y)b_1 + f_\eta(x,y)b_2) \\
&\qquad (e^{2\pi(xu+yv)\mathbf{e}_3\wedge\mathbf{e}_1}b_1 + e^{2\pi(xu-yv)\mathbf{e}_3\wedge\mathbf{e}_1}b_2)\, dx\, dy \\
&= \int_{\mathbb{R}^2} f_\xi(x,y)\, e^{2\pi(xu+yv)\mathbf{e}_3\wedge\mathbf{e}_1}\, dx\, dy\, b_1 \\
&\quad + \int_{\mathbb{R}^2} f_\eta(x,y)\, e^{2\pi(xu-yv)\mathbf{e}_3\wedge\mathbf{e}_1}\, dx\, dy\, b_2 \\
&= F_\xi(u,v)b_1 + F_\eta(u,-v)b_2
\end{aligned}$$

(note that the product in $\mathbb{C}^2$ is evaluated pointwise ($b_1 b_2 = 0$) and that the kernels in (9.18) are the kernels of the complex 2-D FTs $\mathcal{F}\{\cdot\}(u,v)$ and $\mathcal{F}\{\cdot\}(u,-v)$). □

Note that we can obtain the HFT2 of a real signal by the complex spectrum because the spectra F_ξ and F_η are equal in that case. Therefore, the extended (i.e. hypercomplex) representation of a real signal is calculated without increased computational effort!

The simple calculation of the HFT2 spectrum is not the only result of the relation between the complex FT and the HFT2. Using the last theorem, we can state the affine theorem in a straightforward way:

Theorem 9.3.8 (Affine theorem). *Let $F^h(u,v)$ be the HFT2 of a signal $f(x,y)$. Then, the HFT2 of the signal $f'(x,y) = f(x',y')$ reads*

$$F^{h'}(u,v) = \frac{1}{|\det A|}\left(F^h(u',v')b_1 + F^h(u'',v'')b_2\right) \tag{9.19}$$

where $(x',y')^T = A(x,y)^T$, $(u',v')^T = A^{-1}(u,v)^T$, and $(u'',v'')^T = A^{-1}(u,-v)^T$.

Proof. First, we decompose $f(x,y) = f_\xi(x,y)b_1 + f_\eta(x,y)b_2$. According to The. 9.3.7 we have

$$F^h(u,v) = F_\xi(u,v)b_1 + F_\eta(u,-v)b_2$$

where $F_\xi(u,v) = \mathcal{F}\{f_\xi(x,y)\}(u,v)$ and $F_\eta(u,v) = \mathcal{F}\{f_\eta(x,y)\}(u,-v)$. Consider now $f'(x,y) = f(x',y') = f_\xi(x',y')b_1 + f_\eta(x',y')b_2$. The HFT2 of f' is obtained by

$$F^{h'}(u,v) = \mathcal{F}\{f_\xi(x',y')\}(u,v)b_1 + \mathcal{F}\{f_\eta(x',y')\}(u,-v)b_2 \ .$$

According to the affine theorem of the complex FT, we have

$$\mathcal{F}\{f_\xi(x',y')\}(u,v) = \frac{1}{|\det A|}F_\xi(u',v') \qquad \text{and}$$

$$\mathcal{F}\{f_\eta(x',y')\}(u,-v) = \frac{1}{|\det A|}F_\eta(u'',v'')$$

and since $F^h(u,v)b_1 = F_\xi(u,v)b_1$ and $F^h(u,v)b_2 = F_\eta(u,v)b_2$ we obtain (9.19). □

Obviously, the affine theorem is more complicated for the HFT2 than for the complex FT. This results from the fact that the spatial coordinates and the frequency coordinates are not combined by a scalar product. This is some kind of drawback since the recognition of a rotated signal in frequency domain is more complicated. On the other hand, filtering can be performed nearly isotropic (rotation invariant), e.g. the concept of hypercomplex Gabor filters yields a lower energy dependency on the orientation in contrast to complex Gabor filters (see 11).

Last but not least, let us consider the energy of an $\mathcal{H}_2$-valued signal. The magnitude of a multi-vector M is obtained by $|M| = \sqrt{MM^\dagger}$. Consequently, the magnitude of an $\mathcal{H}_2$-valued number

$$h = a + b\mathbf{e}_1 \wedge \mathbf{e}_3 + c\mathbf{e}_2 \wedge \mathbf{e}_4 + d\mathbf{e}_1 \wedge \mathbf{e}_3 \wedge \mathbf{e}_2 \wedge \mathbf{e}_4$$

also reads

$$|h| = \sqrt{a^2 + b^2 + c^2 + d^2} = \sqrt{hh^\dagger} \ .$$

The energy of the HFT2 of a signal $f(x, y)$ is then obtained by

$$\int_{\mathbb{R}^2} F^h(u,v) F^{h^\dagger}(u,v)\, du\, dv \tag{9.20a}$$

$$= \int_{\mathbb{R}^6} f(x,y) e^{2\pi(\boldsymbol{x}\wedge\boldsymbol{u}+\boldsymbol{y}\wedge\boldsymbol{v})} e^{2\pi(\boldsymbol{u}\wedge\boldsymbol{x}'+\boldsymbol{v}\wedge\boldsymbol{y}')} f^\dagger(x',y')\, du\, dv\, dx\, dy\, dx'\, dy'$$

$$= \int_{\mathbb{R}^4} f(x,y)\delta(x-x')\delta(y-y') f^\dagger(x',y')\, dx\, dy\, dx'\, dy' \tag{9.20b}$$

$$= \int_{\mathbb{R}^2} f(x,y) f^\dagger(x,y)\, dx\, dy \tag{9.20c}$$

and therefore, the Parseval equation is satisfied by the FHT2. The energy of the HFT2 spectrum is equal to the energy of the signal.

The last subject of this section is the derivative theorem for the HFT2. It reads analogously to the derivative theorem of the QFT (The. 8.4.7)

$$\frac{\partial}{\partial x} f(x,y) \ \circ\!\!-\!\!\bullet \ 2\pi u \mathbf{e}_1 \wedge \mathbf{e}_3 F^h(u,v) \tag{9.21a}$$

$$\frac{\partial}{\partial y} f(x,y) \ \circ\!\!-\!\!\bullet \ 2\pi v \mathbf{e}_2 \wedge \mathbf{e}_4 F^h(u,v) \ . \tag{9.21b}$$

Finally, we have transferred all global concepts of the QFT to the HFT2.

9.4 The n-Dimensional Hypercomplex Fourier Analysis

In this section we generalize the commutative hypercomplex Fourier transform for arbitrary dimensions (HFTn). Firstly, we have to introduce an algebraic framework which is the systematic extension of $\mathcal{H}_2$: $\mathcal{H}_n$.

9.4.1 The Isomorphism between $\mathcal{H}_n$ and the 2^{n-1}-Fold Cartesian Product of $\mathbb{C}$

Consider the Clifford algebra $\mathbb{R}^+_{2n,0}$ and define a 2^n-dimensional hypercomplex number system based on the space which is induced by

$$i_j = \mathbf{e}_j \wedge \mathbf{e}_{n+j} \ , \quad j = 1, \ldots, n \ .$$

The basis elements of the number system are created by the following rules ($j = 1, \ldots, n$, $s \subseteq \{1, \ldots, n\}$[4])

$$i_j i_j = -1 = -i_\emptyset \tag{9.22a}$$

$$i_j i_s = i_s i_j = i_{\{j\} \cup s} \qquad j \notin s \tag{9.22b}$$

$$i_j i_s = i_s i_j = -i_{s \setminus \{j\}} \qquad j \in s \ . \tag{9.22c}$$

Obviously, span($i_s | \ s \subseteq \{1, \ldots, n\}$) and the Clifford product form a commutative 2^n-dimensional hypercomplex algebra which is denoted $\mathcal{H}_n$ in the sequel. The following lemma identifies this algebra.

Lemma 9.4.1 ($\mathbb{C}^{\otimes n} \cong \mathcal{H}_n$). *The algebra $\mathcal{H}_n \subset \mathbb{R}^+_{2n,0}$ which is formed by (9.22a,b,c) is isomorphic to the n-fold tensor product of the complex algebra.*

Proof. The basis vectors i_j in (9.22a) can be written as

$$i_j = 1 \otimes \ldots \otimes 1 \otimes \underset{\substack{\uparrow \\ j\text{th position}}}{i} \otimes 1 \otimes \ldots \otimes 1 \ . \tag{9.23}$$

Obviously, the basis vectors satisfy (9.22a, b and c) which can be verified by straightforward calculations. □

Consider, for example, $n = 3$. Then, the isomorphism yields the following correspondences of the basis elements:

$$\begin{aligned}
i_\emptyset &= 1 \cong 1 \otimes 1 \otimes 1 \\
i_1 &= \mathbf{e}_1 \wedge \mathbf{e}_4 \cong i \otimes 1 \otimes 1 \\
i_2 &= \mathbf{e}_2 \wedge \mathbf{e}_5 \cong 1 \otimes i \otimes 1 \\
i_3 &= \mathbf{e}_3 \wedge \mathbf{e}_6 \cong 1 \otimes 1 \otimes i \\
i_{12} &= \mathbf{e}_1 \wedge \mathbf{e}_4 \wedge \mathbf{e}_2 \wedge \mathbf{e}_5 \cong i \otimes i \otimes 1 \\
i_{13} &= \mathbf{e}_1 \wedge \mathbf{e}_4 \wedge \mathbf{e}_3 \wedge \mathbf{e}_6 \cong i \otimes 1 \otimes i \\
i_{23} &= \mathbf{e}_2 \wedge \mathbf{e}_5 \wedge \mathbf{e}_3 \wedge \mathbf{e}_6 \cong 1 \otimes i \otimes i \\
i_{123} &= \mathbf{e}_1 \wedge \mathbf{e}_4 \wedge \mathbf{e}_2 \wedge \mathbf{e}_5 \wedge \mathbf{e}_3 \wedge \mathbf{e}_6 \cong i \otimes i \otimes i
\end{aligned}$$

Lemma 9.4.2 (Matrix representation of $\mathbb{C}^{\otimes n}$ and $\mathcal{H}_n$). *The matrices I^n_s, $s \in \mathcal{P}(\{1, \ldots, n\})$ span the matrix representation of $\mathbb{C}^{\otimes n}$ (and therefore for $\mathcal{H}_n$, too). The matrices I^n_s are defined by*

$$I^0_\emptyset = 1 \tag{9.24a}$$

$$I^m_s = \begin{bmatrix} I^{m-1}_s & 0 \\ 0 & I^{m-1}_s \end{bmatrix} \tag{9.24b}$$

$$I^m_{s \cup \{m\}} = \begin{bmatrix} 0 & -I^{m-1}_s \\ I^{m-1}_s & 0 \end{bmatrix} \tag{9.24c}$$

[4] For the sake of short writing, the indices are sometimes denoted as sets in the sequel (e.g., $i_{123} = i_{\{1,2,3\}}$).

with $s \in \mathcal{P}(\{1, \ldots, m-1\})$ *and* $1 \leq m \leq n$.

Proof. Let $\boldsymbol{v} = (v_1, \ldots, v_n)$ be a vector of dimension n and $\operatorname{diag}(\boldsymbol{v})$ denotes the diagonal matrix $\begin{bmatrix} v_1 & 0 & \ldots & \\ 0 & v_2 & 0 & \ldots \\ & & \ddots & \\ \ldots & & 0 & v_n \end{bmatrix}$.

Then, for $s \in \mathcal{P}(\{1, \ldots, n\})$ we have the following identities:

$$I_\emptyset^n = \operatorname{diag}(1, \ldots, 1) \tag{9.25a}$$

$$I_j^n I_j^n = -I_\emptyset^n \qquad \text{with } j \in \{1, \ldots, n\} \tag{9.25b}$$

$$I_s^n I_j^n = I_j^n I_s^n = I_{s \cup \{j\}}^n \qquad \text{with } j \notin s \tag{9.25c}$$

$$I_s^n I_j^n = I_j^n I_s^n = -I_{s \setminus \{j\}}^n \qquad \text{with } j \in s\ . \tag{9.25d}$$

Hence, this matrix algebra follows the same multiplication rules as $\mathcal{H}_n$ and $\mathbb{C}^{\otimes n}$. Since all three algebras are of the same dimension, they are isomorphic as algebras. □

Consider again the case $n = 3$. The matrix representation of $\mathcal{H}_3$ is obtained by:

$$i_\emptyset \cong \begin{bmatrix} 1&0&0&0&0&0&0&0\\ 0&1&0&0&0&0&0&0\\ 0&0&1&0&0&0&0&0\\ 0&0&0&1&0&0&0&0\\ 0&0&0&0&1&0&0&0\\ 0&0&0&0&0&1&0&0\\ 0&0&0&0&0&0&1&0\\ 0&0&0&0&0&0&0&1 \end{bmatrix} \qquad i_{123} \cong \begin{bmatrix} 0&0&0&0&0&0&0&-1\\ 0&0&0&0&0&0&1&0\\ 0&0&0&0&0&1&0&0\\ 0&0&0&0&-1&0&0&0\\ 0&0&0&1&0&0&0&0\\ 0&0&-1&0&0&0&0&0\\ 0&-1&0&0&0&0&0&0\\ 1&0&0&0&0&0&0&0 \end{bmatrix}$$

$$i_1 \cong \begin{bmatrix} 0&-1&0&0&0&0&0&0\\ 1&0&0&0&0&0&0&0\\ 0&0&0&-1&0&0&0&0\\ 0&0&1&0&0&0&0&0\\ 0&0&0&0&0&-1&0&0\\ 0&0&0&0&1&0&0&0\\ 0&0&0&0&0&0&0&-1\\ 0&0&0&0&0&0&1&0 \end{bmatrix} \qquad i_{23} \cong \begin{bmatrix} 0&0&0&0&0&0&1&0\\ 0&0&0&0&0&0&0&1\\ 0&0&0&0&-1&0&0&0\\ 0&0&0&0&0&-1&0&0\\ 0&0&-1&0&0&0&0&0\\ 0&0&0&-1&0&0&0&0\\ 1&0&0&0&0&0&0&0\\ 0&1&0&0&0&0&0&0 \end{bmatrix}$$

$$i_2 \cong \begin{bmatrix} 0&0&-1&0&0&0&0&0\\ 0&0&0&-1&0&0&0&0\\ 1&0&0&0&0&0&0&0\\ 0&1&0&0&0&0&0&0\\ 0&0&0&0&0&0&-1&0\\ 0&0&0&0&0&0&0&-1\\ 0&0&0&0&1&0&0&0\\ 0&0&0&0&0&1&0&0 \end{bmatrix} \qquad i_{13} \cong \begin{bmatrix} 0&0&0&0&0&1&0&0\\ 0&0&0&0&-1&0&0&0\\ 0&0&0&0&0&0&0&1\\ 0&0&0&0&0&0&-1&0\\ 0&-1&0&0&0&0&0&0\\ 1&0&0&0&0&0&0&0\\ 0&0&0&-1&0&0&0&0\\ 0&0&1&0&0&0&0&0 \end{bmatrix}$$

$$i_3 \cong \begin{bmatrix} 0 & 0 & 0 & 0 & -1 & 0 & 0 & 0 \\ 0 & 0 & 0 & 0 & 0 & -1 & 0 & 0 \\ 0 & 0 & 0 & 0 & 0 & 0 & -1 & 0 \\ 0 & 0 & 0 & 0 & 0 & 0 & 0 & -1 \\ 1 & 0 & 0 & 0 & 0 & 0 & 0 & 0 \\ 0 & 1 & 0 & 0 & 0 & 0 & 0 & 0 \\ 0 & 0 & 1 & 0 & 0 & 0 & 0 & 0 \\ 0 & 0 & 0 & 1 & 0 & 0 & 0 & 0 \end{bmatrix} \qquad i_{12} \cong \begin{bmatrix} 0 & 0 & 0 & 1 & 0 & 0 & 0 & 0 \\ 0 & 0 & -1 & 0 & 0 & 0 & 0 & 0 \\ 0 & -1 & 0 & 0 & 0 & 0 & 0 & 0 \\ 1 & 0 & 0 & 0 & 0 & 0 & 0 & 0 \\ 0 & 0 & 0 & 0 & 0 & 0 & 0 & 1 \\ 0 & 0 & 0 & 0 & 0 & 0 & -1 & 0 \\ 0 & 0 & 0 & 0 & 0 & -1 & 0 & 0 \\ 0 & 0 & 0 & 0 & 1 & 0 & 0 & 0 \end{bmatrix}$$

The matrix representation of $\mathcal{H}_n$ has eigenvectors which are *independent* of the coefficients. Therefore, every element of $\mathcal{H}_n$ can be expressed by a diagonal matrix, which is obtained by a fixed eigenvalue transform.

Lemma 9.4.3 (Eigenvectors and eigenvalues of $\mathbb{C}^{\otimes n}$). *Let*

$$A^n = \sum_{s \in \mathcal{P}(\{1,\dots,n\})} k_s I_s^n$$

be the matrix representation of an arbitrary element of $\mathbb{C}^{\otimes n}$. Then, the matrix of eigenvectors (row vectors) of A^n is inductively constructed by

$$E^1 = \begin{bmatrix} 1 & -i \\ 1 & i \end{bmatrix} \quad \text{and} \tag{9.26a}$$

$$E^m = \begin{bmatrix} E^{m-1} & -iE^{m-1} \\ E^{m-1} & iE^{m-1} \end{bmatrix} \quad \text{with } 2 \leq m \leq n \ . \tag{9.26b}$$

The corresponding vector of eigenvalues reads $\boldsymbol{\eta}^n$ where $\mathrm{diag}(\boldsymbol{\eta}^n) = \boldsymbol{\eta}_\emptyset^n$ *and $\boldsymbol{\eta}_t^n$ is inductively defined by*

$$\boldsymbol{\eta}_t^0 = k_t \qquad \text{with } t \in \mathcal{P}(\{1,\dots,n\}) \text{ and} \tag{9.27a}$$

$$\boldsymbol{\eta}_t^m = \begin{bmatrix} \boldsymbol{\eta}_t^{m-1} - i\boldsymbol{\eta}_{\{m\}\cup t}^{m-1} & 0 \\ 0 & \boldsymbol{\eta}_t^{m-1} + i\boldsymbol{\eta}_{\{m\}\cup t}^{m-1} \end{bmatrix} , \tag{9.27b}$$

where $t \in \mathcal{P}(\{m+1,\dots,n\})$ and $1 \leq m \leq n$.

Proof. We prove the lemma by induction over n. Firstly, let $n = 1$. Then we have

$$(E^1)^{-1} = \frac{1}{2} \begin{bmatrix} 1 & 1 \\ i & -i \end{bmatrix}$$

and hence,

$$\begin{aligned} E^1 A^1 (E^1)^{-1} &= E^1 (k_\emptyset I_\emptyset^1 + k_1 I_1^1)(E^1)^{-1} \\ &= \frac{1}{2} \begin{bmatrix} 1 & -i \\ 1 & i \end{bmatrix} \begin{bmatrix} k_\emptyset & -k_1 \\ k_1 & k_\emptyset \end{bmatrix} \begin{bmatrix} 1 & 1 \\ i & -i \end{bmatrix} \\ &= \begin{bmatrix} k_\emptyset - ik_1 & 0 \\ 0 & k_\emptyset + ik_1 \end{bmatrix} = \mathrm{diag}(\boldsymbol{\eta}^1) \end{aligned}$$

Now provide $n > 1$. The induction assumption reads

$$E^{n-1}A^{n-1}(E^{n-1})^{-1} = \mathrm{diag}(\boldsymbol{\eta}^{n-1}) \ , \tag{9.28}$$

for any A^{n-1}. Furthermore,

$$(E^n)^{-1} = \frac{1}{2}\begin{bmatrix} (E^{n-1})^{-1} & (E^{n-1})^{-1} \\ i(E^{n-1})^{-1} & -i(E^{n-1})^{-1} \end{bmatrix} \tag{9.29}$$

and according to Lemma 9.4.2, we obtain

$$A^n = \begin{bmatrix} A_{\emptyset}^{n-1} & -A_{\{n\}}^{n-1} \\ A_{\{n\}}^{n-1} & A_{\emptyset}^{n-1} \end{bmatrix} \ , \tag{9.30}$$

where $A_{\emptyset}^{n-1} = \sum_{s\in\mathcal{P}(\{1,\dots,n-1\})} k_s I_s^{n-1}$ and
$A_{\{n\}}^{n-1} = \sum_{s\in\mathcal{P}(\{1,\dots,n-1\})} k_{s\cup\{n\}} I_s^{n-1}$. Therefore, we have

$$\begin{aligned}
&E^n A^n (E^n)^{-1} \\
&= \tfrac{1}{2}\begin{bmatrix} E^{n-1} & -iE^{n-1} \\ E^{n-1} & iE^{n-1} \end{bmatrix}\begin{bmatrix} A_{\emptyset}^{n-1} & -A_n^{n-1} \\ A_n^{n-1} & A_{\emptyset}^{n-1} \end{bmatrix}\begin{bmatrix} (E^{n-1})^{-1} & (E^{n-1})^{-1} \\ i(E^{n-1})^{-1} & -i(E^{n-1})^{-1} \end{bmatrix} \\
&= \tfrac{1}{2}\begin{bmatrix} E^{n-1}(A_{\emptyset}^{n-1} - iA_n^{n-1}) & E^{n-1}(-A_n^{n-1} - iA_{\emptyset}^{n-1}) \\ E^{n-1}(A_{\emptyset}^{n-1} + iA_n^{n-1}) & E^{n-1}(-A_n^{n-1} + iA_{\emptyset}^{n-1}) \end{bmatrix} \\
&\qquad\qquad\qquad\qquad\qquad\qquad \begin{bmatrix} (E^{n-1})^{-1} & (E^{n-1})^{-1} \\ i(E^{n-1})^{-1} & -i(E^{n-1})^{-1} \end{bmatrix} \\
&= \tfrac{1}{2}\begin{bmatrix} 2E^{n-1}(A_{\emptyset}^{n-1} - iA_n^{n-1})(E^{n-1})^{-1} & 0 \\ 0 & 2E^{n-1}(A_{\emptyset}^{n-1} + iA_n^{n-1})(E^{n-1})^{-1} \end{bmatrix} \\
&= \begin{bmatrix} \boldsymbol{\eta}_{\emptyset}^{n-1} - i\boldsymbol{\eta}_{\{n\}}^{n-1} & 0 \\ 0 & \boldsymbol{\eta}_{\emptyset}^{n-1} + i\boldsymbol{\eta}_{\{n\}}^{n-1} \end{bmatrix} \\
&= \mathrm{diag}(\boldsymbol{\eta}^n) \ ,
\end{aligned}$$

so the induction step is proved and therefore, the lemma is proved, too. □

Again, we present the explicit results for the case $n = 3$:

$$\begin{aligned}
\eta_1 = \eta_8^* &= (k_{\emptyset} - k_{12} - k_{13} - k_{23}) - i(k_1 + k_2 + k_3 - k_{123}) \\
\eta_2 = \eta_7^* &= (k_{\emptyset} + k_{12} + k_{13} - k_{23}) - i(-k_1 + k_2 + k_3 + k_{123}) \\
\eta_3 = \eta_6^* &= (k_{\emptyset} + k_{12} - k_{13} + k_{23}) - i(k_1 - k_2 + k_3 + k_{123}) \\
\eta_4 = \eta_5^* &= (k_{\emptyset} - k_{12} + k_{13} + k_{23}) - i(-k_1 - k_2 + k_3 - k_{123})
\end{aligned}$$

are the eigenvalues of the matrix representation of an arbitrary $\mathcal{H}_3$ element. The matrix of eigenvectors (row-vectors) is given by:

$$E^3 = \begin{bmatrix} 1 & -i & -i & -1 & -i & -1 & -1 & i \\ 1 & i & -i & 1 & -i & 1 & -1 & -i \\ 1 & -i & i & 1 & -i & -1 & 1 & -i \\ 1 & i & i & -1 & -i & 1 & 1 & i \\ 1 & -i & -i & -1 & i & 1 & 1 & -i \\ 1 & i & -i & 1 & i & -1 & 1 & i \\ 1 & -i & i & 1 & i & 1 & -1 & i \\ 1 & i & i & -1 & i & -1 & -1 & -i \end{bmatrix}$$

Theorem 9.4.1 **($\mathbb{C}^{\otimes n} \cong \mathbb{C}^{2^{n-1}}$).** *The n-fold tensor product of $\mathbb{C}$ is isomorphic to the 2^{n-1}-fold Cartesian product of $\mathbb{C}$.*

Proof. The proof follows from Lemma 9.4.3, since the eigenvalues of any matrix representation A^n are complex-valued and the eigenvectors do not depend on the coefficients of A^n. The vector of eigenvalues $\boldsymbol{\eta}^n = (\eta_1, \ldots, \eta_{2^n})$ is Hermite symmetric (i.e. $\eta_i = \eta^*_{2^n - i}$, $i \in \{1, \ldots, 2^n\}$). Hence, $\boldsymbol{\eta}^n$ is uniquely represented by $2^n/2 = 2^{n-1}$ complex values.

From the eigenvectors we obtain the mapping f^n which maps the matrix representation A^n of an arbitrary element of $\mathbb{C}^{\otimes n}$ onto

$$\mathrm{diag}(\boldsymbol{\eta}^n) = f^n(A^n) = E^n A^n (E^n)^{-1}$$

which is the matrix representation of an element of $\mathbb{C}^{2^{n-1}}$.

In order to show that f^n is a vector space isomorphism, we have to show that the kernel of f^n is $\{0^n\}$. Therefore, we must solve

$$f^n(A^n) = E^n A^n (E^n)^{-1} = 0^n \tag{9.31}$$

By multiplication of E^n from the right and $(E^n)^{-1}$ from the left we get the kernel

$$\mathrm{kern}(f^n) = (E^n)^{-1} 0^n E^n = 0^n \ . \tag{9.32}$$

This result already follows from $\mathrm{rank}(E^n) = 2^n$.

In order to show that f^n is an algebra isomorphism, we have to show that $f^n(A^n B^n) = f^n(A^n) f^n(B^n)$:

$$\begin{aligned} f^n(A^n B^n) &= E^n A^n B^n (E^n)^{-1} \\ &= E^n A^n (E^n)^{-1} E^n B^n (E^n)^{-1} \\ &= \mathrm{diag}(\boldsymbol{\eta}^n_A)\mathrm{diag}(\boldsymbol{\eta}^n_A) \\ &= \mathrm{diag}(\eta^A_1 \eta^B_1, \ldots, \eta^A_{2^n-1}\eta^B_{2^n-1}, (\eta^A_{2^n-1}\eta^B_{2^n-1})^*, \ldots, (\eta^A_1\eta^B_1)^*) \\ &= f^n(A^n) f^n(B^n) \ , \end{aligned}$$

where $\boldsymbol{\eta}^n_A = (\eta^A_1, \ldots, \eta^A_{2^n})$ and $\boldsymbol{\eta}^n_B = (\eta^B_1, \ldots, \eta^B_{2^n})$ are the vectors of the eigenvalues of A^n and B^n, respectively. □

Hence, we have identified the algebra $\mathcal{H}_n$ and additionally we have obtained an isomorphism to $\mathbb{C}^{2^{n-1}}$ which will be very useful later on. Using this algebraic framework we introduce now the HFTn.

9.4.2 The n-Dimensional Hypercomplex Fourier Transform

In this section we introduce the HFTn and transfer some theorems from the two-dimensional case. Actually, all but the relation theorem and the affine theorem are formulated for the n-D case. The relation theorem can be formulated for arbitrary but fixed n. A formulation for all n would be *very* technical and therefore hard to understand. The same situation holds with the affine theorem. Additionally, due to their structure these theorems have little practical relevance for high dimensions.

Definition 9.4.1 (n-dimensional HFT). *The n-dimensional commutative hypercomplex Fourier transform $F^h(\boldsymbol{u})$ of an n-dimensional signal $f(\boldsymbol{x})$ is defined by ($\boldsymbol{u} = (u_1, \ldots, u_n)^T, \boldsymbol{x} = (x_1, \ldots, x_n)^T \in \mathbb{R}^n$)*

$$F^h(\boldsymbol{u}) = \int_{\mathbb{R}^n} f(\boldsymbol{x})\, e^{2\pi \sum_{j=1}^n u_j x_j \mathbf{e}_{n+j} \wedge \mathbf{e}_j}\, d^n\boldsymbol{x} \ . \tag{9.33}$$

Note that due to the commutativity of $\mathcal{H}_n$ we can factorize the kernel to n exponential functions.

We do not prove every theorem of Sec. 9.3 for the n-dimensional case, since most proofs are straightforward extensions to the HFT2. Nevertheless, we state the most important ones informally:

- The HFTn is unique and the inverse transform reads

$$f(\boldsymbol{x}) = \int_{\mathbb{R}^n} F^h(\boldsymbol{u})\, e^{2\pi \sum_{j=1}^n u_j x_j \mathbf{e}_j \wedge \mathbf{e}_{n+j}}\, d^n\boldsymbol{u} \ . \tag{9.34}$$

- As in the two-dimensional case, the HFTn of a real signal has the same coefficients as the n-D Clifford spectrum. The reason for this lies in the fact, that all multiplications in the CFT are ordered with respect to the index set. Consequently, no n-blade occurring in the kernel is inverted due to permutations. For non-permuted blades, the multiplication tables of $\mathbb{R}_{0,n}$ and $\mathcal{H}_n$ are identical. If the signal is not real-valued, the spectra differ in general, because there are products of the form $\mathbf{e}'_j\mathbf{e}'_k$ with $j > k$ in the CFT (where $\mathbf{e}'_j$ are the basis one-vectors of $\mathbb{R}_{0,n}$). The result is then of course $-\mathbf{e}'_{kj}$ instead of i_{kj} in the algebra $\mathcal{H}_n$.
- The shift theorem and the symmetry theorem read according to theorems 9.3.3 and 9.3.4, respectively

$$f(\boldsymbol{x} - \boldsymbol{\xi}) \circ\!\!-\!\!\bullet\; e^{2\pi \sum_{j=1}^n u_j \xi_j \mathbf{e}_{n+j} \wedge \mathbf{e}_j} F^h(\boldsymbol{u}) \tag{9.35}$$

and

$$F^{h\dagger}(\boldsymbol{x}) \circ\!\!-\!\!\bullet\; f^\dagger(\boldsymbol{u}) \ , \tag{9.36}$$

(where $f(\boldsymbol{x}) \circ\!\!-\!\!\bullet\; F^h(\boldsymbol{u})$, i.e. $F^h(\boldsymbol{u})$ is the HFTn of $f(\boldsymbol{x})$). Note that there is no such factorized version of the shift theorem possible for the CFT and $n \geq 3$.

– The n-dimensional convolution is defined by

$$(f * g)(\boldsymbol{x}) = \int_{\mathbb{R}^n} f(\boldsymbol{\xi}) g(\boldsymbol{x} - \boldsymbol{\xi})\, d^n \boldsymbol{\xi} \tag{9.37}$$

and the convolution theorem reads

$$(f * g)(\boldsymbol{x}) \circ\!\!-\!\!\bullet\; F^h(\boldsymbol{u})\, G^h(\boldsymbol{u}) \ . \tag{9.38}$$

– The Parseval equation is satisfied by the HFTn which can be verified by a straightforward calculation like in the two-dimensional case.
– The derivative theorem reads

$$\frac{\partial}{\partial x_i} f(\boldsymbol{x}) \circ\!\!-\!\!\bullet\; 2\pi u_i \mathbf{e}_i \wedge \mathbf{e}_{n+i} F^h(\boldsymbol{u}) \ .$$

The only theorems for which we do not have the n-D extensions yet, are theorems 9.3.7 and 9.3.8 (relation FT and HFT and affine theorem). Though we can state the two theorems for any fixed dimension n, we cannot formulate them explicitly for arbitrary n. Nevertheless, we describe how to design the theorems for any n.

Due to The. 9.4.1 we can decompose the (hypercomplex) signal f into 2^{n-1} complex signals. We can do the same with the kernel of the FHTn. The kernel is not only decomposed into 2^{n-1} complex functions but into 2^{n-1} complex *exponential* functions (complex Fourier kernels). In order to understand why this is true, consider the coefficients of the hypercomplex kernel. We obtain

$$k_s = (-1)^{|s|} \prod_{j \in \{1,\dots,n\} \setminus s} \cos(2\pi u_j x_j) \prod_{l \in s} \sin(2\pi u_l x_l) \tag{9.39}$$

for $s \in \mathcal{P}(\{1,\dots,n\})$.

By calculating the eigenvalues (9.27b), one factor changes into an exponential function in each step. Consider, for example, the first step. For $t \in \mathcal{P}(\{2,\dots,n\})$ we obtain

$$\boldsymbol{\eta}_t^1 = \begin{bmatrix} e^{2\pi u_1 x_1} c_t(\boldsymbol{x},\boldsymbol{u}) & 0 \\ 0 & e^{-2\pi u_1 x_1} c_t(\boldsymbol{x},\boldsymbol{u}) \end{bmatrix} \tag{9.40}$$

where $c_t(\boldsymbol{x},\boldsymbol{u}) = (-1)^{|t|} \prod_{j \in \{2,\dots,n\} \setminus t} \cos(2\pi u_j x_j) \prod_{l \in t} \sin(2\pi u_l x_l)$.

The eigenvalues of the kernel of the HFT are the *Fourier kernels for all possible sign-permutations*, i.e. $e^{-i2\pi(\pm x_1 u_1 \pm \dots \pm x_n u_n)}$. Since there are always two exponential functions pairwise conjugated, we have 2^{n-1} different Fourier kernels left.

Now, we have 2^{n-1} signals and 2^{n-1} Fourier transforms. According to the isomorphism we can calculate the HFT by calculating the 2^{n-1} complex transforms and applying the inverse mapping $(f^n)^{-1}$.

Using this knowledge, we can also state an n-dimensional affine theorem by applying the affine theorem for the complex FT to each of the 2^{n-1} transforms. This would result in a sum of 2^{n-1} HFT which would be applied in 2^{n-1} different coordinate systems. Since there is no practical relevance for such a complicated theorem we omit it.

9.5 Conclusion

We have shown in this chapter that the CFT and the QFT can be replaced by the HFTn and the HFT2, respectively. The commutative hypercomplex transforms yield a spectral representation which is as rich as the Clifford spectrum. We have stated several theorems, among those the isomorphism between the n-fold tensor product of $\mathbb{C}$ and the 2^{n-1}-fold Cartesian product of $\mathbb{C}$ is the theoretically most important result. This theorem makes it possible to calculate the hypercomplex spectrum of a signal from the complex spectrum.

The commutative algebra $\mathcal{H}_n$ makes the analytic and numerical calculations easier. We can extend complex 1-D filters to hypercomplex n-D filters by the tensor product. We do not take care of the order of the operation as for the CFT: the shift theorem, the convolution theorem and the affine theorem are easier to formulate. Furthermore, we will be able to state a simple fast algorithm in Chap. 10.

Additionally, the two domains of the HFTn can be visualized by two orthogonal n-D subspaces in a common $2n$-D space. This point of view can lead to further concepts, e.g. a decayed Fourier transform (or Laplace transform), fractional Fourier transforms (similar to [137]), and so on.

Furthermore, the design of hypercomplex filters has to be considered more closely. Our present and future aim is to develop new multi-dimensional concepts which are not only a 'blow-up' of 1-D concepts, but an intrinsic n-D extension, which includes a *new quality* of filter properties.

10. Fast Algorithms of Hypercomplex Fourier Transforms*

Michael Felsberg[1], Thomas Bülow[1], Gerald Sommer[1], and Vladimir M. Chernov[2]

[1] Institute of Computer Science and Applied Mathematics, Christian-Albrechts-University of Kiel
[2] Image Processing System Institute, Russian Academy of Sciences, Samara

10.1 Introduction

In this chapter we consider the computational aspect of the quaternionic Fourier transform (QFT), of the Clifford Fourier transform (CFT), and of the commutative hypercomplex Fourier transform (HFT). We can cover all these transforms with the term *hypercomplex Fourier transforms*, since all mentioned algebras are hypercomplex algebras (see Cha. 9). In order to have a numerical way to evaluate these transforms, we introduce the corresponding discrete transforms by sampling the continuous transforms. Furthermore, we prove the inverse transforms.

The simplest way to create fast algorithms for the discrete transforms (of real n-dimensional signals) is to separate the transforms into $2^n - 1$ transforms so that each transform only effects one spatial coordinate. Therefore, the asymptotic complexity of fast algorithms for n-dimensional transforms

* This work has been supported by German National Merit Foundation and by DFG Grants So-320-2-1, So-320-2-2, and Graduiertenkolleg No. 357.

should not exceed $2^n - 1$ times the complexity of $(2^n - 1)N^n$ FFTs[1], i.e. the complexity should be of order $\mathcal{O}(N^n \log n)$ (or less).

In order to obtain little computational complexities, we formulate several approaches for fast algorithms. We make use of some algebraic properties of hypercomplex algebras in order to optimize the transforms. Some of the algorithms are based on standard algorithms, so that for the implementation there is nearly no additional coding; the standard implementations can be re-used.

All the time and memory complexities of the presented algorithms are estimated, illustrated, and systematically compared. Hence, it should be easy for the reader to choose the algorithm which suits best for his purposes. This decision does not only depend on the purpose but also on the computer environment. We try to stay abreast of this fact by considering cache sizes and memory sizes of today's computers knowing the fact that the usage of swap space (i.e. to swap out data on a hard disc) drives any algorithmetic optimization insane.

10.2 Discrete Quaternionic Fourier Transform and Fast Quaternionic Fourier Transform

In this section we define the discrete quaternionic Fourier transform (DQFT) in a similar way as the discrete complex Fourier transform is defined, i.e. the continuous transform is sampled. We state and prove the inverse discrete QFT, a fast algorithm for the QFT (FQFT) and a fast algorithm for the inverse QFT. Further optimizations of the fast algorithms are presented (and proved) and their complexities are considered in detail.

10.2.1 Derivation of DQFT and FQFT

Starting point for a fast algorithm of the QFT is the discrete quaternionic Fourier transform, of course. It is defined analogously to the discrete Fourier transform by sampling the continuous transform (8.3.4). Formally, the continuous, infinite signal $f(x, y)$ which must be of limited bandwidth (or convolved with a low-pass filter) is convolved with the Shah function (infinite sum of equi-distant Dirac impulses). Afterwards, the new (periodic) signal is sampled so that $f_{m,n} = f(me_x, ne_y)$ where e_x and e_y must be integer divisors of the periods with respect to x and y. Formally, the last step is a pointwise multiplication with a Shah function.

We obtain the discrete transform by two steps. Firstly, we only consider the periodicity. As a result we evaluate the integral of (8.3.4) only in one period (similar to Fourier series). Secondly, we multiply the integral with the Shah function, so the integral changes to a sum:

[1] Note that N indicates the extension of the signal in only one dimension. The total size of the signal is therefore N^n.

Definition 10.2.1 (Discrete QFT). *Let f be a discrete two-dimensional signal of finite size $M \times N$. The discrete quaternionic Fourier transform (DQFT), denoted as $\mathcal{F}_q^D\{f\} = F^q$, is defined by*

$$F^q_{u,v} := \sum_{x=0}^{M-1} \sum_{y=0}^{N-1} e^{-i2\pi uxM^{-1}} f_{x,y} e^{-j2\pi vyN^{-1}} . \tag{10.1}$$

The discrete inverse transform can be derived from the continuous one, too. Due to the fact that a periodic signal in the spatial domain corresponds to a discrete spectrum (Fourier series) and a discrete (infinitely long) signal corresponds to a periodic spectrum, we can simply exchange spatial and frequency domain. This yields the same transform, except for the sign of the exponential term and except for a normalizing factor which must be multiplied in this transform as in the one-dimensional case:

Theorem 10.2.1 (Inverse DQFT). *The inverse discrete quaternionic Fourier transform $\mathcal{F}_q^{D^{-1}}\{F^q\} = f$ reads*

$$f_{x,y} = \frac{1}{MN} \sum_{u=0}^{M-1} \sum_{v=0}^{N-1} e^{i2\pi uxM^{-1}} F^q_{u,v} e^{j2\pi vyN^{-1}} . \tag{10.2}$$

Proof. Due to the fact that we call (10.2) the inverse transform of (10.1), we must prove that the concatenation of both transforms yields the identity.

Firstly, we define abbreviations for the modulation terms:

$$w_i = e^{i2\pi M^{-1}} \tag{10.3a}$$

$$w_j = e^{j2\pi N^{-1}} \tag{10.3b}$$

Applying formula (10.2) to (10.1) yields:

$$\begin{aligned}
&\mathcal{F}_q^{D^{-1}} \left\{ \sum_{x=0}^{M-1} \sum_{y=0}^{N-1} w_i^{-ux} f_{x,y} w_j^{-vy} \right\} \\
&= \frac{1}{MN} \sum_{u=0}^{M-1} \sum_{v=0}^{N-1} w_i^{ux'} \sum_{x=0}^{M-1} \sum_{y=0}^{N-1} w_i^{-ux} f_{x,y} w_j^{-vy} w_j^{vy'} \\
&= \frac{1}{MN} \sum_{x=0}^{M-1} \sum_{y=0}^{N-1} \sum_{u=0}^{M-1} \sum_{v=0}^{N-1} w_i^{u(x'-x)} f_{x,y} w_j^{v(y'-y)} \\
&= \frac{1}{MN} \sum_{x=0}^{M-1} \sum_{y=0}^{N-1} M\delta_{x'-x} f_{x,y} N\delta_{y'-y} \\
&= f_{x',y'}
\end{aligned}$$

So the concatenation of the transforms (10.2) and (10.1) yields the identity and we can call them inverse to each other. Note that the other concatenation need not be proved because the Fourier transform is a 1-1 mapping (it is just a change to another basis of same dimension). □

Now, we are able to calculate the spectrum of a finite, discrete signal. If we use formula (10.1) to implement an algorithm for the DQFT, we obtain a computational complexity of cM^2N^2 (c being a constant). This complexity is quite high. Hence we try to reduce it as it has been done by Cooley and Tuckey when they developed their fast Fourier transform (FFT) algorithm. The idea they used is the *decimation of time* method.

The FFT algorithm has originally been designed for 1-D time signals. The time domain has been divided into two new domains (one consisting of the signal values at even positions and one consisting of the values at odd positions). Hence, it is called the radix-2 method. We will do the same for the DQFT. We apply the decimation method to the spatial domain in order to obtain a recursive algorithm (divide and conquer). If the domain is of the size (1×1) the DQFT is the identity. Otherwise we have the following.

Theorem 10.2.2 (Fast QFT). *Let $(f)_{M\times N}$ be a discrete two-dimensional signal and $M = N = 2^k \geq 2$. Then we have*

$$\mathcal{F}_q^D\{f\} = F^{ee} + w_i^{-u}F^{oe} + F^{eo}w_j^{-v} + w_i^{-u}F^{oo}w_j^{-v} \tag{10.4}$$

where

$$F^{ee} = \mathcal{F}_q^D\{f^{ee}\} = \mathcal{F}_q^D\{f_{2x,2y}\} \tag{10.5a}$$

$$F^{oe} = \mathcal{F}_q^D\{f^{oe}\} = \mathcal{F}_q^D\{f_{2x+1,2y}\} \tag{10.5b}$$

$$F^{eo} = \mathcal{F}_q^D\{f^{eo}\} = \mathcal{F}_q^D\{f_{2x,2y+1}\} \tag{10.5c}$$

$$F^{oo} = \mathcal{F}_q^D\{f^{oo}\} = \mathcal{F}_q^D\{f_{2x+1,2y+1}\} \tag{10.5d}$$

Proof. In order to show that the recursive formula (10.4) is correct we present a constructive proof. We apply the radix-2 method to the definition of the DQFT (10.1) and obtain

$$\mathcal{F}_q^D\{f\} = \sum_{x_1,y_1=0}^{N/2-1} \sum_{x_0,y_0=0}^{1} w_i^{-u(2x_1+x_0)} f_{2x_1+x_0,2y_1+y_0} w_j^{-v(2y_1+y_0)} ,$$

by substituting $u = u_1N/2 + u_0$, $v = v_1N/2 + v_0$

$$\mathcal{F}_q^D\{f\} = \sum_{x_0,y_0=0}^{1} w_i^{-ux_0} \sum_{x_1,y_1=0}^{N/2-1} w_i^{-u_0 2x_1} f_{2x_1+x_0,2y_1+y_0} w_j^{-v_0 2y_1} w_j^{-vy_0}$$

and finally, because $w_i^{-Nu_1x_1} = w_j^{-Nv_1y_1} = 1$

$$\mathcal{F}_q^D\{f\} = \mathcal{F}_q^D\{f_{2x,2y}\} + w_i^{-u}\mathcal{F}_q^D\{f_{2x+1,2y}\} + \mathcal{F}_q^D\{f_{2x,2y+1}\}w_j^{-v} + w_i^{-u}\mathcal{F}_q^D\{f_{2x+1,2y+1}\}w_j^{-v}$$

Note that the signals $f_{2x,2y}$, $f_{2x+1,2y}$, $f_{2x,2y+1}$ and $f_{2x+1,2y+1}$ and their quaternionic Fourier transforms have the size $N/2$. □

In practical applications signals will not always have a size of a power of two. In that case the domain may be filled up by zeros. As a consequence, the period of the signal is changed, which has an effect on the reconstruction of the original signal.

The recursive formula (10.4) and the identity for $M = N = 1$ yield an algorithm for calculating the DQFT with complexity $cN^2 \operatorname{ld} N$ (see section 10.2.3). We need $\operatorname{ld} N = k$ recursive calls of (10.4).

A fast algorithm for the inverse transform can be derived in the same way as for the DQFT except for the normalizing factor and the sign in the exponential function. This method is called decimation of frequency. Nevertheless, it is advantageous to develop another algorithm which calculates the inverse of (10.4) in each step. Consequently, we obtain after step n exactly the same sub-spectra as in the FQFT just before the $(k-n)$th application of (10.4). The reason why this method is advantageous will be given in section 10.2.2.

Using the notations above we state the following.

Theorem 10.2.3 (Inverse FQFT). *The inverse of formula (10.4) reads* $(L = N/2)$

$$F^{ee}_{u,v} = \frac{1}{4}\left(F^q_{u,v} + F^q_{u+L,v} + F^q_{u,v+L,} + F^q_{u+L,v+L)}\right) \tag{10.6a}$$

$$F^{oe}_{u,v} = \frac{1}{4} w_i^u \left(F^q_{u,v} - F^q_{u+L,v} + F^q_{u,v+L)} - F^q_{u+L,v+L)}\right) \tag{10.6b}$$

$$F^{eo}_{u,v} = \frac{1}{4}\left(F^q_{u,v} + F^q_{u+L,v} - F^q_{u,v+L)} - F^q_{u+L,v+L)}\right) w_j^v \tag{10.6c}$$

$$F^{oo}_{u,v} = \frac{1}{4} w_i^u \left(F^q_{u,v} - F^q_{u+L,v} - F^q_{u,v+L)} + F^q_{u+L,v+L)}\right) w_j^v \tag{10.6d}$$

and therefore, the recursive execution of (10.6a-10.6d) reconstructs $f_{x,y}$ *from* $F^q_{u,v}$.

Proof. Since the DQFT (and hence each application of (10.4)) is a 1-1 mapping, we only have to show that the successive application of (10.6a)-(10.6d) and (10.4) yields the identity.

$$\begin{aligned}
&F^{ee} + w_i^{-u} F^{oe} + F^{eo} w_j^{-v} + w_i^{-u} F^{oo} w_j^{-v} \\
&= \frac{1}{4}\Big(\left(F^q_{u,v} + F^q_{u+L,v} + F^q_{u,v+L,} + F^q_{u+L,v+L)}\right) \\
&\quad + w_i^{-u} w_i^u \left(F^q_{u,v} - F^q_{u+L,v} + F^q_{u,v+L)} - F^q_{u+L,v+L)}\right) \\
&\quad + \left(F^q_{u,v} + F^q_{u+L,v} - F^q_{u,v+L)} - F^q_{u+L,v+L)}\right) w_j^v w_j^{-v} \\
&\quad + w_i^{-u} w_i^u \left(F^q_{u,v} - F^q_{u+L,v} - F^q_{u,v+L)} + F^q_{u+L,v+L)}\right) w_j^v w_j^{-v}\Big) \\
&= F^q_{u,v}
\end{aligned}$$

Consequently, (10.6a)-(10.6d) and (10.4) are inverse mappings. If the DQFT of a signal $f_{x,y}$ (size: $N \times N$) is calculated by (10.4), $\operatorname{ld} N$ levels of sub-spectra are created. Eq. (10.4) describes the transition from one level to the next. We have proved that (10.6a)-(10.6d) do the inverse of (10.4) and therefore, they describe the transition from one level to the *previous*. Consequently, recursive execution of (10.6a)-(10.6d) yield the iDQFT. □

Due to this theorem, the equations (10.6a)-(10.6d) can be used to implement an inverse fast quaternionic Fourier transform. In each step one step of the (forward) FQFT is inverted. Hence, after k steps we have reconstructed the original signal. In other words, the recursive calls must stop for $N = 1$. In this case, the iDQFT is the identity. This procedure is illustrated in Fig. 10.1.

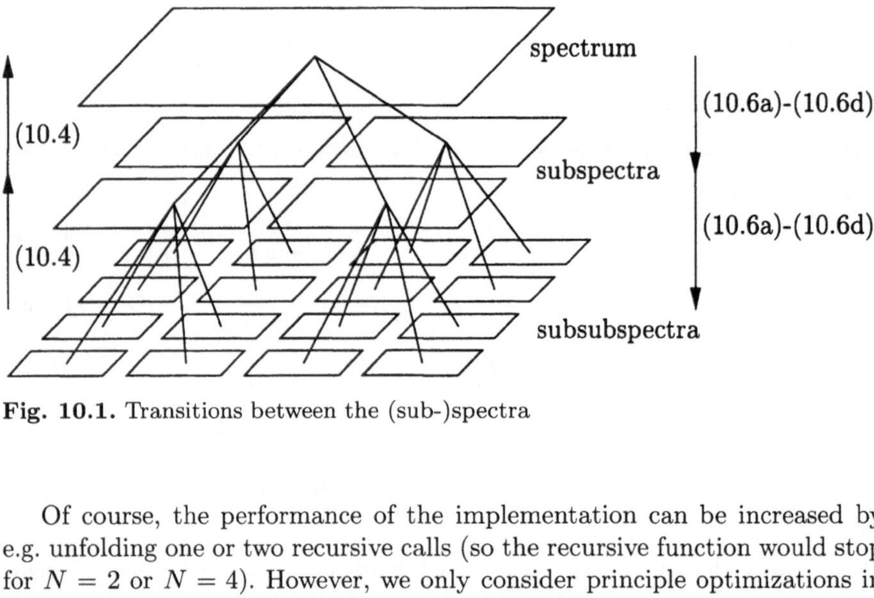

Fig. 10.1. Transitions between the (sub-)spectra

Of course, the performance of the implementation can be increased by e.g. unfolding one or two recursive calls (so the recursive function would stop for $N = 2$ or $N = 4$). However, we only consider principle optimizations in this chapter; further improvements of technical details are left to the programmer. Some general optimizations are given in the following section.

10.2.2 Optimizations by Hermite Symmetry

The first approach for optimizing the fast algorithms does not make use of the specific properties of the spectra of real signals (i.e. Hermite symmetry). The idea is the same as in equations (10.6a)-(10.6d). A phase of π yields only a change of the sign. This property will be called *π-phase* in the following. Hence, we have the same sum (10.4) for $F^q_{u,v}$, $F^q_{u+L,v}$, $F^q_{u,v+L}$ and $F^q_{u+L,v+L}$, except for the signs. Consequently, we only need to perform the multiplications once and can use the products four times. Hence, the multiplicative complexity is divided by four.

The second approach makes use of the quaternionic Hermite symmetry: we can

a) calculate four spectra in one step,
b) evaluate three quarters of each sub-spectrum by copying and applying one of the involutions (8.5).

If we choose a) and use the four spectra from the first decimation-step of (10.4), we make use of a method which is called *overlapping* [40].

This procedure is illustrated in the figure 10.2. The 8×8 real signal is mapped onto a 4×4 quaternionic signal. The real, the i-, the j-, and the k-imaginary part are denoted by r, i, j, and k, respectively.

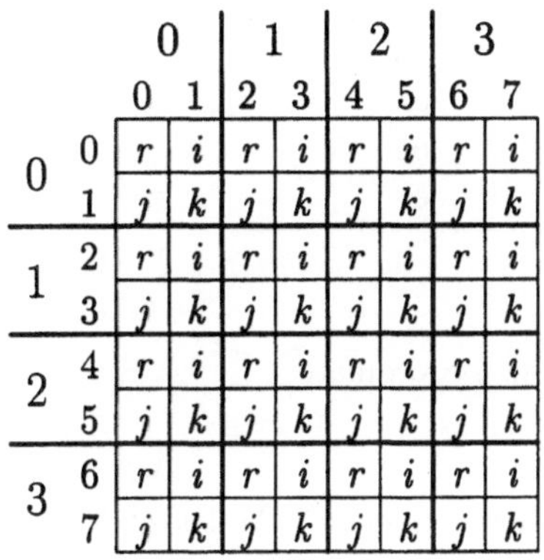

0 1 2 3
0 1 2 3 4 5 6 7
0 0 r i r i r i r i
1 j k j k j k j k
1 2 r i r i r i r i
3 j k j k j k j k
2 4 r i r i r i r i
5 j k j k j k j k
3 6 r i r i r i r i
7 j k j k j k j k

Fig. 10.2. A two-dimensional real signal is mapped onto a quaternionic signal

In (8.5), we have defined three involutions ($q \in \mathbb{H}$):

$$\alpha(q) = -jqj \tag{10.7a}$$
$$\beta(q) = -iqi \tag{10.7b}$$
$$\gamma(q) = -kqk = \alpha(\beta(q)) \tag{10.7c}$$

In order to reconstruct the four spectra from the overlapped one, we simply use the symmetry properties of the signals (see 8.3.6). The symmetries imply the following:

Theorem 10.2.4 (Overlapping). *Let* $\bar{f} = f^{ee} + if^{oe} + jf^{eo} + kf^{oo}$ *and* $\mathcal{F}_q^D\{\bar{f}\} = \bar{F}^q$. *Furthermore let*

$$F^{\alpha}_{u,v} = \alpha(\bar{F}^q_{u,-v}) \tag{10.8a}$$
$$F^{\beta}_{u,v} = \beta(\bar{F}^q_{-u,v}) \tag{10.8b}$$
$$F^{\gamma}_{u,v} = \gamma(\bar{F}^q_{-u,-v}). \tag{10.8c}$$

Then, the partial spectra are obtained by

$$4F^{ee}_{u,v} = \bar{F}^q_{u,v} + F^{\alpha}_{u,v} + F^{\beta}_{u,v} + F^{\gamma}_{u,v} \tag{10.9a}$$
$$4iF^{oe}_{u,v} = \bar{F}^q_{u,v} + F^{\alpha}_{u,v} - F^{\beta}_{u,v} - F^{\gamma}_{u,v} \tag{10.9b}$$
$$4jF^{eo}_{u,v} = \bar{F}^q_{u,v} - F^{\alpha}_{u,v} + F^{\beta}_{u,v} - F^{\gamma}_{u,v} \tag{10.9c}$$
$$4kF^{oo}_{u,v} = \bar{F}^q_{u,v} - F^{\alpha}_{u,v} - F^{\beta}_{u,v} + F^{\gamma}_{u,v}. \tag{10.9d}$$

Proof. Due to linearity of the Fourier transform we obtain the following table of symmetries (table 10.1) for the addends of the overlapped quaternionic spectrum.

Table 10.1. Symmetries of the addends of $\bar{F}^q$

Addend	real part	i-imag. part	j-imag. part	k-imag. part
$\mathcal{F}^D_q\{f^{ee}\}$	ee	oe	eo	oo
$i\mathcal{F}^D_q\{f^{oe}\}$	oe	ee	oo	eo
$j\mathcal{F}^D_q\{f^{eo}\}$	eo	oo	ee	oe
$k\mathcal{F}^D_q\{f^{oo}\}$	oo	eo	oe	ee

e: even
o: odd
f^{eo} is even wrt. the x-coordinate and odd wrt. the y-coordinate

Hence, each partial spectrum can be extracted from $\bar{F}^q$. Without loss of generality we will reconstruct the spectrum F^{ee}. We obtain the real part of F^{ee} by

$$4\,\mathcal{R}(F^{ee}_{u,v}) = \mathcal{R}(\bar{F}^q_{u,v} + \bar{F}^q_{-u,v} + \bar{F}^q_{u,-v} + \bar{F}^q_{-u,-v}),$$

since all odd parts yield zero. Analogously, we obtain the other parts, namely

$$4\,\mathcal{I}(F^{ee}_{u,v}) = \mathcal{I}(\bar{F}^q_{u,v} - \bar{F}^q_{-u,v} + \bar{F}^q_{u,-v} - \bar{F}^q_{-u,-v}),$$
$$4\,\mathcal{J}(F^{ee}_{u,v}) = \mathcal{J}(\bar{F}^q_{u,v} + \bar{F}^q_{-u,v} - \bar{F}^q_{u,-v} - \bar{F}^q_{-u,-v}),$$
$$4\,\mathcal{K}(F^{ee}_{u,v}) = \mathcal{K}(\bar{F}^q_{u,v} - \bar{F}^q_{-u,v} - \bar{F}^q_{u,-v} + \bar{F}^q_{-u,-v}).$$

Due to linearity we can exchange the sums and the selection of the components which yield formula (10.9a). The other three spectra can be reconstructed analogously. □

Now we consider the optimization b) which means direct application of the Hermite symmetry. If we have calculated a spectral value $F^q_{u,v}$ (this might be a spectral value of a sub-spectrum, too), we obtain three more spectral values by

$$F^q_{-u,v} = \beta(F^q_{u,v}) \tag{10.10a}$$
$$F^q_{u,-v} = \alpha(F^q_{u,v}) \tag{10.10b}$$
$$F^q_{-u,-v} = \gamma(F^q_{u,v}). \tag{10.10c}$$

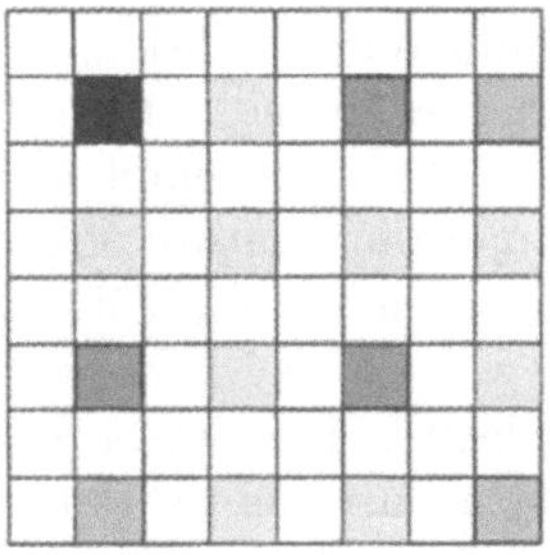

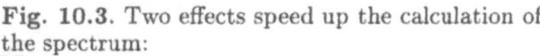

Fig. 10.3. Two effects speed up the calculation of the spectrum:
■ calculated value
■ changed signs
□ Hermite symmetry
□ signs and Hermite symmetry

Note, if u (v) is zero or $N/2$, we have $(-u, v) = (u, v)$ ($(u, -v) = (u, v)$), since F^q is N-periodic.

Both effects (the π-phase and the Hermite symmetry) are illustrated in the following figure 10.2.2.

The black cell (calculated value) is the only cell which has to be *explicitly* calculated. The dark grey cells (changed signs) are obtained without additional multiplications due to the π-phase. The middle grey cells (Hermite symmetry) are obtained from the black cell without any additional arithmetic operations by applying the involutions. Finally, the light grey cells are obtained from the dark grey cells by the Hermite symmetry.

Both methods, a) and b), reduce the whole complexity by a factor of four. Though we only considered the forward transform, we can decrease the complexity of the inverse transform using the same methods. This is possible, since we formulated an inverse transform which reconstructs the Hermite-symmetric sub-spectra (a simple decimation of frequency does not yield such an algorithm).

The overlapping additionally reduces the memory complexity by a factor of four (except for the reconstruction procedure). This is quite important if we consider the execution of the algorithm on a computer.

The memory complexity and the localization of data have to be considered under two aspects: cache size and memory size. If the data exceeds the cache size, we have cache misses (in image processing we have this on most machines for e.g. images of 512×512 pixels). During the execution of the algorithm the data have to be moved between cache and main memory. It is advantageous if the algorithm acts locally, which means that the algorithm uses only a small part of the data (less than the cache size). The latency time for copying data from the main memory to the cache is similar to the latency time of floating point operations. Hence, if we would have a *branch first algorithm* (the recursion is executed layer by layer), the data must be moved in each incarnation of recursion. The resulting complexity of the data handling would be similar to that one of the calculation itself.

Fortunately, we have a *depth first algorithm* which means that the recursion is first finished completely for the first call, then for the second call and so on. Hence, we have two times a cache miss for each value if the cache size is at least a quarter of the size of the spectrum.

If the data even exceeds the main memory, the computer uses disk space to extend the main memory (*swapping*). Since the access to the hard disk is one thousand times slower than the memory access, the time complexity of the algorithm itself becomes obsolete. The only important feature of the algorithm is to work as locally as possible. This problem might appear if 3-D data or image sequences are transformed.

In the following, we only consider the time complexity, since we have a 2-D transform and we assume that the cache is nearly as large as the data.

10.2.3 Complexities

In this section we consider the complexities of the presented algorithms. Since modern computers calculate floating point multiplications much faster than in former times, we do not restrict to the multiplicative complexities.

In literature, we have found some algorithms which reduce the multiplicative complexity of complex and quaternionic multiplications. Complex multiplications can be performed with three multiplications (instead of four) and quaternionic multiplications can be performed with nine multiplications (instead of 16) [203]. This decrease of the multiplicative complexity can only be achieved by increasing the additive complexity. Since multiplications are performed as fast as additions on today's processors (e.g. SUN UltraSPARC-II [224], SGI R10000 [212], DEC Alpha 21164PC [49]), the fastest algorithm is the one with the least number of operations – multiplications plus additions.

Consequently, we always consider the multiplicative complexity, the additive complexity, and the sum of both.

Although the FQFT is performed in the quaternion algebra, we need not calculate the general quaternionic multiplication. We know that in each multiplication of (10.4) one factor (the exponential term) is the element of a subalgebra which is isomorphic to the complex algebra. Hence, the complexity of such a multiplication can be reduced to eight multiplications and four additions.

Formula (10.4) yields four multiplications and three additions in the quaternion algebra or 32 multiplications and $16 + 12 = 28$ additions in the algebra of real numbers for each frequency. Hence, we have $32N^2$ multiplications and $28N^2$ additions for each recursive call. If the effect of the π-phase is used, we reduce the multiplications in the quaternion algebra by a factor of four, i.e. we need $8N^2$ multiplications and $16N^2$ additions in the algebra of real numbers for evaluating (10.4) for all frequencies. Since we have $\mathrm{ld}\, N$ recursive calls, we have a total complexity of $8N^2 \,\mathrm{ld}\, N$ multiplications and $16N^2 \,\mathrm{ld}\, N$ additions (total: $24N^2 \,\mathrm{ld}\, N$ floating point operations).

If we have a real signal, the complexity can be reduced to approximately one quarter. Firstly, we consider overlapping. Since the size of the new signal domain is reduced to $N^2/4$, the complexity is divided by four and $\mathrm{ld}\, N$

is substituted by $\mathrm{ld}\,\frac{N}{2} = -1 + \mathrm{ld}\,N$ (one recursive call less). The reconstruction of the overlapped spectrum (10.9a-d) increases the quadratic additive complexity by three quaternionic additions (or twelve real additions) and one real multiplication per frequency (the factor four can be eliminated component-wise in $\bar{F}^q$). The last recursive call of (10.4) increases the complexities according to the case without Hermite symmetry (e.g. eight real multiplications and 16 real additions per frequency). Finally, the overall complexity reads $\frac{8}{4}N^2(-1 + \mathrm{ld}\,N) + (1+8)N^2 = (7 + 2\,\mathrm{ld}\,N)N^2$ multiplications and $\frac{16}{4}N^2(-1 + \mathrm{ld}\,N) + (12 + 16)N^2 = (24 + 4\,\mathrm{ld}\,N)N^2$ additions (total: $(31 + 6\,\mathrm{ld}\,N)N^2$ operations).

If the Hermite symmetry is directly used, we only need calculate one quarter of each sub-spectrum. The other three quarters are evaluated by the automorphisms (8.5). Since there is no symmetric frequency if u and v are zero or $N/2$ and there is only one instead of three symmetric frequencies if either u or v are zero or $N/2$, we obtain a high additional quadratic complexity. Furthermore, this algorithm has a four times higher memory complexity than the overlapping algorithm (except for the last step). Hence, it is not advantageous to realize this fast algorithm.

Nevertheless, we can decrease the complexity of the overlapping algorithm if we reconstruct only one quarter of the spectrum by formulae (10.9a-d) and (10.4) and copy the other three quarters. This reduces the complexity of the reconstruction to $1 + \frac{8}{4} = 3$ real multiplications and $\frac{28}{4} = 7$ real additions per frequency. Due to the zero-frequencies (and the $N/2$-frequencies), we have an additional *linear* complexity, which is neglected in these considerations. The total complexity of this algorithm is $(1 + 2\,\mathrm{ld}\,N)N^2$ multiplications and $(3 + 4\,\mathrm{ld}\,N)N^2$ additions (total complexity: $(4 + 6\,\mathrm{ld}\,N)N^2$ operations).

The same complexity as for the last algorithm is obtained if four real signals are transformed at the same time. The four signals are mapped to the four parts of a quaternionic signal which is transformed with a complexity of $24N^2\,\mathrm{ld}\,N$ operations (i.e. $6N^2\,\mathrm{ld}\,N$ per spectrum). Afterwards, we reconstruct the four spectra similarly to overlapping (using the Hermite-symmetry), which yields the same additional complexity (N^2 multiplications and $\frac{12}{4}N^2$ additions per spectrum). All results are summarized in table 10.2.

Table 10.2. Complexities of the considered algorithms

Algorithm	multiplications	additions	operations
FQFT (π-phase)	$8N^2\,\mathrm{ld}\,N$	$16N^2\,\mathrm{ld}\,N$	$24N^2\,\mathrm{ld}\,N$
FQFT/overlapping	$N^2(7 + 2\,\mathrm{ld}\,N)$	$N^2(24 + 4\,\mathrm{ld}\,N)$	$N^2(31 + 6\,\mathrm{ld}\,N)$
FQFT/overl.+sym.	$N^2(1 + 2\,\mathrm{ld}\,N)$	$N^2(3 + 4\,\mathrm{ld}\,N)$	$N^2(4 + 6\,\mathrm{ld}\,N)$
FQFT/four spectra	$N^2(1 + 2\,\mathrm{ld}\,N)$	$N^2(3 + 4\,\mathrm{ld}\,N)$	$N^2(4 + 6\,\mathrm{ld}\,N)$

Note, that the algorithm using overlapping with Hermite symmetry in the last step is very complicated to implement. The enormous code length of the implementation could possibly slow down the algorithm more than it is sped up by the Hermite symmetry. Additionally, there is an alternative way to calculate the DQFT which is even faster than all algorithms above (see 10.4.2).

10.3 Discrete and Fast n-Dimensional Transforms

In this section we extend the definition of the discrete QFT for the n-dimensional commutative hypercomplex Fourier transform (HFT). We state the inverse discrete HFT and a fast algorithm (FHFT). Further on, we generalize the π-phase and the overlapping for n dimensions and consider the complexities.

10.3.1 Discrete Commutative Hypercomplex Fourier Transform and Fast Commutative Hypercomplex Fourier Transform

Recall the correspondence of the QFT and the HFT2 (9.3.1): both transforms yield the same coefficient in the case of real signals. We have extended that concept to n dimensions in Sec. 9.4.2, i.e. the coefficients of the HFTn are the same as those of the n-D CFT of a real signal. For the discrete n-D CFT and the discrete HFTn we have the same correspondence as for the continuous transforms, since we have not used the properties of the domain (i.e. if it is infinite and continuous or finite and discrete) in the proof. Therefore, we can calculate the Clifford spectrum via the HFTn and it is sufficient to give the definition of the discrete HFTn in this section.

Additionally, it is not possible to develop directly a fast algorithm for the CFT in the same way as we did for the QFT. We can apply the decimation method in a straightforward way only for commutative algebras because we have to exchange some factors. The QFT is an exception since we have *two* ways of multiplying and therefore we can extract one factor of the QFT kernel to the left (the i-term) and one to the right (the j-term). If $n \geq 3$ we have not enough ways of multiplying (e.g. from the top), so we must really permute some factors. Since the hypercomplex algebra introduced in section 9.4.2 is commutative, we use the commutative hypercomplex Fourier transform for developing a fast algorithm. In this respect, we want to annotate that we have originally introduced the commutative hypercomplex algebra for the purpose of developing fast algorithms. Therefore, the whole theory presented in Cha. 9 has been motivated by the mathematical properties which we needed for the algorithm.

We have shown in The. 9.3.1 that the coefficients of both transforms are the same if the signal is real-valued and due to linearity of the HFT, we can even calculate the Clifford spectrum of a Clifford-valued signal (see 10.4.2). Thus, by developing the discrete and fast HFT, we indirectly obtain a discrete and fast CFT, respectively.

To start with, we now define the discrete HFT in analogy to the DQFT by sampling the continuous transform. The considerations we made for the QFT in section 10.2.1 (about the formal construction of discrete signals) are the same for the HFT, of course.

Definition 10.3.1 (Discrete HFT). *Let f be an n-dimensional discrete signal of the size N^n and let I_n be defined as*

$$I_n = \begin{bmatrix} \mathbf{e}_1 \wedge \mathbf{e}_{1+n} & 0 & \dots & \dots \\ 0 & \mathbf{e}_2 \wedge \mathbf{e}_{2+n} & 0 & \dots \\ \dots & 0 & \ddots & 0 \\ \dots & \dots & 0 & \mathbf{e}_n \wedge \mathbf{e}_{2n} \end{bmatrix} .$$

Then, the discrete HFT of f is defined by

$$F^h_{\boldsymbol{u}} = \sum_{\boldsymbol{x} \in \{0,\dots,N-1\}^n} f_{\boldsymbol{x}} e^{-2\pi \boldsymbol{x} I_n \boldsymbol{u}^T N^{-1}} . \tag{10.11}$$

Note that the discrete signal can have different lengths in each coordinate, i.e. a 3-D signal can have the size $M \times N \times K$ with arbitrary (true positive) M, N, K. We formulated the DHFT for signals with equal length in each dimension, since the formula is more compact and the FHFT (see (10.13)) can only be applied to such signals. Nevertheless, every signal can be embedded in a larger signal of that form (with a changed period, of course).

In analogy to the iDQFT we state a theorem for the inverse transform in the following. The transform and its proof are extensions of the two-dimensional case (10.2). Note that there is one important difference between the 2-D DHFT and the DQFT. The latter is formulated using a special order of multiplications (i-term from the left and j-term from the right). This order implies that the iDQFT must have the same sequence of factors, since the two i-terms and the two j-terms must compensate *directly* each other, respectively.

In contrast to this, the sequence of factors is irrelevant in the case of the DHFT and the iDHFT, since they are commutative. Hence, they can be formulated putting all exponential functions in one exponential function at the end. Additionally, we need not take care of the order of the multiplications in the proof of the theorem.

Theorem 10.3.1 (Inverse discrete HFT). *Let f be an n-dimensional discrete signal of the size N^n and F^h its DHFT. Then, the following equation holds true*

$$f_{\boldsymbol{x}} = \frac{1}{N^n} \sum_{\boldsymbol{u} \in \{0,\dots,N-1\}^n} F^h_{\boldsymbol{u}} e^{2\pi \boldsymbol{x} I_n \boldsymbol{u}^T N^{-1}} . \tag{10.12}$$

Proof. Applying (10.12) to the discrete HFT of f yields:

$$\begin{aligned}
&\frac{1}{N^n} \sum_{\boldsymbol{u}\in\{0,\dots,N-1\}^n} F^h_{\boldsymbol{u}} e^{2\pi \boldsymbol{x}' I_n \boldsymbol{u}^T N^{-1}} \\
&= \frac{1}{N^n} \sum_{\boldsymbol{u}\in\{0,\dots,N-1\}^n} \sum_{\boldsymbol{x}\in\{0,\dots,N-1\}^n} f_{\boldsymbol{x}} e^{-2\pi \boldsymbol{x} I_n \boldsymbol{u}^T N^{-1}} e^{2\pi \boldsymbol{x}' I_n \boldsymbol{u}^T N^{-1}} \\
&= \frac{1}{N^n} \sum_{\boldsymbol{x}\in\{0,\dots,N-1\}^n} \sum_{\boldsymbol{u}\in\{0,\dots,N-1\}^n} e^{2\pi (\boldsymbol{x}'-\boldsymbol{x}) I_n \boldsymbol{u}^T N^{-1}} f_{\boldsymbol{x}} \\
&= \frac{1}{N^n} \sum_{\boldsymbol{x}\in\{0,\dots,N-1\}^n} f_{\boldsymbol{x}} N^n \delta_{\boldsymbol{x}'-\boldsymbol{x}} \\
&= f_{\boldsymbol{x}'} .
\end{aligned}$$

Hence, (10.12) is the inverse DHFT. □

Now, having a discrete transform in a commutative algebra, we are able to state a fast algorithm for the HFT, since we can apply the decimation method. The following theorem is the extension of the FQFT (10.4). The notes we made for the DQFT concerning the order of multiplications are valid for the FHFT, too. Therefore, the recursive formula for the 2-D FHFT differs from that one of the FQFT. Nevertheless, both formulae yield the same coefficients for a real signal. Besides, the proof is simplified by the commutativity.

Theorem 10.3.2 (Fast HFT). *Let $f_{\boldsymbol{x}}$ be a discrete n-dimensional signal of the size N^n with $N = 2^k$. Then we have*

$$F^h_{\boldsymbol{u}} = \sum_{\substack{x_{l0}=0 \\ 1\le l\le n}}^{1} F_{\boldsymbol{x}_0 \boldsymbol{u}} e^{-2\pi \boldsymbol{x}_0 I_n \boldsymbol{u}/N} \tag{10.13}$$

where

$$F_{\boldsymbol{x}_0 \boldsymbol{u}} = \sum_{\substack{x_{l1}=0 \\ 1\le l\le n}}^{N/2-1} f_{2\boldsymbol{x}_1+\boldsymbol{x}_0} e^{-2\pi 2\boldsymbol{x}_1 I_n \boldsymbol{u}/N} . \tag{10.14}$$

Proof. Since the proof of equation (10.13) is the same as the one for the FQFT (10.4) except for the dimension and the notation of even / odd signal-components, it is omitted. Instead of using "e" for even and "o" for odd signal-components, they are denoted by "0" and "1", respectively. Accumulating the indices yields an index-collection, which is an n-dimensional vector. This vector is identical to $\boldsymbol{x}_0 = \boldsymbol{x} \bmod 2$. Again, the order of multiplications is irrelevant. □

The notes which have been made for the FQFT concerning the signal length are valid for the n-D FHFT as well. Obviously, the restriction to the signal size becomes a drawback for higher dimensions, since the area filled up by zeros might be a multiple of the original signal size. For those cases the row-column algorithm in section 10.4.1 evaluates the spectrum with less complexity.

The inverse transform can be calculated by the same algorithm as the FHFT using positive exponential terms and a norming factor. Alternatively, an iDHFT can be developed by reconstructing the sub-spectra as it has been done for the iFQFT. Since such an iFHFT does not include any new ideas which have not been mentioned so far, we omit the explicit formulation of the algorithm.

10.3.2 Optimizations and Complexities

For the FHFT we have the same optimizations as for the FQFT. Essentially, we can apply the π-phase method and overlapping. We will consider only these two methods in the following.

Analogously to the FQFT, we exclusively have multiplications by complex factors. Note that for an implementation of the FQFT or FHFT, we reduce the complex exponential functions to sine and cosine. Therefore, we have to decompose the compact exponential function in (10.13) into n products of the sums of sines and cosines.

The multiplication of a general element U of the 2^n dimensional commutative algebra by a complex factor yields 2^{n+1} real multiplications and 2^n additions. The addition of two general elements U and V yields 2^n real additions.

One application of (10.13) yields

$$\sum_{i=0}^{n} \binom{n}{i} i = n \sum_{i=0}^{n-1} \binom{n-1}{i} = n2^{n-1}$$

of these special multiplications and $2^n - 1$ additions per frequency. Hence, we obtain $n2^{2n}$ real multiplications and

$$n2^{2n-1} + 2^{2n} - 2^n = \left(\frac{n}{2} + 1 - 2^{-n}\right) 2^{2n}$$

real additions. Therefore, we have $n2^{2n}N^n$ multiplications and $(\frac{n}{2} + 1 - 2^{-n})2^{2n}N^n$ additions for the whole spectrum. Since the algorithm is performed $\operatorname{ld} N$ times, we must multiply these complexities by $\operatorname{ld} N$ for obtaining the complexity of the whole algorithm.

Using the effect of the π-phase, we reduce the number of multiplications by 2^n. Consequently we need $n2^n N^n \operatorname{ld} N$ real multiplications and $(\frac{n}{2} + 2^n - 1)2^n N^n \operatorname{ld} N$ real additions for the whole algorithm.

Now, we describe the overlapping for the n-dimensional case: firstly, the domain of the signal $f_{\boldsymbol{x}}$ of the size N^n is divided into 2^n parts by the substitution $\boldsymbol{x} = 2\boldsymbol{x}_1 + \boldsymbol{x}_0$ where $x_{0j} = x_j \bmod 2$ and $x_{1j} = \lfloor x_j/2 \rfloor$. Each signal value $f_{2\boldsymbol{x}_1+\boldsymbol{x}_0}$ is mapped onto one component of a new hypercomplex-valued signal $\bar{f}_{\boldsymbol{x}_1}$. Which value is mapped onto which component is determined by $\boldsymbol{x}_0$.
We define a coding function

$$\boldsymbol{C} : \mathcal{P}(\{1,\ldots,n\}) \longrightarrow \{0,1\}^n$$

with

$$C_k(j) = \begin{cases} 1 & \text{if} \quad k \in j \\ 0 & \text{else} \end{cases} \tag{10.15}$$

and use the following mapping:

$$\mathcal{I}_j\{\bar{f}_{\boldsymbol{x}_1}\} = f_{2\boldsymbol{x}_1+\boldsymbol{C}(j)} \tag{10.16}$$

for all $j \in \mathcal{P}(\{1,\ldots,n\}) \setminus \emptyset$ and

$$\mathcal{R}\{\bar{f}_{\boldsymbol{x}_1}\} = f_{2\boldsymbol{x}_1}. \tag{10.16'}$$

If the set j is empty, the real part is taken, otherwise $\mathcal{I}_j\{\bar{f}_{\boldsymbol{x}_1}\}$. Consequently, a value at a position which is odd with respect to coordinate k is mapped onto an imaginary part which contains i_k.

Calculating the spectrum of $\bar{f}_{\boldsymbol{x}_1}$ yields

$$\bar{F}_{\boldsymbol{u}_0} = \sum_{j \in \mathcal{P}(\{1,\ldots,n\})} i_j F_{\boldsymbol{C}(j)\boldsymbol{u}_0} \tag{10.17}$$

where $F_{\boldsymbol{x}_0\boldsymbol{u}_0}$ are the sub-spectra from (10.13).

The sub-spectra can be extracted from (10.17) by inverting the involutions (8.5):

$$2^n i_j F_{\boldsymbol{C}(j)\boldsymbol{u}_0} = \sum_{k \in \mathcal{P}(\{1,\ldots,n\})} \alpha_k(\bar{F}_{\boldsymbol{u}_0^k})(-1)^{\mathrm{card}(j \cap k)} \tag{10.18}$$

where $u_{0i}^k = \begin{cases} -u_{0i} & \text{if } i \in k \\ u_{0i} & \text{else} \end{cases}$ and $\mathrm{card}(M)$ is the cardinality of M. Finally, the spectrum $F_{\boldsymbol{u}}$ is calculated by use of formula (10.13).

Roughly speaking, overlapping reduces the complexity by a factor of four. The total complexity is a little bit worse, since we have an additional N^n-complexity for the reconstruction. This additional complexity can be calculated analogously to that one in section 10.2.3. Finally, we obtain the following table 10.3:

We want to justify the neglect of the exact evaluation of the N^n-complexities by the fact that we will present a faster approach to calculate the DHFT in section 10.4.

Table 10.3. Complexities of the considered algorithms

Complexity	FHFT (π-phase)	FHFT/overlapping
multiplications	$n2^n N^n \operatorname{ld} N$	$nN^n \operatorname{ld} N + \mathcal{O}(N^n)$
additions	$(\frac{n}{2} + 2^n - 1)2^n N^n \operatorname{ld} N$	$(\frac{n}{2} + 2^n - 1)N^n \operatorname{ld} N + \mathcal{O}(N^n)$
operations	$(\frac{3n}{2} + 2^n - 1)2^n N^n \operatorname{ld} N$	$(\frac{3n}{2} + 2^n - 1)N^n \operatorname{ld} N + \mathcal{O}(N^n)$

The memory complexity is very crucial in the n-dimensional case, since the signal size increases exponentially with the dimension. The presented algorithm gets very slow, if the signal size is greater than the main memory. The reason for this is the global data access of the algorithm in the first recursive steps (the whole domain, 2^{-n}th of the domain, ...). In the section 10.4.1 we present an algorithm which only acts on 1-D sub-signals. For very big signal sizes this algorithm needs less swapping of the data.

10.4 Fast Algorithms by FFT

In this section we describe two methods of evaluating the DHFT and the DCFT by applying complex-valued FFT algorithms. The first method which is called *row-column method* cascades 1-D FFTs in order to calculate the spectrum. The second method uses the isomorphism between the commutative hypercomplex algebra and the indirect product of complex algebras. This isomorphism maps the HFT of a signal onto two complex spectra of the signal and vice versa. We obtain a method to calculate the hypercomplex spectrum from the complex spectrum. The last paragraph deals with the complexities of the presented algorithms.

10.4.1 Cascading 1-D FFTs

There is one simple approach to calculate the DCFT (or the DHFT) by the 1-D FFT algorithm. The method is called *row-column algorithm* and we will firstly introduce it for the DQFT. Of course, it can be generalized for arbitrary dimensions.

The idea is as follows. We pick up one coordinate, without loss of generality we will take the first one (i.e. the x-coordinate). Then, we calculate the 1-D FFT of each row and put the results in two domains of the same size, one for the real part and one for the imaginary part, denoted by f^R and f^I, respectively.

Next, each column of these two signals f^R and f^I is transformed by the 1-D FFT. Finally, each part of both spectra is mapped to one part of the quaternionic spectrum. The real spectrum of f^R is the real part of F^q, the real spectrum of f^I is the i-imaginary part of F^q, the imaginary spectrum of f^R

is the j-imaginary part of F^q and the remaining spectrum is the k-imaginary part of F^q. The method is illustrated in the following figure 10.4.

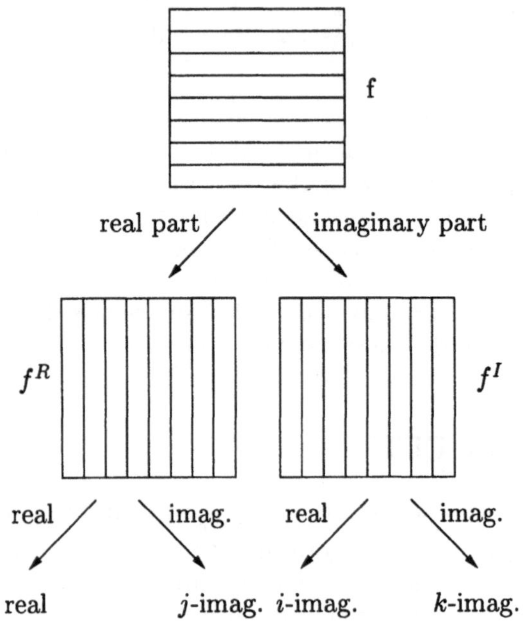

Fig. 10.4. The row-column algorithm for a 2-D signal (the arrow means application of the 1-D FFT)

Theorem 10.4.1 (Row-Column algorithm). *The application of the row-column algorithm to a signal yields its DQFT in the 2-D case.*

Proof. We start this proof with the DQFT of a signal f.

$$F^q_{u,v} = \sum_{x=0}^{M-1} \sum_{y=0}^{N-1} e^{-i2\pi uxM^{-1}} f_{x,y} e^{-j2\pi vyN^{-1}}$$

Now, we will set some parenthesis to define N new 1-D functions $f^y_x = f_{x,y}$. The 1-D Fourier transform of f^y_x is denoted by F^y_u (complex-valued):

$$\begin{aligned} F^q_{u,v} &= \sum_{y=0}^{N-1} \left(\sum_{x=0}^{M-1} e^{-i2\pi uxM^{-1}} f^y_x \right) e^{-j2\pi vyN^{-1}} \\ &= \sum_{y=0}^{N-1} F^y_u e^{-j2\pi vyN^{-1}} \\ &= \sum_{y=0}^{N-1} \mathcal{R}\{F^y_u\} e^{-j2\pi vyN^{-1}} + i \sum_{y=0}^{N-1} \mathcal{I}\{F^y_u\} e^{-j2\pi vyN^{-1}} \end{aligned}$$

The 1-D Fourier transform of $\mathcal{R}\{F^y_u\}$ and $\mathcal{I}\{F^y_u\}$ are denoted by $F^R_{u,v}$ and

$F^I_{u,v}$, respectively. Note that these transforms use y as the spatial coordinate and v as the frequency.

$$
\begin{aligned}
F^q_{u,v} &= \sum_{y=0}^{N-1} \mathcal{R}\{F^y_u\}(\mathcal{R}\{e^{-i2\pi vyN^{-1}}\} + j\mathcal{I}\{e^{-i2\pi vyN^{-1}}\}) \\
&\quad + i\sum_{y=0}^{N-1} \mathcal{I}\{F^y_u\}(\mathcal{R}\{e^{-i2\pi vyN^{-1}}\} + j\mathcal{I}\{e^{-i2\pi vyN^{-1}}\}) \\
&= \mathcal{R}\{F^R_{u,v}\} + j\mathcal{I}\{F^R_{u,v}\} + i\mathcal{R}\{F^I_{u,v}\} + k\mathcal{I}\{F^I_{u,v}\}
\end{aligned}
$$

Hence, we obtain exactly the coefficients as described in the row-column algorithm and due to the fact that two quaternions are equal if and only if all coefficients are equal, the theorem is proved. □

For the DHFT we obtain an analogous algorithm. In each step the number of transforms is doubled, since the real and imaginary parts are transformed separately. Every exponential factor $e^{-i_j 2\pi x_j u_j N_j^{-1}}$ is rewritten as $\mathcal{R}\{e^{-i2\pi x_j u_j N_j^{-1}}\} + i_j\mathcal{I}\{e^{-i2\pi x_j u_j N_j^{-1}}\}$. Hence, we double the number of Fourier transforms $n-1$ times and consequently we obtain $2 \cdot 2^{n-1} = 2^n$ coefficients which we need for the DHFT. Furthermore, we have all imaginary units: $\Pi_{j=1}^{n}(1 + i_j)$.

Since the order of the imaginary units in the product is ascending, we can perform the multiplications as well in the Clifford algebra as in the commutative hypercomplex algebra. This is an improvement, compared to the FHFT (see below). Besides, the drawback of all radix-2 methods (i.e. that the signal size must be a power of two) is less serious for the row-column algorithm. For the FHFT the signal length in each coordinate must be filled up to the same (i.e. the greatest) power of two. The row-column algorithm can handle different signal lengths in each coordinate.

The algorithm must be modified, if the spatial signal f is not real-valued. Consider the two-dimensional case. If the QFT is defined with one exponential factor to the left and one to the right, we can split the signal f into one complex signal $f^1 = \mathcal{R}\{f\} + i\mathcal{I}\{f\}$ and one complex signal $f^2 = \mathcal{J}\{f\} + i\mathcal{K}\{f\}$. Obviously $f = f^1 + f^2 j$. The 1-D Fourier transform (row-wise) of f^1 and f^2 are denoted F^1 and F^2, respectively. Now, we calculate the 1-D Fourier transform (column wise) of $\mathcal{R}\{F^1\} + i\mathcal{R}\{F^2\}$ and $\mathcal{I}\{F^1\} + i\mathcal{I}\{F^2\}$, denoted by F^{RJ} and F^{IK}, respectively. They are identical to $\mathcal{R}\{F^q\}+i\mathcal{J}\{F^q\}$ and $\mathcal{I}\{F^q\} + i\mathcal{K}\{F^q\}$. Hence, we obtain

$$
F^q = \mathcal{R}\{F^{RJ}\} + j\mathcal{I}\{F^{RJ}\} + i(\mathcal{R}\{F^{IK}\} + j\mathcal{I}\{F^{IK}\}). \tag{10.19}
$$

Considering the CFT, a new problem occurs. As we can see in the case of the 2-D CFT, the spectrum is different depending on the i-exponential term standing on the left or on the right of the signal. If we have more than two dimensions, we must use the CFT. Therefore, in order to compensate the exchanged imaginary units, we have to alter some signs. These signs can be evaluated by formally splitting the CFT into 1-D FTs. The formulation of the specific rules for each dimension is omitted here, since the rules can be derivated from the eigenvalues in Lem. 9.4.3 and by the different signs in the multiplication tables of the algebras $\mathbb{R}_{0,n}$ and $\mathcal{H}_n$. Furthermore, we are mostly interested in image analysis and for that purpose the 2-D algorithm is sufficient.

10.4.2 HFT by Complex Fourier Transform

We have proved in section 9.4.2 that the commutative hypercomplex algebra is isomorphic to the Cartesian product of complex algebras. Let us consider this isomorphism more closely with respect to the Fourier transform. Firstly, we take the two-dimensional case.

We obtain the two complex coefficients η, ξ from the hypercomplex value F by the formulae

$$\eta = \mathcal{R}\{F\} - \mathcal{K}\{F\} + i(\mathcal{I}\{F\} + \mathcal{J}\{F\}) \tag{10.20a}$$

$$\xi = \mathcal{R}\{F\} + \mathcal{K}\{F\} + i(\mathcal{I}\{F\} - \mathcal{J}\{F\}) \tag{10.20b}$$

Assume now F being a hypercomplex spectrum. If we consider η, we can see that it is equal to the complex spectrum of the same signal. It is an amazing fact, that two totally different mappings (one between two algebras and one between two signal-theoretic concepts) are equal.

Even more amazing is the correspondence between the second complex component ξ and the complex spectrum. If F is the spectrum of a real signal, i.e. it is Hermite symmetric, we obtain ξ from η by inverting the v-coordinate (the reversion in v-direction yields a changed sign in the j- and k-components of the spectrum).

Using this knowledge, we can develop another fast algorithm for the HFT2 based on a complex 2-D FFT. Assume that the signal is real-valued. Now, calculate its complex spectrum, using an ordinary 2-D FFT algorithm. This spectrum is equal to η. Next, invert the v-axis in order to obtain the ξ component. Last, use the inverse of (10.20a and b) in order to reconstruct the hypercomplex spectrum.

The whole procedure can be shortened by writing

$$F^h = \mathcal{R}\{F^e\} + i\mathcal{I}\{F^e\} + j\mathcal{I}\{F^o\} - k\mathcal{R}\{F^o\} \tag{10.21}$$

where F is the complex spectrum, $F^e_{u,v} = 1/2(F_{u,v} + F_{u,-v})$ and $F^o_{u,v} = 1/2(F_{u,v} - F_{u,-v})$.

Up to now, we have exclusively considered spectra of real signals. In section 10.3.1 we have already mentioned that the DCFT of a Clifford-valued signal can be calculated using the DHFT. In order to develop such an algorithm we use the following derivation starting with the 2-D DCFT (not the DQFT):

$$\begin{aligned}
F^q &= \sum_{x=0}^{M-1}\sum_{y=0}^{N-1}(\mathcal{R}\{f\}+i\mathcal{I}\{f\}+j\mathcal{J}\{f\}+k\mathcal{K}\{f\})e^{-i2\pi ux/M}e^{-j2\pi vy/N} \\
&= \sum_{x=0}^{M-1}\sum_{y=0}^{N-1}\mathcal{R}\{f\}e^{-i2\pi ux/M}e^{-j2\pi vy/N} \\
&\quad +i\sum_{x=0}^{M-1}\sum_{y=0}^{N-1}\mathcal{I}\{f\}e^{-i2\pi ux/M}e^{-j2\pi vy/N} \\
&\quad +j\sum_{x=0}^{M-1}\sum_{y=0}^{N-1}\mathcal{J}\{f\}e^{-i2\pi ux/M}e^{-j2\pi vy/N} \\
&\quad +k\sum_{x=0}^{M-1}\sum_{y=0}^{N-1}\mathcal{K}\{f\}e^{-i2\pi ux/M}e^{-j2\pi vy/N} \\
&= F^R + iF^I + jF^J + kF^K
\end{aligned}$$

where F^R, F^I, F^J, and F^K are spectra of the real signals $\mathcal{R}\{f\}, \mathcal{I}\{f\}, \mathcal{J}\{f\}$, and $\mathcal{K}\{f\}$, respectively. Hence, they can be evaluated either in the Clifford algebra or in the commutative hypercomplex algebra. Therefore, they can be calculated via the complex spectra. Only the last step (the multiplication of the partial spectra by the imaginary units) must be calculated in the Clifford algebra.

Since the isomorphism is proved for any dimension, every hypercomplex spectrum can be calculated using the complex Fourier transform. If the signal is real, the method is straightforward. The first complex coefficient is obtained by the complex Fourier transform. The other coefficient are calculated by inverting each coordinate axis except for the first. If the signal is Clifford-valued, each component must be transformed separately and afterwards the imaginary units are multiplied to the partial spectra using the Clifford algebra multiplication.

10.4.3 Complexities

In this section we consider some complexities of the presented algorithms. Though the row-column algorithm is most advantageous in the case where the signal length varies widely with respect to the different coordinates, we consider the case where all signal lengths are the same, in order to compare the row-column algorithm to the other ones.

We start with the complexities for the two-dimensional case. We already said that for real signals the row-column algorithm doubles the number of 1-D transforms for each dimension. Hence, we need three transforms in the 2-D case. Each transform has to pass N rows (columns) and its complexity is given according to table 10.3 by $N \operatorname{ld} N$ multiplications and $3/2N \operatorname{ld} N$

additions ($5/2N \operatorname{ld} N$ operations). Hence, we obtain $3N^2 \operatorname{ld} N$ multiplications and $9/2N^2 \operatorname{ld} N$ additions for the whole spectrum ($15/2N^2 \operatorname{ld} N$ operations).

For quaternionic signals the row-column algorithm performs two times two 1-D FFTs. Since the signals are not real, the 1-D FFT itself needs twice the number of operations. That yields a complexity of $8N^2 \operatorname{ld} N$ multiplications and $12N^2 \operatorname{ld} N$ additions ($20N^2 \operatorname{ld} N$ operations).

If we use the isomorphism to calculate the Clifford spectrum of a signal, the complexity depends on the 2-D FFT algorithm (up to a quadratic additive complexity). Assuming that a 2-D FFT algorithm can be performed with $3/2N^2 \operatorname{ld} N$ multiplications and $(3/4+2)N^2 \operatorname{ld} N$ additions ($17/4N^2 \operatorname{ld} N$ operations)[121], an FHFT algorithm using the isomorphism has the same complexity if the signal is real.

If the signal is Clifford-valued, the complexity is four times the complexity of the algorithm for a real signal (up to a quadratic additive complexity for the combination of the four spectra). All the complexities calculated above are summarized in table 10.4.

Table 10.4. Complexities of the considered algorithms

algorithm	multiplications	additions	operations
row-column (real)	$3N^2 \operatorname{ld} N$	$9/2N^2 \operatorname{ld} N$	$15/2N^2 \operatorname{ld} N$
row-column (quat.)	$8N^2 \operatorname{ld} N$	$12N^2 \operatorname{ld} N$	$20N^2 \operatorname{ld} N$
isomorphism (real)	$\frac{3}{2}N^2 \operatorname{ld} N$	$\frac{11}{4}N^2 \operatorname{ld} N$	$\frac{17}{4}N^2 \operatorname{ld} N$
isomorphism (quat.)	$6N^2 \operatorname{ld} N$	$11N^2 \operatorname{ld} N$	$17N^2 \operatorname{ld} N$

Finally, we will roughly consider some complexities for the n-dimensional case. The row-column algorithm for the real case uses $\sum_{i=0}^{n-1} 2^i = 2^n - 1$ 1-D FFTs N^{n-1} times. That yields a complexity of $(2^n - 1)N^n \operatorname{ld} N$ multiplications and $(2^n - 1)3/2N^n \operatorname{ld} N$ additions ($(2^n - 1)5/2N^n \operatorname{ld} N$ operations).

The row-column algorithm for Clifford-valued signals needs $n2^{n-1}$ 1-D FFTs (with double complexity) N^{n-1} times. Hence, we have a complexity of $n2^n N^n \operatorname{ld} N$ multiplications and $n2^n 3/2N^n \operatorname{ld} N$ additions ($n2^n 5/2N^n \operatorname{ld} N$ operations).

The algorithm using the isomorphism still has the same complexity as the complex nD FFT in the case of real signals and it has a 2^n times higher complexity in the case of Clifford-valued signals.

If we consider the memory complexity, it becomes obvious that the row-column algorithm applies FFTs only on 1-D sub-signals. Therefore, it works nearly independently on the dimension. If the signal does not fit into the main memory, the data is swapped as often as the coordinate for the 1-D FFT is changed (i.e. n times). Hence, the number of swapped data is nN^n. Note that we provided implicitly, that one row always fits in the main memory, which is quite realistic.

For the n-D FFT (or FHFT) we have one crucial point in the recursion where the data begins to swap for each higher level. The number of the swapping levels s can be calculated from the size of the main memory M and the signal size N^n by the formula

$$s = \lceil \frac{1}{n} \operatorname{ld} \frac{N^n}{M} \rceil = \lceil \operatorname{ld} N - \frac{1}{n} \operatorname{ld} M \rceil = \operatorname{ld} N - \lfloor \operatorname{ld} \sqrt[n]{M} \rfloor . \tag{10.22}$$

The algorithm must swap $(\operatorname{ld} N - \lfloor \operatorname{ld} \sqrt[n]{M} \rfloor) N^n$ values. Since this formula is not as easy to understand as the one for the row-column algorithm, we will give an example.

Firstly, consider that we want to calculate the DQFT ($n = 2$). We know that each quaternion uses 32 bytes (four double floats). Assuming that the main memory consists of 64 megabyte, we obtain $M = 2^{21}$. Hence $\lfloor \operatorname{ld} \sqrt[2]{M} \rfloor = 10$. That means, if the image is greater than 1024×1024 the FHFT begins to swap. The number of swappings is linear with the power of two. The row-column algorithm swaps two times, if the image is greater that 1024×1024.

Now we have $n = 3$. We obtain $M = 2^{20}$ and $\lfloor \operatorname{ld} \sqrt[3]{M} \rfloor = 6$. Again, both algorithms begin to swap if the signal is greater than $64 \times 64 \times 64$. The row-column algorithm swaps three times, the FHFT algorithm needs a number of swappings linear with the power of two. Both examples are illustrated in figure 10.5. The x-axis indicates the exponent of the signal size and the y-axis indicates, how often the whole data is swapped.

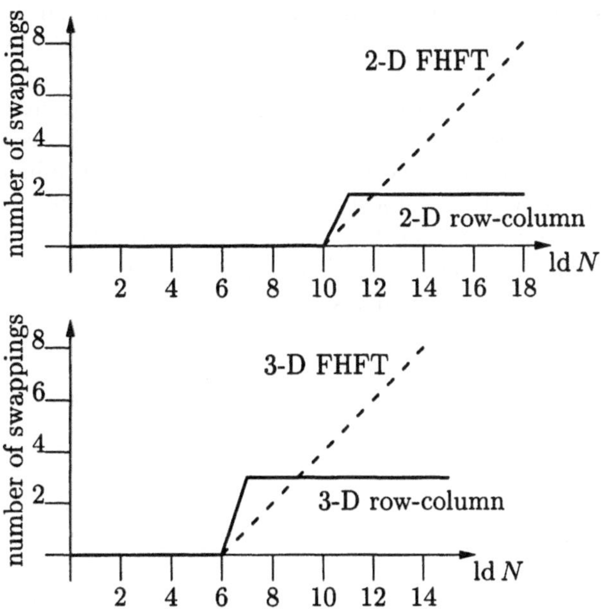

Fig. 10.5. The number of swapping operations for the 2-D and 3-D FHFT and row-column algorithm

Obviously, it depends on the size of the data which algorithm is the best choice.

10.5 Conclusion and Summary

We have considered several fast algorithms for multidimensional hypercomplex Fourier transforms. We can divide these algorithms into three classes: n-D decimation algorithms, n-D algorithms which use the complex FFTn, and row-column algorithms which apply 1-D FFT1s to each coordinate (separability). The last two algorithms use standard transforms and a simple mapping of the data is all to be done. Therefore, these algorithms are to be preferred if one wants to make only some experiments or if the dimension of the signals often changes because they can be implemented fast and easily.

Although the asymptotic complexity of the n-D decimation algorithms are the least and therefore these algorithms seem to be superior to the algorithms using the isomorphism or the row-column method, we cannot recommend using them, because their implementation is slower due to all the array accesses and the exhaustive length of the code. We ourselves recommend using the row-column method since the complex FFTn algorithms are mostly implemented in this way and one can merge the steps from 1-D FFTs to n-D FFT and from n-D FFT to n-D FHFT in one step. Additionally, it is possible to adapt the algorithm to signal sizes which differ for different coordinates.

For very large signals (i.e. greater than the actual memory size) the row-column method is superior as well, since the swapping is reduced to a minimum. Furthermore, it is easier to optimize the 1-D FFT algorithm than to speed up the n-D algorithms. These optimizations of the FFT1 automatically lead to an optimized n-D algorithm.

The most interesting result which we have presented is the fact that a theoretical algebraic result yields a practically optimal algorithm. The isomorphism of the n-fold tensor product and the 2^{n-1}-fold Cartesian product of the complex algebra leads to a decomposition technic for an algorithm. This result emphasizes the importance of deep mathematical knowledge for signal processing. Thus, geometric algebra has been shown to be a powerful embedding for multidimensional signal analysis.

11. Local Hypercomplex Signal Representations and Applications*

Thomas Bülow and Gerald Sommer

Institute of Computer Science and Applied Mathematics,
Christian-Albrechts-University of Kiel

11.1 Introduction

The concept of the analytic signal is an important concept in one-dimensional signal theory since it makes the instantaneous amplitude and phase of a real signal directly accessible. Regrettably, there is no straightforward extension of this concept to multidimensional signals, yet. There are rather different approaches to an extension which have different drawbacks. In the first part of this chapter we will review the main approaches and introduce a new one which overcomes some of the problems of the older approaches. The new definition is easily described in the frequency domain. However, in contrast to the 1-D analytic signal we will use the quaternionic frequency domain instead of the complex Fourier domain. Based on the so defined quaternionic analytic signal [36] the instantaneous amplitude and quaternionic phase of a 2-D signal can be defined [34].

In one-dimensional signal theory it is often useful not to calculate the analytic signal but a bandpass filtered version of the analytic signal. This is done by applying so called quadrature filters. Here, we will use Gabor filters which are good approximations to quadrature filters. Corresponding to the evaluation of the quaternionic analytic signal is the application of quaternionic

* This work has been supported by German National Merit Foundation and by DFG Grants So-320-2-1 and So-320-2-2.

Gabor filters which we introduce based on the quaternionic Fourier transform. As a practical example we demonstrate the application of quaternionic Gabor filters in texture segmentation.

11.2 The Analytic Signal

The notion of the analytic signal of a real one-dimensional signal was introduced in 1946 by Gabor [86]. Before going into technical details we will give a vivid explanation of the meaning of the analytic signal. If we regard a real one-dimensional signal f as varying with time, it can be represented by the oscillating vector from the origin to $f(t)$ on the real line. Taking a snapshot of the vector at time t_0 as shown in figure 11.2 reveals no information about the amplitude or the instantaneous phase of the oscillation. I.e. it is invisible whether f is still growing to the right or already on the returning way and where the extrema of the oscillation lie. The analytic signal of f is a complex-valued signal, denoted by f_A. Thus, f_A can be visualized as a rotating vector in the complex plane. This vector has the property that its projection to the real axis is identical to the vector given by f. Moreover, if a snapshot is taken, the length of the vector, and its angle against the real axis give the instantaneous amplitude and the instantaneous phase of f, respectively. The analytic signal is constructed by adding to the real signal f a signal which is shifted by $-\pi/2$ in phase against f.

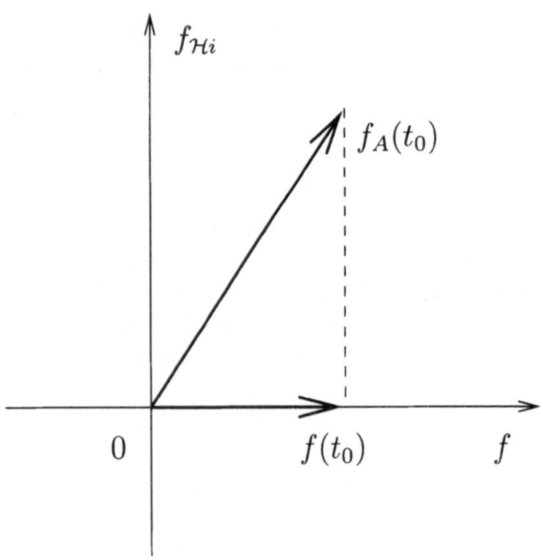

Fig. 11.1. Snapshot of the oscillating vector to f and the rotating vector to f_A at time t_0

In this section we will shortly review the analytic signal in 1-D and four approaches to the analytic signal in 2-D which have occurred in the literature [96, 97, 88, 219]. We investigate the different principles which lie at the basis of the definitions and conclude with a set of desirable properties of the 2-D analytic signal. Based on the QFT it is possible to introduce a novel definition of the analytic signal which fulfills most of the desired properties.

11.2.1 The One-Dimensional Analytic Signal

As mentioned above, the analytic signal f_A of a real one-dimensional signal f is defined as the sum of f and a $-\pi/2$-shifted version of f as imaginary part. The shifted version of f is the Hilbert transform $f_{\mathcal{H}i}$ of f. Thus, the analytic signal can be written as $f_A = f + i f_{\mathcal{H}i}$. This is the generalization of the complex notation of harmonic signals given by Euler's equation $\exp(i2\pi ux) = \cos(2\pi ux) + i\sin(2\pi ux)$.

A phase shift by $-\pi/2$ – which is expected to be done by the Hilbert transform – can be realized by taking the negative derivative of a function. E.g. we have

$$-\frac{\partial}{\partial x}\cos(2\pi ux) = 2\pi u \sin(2\pi ux),$$

which shifts the cosine-function and additionally scales the amplitude with the angular frequency $\omega = 2\pi u$. In order to avoid this extra scaling we divide each frequency component by the absolute value of the angular frequency. This procedure can easily be described in the Fourier domain: Taking the negative derivative results in multiplication by $-i2\pi u$. Dividing by $|2\pi u|$ results in the following procedure in the frequency domain:

$$F(u) \mapsto -i\frac{u}{|u|}F(u) = -i\,\mathrm{sign}(u)F(u),$$

which makes plausible the definition of the Hilbert transform.

The formal definitions of the Hilbert transform and of the analytic signal are as follows:

Definition 11.2.1 (Hilbert transform). *Let f be a real 1-D signal and F its Fourier transform. The Hilbert transform of f is then defined in the frequency domain by*

$$F_{\mathcal{H}i}(u) = -i sign(u)F(u) \quad with \quad sign(u) = \begin{cases} 1 & if\, u > 0 \\ 0 & if\, u = 0 \\ -1 & if\, u < 0 \end{cases}. \tag{11.1}$$

In spatial domain this reads

$$f_{\mathcal{H}i}(x) = f(x) * \frac{1}{\pi x}, \tag{11.2}$$

where $*$ denotes the convolution operation.

The convolution integral in (11.2), namely

$$f_{\mathcal{H}i}(x) = \frac{1}{\pi}\int_{\mathbb{R}} \frac{f(\xi)}{x-\xi}d\xi$$

contains a singularity at $x = \xi$. This is handled by evaluating Cauchy's principle value, i.e.

$$f_{\mathcal{H}i}(x) = \frac{1}{\pi}P\int_{\mathbb{R}} \frac{f(\xi)}{x-\xi}d\xi \tag{11.3}$$

$$= \frac{1}{\pi}\lim_{\epsilon\to 0}\left(\int_{-\infty}^{x-\epsilon} \frac{f(\xi)}{x-\xi}d\xi + \int_{x+\epsilon}^{\infty} \frac{f(\xi)}{x-\xi}d\xi\right) \tag{11.4}$$

Definition 11.2.2 (Analytic signal). *Let f be a real 1-D signal and F its Fourier transform. Its analytic signal in the Fourier domain is then given by*

$$F_A(u) = F(u) + iF_{\mathcal{H}i}(u) \tag{11.5}$$
$$= F(u)(1 + sign(u)).$$

In the spatial domain this definition reads:

$$f_A(x) = f(x) + if_{\mathcal{H}i}(x) = f(x) * \left(\delta(x) + \frac{i}{\pi x}\right). \tag{11.6}$$

Thus, the analytic signal of f is constructed by taking the Fourier transform F of f, suppressing the negative frequencies and multiplying the positive frequencies by two. Note that, applying this procedure, we do not lose any information about f because of the Hermite symmetry of the spectrum of a real function.

The analytic signal enables us to define the notions of the instantaneous amplitude and the instantaneous phase of a signal [88].

Definition 11.2.3 (Instantaneous amplitude and phase). *Let f be a real 1-D signal and f_A its analytic signal. The instantaneous amplitude and and phase of f are then defined by*

$$\textit{instantaneous amplitude of } f(x) = |f_A(x)| \tag{11.7}$$

$$\textit{instantaneous phase of } f(x) = \operatorname{atan2}(\mathcal{I}f_A(x), \mathcal{R}f_A(x)). \tag{11.8}$$

For later use we introduce the notion of a Hilbert pair.

Definition 11.2.4 (Hilbert pair). *Two real one-dimensional functions f and g are called a Hilbert pair if one is the Hilbert transform of the other, i.e.*

$$f_{\mathcal{H}i} = g \quad or \quad g_{\mathcal{H}i} = f.$$

If $f_{\mathcal{H}i} = g$ it follows that $g_{\mathcal{H}i} = -f$.

We illustrate the above definitions by a simple example: The analytic signal of $f(x) = a\cos(\omega x)$ which is $a\cos_A(\omega x) = a\cos(\omega x) + ia\sin(\omega x) = a\exp(i\omega x)$. The instantaneous amplitude of f is given by $|f_A(x)| = a$ while the instantaneous phase is $\operatorname{atan2}(\mathcal{I}f_A(x), \mathcal{R}f_A(x)) = \omega x$. Thus, the instantaneous amplitude and phase of the cosine-function are exactly equal to the expected values a and ωx, respectively. Furthermore, cos and sin constitute a *Hilbert pair*. Figure 11.2 shows another example of an oscillating signal together with its instantaneous amplitude and its instantaneous phase.

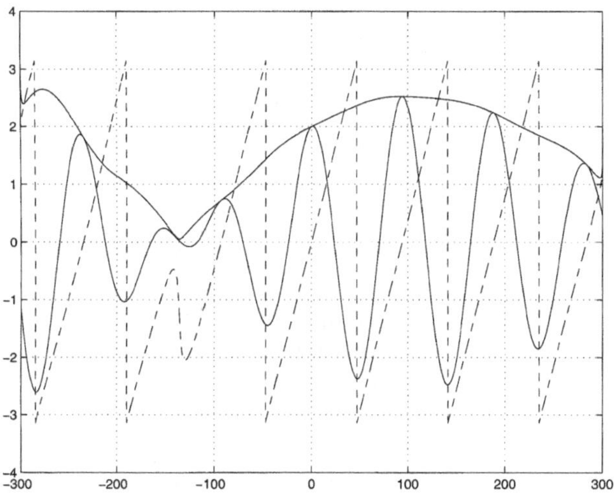

Fig. 11.2. An oscillating signal, its instantaneous amplitude (signal envelope) and its instantaneous phase (dashed)

However, the close relation of the instantaneous amplitude and phase to the local structure of the signal gets lost if the signal has no well defined angular frequency. Most of the time it is sufficient to require the signal to be of narrow bandwidth ([88], p. 171).

For this reason later (in section 11.3.1) Gabor filters will be introduced which establish a relation between the local structure and the local phase of a broader class of signals.

11.2.2 Complex Approaches to the Two-Dimensional Analytic Signal

The construction of the analytic signal is of interest not only in one-dimensional signal processing but in image processing and multidimensional signal processing as well. So far, however, we have merely presented a definition of the one-dimensional analytic signal. Thus, an extension to higher dimensions is needed. There have appeared different approaches to a 2-D analytic signal in the literature. All of these approaches use a combination of the original signal and its Hilbert transform. In this section we will present and discuss these approaches. A novel approach which is based on the quaternionic Fourier transform (see chap. 8, Def. 8.3.4) is introduced in section 11.2.3.

In order to evaluate the different approaches to the analytic signal to 2-D we need some guidelines. As such a guideline we give a list of the main properties of the analytic signal in 1-D. Any new definition will be measured according to the degree to which it extends these properties to higher dimensions.

Table 11.1. Four properties of the analytic signal

1.	The spectrum of an analytic signal is right-sided ($F_A(u) = 0$ for $u < 0$).
2.	Hilbert pairs are orthogonal.
3.	The real part of the analytic signal f_A is equal to the original signal f.
4.	The analytic signal is compatible with the associated harmonic transform (in case of the 1-D analytic signal with the Fourier transform.)

We will explain the forth point. The analytic signal is called compatible with the associated harmonic transform with transformation kernel K if $\mathcal{R}K$ and $\mathcal{I}K$ are a Hilbert pair. In case of the one-dimensional Fourier transform this property is fulfilled, since the real part of the Fourier kernel, i.e. $\mathcal{R}(\exp(-i2\pi\, ux)) = \cos(-2\pi\, ux)$ is the Hilbert transform of $\sin(-2\pi\, ux)$, as was shown above.

The first definition is based on the 2-D Hilbert transform [219]:

Definition 11.2.5 (Total 2-D Hilbert transform). *Let f be a real two-dimensional signal. Its Hilbert transform is given by*

$$f_{\mathcal{H}i}(\boldsymbol{x}) = f(\boldsymbol{x}) * \left(\frac{1}{\pi^2 xy}\right), \tag{11.9}$$

where $*$ denotes the 2-D convolution. In the frequency domain this reads

$$F_{\mathcal{H}i}(\boldsymbol{u}) = -F(\boldsymbol{u})\, sign(u)\, sign(v).$$

Sometimes $f_{\mathcal{H}i}$ is called the total Hilbert transform of f [97].

For later use, we define also the partial Hilbert transforms of a 2-D signal.

Definition 11.2.6 (Partial Hilbert transform).
Let f be a real two-dimensional signal. Its partial Hilbert transforms in x- and y-direction are given by

$$f_{\mathcal{H}i_1}(\boldsymbol{x}) = f(\boldsymbol{x}) * \left(\frac{\delta(y)}{\pi x}\right), \quad \textit{and} \tag{11.10}$$

$$f_{\mathcal{H}i_2}(\boldsymbol{x}) = f(\boldsymbol{x}) * \left(\frac{\delta(x)}{\pi y}\right), \tag{11.11}$$

respectively. In the frequency domain this reads

$$F_{\mathcal{H}i_1}(\boldsymbol{u}) = -iF(\boldsymbol{u})sign(u) \quad \textit{and} \quad F_{\mathcal{H}i_2}(\boldsymbol{u}) = -iF(\boldsymbol{u})sign(v).$$

The partial Hilbert transform of a 2-D signal can of course be defined with respect to any orientation.

In analogy to 1-D an extension of the analytic signal can be defined as follows:

Definition 11.2.7 (Total analytic signal). *The analytic signal of a real 2-D signal f is defined as*

$$f_A(\boldsymbol{x}) = f(\boldsymbol{x}) * (\delta^2(\boldsymbol{x}) + \frac{i}{\pi^2 xy}) \tag{11.12}$$

$$= f(\boldsymbol{x}) + if_{\mathcal{H}i}(\boldsymbol{x}), \tag{11.13}$$

where $f_{\mathcal{H}i}$ is given by (11.9). In the frequency domain this definition reads

$$F_A(\boldsymbol{u}) = F(\boldsymbol{u})(1 - i\, sign(u) sign(v)).$$

The spectrum of f_A according to definition 11.2.7 is shown in figure 11.3. It does not vanish anywhere in the frequency domain. Hence, there is no analogy to the causality property of an analytic signal's spectrum in 1-D. Secondly, Hilbert pairs according to this definition are only orthogonal if the functions are separable [97]. Furthermore, the above definition of the analytic signal is not compatible with the two-dimensional Fourier transform, since $\sin(2\pi\boldsymbol{ux})$ is not the total Hilbert transform of $\cos(2\pi\boldsymbol{ux})$. Thus, the properties 1, 2 and 4 from table 11.1 are not satisfied by this definition. A

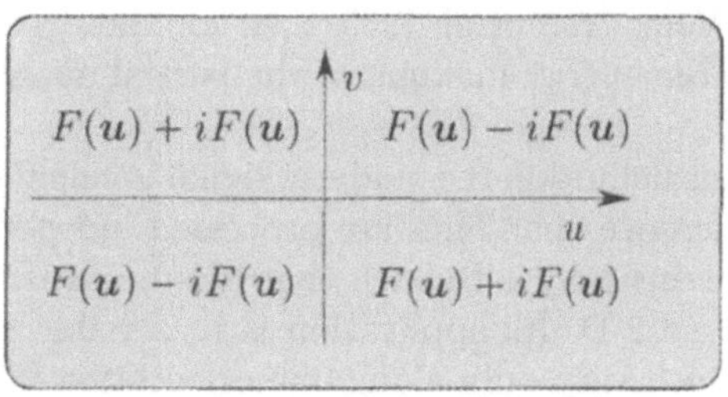

Fig. 11.3. The spectrum of the analytic signal according to definition 11.2.7

common approach to overcome this fact can be found e.g. by Granlund [88]. This definition starts with the construction in the frequency domain. While in 1-D the analytic signal is achieved by suppressing the negative frequency components, in 2-D one half-plane of the frequency domain is set to zero in order to fulfill the causality constraint (property no. 1 in table 11.1). It is not immediately clear how negative frequencies can be defined in 2-D. However, it is possible to introduce a direction of reference defined by the unit vector $\hat{\mathbf{e}} = (\cos(\theta), \sin(\theta))$. A frequency $\boldsymbol{u}$ with $\hat{\mathbf{e}} \cdot \boldsymbol{u} > 0$ is called positive while a frequency with $\hat{\mathbf{e}} \cdot \boldsymbol{u} < 0$ is called negative. The 2-D analytic signal can then be defined in the frequency domain.

Definition 11.2.8 (Partial analytic signal). *Let f be a real 2-D signal and F its Fourier transform. The Fourier transform of the analytic signal is defined by:*

$$F_A(\boldsymbol{u}) = \left\{ \begin{array}{cl} 2F(\boldsymbol{u}) & \textit{if } \boldsymbol{u} \cdot \hat{\mathbf{e}} > 0 \\ F(\boldsymbol{u}) & \textit{if } \boldsymbol{u} \cdot \hat{\mathbf{e}} = 0 \\ 0 & \textit{if } \boldsymbol{u} \cdot \hat{\mathbf{e}} < 0 \end{array} \right\} = F(\boldsymbol{u})(1 + \textit{sign}(\boldsymbol{u} \cdot \hat{\mathbf{e}})). \tag{11.14}$$

In the spatial domain (11.14) reads

$$f_A(\boldsymbol{x}) = f(\boldsymbol{x}) * \left(\delta(\boldsymbol{x} \cdot \hat{\mathbf{e}}) + \frac{i}{\pi \boldsymbol{x} \cdot \hat{\mathbf{e}}} \right) \delta(\boldsymbol{x} \cdot \hat{\mathbf{e}}_\perp). \tag{11.15}$$

The vector $\hat{\mathbf{e}}_\perp$ is a unit vector which is orthogonal to $\hat{\mathbf{e}}$: $\hat{\mathbf{e}} \cdot \hat{\mathbf{e}}_\perp = 0$.

Please note the similarity of this definition with the one-dimensional definition (11.5). For $\hat{\mathbf{e}}^\top = (1,\, 0)$ (11.15) takes the form

$$f_A(\boldsymbol{x}) = f(\boldsymbol{x}) * \left(\delta(x) + \frac{i}{\pi x} \right) \delta(y) \tag{11.16}$$

$$= f(\boldsymbol{x}) + i f_{\mathcal{H}i_1}. \tag{11.17}$$

Thus, the reason for the name *partial analytic signal* lies in the fact that it is the sum of the original signal and the partial Hilbert transform as imaginary part. The partial analytic signal with respect to the two coordinate axes has been used by Venkatesh et al. [235, 234] for the detection of image features. They define the energy maxima of the partial analytic signal as image features.

According to this definition the analytic signal is calculated line-wise along the direction of reference. The lines are processed independently. Hence, definition 11.2.8 is intrinsically 1-D, such that it is no satisfactory extension of the analytic signal to 2-D. Its application is reasonable only for simple signals, i.e. signals which vary only along one orientation [88]. The orientation $\hat{\mathbf{e}}$ can then be chosen according to the direction of variation of the image.

If negative frequencies are defined in the way indicated above, we can say that property 1 of table 11.1 is fulfilled. Properties 2 and 3 are valid as well.

This follows from the fact that merely the 1-D analytic signal is evaluated line-wise, which leads to a trivial extension of these properties. Even property 4 is "almost" valid: $\sin(ux+vy)$ is the partial Hermite transform (i.e. with respect to the x direction) of $\cos(ux+vy)$ for all frequencies $\boldsymbol{u}$ with $u \neq 0$. However, the main drawback of definition 11.2.8 is the intrinsic one-dimensionality of the definition and the non-uniqueness with regard to the orientation of reference $\hat{\boldsymbol{e}}$.

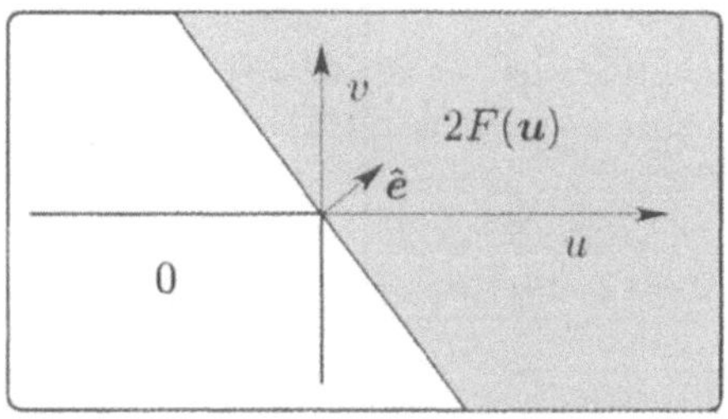

Fig. 11.4. The spectrum of the analytic signal according to definition 11.2.8

The both definitions presented so far seem to establish the following dilemma: Either an intrinsically two-dimensional definition of the analytic signal based on the total Hilbert transform can be introduced, which does not extend the main properties of the 1-D analytic signal, or these properties are extended by an intrinsically one-dimensional definition based on the partial Hilbert transform.

An alternative to these approaches was recently introduced by Hahn [96, 97]. Hahn avoids the term "analytic signal" and uses Gabor's original term "complex signal" instead.

Definition 11.2.9. *Let f be a real, two-dimensional function and F its Fourier transform. The 2-D complex signal (according to Hahn [97]) is defined in the frequency domain by*

$$F_A(\boldsymbol{u}) = (1 + sign(u))(1 + sign(v))F(\boldsymbol{u}).$$

In the spatial domain this reads

$$f_A(\boldsymbol{x}) = f(\boldsymbol{x}) * \left(\delta(x) + \frac{i}{\pi x}\right)\left(\delta(y) + \frac{i}{\pi y}\right) \tag{11.18}$$

$$= f(\boldsymbol{x}) - f_{\mathcal{H}i}(\boldsymbol{x}) + i(f_{\mathcal{H}i_1}(\boldsymbol{x}) + f_{\mathcal{H}i_2}(\boldsymbol{x})), \tag{11.19}$$

where $f_{\mathcal{H}i}$ is the total Hilbert transform, and $f_{\mathcal{H}i_1}$ and $f_{\mathcal{H}i_2}$ are the partial Hilbert transforms.

The meaning of definition 11.2.9 becomes clear in the frequency domain: Only the frequency components with $u > 0$ and $v > 0$ are kept, while the components in the three other quadrants are suppressed (see figure 11.5):

$$F_A(\boldsymbol{u}) = (1 + \mathrm{sign}(u))(1 + \mathrm{sign}(v))F(\boldsymbol{u}).$$

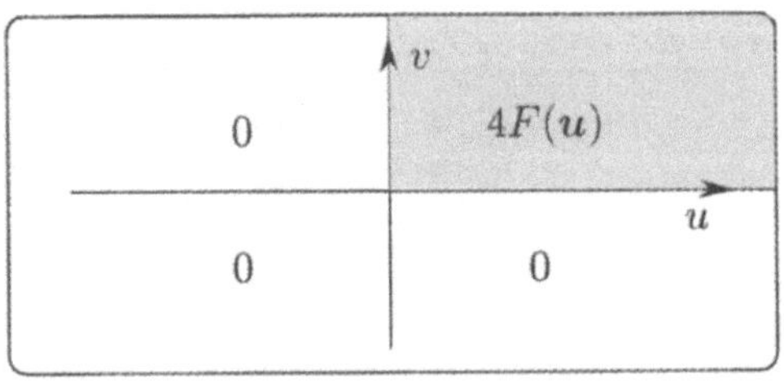

Fig. 11.5. The spectrum of the analytic signal according to Hahn [96] (definition 11.2.9)

Thus, the problem of defining positive frequencies is solved in another way then in definition 11.2.8.

A main problem of definition 11.2.9 is the fact that the original signal is not reconstructible from the analytic signal, since due to the Hermite symmetry only one half-plane of the frequency domain of a real signal is redundant. For this reason Hahn proposes to calculate not only the analytic signal with the spectrum in the upper right quadrant but also another analytic signal with its spectrum in the upper left quadrant. It can be shown that these two analytic signals together contain all the information of the original signal [97]. When necessary we distinguish the two analytic or complex signals by referring to them as definition 11.2.9a and 11.2.9b, respectively.

Thus, the complete analytic signal according to definition 11.2.9 consists of two complex signals, i.e. two real parts and two imaginary parts or, in polar representation, of two amplitude- and two phase-components which makes the interpretation, especially of the amplitude, difficult. Furthermore, it would be more elegant to express the analytic signal with only one function instead of two. Definition 11.2.9 fulfills properties 1 and 2 from table 11.1. The very important property that the signal should be reconstructible from its analytic signal is only fulfilled if two different complex signals are calculated using two neighbored quadrants of the frequency domain. Hahn [97] mentions that his definition of the 2-D analytic signal is compatible with the 2-D Fourier transform for the following reason: The 2-D Fourier kernel can be written in the form

$$\exp(i2\pi\boldsymbol{u}\boldsymbol{x}) = \cos(2\pi ux)\cos(2\pi vy) - \sin(2\pi ux)\sin(2\pi vy) \quad (11.20)$$
$$+\, i(\cos(2\pi ux)\sin(2\pi vy) + \sin(2\pi ux)\cos(2\pi vy)) \quad (11.21)$$

where for convenience we have omitted the minus sign in the exponential. According to definition 11.2.9 this is exactly the complex signal of $f(\boldsymbol{x}) = \cos(2\pi ux)\cos(2\pi vy)$. However, this fulfills only a weak kind of compatibility and not the one defined by us above. This would require that the analytic signal of $\mathcal{R}\exp(i2\pi\boldsymbol{u}\boldsymbol{x})$ would equal $\exp(i2\pi\boldsymbol{u}\boldsymbol{x})$.

The remaining problems can be summarized as follows. The original signal cannot be recovered from Hahn's analytic signal. This restriction can only be overcome by introducing two complex signals for each real signal, which is not a satisfactory solution. Furthermore, Hahn's analytic signal is not compatible with the 2-D Fourier transform in the strong sense.

Apart from these disadvantages, it is clear from the above analysis, that, among the definitions introduced so far, Hahn's definition is closest to a satisfactory 2-D extension of the analytic signal. In the following section we will show how Hahn's frequency domain construction can be applied to the construction of a quaternionic analytic signal, which overcomes the remaining problems.

11.2.3 The 2-D Quaternionic Analytic Signal

Hahn's approach to the analytic signal faces the problem that a two-dimensional complex hermitian signal can not be recovered from one quadrant of its domain. For this reason Hahn introduced two complex signals to each real two-dimensional signal. We will show how this problem is solved using the QFT.

Since the QFT of a real signal is quaternionic hermitian (see chapter 8, theorem 8.4.8) we do not lose any information about the signal in this case. This fact is visualized in figure 11.6.

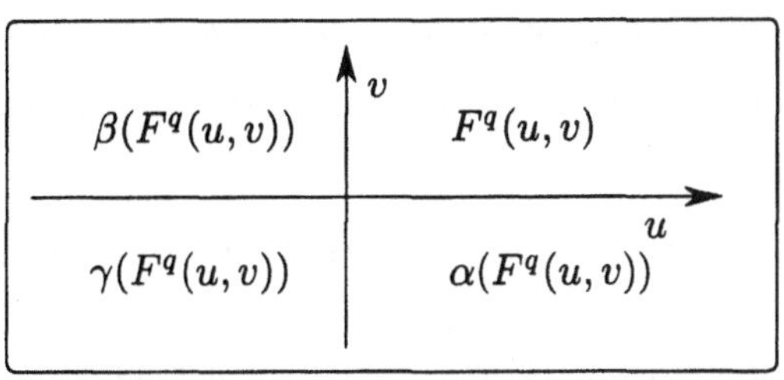

Fig. 11.6. The quaternionic spectrum of a real signal can be reconstructed from only one quadrant

Thus, we define the quaternionic analytic signal in the frequency domain as in definition 11.2.9, with the only difference that we use the quaternionic frequency domain defined by the QFT instead of the complex frequency domain.

Definition 11.2.10 (Quaternionic analytic signal). *Let f be a real two-dimensional signal and F^q its QFT. In the quaternionic frequency domain we define the quaternionic analytic signal of a real signal as*

$$F^q_A(\boldsymbol{u}) = (1 + sign(u))(1 + sign(v))F^q(\boldsymbol{u}),$$

where $\boldsymbol{x} = (x, y)$ and $\boldsymbol{u} = (u, v)$. Definition 11.2.10 can be expressed in the spatial domain as follows:

$$f^q_A(\boldsymbol{x}) = f(\boldsymbol{x}) + \boldsymbol{n} \cdot \boldsymbol{f}_{\mathcal{H}i}(\boldsymbol{x}), \tag{11.22}$$

where $\boldsymbol{n} = (i, j, k)^\top$ and $\boldsymbol{f}_{\mathcal{H}i}$ is a vector which consists of the total and the partial Hilbert transforms of f according to definitions 11.2.5 and 11.2.6:

$$\boldsymbol{f}_{\mathcal{H}i}(\boldsymbol{x}) = \begin{pmatrix} f_{\mathcal{H}i_1}(\boldsymbol{x}) \\ f_{\mathcal{H}i_2}(\boldsymbol{x}) \\ f_{\mathcal{H}i}(\boldsymbol{x}) \end{pmatrix}. \tag{11.23}$$

Note that, formally, (11.22) resembles the definition of the one-dimensional analytic signal (11.6). Since the quaternionic analytic signal consists of four components we replace the notion of a Hilbert pair (definition 11.2.4) by the notion of a *Hilbert quadruple.*

Definition 11.2.11 (Hilbert quadruple). *Four real two-dimensional functions f_i, $i \in \{1, \ldots 4\}$ are called a Hilbert quadruple if*

$$\mathcal{I}(f_k)^q_A = f_l \tag{11.24}$$
$$\mathcal{J}(f_k)^q_A = f_m \tag{11.25}$$
$$\mathcal{K}(f_k)^q_A = f_n \tag{11.26}$$

for some permutation of pairwise different $k, l, m, n \in \{1, \ldots 4\}$.

Theorem 11.2.1. *The four components of the QFT-kernel build a Hilbert quadruple.*

Proof. Since the quaternionic analytic signal of $f(\boldsymbol{x}) = \cos(\omega_x x)\cos(\omega_y y)$ is given by $f^q_A(\boldsymbol{x}) = \exp(i\omega_x x)\exp(j\omega_y y)$, which is the QFT-kernel, we have

$$\mathcal{I}(\mathcal{R}\exp(i\omega_x x)\exp(j\omega_y y))^q_A = \mathcal{I}\exp(i\omega_x x)\exp(j\omega_y y) \tag{11.27}$$
$$\mathcal{J}(\mathcal{R}\exp(i\omega_x x)\exp(j\omega_y y))^q_A = \mathcal{J}\exp(i\omega_x x)\exp(j\omega_y y) \tag{11.28}$$
$$\mathcal{K}(\mathcal{R}\exp(i\omega_x x)\exp(j\omega_y y))^q_A = \mathcal{K}\exp(i\omega_x x)\exp(j\omega_y y). \tag{11.29}$$

which concludes the proof. □

11.2.4 Instantaneous Amplitude

One main feature of the analytic signal is that it makes accessible instantaneous phase and amplitude information directly. In the following we define the instantaneous amplitude of real 2-D signal as the absolute value of its analytic signal. Clearly, the different definitions of the analytic signal given in the last section result in different definitions of the instantaneous amplitude of a signal. We summarize these definitions in table 11.2. Figure 11.7 shows

analytic signal	instantaneous amplitude
Def. 11.2.7	$\sqrt{f^2(\boldsymbol{x}) + f^2_{\mathcal{H}i}(\boldsymbol{x})}$
Def. 11.2.8	$\sqrt{f^2(\boldsymbol{x}) + f^2_{\mathcal{H}i_1}(\boldsymbol{x})}$
Def. 11.2.9	$\sqrt{[f(\boldsymbol{x}) - f_{\mathcal{H}i}(\boldsymbol{x})]^2 + [f_{\mathcal{H}i_1}(\boldsymbol{x}) + f_{\mathcal{H}i_2}(\boldsymbol{x})]^2}$
Def. 11.2.10	$\sqrt{f^2(\boldsymbol{x}) + f^2_{\mathcal{H}i_1}(\boldsymbol{x}) + f^2_{\mathcal{H}i_2}(\boldsymbol{x}) + f^2_{\mathcal{H}i}(\boldsymbol{x})}$

Table 11.2. The tabular shows the different possible definitions of the instantaneous magnitude in 2-D. On the right hand side the instantaneous amplitude of the 2-D signal f is given according to the definition of the analytic signal indicated on the left hand side

an image of D. Hilbert and the instantaneous amplitude of this image. The instantaneous amplitude is expected to take high values wherever the image has considerable contrast. From this point of view only the instantaneous amplitude constructed via the partial analytic signal and the quaternionic analytic signal yield acceptable results. However, at positions where the local structure is intrinsically 2-D the quaternionic analytic signal yields better results.

11.2.5 The n-Dimensional Analytic Signal

All approaches to the 2-D analytic signal can easily be extended to n-dimensional signals. We merely give the definitions here and forego a detailed discussion, since the main properties of and differences between the different approaches remain the same in n-D as in 2-D.

Definition 11.2.12 (Total analytic signal). *The analytic signal of a real n-D signal f is defined as*

Fig. 11.7. An image of Hilbert and its instantaneous amplitude according to the different definitions of the 2-D analytic signal given in section 11.2.3. From top left to bottom right: The original image, the instantaneous amplitude (IA) according to the total analytic signal, the partial analytic signal (with respect to the x-direction), the definition of Hahn (maintaining the upper right quadrant and the upper left quadrant, respectively), and the IA with respect to the quaternionic analytic signal

$$f_A(\boldsymbol{x}) = f(\boldsymbol{x}) * (\delta^n(\boldsymbol{x}) + \frac{i}{\pi^n \prod_{j=1}^n x_j}) \tag{11.30}$$

$$=: f(\boldsymbol{x}) + i f_{\mathcal{H}i}(\boldsymbol{x}), \tag{11.31}$$

where $f_{\mathcal{H}i}$ is the n-D total Hilbert transform of f. In the frequency domain this definition reads

$$F_A(\boldsymbol{u}) = F(\boldsymbol{u})(1 - i \prod_{j=1}^{n} sign(u_j)).$$

Definition 11.2.13 (Partial analytic signal). *Let f be a real n-D signal and F its Fourier transform. The Fourier transform of the analytic signal with respect to some n-D unit vector $\hat{\mathbf{e}}$ is defined by:*

$$F_A(\boldsymbol{u}) = \begin{Bmatrix} 2F(\boldsymbol{u}) & \text{if } \boldsymbol{u}\cdot\hat{\mathbf{e}} > 0 \\ F(\boldsymbol{u}) & \text{if } \boldsymbol{u}\cdot\hat{\mathbf{e}} = 0 \\ 0 & \text{if } \boldsymbol{u}\cdot\hat{\mathbf{e}} < 0 \end{Bmatrix} = F(\boldsymbol{u})(1 + sign(\boldsymbol{u}\cdot\hat{\mathbf{e}})). \tag{11.32}$$

Definition 11.2.14. *Let f be a real, n-dimensional function and F its Fourier transform. The n-D complex signal (according to Hahn [97]) is defined in the frequency domain by*

$$F_A(\boldsymbol{u}) = \prod_{j=1}^{n}(1 + sign(u_j))F(\boldsymbol{u}).$$

In the spatial domain this reads

$$f_A(\boldsymbol{x}) = f(\boldsymbol{x}) * \prod_{j=1}^{n}\left(\delta(x_j) + \frac{i}{\pi x_j}\right). \tag{11.33}$$

Finally we define the n-dimensional version of the quaternionic analytic signal, namely the Clifford analytic signal.

Definition 11.2.15 (Clifford analytic signal).
Let f be a real, n-dimensional function and F^c its Clifford Fourier transform. The n-D Clifford analytic signal is defined in the frequency domain by

$$F_A^c(\boldsymbol{u}) = \prod_{j=1}^{n}(1 + sign(u_j))F^c(\boldsymbol{u}).$$

In the spatial domain this reads

$$f_A^c(\boldsymbol{x}) = f(\boldsymbol{x}) * \prod_{j=1}^{n}\left(\delta(x_j) + \frac{e_j}{\pi x_j}\right). \tag{11.34}$$

11.3 Local Phase in Image Processing

We have shown how the instantaneous phase can be evaluated using the analytic signal. However, the instantaneous phase loses its direct relation to the local signal structure, when the signal is not of narrow bandwidth [88]. In order to overcome this restriction, bandpass-filters with a one-sided transfer function can be applied to a signal. According to the definition of the 1-D analytic signal the impulse responses of these filters, and the filter responses to any real signal as well, are analytic signals. Filters of this kind are called *quadrature filters.* The angular phase of the quadrature filter response to a real signal is called the *local phase.* In the following we will introduce complex Gabor filters as approximations to quadrature filters. Using these filters we will define the local complex phase of an n-D signal. Since the local complex phase is an intrinsically 1-D concept it is a reasonable concept merely for simple or locally intrinsically 1-D signals. In section 11.3.2 we introduce quaternionic Gabor filters based on the quaternionic Fourier transform. Using these filters the concept of local phase of 2-D signals is extended.

11.3.1 Local Complex Phase

Complex Gabor filters are defined as linear shift-invariant filters with the Gaussian windowed basis functions of the Fourier transform as their basis functions.

Definition 11.3.1 (1-D Complex Gabor filter). *A one-dimensional complex Gabor filter is a linear shift-invariant filter with the impulse response*

$$h(x; N, u_0, \sigma) = g(x; N, \sigma) \exp(i2\pi u_0 x), \tag{11.35}$$

where $g(x; N, \sigma)$ is the Gauss function

$$g(x; N, \sigma) = N \exp\left(-\frac{x^2}{2\sigma^2}\right).$$

The Gabor filters have as parameters the normalization constant N, the *center frequency* u_0 and the variance σ of the Gauss function. However, most of the time we will not write down these arguments explicitly. Where no confusion is possible we use the notation $h(x)$ and $g(x)$ for the Gabor filter and the Gaussian function at position x, respectively.

We will use the normalization $N = (\sqrt{2\pi\sigma^2})^{-1}$ such that $\int_{\mathbb{R}} g(x)dx = 1$ in the following. Analogously the definition of 2-D complex Gabor filters is based on the 2-D Fourier transform:

Definition 11.3.2 (2-D Complex Gabor filter). *A two-dimensional complex Gabor filter is a linear shift-invariant filter with the impulse response*

$$h(\boldsymbol{x}; \boldsymbol{u}_0, \sigma, \epsilon, \phi) = g(x', y') \exp(2\pi i(u_0 x + v_0 y)) \tag{11.36}$$

with

$$g(x, y) = N \exp\left(-\frac{x^2 + (\epsilon y)^2}{\sigma^2}\right)$$

where ϵ is the aspect ratio. The coordinates (x', y') are derived from (x, y) by a rotation about the origin through the angle ϕ:

$$\begin{pmatrix} x' \\ y' \end{pmatrix} = \begin{pmatrix} \cos\phi & \sin\phi \\ -\sin\phi & \cos\phi \end{pmatrix} \begin{pmatrix} x \\ y \end{pmatrix}. \qquad (11.37)$$

Again, we will choose the normalization such that $\int_{\mathbb{R}} g(x, y) dx\, dy = 1$, i.e. $N = \frac{\epsilon}{2\pi\sigma^2}$. In frequency domain the 1-D Gabor filters take the following form:

$$h(x; u_0, \sigma) \circ\!\!-\!\!\bullet\; H(u; u_0, \sigma) = \exp(-2\pi^2\sigma^2(u - u_0)^2).$$

The transfer function of a 2-D Gabor filter is given by

$$h(\boldsymbol{x}; \boldsymbol{u}_0, \sigma, \epsilon, \phi) \circ\!\!-\!\!\bullet\; H(\boldsymbol{u}; \boldsymbol{u}_0, \sigma, \epsilon, \phi) = \exp(-2\pi^2\sigma^2[|(\boldsymbol{u}' - \boldsymbol{u}_0')|^2/\epsilon]).$$

Thus, Gabor filters are bandpass filters. The radial center frequency of the 2-D Gabor filter is given by $F = \sqrt{u_0^2 + v_0^2}$ and its orientation is $\theta = \mathrm{atan}(v_0/u_0)$. In most cases it is convenient to choose $\theta = \phi$ such that the orientation of the complex sine gratings is identical with the orientation of one of the principle axes of the Gauss function. Figure 11.8 shows the transfer function of a one-dimensional complex Gabor filter. Figure 11.8 shows that

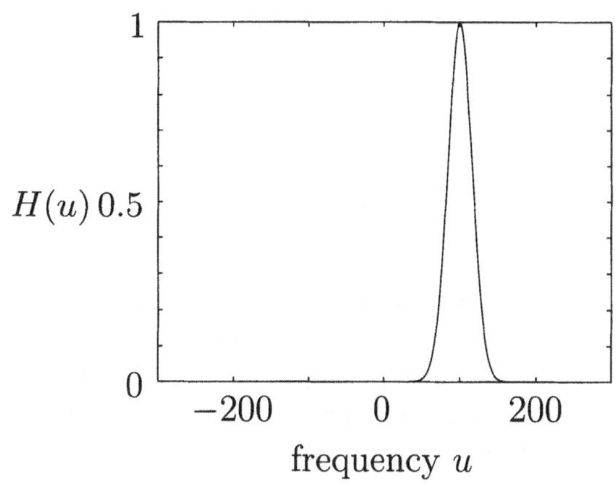

Fig. 11.8. The transfer function of a one-dimensional Gabor filter with $u_0 = 100$ and $\sigma = 0.01$

the main amount of energy of the Gabor filter is centered around the center frequency u_0 in the positive half of the frequency domain. However, the energy in the negative half is not equal to zero. Because of this property, the filter response of the Gabor filter to a real signal is only an approximation to an analytic signal (which is only one-sided in the frequency domain). The error of this approximation decreases with increasing u and with increasing σ.

The *local phase* of a signal is defined as the angular phase of its complex Gabor filter response. The relation to the local structure of the signal becomes clear in the following way. At a signal position with locally even symmetry only the even part of the Gabor filter, which is real-valued matches. The angular phase of a real number is either 0 for a positive number or π for a negative one. Thus, if the even filter component matches the signal positive, the local phase is 0, if it matches negative, the local phase is π. A similar reflection clarifies the case of a locally odd structure. In this case only the odd, and thus imaginary, filter component matches the signal. Since the angular phase of a pure imaginary number is $\pi/2$ for a positive imaginary part and $-\pi/2$ otherwise, these values represent odd local structures. Figure 11.9 sketches the relation between structure and phase: the orientation in the circle indicates the value of the local phase. At the values 0, $\pi/2$, π and $-\pi/2$ the related structure is shown. An important feature of the **local phase** is

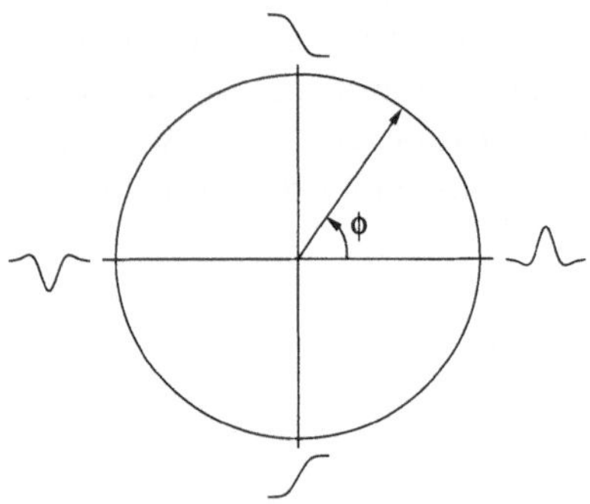

Fig. 11.9. The relation between local signal structure and local phase (See [88].)

that it **is independent of the signal energy**. This makes the local phase very robust against changing lighting conditions.

It should be mentioned here that the value of the local phase at a certain signal position depends on the chosen filter parameters. I.e. Gabor filters will only detect features at the scale to which they are tuned.

11.3.2 Quaternionic Gabor Filters

In analogy to the complex Gabor filters we introduce quaternionic Gabor filters.

Definition 11.3.3 (Quaternionic Gabor filter). *The impulse response of a quaternionic Gabor filter is a Gaussian windowed basis function of the QFT:*

$$h^q(\boldsymbol{x}; \boldsymbol{u}_0, \sigma, \epsilon) = g(\boldsymbol{x}; \sigma, \epsilon) \exp(i2\pi u_0 x) \exp(j2\pi v_0 y). \tag{11.38}$$

Note that we do not use rotated Gaussian windows here.

It follows from the modulation theorem of the Fourier transform that complex Gabor filters are shifted Gaussians in the frequency domain. In section 8.4.2 of chapter 8 we showed that there exists a modulation theorem for the QFT as well. Consequently, quaternionic Gabor filters are shifted Gaussian functions in the quaternionic frequency domain. Quaternionic Gabor filters thus belong to the "world" of the QFT rather than to the "complex Fourier world". The QFT of a quaternionic Gabor filter is given by

$$h^q(\boldsymbol{x}; \boldsymbol{u}_0, \sigma, \epsilon) \overset{\mathbb{H}}{\circ\!\!-\!\!\bullet} H^q(\boldsymbol{u}; \boldsymbol{u}_0, \sigma, \epsilon) = \exp(-2\pi^2\sigma^2[|\boldsymbol{u} - \boldsymbol{u}_0|^2/\epsilon^2])$$

Thus, for positive frequencies u_0 and v_0 the main amount of the Gabor filter's energy lies in the upper right quadrant. Therefore, convolving a real signal with a quaternionic Gabor filter yields an approximation to a quaternionic analytic signal.

A typical quaternionic Gabor filter is shown in figure 11.10.

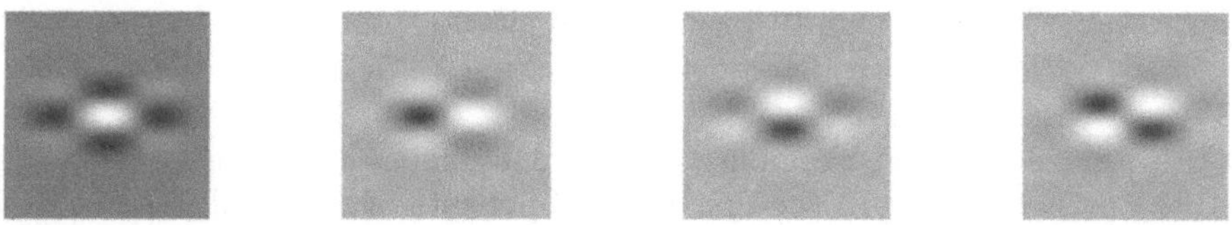

Fig. 11.10. A quaternionic Gabor filter with parameters $\sigma_1 = 20$, $\sigma_2 = 10$, $2\pi u_0 \sigma_1 = 2\pi v_0 \sigma_2 = 2$. The size of the filter mask is 100×100

11.3.3 Local Quaternionic Phase

We now define the local quaternionic phase of a real two-dimensional signal as the angular phase of the filter response to a quaternionic Gabor filter. The angular phase is evaluated according to the rules given in table 8.1. If k^q is the quaternionic Gabor filter response of some image f the local quaternionic phase $(\phi(\boldsymbol{x}), \theta(\boldsymbol{x}), \psi(\boldsymbol{x})$ is defined by

$$k^q(\boldsymbol{x}) = |k^q(\boldsymbol{x})| e^{i\phi(\boldsymbol{x})} e^{k\psi(\boldsymbol{x})} e^{j\theta(\boldsymbol{x})}$$

according to def. 8.3.1 given in chapter 8.

In 1-D we can make the statement: **The local phase estimates and spatial position are equivariant** [88]. I.e. generally the local phase of a signal varies monotonically up to 2π-wrap-arounds. There are only singular points with low or zero signal energy where this equivariance cannot be found anymore. A simple example is the cosine function $\cos(x)$. If we apply a well tuned Gabor filter for estimating the local phase ϕ of this function, we find that it is almost equal to the spatial position: $\phi(x) \approx x$ for $x \in [-\pi, \pi[$ (see figure 11.11). This leads us to an interpretation of the local quaternionic phase.

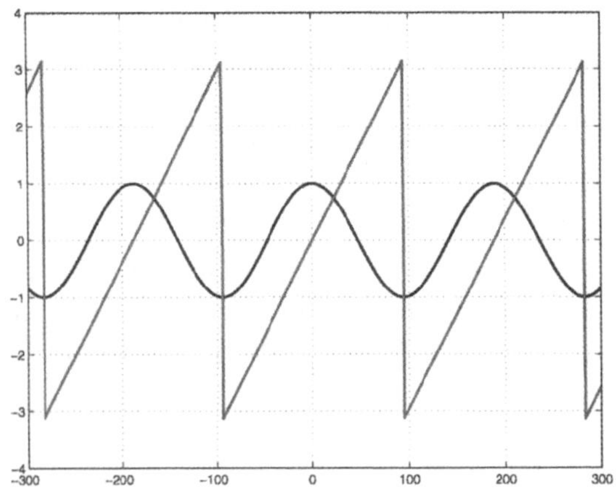

Fig. 11.11. The cosine function and its local phase

We make a similar example as in the one-dimensional case by replacing $\cos(x)$ by $\cos(x)\cos(y)$. The first two components of the local phase ϕ and θ turn out to approximate the spatial position: $\phi(\boldsymbol{x}) \approx x$ and $\theta(\boldsymbol{x}) \approx y$ for $(x, y) \in [0, 2\pi[\times[0, \pi[$. In general it turns out that these two components of the local phase are equivariant with spatial position. The reason for the interval $[0, 2\pi[\times[0, \pi[$, which follows mathematically from the definition of the angular phase of unit quaternions, can be understood from figure 11.12.

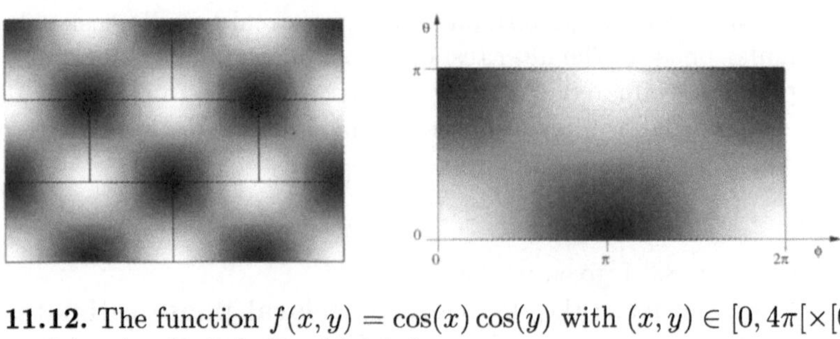

Fig. 11.12. The function $f(x, y) = \cos(x)\cos(y)$ with $(x, y) \in [0, 4\pi[\times[0, 3\pi[$ (left) and $(x, y) \in [0, 2\pi[\times[0, \pi[$ (right)

While the spatial position can be recovered uniquely from the local signal structure within the interval $[0, 2\pi[\times[0, \pi[$, there will occur ambiguities if the interval is extended. The whole function $\cos(x)\cos(y)$ can be build from patches of the size $2\pi \times \pi$. Considering this example the third component of the local phase is always zero: $\psi = 0$. The meaning of this phase component becomes obvious if we vary the structure of the test signal in the following way. The function $\cos(x)\cos(y)$ can be written as the sum

$$\cos(x)\cos(y) = \frac{1}{2}(\cos(x+y) + \cos(x-y)).$$

If we consider linear combinations of the form

$$f(\boldsymbol{x}) = (1-\lambda)\cos(x+y) + \lambda\cos(x-y)$$

we find that ψ varies monotonically with the value of $\lambda \in [0, 1]$. This behavior is shown in figure 11.13.

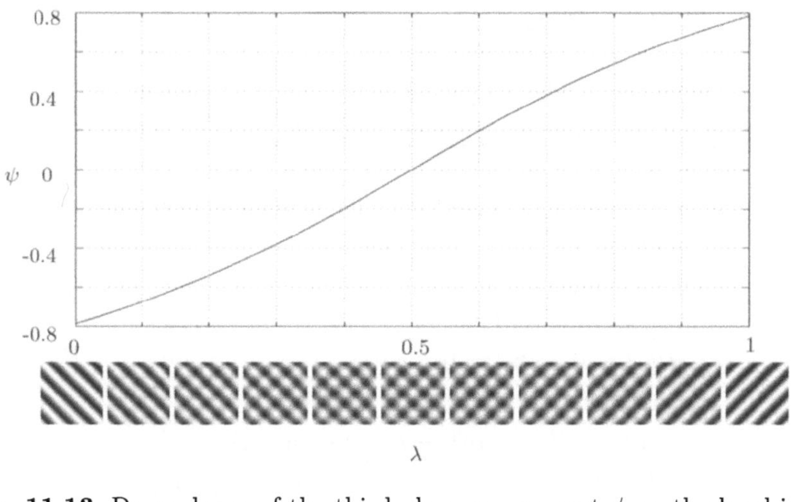

Fig. 11.13. Dependence of the third phase component ψ on the local image structure

The first two phase components, namely ϕ and θ do not change their meaning, while λ varies. Only for the values $\lambda = 0$ and $\lambda = 1$, i.e. $\psi = \mp\frac{\pi}{4}$ the structure degenerates into an intrinsically one-dimensional structure. Hence, the spatial position cannot any longer be recovered from the local structure. This corresponds to the singularity in the evaluation the angular phase of a quaternion when $\psi = \pm\frac{\pi}{4}$. In this case only $\phi \mp \theta$ can be evaluated. The remaining degree of freedom can be eliminated by setting $\theta = 0$.

11.3.4 Relations between Complex and Quaternionic Gabor Filters

There is a simple relation between complex and quaternionic Gabor filters. Each component of a complex Gabor filter with aspect ratio $\epsilon = 1$ may be written as the sum of two quaternionic Gabor filter components:

$$\begin{aligned} h_e(x,y) &= g(x,y)\cos(\omega_1 x + \omega_2 y) \\ &= g(x,y)(\cos(\omega_1 x)\cos(\omega_2 y) - \sin(\omega_1 x)\sin(\omega_2 y)) \\ &= h^q_{ee}(x,y) - h^q_{oo}(x,y) \end{aligned} \tag{11.39}$$

$$\begin{aligned} h_o(x,y) &= g(x,y)\sin(\omega_1 x + \omega_2 y) \\ &= g(x,y)(\cos(\omega_1 x)\sin(\omega_2 y) + \sin(\omega_1 x)\cos(\omega_2 y)) \\ &= h^q_{eo}(x,y) + h^q_{oe}(x,y). \end{aligned} \tag{11.40}$$

From the same quaternionic Gabor filter a second complex Gabor filter can be generated by

$$\begin{aligned} h_e(x,y) &= g(x,y)\cos(\omega_1 x - \omega_2 y) \\ &= h^q_{ee}(x,y) + h^q_{oo}(x,y) \end{aligned} \tag{11.41}$$

$$\begin{aligned} h_o(x,y) &= g(x,y)\sin(\omega_1 x - \omega_2 y) \\ &= h^q_{oe}(x,y) - h^q_{eo}(x,y). \end{aligned} \tag{11.42}$$

Thus, each quaternionic Gabor filter corresponds to *two* complex Gabor filters. Sometimes these two complex filters are denoted by h^+ (11.39, 11.40) and h^- (11.41, 11.42), respectively. The response of a signal $f(x,y)$ to a Gabor filter will be denoted by $k(x,y)$ for a complex Gabor filter and $k^q(x,y)$ for a quaternionic Gabor filter:

$$\begin{aligned} k(x,y) &= (h * f)(x,y) \\ &= ((h_e + ih_o) * f)(x,y) \\ &= k_e(x,y) + ik_o(x,y) \end{aligned} \tag{11.43}$$

$$\begin{aligned} k^q(x,y) &= (h^q * f)(x,y) \\ &= ((h^q_{ee} + ih^q_{oe} + jh^q_{eo} + kh^q_{oo}) * f)(x,y) \\ &= k^q_{ee}(x,y) + ik^q_{oe}(x,y) + jk^q_{eo}(x,y) + kk^q_{oo}(x,y). \end{aligned} \tag{11.44}$$

Theorem 11.3.1. *The filter responses of the complex Gabor filters h^+ and h^- can be obtained from k^q by*

$$k^+(\boldsymbol{x}) = (k^q_{ee} - k^q_{oo}) + i(k^q_{oe} + k^q_{eo}) \tag{11.45}$$

$$k^-(\boldsymbol{x}) = (k^q_{ee} + k^q_{oo}) + i(k^q_{oe} - k^q_{eo}). \tag{11.46}$$

Proof. The theorem follows from the definition of h^+ and h^- and the fact that h^q is an LSI-filter. □

Algebraically, the relation between quaternionic and complex Gabor filters can be illuminated if we apply a mapping from the algebra $\mathbb{H}$ to the four-

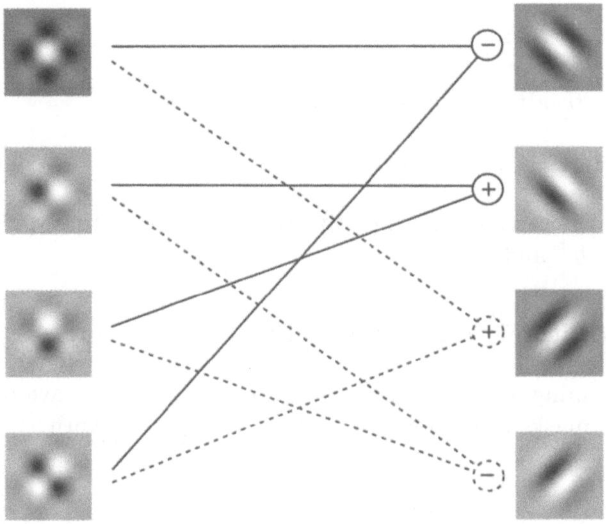

Fig. 11.14. Relation between quaternionic and complex Gabor filters

dimensional commutative hypercomplex algebra $\mathcal{H}_2$ introduced in chapter 9 called *switching*.

Definition 11.3.4 (Switching). *The one-to-one mapping $S_2 : \mathbb{H} \to \mathcal{H}_2$ is defined by*

$$S_2(a + bi + cj + dk) = a + bi_1 + ci_2 + di_3.$$

The multiplication table of $\mathcal{H}_2$ is given in table 11.3 (see also table 9.2)

Table 11.3. Multiplication table of $\mathcal{H}_2$

	i_1	i_2	i_3
i_1	-1	i_3	$-i_2$
i_2	i_3	-1	$-i_1$
i_3	$-i_2$	$-i_1$	1

Theorem 11.3.2. *Let h^q be a quaternionic Gabor filter. Then*

$$\eta(S_2(h^q(\boldsymbol{x}))) = (h^+(\boldsymbol{x}), h^-(\boldsymbol{x})) \in \mathbb{C}^2,$$

where η establishes the isomorphism between $\mathcal{H}_2$ and $\mathbb{C}^2$:

$$\eta : \mathcal{H}_2 \to \mathbb{C}^2 \tag{11.47}$$
$$(\alpha + \beta i_1 + \gamma i_2 + \delta i_3) \mapsto ((\alpha - \delta) + i(\beta + \gamma), (\alpha + \delta) + i(\beta - \gamma)) \tag{11.48}$$

The same is true for the filter responses to real images

$$\eta(S_2(k^q(\boldsymbol{x}))) = (k^+(\boldsymbol{x}), k^-(\boldsymbol{x})) \in \mathbb{C}^2.$$

Proof. The theorem follows directly from applying η to $S_2(h^q(\boldsymbol{x}))$ and the definition of h^+ and h^-. □

11.3.5 Algorithmic Complexity of Gabor Filtering

When performing a Gabor filtering on the computer we have to use discrete Gabor filter masks of the form: $h = [h_{m,n}]_{m,n \in \{1,...M\}}$ with

$$h_{m,n} = h(m - \frac{M-1}{2}, n - \frac{N-1}{2}), \tag{11.49}$$

where the right hand side is the continuous Gabor filter as in Def. 11.3.2.

Using this convention the Gabor filter mask is an $M \times M$ matrix. The origin is located at the center of the matrix, therefore it is advantageous to choose M odd, in order to have a pixel in the center of the filter mask. The frequencies u and v count how many periods fit into the filter mask in horizontal and vertical direction, respectively.

The number of multiplications required by the convolution of an $N \times N$ image with an $M \times M$ filter mask in a direct manner is $O(N^2M^2)$. When the filter mask h is separable ($h = h_c * h_r$), where h_c and h_r are a column vector and a row vector of length M, respectively, the filtering operation is of linear asymptotic complexity. Since the convolution operation is associative we can write the filtering as

$$F = f * (h_c * h_r) = (f * h_c) * h_r. \tag{11.50}$$

Thus, the number of required multiplications reduces to $O(N^2M)$. It has been shown how complex Gabor filter components can be constructed as the sum of components of a quaternionic Gabor filter. Since quaternionic Gabor filters are separable, **this opens the possibility of implementing the convolution with complex Gabor filters in a separable way.** This result is especially important since Gabor filters are known to be not exactly steerable [169]. Figure 11.15 clarifies this result in "image notation".

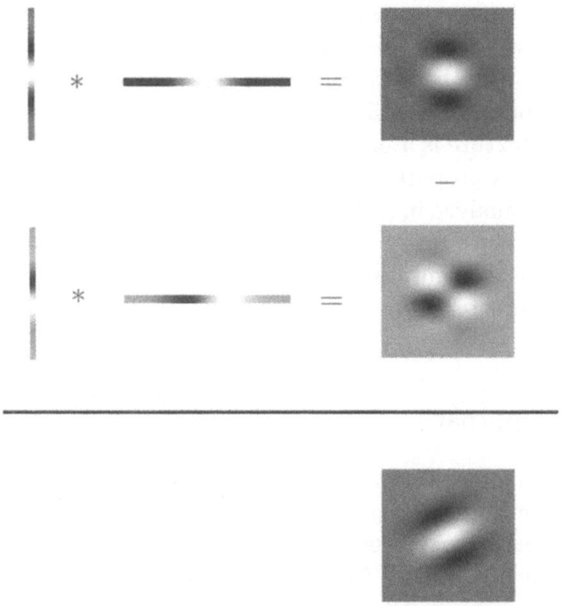

Fig. 11.15. The real part of a complex Gabor filter as linear combination of separable quaternionic Gabor filter components

11.4 Texture Segmentation Using the Quaternionic Phase

The task addressed in this section is: *Segment a given image into uniformly textured regions.* This so-called *texture segmentation* problem is one branch of the general problem of *image segmentation* which is one important step in many computer vision tasks. Regarding global variations of gray values or mean gray values over some neighborhood is in most cases not sufficient for a correct segmentation. For this reason rather the global variations of local measures characterizing the texture have to be regarded.

The posed problem is rather vague since the term *texture* is not well defined and there is no unique way of characterizing mathematically the local gray-value variations perceived as texture by human observers. For this reason very different approaches to texture segmentation have been taken. As local measure for the characterization of texture local statistical properties [99, 121] and local geometric building blocks (textons) [123] have been used among others. Another whole branch in texture segmentation research is based on the local spatial frequency for characterizing texture. On the one hand the Gabor filter based approaches to texture analysis are motivated by psychophysical research since 2-D Gabor filters have proven to be a good model for the cortical receptive field profiles [57] while on the other hand they are supported by the observation that a whole class of textures (so-called *deterministic textures*) give rise to periodic gray value structures. We will restrict ourselves to the Gabor filter based approaches here. In the following the term texture will always be understood as *image texture* in contrast to *surface texture.* While surface texture is a property of a 3D real-world object, image texture in this context is a property of a 2-D intensity image.

In the following sections we analyze in detail the pioneering work of Bovik et al. [26] and in parallel introduce the corresponding quaternionic Gabor filter based approach to texture segmentation. In the final section we discuss our result and make some remarks on other texture segmentation approaches based on Gabor filters.

11.4.1 The Gabor Filter Approach

Bovik et al. [26] introduced a Gabor filter based approach to texture segmentation. As mentioned above, texture segmentation is the task of segmenting an image into uniformly textured regions. According to Bovik's approach a uniform texture is described by a dominant frequency and orientation. Thus, different textures occurring in a given image are supposed to differ significantly at least in either the dominant frequency or the dominant orientation.

This assumption leads to the following simple *texture model.* An image containing only one homogeneous texture is modeled as

$$\begin{aligned} f_i(\boldsymbol{x}) &= c_i(\boldsymbol{x})\cos(2\pi(u_i x + v_i y)) + s_i(\boldsymbol{x})\sin(2\pi(u_i x + v_i y)) \\ &= a_i(\boldsymbol{x})\cos(2\pi(u_i x + v_i y) + p_i(\boldsymbol{x})), \end{aligned} \tag{11.51}$$

where the amplitude $a_i = \sqrt{c_i^2 + s_i^2}$ and the phase $p_i = -\tan^{-1}\left(\frac{s_i}{c_i}\right)$ are assumed to vary slowly, i.e. in such a way that the dominant frequency component is always well approximated by (u_i, v_i). The characterizing dominant frequency and orientation of the texture f_i are $|\boldsymbol{u}_i| = \sqrt{u_i^2 + v_i^2}$ and $\alpha_i = -\tan^{-1}(\frac{v_i}{u_i})$, respectively.

A textured image containing n different textures f_i is then given by n textures of the form (11.51) each of which occurs in exactly one connected region $\mathcal{R}_i$ of the image. Defining the characteristic functions z_i of the regions

$$z_i(\boldsymbol{x}) = \begin{cases} 1 & \text{if } \boldsymbol{x} \in \mathcal{R}_i \\ 0 & \text{else,} \end{cases}$$

we can write the texture image f as

$$f(\boldsymbol{x}) = \sum_{i=1}^{n} f_i(\boldsymbol{x}) z_i(\boldsymbol{x}). \tag{11.52}$$

The regions $\mathcal{R}_i$ are assumed to define a partitioning of the domain of f, i.e. $\sum_{i=1}^{n} z_i(\boldsymbol{x}) \equiv 1$ and $z_i(\boldsymbol{x}) z_j(\boldsymbol{x}) \equiv 0$ if $i \neq j$. The set of all possible textures f will be denoted by $\mathcal{T}$. This texture model fits optimally the texture segmentation technique applied by Bovik et al.

The first step in the segmentation procedure is devoted to *filter selection*. In this stage the parameters of a number of Gabor filters that will be used for the segmentation are chosen. For a review of possible methods we refer to Bovik's article [26]. The image f is convolved with the set of selected Gabor filters h_i which yields n filtered images k_i, where n is the number of selected filters. The complex filtered images are transformed into the amplitude/phase-representation according to

$$m_i = |k_i|, \quad \phi = -\tan^{-1}\left(\frac{\mathcal{I}(k_i)}{\mathcal{R}(k_i)}\right). \tag{11.53}$$

The first level of segmentation is based on the comparison of the channel amplitudes. At this stage each pixel of the image is assigned to one channel. We will denote the region of pixels belonging to channel i by $\mathcal{R}_i$. The classification is simply based on the comparison of the amplitudes m_i at each position in the image:

$$\boldsymbol{x} \in \mathcal{R}_i \Longleftrightarrow arg\left(\max_{j\in\{1,\dots,n\}}(m_j(\boldsymbol{x}))\right) = i, \tag{11.54}$$

where the function arg returns the index of m. A second segmentation step is based on phase discontinuities. In this step regions which contain the same texture but which are shifted against each other are separated.

11.4.2 Quaternionic Extension of Bovik's Approach

The extension of Bovik's approach to texture segmentation using quaternionic Gabor filters is straightforward. Before outlining the segmentation procedure in the quaternionic case we modify the texture model given above. If quaternionic Gabor filters are applied instead of complex filters the following texture model is more appropriate. A textured image is assumed to consist of homogeneously textured regions

$$f^q(\boldsymbol{x}) = \sum_{i=1}^{n} f_i^q(\boldsymbol{x}) z_i(\boldsymbol{x}), \tag{11.55}$$

where this time the homogeneous textures are of the form

$$\begin{aligned} f_i^q(\boldsymbol{x}) &= cc_i(\boldsymbol{x})\cos(2\pi u_i x)\cos(2\pi v_i y) \\ &+ sc_i(\boldsymbol{x})\sin(2\pi u_i x)\cos(2\pi v_i y) \\ &+ cs_i(\boldsymbol{x})\cos(2\pi u_i x)\sin(2\pi v_i y) \\ &+ ss_i(\boldsymbol{x})\sin(2\pi u_i x)\sin(2\pi v_i y). \end{aligned}$$

Again, the functions cc_i, sc_i, cs_i and ss_i are assumed to vary slowly. The set of all possible textures f^q will be denoted by $\mathcal{T}^q$. Obviously, this model is most appropriate for the use of quaternionic filters, since the four terms exactly correspond to the modulation functions of the components of a quaternionic Gabor filter. In figure 11.16 two model textures are shown which demonstrate the difference between the two models.

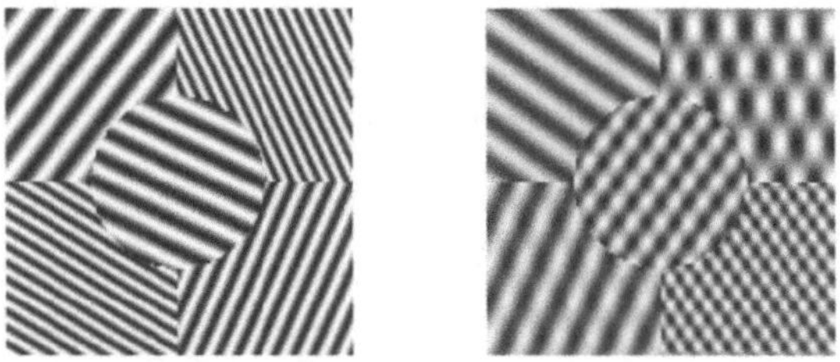

Fig. 11.16. Two examples of textured images. Left: A textured image fitting Bovik's texture model (11.52). Right: An image fitting the extended texture model (11.55). For simplicity, in both examples constant coefficients have been chosen

Note that the quaternionic texture model comprises Bovik's model as a special case, i.e. $\mathcal{T} \subset \mathcal{T}^q$.

The first stages of the segmentation procedure stay basically the same as described in the previous section. Only slight modifications have to be made. The filter selection stage is performed by a peak-finding algorithm in the quaternionic power spectrum. The difference is that here the peak finding is only performed over one quadrant of the frequency domain instead of one half in the complex approach. As we have shown when introducing the quaternionic analytic signal in section 11.2.3, one quadrant of the quaternionic frequency domain contains the complete information about the image.

Having selected a set of n quaternionic Gabor filters h_i^q the textured image is convolved with these filters, which yields the filtered images k_i^q. These image values are transformed into the polar representation of quaternions introduced in section 8.3.1. This leads to an amplitude/phase-representation $(m_i, \phi_i, \theta_i, \psi_i)$ of the filtered images.

Since we have shown that complex Gabor filters are contained in the quaternionic Gabor filters, the first levels of Bovik's approach, i.e. channel assignment and detecting phase discontinuities, can as well be performed using quaternionic Gabor filters. Thus, we do not go into details on these steps but show which additional information is contained in the quaternionic Gabor filter response, which can be used for segmentation purposes.

As shown in Fig. 11.13 the ψ-component of the phase holds the information about the mixture of two superimposed frequency components, i.e. f_1 and f_2. Denoting the mixed texture by $f = (1 - \lambda) f_1 + \lambda f_2$ there is a one-to-one mapping of ψ to λ. Thus, it is possible to use the ψ-component of the local quaternionic phase in order to separate regions belonging to the same frequency channel but having different structure according to the continuum of structures shown in Fig. 11.13.

11.4.3 Experimental Results

We demonstrate the segmentation power of the ψ-component of the local quaternionic phase first on a synthetic texture consisting of three different textures (figure 11.17). This image resembles an image used by Bovik ([26] p. 64, fig. 6), with the difference that in [26] only two different regions are used. The third region (upper right and lower left region), which is the superposition of the two orthogonally oriented sinusoidals, can not be segmented using the complex approach. In contrast, the ψ-component of the quaternionic phase distinguishes not only local frequency and orientation but also local structure as explained in the last section. See also figure 11.19 for clarification.

We tested the robustness of ψ for segmentation by adding Gaussian noise to the synthetic texture in figure 11.17. The result is shown in figure 11.18.

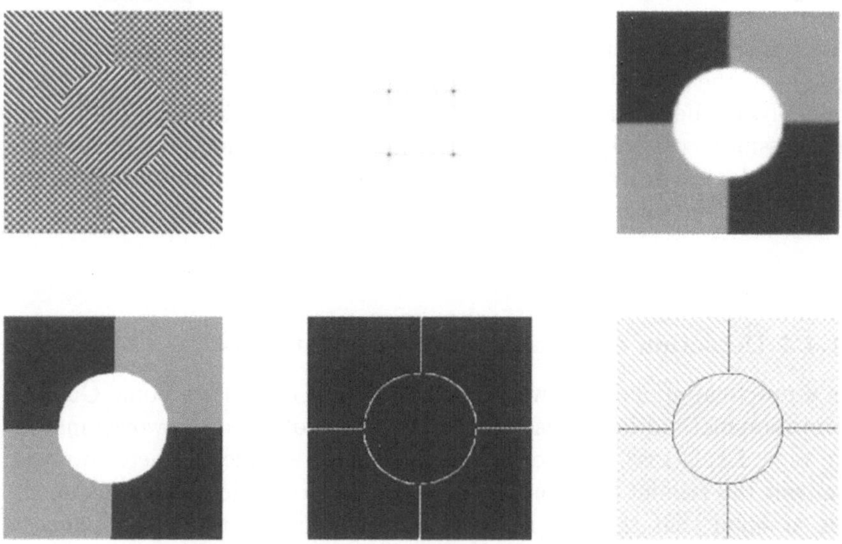

Fig. 11.17. The textured image, its QFT-magnitude spectrum, and the ψ-component of the local phase (top), and the segmentation result, the pixels which were misclassified (1.22%) and the edges of the ψ-component found by a Sobel filter superimposed to the original texture (bottom)

We added noise with zero mean and variance 1.5 and 5, respectively. The texture itself has zero mean and takes values between -1 and 1. The SNR is -2.7 dB and -13.2 dB, respectively. Although it is almost impossible for a human observer to segment the image with the strongest noise, by means of ψ more than 78% of the pixels are correctly classified.

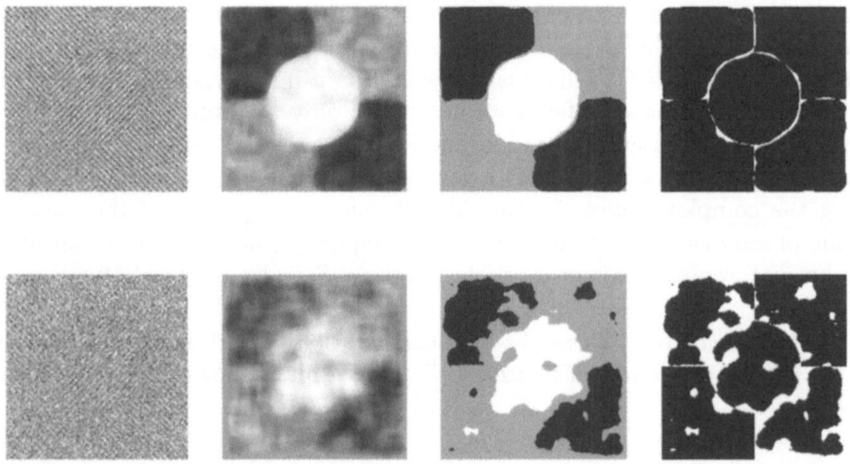

Fig. 11.18. The texture from figure 11.17 with added Gaussian noise. In the upper row the SNR is -2.7 dB, and more than 97% of the pixels are classified correctly. In the lower row the SNR is -13.2 dB and about 78% of the pixels are classified correctly. From left to right the rows show the contaminated texture, the median filtered ψ-component of the local phase, the segmented texture and the false classified pixels

11.4.4 Detection of Defects in Woven Materials

As a practical application we demonstrate how the quaternionic Gabor segmentation method can be used for the detection of defects in woven materials. We regard this task as a texture segmentation problem, where we want to segment the regular texture from defective regions. However, defects are often so small that they do not exhibit periodic structure. That makes the defect detection not feasible for a channel assignment method — complex or quaternionic — based on the magnitude of response to a certain channel filter. We test the following method here. Given a homogeneous woven texture we extract the dominant quaternionic frequency component. The image is convolved with the corresponding quaternionic Gabor filter (where the remaining parameters are chosen as $c_h = c_v = 3$) and the ψ-component of the local phase is extracted. A flaw in the texture manifests itself in a change of the lo-

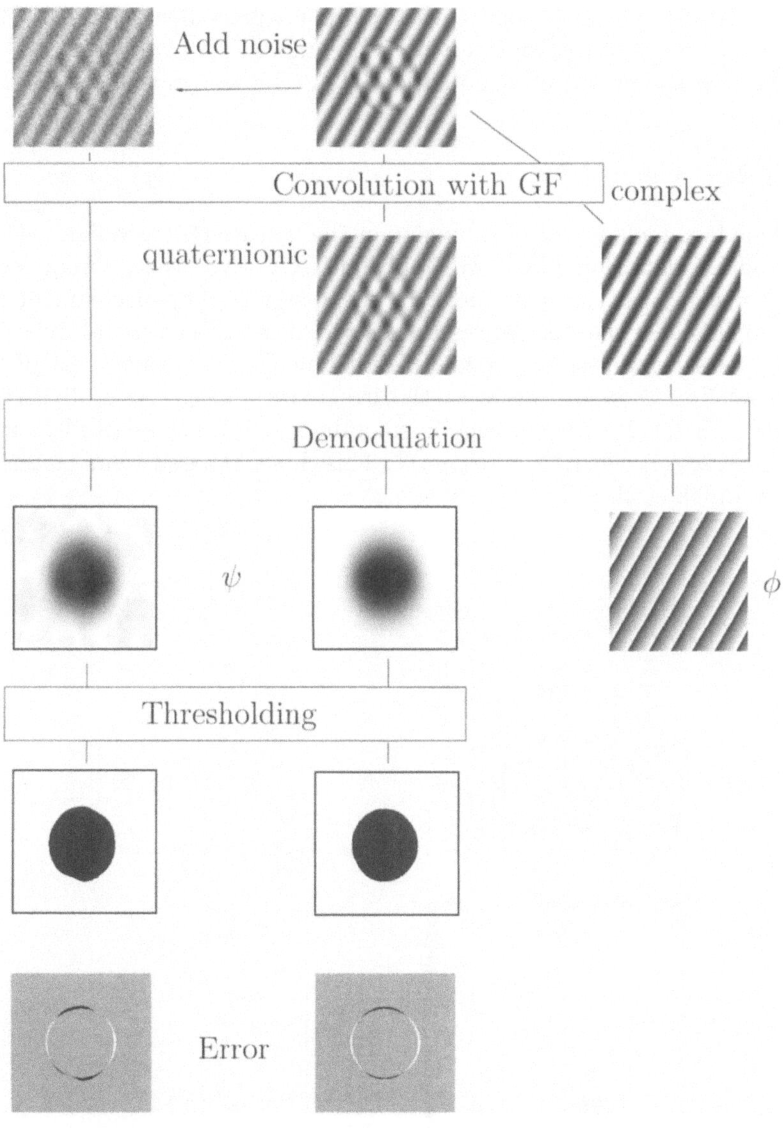

Fig. 11.19. Comparison of the complex and the quaternionic segmentation approach. The input image (top) is convolved with an optimally tuned complex (right column) and quaternionic (middle column) Gabor filter. In the second row the real parts of the filter responses are shown. The filtered images are transformed into amplitude/phase-representation. In the complex case the magnitude (not shown) is constant, and the phase ϕ is varying monotonically. No segmentation is possible. In the quaternionic case segmentation based on the ψ-component (magnitude and other phase-components are not shown) is possible. The left column is like the middle column, but with added noise (SNR=0 dB)

cal structure, which is what is measured by the ψ-phase. As the experiments show, ψ varies only very modestly within a homogeneously textured region. The mean ψ-value of a homogeneous texture f will be denoted as ψ_f. For the segmentation we chose an interval of acceptance $I_{Texture} = [\psi_f - \delta, \psi_f + \epsilon]$. The defective region will be denoted by $\mathcal{R}_{Flaw}$. The assignment rule is then given by

$$\boldsymbol{x} \in \mathcal{R}_{Flaw} \Leftrightarrow \psi(\boldsymbol{x}) \notin I_{Texture}.$$

As a second example we use a subregion of the texture D77 (see figure 11.20 taken from Brodatz album [32]). We apply one QGF whose central frequencies have been tuned to the main peak in the power (QFT)-spectrum of the image. In this case the frequencies are 21 cycles/image in vertical direction and 12 cycles per image in horizontal direction. In the regular part of the texture we find $\psi \approx 0.5$ while at the irregularity we get $\psi \leq 0$. Before applying a threshold, the ψ-image is smoothed with a Gaussian filter with $\boldsymbol{\sigma}_{Gauss} = 1.5\boldsymbol{\sigma}_{QGF}$ where $\boldsymbol{\sigma} = (\sigma_1, \sigma_2)^\top$. This choice is based on an empirical result by Bovik et al. [26].

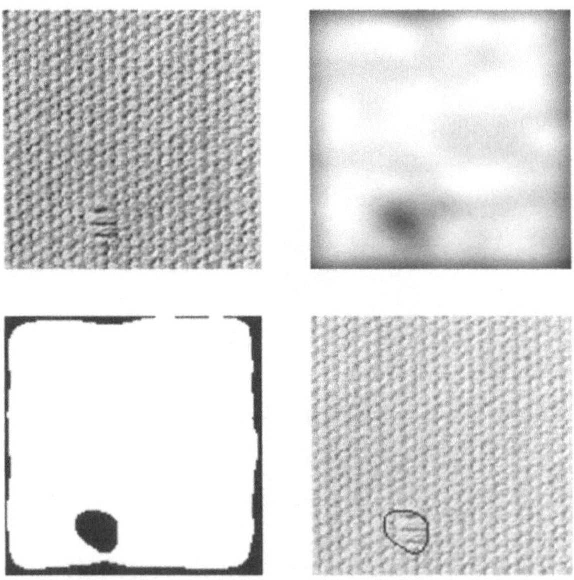

Fig. 11.20. A subregion of Brodatz texture D77 (top, left). The smoothed ψ-component of the local quaternionic phase as intensity image (top, right) and after applying a threshold (bottom, left). The edges of the thresholded ψ-phase superimposed to the input image (bottom, right)

Since at the flaw the applied filters do not match optimally, also the amplitude of the filter output yields a hint for the defect searched for. However, the amplitude is very sensitive to changing lightning conditions as shown in the following experiments. However, ψ is insensitive to changes in contrast. This is important, because of the fact that the lighting conditions are not necessarily optimal (e.g. not homogeneous) in practical applications [74].

We simulate changing lighting conditions by adding a gray-value ramp with constant slope (figure 11.21) and by changing the contrast inhomogeneously (figure 11.22). In figure 11.23 the amplitude of the filter responses are shown for the different lighting conditions. A segmentation on the basis of the amplitude envelopes is not possible by a thresholding procedure.

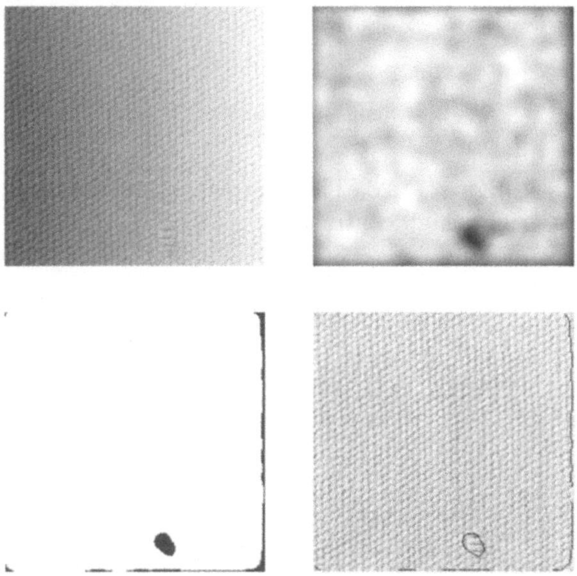

Fig. 11.21. As in figure 11.20. To the original image a gray value ramp with constant slope is added

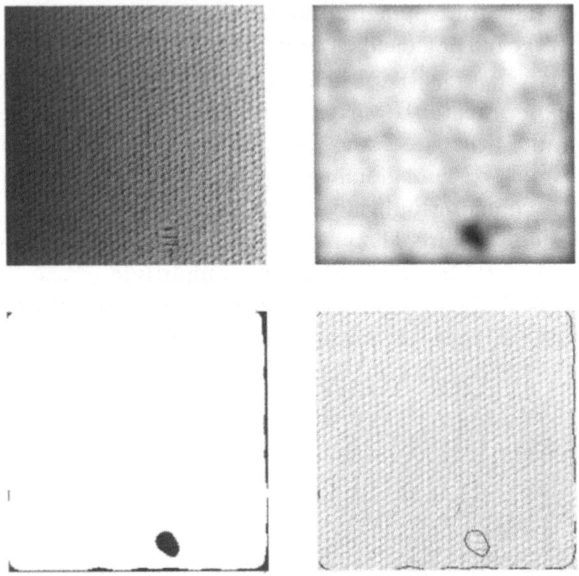

Fig. 11.22. As in figure 11.20. The contrast is modified to vary from left (low contrast) to right (high contrast)

Fig. 11.23. The amplitude envelopes of the quaternionic Gabor filter response to the texture D77 under different lighting conditions. Left: Original illumination. Middle: A gray value ramp added. Right: Changing contrast

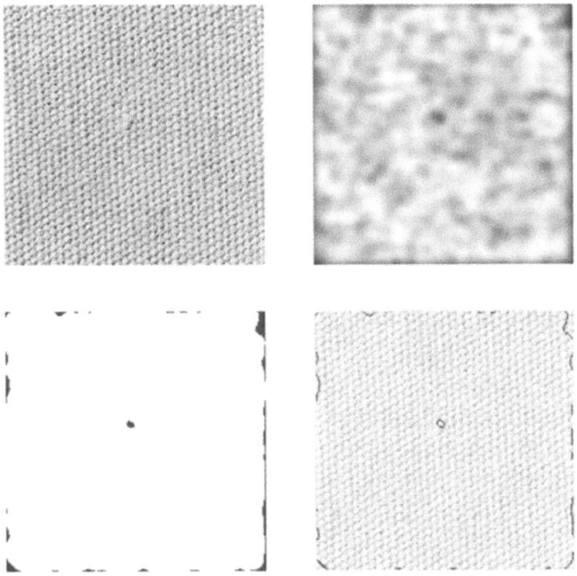

Fig. 11.24. Another subregion of D77. As in figure 11.20

The flaw detection method presented here has the advantage of being fast, since only separable convolutions have to be performed and only the ψ-component of the local phase has to be evaluated which is a pointwise nonlinear operation. The method is robust to changing lighting conditions.

11.5 Conclusion

In this chapter the quaternionic Fourier transform has been used in order to generalize the concept of the analytic signal which is well-known in 1-D signal theory to 2-D in a novel manner. Based on the quaternionic analytic signal the instantaneous quaternionic phase has been introduced. The local phase concept as introduced in this chapter is based on the approximation of an analytic signal by a Gabor filtered image. In order to introduce a local quaternionic phase, quaternionic Gabor filters have been introduced as windowed basisfunctions of the quaternionic Fourier transform. The local quaternionic phase has been used for texture segmentation where it could be shown that the ψ-component of the quaternionic phase yields a novel feature and provides useful information for the segmentation.

12. Introduction to Neural Computation in Clifford Algebra*

Sven Buchholz and Gerald Sommer

Institute of Computer Science and Applied Mathematics,
Christian-Albrechts-University of Kiel

12.1 Introduction

This is the first of two chapters on neural computation in Clifford algebra. The name Clifford algebra refers to its inventor WILLIAM K. CLIFFORD (1845-1879). We will restrict ourselves on Clifford algebras generated by non–degenerate quadratic forms. Thus, Clifford algebras are non–degenerated geometric algebras hereafter.

Any degenerate geometric algebra can be embedded in a larger dimensional non-degenerate geometric (Clifford) algebra as stated in the first chapter of this book. In that sense, our restriction will not be significant. However, it is necessary because it is not possible to derive learning algorithms for neural networks in degenerate algebras. We will explain this in full detail in the second chapter on Clifford multilayer perceptrons (Clifford MLPs).

The idea of developing neural networks in other than the real domain is not new. Complex valued networks were already introduced at the beginning of our decade, see e.g. [87]. Recently, [7] proposed a quaternionic valued neural network. A first attempt of developing neural networks in Clifford algebras was made by [182]. Unfortunately, the approach proposed there had many drawbacks. We showed this in [33], [15], where we also sketched an alternative that leads to correct learning algorithms for Clifford–valued neural networks.

* This work has been supported by DFG Grants So-320-2-1 and So-320-2-2.

In this chapter we will not speak of Clifford neural networks yet. Instead, the center of our studies will be the Clifford neuron itself. Thus, we will develop Clifford neural networks directly from their building blocks. At the level of neurons it is much easier to understand the principles of neural computation in Clifford algebras. The main principle is, that neural computation in Clifford algebras can be seen as model-based in comparison to that in the real case. To show this theoretically and experimentally is the main goal of this chapter.

The model–based approach will allow us to interpret the non–commutativity of Clifford algebras in general as a feature of Clifford neural computation. That aspect is not mentioned in the previous work of [7] and [182]. This will lead us to a special type of Clifford neuron called spinor neuron.

The split–up of the discussion of Clifford neural computation in one chapter on Clifford neurons and one other on Clifford MLPs follows also a classical road. That is the design from linear units to non–linear networks. The last section of this chapter is therefore dedicated to linearization with Clifford neurons. We will work out this completely on the example of Möbius transformations.

We start with an outline of Clifford algebra now.

12.2 An Outline of Clifford Algebra

A compact and comprehensive introduction on geometric algebras (and therefore also on Clifford algebras) has already been given by the first chapter of this book. Here, we will only review those facts needed in the following.

In addition, we will put more emphasis to the direct generation of algebras by non-degenerate quadratic forms. This gives the signature of a vector space in a very natural way.

So let be Q a non–degenerate quadratic form on $\mathbb{R}^n$. For shortness, we will call $(\mathbb{R}^n, Q)$ a quadratic space.

By a theorem of linear algebra there exists a basis of $\mathbb{R}^n$ such that

$$Q(v) = -v_1^2 - v_2^2 - \ldots - v_p^2 + v_{p+1}^2 \cdots + v_{p+q}^2 \tag{12.1}$$

for all $v = (v_1, \ldots, v_n) \in \mathbb{R}^n$, $p, q \in \mathbb{N}$ and $p + q = n$. This allows us to work with quadratic forms in an abstract fashion. In that sense, a quadratic form is already determined by (p, q). Then, we will denote a quadratic space by $\mathbb{R}^{p,q}$ hereafter. For the vectors of an orthonormal basis $\{e_1, \ldots, e_n\}$ of $\mathbb{R}^n$ we get from (12.1)

$$-Q(e_i) = +1 \quad \text{if} \quad i \leq p \tag{12.2}$$

$$-Q(e_i) = -1 \quad \text{if} \quad i > p\,. \tag{12.3}$$

With the corresponding scalar product to Q in mind we can also speak of $\mathbb{R}^{0,q}$ as an Euclidean space, $\mathbb{R}^{p,0}$ as an anti–Euclidean space, and $\mathbb{R}^{p,q}$ $(p \neq 0 \wedge q \neq 0)$ as an indefinite space, respectively.

Equations (12.1)-(12.3) together now allow the following definition of a Clifford algebra [154].

Definition 12.2.1. *An associative algebra with unity 1 over* $\mathbb{R}^{p,q}$ *containing* $\mathbb{R}^{p,q}$ *and* $\mathbb{R}$ *as distinct subspaces is called the Clifford algebra* $\mathcal{C}_{p,q}$ *of* $\mathbb{R}^{p,q}$*, iff*

(a) $v \otimes_{p,q} v = -Q(v)\,, \quad v \in \mathbb{R}^{p,q}$

(b) $\mathcal{C}_{p,q}$ *is generated as an algebra by* $\mathbb{R}^{p,q}$

(c) $\mathcal{C}_{p,q}$ *is not generated by any proper subspace of* $\mathbb{R}^{p,q}$.

Examples of Clifford algebras are the real numbers $\mathbb{R}$ corresponding to $\mathcal{C}_{0,0}$, the complex numbers $\mathbb{C}$ corresponding to $\mathcal{C}_{0,1}$, and the quaternions $\mathbb{H}$ corresponding to $\mathcal{C}_{0,2}$, respectively.

An element of a Clifford algebra is called a *multivector*, due to the fact that it consists of objects of different types by definition. The algebra multiplication $\otimes_{p,q}$ of a Clifford algebra $\mathcal{C}_{p,q}$ is called the *geometric product.*

Condition (a) of the above definition implies equations (12.2),(12.3) and further for all $i, j \in \{1, \ldots, p+q\}$

$$e_i \otimes_{p,q} e_j = -e_j \otimes_{p,q} e_i\,. \tag{12.4}$$

Clearly, 2^{p+q} is then an upper bound for the dimension of a Clifford algebra. Condition (c) guarantees that no lower dimensional algebras are generated. For a complete proof see e.g. [189]. Hence, the dimension of a Clifford algebra is 2^{p+q}.

A further very important consequence of equation (12.4) is, that only Clifford algebras up to dimension 2 are commutative ones.

We will now give explicitly a basis of a Clifford algebra in terms of the basis vectors $\{e_1, \ldots, e_n\}$ of the underlying quadratic space. Using the canonical order of the power set $\mathcal{P}(\{1, \ldots, n\})$ to derive the index set

$$\mathcal{A} := \{\{a_1, \ldots, a_r\} \in \mathcal{P}(\{1, \ldots, n\}) \mid 1 \leq a_1 \leq \ldots \leq a_r \leq n\} \tag{12.5}$$

and then defining for all $A \in \mathcal{A}$

$$e_A := e_{a_1} \ldots e_{a_r}\,, \tag{12.6}$$

we achieve a basis $\{e_A \mid A \in \mathcal{A}\}$ of the Clifford algebra $\mathcal{C}_{p,q}$. Every $x \in \mathcal{C}_{p,q}$ can then be written as

$$x = \sum_{A \in \mathcal{A}} x_A e_A\,. \tag{12.7}$$

For every $r \in \{0, \ldots, 2^{p+q}-1\}$ the set $\{e_A \mid A \in \mathcal{A},\ |A| = r\}$ is spanning a linear subspace of $\mathcal{C}_{p,q}$. This linear subspace is called the *r–vector part* of the Clifford algebra $\mathcal{C}_{p,q}$. An element of such a subspace is then called an r–vector. For r running from 0 to 3 an r–vector is also often called a scalar, vector, bivector, or trivector, respectively.

The vector part of a Clifford algebra $\mathcal{C}_{p,q}$ should get its own notation $\mathbb{R}_{p,q}$ to be distinguished from the quadratic space $\mathbb{R}^{p,q}$ itself.

All even r-vectors form the even part $\mathcal{C}^+_{p,q}$ of the Clifford algebra $\mathcal{C}_{p,q}$. $\mathcal{C}^+_{p,q}$ is a subalgebra of $\mathcal{C}_{p,q}$ isomorphic to $\mathcal{C}_{p,q-1}$. Whereas the odd part $\mathcal{C}^-_{p,q}$ formed by all odd r-vectors is not a subalgebra.

Clifford algebras are $\mathbb{R}$–linear

$$\forall \lambda \in \mathbb{R}\, \forall x, y \in \mathcal{C}_{p,q} : \; (\lambda\, x)y = x(\lambda\, y) = \lambda\, (xy)\,, \tag{12.8}$$

so every Clifford algebra is isomorphic to some matrix algebra. The matrix representations of Clifford algebras $\mathcal{C}_{p,q}$ up to dimension 16 are given in Table 12.1. As we can see, there are many isomorphic Clifford algebras.

Table 12.1. Matrix representations of Clifford algebras up to dimension 16

$p \backslash q$	0	1	2	3	4
0	$\mathbb{R}$	$\mathbb{C}$	$\mathbb{H}$	${}^2\mathbb{H}$	$\mathbb{H}(2)$
1	${}^2\mathbb{R}$	$\mathbb{R}(2)$	$\mathbb{C}(2)$	$\mathbb{H}(2)$	${}^2\mathbb{H}(2)$
2	$\mathbb{R}(2)$	${}^2\mathbb{R}(2)$	$\mathbb{R}(4)$	$\mathbb{C}(4)$	$\mathbb{H}(4)$
3	$\mathbb{C}(2)$	$\mathbb{R}(4)$	${}^2\mathbb{R}(4)$	$\mathbb{R}(8)$	$\mathbb{C}(8)$
4	$\mathbb{H}(2)$	$\mathbb{C}(4)$	$\mathbb{R}(8)$	${}^2\mathbb{R}(8)$	$\mathbb{R}(16)$

Next, we will deal with involutions of Clifford algebras. An involution is an algebra mapping of order 2. Thus the set of all involutions of a Clifford algebra is given by

$$In(\mathcal{C}_{p,q}) := \{f : \mathcal{C}_{p,q} \to \mathcal{C}_{p,q} \mid f^2 = id\}\,. \tag{12.9}$$

The most important involutions of a Clifford algebra are the following ones. The first, called *inversion*

$$\hat{x} = \sum_{A \in \mathcal{A}} (-1)^{|A|}\, x_A e_A \tag{12.10}$$

is an automorphism ($\hat{x}\hat{y} = \widehat{xy}$), whereas *reversion*

$$\tilde{x} = \sum_{A \in \mathcal{A}} (-1)^{\frac{|A|\,(|A|-1)}{2}} x_A e_A\,, \tag{12.11}$$

and *conjugation*

$$\bar{x} = \sum_{A \in \mathcal{A}} (-1)^{\frac{|A|\,(|A|+1)}{2}} x_A e_A \tag{12.12}$$

are anti–automorphisms ($\tilde{x}\tilde{y} = \widetilde{yx}$, $\bar{x}\bar{y} = \overline{yx}$). Conjugation is obviously a composition of inversion and reversion. The conjugation of complex numbers results a special case of (12.12).

Finally, we want to analyse which Clifford algebras are division algebras. The answer is given by the famous FROBENIUS theorem. That theorem states, that there are no other real division algebras despite of $\mathbb{R}$, $\mathbb{C}$, and $\mathbb{H}$. A finite-dimensional associative algebra $\mathcal{A}$ is a division algebra, iff it contains no divisors of zero. Therefore, any other Clifford algebras except the ones mentioned above will contain divisors of zero. Thereby, an element $a \in \mathcal{A}$ is a divisor of zero, iff there exists an element $b \in \mathcal{A}\backslash\{0\}$ with $ab = 0$ or $ba = 0$.

The existence of divisors of zero can cause many problems in the design of neural algorithms in the frame of Clifford algebras. We will see this already in outlines in the next section.

12.3 The Clifford Neuron

In this section we will start with a generic neuron as computational unit. From this, a standard real valued neuron is then derived. Finally, we will introduce the Clifford neuron based on the geometric product. Through this way, we will also introduce some basics of neural computation in general very briefly. To characterize the computation with Clifford neurons as model–based in relation to that with real neurons is the main goal of this section.

A generic neuron is a computational unit of the form shown in Figure 12.1. The computation within such a neuron is performed in two steps. Firstly, a propagation function f associates the input vector x with the parameters of the neuron comprised in the weight vector w. Then, the application of a activation function g follows. Thus, the output of a generic neuron is given by

$$y = g(f(x; w)) . \tag{12.13}$$

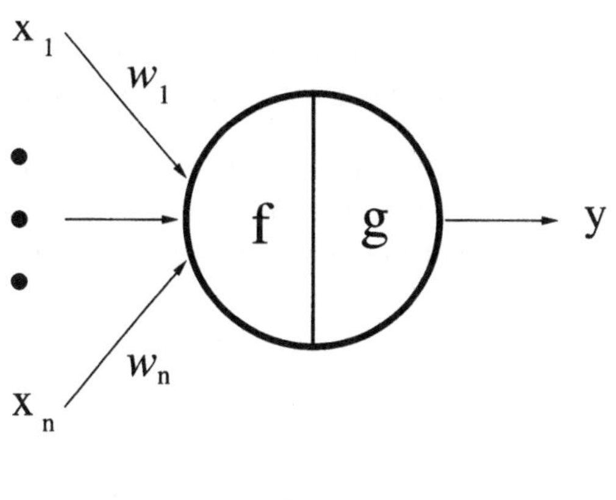

Fig. 12.1. Generic neuron

In general, the propagation function f is a mapping

$$f : D^n \to D \tag{12.14}$$

for a domain D. The activation function g is a mapping

$$g : D \rightarrow D' \tag{12.15}$$

to a domain D'. Mostly, D is a continuous domain. In this case, we have usually $D' = D$ for function approximation. On the other hand, the neuron computes a classification if D' is discrete.

From now on, we will assume if no other statement is made, that g is set to be the identity. We will also speak of a neuron with this in mind.

12.3.1 The Real Neuron

For a real neuron we have with our previous notation $D = \mathbb{R}$ and $w, x \in \mathbb{R}^n$. The most common propagation function for such a neuron simply computes a weighted sum of the inputs of a real neuron

$$f(x) = \sum_{i=1}^{n} w_i x_i + \theta\,, \tag{12.16}$$

with an additional parameter $\theta \in \mathbb{R}$, that works as a bias. By extending the domain by one dimension and then using an extended input vector $x^+ := (x, 1)$ and an extended weight vector $w^+ := (w, \theta)$ we can rewrite (12.16) in the form

$$f(x^+) = \sum_{i=1}^{n+1} w_i^+ x_i^+\,. \tag{12.17}$$

A real neuron with the above propagation function is therefore a linear associator. Non-linearity of the neuron could be achieved by applying a non-linear activation function g.

As a linear associator we can use the real neuron for linear regression. This (neural computation) is done by formulating linear regression as a learning problem.

So let us consider a training set $T := \{(x^1, t^1), \ldots, (x^m, t^m)\}$ consisting of input–output pairs (x^i, t^i) with $x^i \in \mathbb{R}^n$, $t^i \in \mathbb{R}$. The aim of learning is to find a weight vector $w = (w_1, \ldots, w_n)$ that minimizes the sum-of-squared error (SSE)

$$E = \frac{1}{2} \sum_{i=1}^{m} (t^i - \sum_{j=1}^{n} w_j x_j^i)^2 \tag{12.18}$$

iteratively. A well known method to do so is using gradient descent. Then, at each step the following correction of the weights

$$\Delta w_j = -\frac{\partial E}{\partial w_j}\,. \tag{12.19}$$

has to be made. In terms of neural networks this is called *back–propagation*, due to the fact that the error is propagated back from the output.

Since the error function (12.18) is convex, back–propagation will always find the global minimum.

Provided with the above basic knowledge about generic and real neurons we are now able to study Clifford neurons in detail.

12.3.2 The Clifford Neuron

An abstract Clifford neuron is easily derived as a special case of a generic neuron by taking in (12.13) a Clifford algebra as domain. However, some care has to be taken already. The propagation function of a generic Clifford neuron should obviously be a mapping of the form

$$f : \mathcal{C}_{p,q} \to \mathcal{C}_{p,q}. \tag{12.20}$$

The above function is then just a special case of (12.14) with $D = \mathcal{C}_{p,q}$ and $n = 1$. In that case the illustration of a generic neuron in Figure 12.1 has no great strength anymore, because we have just one input and one weight. But through that, we can also see immediately that f has lost its independent function. More precisely, it is fully determined by the way the association of the one input with the one weight is done. Clearly, there is only one intented way of association — the geometric product.

The propagation function f of a Clifford neuron is given either by

$$f(x) = w \otimes_{p,q} x + \theta \tag{12.21}$$

or by

$$f(x) = x \otimes_{p,q} w + \theta\,. \tag{12.22}$$

All the entities are now multivectors, i.e. $x, w, \theta \in \mathcal{C}_{p,q}$.

Of course, we have to distinguish left–sided and right–sided weight multiplication in the general case of a non-commutative Clifford algebra.

Formally, we have just replaced the scalar product by the geometric product. As in the real case, we can interpret the parameter θ as a bias. However, now an extension of the form (12.17) is possible to treat θ as a normal weight.

The input-weight association of a Clifford neuron should now be made concretely. For the sake of simplicity let us choose the complex numbers $\mathcal{C}_{0,1}$ as an example. A complex neuron computes just a complex multiplication, say $xw = y$. Further, let be $x = x_1 + x_2\,\mathrm{i}$ and $y = y_1 + y_2\,\mathrm{i}$.

Now assume we want to compute a complex multiplication with real neurons. Clearly, this requires 2 real input and 2 real output neurons. We are looking then for a weight matrix $W \in \mathbb{R}(2)$ that fulfills $(x_1, x_2)\, W = (y_1, y_2)$. This is achieved by setting $w_{11} = w_{22}$ and $w_{12} = -w_{21}$, which just results in

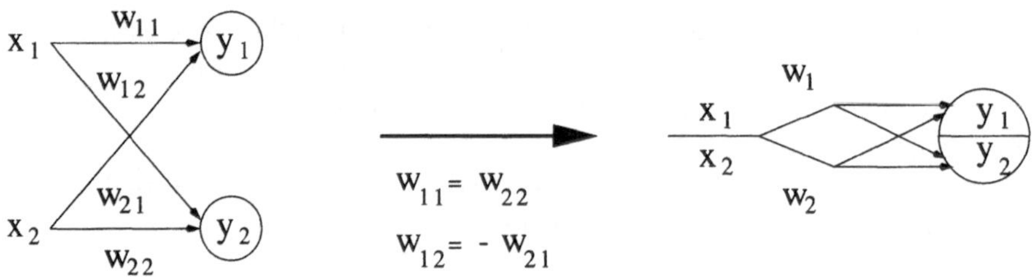

Fig. 12.2. Computation of a complex neuron (right) and simulation with real neurons (left)

the well-known matrix representation of complex numbers. Figure 12.2 gives an illustration of the situation.

Thus, complex multiplication is just a certain linear transformation, namely a dilatation-rotation, which easily follows from the polar form of complex numbers. This means in terms of neural computation, a complex neuron can be seen as model-based. Instead of an unspecified linear function (real neurons) we use a dilatation-rotation (complex neuron). If this model is applicable to given data, we would only need half of the parameters (see again Figure 12.2) for computation. Furthermore, the real neurons have to imitate the model "by finding the given weight constraints" with independent weights. This approach should then also be less efficient with respect to time complexity or less accurate.

To be able to verify this experimentally, we now need a correct learning algorithm for a complex neuron. Yet, we will give here the rule for updating the weight in the general case of an arbitrary Clifford neuron. So let be $T := \{(x^1, t^1), \ldots, (x^m, t^m)\}$ the training set consisting of input–output pairs (x^i, t^i) with $x^i, t^i \in \mathcal{C}_{p,q}$. The SSE defined analogously to (12.18) is then minimized by applying the correction step

$$\Delta w = \bar{x^i} \otimes_{p,q} (t^i - w \otimes_{p,q} x^i)\,. \tag{12.23}$$

for left–sided weight multiplication and

$$\Delta w = (t^i - x^i \otimes_{p,q} w) \otimes_{p,q} \bar{x^i}\,. \tag{12.24}$$

for right–sided weight multiplication, respectively. Here, the function $^-$ stands for that univocally determined involution yielding

$$x \otimes_{p,q} \overline{y} = \sum_i x_i y_i\,. \tag{12.25}$$

Using this function avoids the appearance of divisors of zeros during back–propagation. This is necessary, otherwise learning could stop for an non–zero error. The proof of correctness of the algorithm will be postponed to the next chapter.

Now, we can perform our first intented experiment.

Experiment 1 (Complex multiplication).
The task for a complex neuron and for real neurons as in Figure 12.2 was simply to learn the complex multiplication with $2 - 4\mathrm{i}$. As training set $T = \{(-0.3, 0), (-0.5, -0.3), (-0.6, 0)\}$ was used. After 116 epochs (which means after applying the training patterns 116 times) the SSE of the complex neuron where dropped under 0.000001. The learned weight of the complex neuron was $w = 2.0000 - 4.0000\,\mathrm{i}$. In contrast, the SSE of the real neurons dropped under 0.000001 after 246 steps but the weight matrix was

$$W = \begin{pmatrix} 1.99741 & -3.99738 \\ 4.00252 & 1.99700 \end{pmatrix}.$$

Thus, our very plain considerations are right. Simulation of a model seems worse than using a model directly.

In the case of complex numbers we have identified the input–weight association by the geometric (complex) product completely and characterized it as model-based. The generalization of this is quiet easy. Due to the $\mathbb{R}$-linearity of Clifford algebras (12.8), any geometric product can be expressed as a special matrix multiplication. This means that the computation of an arbitrary single Clifford neuron can also be performed by the corresponding number of real neurons. However, this point of view on the neuron level is too artificial. In practice we have to deal with *real* data of *any* dimension.

We have introduced Clifford algebras as the algebras of quadratic spaces in section 2. Therefore, a natural computation of a Clifford neuron should process (real) data of the underlying quadratic space. In fact, the complex multiplication of a Clifford neuron should also be seen in this way. As we know already, a complex neuron computes a dilatation–rotation. More precisely, it computes a transformation of vectors of $\mathbb{R}^2$ in such a manner. As real vector spaces of the same dimension $\mathbb{R}^2$ and $\mathbb{C}$ are isomorphic. In that sense a complex neuron processes also indeed points of $\mathbb{R}^2$. However, complex numbers are no vectors.

This interpretation problem will be easily resolved in the next section. That section will be fully dedicated to the processing of data drawn from quadratic spaces with Clifford neurons in a formally consistent manner. By doing so we will also get a better understanding of the model–based nature of Clifford neurons.

12.4 Clifford Neurons as Linear Operators

Following the ideas developed at the end of the previous section, we are now interested how a linear transformation of the form

$$f : \mathbb{R}_{p,q} \to \mathbb{R}_{p,q} \tag{12.26}$$

can be computed with Clifford neurons. To be able to do so, we need a theoretical method to describe such transformation in Clifford algebras.

Fortunately, any multivector that has a multiplicative inverse defines such a transformation already. Thus, the mathematical object we have to look at is the group formed by these multivectors. This group is called the Clifford group.

Applying a group to the elements of a set is generally formalized in the following way.

Definition 12.4.1. *Let G be a group and M be a non-empty set. The map*

$$\star : G \times M \to M; \quad (a, x) \mapsto a \star x \tag{12.27}$$

is called the operation of G on M, if $1_G \star x = x$ and $a \star (b \star x) = (a \star b) \star x$ for all $x \in M,\ a, b \in G$.

For example, the general linear group $GL(n, \mathbb{R})$ of $\mathbb{R}^n$ operates on (column) vectors by matrix multiplication

$$\cdot : GL(n, \mathbb{R}) \times \mathbb{R}^n \to \mathbb{R}^n; \quad (A, x) \mapsto Ax\,. \tag{12.28}$$

The Clifford case is more complicated than that. It will be studied in detail in the next subsection. The results of this study will then be transposed to the level of Clifford neurons and will be verified there experimentally.

12.4.1 The Clifford Group

Let us start directly with the definition of the Clifford group.

Definition 12.4.2. *The Clifford group $\Gamma_{\mathrm{p,q}}$ of a Clifford algebra $\mathcal{C}_{p,q}$ is defined as*

$$\Gamma_{\mathrm{p,q}} := \{s \in \mathcal{C}_{p,q} \mid \forall x \in \mathbb{R}_{p,q} :\ sx\hat{s}^{-1} \in \mathbb{R}_{p,q}\}\,. \tag{12.29}$$

From that definition we get immediately

$$\Gamma_{\mathrm{p,q}} \times \mathbb{R}_{p,q} \to \mathbb{R}_{p,q}; \quad (s, x) \mapsto sx\hat{s}^{-1} \tag{12.30}$$

as the operation of the Clifford group $\Gamma_{\mathrm{p,q}}$ on $\mathbb{R}_{p,q}$. Thus, the operation of $\Gamma_{\mathrm{p,q}}$ is not one single primitive operation, as it was the case in the example of $GL(n, \mathbb{R})$ (12.28). Another important difference to that case is, that the elements of the group are of the same type as the elements of the set on which the group is operating. Actually, this is one of the great advantages of Clifford algebra. We shall call an element of $\Gamma_{\mathrm{p,q}}$ a linear operator to distinguish it from an ordinary multivector. It is indeed a linear operator since the Clifford group $\Gamma_{\mathrm{p,q}}$ consists of linear transformations of $\mathbb{R}_{p,q}$ by definition (12.26).

Hence, $\Gamma_{\mathrm{p,q}}$ is isomorphic to a general linear group or one of its subgroups. The relation of $\Gamma_{\mathrm{p,q}}$ to those classical groups can be concluded from the map

$$\psi_s : \mathbb{R}^{p,q} \longrightarrow \mathbb{R}^{p,q};\ x \mapsto s x \hat{s}^{-1}\,. \tag{12.31}$$

For all $x \in \mathbb{R}_{p,q}$, $s \in \Gamma_{\mathrm{p,q}}$ we have

$$Q(\psi_s(x)) = (\widehat{sx\hat{s}^{-1}})sx\hat{s}^{-1} = \hat{s}\hat{x}s^{-1}sx\hat{s}^{-1} = \hat{x}x = Q(x)\,, \tag{12.32}$$

so ψ_s is an orthogonal map. In fact, it is easy to see that ψ_s is even an orthogonal automorphism of $\mathbb{R}_{p,q}$. Thereby, we have proofed the following theorem in principle.

Theorem 12.4.1. *The map $\Psi_s : \Gamma_{\mathrm{p,q}} \rightarrow O(p,q);\ s \mapsto \psi_s$ is a group epimorphism.*

Indeed, $\Gamma_{\mathrm{p,q}}$ is a multiple cover of the orthogonal group $O(p,q)$ since the kernel of Ψ_s is $\mathbb{R}\setminus\{0\}$.

Altogether, we know now that the Clifford group $\Gamma_{\mathrm{p,q}}$ is an orthogonal transformation group. However, it is still unnecessarily large. Therefore, we first reduce $\Gamma_{\mathrm{p,q}}$ to a two-fold cover of $O(p,q)$ by defining the so-called Pin group

$$\mathrm{Pin(p,q)} := \{s \in \Gamma_{\mathrm{p,q}} \mid s\tilde{s} = \pm 1\}\,. \tag{12.33}$$

The even elements of Pin(p, q) form the spin group

$$\mathrm{Spin(p,q)} := \mathrm{Pin(p,q)} \cap \mathcal{C}^{+}_{p,q} \tag{12.34}$$

which is a double cover of the special orthogonal group $SO(p,q)$. Finally, those elements of Spin(p, q) with Clifford norm equal 1 form a further subgroup

$$\mathrm{Spin_+(p,q)} := \{s \in \mathrm{Spin(p,q)} \mid s\tilde{s} = 1\} \tag{12.35}$$

that covers $SO_+(p,q)$ twice. Thereby, $SO_+(p,q)$ is the connected component of the identity of $O(p,q)$.

As usual, we write Pin(p) for Pin(p, q) and so on. We shall remark here, that $\mathrm{Spin(p,q)} \simeq \mathrm{Spin(q,p)}$ and $\mathrm{Spin(p)} = \mathrm{Spin_+(p)}$. Both follows easily from the properties of the orthogonal groups together with $\mathcal{C}^{+}_{p,q} \simeq \mathcal{C}^{+}_{q,p}$.

For the spin group Spin(p, q) there exists another way besides the standard one (12.30) of operating as a dilatation–rotation operator. This way will allow the reduction of the dimension of the Clifford algebra in use. Also it will resolve the interpretation problem regarding complex multiplication noticed earlier in section 3.

Spin(p, q) consists by definition only of even elements. Remembering $\mathcal{C}^{+}_{p,q} \simeq \mathcal{C}_{p,q-1}$, we can interpret a spinor also as an element of $\mathcal{C}_{p,q-1}$.

Let us denote by $\lambda\mathbb{R}_{p,q-1}$ both the scalar and vector part of $\mathcal{C}_{p,q-1}$. This space is called the space of paravectors. Then the operation of Spin(p, q) on $\mathbb{R}_{p,q-1}$ is the same as on $\mathbb{R}_{p,q}$ [189]. More precisely, for every $s \in \mathrm{Spin(p,q)}$ the map

$$\phi_s : \lambda\mathbb{R}_{p,q-1} \to \lambda\mathbb{R}_{p,q-1};\ x \mapsto sx\hat{s}^{-1} \tag{12.36}$$

is a dilatation–rotation of $\mathbb{R}_{p,q}$. If the underlained Clifford algebra in (12.36) is commutative in addition we have

$$\phi_s(x) = sx\hat{s}^{-1} = xs\hat{s}^{-1} = xs' \quad (s' := s\hat{s}^{-1} \in \mathrm{Spin(p,q)})\,. \tag{12.37}$$

In the special case of complex numbers the above relations together with $\mathcal{C}_{0,1} = \lambda\mathbb{R}_{0,1}$ implies that any complex multiplication is indeed a dilatation–rotation.

All the obtained results will be transposed to the computational level of Clifford neurons now.

12.4.2 Spinor Neurons

In the previous section we have studied the group of linear transformations $\Gamma_{\mathrm{p,q}}$ of a Clifford algebra. Actually, we have found out that $\Gamma_{\mathrm{p,q}}$ consists of orthogonal transformations only. The operation of $\Gamma_{\mathrm{p,q}}$ can be simulated by concatenation of a left–sided and a right-sided (or vice versa) Clifford neuron. This architecture is shown in Figure 12.3.

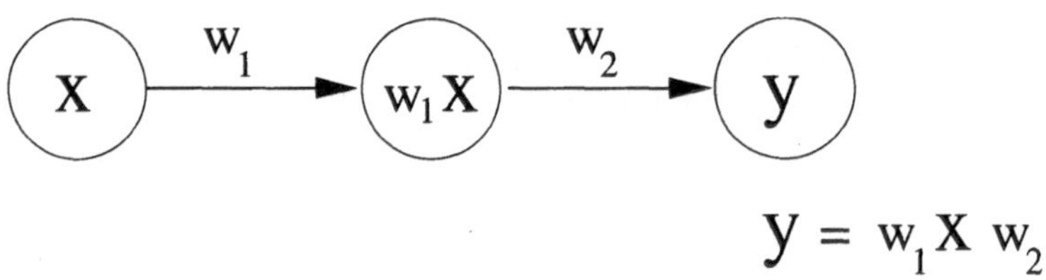

Fig. 12.3. Simulation of the operation of $\Gamma_{\mathrm{p,q}}$ with Clifford neurons

Every orthogonal transformation is computable by this architecture. This is just done by using the vector part of the input and output neuron to process the data. However, there might exist other suitable ways of representing the data. In general there are $\binom{n}{k}^2$ possibilities of input-output representations of k–dimensional data in n dimensions. Therefore, we will only study the case of plane transformations in Clifford algebras of dimension 4. In that case there are 36 possibilities of data representation.

Table 12.2. Used codes for 2 dimensional data in 4 dimension

	Representation
1	0xx0
2	0x0x
3	00xx
4	xx00
5	x0x0
6	x00x

Using the notations of Table 12.2 the number 11 then denotes input representation 1 and output representation 1 and thus input–output representation $0xx0 - 0xx0$. This is the representation corresponding directly to the definition of $\Gamma_{\mathrm{p,q}}$. The results for the computation of 2-D Euclidean transformations are listed in Table 12.3.

Table 12.3. Suitable data representations for SO(2) and O(2) computation

Algebra	Weight multiplication	Data representation
$\mathcal{C}_{0,2}$	left-right, right-left	all
	left	11,22,33,44,55,66,25,52
	right	11,22,33,44,55,66,16,61,34,43
$\mathcal{C}_{1,1}$	left, left-right, right-left	22,55,25,52
	right	55,22
$\mathcal{C}_{2,0}$	left, left-right, right-left	11,66,16,61
	right	11,66

As we can see, there is no difference between the computation of SO(2) and O(2). Remarkable, all representations with two weights in $\mathcal{C}_{0,2}$ are suitable. Due to the existence of complex number representations we get also representations that work with only one weight. In the case of an anti–Euclidean transformation we have to distinguish SO(1,1) and O(1,1). The suitable data representations can be found in Table 12.4 and Table 12.5, respectively.

Before starting to discuss the above listed results we should re-think the situation in general. All the reported results were obtained by applying data of a transformation of one of the mentioned types. So we actually just checked which representation will not work. Having in mind that all the transformations could also be computed by real neurons as in Fig. 12.2, we should extend our point of view again.A main idea of this introductory chapter is to develop interpretations of the computation of Clifford neurons. This should always be done by characterizing Clifford neurons as model–based ones as in section 3.

Table 12.4. Suitable data representations for SO(1,1) computation

Algebra	Weight multiplication	Data representation
$\mathcal{C}_{0,2}$	left, right, left-right, right-left	none
$\mathcal{C}_{1,1}$	left-right, right-left	44,64,14,34,46,66,16,36, 41,61,11,31,53,63,13,33
	left	44,66,16,61,11,33
	right	44,34,66,11,43,33
$\mathcal{C}_{2,0}$	left-right, right-left	44,54,24,34,45,55,25,35, 42,52,22,32,43,53,23,33
	left	44,55,25,52,22,33
	right	44,34,55,22,43,33

Table 12.5. Suitable data representations for O(1,1) computation

Algebra	Weight multiplication	Data representation
$\mathcal{C}_{0,2}$	left, right, left-right, right-left	none
$\mathcal{C}_{1,1}$	left-right, right-left	44,64,14,34,46,66,16,36 41,61,11,31,53,63,13,33
	left	34,43
	right	16,61
$\mathcal{C}_{2,0}$	left-right, right-left	44,54,24,34,45,55,25,35, 42,52,22,32,43,53,23,33
	left	34,43
	right	25,52

This step has to be made for the computation of orthogonal transformations with Clifford neurons now. That is, we have to determine the conditions so that the computation of Clifford neurons as in Figure 12.3 can be forced to be an orthogonal computation. In that case we would apply this model independent of the processed data. To be able to do so, we have to constrain the weights of the left–sided and right–sided Clifford neurons in Figure 12.3 together. This should result in one neuron with one weight that is multiplied from the left and from the right. But this will be not possible for an orthogonal transformation in general. However, it is possible for the operation of a spinor. The corresponding neuron is then named a *spinor neuron*. Computation with such neurons is always model–based. In the case of a 2–dimensional Clifford algebra this is also valid for any orthogonal transformation due to the commutativity of the algebra. However, this could require a special data representation as shown in Tables 12.3-12.5. So we use the notion of a spinor neuron in that sense that the operation of the neuron as a linear operator is performed by one weight. After we have reflected that spinor neurons are model–based, we will now perform simulations to compare them with real neurons.

12.4.3 Simulations with Spinor Neurons

With the following two experiments we want to test the strength of the model of single spinor neurons in comparison with multiple real neurons, especially in the presence of noise. We will just speak of real neurons, since the number of real input and output neurons to compute a linear transformation of $\mathbb{R}^n$ is always n. See again Figure 12.2.

Experiment 2 (Euclidean 2D similarity transformation).
The transformation that should be learned was a composition of a Euclidean 2D rotation about -55°, a translation of [+1, -0.8], and a scaling of factor 1.5. The training and test data is shown in Figure 12.4.

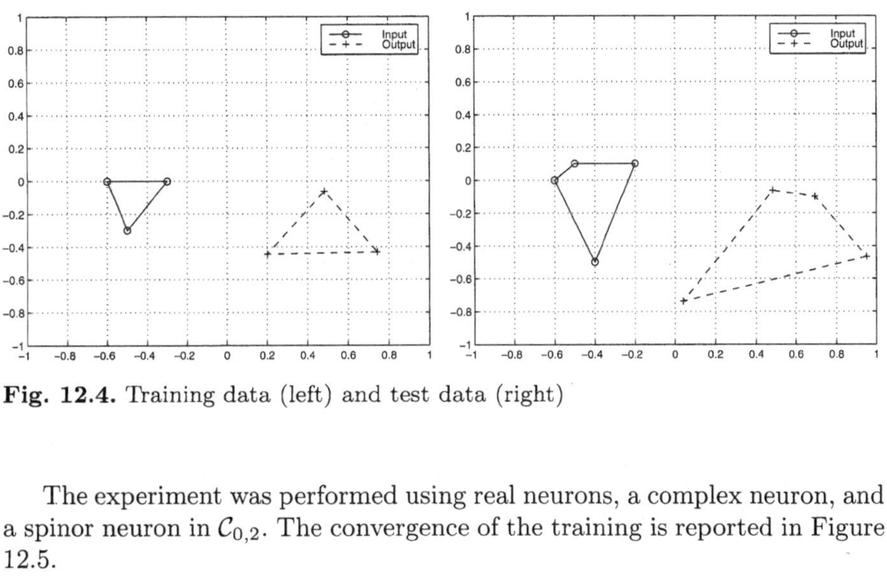

Fig. 12.4. Training data (left) and test data (right)

The experiment was performed using real neurons, a complex neuron, and a spinor neuron in $\mathcal{C}_{0,2}$. The convergence of the training is reported in Figure 12.5.

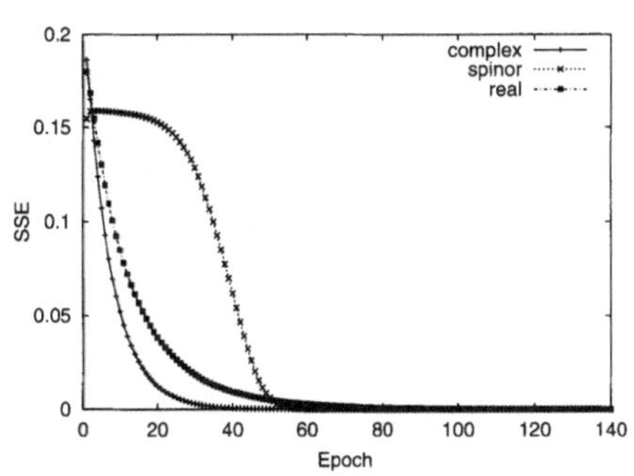

Fig. 12.5. Convergence of the learning

The spinor neuron learned indeed a spinor representation. Its weights in the odd components were zero. This required some more epochs of learning in comparison with the complex neuron. But it learned the task still faster than the real neurons. Besides the qualitative difference of the learning curve of the spinor neuron to the curves of the other neurons, no great quantitative difference could be noticed.

To test the generalization performance of the different neurons we also made simulations with noisy training data, by adding median-free uniform noise up to a level of 20%. The obtained results are shown in Figure 12.6.

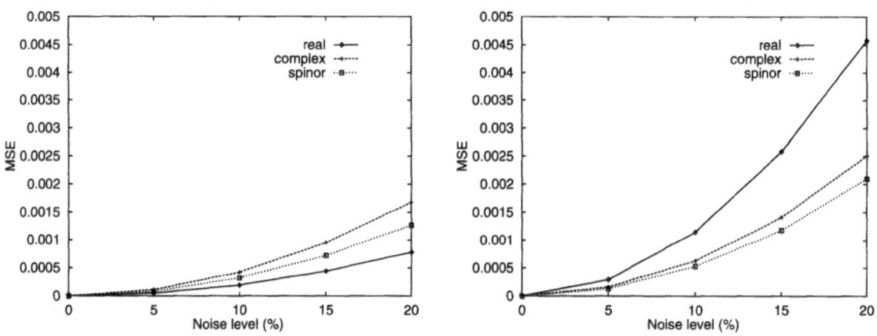

Fig. 12.6. Training errors (left) and generalization errors (right) by different noise levels

Due to the fact that the real neurons compute a general linear transformation, they have learned the noise better than the Clifford neurons. As a consequence, the generalization was then much worse in comparison with the Clifford neurons. There was no significant difference in generalization between the both Clifford neurons. The output on the test data of the real neurons and the complex neuron is shown in Figure 12.7.

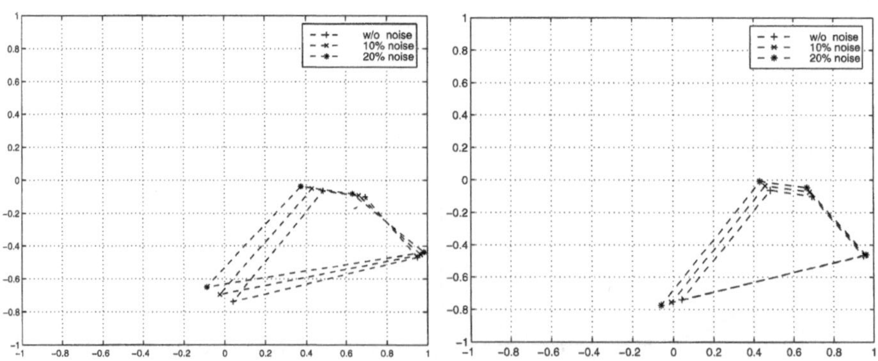

Fig. 12.7. Generalization obtained by the real neurons (left) and the complex neuron (right) by different noise levels

Using the model–based Clifford neurons for data fitting this model gave better results than using real neurons, especially on training with noisy data.

Experiment 3 (Euclidean 3D similarity transformation).
The only 4–dimensional Clifford algebra in which a Euclidean 3D rotation can be computed is $\mathcal{C}_{0,2}$. Thus, we can only compare experimentally the quaternionic and the real way of neural computation of such transformations. Actually, a quaternionic spinor neuron with any input-output representation can compute such a transformation. For the following experiment we use the standard spinor representation. That is we used the input–output representation $xxx0 - xxx0$. For this single quaternionic spinor neuron and a network of real neurons the task was to learn a rotation of -60° about the axis [0.5, $\sqrt{0.5}$, 0.5] with translation about [0.2,-0.2,0.3]. The data for training is shown in Figure 12.8.

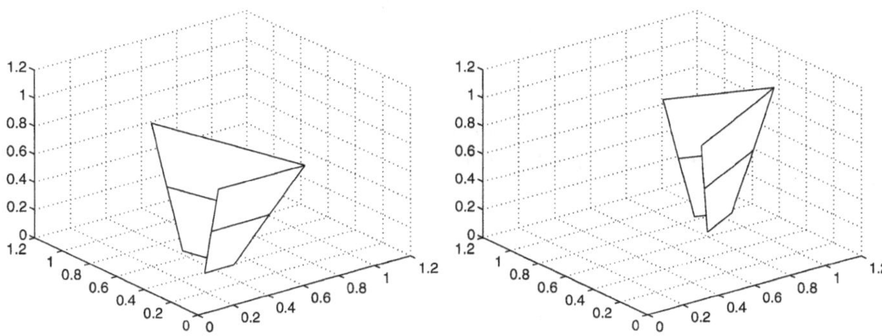

Fig. 12.8. Training data input (left) and training data output(right)

As test set we use a transformed version of the training data as shown in Figure 12.9.

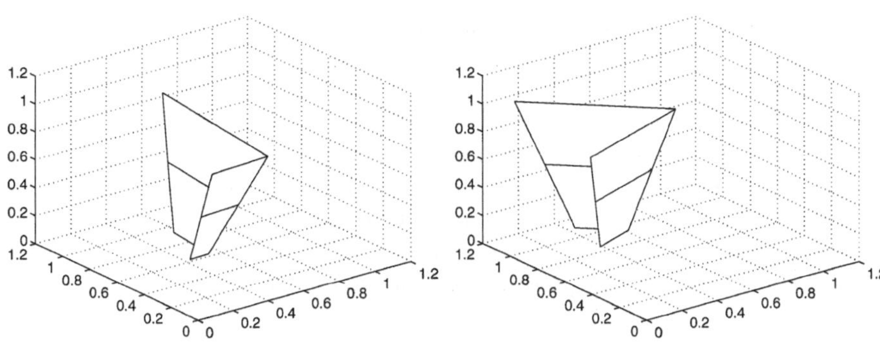

Fig. 12.9. Test data input (left) and test data output (right)

The convergence of the training is shown in Figure 12.10. As we can see, the quaternionic spinor neuron converges much faster than the real neurons. The real neurons have to learn the matrix representation of the quaternionic multiplication. Due to that fact, it was impossible to drop the SSE < 0.00001 for the real neurons. Thus, there exists already a numerical boundary value of reachable accuracy for the computation with real neurons.

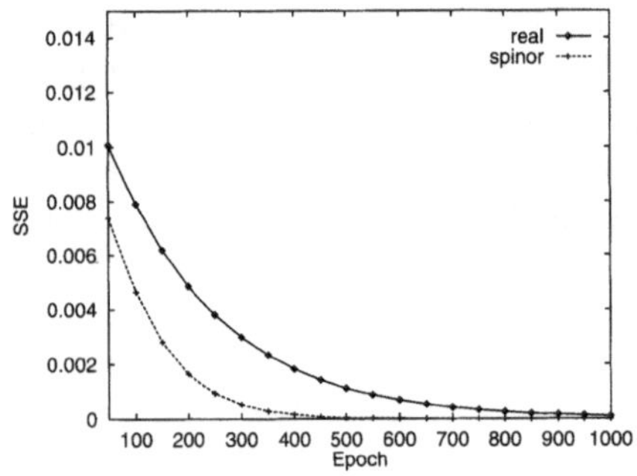

Fig. 12.10. Convergence of the learning

Clearly, this effects the performance of the real neuron on noisy training data in a quiet negative way. The errors for different noise level are shown in Figure 12.11.

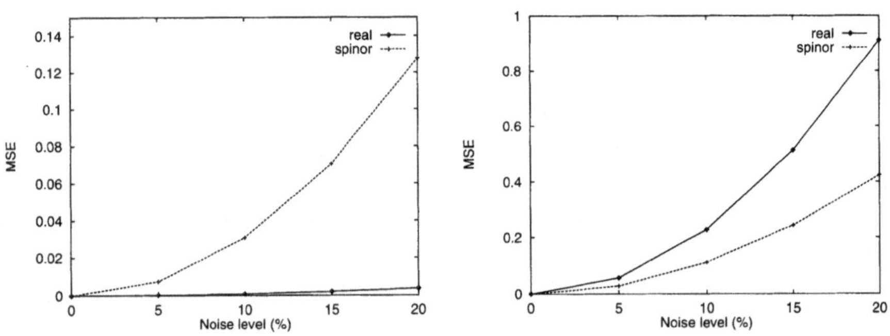

Fig. 12.11. Training errors (left) and generalization errors (right) by different noise levels

The real neurons simply learned the noise. Therefore, their generalization is worse than that of the Clifford neuron by a factor two. Actually, the real neurons performed much worse than indicated by that, as it can be seen by looking at the obtained generalization results shown in Figure 12.12 and Figure 12.13, where crosses indicate the desired output.

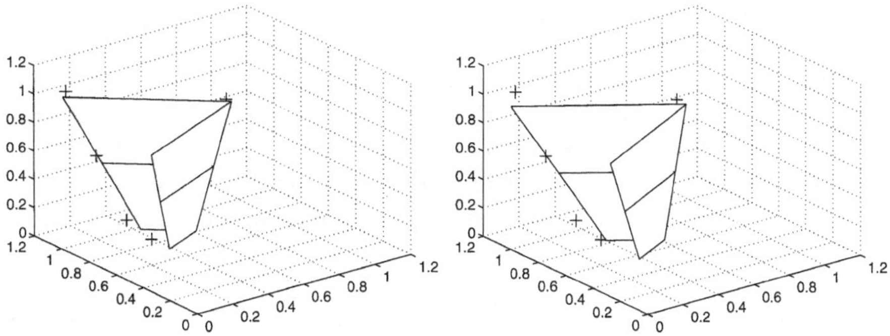

Fig. 12.12. Generalization obtained with spinor neurons by 10% noise (left) and by 20% noise (right)

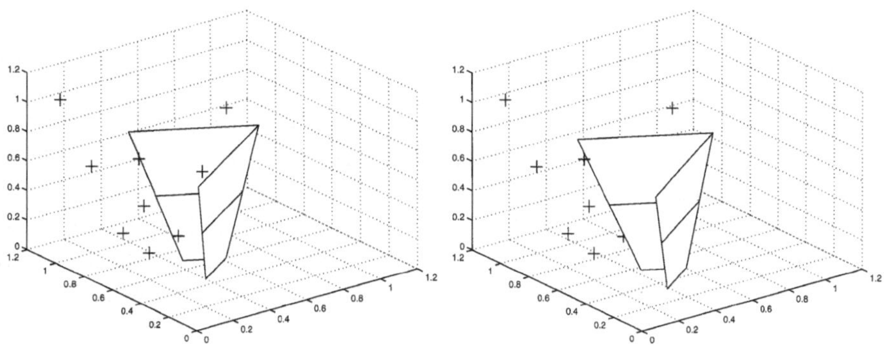

Fig. 12.13. Generalization obtained with real neurons by 10% noise (left) and by 20% noise (right)

The pose of the object generalized by the real neurons is completely wrong. In fact, this is already caused by the way the real neurons learned the task. The model applied is that of the noise. Instead of still separating the problem in a rotation part (weight matrix) and a translation part (biases) they will always use the biases strongly to fit the noise. Instead, a Clifford neuron applies this model of separation.

With this simulation we will finish our study of Clifford neurons as linear operators.

12.5 Möbius Transformations

In this section we will demonstrate, that Clifford neurons are able to learn transformations that are not learnable with real neurons. Clearly, this will require to linearize a non-linear transformation in a unique way in the framework of Clifford algebras.

The transformations in mind are the plane projective transformations. These are the most general transformations mapping lines to lines.

Of course, some theoretical preparations have to be made first. The idea is to relate the projective transformation groups to Möbius transformation groups. So let therefore the *complex general projective group*, denoted by PGL(2,$\mathbb{C}$), $\widehat{\mathbb{C}} := \mathbb{C} \cup \{\infty\}$, be the one-point compactification of $\mathbb{C}$.

The biholomorphic functions of $\widehat{\mathbb{C}}$ in itself are isomorphic to the group of the fractional-linear transformations

$$z \mapsto \tau_A(z) := \frac{az+b}{cz+d}\,, \; A = \begin{pmatrix} a & b \\ c & d \end{pmatrix} \in GL(2,\mathbb{C})\,. \tag{12.38}$$

This group is also called the Möbius group of $\widehat{\mathbb{C}}$, denoted by $M(\widehat{\mathbb{C}})$. Further then, the map

$$GL(2,\mathbb{C}) \to M(\widehat{\mathbb{C}})\,, \; A \mapsto \tau_A \tag{12.39}$$

is a group isomorphism with kernel identical to the center of $GL(2,\mathbb{C})$. Due to the fact, that PGL(2,$\mathbb{C}$) is GL(2,$\mathbb{C}$) factorized to its center, we then have PGL(2,$\mathbb{C}$) $\simeq M(\widehat{\mathbb{C}})$.

The definition of Möbius transformations of the complex plane $\mathbb{C}$, can be easily generalized to the general case of a quadratic space $\mathbb{R}^{p,q}$, where no explicit notion of the corresponding one-point compactification will be made anymore.

Definition 12.5.1. *The map*

$$\mathcal{C}_{p,q} \to \mathcal{C}_{p,q}, x \mapsto (ax+b)(cx+d)^{-1} \quad a,b,c,d \in \mathcal{C}_{p,q}, \quad (cx+d) \in \Gamma_{\mathrm{p,q}}$$

is called a Möbius transformation of $\mathbb{R}^{p,q}$.

Again, the group formed by Möbius transformations of $\mathbb{R}^{p,q}$ is called the *Möbius group* and will be denoted by $M(p,q)$, that is, $M(\widehat{\mathbb{C}})$ is now denoted by $M(0,1)$. The Möbius group $M(p,q)$ is covered by the orthogonal group $O(p+1,q+1)$, and is therefore (section 3.2) four times covered by $\mathrm{Pin(p+1,q+1)}$. Clearly then, we have immediately $M(p,q) \simeq M(q,p)$. However, $\mathrm{Pin(p+1,q+1)}$ acts not directly on elements of $\mathcal{C}_{p,q}$ in $\mathcal{C}_{p+1,q+1}$.

To be able to achieve the intented embedding of $\mathcal{C}_{p,q}$ in $\mathcal{C}_{p+1,q+1}$, i.e. to find a way to let $M(p,q)$ (or $\mathrm{Pin(p+1,q+1)}$, respectively) operate on $\mathcal{C}_{p,q}$, we must proceed our study of Möbius transformations in terms of matrix groups. We will restrict ourselves thereby essentially to the case of interest, that is Möbius transformations of anti–Euclidean spaces $\mathbb{R}^{0,n}$. The following characterization theorem for that was given already by *Vahlen, 1902.*

Theorem 12.5.1. *A matrix* $A = \begin{pmatrix} a & b \\ c & d \end{pmatrix}$ *with entries in* $\mathcal{C}_{0,n}$ *represents a Möbius transformation of* $\mathbb{R}^{0,n}$, *iff*

(a) $a, b, c, d \in \Gamma_{0,n} \cup \{0\}$
(b) $\bar{a}b, b\bar{d}, \bar{d}c, c\bar{a} \in \mathbb{R}^{0,n}$
(c) $a\tilde{d} - b\tilde{c} \in \mathbb{R}\backslash\{0\}$.

Matrices fulfilling these conditions are called *Vahlen matrices.*

A characterization of Möbius transformations of Euclidean spaces is easily obtained by switching the signature $(0, n)$ to $(n, 0)$ in the above theorem. For the general case of a quadratic space with an arbitrary signature one has to allow all products of vectors (not only invertible) in $\mathbb{R}^{p,q}$ in condition (a) of Theorem 12.5.1, which is just the same if $p = 0$ or $q = 0$.

We will now develop the representation of Möbius transformations of the complex plane in detail. Due to the fact that $\mathcal{C}_{p,q}(2) \simeq \mathcal{C}_{p+1,q+1}$, the algebra to concern is $\mathcal{C}_{1,2}$, for which we need a matrix representation firstly. This is given by defining the following basis

$$e_0 := \begin{pmatrix} 1 & 0 \\ 0 & 1 \end{pmatrix} \; e_1 := \begin{pmatrix} 0 & 1 \\ 1 & 0 \end{pmatrix} \; e_2 := \begin{pmatrix} i & 0 \\ 0 & -i \end{pmatrix} \; e_3 := \begin{pmatrix} 0 & -1 \\ 1 & 0 \end{pmatrix}$$

and the remaining basis vectors are easily obtained by matrix multiplication, e.g.

$$e_{123} = \begin{pmatrix} -i & 0 \\ 0 & -i \end{pmatrix}.$$

A complex number z can therefore be represented as a matrix in an obvious way either by

$$Z' := \begin{pmatrix} z & 0 \\ 0 & z \end{pmatrix}$$

or equivalently by

$$Z'' := \begin{pmatrix} z & 0 \\ 0 & \bar{z} \end{pmatrix}$$

with the corresponding multivectors $(Re(z), 0, 0, 0, 0, 0, 0, -Im(z))$ and $(Re(z), 0, Im(z), 0, 0, 0, 0, 0)$, respectively. Although outside our main focus, we should remark as a warning, that none of them gives a multivector representation of complex numbers in $\mathcal{C}_{1,2}$, because complex multiplication is not preserved. For a complex Vahlen matrix V neither $VZ'V^{\tilde{}}$ nor $VZ''V^{\tilde{}}$ represent a Möbius transformation in general.

The right embedding to choose is

$$Z := \begin{pmatrix} z & z\bar{z} \\ 1 & \bar{z} \end{pmatrix}, \tag{12.40}$$

which can be deduced by using the concept of paravectors, mentioned earlier in section 3.2.
The corresponding multivector is then given by

$$(Re(z), \frac{1}{2}(1 + z\bar{z}), Im(z), \frac{1}{2}(1 - z\bar{z}), 0, 0, 0, 0)\,.$$

Applying now a complex Vahlen matrix $\begin{pmatrix} a & b \\ c & d \end{pmatrix}$ as a spinor to Z one obtains

$$\begin{aligned} &\begin{pmatrix} a & b \\ c & d \end{pmatrix} \begin{pmatrix} z & z\bar{z} \\ 1 & \bar{z} \end{pmatrix} \begin{pmatrix} a & b \\ c & d \end{pmatrix} \\ &= \begin{pmatrix} a & b \\ c & d \end{pmatrix} \begin{pmatrix} z & z\bar{z} \\ 1 & \bar{z} \end{pmatrix} \begin{pmatrix} \bar{d} & \bar{b} \\ \bar{c} & \bar{a} \end{pmatrix} \\ &= \lambda \begin{pmatrix} z' & z'\bar{z}' \\ 1 & \bar{z}' \end{pmatrix} \end{aligned}$$

where $\lambda = |bz + d|^2$ and $z' = (az + b)(cz + d)^{-1}$. Thus, we have found the spinor representation of a complex Möbius transformation in $\mathcal{C}_{1,2}$. With some effort λ could be expressed only in terms of the parameters of the Möbius transformation. Therefore, we can speak of it as a scaling factor.

Experiment 4 (Möbius transformation).
As an example, we will now study how the Möbius transformation

$$z \mapsto \frac{0.5(1 + i)z + 0.5(1 - i)}{-0.5(1 + i)z + 0.5(1 - i)}$$

can be learned by $\mathcal{C}_{1,2}$-Neurons, but not by real valued neurons.

The training and test data used for this task is shown below in Fig. 12.14.

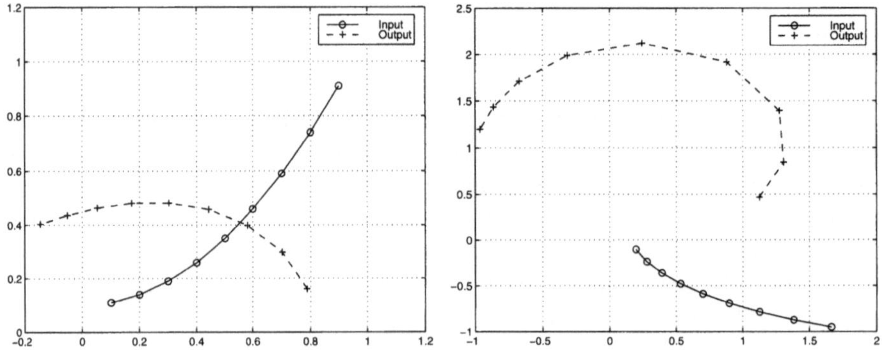

Fig. 12.14. Training data (left) and test data (right)

Neither on the pure complex input-output pairs nor on the coded data (12.40) an MLP can generalize the transformation. In the first case a training SSE of 0.00326 results in a generalization SSE of 6.15109 on the test set, in the second case a generalization SSE of 2.31305 was reached, although the training SSE was only 0.00015. So the MLP has in both cases just memorized the training examples, but not learned the structure within the data, because of its missing abilities to do so, namely the to embed the 4-dimensional data correctly in the required 8-dimensional space. This was done, as theoretically derived, by the Clifford neurons with nice robustness with respect to noise as shown in Figs. 12.15 and 12.16, respectively.

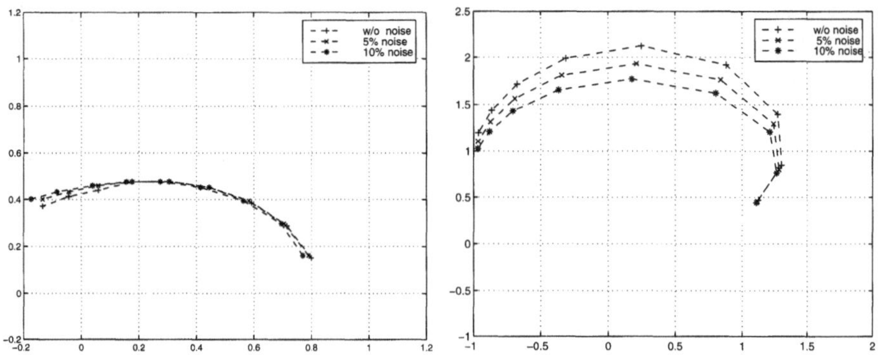

Fig. 12.15. Learned transformation on training data (left) and test data(right)

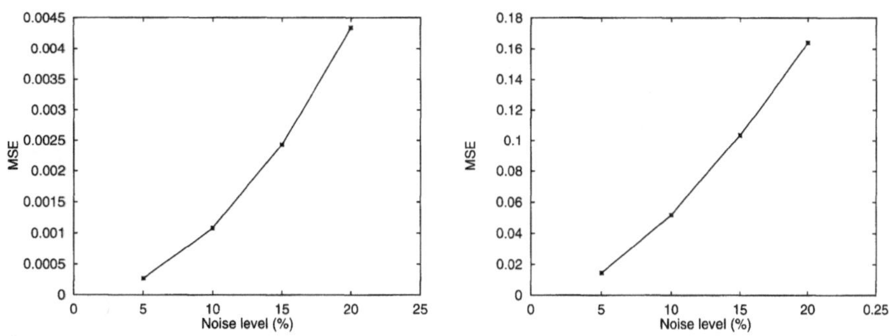

Fig. 12.16. MSE vs. noise level for training (left) and testing (right)

12.6 Summary

In this first of two chapters on Clifford neural computation we discussed the Clifford neuron in detail. We showed how the geometric product can be used with an associator. We introduced the special neuron model of spinor neurons that allows to compute orthogonal transformations very elegantly. This way of computation was proven to be faster and much more robust against noise as real single–layer neural networks. Moreover, we were able to show on the example of Möbius transformations that there exist geometric transformations that are only exclusively computable by Clifford neurons. This was done by using a non–linear coding of the data which resulted in a linearization in Clifford algebras.

We now will make the transition from the Clifford neuron and linearity to Clifford neural networks and non–linearity in the subsequent chapter.

13. Clifford Algebra Multilayer Perceptrons

Sven Buchholz and Gerald Sommer*

Institute of Computer Science and Applied Mathematics,
Christian-Albrechts-University of Kiel

13.1 Introduction and Preliminaries

Multilayer Perceptrons (MLPs) are one of the most common and popular neural architectures. They are widely used in many different areas like handwriting recognition, speech recognition, and time series prediction for instance. In this chapter, we will extend MLPs from the domain of real numbers to Clifford algebra domains.

MLPs consist of Perceptron–type neurons as processing units grouped together in layers. The computation in an MLP is feed forward only. The neurons processing the input to the net are grouped in the *input layer*. The output of the net is taken from the output neurons grouped in the output layer. Usually, there are also one or more layers between input and output layer called *hidden layers*, since they are not visible from the outside. Input neurons are just for making the data available to the net, they do not perform a computation. Any other single neuron computes as its so–called propagation function a weighted sum of its received inputs. Thus, the association of the weights and the inputs is linear. Nonlinearity is achieved by applying a so-called activation function g. The computation of such a neuron is therefore given by

* This work has been supported by DFG Grants So-320-2-1 and So-320-2-2.

$$y = g(\sum_{i=1}^{n} w_i x_i + \theta) \tag{13.1}$$

in the real case ($w, x \in \mathbb{R}^n, \theta \in \mathbb{R}$) and by either

$$\boldsymbol{y} = \boldsymbol{g}(w \otimes_{p,q} x + \theta) \tag{13.2}$$

or by

$$\boldsymbol{y} = \boldsymbol{g}(x \otimes_{p,q} w + \theta) \tag{13.3}$$

in the general case of a Clifford algebra ($w, x, \theta \in \mathcal{C}_{p,q}$) using the *geometric product* $\otimes$ as associator. The Clifford neuron in comparison with the real neuron was fully discussed in chapter 12. There we assumed g to be the identity to discuss propagation functions and linear aspects exclusively.

A suitable nonlinear activation function g allows to built powerful neural networks out of real neurons by using the superposition principle

$$y = \sum_i \lambda_i \, g_i(x) \,. \tag{13.4}$$

Thus, one hidden layer may be sufficient and no activation function in the output neurons is needed. This transforms directly in the MLP architecture shown below in Figure 13.1.

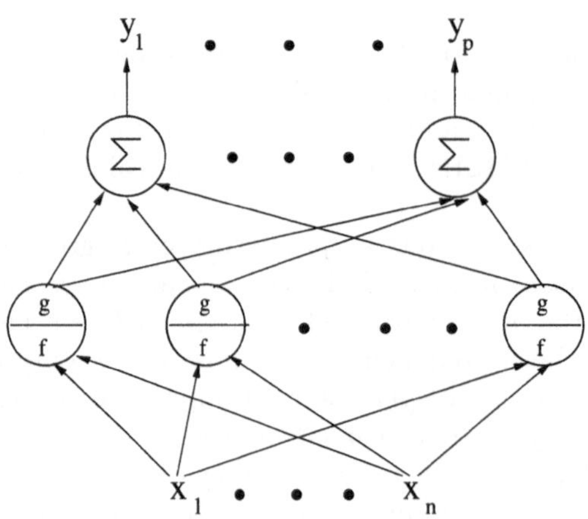

Fig. 13.1. *MLP with one hidden layer*

Cybenko proved in [53] that for so–called sigmoidal activation functions MLPs are universal approximators of continuous functions. In [117] these results were extended to the class of Borel measurable functions. Universal approximation in that sense means, that for any required approximation accuracy an MLP with one hidden layer with finite number of neurons is sufficient.

The graph representation (Figure 13.1) of a neural network is called its *topology*. Since an MLP is fully connected, its topology is therefore fully determined by the sequence of the number of nodes in any layer starting from the input layer.

Throughout this chapter let N denote the number of neurons in the input layer, M denote the number of neurons in the hidden layer, and P denote the number of neurons in the output layer, respectively. Hence, we can speak of an (N, M, P)–MLP to denote the topology completely.

Let us denote the other parameters of the network according to Table 13.1.

Table 13.1. *Notations of MLP parameters*

- w^1_{nm} weight connecting the n–th input neuron to the m–th hidden neuron
- w^2_{mp} weight connecting the m–th hidden neuron to the p–th output neuron
- θ^1_m bias of the m–th hidden neuron
- θ^2_p bias of the p–th output neuron

The detailed structure of the chapter is as follows. Starting with a mathematically precise formulation of required notions of approximation theory we will derive in section 2 a sufficient criterion on activation functions to make Clifford MLPs universal approximators as well. In section 3, we will study activation functions of Clifford MLPs in detail. Reviewing the real, complex, and quaternionic special cases and proposed activation functions in the literature, we will get a systematic survey of the topic. We will prove therein that Clifford MLPs with sigmoid activation functions in every multivector component are universal approximators. After this we will develop in the subsequent section a backpropagation algorithm for such Clifford MLPs. In the final section we will report experimental results.

13.2 Universal Approximation by Clifford MLPs

In the introduction we gave already an informal characterization of the universal approximation property. Let us start with the formalization thereof by introducing the notion of "denseness". Thereby we will use $\mathcal{N}$ to indicate the class of functions realizable by a certain neural network and $\mathcal{F}$ the class of function that shall be approximated by $\mathcal{N}$. Then, the concept of denseness can be defined as follows.

Definition 13.2.1. *Let F and N be sets of functions of a normed space (X, p). Let d be the metric induced by p. That is for all $x \in \mathcal{N}, y \in \mathcal{F}$ the distance is defined by $d(x, y) = p(x - y)$. Then N is dense under the norm p in F if, for any $f \in \mathcal{F}$ and any $\epsilon > 0$, there exists some $n \in \mathcal{N}$ with*

$$d(f, n) < \epsilon \,. \tag{13.5}$$

Thus, denseness is always measured with respect to some norm. Typical norms are the L_p norms $(1 < p < \infty)$

$$\|f\|_p = \left(\int_X |f(x)|^p dx \right)^{1/p} . \tag{13.6}$$

However, the most relevant norm in our case will be the supremum norm L_∞

$$\|f\|_\infty = \sup_{x \in X} |f(x)| \,. \tag{13.7}$$

A well known density theorem is the famous Weierstrass theorem of real analysis. It states that polynomials of one real variable are dense in the set $C^0([a, b], \mathbb{R})$. A generalization of this, the Stone-Weierstrass theorem, was used in [117] to prove the universal approximation capability of real valued MLPs. However, we cannot profit from these results in Clifford algebras. Moreover, these do not lead to a general density criterion of Clifford MLPs which is what we are looking for.

To reach this goal, we need functional analysis in Clifford algebras, especially an appropriate version of the Hahn-Banach theorem. The real and complex Hahn-Banach theorem has already been used in such a manner in [53] and accordingly in [6].

Let us first have an informal look at the Hahn-Banach theorem before going into technical details. In its dominated extended version it states the following.

Let M be a subspace of a linear space X over $\mathbb{R}$, let p be a sublinear functional defined on X and let f be a linear form defined on M dominated by p. The theorem asserts the existence of a linear extension F of f to X dominated by p everywhere. The diagram below gives an illustration of the statement of the Hahn-Banach theorem.

$$\begin{array}{lll} F : X & & F \le p \\ \quad | \searrow & & \\ f : M \to \mathbb{R} & & f \le p \end{array}$$

The basic idea now is the following. The neural architecture of f above is $\mathcal{N}$ and F is the class of functions it should be dense in. With this in mind we will get a nice criterion of denseness as a corollary from that theorem soon. To go ahead in this direction we have to formulate the theorem in terms of Clifford algebra. The functionality of a generic Clifford MLP is

$$\mathcal{C}^n_{p,q} \longrightarrow \mathcal{C}^m_{p,q} \,. \tag{13.8}$$

In general, $\mathcal{C}^n_{p,q}$ cannot be a linear space, since $\mathcal{C}_{p,q}$ itself is not a skew field in general. Thus, we have to replace the concept of a linear space with the algebraic weaker one of a module.

Definition 13.2.2. *Let R be a ring with 1. A left R-module $\mathcal{G}_l$ is an abelian group $G = (G, +)$ together with a mapping $R \times G_l \to G_l : (r, g) \mapsto rg$ in such a way, that*

$$\begin{array}{lll}
(a) & \forall g_1, g_2 \in G \; \forall r \in R : & r(g_1 + g_2) = rg_1 + rg_2 \\
(b) & \forall g \in G \; \forall r_1, r_2 \in R : & (r_1 + r_2)g = r_1 g + r_2 g \\
(c) & \forall g \in G \; \forall r_1, r_2 \in R : & (r_1 r_2)g = r_1(r_2 g) \\
(d) & \forall g \in G : & 1g = g
\end{array}$$

are fulfilled.

The corresponding definition of right modules is obvious. However, we only have to choose one version to be formally consistent without loss of generality. From now on we will always use left modules. To bound a function as required by the HAHN-BANACH theorem we next introduce the notion of a seminorm.

Definition 13.2.3. *Let X be a $\mathcal{C}_{p,q}$-module. A function $p : X \longrightarrow \mathbb{R}$ is called a seminorm on X if it fulfills for all $f, g \in X, \lambda \in \mathcal{C}_{p,q}$ and $\kappa \in \mathbb{R}$*

$$\begin{array}{ll}
(a) & p(f + g) \leq p(f) + p(g) \\
(b) & p(f) = 0 \Rightarrow f = 0 \\
(c) & p(\lambda f) \leq C\|\lambda\| p(f) \\
 & p(\kappa f) = |\kappa| p(f) \,.
\end{array}$$

The next definition gives then complete access to the needed concept of boundness.

Definition 13.2.4. *Let X be a $\mathcal{C}_{p,q}$-module. A family P of seminorms $p : X \longrightarrow \mathbb{R}$ is called a proper system of seminorms on X if for any finite sequence $p_1, p_2, \ldots, p_k \in P$ there exist $p \in P$ and $C > 0$ such that for all $f \in X$*

$$\sup_{j=1,\ldots,k} p_j(f) \leq C p(f) \,. \tag{13.9}$$

Hereafter, we will speak of a module equipped with a proper system of seminorms as a proper module for shortness. We are now in the position to formulate the HAHN-BANACH theorem of Clifford analysis.

Theorem 13.2.1. *Let X be a proper $C_{p,q}$-module, let Y be a submodule of X and let T be a bounded left $C_{p,q}$-functional on Y. Then there exists a bounded left $C_{p,q}$-functional T^* on X such that*

$$T^*_{|Y} = T\,. \tag{13.10}$$

For a proof of this theorem and the following corollary see again [30]. This corollary will give us now the desired density criterion.

Corollary 13.2.1. *Let X be a left proper $C_{p,q}$-module and Y a submodule of X. Then Y is dense in X, iff for all $T \in X^*$ with $T_{|Y} = 0$ follows $T = 0$ on X.*

Let us now return to our neural architecture $\mathcal{N}$ that should be dense in the function class $\mathcal{F}$. If it is not, then the closure $\overline{\mathcal{N}}$ is not completely $\mathcal{F}$. By corollary 13.2.1 of the Clifford HAHN-BANACH theorem there exists a bounded linear functional $L : \mathcal{F} \to C_{p,q}$, with $L(\overline{\mathcal{N}}) = L(\mathcal{N})$ and $L \neq 0$. Then furthermore, by the Clifford RIESZ theorem [30] there exists a unique Clifford measure μ on X such that for all $g \in C^0(X, C_{p,q})$

$$L(g) = \int_X g d\mu(x)\,. \tag{13.11}$$

Let us assume that the function g has the special property to be *discriminatory.*

Definition 13.2.5. *A function $g : C_{p,q} \to C_{p,q}$ is said to be discriminatory if*

$$\int_{I^{2^{p+q}}} g(w \otimes x + \theta) d\mu(x) = 0 \tag{13.12}$$

implies that $\mu(x) = 0$ for any finite regular Clifford measure μ with support $I^{2^{p+q}} := [0,1]^{2^{p+q}}$.

If g is discriminatory, then follows immediately by definition that $\mu(x) = 0$. But this is a contradiction to $L \neq 0$, which was a consequence of the assumption that $\mathcal{N}$ is dense in $\mathcal{F}$. Thus, we can conclude, that the use of a discriminatory activation function is sufficient to make Clifford MLPs universal approximators of $C^0(I^{2^{p+q}}, C_{p,q})$ functions.

13.3 Activation Functions

With the discriminatory property of the preceding section we have already a criteria on hand regarding the approximation capabilities of activation functions of Clifford MLPs. We will now turn our attention to properties necessary from the algorithmic point of view. We will start with the real case and then proceed to the multi–dimensional Clifford algebra case. We can thereby make extensive use of previous work by other authors in the complex and quaternionic case.

13.3.1 Real Activation Functions

Let us start with a general property an activation function has to fulfill independent of the concrete training algorithm. It is simply due to implementation aspects. The property in mind is boundness to avoid overflows during simulation on computers.

In the real case, this property is easy to check and expressed by so–called *squashing functions.* A function $g : \mathbb{R} \to \mathbb{R}$ is called a squashing function if $\lim_{x\to-\infty} g(x) = a$ and $\lim_{x\to\infty} g(x) = b$ for $a, b \in \mathbb{R}, a < b$.

Since backpropagation is gradient descent in the weight space of the MLP, the activation function has to be differentiable. A class of squashing functions with this property are the *sigmoid* functions.

Definition 13.3.1. *The function*

$$\sigma_\beta(x) : \mathbb{R} \to \mathbb{R}; \quad x \mapsto \frac{1}{1 + \exp(-\beta x)} \tag{13.13}$$

is called a sigmoid function.

Also the widely used hyperbolic tangent in real MLPs is only a slight modification of a sigmoid function, since

$$\tanh = 2\sigma_2 - 1\,. \tag{13.14}$$

In the following, we will only proceed with the most used activation function of real MLPs which is the so–called *logistic* function $\sigma := \sigma_1$. It has a very simple derivative $\dot{\sigma} = \sigma(1 - \sigma)$. Figure 13.2 shows a plot of the logistic function.

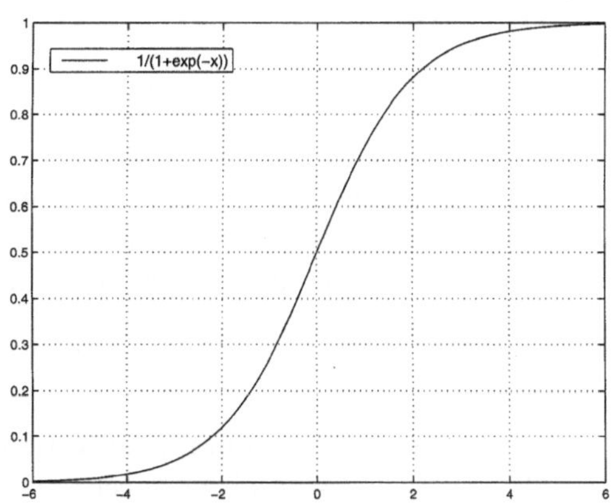

Fig. 13.2. *Logistic activation function*

The logistic function has approximately linear behaviour in $[-1, 1]$ and saturation is reached quickly outside of this region. To be complete at this point, we mention again that the universal approximator capability of real–valued MLPs with sigmoid activation function was first proven by Cybenko [53].

13.3.2 Activation Function of Clifford MLPs

There are two possible ways of generalization of real activation functions to Clifford algebra activation functions. One is to use the corresponding Clifford algebra formulation of such a function, the other is to use the real activation function in every multivector component separately. A formal characterization can be made in the following way.

Definition 13.3.2. *Let be $G : \mathcal{C}_{p,q} \to \mathcal{C}_{p,q}$ $(p+q>1)$, $n := 2^{p+q}$. G is called a component–wise activation function if*

$$\forall i \in \{1, \dots, n\}\ \exists g_i \in \mathbb{R} \to \mathbb{R}\ \forall x = (x_1, \dots, x_n) \in \mathcal{C}_{p,q} : [G(x)]_i = g_i(x_i)\,,$$

otherwise a multivector activation function.

Thus, in the complex case ($G : \mathbb{C} \to \mathbb{C}$) a multivector activation function has the generic form

$$G(z) = u\,(x, y) + u\,(x, y)\,\mathrm{i}\,. \tag{13.15}$$

On the other hand, the generic form of a component–wise activation function is given by

$$G(z) = v\,(x) + v\,(y)\,\mathrm{i}\,. \tag{13.16}$$

The use of a multivector activation function seems to be more natural and quiet more sophisticated in comparison to component activation functions. Hence, as the complex MLP was introduced, the first proposed activation function (see e.g. [147]) was the extension of the real logistic function σ to the complex domain

$$\sigma_{\mathbb{C}} : \mathbb{C} \to \mathbb{C};\ z \mapsto \frac{1}{1 + \exp(-z)}\,. \tag{13.17}$$

Later, it was pointed out by Georgiou and Koutsougeras [87] that $\sigma_{\mathbb{C}}$ is not bounded, since it has singularities with value $+\infty$ at $0 + \pi(2n+1)\,\mathrm{i}$ $(n \in \mathbb{N})$. Due to that fact, these authors proposed as alternative the activation function

$$G_{c,r} : \mathbb{C} \to \mathbb{C};\ z \mapsto \frac{z}{c + \frac{1}{r}\,|z|} \qquad (c, r \in \mathbb{R})\,. \tag{13.18}$$

They also gave a complete list of necessary properties that complex activation functions of the form (13.15) have to fulfill. One of these is with respect to the backpropagation algorithm, that all partial derivatives have to exist together with further conditions on them. Clearly, these requirements are also valid in the general case of an arbitrary Clifford algebra.

The first Clifford MLP introduced by Pearson [182] used the straightforward Clifford algebra domain extension of the above function (13.18). However, he has not paid any attention to the fact that Clifford algebras are not division algebras in general. The consequence (impossibility) for the formulation of a correct backpropagation algorithm is the subject of the next section. Also, no universal approximation theorem for such networks could yet be proven.

Instead, we will use in our Clifford MLP the real logistic activation function in any component. Let us denote the function derived in this manner by $\boldsymbol{\sigma}$. This function was first introduced by Arena et al. [7] in their work on the quaternionic MLP (QMLP). For this QMLP they further derived a quaternionic version of the backpropagation algorithm and proved its universal approximation capability. To give a proof sketch for the universal approximation capability of Clifford MLPs with activation function $\boldsymbol{\sigma}$ is the final part of this section.

Theorem 13.3.1. *The function*

$$\boldsymbol{\sigma}(w;\theta) : \mathcal{C}_{p,q} \to \mathcal{C}_{p,q}; \quad x \mapsto \sum_{A\in\mathcal{A}} \sigma([w \otimes x + \theta])_A e_A \tag{13.19}$$

is discriminatory.

Proof (Sketch). Let $\mu(x)$ be a finite regular Clifford measure on the set $C^0(I^{2^{p+q}}, \mathcal{C}_{p,q})$ such that

$$\int_{I^{2^{p+q}}} \boldsymbol{\sigma}(w \otimes x + \theta) d\mu(x) = 0\,, \tag{13.20}$$

for all $w, x \in \mathcal{C}^n_{p,q}, \theta \in \mathcal{C}_{p,q}$. According to the definition of $\boldsymbol{\sigma}$, we have for all $i \in \ 1, \ldots, 2^{p+q}\}$

$$[\boldsymbol{\sigma}(w \otimes x + \theta)]_i = \sigma([w \otimes x + \theta]_i)\,. \tag{13.21}$$

Let us now consider the pointwise limit

$$\phi_i(w \otimes x + \theta) := \lim_{\lambda\to\infty} \sigma(\lambda[w \otimes x + \theta]_i)\,, \tag{13.22}$$

with $\lambda \in \mathbb{R}$. This limit evaluates to

$$\phi_i(w \otimes x + \theta) = \begin{cases} 1 & : \; if \; [w \otimes x + \theta]_i > 0 \\ 0 & : \; if \; [w \otimes x + \theta]_i \leq 0 \end{cases} \tag{13.23}$$

With the *Lesbesgue*-dominated convergence theorem of Clifford analysis follows

$$\begin{aligned} 0 &= \int_{I^{2^{p+q}}} \sigma(w \otimes x + \theta) d\mu(x) \\ &= \int_{I^{2^{p+q}}} \Big(\sum_{A \in \mathcal{A}} \phi_A(w \otimes x + \theta) e_A\Big)\, d\mu(x) \\ &= \lim_{\lambda \to \infty} \sigma(\lambda [w \otimes x + \theta]_i)\,. \end{aligned}$$

For all $j \in \{0,1\}^{2^{p+q}}$ define the following sets

$$H_j := \bigcap_{\substack{i \in \{1,\dots,2^{p+q}\} \\ j[i]=1}} \{[w \otimes x + \theta]_i > 0\} \cap \bigcap_{\substack{i \in \{1,\dots,2^{p+q}\} \\ j[i]=0}} \{[w \otimes x + \theta]_i \leq 0\}\,. \tag{13.24}$$

Thus, the H_j sets give us a partition of $I^{2^{p+q}}$. Therefore, we have with (13.23), (13.24)

$$\mu(\cup H_j) = 0\,. \tag{13.25}$$

Unfortunately, no assumptions on μ can be made. Therefore, one has to prove that for all $j \in \{0,1\}^{2^{p+q}}$

$$\mu(H_j) = 0\,. \tag{13.26}$$

By the real theorem of Cybenko [53] we only know $\mu(H_{10\dots0}) = 0$. This can be extended with some effort to show that

$$\mu(\{H_j \mid \sum_{i=1}^{2^{p+q}} j[i] = 1\}) = 0 \tag{13.27}$$

However, the other cases remain open problems to be proved.

13.4 Clifford Back–Propagation Algorithm

In this section we will derive the Clifford back–propagation algorithm. For the sake of simplicity, we only deal with a Clifford MLP with one hidden layer, reminding the reader that this structure is already a universal approximator. Let N, M and P denote the number of input, hidden and output nodes, respectively. Furthermore, let be w^1_{nm} the multivector weight connecting the n-th input node with the m-th hidden node, and w^2_{mp} the one connecting the m-th hidden node with the p-th output node. Analogously, the bias nodes are denoted by θ^1_m and θ^2_p, respectively.

Using the above nomenclature the feed-forward phase is given as follows:

- hidden node activation and output value

$$S_m^1 := \sum_{n=1}^{N} w_{nm}^1 \otimes x_n + \theta_m^1 \tag{13.28}$$

$$h_m := \boldsymbol{\sigma}(S_m^1) \tag{13.29}$$

- output node activation and output value

$$S_p^2 := \sum_{m=1}^{M} w_{mp}^2 \otimes h_m + \theta_p^2 \tag{13.30}$$

$$o_p := \boldsymbol{\sigma}(S_p^2)\,. \tag{13.31}$$

We will now apply gradient descent with respect to the weights to minimize the common sum–of–squared error function

$$E = \frac{1}{2}\sum_{p=1}^{P}(y_p - o_p)^2\,, \tag{13.32}$$

whereby $y = (y_1, \ldots, y_P)$ stands for the expected output value. First, we have to compute the weights of the output layer according to

$$\nabla E_{w_{mp}^2} = \sum_{A \in \mathcal{A}} \frac{\partial E}{\partial [w_{mp}^2]_A}\, e_A\,. \tag{13.33}$$

Applying the chain rule to each term of (13.33) gives

$$\frac{\partial E}{\partial [w_{mp}^2]_A} = \sum_{B \in \mathcal{A}} \frac{\partial E}{\partial [S_p^2]_B} \frac{\partial [S_p^2]_B}{\partial [w_{mp}^2]_A}\,. \tag{13.34}$$

For the partial derivatives of the error function wrt. the output node activation S_p^2 we obtain

$$\frac{\partial E}{\partial [S_p^2]_B} = \frac{\partial E}{\partial [y_p]_B} \frac{\partial [y_p]_B}{\partial [S_p^2]_B} = ([y_p]_B - [o_p]_B)\, \dot{\boldsymbol{\sigma}}([S_p^2]_B)\,. \tag{13.35}$$

The computation of the partial derivatives of the output node activation wrt. the output layer weights is as easy to compute. However, some effort has to be made to get one single compact formula for it. Let us take a look at the case of 4–dimensional Clifford algebras. Table 13.2 shows exemplarily the partial derivatives $\frac{\partial [S_p^2]}{\partial [w_{mp}^2]_{e_1}}$.

Table 13.2. *Partial derivatives* $\frac{\partial[S_p^2]}{\partial[w_{mp}^2]_{e_1}}$

	$\mathcal{C}_{0,2}$	$\mathcal{C}_{1,1}$	$\mathcal{C}_{2,0}$
$\frac{\partial[S_p^2]_0}{\partial[w_{mp}^2]_{e_1}}$	$-[h_m]_{e_1}$	$+[h_m]_{e_1}$	$+[h_m]_{e_1}$
$\frac{\partial[S_p^2]_{e_1}}{\partial[w_{mp}^2]_{e_1}}$	$+[h_m]_0$	$+[h_m]_0$	$+[h_m]_0$
$\frac{\partial[S_p^2]_{e_2}}{\partial[w_{mp}^2]_{e_1}}$	$+[h_m]_{e_{12}}$	$-[h_m]_{e_{12}}$	$+[h_m]_{e_{12}}$
$\frac{\partial[S_p^2]_{e_{12}}}{\partial[w_{mp}^2]_{e_1}}$	$-[h_m]_{e_2}$	$-[h_m]_{e_2}$	$+[h_m]_{e_2}$

It is easy to conclude from the above example (and also easy to verify directly), that

$$\frac{\partial[S_p^2]}{\partial w_{mp}^2} = h_m^* \tag{13.36}$$

for some involution * dependent on the underlying Clifford algebra. Clearly, this involution is already determined uniquely by any partial derivative $\frac{\partial[S_p^2]}{\partial[w_{mp}^2]_A}$. For Clifford algebras of the type $\mathcal{C}_{(0,q)}$ this involution is just conjugation, i.e. we have $h_m^* = \overline{h_m}$ in (13.36). Due to the fact that the geometric product of a multivector with a scalar is ordinary component–wise real multiplication we can get a very elegant description of the involution * via the scalar component. We can then use the fact

$$[x \otimes \bar{y}]_0 = xy^T \tag{13.37}$$

to describe * as the unique involution yielding

$$[x \otimes y^*]_0 = xy^T\,. \tag{13.38}$$

Putting now all the derived results together and using the symbol $\odot$ to denote component–wise multiplication, we get the following update rule for the weights of the output layer

$$\Delta w_{mp}^2 = [\,\underbrace{(y_p - o_p) \odot \dot{\sigma}(S_p^2)}_{\delta_p^2}\,] \otimes \overline{h}_m\,. \tag{13.39}$$

The derivation of the updating rule for the weights of the hidden layer is analog, resulting in

$$\Delta w_{nm}^1 = \underbrace{[\,(\sum_{p=1}^{P} \overline{w}_{mp}^2 \otimes \delta_p^2) \odot \dot{\sigma}(S_m^1)\,]}_{\delta_m^1}\,.\otimes \overline{x}_n \tag{13.40}$$

Finally, the update rule for the biases is then given by

$$\Delta \theta_p^2 = \delta_p^2 \quad \text{and} \quad \Delta \theta_m^1 = \delta_m^1 \,. \tag{13.41}$$

Let us now verify briefly our claim made in section 3 regarding the impossibility of a general correctly Clifford back–propagation algorithm for not component–wise activation functions. This is simply due to the existence of divisors of zero in general Clifford algebras. Using non component–wise activation functions results in a geometric product $\otimes$ instead of a component–wise product $\odot$ in (13.39), respectively (13.40). Thus, δ_m^1 and δ_p^2 could then always be zero even if a non–zero error occurred. Due to the definition of * this can never be the case for the geometric products involving *.

The above derived Clifford back-propagation rule therefore avoids problems with divisors of zero completely.

13.5 Experimental Results

Both real MLPs and Clifford MLPs (CMLPs) are universal approximators of continuous functions in several variables as we know from the previous sections. Thus, they have the same theoretical strength in principle. Moreover, they use the "same" activation function, since our Clifford MLP uses the logistic function σ in every component. However, alternative activation functions are rare as argued before. Thus, a potential advantage of CMLPs versus MLPs seems to be based on the propagation function, i.e. on the involved geometric product. The propagation function was fully discussed in chapter 12, however only in the case of a single linear neuron. As we know from that chapter, in a Clifford MLP real vector data can be presented in many arbitrary different ways. But it is difficult to give general advises for Clifford MLPs for the optimal choice, especially due to the incorporated non–linearity.

However, this is only valid in a theoretically provable sense. In this section instead, we try to conclude from an experimental approach. Thereby, we will compare the space and time complexity of the real MLP and the CMLP with respect to their generalization performance. Time complexity is used in the loose sense of convergence time.

Space complexity is measured by the amount of real parameters. This is given for an MLP with one hidden layer by the formula

$$\#MLP := M \cdot (N+1) + P \cdot (M+1) \,. \tag{13.42}$$

The weights of an CMLP can be easily converted into real parameters by counting them component by component. Thus, one obtains for the number of real parameters of an CMLP with one hidden layer

$$\#CMLP_{(p,q)} := 2^{p+q} \cdot M \cdot (N+1) + 2^{p+q} \cdot P \cdot (M+1) \,. \tag{13.43}$$

An MLP with the same number of component activation functions as an CMLP would have 2^{p+q} times of real parameters, which follows easily from (13.42), (13.43). Clearly, the assumption that an MLP would require the same amount of activation functions as an CMLP to achieve the same performance is not realistic. However, about 20–25% fewer parameters of an CMLP in comparison to an MLP where reported by Arena et al. [6], [7] and this result was also obtained in earlier work of ours [15]. However, it is not easy to find another reason for this phenomenon than the more compact weight structure of CMLPs, especially in the case of processing real data. Thus, we will not make a simulation of the approximation of a real vector function, but one of a Clifford–valued function. The main goal of the simulations is to check, whether or not there are indications of algebraic model–based behavior of Clifford MLPs.

Let us study only one very simple example. The considered function is

$$f : \mathbb{C} \to \mathbb{C};\ (x + y\,\mathrm{i}) \mapsto (x^2 + y^2 + 2xy\,\mathrm{i})\,, \tag{13.44}$$

which is plotted in Figure 13.3.

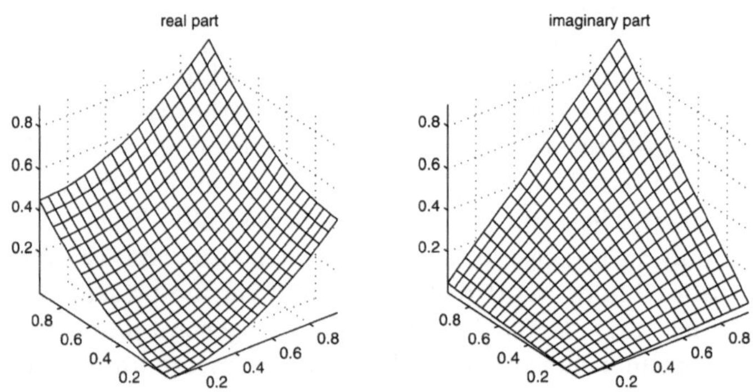

Fig. 13.3. *Plot of the real and the imaginary parts of f*

There is a very good reason to choose a low dimensional example. Namely, everything can be visualized. This is clearly helpful, to get a real evaluation of the performance of different networks. As outlined, we want to investigate whether or not there might be indications of algebraic interpretations of the approximation capability of the $\mathrm{CMLP}_{(0,1)}$. If on the other hand a $\mathrm{CMLP}_{(1,0)}$ achieved an equal or better performance then an algebraic reason would have to be rejected. Clearly, also if this were true for a real MLP. If both Clifford networks showed equivalent but better performance than the real MLP, this would then only be due to their more compact weight structure. As in the simulations of chapter 12 we will also have a closer look at the performance of the networks in the presence of noise.

Next, we report the obtained results in detail.

The training data consisted of 100 randomly drawn points from $[0,1] \times [0,1]$ with uniform distribution. For the test set we sampled this domain with a regular grid of size 20×20. Thus, we got 100 training points and 400 test points. This 20%/80% ratio of samples is well established and therefore often used in neural network simulations.

The number of hidden nodes of the Clifford networks was easy to determine. Two hidden nodes already gave results, which could not be improved significantly by the use of more hidden nodes. The performance of an MLP with two hidden nodes was not sufficient. The convergence of the training of the two CMLPs, a (2,3,2)-MLP and a (2,4,2)-MLP are shown in Figure 13.4.

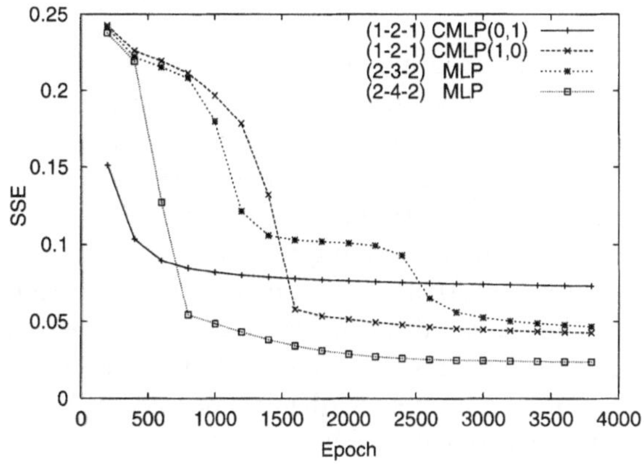

Fig. 13.4. *Convergence of the learning*

A first look at these results seems to be quite astonishing. The complex MLP shows the worst performance of all networks, followed by the $\text{CMLP}_{(1,0)}$. Taking into account that the number of parameters of the (2,3,2)-MLP is 17 ((2,4,2)-MLP: 22) and that of both CMLPs is 14, no advantage seems visible from the space complexity point of view either. We should remark, that all networks have reached a stable and optimal error plateau.

As already observed in the simulations on Clifford neurons in chapter 12, generalization is the measure that counts indeed. The obtained sum-of-squared errors (SSE) in training and testing are reported together in Table 13.3.

Table 13.3. *SSEs by noise free training*

network	SSE Training	SSE Testing
(1,2,1)-CMLP$_{(0,1)}$	0.07302	0.00040
(1,2,1)-CMLP$_{(1,0)}$	0.04175	0.00067
(2,4,2)-MLP	0.00971	0.03298
(2,3,2)-MLP	0.04315	0.15539

As we can see there, the complex MLP showed the best generalization performance of all nets, followed closely by the CMLP$_{(1,0)}$. Both generalization errors are very low, and remarkable orders smaller than the trainings errors. The MLPs on the other hand have both approximately 4-times higher generalization errors than training errors. With the generalization errors the situation has changed completely.

The complex MLP has reached the training error level corresponding to this superb generalization error very quickly within about 1000 epochs. Thus, we could say that it converges fastest. Another interesting fact observable from Figure 13.4 is that all other nets except the complex MLP show the same behavior during the beginning of the training. Hence, we could argue that the complex CMLP could match the underlying model of the data very early in the training, whereas both real MLPs have not observed the right model as can be concluded from their generalization error. The rapid descent of the MLP with 4 hidden nodes in comparison to that with 3 hidden nodes is clearly due to its greater amount of parameters (degrees of freedom).

Let us now have a closer look at the output of the networks shown in Figure 13.5. Especially compare the numerically nearly identical generalization of both CMLPs. Thereby, areas of high approximation errors are indicated by light shading in Figure 13.5.

Between them there are no great differences visible in fact. Both have learned the two component functions indeed, with less accuracy on the imaginary one. However, the learned real and imaginary functions of the MLPs are similar and (thus) far away from the expected shapes. Obviously, the MLPs have applied a global numerically concept to match the data, without notice to the structure of the data. The effect of the 4–th MLP hidden node is also easy to interpret. With only 3 hidden nodes the MLP learned a similar concept as with 4 hidden nodes. However, it decided to not descend down to zero height to approximate the range $[0, 0.3]^2$ accurately. Since the values of the function are low in this area an error is "cheap" with respect to the other areas.

Thus, an MLP is not able to detect the algebraic structure of the data. But the CMLP$_{(1,0)}$ seems to be able to do so. We should remember that there is only one sign in the multiplication tables that makes both 2-dimensional Clifford algebras different.

In the following we discuss how things changed in the presence of noise. We therefore added 20% mean–free noise to the training data. The obtained errors are presented in Table 13.4.

Table 13.4. *SSEs by 20% noise in training data*

network	SSE Training	SSE Testing
(1,2,1)-CMLP$_{(0,1)}$	1.32612	0.72041
(1,2,1)-CMLP$_{(1,0)}$	1.57850	1.15853
(2,4,2)-MLP	1.24180	0.74325
(2,3,2)-MLP	1.51760	1.12471

The errors of the (2,4,2)-MLP and the (1,2,1)-CMLP$_{(0,1)}$ are nearly equal. The same can be said about the errors of the (2,3,2)-MLP and the (1,2,1)-CMLP$_{(1,0)}$. The (1,2,1)-CMLP$_{(0,1)}$ shows now a better performance than the (1,2,1)-CMLP$_{(1,0)}$. Clearly, that of the MLP with 4 hidden nodes is better than that with 3 hidden nodes again.

Let us study the outputs of the networks shown in Figure 13.6 beginning with that of the real MLPs. We can see a clear negative effect of the 4 hidden nodes there. In the presence of noise, this "additional" degree of freedom is just used to learn the noise. Actually, any simulation of an (2,4,2)-MLP with high noisy training data can produce an arbitrary output scheme. The applied concept leads then no longer to equally well learned component functions. This is still the case for the (2,3,2)-MLP. The real part in the case of the (2,4,2)-MLP shows an additional scaling error, while its imaginary part fits randomly the imaginary component function.

An important but well known conclusion from the simulations is that more degrees of freedom in an MLP (which cannot match the model of data) are only good for memorization. Clearly, things then get worse very quickly in the presence of noise.

The difference between both CMLPs does not seem too large again. However, it is significant as seen in Table 13.4. The imaginary part of the function in the range $[0, 0.2] \times [0, 0.8]$ is better approximated by the complex CMLP.

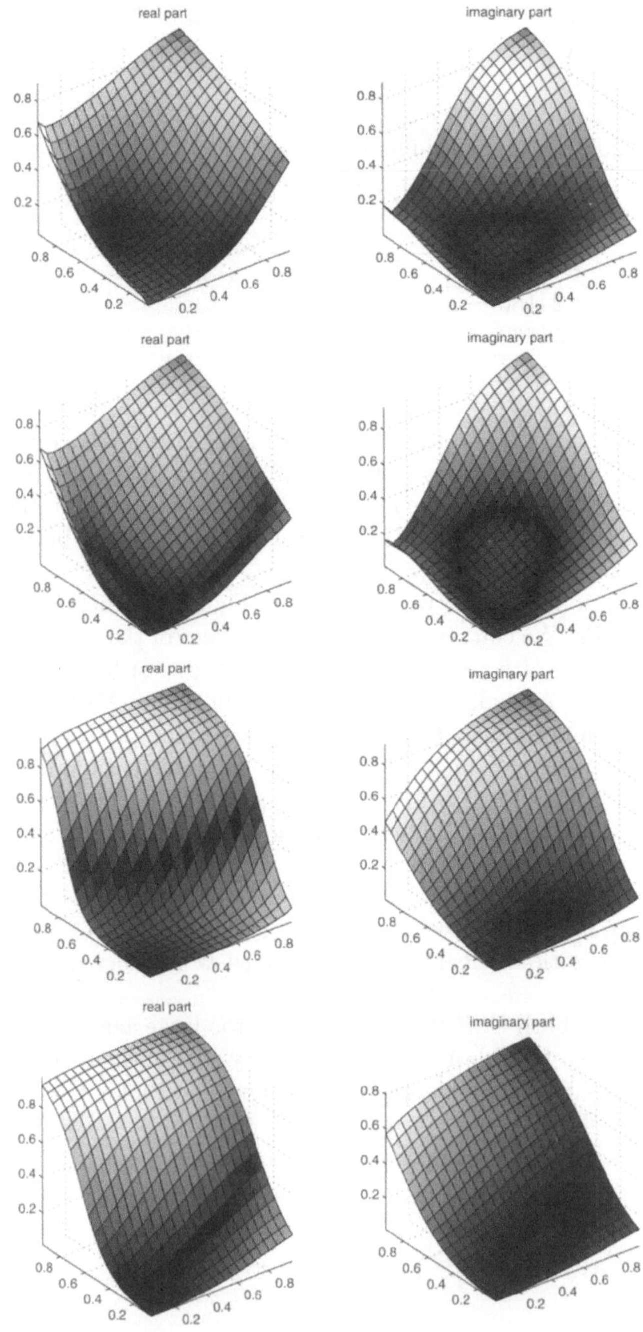

Fig. 13.5. *Approximation results (from top to bottom):* *(1,2,1)-CMLP*$_{(0,1)}$, *(1,2,1)-CMLP*$_{(1,0)}$, *(2,4,2)-MLP, (2,3,2)-MLP*

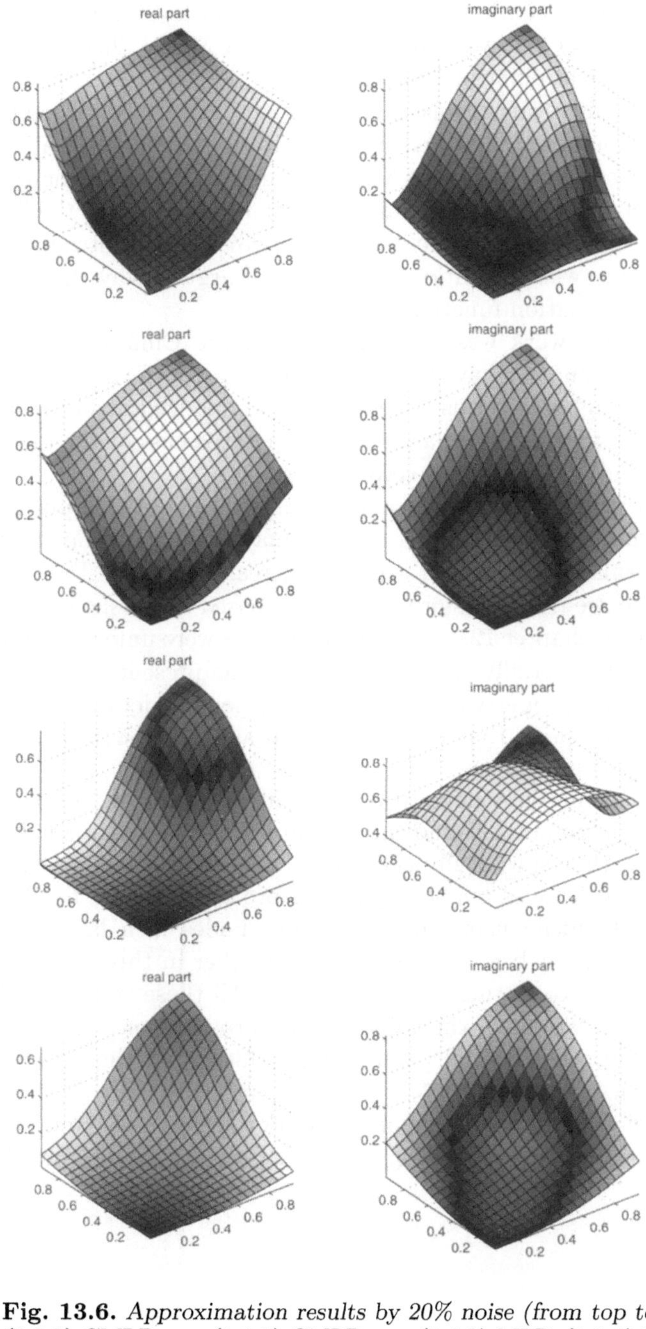

Fig. 13.6. *Approximation results by 20% noise (from top to bottom): (1,2,1)-CMLP$_{(0,1)}$, (1,2,1)-CMLP$_{(1,0)}$, (2,4,2)-MLP, (2,3,2)-MLP*

13.6 Conclusions and Outlook

In this chapter we introduced the Clifford Algebra Multilayer Perceptron (CMLP) as an extension of the well known real–valued MLP. Thereby, we applied mainly a theoretical point of view.

We discussed questions regarding the theory of function approximation in Clifford algebras in some detail. This led us to a criterion on the activation function that guarantees that CMLPs as MLPs are universal approximators too. We then reviewed basic facts on activation functions known from the literature in the complex and quaternionic special cases [87] [7]. We introduced the notion of component–wise activation functions and argued why this is a necessary property of activation functions for CMLPs.

A central part of this work was the derivation of the Clifford algebra back–propagation algorithm. We have found an elegant way to formulate the updating rules of the weights in terms of generally characterized involutions. The properties of these involutions guarantee the operativeness of the algorithm because excluding problems with zero divisors.

Although, concentrating on theoretical and technical aspects of computing with Clifford MLPs throughout this paper, we also made a simple simulation to compare the performance of Clifford MLPs with real–valued MLPs. Our interest was thereby to see if CMLPs are also model–based as we showed for single Clifford neurons in chapter 12. The obtained results were unfortunately weaker in that sense and partially showed up only in the presence of noise. However, the model–based property of CMLPs is not in general doubt. On the other hand it is also very clear, that non–linearity in any MLP architecture makes things less easy to interpret.

Thus, many more simulations have to be done in the future to get empirical confidence at this points. It is indicated, that such simulations should be biased in that way, that specifically geometric tasks are chosen for Clifford MLPs. However, such tasks might require a more suitable and flexible architecture. This could mean using inhomogeneous nodes in a layer, i.e. nodes of different Clifford algebra type. A step still further in this direction would be the use of nodes operating on single blades. All these steps would only require small modifications of the Clifford back–propagation algorithm as derived herein. From a conceptual point of view this would then invoke questions of self–organization.

It is our strong belief that a way based on this work towards more complex Clifford neural computation is worth being considered and would be fruitful.

Part III

Geometric Algebra for Computer Vision and Robotics

14. A Unified Description of Multiple View Geometry

Christian B.U. Perwass[1] and Joan Lasenby[2]

[1] Cavendish Laboratory, Cambridge
[2] C. U. Engineering Department, Cambridge

14.1 Introduction

Multiple view tensors play a central role in many areas of Computer Vision. The *Fundamental Matrix*, *Trifocal Tensor* and *Quadfocal Tensor* have been investigated by many researchers using a number of different formalisms. For example, standard matrix analysis has been used in [102] and [210]. An analysis of multiple view tensors in terms of Grassmann-Cayley (GC) algebra can be found in [82], [179], [80]. Geometric Algebra (GA) has also been applied to the problem [184], [185], [142], [141].

In this article we will show how Geometric Algebra can be used to give a unified *geometric* picture of multiple view tensors. It will be seen that with the GA approach multiple view tensors can be derived from simple geometric considerations. In particular, constraints on the internal structure of multiple view tensors will all be derived from the trivial fact that the intersection points of a line with three planes, all lie along a line. Our analysis will also show how closely linked the numerous different expressions for multiple view tensors are.

The structure of this article will be as follows. First we give a short introduction to projective geometry, mainly to introduce our notation. We then describe the Fundamental Matrix, the Trifocal Tensor and the Quadfocal Tensor in detail, investigating their derivations, inter-relations and other properties. Following on from these analytical investigations, we show how the self-consistency of a trifocal tensor influences its reconstruction quality. We end this article with some conclusions and a table summarising the main properties of the three multiple view tensors described here.

14.2 Projective Geometry

In this section we will outline the GA framework for projective geometry. We assume that the reader is familiar with the basic ideas of GA and is able to manipulate GA expressions.

We define a set of 4 orthonormal basis vectors $\{e_1, e_2, e_3, e_4\}$ with signature $\{---+\}$. The pseudoscalar of this space is defined as $I = e_1{\wedge}e_2{\wedge}e_3{\wedge}e_4$. A vector in this 4D-space ($\mathbb{P}^3$), which will be called a *homogeneous* vector, can then be regarded as a projective line which describes a point in the corresponding 3D-space ($\mathbb{E}^3$). Also, a line in $\mathbb{E}^3$ is represented in $\mathbb{P}^3$ by the outer product of two homogeneous vectors, and a plane in $\mathbb{E}^3$ is given by the outer product of three homogeneous vectors in $\mathbb{P}^3$. In the following, homogeneous vectors in $\mathbb{P}^3$ will be written as capital letters, and their corresponding 3D-vectors in $\mathbb{E}^3$ as lower case letters in bold face.

Note that the set of points $\{X\}$ that lie on a line $(A{\wedge}B)$ are those that satisfy $X{\wedge}(A{\wedge}B) = 0$. Similarly, a plane is defined through the set of points $\{X\}$ that satisfy $X{\wedge}(A{\wedge}B{\wedge}C) = 0$. Therefore, it is clear that if two lines, or a line and a plane intersect, their outer product is zero.

The projection of a 4D vector A into $\mathbb{E}^3$ is given by,

$$\boldsymbol{a} = \frac{A{\wedge}e_4}{A{\cdot}e_4}$$

This is called the *projective split*. Note that a homogeneous vector with no e_4 component will be projected onto the plane at infinity.

A set $\{A_\mu\}$ of four homogeneous vectors forms a basis or *frame* of $\mathbb{P}^3$ if and only if $(A_1 \wedge A_2 \wedge A_3 \wedge A_4) \neq 0$. The *characteristic pseudoscalar* of this frame for 4 such vectors is defined as $I_a = A_1{\wedge}A_2{\wedge}A_3{\wedge}A_4$. Note that $I_a = \rho_a I$, where ρ_a is a scalar. This and results relating the inner products of multivectors with the pseudoscalars of the space are given in [185].

Another concept which is very important in the analysis to be presented is that of the dual of a multivector X. This is written as X^* and is defined as $X^* = XI^{-1}$. It will be extremely useful to introduce the *dual bracket* and the *inverse dual bracket.* They are related to the bracket notation as used in GC algebra and GA, [141]. The bracket of a pseudoscalar P is a scalar, defined as the dual of P in GA. That is, $[P] = PI^{-1}$. The dual and inverse dual brackets are defined as

$$[\![A_{\mu_1} \cdots A_{\mu_n}]\!]_a \equiv (A_{\mu_1}{\wedge}\ldots{\wedge}A_{\mu_n})I_a^{-1} \tag{14.1a}$$

$$[\![A_{\mu_1} \cdots A_{\mu_n}]\!] \equiv (A_{\mu_1}{\wedge}\ldots{\wedge}A_{\mu_n})I^{-1} \tag{14.1b}$$

$$\langle\!\langle A_{\mu_1} \cdots A_{\mu_n} \rangle\!\rangle_a \equiv (A_{\mu_1} \wedge \ldots \wedge A_{\mu_n}) I_a \tag{14.2a}$$
$$\langle\!\langle A_{\mu_1} \cdots A_{\mu_n} \rangle\!\rangle \equiv (A_{\mu_1} \wedge \ldots \wedge A_{\mu_n}) I \tag{14.2b}$$

with $n \in \{0, 1, 2, 3, 4\}$. The range given here for n means that in $\mathbb{P}^3$ none, one, two, three or four homogeneous vectors can be bracketed with a dual or inverse dual bracket. For example, if $P = A_1 \wedge A_2 \wedge A_3 \wedge A_4$, then $[\![A_1 A_2 A_3 A_4]\!] = [\![P]\!] = [P] = \rho_a$.

Using this bracket notation the *normalized reciprocal A-frame*, written $\{A_a^{\mu}\}$, is defined as $A_a^{\mu_1} = [\![A_{\mu_2} A_{\mu_3} A_{\mu_4}]\!]_a$. It is also useful to define a *standard reciprocal A-frame*: $A^{\mu_1} = [\![A_{\mu_2} A_{\mu_3} A_{\mu_4}]\!]$. Then, $A_\mu \cdot A_a^\nu = \delta_\mu^\nu$ and $A_\mu \cdot A^\nu = \rho_a \delta_\mu^\nu$, where δ_μ^ν is the Kronecker delta. That is, *a reciprocal frame vector is nothing else but the dual of a plane.* In the GC algebra these reciprocal vectors would be defined as elements of a *dual space*, which is indeed what is done in [80]. However, because GC algebra does not have an explicit inner product, elements of this dual space cannot operate on elements of the "normal" space. Hence, the concept of reciprocal frames cannot be defined in the GC algebra.

A reciprocal frame can be used to transform a vector from one frame into another. That is, $X = (X \cdot A_a^\mu) A_\mu = (X \cdot A_\nu) A_a^\nu$. Note that in general we will use greek indices to count from 1 to 4 and latin indices to count from 1 to 3. We also adopt the convention that if a subscript index is repeated as a superscript, or vice versa, it is summed over its implicit range, unless stated otherwise. That is, $\sum_{\mu=1}^{4} (X \cdot A_a^\mu) A_\mu \equiv (X \cdot A_a^\mu) A_\mu$.

It will be important later not only to consider vector frames but also line frames. The A-line frame $\{L_a^i\}$ is defined as $L_a^{i_1} = A_{i_2} \wedge A_{i_3}$. The $\{i_1, i_2, i_3\}$ are assumed to be an even permutation of $\{1, 2, 3\}$. The *normalised reciprocal A-line frame* $\{\bar{L}_i^a\}$ and the *standard reciprocal A-line frame* $\{L_i^a\}$ are given by $\bar{L}_i^a = [\![A_i A_4]\!]_a$ and $L_i^a = [\![A_i A_4]\!]$, respectively. Hence, $L_a^i \cdot \bar{L}_j^a = \delta_j^i$ and $L_a^i \cdot L_j^a = \rho_a \delta_j^i$. Again, this shows the universality of the inner product: bivectors can be treated in the same fashion as vectors.

The *meet* and *join* are the two operations needed to calculate intersections between two lines, two planes or a line and a plane – these are discussed in more detail in [185], [141] and [114]; here we will give just the most relevant expression for the meet. If A and B represent two planes or a plane and a line in $\mathbb{P}^3$ their meet may be written as

$$A \vee B = \langle\!\langle [\![A]\!] [\![B]\!] \rangle\!\rangle = [\![A]\!] \cdot B \equiv (A I^{-1}) \cdot B \tag{14.3}$$

From this equation it also follows that

$$\langle\!\langle A \rangle\!\rangle \vee \langle\!\langle B \rangle\!\rangle = \langle\!\langle AB \rangle\!\rangle \tag{14.4}$$

Later on we will need the *dual representations* of points and lines. For lines they are given by,

$$L_a^{i_1} = A_{i_2} \wedge A_{i_3} \simeq \langle\!\langle A_a^{i_1} A_a^4 \rangle\!\rangle \quad \text{and} \quad A_{i_1} \wedge A_4 \simeq \langle\!\langle A_a^{i_2} A_a^{i_3} \rangle\!\rangle \tag{14.5}$$

The symbol $\simeq$ denotes equality up to a scalar factor. This shows that a line can either be expressed as the outer product of two vectors or by the intersection of two planes, since $\langle\!\langle A_a^{i_1} A_a^4 \rangle\!\rangle = \langle\!\langle A_a^{i_1} \rangle\!\rangle \vee \langle\!\langle A_a^4 \rangle\!\rangle$. Similarly, for points we have

$$A_{\mu_1} \simeq \langle\!\langle A_a^{\mu_2} A_a^{\mu_3} A_a^4 \rangle\!\rangle \tag{14.6}$$

That is, a point can also be described as the intersection of three planes.

A pinhole camera can be defined by 4 homogeneous vectors in $\mathbb{P}^3$: one vector gives the optical centre and the other three define the image plane [142], [141]. Thus, the vectors needed to define a pinhole camera also define a frame for $\mathbb{P}^3$. Conventionally the fourth vector of a frame, eg. A_4, defines the optical centre, and the outer product of the other three defines the image plane.

Suppose that X is given in some frame $\{Z_\mu\}$ as $X = \zeta^\mu Z_\mu$, it can be shown [185] that the projection of some point X onto image plane A can be written as

$$X_a = (X \cdot A^i) A_i = (\zeta^\mu Z_\mu \cdot A^i) A_i = \zeta^\mu K_{\underset{\mu}{i}} A_i ; \quad K_{\underset{\mu}{i}} \equiv Z_\mu \cdot A^i \tag{14.7}$$

The matrix $K_{\underset{\mu}{i}}$ is the *camera matrix* of camera A, for projecting points given in the Z-frame onto image plane[1] A. In general we will write the projection of some point X onto image plane P as $X \xrightarrow{P} X_p$.

In [80] the derivations begin with the camera matrices by noting that the row vectors *refer* to planes. As was shown here, the row vectors of a camera matrix are the reciprocal frame vectors $\{A^i\}$, whose dual *is* a plane.

With the same method as before, lines can be projected onto an image plane. For example, let L be some line in $\mathbb{P}^3$, then its projection onto image plane A is: $(L \wedge A_4) \vee (A_1 \wedge A_2 \wedge A_3) = (L \cdot L_i^a) L_a^i$.

An *epipole* is the projection of the optical centre of one camera onto the image plane of another. Therefore epipoles contain important information about the relative placements of cameras.

As an example consider two cameras A and B represented by frames $\{A_i\}$ and $\{B_i\}$, respectively. The projection of the optical centre of camera B onto image plane A will be denoted E_{ab}. That is, $E_{ab} = B_4 \cdot A^i A_i$ or simply $E_{ab} = \varepsilon_{ab}^i A_i$, with $\varepsilon_{ab}^i \equiv B_4 \cdot A^i$. Note, that we adopted the general GA convention that the inner product takes precedence over the geometric product[2].The only other epipole in this two camera set-up is E_{ba} given by

[1] Note that the indices of K are not given as super- and subscripts of K but are raised (or lowered) relative to each other. This notation was adopted since it leaves the superscript position of K free for other usages.

[2] Also, the outer product has precedence over the inner product. That is, $A \cdot B \wedge C = A \cdot (B \wedge C)$.

$E_{ba} = A_4 \cdot B^i B_i$. This may also be written as $E_{ba} = \varepsilon^i_{ba} B_i$, with $\varepsilon^i_{ba} \equiv A_4 \cdot B^i$. If there are three cameras then each image plane contains two epipoles. With four cameras each image plane contains three epipoles. In general the total number of epipoles is $N(N-1)$ where N is number of cameras present.

Let $\{B_\mu\}$ define a camera in $\mathbb{P}^3$ and $\{A_\mu\}$ be some other frame of the same projective space. Also, define A_4 to be the origin of $\mathbb{P}^3$. Then E_{ba} contains some information about the placement of camera B relative to the origin. Therefore, $A_4 \cdot B^j$ may be regarded as a *unifocal* tensor U_b.

$$U^i_b \equiv \varepsilon^i_{ba} = A_4 \cdot B^i = K^b_{i_4} \simeq \langle\!\langle A^1 A^2 A^3 B^i \rangle\!\rangle \tag{14.8}$$

Obviously the unifocal tensor is of rank 1. The definition of a unifocal tensor is only done for completeness and is not strictly necessary since every unifocal tensor is also an epipole vector.

Later on we will have to deal with determinants of various 3×3 matrices. Such a determinant can be written in terms of the ϵ_{ijk} operator, which is defined as

$$\epsilon_{ijk} = \begin{cases} +1 \text{ if the } \{ijk\} \text{ form an even permutation of } \{123\} \\ 0 \quad \text{if any two indices of } \{ijk\} \text{ are equal} \\ -1 \text{ if the } \{ijk\} \text{ form an odd permutation of } \{123\} \end{cases} \tag{14.9}$$

Let $\alpha_1^{i_a}$, $\alpha_2^{i_b}$ and $\alpha_3^{i_c}$ give the three rows of a 3×3 matrix M. Then the determinant of M is $\det(M) = \epsilon_{i_a i_b i_c} \alpha_1^{i_a} \alpha_2^{i_b} \alpha_3^{i_c}$. Note that there is an implicit summation over all indices. It will simplify the notation later on if we define

$$\det(\alpha_1^{i_a}, \alpha_2^{i_b}, \alpha_3^{i_c})_{i_a i_b i_c} = \det(\alpha_j^i)_{ij} \equiv \epsilon_{i_a i_b i_c} \alpha_1^{i_a} \alpha_2^{i_b} \alpha_3^{i_c} = \det(M) \tag{14.10}$$

Furthermore, if the rows of the matrix M are written as vectors $\boldsymbol{a}_j = \alpha_j^i e_i$, then we can also adopt the notation

$$\det(\boldsymbol{a}_1, \boldsymbol{a}_2, \boldsymbol{a}_3) = |\boldsymbol{a}_1 \boldsymbol{a}_2 \boldsymbol{a}_3| \equiv \det(M) \tag{14.11}$$

As an example, let the $\{A_\mu\}$ form a frame of $\mathbb{P}^3$, with reciprocal frame $\{A^\mu\}$. Then from the definition of the square and angle brackets, it follows that

$$\epsilon_{i_a i_b i_c} = [\![A_{i_a} A_{i_b} A_{i_c} A_4]\!]_a \quad \text{and} \quad \epsilon^{i_a i_b i_c} = \langle\!\langle A^{i_a} A^{i_b} A^{i_c} A^4 \rangle\!\rangle_a \tag{14.12}$$

Therefore, we may, for example, express a determinant as $\det(\alpha_j^i)_{ij} = \alpha_1^{i_a} \alpha_2^{i_b} \alpha_3^{i_c} [\![A_{i_a} A_{i_b} A_{i_c} A_4]\!]_a$.

14.3 The Fundamental Matrix

14.3.1 Derivation

Let $\{A_\mu\}$ and $\{B_\mu\}$ define two cameras in $\mathbb{P}^3$. A point X in $\mathbb{P}^3$ may be transformed into the A and B frames via

$$X = X\cdot A_a^{\mu} A_{\mu} = X\cdot B_b^{\nu} B_{\nu} \tag{14.13}$$

Recall that there is an implicit summation over μ and ν. From that follows that the line $A_4 \wedge X$ can also be written as

$$\begin{aligned} A_4 \wedge X &= X\cdot A_a^{i}\; A_4 \wedge A_i \\ &= \rho_a^{-1}\, A_4 \wedge X_a \end{aligned} \tag{14.14}$$

where $X_a = X\cdot A^i A_i$. Let X_a and X_b be the images of some point $X \in \mathbb{P}^3$ taken by cameras A and B, respectively. Then, since the lines from A and B to X intersect at X

$$\begin{aligned} 0 &= (\, A_4 \wedge X \wedge B_4 \wedge X\,) I^{-1} \\ &\simeq (\, A_4 \wedge X_a \wedge B_4 \wedge X_b\,) I^{-1} \\ &= \alpha^i \beta^j [\![A_4 A_i B_4 B_j]\!] \end{aligned} \tag{14.15}$$

where $\alpha^i \equiv X\cdot A^i$ and $\beta^j \equiv X\cdot B^j$ are the image point coordinates of X_a and X_b, respectively. Therefore, for a *Fundamental Matrix* defined as

$$F_{ij} \equiv [\![A_4 A_i B_4 B_j]\!] \tag{14.16}$$

we have

$$\alpha^i \beta^j F_{ij} = 0 \tag{14.17}$$

if the image points given by $\{\alpha^i\}$ and $\{\beta^j\}$ are images of the same point in space. Note, however, that equation (14.17) holds as long as X_a is the image of any point along $A_4 \wedge X_a$ and X_b is the image of any point along $B_4 \wedge X_b$. In other words, the condition in equation (14.17) only ensures that lines $A_4 \wedge X_a$ and $B_4 \wedge X_b$ are co-planar.

In the following let any set of indices of the type $\{i_1, i_2, i_3\}$ be an even permutation of $\{1, 2, 3\}$. It may be shown that

$$[\![B_4 B_{j_1}]\!] \simeq B^{j_2} \wedge B^{j_3} \tag{14.18}$$

Thus, equation (14.16) can also be written as

$$F_{ij_1} \simeq (A_i \wedge A_4)\cdot(B^{j_2} \wedge B^{j_3}) \tag{14.19}$$

This may be expanded to

$$\begin{aligned} F_{ij_1} &= (A_4\cdot B^{j_2})(A_i\cdot B^{j_3}) - (A_4\cdot B^{j_3})(A_i\cdot B^{j_2}) \\ &= U_b^{j_2} K_{j_3 i}^{b} - U_b^{j_3} K_{j_2 i}^{b} \end{aligned} \tag{14.20}$$

That is, the Fundamental Matrix is just the standard cross product between the epipole[3] $U_b^{\bullet}$ and the column vectors $K_{\bullet_i}^{b}$.

$$F_{i\bullet} \simeq U_b^{\bullet} \times K_{\bullet_i}^{b} \tag{14.21}$$

In order to have a unified naming convention the Fundamental Matrix will be refered to as the *bifocal tensor*.

[3] Recall that $U_b \equiv E_{ba}$.

14.3.2 Rank of F

Note that we use the term "rank" in relation to tensors in order to generalise the notion of rank as used for matrices. That is, we would describe a rank 2 matrix as a rank 2, 2-valence tensor.

In general a tensor may be decomposed into a linear combination of rank 1 tensors. The minimum number of terms necessary for such a decomposition gives the *rank* of the tensor. For example, a rank 1, 2-valence tensor M is created by combining the components $\{\alpha^i\}$, $\{\beta^i\}$ of two vectors as $M^{ij} = \alpha^i \beta^j$.

The rank of F can be found quite easily from geometric considerations. Equation (14.16) can also be written as

$$F_{ij} \simeq A_i \cdot [\![A_4 B_4 B_j]\!] \tag{14.22}$$

The expression $[\![A_4 B_4 B_j]\!]$ gives the normal to the plane $(A_4 \wedge B_4 \wedge B_j)$. This defines three planes, one for each value of j, all of which contain the line $A_4 \wedge B_4$. Hence, all three normals lie in a plane. Furthermore, no two normals are identical since the $\{B_j\}$ are linearly independent by definition. It follows directly that at most two columns of F_{ij} can be linearly independent. Therefore, F is of rank 2.

The rank of the bifocal tensor F can also be arrived at through a *minimal* decomposition of F into rank 1 tensors. To achieve this we first define a new A-image plane frame $\{A'_i\}$ as

$$A'_i \equiv s(A_i + t_i A_4) \tag{14.23}$$

where s and the $\{t_i\}$ are some scalar components. Thus we have

$$\begin{aligned} A_4 \wedge A'_i &= s A_4 \wedge (A_i + t_i A_4) \\ &= s A_4 \wedge A_i \end{aligned} \tag{14.24}$$

Hence, F is left unchanged up to an overall scale factor under the transformation $A_i \longrightarrow A'_i$. In other words, the image plane bases $\{A_i\}$ and $\{B_j\}$ can be changed along the projective rays $\{A_4 \wedge A_i\}$ and $\{B_4 \wedge B_j\}$, respectively, without changing the bifocal tensor relating the two cameras. This fact limits the use of the bifocal tensor, since it cannot give any information about the actual placement of the image planes.

Define two bifocal tensors F and F' as

$$F_{ij} = [\![A_4 A_i B_4 B_j]\!] \tag{14.25a}$$

$$F'_{ij} = [\![A_4 A'_i B_4 B_j]\!] \tag{14.25b}$$

From equation (14.24) it follows directly that $F_{ij} \simeq F'_{ij}$. Since the $\{A'_i\}$ can be chosen arbitrarily along the line $A_4 \wedge A_i$ we may write

$$A'_i = (A_4 \wedge A_i) \vee P \tag{14.26}$$

where P is some plane in $\mathbb{P}^3$. $P = (B_4 \wedge B_1 \wedge B_2)$ seems a good choice, since then the $\{A'_i\}$ all lie in a plane together with B_4. The effect of this is that the projections of the $\{A'_i\}$ on image plane B will all lie along a line. The matrix $A'_i \cdot B^j$ therefore only has two linearly independent columns because the column vectors *are* the projections of the $\{A'_i\}$ onto image plane B. That is, $A'_i \cdot B^j$, which is the 3×3 minor of K^b, is of rank 2.

This matrix could only be of rank 1, if the $\{A'_i\}$ were to project to a single point on image plane B, which is only possible if they lie along a line in $\mathbb{P}^3$. However, then they could not form a basis for image plane A which they were defined to be.

Thus $A'_i \cdot B^j$ can minimally be of rank 2. Such a minimal form is what we need to find a minimal decomposition of F into rank 1 tensors using equation (14.20). Substituting $P = (B_4 \wedge B_1 \wedge B_2)$ into equation (14.26) gives

$$\begin{aligned} A'_i &= (A_4 \wedge A_i) \vee (B_4 \wedge B_1 \wedge B_2) \\ &= [\![A_4 A_i]\!] \cdot (B_4 \wedge B_1 \wedge B_2) \\ &= [\![A_4 A_i B_4 B_1]\!] B_2 - [\![A_4 A_i B_4 B_2]\!] B_1 + [\![A_4 A_i B_1 B_2]\!] B_4 \\ &= F_{i1} B_2 - F_{i2} B_1 + [\![A_4 A_i B_1 B_2]\!] B_4 \end{aligned} \tag{14.27}$$

Expanding F' in the same way as F in equation (14.20) and substituting the above expressions for the $\{A'_i\}$ gives

$$\begin{aligned} F'_{ij_1} &= (A_4 \cdot B^{j_2})(A'_i \cdot B^{j_3}) - (A_4 \cdot B^{j_3})(A'_i \cdot B^{j_2}) \\ &= \quad (A_4 \cdot B^{j_2}) \Big[- F_{i2}(B_1 \cdot B^{j_3}) + F_{i1}(B_2 \cdot B^{j_3}) \Big] \\ &\quad - (A_4 \cdot B^{j_3}) \Big[- F_{i2}(B_1 \cdot B^{j_2}) + F_{i1}(B_2 \cdot B^{j_2}) \Big] \\ &= \quad \varepsilon^{j_2}_{ba} \Big[- F_{i2} \delta^{j_3}_1 + F_{i1} \delta^{j_3}_2 \Big] \\ &\quad - \varepsilon^{j_3}_{ba} \Big[- F_{i2} \delta^{j_2}_1 + F_{i1} \delta^{j_2}_2 \Big] \\ &= \quad F_{i1} \Big[\varepsilon^{j_2}_{ba} \delta^{j_3}_2 - \varepsilon^{j_3}_{ba} \delta^{j_2}_2 \Big] \\ &\quad - F_{i2} \Big[\varepsilon^{j_2}_{ba} \delta^{j_3}_1 - \varepsilon^{j_3}_{ba} \delta^{j_2}_1 \Big] \end{aligned} \tag{14.28}$$

where we used the fact that $B_4 \cdot B^j = 0$. Clearly, F_{i1}, F_{i2} and the expressions in the square brackets all represent vectors. Therefore, equation (14.28) expresses F' as a linear combination of two rank 1 tensors (matrices). This shows again that the bifocal tensor is of rank 2.

But why should we do all this work of finding a minimal decomposition of F if its rank can be found so much more easily from geometric considerations? There are two good reasons:

1. for the trifocal and quadfocal tensor, a minimal decomposition will be the easiest way to find the rank, and

2. such a decomposition is useful for evaluating F with a non-linear algorithm, since the self-consistency constraints on F are automatically satisfied.

14.3.3 Degrees of Freedom of F

Equation (14.28) is in fact a minimal parameterisation of the bifocal tensor. This can be seen by writing out the columns of F'.

$$F'_{i1} = -\varepsilon^3_{ba} F_{i1}\,; \quad F'_{i2} = -\varepsilon^3_{ba} F_{i2}\,; \quad F'_{i3} = \varepsilon^1_{ba} F_{i1} + \varepsilon^2_{ba} F_{i2} \qquad (14.29)$$

As expected, the third column (F_{i3}) is a linear combination of the first two. Since an overall scale is not important we can also write

$$F'_{i1} = F_{i1}\,; \quad F'_{i2} = F_{i2}\,; \quad F'_{i3} = -\bar{\varepsilon}^1_{ba} F_{i1} - \bar{\varepsilon}^2_{ba} F_{i2} \qquad (14.30)$$

where $\bar{\varepsilon}^i_{ba} \equiv \varepsilon^i_{ba}/\varepsilon^3_{ba}$. This is the most general form of a rank 2, 3×3 matrix. Furthermore, since there are no more constraints on F_{i1} and F_{i2} this is also a minimal parameterisation of the bifocal tensor. That is, eight parameters are minimally necessary to form the bifocal tensor. It follows that since an overall scale is not important the bifocal tensor has *seven* degrees of freedom (DOF).

This DOF count can also be arrived at from more general considerations: each camera matrix has 12 components. However, since an overall scale is not important, each camera matrix adds only 11 DOF. Furthermore, the bifocal tensor is independent of the choice of basis. Therefore, it is invariant under a projective transformation, which has 16 components. But again, an overall scale is not important. Thus only 15 DOF can be subtracted from the DOF count due to the camera matrices. For two cameras we therefore have $2 \times 11 - 15 = 7$ DOF.

14.3.4 Transferring Points with F

The bifocal tensor can also be used to transfer a point in one image to a line in the other. Starting again from equation (14.16) the bifocal tensor can be written as

$$\begin{aligned} F_{ij} &= [\![A_i A_4 B_j B_4]\!] \\ &= (A_i \wedge A_4) \cdot [\![B_j B_4]\!] \\ &= (A_i \wedge A_4) \cdot L^b_j \end{aligned} \qquad (14.31)$$

This shows that F_{ij} gives the components of the projection of line $(A_i \wedge A_4)$ onto image plane B. Therefore,

$$(A_i \wedge A_4) \xrightarrow{B} F_{ij} L^j_b. \qquad (14.32)$$

Since $A_4 \xrightarrow{B} E_{ba}$ (the epipole on image plane B), $F_{ij}L^j_b$ defines an epipolar line.

Thus, contracting F with the coordinates of a point on image plane A, results in the homogeneous line coordinates of a line passing through the corresponding point on image plane B and the epipole E_{ba}.

$$\alpha^i F_{ij} = \lambda^b_j \tag{14.33}$$

where the $\{\alpha^i\}$ are some point coordinates and the $\{\lambda^b_j\}$ are the homogeneous line coordinates of an epipolar line.

14.3.5 Epipoles of *F*

Recall that if there are two cameras then two epipoles are defined;

$$E_{ab} \equiv B_4 \cdot A^i A_i = \varepsilon^i_{ab} A_i \tag{14.34a}$$

$$E_{ba} \equiv A_4 \cdot B^i B_i = \varepsilon^i_{ba} B_i \tag{14.34b}$$

Contracting F_{ij} with ε^i_{ab} gives

$$\begin{aligned} \varepsilon^i_{ab} F_{ij} &= \varepsilon^i_{ab} [\![A_4 A_i B_4 B_j]\!] \\ &= \rho_a [\![A_4 (\, B_4 \cdot A^i_a \; A_i) B_4 B_j]\!] \\ &= \rho_a [\![A_4 B_4 B_4 B_j]\!] \; ; \quad \text{from equation (14.14)} \\ &= 0 \end{aligned} \tag{14.35}$$

Similarly,

$$\varepsilon^j_{ba} F_{ij} = 0 \tag{14.36}$$

Therefore, vectors $\{\varepsilon^i_{ab}\}$ and $\{\varepsilon^j_{ba}\}$ can be regarded respectively as the left and right null spaces of matrix F. Given a bifocal tensor F, its epipoles can therefore easily be found using, for example, a singular value decomposition (SVD).

14.4 The Trifocal Tensor

14.4.1 Derivation

Let the frames $\{A_\mu\}$, $\{B_\mu\}$ and $\{C_\mu\}$ define three distinct cameras. Also, let $L = X \wedge Y$ be some line in P^3. The plane $L \wedge B_4$ is then the same as the plane $\lambda^b_i L^i_b \wedge B_4$, up to a scalar factor, where $\lambda^b_i = L \cdot L^b_i$. But,

$$L^{i_1}_b \wedge B_4 = B_{i_2} \wedge B_{i_3} \wedge B_4 = \langle\!\langle B^{i_1} \rangle\!\rangle$$

Intersecting planes $L \wedge B_4$ and $L \wedge C_4$ has to give L. Therefore, $(\lambda^b_i \langle\!\langle B^i \rangle\!\rangle) \vee (\lambda^c_j \langle\!\langle C^j \rangle\!\rangle)$ has to give L up to a scalar factor. Now, if two lines intersect, their outer product is zero. Thus, the outer product of lines $X \wedge A_4$ (or $Y \wedge A_4$) and L has to be zero. Note that $X \wedge A_4$ defines the same line as $(\alpha^i A_i) \wedge A_4$, up to a scalar factor, where $\alpha^i = X \cdot A^i$. Figure 14.1 shows this construction.

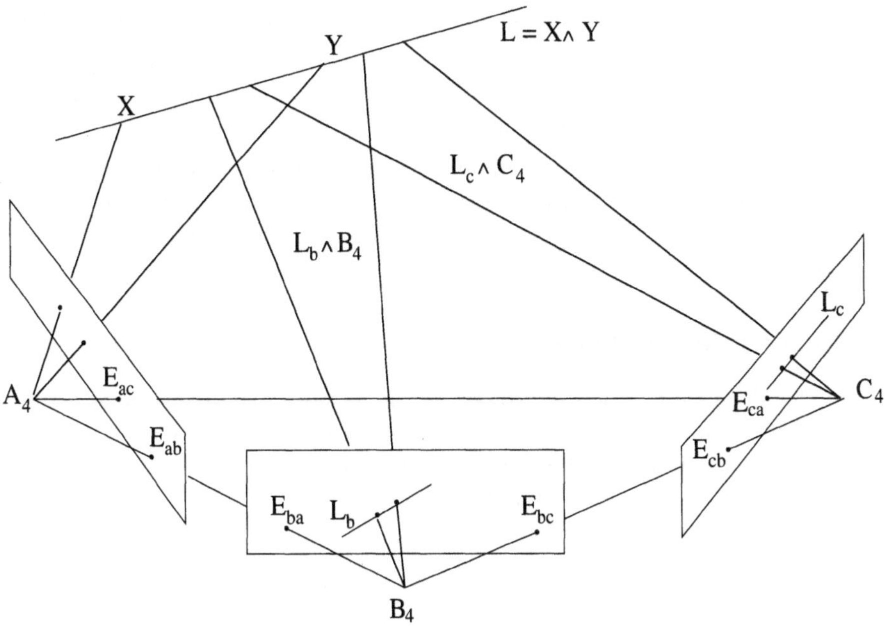

Fig. 14.1. Line projected onto three image planes. Note that although the figure is drawn in $\mathbb{E}^3$, lines and points are denoted by their corresponding vectors in $\mathbb{P}^3$

Combining all these expressions gives

$$\begin{aligned} 0 &= (X \wedge A_4 \wedge L) I^{-1} \\ &= \alpha^i \lambda^b_j \lambda^c_k \left[\!\left[(A_i \wedge A_4)(\langle\!\langle B^j \rangle\!\rangle \vee \langle\!\langle C^k \rangle\!\rangle) \right]\!\right] \\ &= \alpha^i \lambda^b_j \lambda^c_k \left[\!\left[(A_i \wedge A_4) \langle\!\langle B^j C^k \rangle\!\rangle \right]\!\right] \end{aligned} \tag{14.37}$$

where the identity from equation (14.4) was used. If the trifocal tensor $T_i{}^{jk}$ is defined as

$$T_i{}^{jk} = \left[\!\left[(A_i \wedge A_4) \langle\!\langle B^j C^k \rangle\!\rangle \right]\!\right] \tag{14.38}$$

then, from equation (14.37) it follows that it has to satisfy $\alpha^i \lambda^b_j \lambda^c_k T_i{}^{jk} = 0$. This expression for the trifocal tensor can be expanded in a number of different ways. One of them is,

$$\begin{aligned} T_i{}^{jk} &= (A_i \wedge A_4) \cdot \left[\!\left[\langle\!\langle B^j C^k \rangle\!\rangle \right]\!\right] \\ &= (A_i \wedge A_4) \cdot (B^j \wedge C^k) \\ &= (A_4 \cdot B^j)(A_i \cdot C^k) - (A_4 \cdot C^k)(A_i \cdot B^j) \\ &= U^j_b K^c_{k_i} - U^k_c K^b_{j_i} \end{aligned} \tag{14.39}$$

where $K^b_{j_i} \equiv A_i \cdot B^j$ and $K^c_{k_i} \equiv A_i \cdot C^k$ are the camera matrix minors for cameras B and C, respectively, relative to camera A. This is the expression for the trifocal tensor given by Hartley in [102]. Note that the camera matrix for camera A would be written as $K^a_{j_\mu} \equiv A_\mu \cdot A^j \simeq \delta^j_i$. That is, $K^a = [I|0]$ in standard matrix notation. In many other derivations of the trifocal tensor (eg. [102]) this form of the camera matrices is assumed at the beginning. Here, however, the trifocal tensor is defined first geometrically and we then find that it implies this particular form for the camera matrices.

14.4.2 Transferring Lines

The trifocal tensor can be used to transfer lines from two images to the third. That is, if the image of a line in $\mathbb{P}^3$ is known on two image planes, then its image on the third image plane can be found. This can be seen by expanding equation (14.38) in the following way,

$$\begin{aligned} T_{i^{jk}} &= [\![A_i A_4]\!] \cdot \langle\!\langle B^j C^k \rangle\!\rangle \\ &= L^a_i \cdot \langle\!\langle B^j C^k \rangle\!\rangle \end{aligned} \tag{14.40}$$

This shows that the trifocal tensor gives the homogeneous line components of the projection of line $\langle\!\langle B^j C^k \rangle\!\rangle$ onto image plane A. That is,

$$\langle\!\langle B^j C^k \rangle\!\rangle \xrightarrow{A} T_{i^{jk}} L^i_a \tag{14.41}$$

It will be helpful later on to define the following two lines.

$$T^{jk} \equiv \langle\!\langle B^j C^k \rangle\!\rangle \tag{14.42a}$$

$$T^{jk}_a \equiv T_{i^{jk}} L^i_a \tag{14.42b}$$

such that $T^{jk} \xrightarrow{A} T^{jk}_a$. Let the $\{\lambda^b_j\}$ and $\{\lambda^c_k\}$ be the homogeneous line coordinates of the projection of some line $L \in \mathbb{P}^3$ onto image planes B and C, respectively. Then recall that $\lambda^b_j \lambda^c_k \langle\!\langle B^j C^k \rangle\!\rangle$ gives L up to an overall scalar factor, i.e.

$$L \simeq \lambda^b_j \lambda^c_k \langle\!\langle B^j C^k \rangle\!\rangle \, ; \qquad \lambda^b_j \equiv L \cdot L^b_j \text{ and } \lambda^c_k \equiv L \cdot L^c_k \tag{14.43}$$

The image of L on image plane A, L_a, can therefore be found via

$$\begin{aligned} L_a &= L \cdot L^a_i L^i_a \\ &\simeq \lambda^b_j \lambda^c_k \langle\!\langle B^j C^k \rangle\!\rangle \cdot L^a_i L^i_a \\ &= \lambda^b_j \lambda^c_k T_{i^{jk}} L^i_a \end{aligned} \tag{14.44}$$

Thus, we have

$$\lambda^a_i \simeq \lambda^b_j \lambda^c_k T_{i^{jk}} \tag{14.45}$$

14.4.3 Transferring Points

It is also possible to find the image of a point on one image plane if its image is known on the other two. To see this, the expression for the trifocal tensor needs to be expanded in yet another way. Substituting the dual representation of line $A_{i_1} \wedge A_4$, i.e. $\langle\!\langle A_a^{i_2} A_a^{i_3} \rangle\!\rangle$ into equation (14.38) gives

$$\begin{aligned} T_{i_1}{}^{jk} &= \left[\!\left[(A_{i_1} \wedge A_4) \langle\!\langle B^j C^k \rangle\!\rangle \right]\!\right] \\ &= \left[\!\left[\langle\!\langle A_a^{i_2} A_a^{i_3} \rangle\!\rangle \langle\!\langle B^j C^k \rangle\!\rangle \right]\!\right] \\ &= \langle\!\langle A_a^{i_2} A_a^{i_3} \rangle\!\rangle \cdot (B^j \wedge C^k) \\ &= \langle\!\langle A_a^{i_2} A_a^{i_3} B^j C^k \rangle\!\rangle \end{aligned} \tag{14.46}$$

It can be shown that this form of the trifocal tensor is equivalent to the determinant form given by Heyden in [116]. Now only one more step is needed to see how the trifocal tensor may be used to transfer points.

$$\begin{aligned} T_{i_1}{}^{jk} &= \langle\!\langle A_a^{i_2} A_a^{i_3} B^j C^k \rangle\!\rangle \\ &= \langle\!\langle A_a^{i_2} A_a^{i_3} B^j \rangle\!\rangle \cdot C^k \\ &= X^T_{i_1}{}^j \cdot C^k \; ; \quad X^T_{i_1}{}^j \equiv \langle\!\langle A_a^{i_2} A_a^{i_3} B^j \rangle\!\rangle \end{aligned} \tag{14.47}$$

Note that the points $\{X^T_{i_1}{}^j\}$ are defined through their dual representation as the set of intersection points of lines $\{A_{i_1} \wedge A_4\}$ ($\simeq \{\langle\!\langle A_a^{i_2} A_a^{i_3} \rangle\!\rangle\}$) and planes $\{\langle\!\langle B^j \rangle\!\rangle\}$ ($\simeq \{L_b^j \wedge B_4\}$). Let $L = X \wedge Y$ be a line in $\mathbb{P}^3$. Then

$$X \xrightarrow{A} X_a = \alpha^i A_i \tag{14.48a}$$

$$L \xrightarrow{B} L_B = \lambda_j^b L_b^j \tag{14.48b}$$

Hence

$$\begin{aligned} X &\simeq (\alpha^{i_1} \underbrace{A_{i_1} \wedge A_4}_{\langle A^{i_2} A^{i_3} \rangle}) \vee (\lambda_j^b \underbrace{L_b^j \wedge B_4}_{\langle B^j \rangle}) \\ &= \textstyle\sum_{i_1} \alpha^{i_1} \lambda_j^b \langle\!\langle A^{i_2} A^{i_3} B^j \rangle\!\rangle \\ &= \alpha^{i_1} \lambda_j^b X^T_{i_1}{}^j \end{aligned} \tag{14.49}$$

Now, the projection of X onto image plane C is simply

$$\begin{aligned} X_c &= X \cdot C^k C_k \\ &\simeq \alpha^i \lambda_j^b X^T_{i_1}{}^j \cdot C^k C_k \\ &= \alpha^i \lambda_j^b T_i{}^{jk} C_k \end{aligned} \tag{14.50}$$

That is,

$$\eta^k \simeq \alpha^i \lambda^b_j T_{\underset{i}{\,}jk} \tag{14.51}$$

with $\eta^k \equiv X \cdot C^k$. Similarly we also have,

$$\beta^k \simeq \alpha^i \lambda^c_k T_{\underset{i}{\,}jk} \tag{14.52}$$

Therefore, if the image of a point and a line through that point are known on two image planes, respectively, then the image of the point on the third image plane can be calculated. Note that the line defined by the $\{\lambda^b_j\}$ can be any line that passes through the image of X on image plane B. That is, we may choose the point $(0, 0, 1)$ as the other point the line passes through. Then we have

$$\lambda^b_1 = \beta^2 ; \quad \lambda^b_2 = -\beta^1 ; \quad \lambda^b_3 = 0 \tag{14.53}$$

Hence, equation (14.51) becomes

$$\eta^k \simeq \alpha^i (\beta^2 T_{\underset{i}{\,}1k} - \beta^1 T_{\underset{i}{\,}2k}) \tag{14.54}$$

and equation (14.52) becomes

$$\beta^k \simeq \alpha^i (\eta^2 T_{\underset{i}{\,}j1} - \eta^1 T_{\underset{i}{\,}j2}) \tag{14.55}$$

14.4.4 Rank of T

Finding the rank of T is somewhat harder than for the bifocal tensor, mainly because there is no simple geometric construction which yields its rank. As was mentioned before the rank of a tensor is given by the minimum number of terms necessary for a linear decomposition of it in terms of rank 1 tensors[4]. As for the bifocal tensor, the transformation $A_i \rightarrow A'_i = s(A_i + t_i A_4)$ leaves the trifocal tensor unchanged up to an overall scale. A good choice for the $\{A'_i\}$ seems to be

$$A'_i = (A_i \wedge A_4) \vee (B_3 \wedge B_4 \wedge C_4) \tag{14.56}$$

since then all the $\{A'_i\}$ lie in a plane together with B_4 and C_4. Therefore, the camera matrix minors $K^b_{\underset{i}{j}} = A'_i \cdot B^j$ and $K^c_{\underset{i}{k}} = A'_i \cdot C^k$ are of rank 2. As was shown before, this is the minimal rank camera matrix minors can have. To see how this may help to find a minimal decomposition of T recall equation (14.39);

$$T_{\underset{i}{\,}jk} = U^j_b K^c_{\underset{i}{k}} - U^k_c K^b_{\underset{i}{j}}$$

[4] For example, a rank 1 3-valence tensor is created by combining the components $\{\alpha^i\}$, $\{\beta^i\}$, $\{\eta^i\}$ of three vectors as $T^{ijk} = \alpha^i \beta^j \eta^k$.

This decomposition of T shows that its rank is at most 6, since U_b and U_c are vectors, and K^c and K^b cannot be of rank higher than 3. Using the above choice for K^b and K^c however shows that the rank of T is 4, since then the rank of the camera matrices is minimal, and we thus have a minimal linear decomposition of T.

14.4.5 Degrees of Freedom of T

As for the bifocal tensor we can also write down an explicit parameterisation for the trifocal tensor. Starting with equation (14.56) we get

$$\begin{aligned} A'_i &= (A_i \wedge A_4) \vee (B_3 \wedge B_4 \wedge C_4) \\ &= [\![A_i A_4]\!] \cdot (B_3 \wedge B_4 \wedge C_4) \\ &= [\![A_i A_4 B_4 C_4]\!] B_3 - [\![A_i A_4 B_3 C_4]\!] B_4 + [\![A_i A_4 B_3 B_4]\!] C_4 \\ &= \alpha_i^1 B_3 + \alpha_i^2 B_4 + \alpha_i^3 C_4 \end{aligned} \tag{14.57}$$

where α_i^1, α_i^2 and α_i^3 are defined appropriately. The trifocal tensor may be expressed in terms of the $\{A'_i\}$ as follows (see equation (14.39)).

$$\begin{aligned} T_i{}^{jk} &= (A_4 \cdot B^j)(A'_i \cdot C^k) - (A_4 \cdot C^k)(A'_i \cdot B^j) \\ &= \quad (A_4 \cdot B^j)\left[\alpha_i^1 B_3 \cdot C^k + \alpha_i^2 B_4 \cdot C^k\right] \\ &\quad - (A_4 \cdot C^k)\left[\alpha_i^1 B_3 \cdot B^j + \alpha_i^3 C_4 \cdot B^j\right] \\ &= \quad \varepsilon^j_{ba}\left[\alpha_i^1 B_3 \cdot C^k + \alpha_i^2 \varepsilon^k_{cb}\right] \\ &\quad - \varepsilon^k_{ca}\left[\alpha_i^1 \delta^j_3 + \alpha_i^3 \varepsilon^j_{bc}\right] \end{aligned} \tag{14.58}$$

This decomposition of T has $5 \times 3 + 3 \times 3 - 1 = 23$ DOF. The general formula for finding the DOF of T gives $3 \times 11 - 15 = 18$ DOF. Therefore, equation (14.58) is an overdetermined parameterisation of T. However, it will still satisfy the self-consistency constraints of T.

14.4.6 Constraints on T

To understand the structure of T further, we will derive self-consistency constraints for T. Heyden derives the constraints on T using the "quadratic p-relations" [116]. In GA these relations can easily be established from geometric considerations.

The simplest constraint on T may be found as follows. Recall equation (14.47), where the trifocal tensor was expressed in terms of the projection of points $X^T_{i_1}{}^j = \langle\!\langle A_a^{i_2} A_a^{i_3} B^j \rangle\!\rangle$ onto image plane C, i.e.

$$T_{i_1}{}^{jk} = X^T_{i_1}{}^j \cdot C^k$$

Now consider the following trivector.

$$
\begin{aligned}
& X^T_{i_1}{}^{j_a} \wedge X^T_{i_1}{}^{j_b} \wedge X^T_{i_1}{}^{j_c} \\
&= \left(\langle\!\langle A^{i_2}_a A^{i_3}_a \rangle\!\rangle \vee \langle\!\langle B^{j_a} \rangle\!\rangle \right) \\
&\quad \wedge \left(\langle\!\langle A^{i_2}_a A^{i_3}_a \rangle\!\rangle \vee \langle\!\langle B^{j_b} \rangle\!\rangle \right) \wedge \left(\langle\!\langle A^{i_2}_a A^{i_3}_a \rangle\!\rangle \vee \langle\!\langle B^{j_c} \rangle\!\rangle \right) \\
&= 0
\end{aligned}
\tag{14.59}
$$

The first step follows from equation (14.4). It is clear that this expression is zero because we take the outer product of the intersection points of line $\langle\!\langle A^{i_2}_a A^{i_3}_a \rangle\!\rangle$ with the planes $\langle\!\langle B^{j_1} \rangle\!\rangle$, $\langle\!\langle B^{j_2} \rangle\!\rangle$ and $\langle\!\langle B^{j_3} \rangle\!\rangle$. In other words, this equation says that the intersection points of a line with three planes all lie along a line (see figure 14.2).

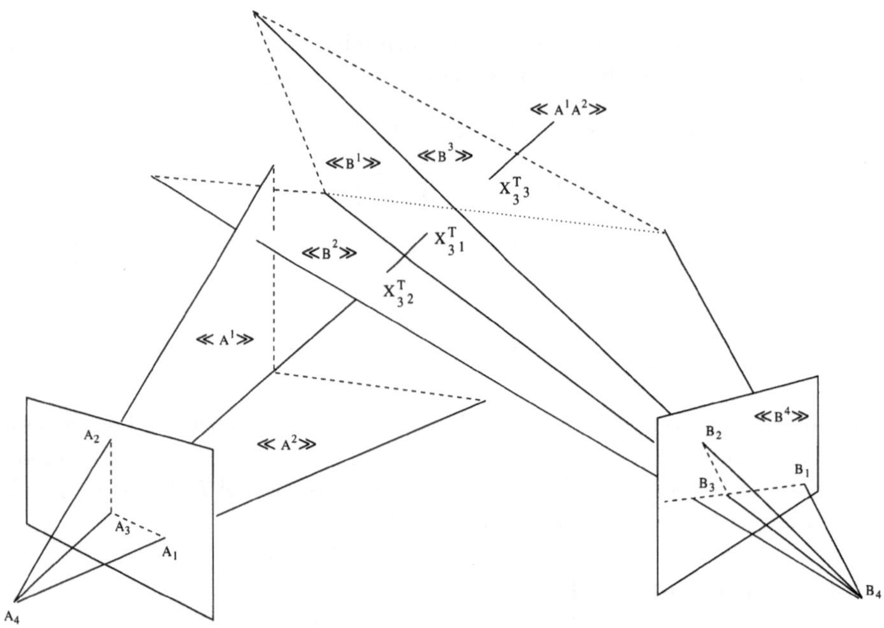

Fig. 14.2. This demonstrates the constraint from equation (14.59) for $i_2 = 1$, $i_3 = 2$ and $j_a = 1$, $j_b = 2$, $j_c = 3$. The figure also visualises the use of the inverse dual bracket to describe planes and lines

When projecting the three intersection points onto image plane C they still have to lie along a line. That is,

$$
\begin{aligned}
0 &= (X^T_{\substack{j_a\\i}} \cdot C^{k_a})(X^T_{\substack{j_b\\i}} \cdot C^{k_b})(X^T_{\substack{j_c\\i}} \cdot C^{k_c}) C_{k_a} \wedge C_{k_b} \wedge C_{k_c} \\
\iff 0 &= T_{\substack{j_a k_a\\i}} T_{\substack{j_b k_b\\i}} T_{\substack{j_c k_c\\i}} [\![C_{k_a} C_{k_b} C_{k_c} C_4]\!]_c \\
&= \epsilon_{k_a k_b k_c} T_{\substack{j_a k_a\\i}} T_{\substack{j_b k_b\\i}} T_{\substack{j_c k_c\\i}} \\
&= \det(T_{\substack{jk\\i}})_{jk}
\end{aligned}
\tag{14.60}
$$

14.4.7 Relation between T and F

We mentioned before that the quadratic p-relations can be used to find constraints on T [116]. The equivalent expressions in GA are of the form

$$
\langle\!\langle B^1 B^2 \rangle\!\rangle \wedge \langle\!\langle A^1 A^2 A^3 \rangle\!\rangle \wedge \langle\!\langle B^1 B^2 C^1 \rangle\!\rangle = 0 \tag{14.61}
$$

This expression is zero because $\langle\!\langle B^1 B^2 \rangle\!\rangle \wedge \langle\!\langle B^1 B^2 C^1 \rangle\!\rangle = 0$. This becomes obvious immediately from a geometric point of view: the intersection point of line $\langle\!\langle B^1 B^2 \rangle\!\rangle$ with plane $\langle\!\langle C^1 \rangle\!\rangle$ clearly lies on line $\langle\!\langle B^1 B^2 \rangle\!\rangle$.

In the following we will write $T^{XYZ}_{\substack{jk\\i_1}}$ to denote the trifocal tensor

$$
T^{XYZ}_{\substack{jk\\i_1}} = \langle\!\langle X^{i_2} X^{i_3} Y^j Z^k \rangle\!\rangle
$$

We will similarly write $F^{XY}_{i_1 j_1}$ to denote the bifocal tensor

$$
F^{XY}_{i_1 j_1} = \langle\!\langle X^{i_2} X^{i_3} Y^{j_2} Y^{j_3} \rangle\!\rangle
$$

If no superscripts are given then $T_{\substack{jk\\i}}$ and F_{ij} take on the same meaning as before. That is,

$$
T_{\substack{jk\\i}} \equiv T^{ABC}_{\substack{jk\\i}} \tag{14.62a}
$$

$$
F_{ij} \equiv F^{AB}_{ij} \tag{14.62b}
$$

We can obtain a constraint on T by expanding equation (14.61).

$$
\begin{aligned}
0 &= \langle\!\langle B^1 B^2 \rangle\!\rangle \wedge \langle\!\langle A^1 A^2 A^3 \rangle\!\rangle \wedge \langle\!\langle B^1 B^2 C^1 \rangle\!\rangle \\
&= \quad \langle\!\langle A^1 A^2 B^1 B^2 \rangle\!\rangle \langle\!\langle B^1 B^2 A^3 C^1 \rangle\!\rangle \\
&\quad + \langle\!\langle A^3 A^1 B^1 B^2 \rangle\!\rangle \langle\!\langle B^1 B^2 A^2 C^1 \rangle\!\rangle \\
&\quad + \langle\!\langle A^2 A^3 B^1 B^2 \rangle\!\rangle \langle\!\langle B^1 B^2 A^1 C^1 \rangle\!\rangle \\
&= F_{33} T^{BAC}_{\substack{31\\3}} + F_{23} T^{BAC}_{\substack{21\\3}} + F_{13} T^{BAC}_{\substack{11\\3}} \\
&= F_{i3} T^{BAC}_{\substack{i1\\3}}
\end{aligned}
\tag{14.63}
$$

Note that there is an implicit summation over i, because it is repeated as a (relative) superscript. Of course, we could have chosen different indices for the reciprocal B vectors and the reciprocal C vector. Therefore, we can obtain the following relation between the trifocal tensor and the bifocal tensor.

$$F_{ij} T^{BAC}_{\underset{j}{\cdot} ik} = 0 \tag{14.64}$$

Again there is an implicit summation over the i index but not over the j index. From this equation it follows that the three column vectors of the bifocal tensor give the three "left" null vectors of the three matrices $T_{\underset{i}{\cdot}\bullet\bullet}$, respectively. Equation (14.64) has two main uses: it can be used to find some epipoles of the trifocal tensor via equations (14.35) and (14.36), but it also serves to give more constraints on T since $\det F = 0$.

The columns of F may be found from equation (14.64) using, for example, an SVD. However, since the columns are found separately they will not in general be scaled consistently. Therefore, F found from equation (14.64) has only a limited use. Nonetheless, we can still find the correct left null vector of F, i.e. ε^{i}_{ab}, because each column is consistent in itself. Note also that, the determinant of F is still zero, since the rank F cannot be changed by scaling its columns separately. We cannot use this F, though, to find the right null vector, i.e. ε^{i}_{ba}, or to check whether image points on planes A and B are images of the same world point. Finding a consistent F is not necessary to find the right null vector of F, as will be shown later on. Therefore, unless we need to find a bifocal tensor from T which we can use to check image point pair matches, a fully consistent F is not necessary. A consistent F can, however, be found as shown in the following.

We can find the bifocal tensor row-wise in the following way.

$$\begin{aligned} 0 &= \langle\!\langle A^{i_2} A^{i_3} \rangle\!\rangle \wedge \langle\!\langle B^1 B^2 B^3 \rangle\!\rangle \wedge \langle\!\langle A^{i_2} A^{i_3} C^k \rangle\!\rangle \\ &= F_{i_1 j} T_{\underset{i_1}{\cdot} jk} \end{aligned} \tag{14.65}$$

Knowing F row-wise and column-wise we can find a consistently scaled bifocal tensor. What remains is to find T^{BAC} from T. To do so we define the following intersection points in terms of the lines $T^{i_a j_a} \equiv \langle\!\langle B^{i_a} C^{j_a} \rangle\!\rangle$ (see equation (14.42a)).

$$\begin{aligned} p(i_a j_a, i_b j_b) &\equiv (A_4 \wedge T^{i_a j_a}) \vee T^{i_b j_b} \\ &= \Big\langle\!\!\Big\langle \Big[\!\!\Big[A_4 \langle\!\langle B^{i_a} C^{j_a} \rangle\!\rangle \Big]\!\!\Big] \Big[\!\!\Big[\langle\!\langle B^{i_b} C^{j_b} \rangle\!\rangle \Big]\!\!\Big] \Big\rangle\!\!\Big\rangle \\ &= \Big\langle\!\!\Big\langle \Big(A_4 \cdot \Big[\!\!\Big[\langle\!\langle B^{i_a} C^{j_a} \rangle\!\rangle \Big]\!\!\Big] \Big) B^{i_b} C^{j_b} \Big\rangle\!\!\Big\rangle \\ &= \Big\langle\!\!\Big\langle \Big(A_4 \cdot (B^{i_a} \wedge C^{j_a}) \Big) B^{i_b} C^{j_b} \Big\rangle\!\!\Big\rangle \\ &= \Big\langle\!\!\Big\langle \; (A_4 \cdot B^{i_a}) C^{j_a} B^{i_b} C^{j_b} \\ &\qquad - (A_4 \cdot C^{j_a}) B^{i_a} B^{i_b} C^{j_b} \Big\rangle\!\!\Big\rangle \\ &= \varepsilon^{i_a}_{ba} \langle\!\langle C^{j_a} B^{i_b} C^{j_b} \rangle\!\rangle + \varepsilon^{j_a}_{ca} \langle\!\langle B^{i_a} C^{j_b} B^{i_b} \rangle\!\rangle \end{aligned} \tag{14.66}$$

Two useful special cases are

$$p(i_1 j, i_2 j) = \varepsilon^{j}_{ca} \langle\!\langle B^{i_1} C^{j} B^{i_2} \rangle\!\rangle \tag{14.67a}$$

$$p(i j_1, i j_2) = \varepsilon^{i}_{ba} \langle\!\langle C^{j_1} B^{i} C^{j_2} \rangle\!\rangle \tag{14.67b}$$

The projection of $p(i_1 j, i_2 j)$ onto image plane A, denoted by $p_a(i_1 j, i_2 j)$ gives

$$\begin{aligned} p_a(i_2 k, i_3 k) &= \varepsilon^{k}_{ca} \Big(A^{j} \cdot \langle\!\langle B^{i_2} C^{k} B^{i_3} \rangle\!\rangle \Big) A_j \\ &= \varepsilon^{k}_{ca} \langle\!\langle A^{j} B^{i_2} C^{k} B^{i_3} \rangle\!\rangle A_j \\ &= -\varepsilon^{k}_{ca} \langle\!\langle B^{i_2} B^{i_3} A^{j} C^{k} \rangle\!\rangle A_j \\ &= -\varepsilon^{k}_{ca} T^{BAC}_{\underset{i_1}{\cdot} jk} A_j \end{aligned} \tag{14.68}$$

We can also calculate $p_a(j_a k_a, j_b k_b)$ by immediately using the projections of the T^{jk} onto image plane A (see equation (14.42b)). That is,

$$\begin{aligned} p_a(j_a k_a, j_b k_b) &= (A_4 \wedge T^{i_a j_a}_{a}) \vee T^{i_b j_b}_{a} \\ &= T_{\underset{i_a}{\cdot} j_a k_a} T_{\underset{i_b}{\cdot} j_b k_b} (A_4 \wedge L^{i_a}_{a}) \vee L^{i_b}_{a} \\ &= T_{\underset{i_a}{\cdot} j_a k_a} T_{\underset{i_b}{\cdot} j_b k_b} (A_4 \wedge \langle\!\langle A^{i_a}_{a} A^{4}_{a} \rangle\!\rangle) \vee \langle\!\langle A^{i_b}_{a} A^{4}_{a} \rangle\!\rangle \\ &= T_{\underset{i_a}{\cdot} j_a k_a} T_{\underset{i_b}{\cdot} j_b k_b} \Big\langle\!\!\Big\langle \Big(A_4 \cdot (A^{i_a}_{a} \wedge A^{4}_{a}) \Big) A^{i_b}_{a} A^{4}_{a} \Big\rangle\!\!\Big\rangle \\ &\simeq T_{\underset{i_a}{\cdot} j_a k_a} T_{\underset{i_b}{\cdot} j_b k_b} \langle\!\langle A^{i_a}_{a} A^{i_b}_{a} A^{4}_{a} \rangle\!\rangle \end{aligned} \tag{14.69}$$

From the definition of the inverse dual bracket we have

$$A_{i_3} = \langle\!\langle A^{i_1}_{a} A^{i_2}_{a} A^{4}_{a} \rangle\!\rangle_a$$

Therefore, from equation (14.69) we find

$$p_a(j_1 k, j_2 k) \simeq (T_{\underset{i_1}{\cdot} j_1 k} T_{\underset{i_2}{\cdot} j_2 k} - T_{\underset{i_2}{\cdot} j_1 k} T_{\underset{i_1}{\cdot} j_2 k}) A_{i_3} \tag{14.70}$$

Equating this with equation (14.68) gives

$$T^{BAC}_{\underset{j_3}{\cdot} i_3 k} \simeq (\varepsilon^{k}_{ca})^{-1} (T_{\underset{i_1}{\cdot} j_1 k} T_{\underset{i_2}{\cdot} j_2 k} - T_{\underset{i_2}{\cdot} j_1 k} T_{\underset{i_1}{\cdot} j_2 k}) \tag{14.71}$$

Since ε^{k}_{ca} can be found from T (as will be shown later) we can find T^{BAC} from T up to an overall scale. Equation (14.71) may also be written in terms of the standard cross product.

$$T^{BAC}_{\underset{j_3}{\cdot} \bullet k} \simeq (\varepsilon^{k}_{ca})^{-1} (T_{\underset{\bullet}{\cdot} j_1 k} \times T_{\underset{\bullet}{\cdot} j_2 k}) \tag{14.72}$$

Had we used equation (14.67b) instead of equation (14.67a) in the previous calculation, we would have obtained the following relation.

$$T^{CBA}_{k_3\ \, ji_3} \simeq (\varepsilon^{j}_{ba})^{-1}(T_{i_1}^{\ \, jk_1} T_{i_2}^{\ \, jk_2} - T_{i_2}^{\ \, jk_1} T_{i_1}^{\ \, jk_2}) \tag{14.73}$$

Or, in terms of the standard cross product

$$T^{CBA}_{k_3\ \, j\bullet} \simeq (\varepsilon^{j}_{ba})^{-1}(T_{\bullet}^{\ \, jk_1} \times T_{\bullet}^{\ \, jk_2}) \tag{14.74}$$

Hence, we can also obtain T^{CBA} from T up to an overall scale. Note that since

$$\begin{aligned} T^{ABC}_{i_1\ \, jk} &= \langle\!\langle A^{i_2} A^{i_3} B^j C^k \rangle\!\rangle \\ &= -\langle\!\langle A^{i_2} A^{i_3} C^k B^j \rangle\!\rangle \\ &= -T^{ACB}_{i_1\ \, kj} \end{aligned} \tag{14.75}$$

we have found *all* possible trifocal tensors for a particular camera setup from T.

Equations (14.72) and (14.74) simply express that the projections of the intersection points between some lines onto image plane A are the same as the intersection points between the projections of the same lines onto image plane A. This implies that independent of the intersection points, i.e. the components of $T_{i}^{\ \, jk}$, equations (14.72) and (14.74) will always give a self-consistent tensor, albeit not necessarily one that expresses the correct camera geometry.

14.4.8 Second Order Constraints

There are more constraints on T which we will call "second order" because they are products of determinants of components of T. Their derivation is more involved and can be found in [184] and [185]. Here we will only state the results. These constraints may be used to check the self-consistency of T when it is calculated via a non-linear method.

$$\begin{aligned} 0 = \ & |T_a^{j_a k_a} T_a^{j_b k_a} T_a^{j_a k_b}|\ |T_a^{j_b k_b} T_a^{j_a k_c} T_a^{j_b k_c}| \\ - & |T_a^{j_a k_a} T_a^{j_b k_a} T_a^{j_b k_b}|\ |T_a^{j_a k_b} T_a^{j_a k_c} T_a^{j_b k_c}| \end{aligned} \tag{14.76}$$

$$\begin{aligned} 0 = \ & |T_a^{j_a k_a} T_a^{j_a k_b} T_a^{j_b k_a}|\ |T_a^{j_b k_b} T_a^{j_c k_a} T_a^{j_c k_b}| \\ - & |T_a^{j_a k_a} T_a^{j_a k_b} T_a^{j_b k_b}|\ |T_a^{j_b k_a} T_a^{j_c k_a} T_a^{j_c k_b}| \end{aligned} \tag{14.77}$$

$$\begin{aligned} 0 = \ & |T_a^{i_a j_a} T_a^{i_b j_a} T_a^{i_a j_b}|\ |T_a^{i_b j_b} T_a^{i_a j_b} T_a^{i_b j_c}| \\ - & |T_a^{i_a j_a} T_a^{i_b j_a} T_a^{i_b j_b}|\ |T_a^{i_a j_b} T_a^{i_a j_c} T_a^{i_b j_b}| \end{aligned} \tag{14.78}$$

Where the determinants are to be be interpreted as

$$|T_a^{j_a k_a} T_a^{j_b k_a} T_a^{j_a k_b}| = \det(T_{i_a}^{\ \, j_a k_a}, T_{i_b}^{\ \, j_b k_a}, T_{i_c}^{\ \, j_a k_b})_{i_a i_b i_c}$$

14.4.9 Epipoles

The epipoles of T can be found indirectly via the relation of bifocal tensors to T (e.g. equation (14.64)). Also recall that the right null vector of some F_{ij}^{XY} is ε_{yx}^{j}, whereas the left null vector is ε_{xy}^{i} (equations (14.35) and (14.36)). From equation (14.65) we know that

$$F_{ij}T_{i}{}^{jk} = 0$$

When calculating F from this equation, we cannot guarantee that the rows are scaled consistently. Nevertheless, this does not affect the right null space of F. Hence, we can find ε_{ba}^{j} from this F. In the following we will list the necessary relations to find all epipoles of T.

$$\begin{aligned} 0 &= \langle\!\langle A^{i_2}A^{i_3}\rangle\!\rangle \wedge \langle\!\langle B^1B^2B^3\rangle\!\rangle \wedge \langle\!\langle A^{i_2}A^{i_3}C^k\rangle\!\rangle \\ &= F_{i_1 j}T_{i_1}{}^{jk} \qquad \rightarrow \varepsilon_{ba}^{j} \end{aligned} \tag{14.79a}$$

$$\begin{aligned} 0 &= \langle\!\langle A^{i_2}A^{i_3}\rangle\!\rangle \wedge \langle\!\langle C^1C^2C^3\rangle\!\rangle \wedge \langle\!\langle A^{i_2}A^{i_3}B^j\rangle\!\rangle \\ &= F_{i_1 k}^{AC}T_{i_1}{}^{jk} \qquad \rightarrow \varepsilon_{ca}^{k} \end{aligned} \tag{14.79b}$$

$$\begin{aligned} 0 &= \langle\!\langle B^{i_2}B^{i_3}\rangle\!\rangle \wedge \langle\!\langle A^1A^2A^3\rangle\!\rangle \wedge \langle\!\langle B^{i_2}B^{i_3}C^k\rangle\!\rangle \\ &= F_{i_1 j}^{BA}T_{i_1}^{BAC}{}^{jk} \qquad \rightarrow \varepsilon_{ab}^{j} \end{aligned} \tag{14.80a}$$

$$\begin{aligned} 0 &= \langle\!\langle B^{i_2}B^{i_3}\rangle\!\rangle \wedge \langle\!\langle C^1C^2C^3\rangle\!\rangle \wedge \langle\!\langle B^{i_2}B^{i_3}A^j\rangle\!\rangle \\ &= F_{i_1 k}^{BC}T_{i_1}^{BAC}{}^{jk} \qquad \rightarrow \varepsilon_{cb}^{k} \end{aligned} \tag{14.80b}$$

$$\begin{aligned} 0 &= \langle\!\langle C^{i_2}C^{i_3}\rangle\!\rangle \wedge \langle\!\langle A^1A^2A^3\rangle\!\rangle \wedge \langle\!\langle C^{i_2}C^{i_3}B^j\rangle\!\rangle \\ &= F_{i_1 k}^{CA}T_{i_1}^{CBA}{}^{jk} \qquad \rightarrow \varepsilon_{ac}^{k} \end{aligned} \tag{14.81a}$$

$$\begin{aligned} 0 &= \langle\!\langle C^{i_2}C^{i_3}\rangle\!\rangle \wedge \langle\!\langle B^1B^2B^3\rangle\!\rangle \wedge \langle\!\langle C^{i_2}C^{i_3}A^k\rangle\!\rangle \\ &= F_{i_1 j}^{CB}T_{i_1}^{CBA}{}^{jk} \qquad \rightarrow \varepsilon_{bc}^{j} \end{aligned} \tag{14.81b}$$

By $\rightarrow \varepsilon_{xy}^{j}$ we denote the epipole that can be found from the respective relation[5] . Note that since

$$\begin{aligned} F_{i_1 j_1}^{XY} &= \langle\!\langle X^{i_2}X^{i_3}Y^{j_2}Y^{j_3}\rangle\!\rangle \\ &= \langle\!\langle Y^{j_2}Y^{j_3}X^{i_2}X^{i_3}\rangle\!\rangle \\ &= F_{j_1 i_1}^{YX} \end{aligned} \tag{14.82}$$

we have also found all fundamental matrices.

[5] Initial computations evaluating the quality of the epipoles found via this method indicate that this may not be the best way to calculate the epipoles. It seems that better results can be obtained when the epipoles are found directly from T^{ABC}.

14.5 The Quadfocal Tensor

14.5.1 Derivation

Let L be a line in $\mathbb{P}^3$ and let $\{A_\mu\}$, $\{B_\mu\}$, $\{C_\mu\}$ and $\{D_\mu\}$ define four cameras A, B, C and D, respectively. The projection of L onto the image planes of these four cameras is

$$L \xrightarrow{A} L_A = L\cdot L_i^a L_a^i = \lambda_i^a L_a^i \tag{14.83a}$$

$$L \xrightarrow{B} L_B = L\cdot L_i^b L_b^i = \lambda_i^b L_b^i \tag{14.83b}$$

$$L \xrightarrow{C} L_C = L\cdot L_i^c L_c^i = \lambda_i^c L_c^i \tag{14.83c}$$

$$L \xrightarrow{D} L_D = L\cdot L_i^d L_d^i = \lambda_i^d L_d^i \tag{14.83d}$$

The intial line L can be recovered from these projections by intersecting any two of the planes $(L_A \wedge A_4)$, $(L_B \wedge B_4)$, $(L_C \wedge C_4)$ and $(L_D \wedge D_4)$. For example,

$$L \simeq (L_A \wedge A_4) \vee (L_B \wedge B_4) \simeq (L_C \wedge C_4) \vee (L_D \wedge D_4) \tag{14.84}$$

Therefore,

$$\begin{aligned}
0 &= \Big[\!\!\Big[\Big((L_A \wedge A_4) \vee (L_B \wedge B_4) \Big) \wedge \Big((L_C \wedge C_4) \vee (L_D \wedge D_4) \Big) \Big]\!\!\Big] \\
&= \lambda_i^a \lambda_j^b \lambda_k^c \lambda_l^d \Big[\!\!\Big[\Big((L_a^i \wedge A_4) \vee (L_b^j \wedge B_4) \Big) \\
&\qquad\qquad \Big((L_c^k \wedge C_4) \vee (L_d^l \wedge D_4) \Big) \Big]\!\!\Big] \\
&= \lambda_i^a \lambda_j^b \lambda_k^c \lambda_l^d \Big[\!\!\Big[\Big(\langle\!\langle A^i \rangle\!\rangle \vee \langle\!\langle B^j \rangle\!\rangle \Big) \Big(\langle\!\langle C^k \rangle\!\rangle \vee \langle\!\langle D^l \rangle\!\rangle \Big) \Big]\!\!\Big] \\
&= \lambda_i^a \lambda_j^b \lambda_k^c \lambda_l^d \langle\!\langle A^i B^j C^k D^l \rangle\!\rangle
\end{aligned} \tag{14.85}$$

Therefore, a *quadfocal tensor* may be defined as

$$Q^{ijkl} = \langle\!\langle A^i B^j C^k D^l \rangle\!\rangle \tag{14.86}$$

If the quadfocal tensor is contracted with the homogeneous line coordinates of the projections of one line onto the four camera image planes, the result is zero. In this way the quadfocal tensor encodes the relative orientation of the four camera image planes. However, note that contracting the quadfocal tensor with the line coordinates of the projection of one line onto only three image planes gives a zero vector. This follows directly from geometric considerations. For example,

$$\begin{aligned}
\lambda_i^a \lambda_j^b \lambda_k^c Q^{ijkl} &= \lambda_i^a \lambda_j^b \lambda_k^c \langle\!\langle A^i B^j C^k \rangle\!\rangle \cdot D^l \\
&\simeq \Big(L \vee (\lambda_k^c C^k) \Big) \cdot D^l
\end{aligned} \tag{14.87}$$

where L is the line whose images on image planes A, B and C have coordinates $\{\lambda_i^a\}$, $\{\lambda_j^b\}$ and $\{\lambda_k^c\}$, respectively. Hence, L lies on plane $\lambda_k^c C^k$, and thus their meet is zero. This also shows that the quadfocal tensor does not add any new information to what can be known from the trifocal tensor, since the quadfocal tensor simply relates any three image planes out of a group of four.

The form for Q given in equation (14.86) can be shown to be equivalent to the form given by Heyden in [116]. In this form it is also immediately clear that changing the order of the reciprocal vectors in equation (14.86) at most changes the overall sign of Q.

14.5.2 Transferring Lines

If the image of a line is known on two image planes, then the quadfocal tensor can be used to find its image on the other two image planes. This can be achieved through a somewhat indirect route. Let L be a line projected onto image planes A and B with coordinates $\{\lambda_i^a\}$ and $\{\lambda_j^b\}$, respectively. Then we know that

$$L \simeq \lambda_i^a \lambda_j^b \langle\!\langle A^i B^j \rangle\!\rangle \tag{14.88}$$

Therefore, we can define three points $\{X_L^k\}$ that lie on L as

$$\begin{aligned} X_L^k &\equiv \lambda_i^a \lambda_j^b (\langle\!\langle A^i B^j \rangle\!\rangle \vee \langle\!\langle C^k \rangle\!\rangle) \\ &= \lambda_i^a \lambda_j^b \langle\!\langle A^i B^j C^k \rangle\!\rangle \end{aligned} \tag{14.89}$$

The projections of the $\{X_L^k\}$ onto image plane D, denoted by $\{X_{L_d}^k\}$ are given by

$$\begin{aligned} X_{L_d}^k &\equiv X_L^k \cdot D^l \\ &= \lambda_i^a \lambda_j^b \langle\!\langle A^i B^j C^k \rangle\!\rangle \cdot D^l \\ &= \lambda_i^a \lambda_j^b \langle\!\langle A^i B^j C^k D^l \rangle\!\rangle \\ &= \lambda_i^a \lambda_j^b Q^{ijkl} \end{aligned} \tag{14.90}$$

From the points $\{X_{L_d}^k\}$ the projection of line L onto image plane D can be recovered.

14.5.3 Rank of Q

The form for the quadfocal tensor as given in equation (14.86) may be expanded in a number of ways. For example,

$$\begin{aligned} Q^{i_1 jkl} &= (A_{i_2} \wedge A_{i_3} \wedge A_4) \cdot (B^j \wedge C^k \wedge D^l) \\ &= \quad U_b^j \Big[K_{k_{i_3}}^c K_{l_{i_2}}^d - K_{l_{i_3}}^d K_{k_{i_2}}^c \Big] \\ &\quad - U_c^k \Big[K_{j_{i_3}}^b K_{l_{i_2}}^d - K_{l_{i_3}}^d K_{j_{i_2}}^b \Big] \\ &\quad + U_d^l \Big[K_{j_{i_3}}^b K_{k_{i_2}}^c - K_{k_{i_3}}^c K_{j_{i_2}}^b \Big] \end{aligned} \tag{14.91}$$

In terms of the standard cross product this may be written as

$$Q^{\bullet jkl} = U_b^j (K_{k_\bullet}^c \times K_{l_\bullet}^d) - U_c^k (K_{j_\bullet}^b \times K_{l_\bullet}^d) + U_d^l (K_{j_\bullet}^b \times K_{k_\bullet}^c) \tag{14.92}$$

This decomposition of Q shows that the quadfocal tensor can be at most of rank 9. From equation (14.91) it becomes clear that, as for the trifocal tensor, the transformation $A_i \mapsto s(A_i + t_i A_4)$ leaves Q unchanged up to an overall scale.

Let $P = B_4 \wedge C_4 \wedge D_4$. As for the trifocal tensor case, define a basis $\{A'_i\}$ for image plane A by

$$A'_i = (A_i \wedge A_4) \vee P \tag{14.93}$$

All the $\{A'_i\}$ lie on plane P, that is they lie on the plane formed by B_4, C_4 and D_4. Therefore, $K_{j_i}^{b\prime} = A'_i \cdot B^j$, $K^{c\prime} = A'_i \cdot C^k$ and $K^{d\prime} = A'_i \cdot D^l$ are of rank 2. As was shown previously, this is the minimum rank the camera matrices can have. Hence, forming Q with the $\{A'_i\}$ should yield its rank. However, it is not immediately obvious from equation (14.91) what the rank of Q is when substituting the $\{A'_i\}$ for the $\{A_i\}$. A more yielding decomposition of Q is achieved by expanding equation (14.93).

$$\begin{aligned} A'_i &= (A_i \wedge A_4) \vee P \\ &\simeq [\![A_i A_4]\!] \cdot (B_4 \wedge C_4 \wedge D_4) \\ &= [\![A_i A_4 B_4 C_4]\!] D_4 - [\![A_i A_4 B_4 D_4]\!] C_4 + [\![A_i A_4 C_4 D_4]\!] B_4 \\ &= \alpha_i^1 B_4 + \alpha_i^2 C_4 + \alpha_i^3 D_4 \end{aligned} \tag{14.94}$$

where the $\{\alpha_i^j\}$ are defined accordingly. Furthermore,

$$A'_{i_1} \wedge A'_{i_2} = \lambda_{i_3}^1 \, C_4 \wedge D_4 + \lambda_{i_3}^2 \, D_4 \wedge B_4 + \lambda_{i_3}^3 \, B_4 \wedge C_4 \tag{14.95}$$

with $\lambda_{i_3}^{j_3} = \alpha_{i_1}^{j_1} \alpha_{i_2}^{j_2} - \alpha_{i_2}^{j_1} \alpha_{i_1}^{j_2}$. Equation (14.91) may also be written as

$$\begin{aligned} Q^{i_1 jkl} &= (A_{i_2} \wedge A_{i_3} \wedge A_4) \cdot (B^j \wedge C^k \wedge D^l) \\ &= \quad U_b^j \Big[(A'_{i_2} \wedge A'_{i_3}) \cdot (C^k \wedge D^l) \Big] \\ &\quad - U_c^k \Big[(A'_{i_2} \wedge A'_{i_3}) \cdot (B^j \wedge D^l) \Big] \\ &\quad + U_d^l \Big[(A'_{i_2} \wedge A'_{i_3}) \cdot (B^j \wedge C^k) \Big] \end{aligned} \tag{14.96}$$

From equation (14.95) it then follows

$$\begin{aligned}(A'_{i_2}\wedge A'_{i_3})\cdot(C^k\wedge D^l) = {} & \lambda^1_{i_1}\, D_4\cdot C^k\; C_4\cdot D^l \\ & - \lambda^2_{i_1}\, D_4\cdot C^k\; B_4\cdot D^l \\ & - \lambda^3_{i_1}\, B_4\cdot C^k\; C_4\cdot D^l\end{aligned} \tag{14.97a}$$

$$\begin{aligned}(A'_{i_2}\wedge A'_{i_3})\cdot(B^j\wedge D^l) = {} & \lambda^1_{i_1}\, D_4\cdot B^j\; C_4\cdot D^l \\ & - \lambda^2_{i_1}\, D_4\cdot B^j\; B_4\cdot D^l \\ & + \lambda^3_{i_1}\, C_4\cdot B^j\; B_4\cdot D^l\end{aligned} \tag{14.97b}$$

$$\begin{aligned}(A'_{i_2}\wedge A'_{i_3})\cdot(B^j\wedge C^k) = {} & - \lambda^1_{i_1}\, C_4\cdot B^j\; D_4\cdot C^k \\ & - \lambda^2_{i_1}\, D_4\cdot B^j\; B_4\cdot C^k \\ & + \lambda^3_{i_1}\, C_4\cdot B^j\; B_4\cdot C^k\end{aligned} \tag{14.97c}$$

Each of these three equations has a linear combination of three rank 1, 3-valence tensors on its right hand side. Furthermore, none of the rank 1, 3-valence tensors from one equation is repeated in any of the others. Therefore, substituting equations (14.97) into equation (14.96) gives a decomposition of Q in terms of 9 rank 1 tensors. Since this is a minimal decomposition, Q is of rank 9.

14.5.4 Degrees of Freedom of Q

Substituting equations (14.97) back into equation (14.96) gives

$$\begin{aligned}Q^{ijkl} = {} & \varepsilon^j_{ba}\left[\lambda^1_i \varepsilon^k_{cd}\varepsilon^l_{dc} - \lambda^2_i \varepsilon^k_{cd}\varepsilon^l_{db} + \lambda^3_i \varepsilon^k_{cb}\varepsilon^l_{dc}\right] \\ & - \varepsilon^k_{ca}\left[\lambda^1_i \varepsilon^j_{bd}\varepsilon^l_{dc} - \lambda^2_i \varepsilon^j_{bd}\varepsilon^l_{db} + \lambda^3_i \varepsilon^j_{bc}\varepsilon^l_{db}\right] \\ & + \varepsilon^l_{da}\left[\lambda^1_i \varepsilon^j_{bc}\varepsilon^k_{cd} - \lambda^2_i \varepsilon^j_{bd}\varepsilon^k_{cb} + \lambda^3_i \varepsilon^j_{bc}\varepsilon^k_{cb}\right]\end{aligned} \tag{14.98}$$

This decomposition of Q has $9\times3+3\times3-1 = 35$ DOF. The general formula for the DOF of Q gives $4\times11-15 = 29$ DOF. Therefore the parameterisation of Q in equation (14.98) is overdetermined. However, it will still give a self-consistent Q.

14.5.5 Constraints on Q

The constraints on Q can again be found very easily through geometric considerations. Let the points $\{X^{ijk}_Q\}$ be defined as

$$X^{ijk}_Q \equiv \langle\!\langle A^i B^j C^k \rangle\!\rangle \tag{14.99}$$

A point X^{ijk}_Q can be interpreted as the intersection of line $\langle\!\langle A^i B^j \rangle\!\rangle$ with plane $\langle\!\langle C^k \rangle\!\rangle$. Therefore,

$$X^{ijk_a}_Q \wedge X^{ijk_b}_Q \wedge X^{ijk_c}_Q = 0 \tag{14.100}$$

because the three intersection points $X_Q^{ijk_a}$, $X_Q^{ijk_b}$ and $X_Q^{ijk_c}$ lie along line $\langle\!\langle A^i B^j \rangle\!\rangle$. Hence, also their projections onto an image plane have to lie along a line. Thus, projecting the intersection points onto an image plane D we have

$$\begin{aligned} 0 &= (X_Q^{ijk_a} \cdot D^{l_a})\,(X_Q^{ijk_b} \cdot D^{l_b})\,(X_Q^{ijk_c} \cdot D^{l_c}) \\ &\quad (D_{l_a} \wedge D_{l_b} \wedge D_{l_c}) \\ \Longleftrightarrow\; 0 &= Q^{ijk_a l_a}\, Q^{ijk_b l_b}\, Q^{ijk_c l_c} [\![D_{l_a} D_{l_b} D_{l_c} D_4]\!]_d \\ &= \epsilon_{l_a l_b l_c}\, Q^{ijk_a l_a}\, Q^{ijk_b l_b}\, Q^{ijk_c l_c} \\ &= \det(Q^{ijkl})_{kl} \end{aligned} \tag{14.101}$$

Similarly, this type of constraint may be shown for every pair of indices. We therefore get the following constraints on Q.

$$\begin{aligned} &\det(Q^{ijkl})_{ij} = 0;\; \det(Q^{ijkl})_{ik} = 0;\; \det(Q^{ijkl})_{il} = 0 \\ &\det(Q^{ijkl})_{jk} = 0;\; \det(Q^{ijkl})_{jl} = 0;\; \det(Q^{ijkl})_{kl} = 0 \end{aligned} \tag{14.102}$$

14.5.6 Relation between Q and T

We can find the relation between Q and T via the method employed to find the relation between T and F. For example,

$$\begin{aligned} 0 &= \langle\!\langle A^1 A^2 A^3 \rangle\!\rangle \wedge \langle\!\langle B^j C^k D^l \rangle\!\rangle \wedge \langle\!\langle B^j C^k \rangle\!\rangle \\ &= \textstyle\sum_{i_1} \left(\langle\!\langle A^{i_1} B^j C^k D^l \rangle\!\rangle \langle\!\langle A^{i_2} A^{i_3} B^j C^k \rangle\!\rangle \right) \\ &= Q^{ijkl} T_i{}^{jk} \end{aligned} \tag{14.103}$$

Similarly, equations for the other possible trifocal tensors can be found. Because of the trifocal tensor symmetry detailed in equation (14.75) all trifocal tensors may be evaluated from the following set of equations.

$$\begin{aligned} &Q^{ijkl}\, T^{ABC}_{i\;\,jk} = 0;\; Q^{ijkl}\, T^{ABD}_{i\;\,jl} = 0;\; Q^{ijkl}\, T^{ACD}_{i\;\,kl} = 0 \\ &Q^{ijkl}\, T^{BAC}_{j\;\,ik} = 0;\; Q^{ijkl}\, T^{BAD}_{j\;\,il} = 0;\; Q^{ijkl}\, T^{BCD}_{j\;\,kl} = 0 \\ &Q^{ijkl}\, T^{CAB}_{k\;\,ij} = 0;\; Q^{ijkl}\, T^{CAD}_{k\;\,il} = 0;\; Q^{ijkl}\, T^{CBD}_{k\;\,jl} = 0 \\ &Q^{ijkl}\, T^{DAB}_{l\;\,ij} = 0;\; Q^{ijkl}\, T^{DAC}_{l\;\,ik} = 0;\; Q^{ijkl}\, T^{DBC}_{l\;\,jk} = 0 \end{aligned} \tag{14.104}$$

Note that the trifocal tensors found in this way will not be of consistent scale. To fix the scale we start by defining intersection points

$$\begin{aligned} X^{jkl}_{BCD} &\equiv \left[A_4 \wedge \langle\!\langle B^j C^k \rangle\!\rangle \right] \vee \langle\!\langle C^k D^l \rangle\!\rangle \\ &\simeq \varepsilon^{k}_{ca} \langle\!\langle B^j C^k D^l \rangle\!\rangle \end{aligned} \tag{14.105}$$

Projecting these points onto image plane A gives

$$\begin{aligned} X^{jkl}_{BCD_a} &\equiv X^{jkl}_{BCD} \cdot A^i A_i \\ &\simeq \varepsilon^k_{ca} \langle\!\langle B^j C^k D^l \rangle\!\rangle \cdot A^i A_i \\ &\simeq \varepsilon^k_{ca} \langle\!\langle A^i B^j C^k D^l \rangle\!\rangle A_i \\ &= \varepsilon^k_{ca} Q^{ijkl} A_i \end{aligned} \tag{14.106}$$

But we could have also arrived at an expression for $X^{jkl}_{BCD_a}$ via

$$\begin{aligned} X^{jkl}_{BCD_a} &\simeq \left(\langle\!\langle B^j C^k \rangle\!\rangle \cdot L^a_{i_a}\right)\left(\langle\!\langle C^k D^l \rangle\!\rangle \cdot L^a_{i_b}\right)\left[A_4 \wedge L^{i_a}_a\right] \vee L^{i_b}_a \\ &\simeq \left(T^{ABC}_{i_1 jk} T^{ACD}_{i_2 kl} - T^{ABC}_{i_2 jk} T^{ACD}_{i_1 kl}\right) A_{i_3} \end{aligned} \tag{14.107}$$

Equating this with equation (14.106) gives

$$T^{ABC}_{i_1 jk} T^{ACD}_{i_2 kl} - T^{ABC}_{i_2 jk} T^{ACD}_{i_1 kl} \simeq \varepsilon^k_{ca} Q^{i_3 jkl} \tag{14.108}$$

This equation may be expressed more concisely in terms of the standard cross product.

$$T^{ABC}_{\bullet jk} \times T^{ACD}_{\bullet kl} \simeq \varepsilon^k_{ca} Q^{\bullet jkl} \tag{14.109}$$

Furthermore, from the intersection points

$$X^{kjl}_{CBD} \equiv \left[A_4 \wedge \langle\!\langle C^k B^j \rangle\!\rangle\right] \vee \langle\!\langle B^j D^l \rangle\!\rangle$$

and their projections onto image plane A we get

$$T^{ABC}_{\bullet jk} \times T^{ABD}_{\bullet jl} \simeq \varepsilon^j_{ba} Q^{\bullet jkl} \tag{14.110}$$

We can now find the correct scales for T^{ABC} by demanding that

$$\frac{T^{ABC}_{i_1 jk} T^{ACD}_{i_2 kl} - T^{ABC}_{i_2 jk} T^{ACD}_{i_1 kl}}{Q^{i_3 jkl}} = \phi \tag{14.111}$$

for all j while keeping i_1, k and l constant, where ϕ is some scalar. Furthermore, we know that

$$\frac{T^{ABC}_{i_1 jk} T^{ABD}_{i_2 jl} - T^{ABC}_{i_2 jk} T^{ABD}_{i_1 jl}}{Q^{i_3 jkl}} = \phi \tag{14.112}$$

for all k while keeping i_1, k and l constant, where ϕ is some different scalar. Equations (14.111) and (14.112) together fix the scales of T^{ABC} completely. Note that we do not have to know the epipoles ε^k_{ca} and ε^j_{ba}.

Similarly, all the other trifocal tensors can be found. These in turn can be used to find the fundamental matrices and the epipoles.

14.6 Reconstruction and the Trifocal Tensor

In the following we will investigate a computational aspect of the trifocal tensor. In particular we are interested in the effect the determinant constraints have on the "quality" of a trifocal tensor. That is, a trifocal tensor calculated only from point matches has to be compared with a trifocal tensor calculated form point matches while enforcing the determinant constraints.

For the calculation of the former a simple linear algorithm is used that employs the trilinearity relationships, as, for example, given by Hartley in [102]. In the following this algorithm will be called the "7pt algorithm".

To enforce all the determinant constraints, an estimate of the trifocal tensor is first found using the 7pt algorithm. From this tensor the epipoles are estimated. Using these epipoles the image points are transformed into the epipolar frame. With these transformed point matches the trifocal tensor can then be found in the epipolar basis.

It can be shown [142] that the trifocal tensor in the epipolar basis has only 7 non-zero components[6]. Using the image point matches in the epipolar frame these 7 components can be found linearly. The trifocal tensor in the "normal" basis is then recovered by tranforming the trifocal tensor in the epipolar basis back with the initial estimates of the epipoles. The trifocal tensor found in this way has to be fully self-consistent since it was calculated from the minimal number of parameters. That also means that the determinant constraints have to be fully satisfied. This algorithm will be called the "MinFact" algorithm.

The main problem with the MinFact algorithm is that it depends crucially on the quality of the initial epipole estimates. If these are bad, the trifocal tensor will still be perfectly self-consistent but will not represent the true camera structure particularly well. This is reflected in the fact that typically a trifocal tensor calculated with the MinFact algorithm does not satisfy the trilinearity relationships as well as a trifocal tensor calculated with the 7pt algorithm, which is of course calculated to satisfy these relationships as well as possible.

Unfortunately, there does not seem to be a way to find the epipoles and the trifocal tensor in the epipolar basis simultaneously with a linear method. In fact, the trifocal tensor in a "normal" basis is a *non-linear* combination of the epipoles and the 7 non-zero components of the trifocal tensor in the epipolar basis.

Nevertheless, since the MinFact algorithm produces a fully self-consistent tensor, the camera matrices extracted from it also have to form a self-consistent set. Reconstruction using such a set of camera matrices may be expected to be better than reconstruction using an inconsistent set of camera

---

[6] From this it follows directly that the trifocal tensor has 18 DOF: 12 epipolar components plus 7 non-zero components of the trifocal tensor in the epipolar basis minus 1 for an overall scale.

matrices, as typically found from an inconsistent trifocal tensor. The fact that the trifocal tensor found with the MinFact algorithm may not resemble the true camera structure very closely, might not matter too much, since reconstruction is only exact up to a projective transformation. The question is,

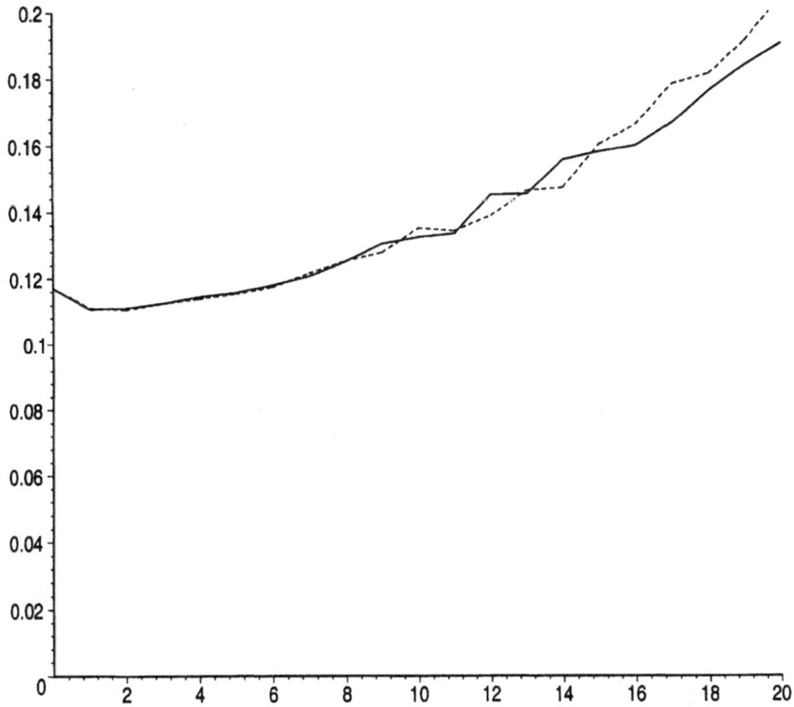

Fig. 14.3. Mean distance between original points and reconstructed points in arbitrary units as a function of mean Gaussian error in pixels introduced by the cameras. The solid line shows the values using the MinFact algorithm, and the dashed line the values for the 7pt algorithm

of course, *how* to measure the quality of the trifocal tensor. Here the quality is measured by how good a reconstruction can be achieved with the trifocal tensor in a geometric sense. This is done as follows:

1. A 3D-object is projected onto the image planes of the three cameras, which subsequently introduce some Gaussian noise into the projected point coordinates. These coordinates are then quantised according to the simulated camera resolution. The magnitude of the applied noise is measured in terms of the mean Gaussian deviation in pixels.

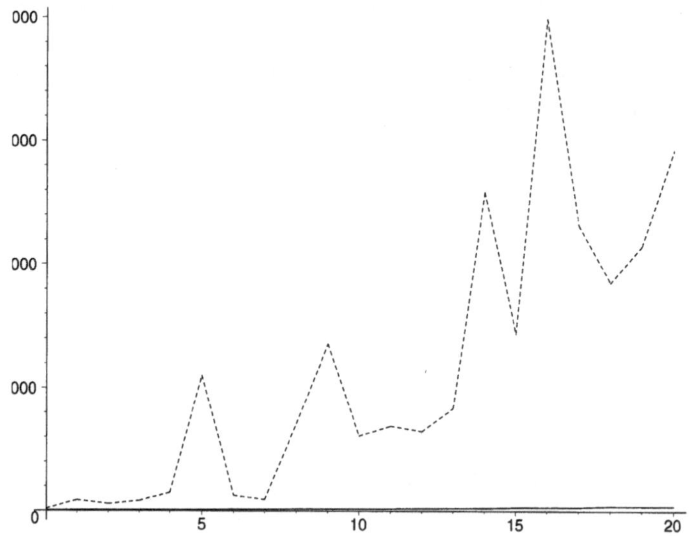

Fig. 14.4. Mean difference between elements of calculated and true tensors in percent. Solid line shows values for trifocal tensor calculated with 7pt algorithm, and dashed line shows values for trifocal tensor calculated with MinFact algorithm

2. The trifocal tensor is calculated in one of two ways from the available point matches:
   a) using the 7pt algorithm, or
   b) using the MinFact algorithm.
3. The epipoles and the camera matrices are extracted from the trifocal tensor. The camera matrices are evaluated using Hartley's recomputation method [102].
4. The points are reconstructed using a version of what is called "Method 3" in [194] and [195] adapted for three views. This uses a SVD to solve for the homogeneous reconstructed point algebraically using a set of camera matrices. In [194] and [195] this algorithm was found to perform best of a number of reconstruction algorithms.
5. This reconstruction still contains an unknown projective transformation. Therefore it cannot be compared directly with the original object. However, since only synthetic data is used here, the 3D-points of the original object are known exactly. Therefore, a projective transformation matrix that best transforms the reconstructed points into the true points can be calculated. Then the reconstruction can be compared with the original 3D-object geometrically.
6. The final measure of "quality" is arrived at by calculating the mean distance in 3D-space between the reconstructed and the true points.

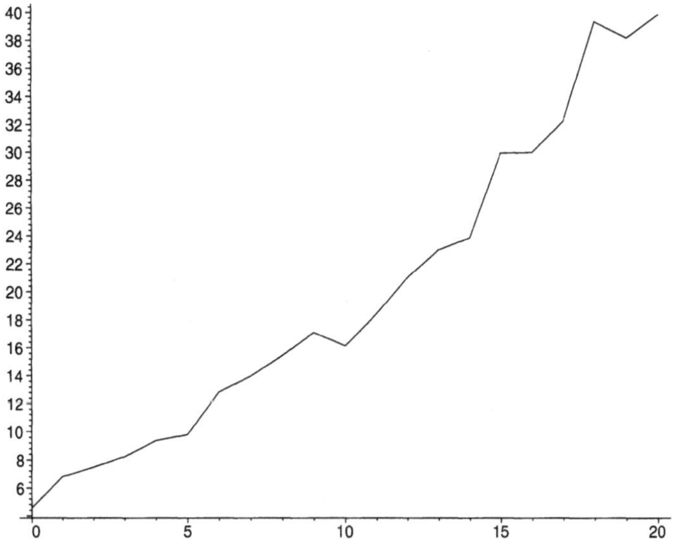

Fig. 14.5. Mean difference between elements of true trifocal tensor and trifocal tensor calculated with 7pt algorithm in percent

These quality values are evaluated for a number of different noise magnitudes. For each particular noise magnitude the above procedure is performed 100 times. The final quality value for a particular noise magnitude is then taken as the average of the 100 trials.

Figure 14.3 shows the mean distance between the original points and the reconstructed points in 3D-space in some arbitrary units[7], as a function of the noise magnitude. The camera resolution was 600 by 600 pixels.

This figure shows that for a noise magnitude of up to approximately 10 pixels both trifocal tensors seem to produce equally good reconstructions. Note that for zero added noise the reconstruction quality is not perfect. This is due to the quantisation noise of the cameras. The small increase in quality for low added noise compared to zero added noise is probably due to the cancellation of the quantisation and the added noise.

Apart from looking at the reconstruction quality it is also interesting to see how close the components of the calculated trifocal tensors are to those of the true trifocal tensor. Figures 14.4 and 14.5 both show the mean of the percentage differences between the components of the true and the calculated trifocal tensors as a function of added noise in pixels. Figure 14.4 compares the trifocal tensors found with the 7pt and the MinFact algorithms. This shows that the trifocal tensor calculated with the MinFact algorithm is indeed very

[7] The particular object used was 2 units wide, 1 unit deep and 1.5 units high in 3D-space. The Y-axis measures in the same units.

different to the true trifocal tensor, much more so than the trifocal tensor calculated with the 7pt algorithm (shown enlarged in figure 14.5).

Table 14.1. Comparison of Multiple View Tensors

Fundamental Matrix	Trifocal Tensor
$F_{i_1 j_1} = \langle\!\langle A^{i_2} A^{i_3} B^{j_2} B^{j_3} \rangle\!\rangle$	$T_{i_1}{}^{jk} = \langle\!\langle A^{i_2} A^{i_3} B^{j} C^{k} \rangle\!\rangle$
$F_{i j_1} = \varepsilon^{j_3}_{ba} {K^b_{j_2}}_i - \varepsilon^{j}_{ba} {K^b_{j_3}}_i$	$T_i{}^{jk} = \varepsilon^{j}_{ba} {K^c_{k}}_i - \varepsilon^{k}_{ca} {K^b_{j}}_i$
$F_{ij} = L^a_i \cdot \underbrace{(B_j \wedge B_4)}_{\text{line}}$	$T_i{}^{jk} = L^a_i \cdot \underbrace{\langle\!\langle B^j C^k \rangle\!\rangle}_{\text{line}}$
$\det F = 0$	$\det(T_i{}^{jk})_{jk} = 0$ for each i
7 DOF	18 DOF
rank 2	rank 4

Quadfocal Tensor
$Q^{ijkl} = \langle\!\langle A^i B^j C^k D^l \rangle\!\rangle$
$\begin{aligned} Q^{i_1 jkl} = \quad & \varepsilon^{j}_{ba} \left[{K^c_k}_{i_3} {K^d_l}_{i_2} - {K^d_l}_{i_3} {K^c_k}_{i_2} \right] \\ - \; & \varepsilon^{k}_{ca} \left[{K^b_j}_{i_3} {K^d_l}_{i_2} - {K^d_l}_{i_3} {K^b_j}_{i_2} \right] \\ + \; & \varepsilon^{l}_{da} \left[{K^b_j}_{i_3} {K^c_k}_{i_2} - {K^c_k}_{i_3} {K^b_j}_{i_2} \right] \end{aligned}$
$Q^{ijkl} = A^i \cdot \underbrace{\langle\!\langle B^j C^k D^l \rangle\!\rangle}_{\text{point}}$
$\det(Q^{ijkl})_{xy} = 0$ where x and y are any pair of $\{ijkl\}$
29 DOF
rank 9

The data presented here seems to indicate that a tensor that obeys the determinant constraints, i.e. is self-consistent, but does not satisfies the trilinearity relationships particularly well is equally as good, in terms of reconstruction ability, as an inconsistent trifocal tensor that satisfies the trilinearity relationships quite well. In particular the fact that the trifocal tensor calculated with the MinFact algorithm is so very much different to the true trifocal tensor (see figure 14.4) does not seem to have a big impact on the final recomputation quality.

14.7 Conclusion

Table 14.1 summarises the expressions for the different tensors, their degrees of freedom, their rank and their main constraints. In particular note the similarities between the expressions for the tensors.

We have demonstrated in this paper how Geometric Algebra can be used to give a unified formalism for multiple view tensors. Almost all properties of the tensors could be arrived at from geometric considerations alone. In this way the Geometric Algebra approach is much more intuitive than traditional tensor methods. We have gained this additional insight into the workings of multiple view tensors because Projective Geometry in terms of Geometric Algebra allows us to describe the geometry on which multiple view tensors are based, directly. Therefore, we can understand their "inner workings" and inter-relations. The best examples of this are probably the derivations of the constraints on T and Q which followed from the fact that the intersection points of a line with three planes all have to lie along a line. It is hard to imagine a more trivial fact.

A similar analysis of multiple view tensors was presented by Heyden in [116]. However, we believe our treatment of the subject is more intuitive due to its geometric nature. In particular the "quadratic p-relations" used by Heyden were here replaced by the geometric fact that the intersection point of a line with a plane lies on that line.

We hope that our unified treatment of multiple view tensors has not just demonstrated the power of Geometric Algebra, but will also give a useful new tool to researchers in the field of Computer Vision.

15. 3D-Reconstruction from Vanishing Points

Christian B.U. Perwass[1] **and Joan Lasenby**[2]

[1] Cavendish Laboratory, Cambridge
[2] C. U. Engineering Department, Cambridge

15.1 Introduction

3D-reconstruction is currently an active field in Computer Vision, not least because of its many applications. It is applicable wherever the "real world" has to be understood by a computer. This may be with regard to control movement (robots), to survey a scene for later interpretation (medicine), or to create and mix artificial with real environments (special effects).

Research on 3D-reconstruction can roughly be separated into three areas:

1. *Reconstruction with calibrated cameras*, [143, 141, 115, 79, 152]. In this case, a set of images is taken of a scene with one or more calibrated cameras. However, the camera positions are unknown. To perform a 3D-reconstruction we therefore first have to reconstruct the camera positions. To do this it is assumed that image point matches are known.
2. *Reconstruction from sequences of images*, [228, 230, 229, 238, 71, 62, 159]. Here a series of monocular, binocular or trinocular images is taken. To perform a reconstruction it is then assumed that point matches between the views in space and over time are known, and that the relative camera geometry and their internal parameters do not change. A popular method in this area is the use of the Kruppa equations [133, 78].
3. *Reconstruction from static views*, [38, 44]. A set of images of a scene taken with unknown cameras, from unknown positions is given. We still

assume that we have point matches over the images. However, note that we cannot assume anymore that the internal parameters of the cameras that took the images are the same.

The least information about a scene is given in point 3. In fact, there is so little information that a correct 3D-reconstruction is *impossible*, as we have shown in chapter 14. Therefore, some additional information is needed. Such information could be the knowledge of lengths, angles or parallel lines.

Our approach to 3D-reconstruction falls into the area of *Reconstruction from static views.* We have two images taken with unknown cameras from unknown positions and assume that apart from the point matches we also know the projections of a number of sets of parallel world lines. The latter are used to find vanishing points but also to constrain the reconstruction. This information allows us to perform an affine reconstruction of the scene. That is, we find the rotation, translation and the internal parameters of the second camera *relative* to the first. If we assume furthermore, that we have three mutually orthogonal sets of parallel lines, we can also find the internal calibration of the first camera and thus obtain a Euclidean 3D-reconstruction.

In the following discussion of our reconstruction algorithm we use the same notation as in chapter 14. We will also assume that the reader is familiar with our description of reciprocal frames, pinhole cameras, camera matrices and the basic form of the fundamental matrix. Of course, all this assumes some familiarity with Geometric Algebra (GA).

15.2 Image Plane Bases

We will be working in projective space ($\mathbb{P}^3$) with basis $\{e_1, e_2, e_3, e_4\}$ which has signature $\{---+\}$. We can project down to the corresponding Euclidean space ($\mathbb{E}^3$) via the projective split. Our general setup is that we have two pinhole cameras described by frames $\{A_\mu\}$ and $\{B_\mu\}$, respectively. The frame $\{A_\mu\}$ is also regarded as the world frame which we use for our reconstruction.

The basic form of our calculation is as follows. We start with the image points obtained from real cameras, i.e. in $\mathbb{E}^3$. These image points are embedded in $\mathbb{P}^3$. All our calculations are then performed in $\mathbb{P}^3$ and the resultant reconstruction is projected back into $\mathbb{E}^3$. This method forces us to take note of two important concepts.

1. **Correct Basis.** The power of GA in this field derives from the fact that we are not working purely with coordinates, but with the *underlying geometric basis.* Therefore, we have to make sure that the basis we are working with is actually appropriate for our problem.
2. **Scale Invariance.** The projection of homogeneous vectors into $\mathbb{E}^3$ is independent of the overall scale of the homogeneous vector. Calculations in $\mathbb{P}^3$ may depend on such an overall scale, though. We have to make sure that all our calculations are *invariant* under a scaling of the homogeneous

vectors, because such a scaling cannot and should not have any influence on our final result. Furthermore, since we are initially embedding vectors from $\mathbb{E}^3$ in $\mathbb{P}^3$ we are not given any particular scale. Any expression that is invariant under a scaling of its component homogeneous vectors will be called *scale invariant*.

As mentioned above, the frames $\{A_\mu\}$ and $\{B_\mu\}$ define two pinhole cameras. Since $\{A_\mu\}$ also serves as our world frame in $\mathbb{P}^3$ we can choose that A_4, the optical centre of camera A, sits at the origin. A_1, A_2 and A_3 define the image plane of camera A. If we want to be true to our previously stated concepts, we need to give some thought as to how we should choose the $\{A_i\}$.

Note here that we use latin indices to count from 1 to 3 and greek indices to count from 1 to 4. We also make use of the Einstein summation convention, i.e. if a superscript index is repeated as a subscript within a product, a summation over the range of the index is implied. Hence, $\alpha^i A_i \equiv \sum_{i=1}^3 \alpha^i A_i$.

The images we obtain from real cameras are 2-dimensional. Therefore, the image point coordinates we get are of the form $\{x, y\}$, which give the displacement in a horizontal and vertical direction[1] in the image coordinate frame. However, in $\mathbb{P}^3$ an image plane is defined by three vectors. Therefore, a point on a plane in $\mathbb{P}^3$ is defined by *three* coordinates. A standard way given in the literature to extend the 2D image point coordinates obtained from a real camera to $\mathbb{P}^3$ is by writing the vector $\{x, y\}$ as $\{x, y, 1\}$. This is a well founded and very practical choice, and if we just worked with matrices and tensors we would not need to do anything else. However, since we want to tap into the power of GA, we need to understand what kind of basis is *implicitly assumed* when we write our image point coordinates in the form $\{x, y, 1\}$.

The best way to proceed is first to describe a 2D-image point in a 3D basis and then to embed this point in $\mathbb{P}^3$. An image point $\{x, y\}$ gives the horizontal and vertical displacement in the 2D-image plane coordinate frame. Let the basis corresponding to this 2D frame in $\mathbb{E}^3$ be $\{\boldsymbol{a}_1, \boldsymbol{a}_2\}$. If we define a third vecor $\boldsymbol{a}_3$ to point to the origin of the 2D frame in $\mathbb{E}^3$, then an image point with coordinates $\{x, y\}$ can be expressed as follows in $\mathbb{E}^3$.

$$\boldsymbol{x}_a = x\,\boldsymbol{a}_1 + y\,\boldsymbol{a}_2 + 1\,\boldsymbol{a}_3 = \hat{\alpha}^i\,\boldsymbol{a}_i, \tag{15.1}$$

with $\{\hat{\alpha}^i\} \equiv \{x, y, 1\}$. The $\{\hat{\alpha}^i\}$ are the image point coordinates corresponding to image point $\{x, y\}$ in $\mathbb{E}^3$. Now we embed the point $\boldsymbol{x}_a$ in $\mathbb{P}^3$.

$$X_a = (\boldsymbol{x}_a \cdot e_4) + e_4 = \hat{\alpha}^i A_i, \tag{15.2}$$

where we defined $A_1 \equiv \boldsymbol{a}_1 \cdot e_4$, $A_2 \equiv \boldsymbol{a}_2 \cdot e_4$ and $A_3 \equiv (\boldsymbol{a}_3 \cdot e_4) + e_4$. That is, A_1 and A_2 are *direction vectors*, or points at infinity, because they have no e_4 component. However, they still lie on image plane A. More precisely, they lie on the intersection line of image plane A with the plane at infinity. Note

[1] Note that although we call these directions horizontal and vertical, they may not be at a 90 degree angle to each other in general.

that A_1 and A_2 do not project to $\boldsymbol{a}_1$ and $\boldsymbol{a}_2$, respectively, when projected back to Euclidean space. For example,

$$\frac{A_1 \wedge e_4}{A_1 \cdot e_4} = \frac{\boldsymbol{a}_1}{0} \longrightarrow \infty. \tag{15.3}$$

Nevertheless, $\{A_i\}$ is still the projective image plane basis we are looking for, as can be seen when we project X_a down to Euclidean space.

$$\boldsymbol{x}_a = \frac{X_a \wedge e_4}{X_a \cdot e_4} = \frac{\hat{\alpha}^i \boldsymbol{a}_i}{\hat{\alpha}^3} = x\,\boldsymbol{a}_1 + y\,\boldsymbol{a}_2 + 1\,\boldsymbol{a}_3\ ; \quad \hat{\alpha}^3 \equiv 1. \tag{15.4}$$

What is important here is that neither $\hat{\alpha}^1$ nor $\hat{\alpha}^2$ appear in the denominator. This shows that by writing our image point coordinates in the form $\{x, y, 1\}$ we have implicitly assumed this type of frame, which we will call a **normalised homogeneous camera frame**. The camera frames we will use in the following are all normalised homogeneous camera frames.

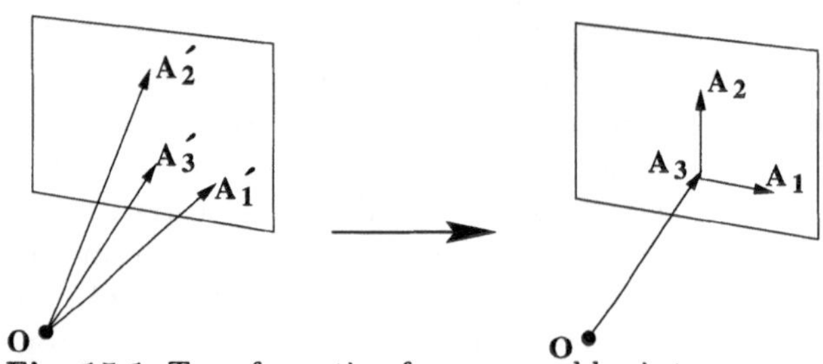

Fig. 15.1. Transformation from general basis to normalised homogeneous camera frame, in which image points have coordinates of the type $\{x, y, 1\}$

Figure 15.1 shows the difference between a general image plane basis in $\mathbb{P}^3$, denoted by $\{A'_i\}$, and a normalised homogeneous camera frame $\{A_i\}$. Note that homogeneous vectors A_1 and A_2 are drawn as lying in the image plane to indicate that they are direction vectors.

It might seem a bit odd that we have devoted so much space to the development of normalised homogeneous camera frames. However, this has far reaching implications later on and is *essential* to understand our derivation.

In $\mathbb{P}^3$ a point on the image plane of camera A can be written as $X_a = \alpha^i A_i$ in general. We can normalise the coordinates without changing the projection of X_a into $\mathbb{E}^3$. That is, $X_a \simeq \bar{\alpha}^i A_i$ with $\bar{\alpha}^i \equiv \alpha^i/\alpha^3$. The symbol $\simeq$ means equality up to a scalar factor. In this case we clearly have $\{\bar{\alpha}^i\} = \{\hat{\alpha}^i\}$.

A general point in $\mathbb{P}^3$ can be written as $X = \alpha^\mu A_\mu$ in the A-frame. We can normalise the coordinates of X_a in the same way as before to obtain $X \simeq \bar{\alpha}^\mu A_\mu$ with $\bar{\alpha}^\mu \equiv \alpha^\mu/\alpha^3$. If we project this point down to $\mathbb{E}^3$ we get[2]

$$\boldsymbol{x} = \frac{X \wedge e_4}{X \cdot e_4} = \frac{\bar{\alpha}^i}{1 + \bar{\alpha}^4} \boldsymbol{a}_i = \hat{\alpha}^i \boldsymbol{a}_i, \tag{15.5}$$

with $\hat{\alpha}^i \equiv \bar{\alpha}^i/(1 + \bar{\alpha}^4)$. That is, if $\bar{\alpha}^4 = 0$, then X is a point on the image plane of camera A. Also, if $\bar{\alpha}^4 = -1$ then X is a point at infinity. We will call $\bar{\alpha}^4$ the **projective depth** of a point in $\mathbb{P}^3$.

15.3 Plane Collineation

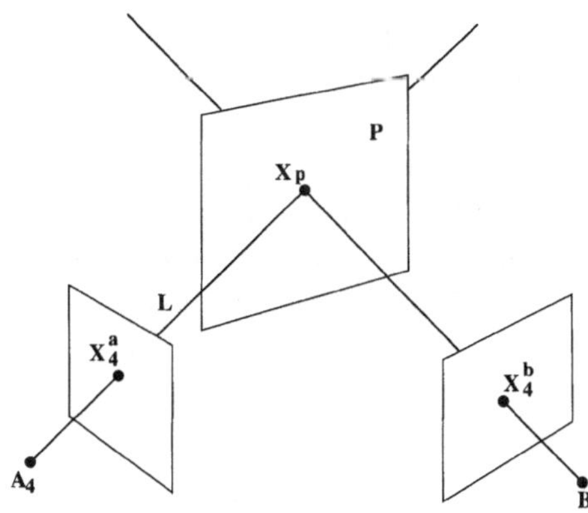

Fig. 15.2. Schematic representation of a plane collineation. Image point X_4^a is projected to X_4^b under the P-collineation

Before we can get started on the actual reconstruction algorithm, we need to derive some more mathematical objects which we will need as tools. The problem we want to solve first is the following. Let us assume we have three image point matches in cameras A and B. That is, if three points in space, $\{X_i\}$, are projected onto image planes A and B to give images $\{X_i^a\}$ and $\{X_i^b\}$ respectively, then we know that the pairs $\{X_i^a, X_i^b\}$ are images of the same point in space. If the three points in space do not lie along a line, they define a plane. This plane induces a collineation, which means that we can transfer image points from camera A to camera B through that plane. For example, let X_4^a be the image point on image plane A which we want to transfer to camera B through the plane. First we have to find the intersection point of line $A_4 \wedge X_4^a$ with the plane[3], and then we project this intersection point onto image plane B (see figure 15.2). This transformation can also be

[2] Recall that $A_4 = e_4$ (the origin of $\mathbb{P}^3$) and that the $\{A_i\}$ are a normalised homogeneous camera frame.

[3] Recall that A_4 is the optical centre of camera A.

represented by a 3×3 matrix, which is called a collineation matrix. Our goal is to find the collineation induced by the plane $P \equiv X_1 \wedge X_2 \wedge X_3$ by knowing the projections of the points $\{X_i\}$ onto image planes A and B, and the fundamental matrix for the two cameras. Since we know the fundamental matrix we can also calculate the epipoles. The epipoles on the two image planes are always projections of a single point in space and thus give us the projections of a fourth point on any plane in space. That is, we have in fact the projections of four points that lie on some plane P. Hence, we can find the collineation matrix directly through a matrix diagonalisation.

However, it is interesting to see what this means geometrically. Faugeras gives a geometrical interpretation[4] in [78]. We will follow his construction method to obtain a $3 \times 3 \times 3$ collineation tensor.

We start by defining three points $X_i = \alpha_i^{\mu} A_{\mu}$. The projections of these three points onto image planes A and B are $X_i^a = \bar{\alpha}_i^j A_j$ and $X_i^b = \bar{\beta}_i^j B_j$, respectively. We know the coordinates $\{\bar{\alpha}_i^j\}$ and $\{\bar{\beta}_i^j\}$, and we know that the pairs $\{\bar{\alpha}_i^j, \bar{\beta}_i^k\}$ are images of the same point in space. Furthermore, we have the fundamental matrix for the two cameras. We find the collineation induced by the plane $P = X_1 \wedge X_2 \wedge X_3$ geometrically through a two step construction.

Step 1:

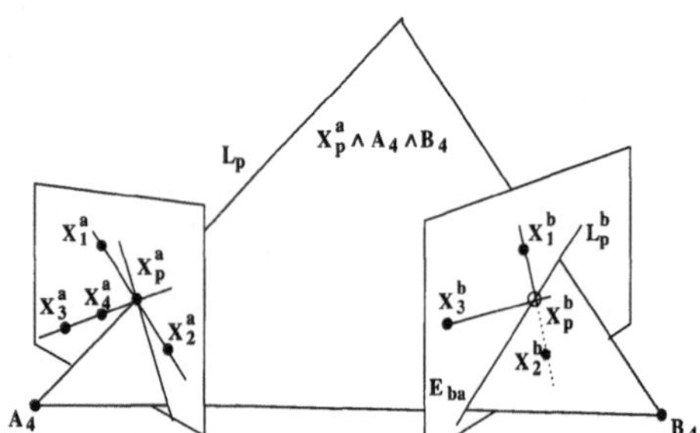

Let $X_4^a = \alpha_4^i A_i$ be the image point we want to project onto image plane B under the P-collineation. Now consider the intersection point X_p^a of lines $X_3^a \wedge X_4^p$ and $X_1^a \wedge X_2^a$. The intersection point of line $L_p \equiv A_4 \wedge X_p^a$ with an arbitrary plane in $\mathbb{P}^3$ obviously lies on L_p. Denote the projection of L_p onto image plane B by L_p^b. Obviously X_p^a can only be projected to some point on L_p^b, independent of the collineation. We also know that X_p^a has to project to some point on the line $X_1^b \wedge X_2^b$ under the specific P-collineation. Hence, X_p^b is the intersection point of lines L_p^b and $X_1^b \wedge X_2^b$. We can also write this as

$$X_p^b = (X_p^a \wedge A_4 \wedge B_4) \vee (X_1^b \wedge X_2^b) \tag{15.6}$$

[4] In [78] this method is called the **Point-Plane** procedure.

Step 2:

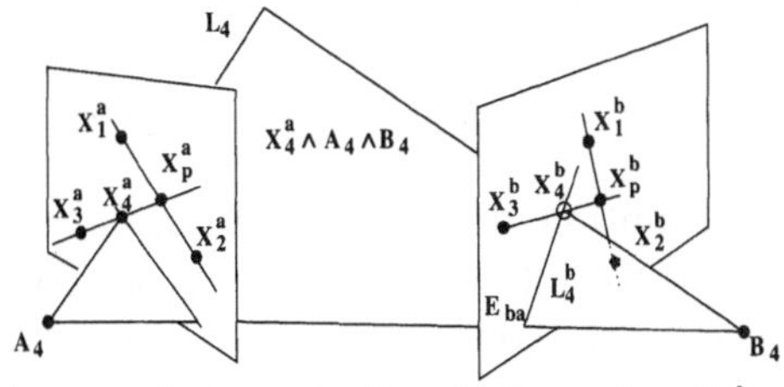

Now that we have calculated the point X_p^b, we can project X_4^a under the P-collineation in an analogue way. We form a line $L_4 = A_4 \wedge X_4^a$ which we project onto image plane B. X_4^b, the projection of X_4^a under the P-collineation, is then the intersection point of L_4^b and line $X_3^b \wedge X_p^b$. This can also be expressed as

$$X_4^b = (X_4^a \wedge A_4 \wedge B_4) \vee (X_3^b \wedge X_p^b) \tag{15.7}$$

By substituting equation (15.6) into equation (15.7) we can find a collineation tensor M_{ij}^k. Details of this calculation can be found in [186]. The resultant expression for M_{ij}^k is

$$M_{ij}^k \equiv \Big[\quad \Big(F(1,2)\, \bar{\lambda}_{a\,i}^{1}\, \bar{\beta}_1^k - F(2,1)\, \bar{\lambda}_{a\,i}^{2}\, \bar{\beta}_2^k\Big)\, f_{j3}^b \tag{15.8}$$
$$- \Big(F(1,2)\, \bar{\lambda}_{a\,i}^{1}\, f_{j1}^b - F(2,1)\, \bar{\lambda}_{a\,i}^{2}\, f_{j2}^b\Big)\, \bar{\beta}_3^k \Big],$$

with

$$F(r,s) \equiv \bar{\alpha}_r^i \bar{\beta}_s^j\, F_{ij}\ ; \quad f_{ir}^b \equiv \bar{\beta}_r^j\, F_{ij}\ ; \quad \bar{\lambda}_{a\,k_1}^{j_1} \equiv (\bar{\alpha}_{j_2}^{k_2} \bar{\alpha}_{j_3}^{k_3} - \bar{\alpha}_{j_2}^{k_3} \bar{\alpha}_{j_3}^{k_2}), \tag{15.9}$$

where F_{ij} is the fundamental matrix for the two cameras. Here, and throughout the rest of this chapter, indices of the type $\{i_1, i_2, i_3\}$ are taken to be an even permutation of $\{1,2,3\}$. Also indices of the type $\{\mu_1, \mu_2, \mu_3, \mu_4\}$ are an even permutation of $\{1,2,3,4\}$.

To project a point $X_4^a = \bar{\alpha}_4^i A_i$ on image plane A, onto image plane B under the collineation described by points $\{X_1, X_2, X_3\}$, we can now simply write

$$\beta_4^k \simeq \bar{\alpha}_4^i \bar{\alpha}_4^j\, M_{ij}^k, \tag{15.10}$$

where the $\{\bar{\beta}_4^j\}$ are the coordinates of the projected point $X_4^b = \bar{\beta}^j B_j$ on image plane B. It can be shown that M_{ij}^k is scale invariant [186].

Equation (15.10) seems to indicate that a collineation is a quadratic relation. However, we know that $\beta_4^k = \alpha_4^i H_i^k$ where H_i^k is the collineation matrix. If we take a closer look at the components of equation (15.8) we find that $\bar{\lambda}_{a\,3}^r$ is linearly dependent on $\bar{\lambda}_{a\,1}^r$ and $\bar{\lambda}_{a\,2}^r$. Therefore, the three matrices in indices i,j of M_{ij}^k are of rank 2. We can write equation (15.10) as

$$\begin{aligned}\beta_4^k \simeq\ & \bar{\alpha}^1\bar{\alpha}^1\ M_{11}^k + \bar{\alpha}^2\bar{\alpha}^2\ M_{22}^k + \bar{\alpha}^1\bar{\alpha}^2\ (M_{12}^k + M_{21}^k) \\ & +\bar{\alpha}^1\ (M_{13}^k + M_{31}^k) + \bar{\alpha}^2\ (M_{23}^k + M_{32}^k) + \bar{\alpha}^3\ M_{33}^k\end{aligned} \tag{15.11}$$

since $\bar{\alpha}^3 = 1$ by definition. Thus, if we perform a set of similarity transforms on M_{ij}^k such that the components $M_{11}^k, M_{22}^k, M_{12}^k, M_{21}^k$ are zero, we can read off the components of the collineation matrix from the transformed M_{ij}^k. Such a similarity transformation on M_{ij}^k is possible because the matrices in indices i, j of M_{ij}^k are of rank 2.

15.4 The Plane at Infinity and Its Collineation

It will be very useful for us to see what the collineation of the plane at infinity looks like. Recall that $A_4 = e_4$ and that the $\{A_i\}$ form a normalised homogeneous camera frame. That is, A_1 and A_2 are direction vectors. Therefore, the plane at infinity P_∞ may be given by

$$P_\infty = A_1 \wedge A_2 \wedge (A_3 - A_4) \tag{15.12}$$

Now that we have the plane at infinity we can also find an expression for the collineation matrix associated with it. More details of the following calculation can be found in [186].

We want to project a point $X^a = \alpha^i A_i$ on image plane A to image plane B under the P_∞-collineation. First we have to find the intersection point X_p of line $L = A_4 \wedge X^a$ with P_∞.

$$X_p = (A_4 \wedge X^a) \vee P_\infty \simeq \alpha^i A_i - \alpha^3 A_4 \tag{15.13}$$

Now we need to find the projection X_p^b of X_p onto image plane B.

$$X_p^b = X_p \cdot B^j\ B_j = \left(\alpha^i K_{j_i}^b - \alpha^3 \varepsilon_{ba}^j\right) B_j \tag{15.14}$$

where $K_{j_i}^b \equiv A_i \cdot B^j$ is the 3×3 camera matrix minor of camera B, and $\varepsilon_{ba}^j \equiv A_4 \cdot B^j$ is the epipole of camera B and also the fourth column of the full camera matrix[5]. Note that we use here a notation of relative super- and subscripts to keep the absolute superscript position free for other uses. From equation (15.14) it follows that we can write the collineation matrix of P_∞ as

$$\Psi_{j_i}^\infty \equiv [K_{j_1}^b, K_{j_2}^b, K_{j_3}^b - \varepsilon_{ba}^j] \tag{15.15}$$

[5] The full camera matrix is given by $K_{j_\mu}^b = A_\mu \cdot B^j$. See chapter 14for details on camera matrices and epipoles.

where i counts the columns. Therefore, if we want to project a point $X_a = \alpha^i A_i$ on image plane A, onto image plane B under the P_∞-collineation we can write

$$\beta^j_\infty \simeq \alpha^i \Psi^\infty_{j_i}. \tag{15.16}$$

What does the P_∞-collineation describe geometrically? If X_a is an image point in camera A and X^∞_b is its projection under the P_∞-collineation, then from the construction of the collineation it follows that the lines $L_a = A_4 \wedge X_a$ and $L_b = B_4 \wedge X^\infty_b$ meet in a point on P_∞. If two lines meet in a point on the plane at infinity, they are parallel. Therefore, the P_∞-collineation tells us which two image points X_a and X^∞_b on image planes A and B, repectively, correspond such that the lines $A_4 \wedge X_a$ and $B_4 \wedge X^\infty_b$ are parallel. Obviously, this tells us something about the relative orientation of the two cameras.

We can use our knowledge of the relation between Ψ^∞ and the camera matrix to find the depths of a set of world points whose projections are known in both cameras, if we also know the projections of at least three pairs of parallel lines. We will assume for the moment that for each point pair $\{\bar{\alpha}^i, \bar{\beta}^j\}$ we also know $\bar{\beta}^j_\infty$, which is the projection of $\bar{\alpha}^i$ under the P_∞-collineation. From the definition of the camera matrix we know that

$$\beta^j = \alpha^i K^b_{j_i} + \alpha^4 \varepsilon^j_{ba}. \tag{15.17}$$

Furthermore, equation (15.16) may be rewritten as

$$\beta^j_\infty \simeq \alpha^i K^b_{j_i} - \alpha^3 \varepsilon^j_{ba} \tag{15.18}$$

We can now combine equations (15.17) and (15.18) to obtain the following expression (see [186] for details).

$$\bar{\alpha}^4 = \bar{\alpha}^i \bar{K}^b_{3_i}\, \zeta^j_1 - \zeta^j_2\ ; \quad j \in \{1,2\}. \tag{15.19}$$

with

$$\zeta^j_1 \equiv \frac{\bar{\beta}^j_\infty - \bar{\beta}^j}{\bar{\beta}^j - \bar{\varepsilon}^j_{ba}}\ ; \quad \zeta^j_2 \equiv \frac{\bar{\beta}^j_\infty - \bar{\varepsilon}^j_{ba}}{\bar{\beta}^j - \bar{\varepsilon}^j_{ba}} \tag{15.20}$$

Since equation (15.19) has to give the same result for both $j = 1$ and $j = 2$ independent of $\bar{K}^b_{3_i}$, it follows that $\zeta^1_1 = \zeta^2_1$ and $\zeta^1_2 = \zeta^2_2$. Therefore, we will discard the superscript of the ζs in the following.

Equation 15.19 by itself is still not useful, since we neither know $\bar{\alpha}^4$ nor $\bar{K}^b_{3_i}$. However, if we had some constraints on the projective depths ($\bar{\alpha}^4$) for a number of points we could find $\bar{K}^b_{3_i}$. Once $\bar{K}^b_{3_i}$ is known for a particular camera setup, we can use it to calculate the depths for any point matches. Before we show how $\bar{K}^b_{3_i}$ can be evaluated, we will take a closer look at how to find the $\{\bar{\beta}^j_\infty\}$.

15.5 Vanishing Points and P_∞

We mentioned earlier that the $\{\beta^j_\infty\}$ are the projections of the $\{\alpha^i\}$ onto image plane B under the P_∞-collineation. We can find the P_∞-collineation Ψ^∞ from the projection pairs of three points on P_∞ and the fundamental matrix.

If two parallel world lines are projected onto an image plane, their projections are only parallel if the image plane is parallel to the world lines. The intersection point of the projections of two parallel world lines is called a *vanishing point.*

Two parallel world lines meet at infinity. In projective space $\mathbb{P}^3$ this may be expressed by saying that the intersection point of two parallel world lines lies on P_∞. Points on P_∞ may also be interpreted as directions. Therefore, intersecting a line with P_∞ gives its direction. In this light, a vanishing point is the projection of the intersection point of two parallel lines. Or, in other words, it is the projection of a direction.

If we knew three vanishing points which are projections of three mutually orthogonal directions, we would know how a basis for the underlying Euclidean space $\mathbb{E}^3$ projects onto the camera used. This information can be used to find the internal camera calibration [44]. Here our initial goal is to find the *relative* camera calibration of the two cameras. We can then find an affine reconstruction. To achieve this, we do not require the vanishing points to relate to orthogonal directions. However, the more mutually orthogonal the directions related to the vanishing points are, the better the reconstruction will work.

15.5.1 Calculating Vanishing Points

Before we go any further with the actual reconstruction algorithm, let us take a look at how to calculate the vanishing points. Suppose we have two image point pairs $\{\bar{\alpha}^i_{u1}, \bar{\alpha}^i_{u2}\}$ and $\{\bar{\alpha}^i_{v1}, \bar{\alpha}^i_{v2}\}$, defining two lines on image plane A, which are projections of two parallel world lines. The vanishing point is the intersection of lines L_u and L_v where

$$L_u = \lambda^u_i L^i_a \quad ; \quad L_v = \lambda^v_i L^i_a, \tag{15.21}$$

and

$$\lambda^u_{i_1} \equiv \bar{\alpha}^{i_2}_{u1}\bar{\alpha}^{i_3}_{u2} - \bar{\alpha}^{i_3}_{u1}\bar{\alpha}^{i_2}_{u2} \quad ; \quad \lambda^v_{i_1} \equiv \bar{\alpha}^{i_2}_{v1}\bar{\alpha}^{i_3}_{v2} - \bar{\alpha}^{i_3}_{v1}\bar{\alpha}^{i_2}_{v2}, \tag{15.22}$$

are the homogeneous line coordinates. Also note that $L^{i_1}_a \equiv A_{i_2} \wedge A_{i_3}$ (see chapter 14). The intersection point X^a_{uv} of lines L_u and L_v is then given by

$$X^a_{uv} = L_u \vee L_v = \alpha^i_{uv} A_i, \tag{15.23}$$

where

$$\alpha_{uv}^{i_1} \equiv (\lambda_{i_2}^{v}\,\lambda_{i_3}^{u} - \lambda_{i_3}^{v}\,\lambda_{i_2}^{u}). \tag{15.24}$$

First of all note that the $\{\alpha_{uv}^{i}\}$ define a point in $\mathbb{P}^2$. Since we defined A_1 and A_2 to be directions, the image point coordinates $\{x, y\}$ in $\mathbb{E}^2$ corresponding to the $\{\alpha_{uv}^{i}\}$, are found to be $\{\bar{\alpha}_{uv}^{1}, \bar{\alpha}_{uv}^{2}\}$ through the projective split, where $\bar{\alpha}_{uv}^{i} \equiv \alpha_{uv}^{i}/\alpha_{uv}^{3}$. Note that points which lie at infinity in $\mathbb{E}^2$ can be expressed in $\mathbb{P}^2$ by points which have a zero third component. Such points will also be called directions.

The fact that points at infinity in $\mathbb{E}^2$ are nothing special in $\mathbb{P}^2$ shows an immediate advantage of using homogeneous coordinates for the intersection points over using 2D-coordinates. Since we are looking for the intersection point of the projections of two parallel world lines, it may so happen, that the projections are also parallel, or nearly parallel. In that case, the 2D image point coordinates of the vanishing point would be very large or tend to infinity. This, however, makes them badly suited for numerical calculations. When using homogeneous coordinates, on the other hand, we do not run into any such problems.

15.5.2 Vanishing Points from Multiple Parallel Lines

Above we described how to find a vanishing point from the projections of two parallel world lines. In practical applications the lines will only be known with a finite precision and will also be subject to a measurment error. Therefore, we could improve on the quality of a vanishing point if sets of more than two parallel lines are known. In particular, the vanishing point quality is improved if these parallel lines are taken from varying depths within in world scene. In [186] we discuss a standard method, which consists of finding the null space of a matrix of the homogeneous line coordinates. This method gives us the best fitting vanishing point in homogeneous coordinates, in the least squares sense.

Note that in [38] vanishing points are found as 2D-image point coordinates, which means that only parallel world lines can be used that are not parallel in the image. In [44] the projections of at least three parallel world lines have to be known to calculate a vanishing point. The implementation of our algorithm switches automatically between finding a vanishing point from two parallel lines, and calculating it from multiple parallel lines, depending on how much information is available.

15.5.3 Ψ^∞ from Vanishing Points

Now we return to our reconstruction algorithm. We discussed vanishing points since they are projections of points on P_∞. If we know three vanishing point matches over cameras A and B and the epipoles, we can calculate the P_∞-collineation matrix Ψ^∞. Once we have Ψ^∞ we can find the projections of some image points $\{\bar{\alpha}_n^i\}$ on image plane A, onto image plane B under the P_∞-collineation. That is,

$$\bar{\beta}_{n\,\infty}^k \simeq \bar{\alpha}_n^i \, \Psi_i^\infty \tag{15.25}$$

We can now use the $\{\bar{\beta}_{n\,\infty}^j\}$ to find the $\{\zeta_n^j\}$ for equation (15.19).

15.6 3D-Reconstruction of Image Points

Now that we have found Ψ^∞ and thus can calculate the $\{\zeta_n\}$ from equation (15.20), we can think about how to find the correct depth values for the image point matches $\{\bar{\alpha}_n^i, \bar{\beta}_n^j\}$.

We will perform an affine reconstruction. That is, we reconstruct in the frame of camera A. When we plot our final reconstructed points we will assume that the A-frame forms an orthonormal frame of $\mathbb{E}^3$, though. However, we do not need to assume anything about the frame of camera B, since we will find the translation, rotation and internal calibration of camera B *relative* to camera A. To find the internal calibration of camera A relative to an orthonormal frame of $\mathbb{E}^3$, we would need to know the projection of this orthonormal set of directions onto camera A [44].

We have already found sets of parallel lines to calculate vanishing points. We can reuse these sets of lines to constrain the depth values found with equation (15.19). In particular, we will regard the $\{\bar{K}_{3_i}^b\}$ as free parameters. If we now take the image point matches that define the projections of two parallel world lines, we can use this extra information to constrain the $\{\bar{K}_{3_i}^b\}$. That is, we vary the free parameters until the reconstructed points define a pair of parallel world lines again.

15.6.1 The Geometry

Before we start developing an algorithm to find the best $\{\bar{K}_{3_i}^b\}$ we will take a quick look at the relevant geometry. In figure 15.3 we have drawn the geometry underlying our reconstruction algorithm.

A_4 and B_4 are the optical centres of cameras A and B, respectively. We have also chosen A_4 to lie at the origin of $\mathbb{E}^3$. Recall that A_1, A_2 and B_1, B_2 are direction vectors in $\mathbb{P}^3$. We have drawn these vectors here as lying on the image planes to indicate this.

A world point X is projected onto image planes A and B giving projections X_a and X_b, respectively. X_b^∞ is the projection of X_a onto image plane B under the P_∞-collineation. Also, E_{ba} is the epipole of camera B.

Now we can see what the $\{\zeta_{1n}, \zeta_{2n}\}$ components from equation (15.19) express.

$$\zeta_{1n} \equiv \frac{\bar{\beta}_{n\,\infty}^j - \bar{\beta}_n^j}{\bar{\beta}_n^j - \bar{\varepsilon}_{ba}^j}$$

gives the ratio of the distance (in x or y direction) between X_b^∞ and X_b, and X_b and E_{ba}.

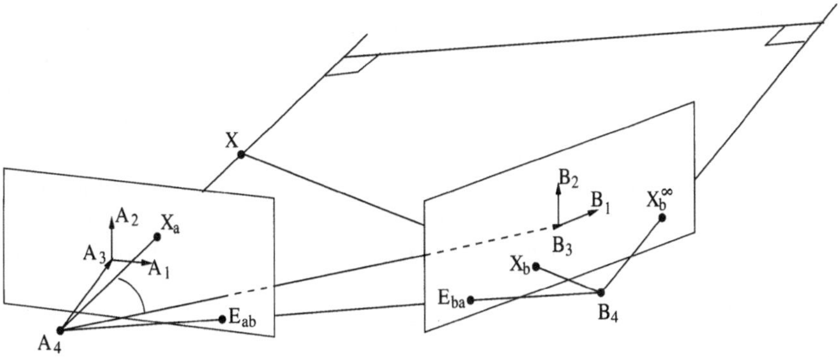

Fig. 15.3. This figure shows the geometry behind equation (15.19). A point X is projected onto cameras A and B, giving images X_a and X_b, respectively. Projecting X_a onto image plane B under the P_∞-collineation gives X_b^∞. We choose A_4 to be the origin of $\mathbb{E}^3$. $K^b_{\underset{i}{3}}$ gives the components of A_1, A_2 and A_3 along B_3

$$\zeta_{2n} \equiv \frac{\bar{\beta}^j_{n\,\infty} - \bar{\varepsilon}^j_{ba}}{\bar{\beta}^j_n - \bar{\varepsilon}^j_{ba}}$$

gives the ratio of the distance (in x or y direction) between X_b^∞ and E_{ba}, and X_b and E_{ba}.

Recall that $K^b_{\underset{i}{3}} = A_i \cdot B^3$, that is, it gives the components of the $\{A_i\}$ along B_3. Therefore, varying the $\{K^b_{\underset{i}{3}}\}$ means that we are moving B_3, which is the principal point on image plane B. Since X_b^∞ cannot change when we vary $K^b_{\underset{i}{3}}$ the relation between B_3 and B_4 is fixed. Thus, changing B_3 means changing B_4. In this respect, finding the correct $\{K^b_{\underset{i}{3}}\}$ means finding the correct translation of camera B relative to camera A. The relative rotation has already been fixed through finding P_∞.

However, it is only the relative sizes of the $\{\bar{K}^b_{\underset{i}{3}}\}$ that are really important. An overall scale factor will only change the depths of all reconstructed points simultaneously. Therefore, we can fix the depth of one image point, to fix the scale of $\bar{K}^b_{\underset{i}{3}}$.

15.6.2 The Minimization Function

We mentioned before that we will use our knowledge of parallel lines once again to constrain the $\{\bar{K}^b_{\underset{i}{3}}\}$ from equation (15.19). Let $L^a_u = X^a_{u1} \wedge X^a_{u2}$ and $L^a_v = X^a_{v1} \wedge X^a_{v2}$ be the projections of two parallel world lines onto image plane A. In general we define world points and image points as

$$\left.\begin{aligned} X_{ur} &\equiv \bar{\alpha}^\mu_{ur}\, A_\mu\ ; \quad X^a_{ur} \equiv \bar{\alpha}^i_{ur}\, A_i \\ X_{vr} &\equiv \bar{\alpha}^\mu_{vr}\, A_\mu\ ; \quad X^a_{vr} \equiv \bar{\alpha}^i_{vr}\, A_i \end{aligned}\right\} r \in \{1, \ldots, n\}. \tag{15.26}$$

Furthermore, if we know the image points on image plane B corresponding to X^a_{u1}, X^a_{u2}, X^a_{v1} and X^a_{v2}, and we have found Ψ^∞, then we can calculate the corresponding ζs from equation (15.20). Equation (15.19) will now allow us to find the projective depths for X^a_{u1}, X^a_{u2}, X^a_{v1} and X^a_{v2}. Therefore, we can calculate the world lines $L_u = X_{u1} \wedge X_{u2}$ and $L_v = X_{v1} \wedge X_{v2}$.

Now, we know that L_u and L_v are supposed to be parallel, which means that they have to intersect P_∞ in the same point. This will be the constraint which we will use to find the correct $\{\bar{K}^b_{3_i}\}$. Let X^∞_u and X^∞_v be defined as

$$X^\infty_u \equiv L_u \vee P_\infty \; ; \quad X^\infty_v \equiv L_v \vee P_\infty . \tag{15.27}$$

Lines L_u and L_v are parallel iff

$$X^\infty_u \wedge X^\infty_v = 0 \tag{15.28}$$

Instead of using this condition we could also project L_u and L_v into $\mathbb{E}^3$, and then check that they are parallel. However, projecting into $\mathbb{E}^3$ means dividing through the projective depth, which means that our free parameters are now in the denominator of a minimisation function. Apart from creating a minimisation surface with singularities, the derivatives of such a minimisation function will be more complicated and thus cost more computing time.

Finding the Minimisation Parameters. The following expression for X^∞_u is derived in more detail in [186].

$$X^\infty_u = L_u \vee P_\infty = \chi^i_u \, A^\infty_i \tag{15.29}$$

where

$$\begin{aligned} &\chi^i_u \equiv (\bar{\lambda}^u_{i3} + \bar{\lambda}^u_{i4}) \; ; \quad \bar{\lambda}^u_{\mu_1\mu_2} \equiv \bar{\alpha}^{\mu_1}_{u1} \, \bar{\alpha}^{\mu_2}_{u2} - \bar{\alpha}^{\mu_2}_{u1} \, \bar{\alpha}^{\mu_1}_{u2} \\ &A^\infty_1 \equiv A_1 \; ; \quad A^\infty_2 \equiv A_2 \; ; \quad A^\infty_3 \equiv A_3 - A_4 \end{aligned} \tag{15.30}$$

The free parameters we have are the $\{\bar{K}^b_{3_i}\}$. To make future equations somewhat clearer we will define $\varphi_i \equiv \bar{K}^b_{3_i}$. Hence, equation (15.19) will be written as

$$\bar{\alpha}^4_n = \bar{\alpha}^i_n \, \zeta_{1n} \, \varphi_i - \zeta_{2n} . \tag{15.31}$$

Recall that lines L_u and L_v are parallel iff $X^\infty_u \wedge X^\infty_v = 0$. We can now write this expression in terms of the $\{\chi^i\}$.

$$X^\infty_u \wedge X^\infty_v = \Lambda^{uv}_i \, L^i_\infty \; ; \quad \Lambda^{uv}_{i_1} \equiv \chi^{i_2}_u \, \chi^{i_3}_v - \chi^{i_3}_u \, \chi^{i_2}_v \tag{15.32}$$

with $L^{i_1}_\infty \equiv A^\infty_{i_2} \wedge A^\infty_{i_3}$. Each of the $\{\Lambda^{uv}_i\}$ has to be zero if $X^\infty_u \wedge X^\infty_v = 0$. Therefore, from an analytical point of view, the expression we should try to minimise for each parallel line pair $\{L_u, L_v\}$ is

$$\Delta^{uv} : \varphi_j \longrightarrow \sum_{i=1}^{3} (\Lambda^{uv}_i)^2 . \tag{15.33}$$

Improving Computational Accuracy. However, for a computer with finite floating point precision, this equation poses a problem. The culprits in this case are the $\{\chi^i\}$. Recall that they give the direction of a line in homogeneous coordinates. Before they are used in equation (15.32) they should be normalised to improve the precision of the equation on a computer.

$$\hat{\chi}_u^i \equiv \frac{\chi_u^i}{\sqrt{\sum_i (\chi_u^i)^2}} \tag{15.34}$$

Therefore, the minimisation function we will use is

$$\Delta^{uv} : \varphi_j \longrightarrow \sum_{i=1}^{3} (\hat{\Lambda}_i^{uv})^2 \; ; \quad \hat{\Lambda}_{i_1}^{uv} \equiv \hat{\chi}_u^{i_2} \hat{\chi}_v^{i_3} - \hat{\chi}_u^{i_3} \hat{\chi}_v^{i_2} \tag{15.35}$$

The Derivatives. The derivative of Δ^{uv} is computationally not a particularly expensive expression. Therefore, we can use a minimisation routine that also uses the derivatives of the minimisation function. This will make the minimisation process more efficient and robust. Details about the derivatives can be found in [186].

Implementing the Depth Constraint. At the moment the minimisation function Δ^{uv} depends on three parameters: the $\{\varphi_j\}$. However, we mentioned earlier that we can fix, the depth of one point. This will reduce the number of free parameters to two. How this is done best is described in [186]. It turns out that constraining the depth of one point is necessary. Otherwise the minimisation routine tries to push the whole scene to infinity.

The Minimisation Routine. We use a modified version of the *conjugate gradient* method to perform the minimisation. This modified version is called *MacOpt* and was developed by David MacKay [160]. It makes a number of improvements over the conjugate gradient method as given in [190]. MacOpt assumes that the minimisation surface is fundamentally convex with no local minima. However, our surface is only of that shape near the absolute minimum[6]. It turns out that the success rate of finding the absolute minimum can be improved if we first use the unnormalised χs to step towards the minimum, and then use the normalised χs to find the minimum with high accuracy. This is because the minimisation surface for the unnormalised χs is of a convex shape, whereas the minimisation surface for the normalised χs has a number of local minima.

---

[6] A number of examples of minimisation surfaces and their corresponding reconstructions are demonstrated by the program **MVT**, which can be downloaded from C.Perwass' home page. This program runs under Windows 95/98 and NT4/5.

Image Point Normalisation. Before we can calculate the collineation tensor for the P_∞-collineation we have to find the fundamental matrix (F) for the two views (see equation (15.8)). For the calculation of the fundamental matrix we cannot use the pixel coordinates directly, because they are typically too large to obtain good accuracy in our numerical calculations. This is also true for all other calculations performed here. Therefore, we need to scale the image point coordinates so that they are of order 1.

In [102] Hartley suggests that the scales and skews applied to the image point coordinates are found in the following way. The skew is given by the coordinates of the centroid of all image points. Then the average distance of the skewed image points from the origin is calculated. The inverse of that distance gives the scale.

This is a good method if we just wanted to calculate F. However, it turns out that for our purposes such a scaling is not suitable. In fact, we found that it is important to conserve the aspect ratio of the images (separately), and to ensure that the origin of the image plane is chosen in the same way in both images.

We choose the image plane origin to be in the centre of each image plane and then scale the image points by dividing their x and y coordinate by the image resolution in the x-direction. This preserves the aspect ratio.

15.7 Experimental Results

We can now outline the structure of our reconstruction algorithm.

Step 1: We find point matches and sets of projections of parallel lines over the two images.

Step 2: We calculate three vanishing points and the fundamental matrix. This allows us to find the P_∞-collineation matrix Ψ^∞.

Step 3: We select a set of parallel lines that we want to use to constrain our minimisation. Note that one pair of parallel lines may be enough. More pairs do not necessarily improve the result, since they may not be consistent due to errors.

Step 4: The image points on image plane A which define the chosen parallel lines are projected onto image plane B under the P_∞-collineation with Ψ^∞.

Step 5: We can now find the $\{K^b_{3_i}\}$ by minimising equation (15.33) or equation (15.35).

Step 6: Once we have found $K^b_{3_i}$ we can use it in conjunction with Ψ^∞ in equation (15.31) to reconstruct any other image point matches for this camera setup.

15.7.1 Synthetic Data

To test the quality of the reconstructions we created synthetic data. The advantage of using synthetic data is that we can get a *geometric* quality measure of the reconstruction. Also if an algorithm fails with synthetic data it is clearly unlikely to work with real data.

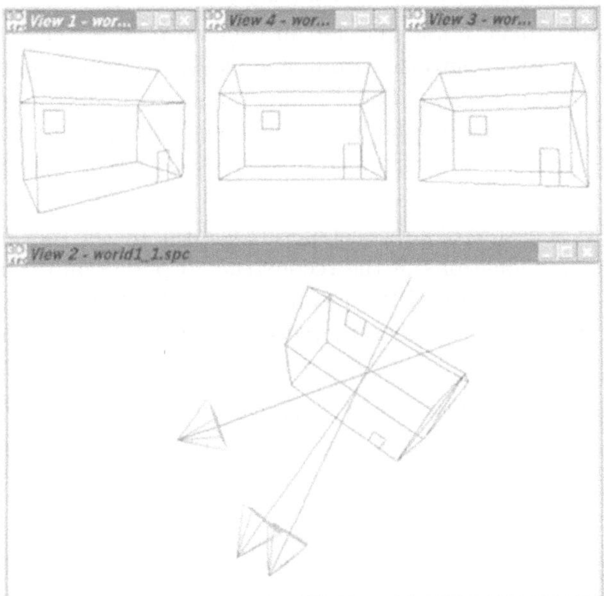

Fig. 15.4. The synthetic data was created from projections of the house onto the cameras

The lower picture in figure 15.4 shows a house with three cameras. The three smaller pictures on top show the projections of the house onto the three image planes. The house consists of 18 vertices, which were all used in our calculations. We performed two trials: trial 1 uses an orthogonal set of vanishing points. Trial 2 uses two orthogonal vanishing points but the third vanishing point is found from the two lines on the roof which are vertically sloping and closest to the camera. In each trial we also tested two camera configurations: the camera to the very left and the very right, and the two cameras which are close together. The former will be called the *far cameras* and the latter the *close cameras* configuration.

Recall that we can and, in fact, have to fix the depth of one point. Since we know the true points we can set this depth to its true value. Also remember that we perform our reconstruction in the frame of one of the cameras. But we also know this frame and can therefore transform our reconstructed points to lie in the appropriate frame. The reconstruction obtained in this way can then be compared *directly* with the true object.

In our experiments we added a Gaussian error with a mean deviation between 0 and 12 pixels to the image points. The camera resolutions were 600×600 pixels. For each setting of the mean deviation of the induced error we calculated the $\{K^b_{3_i}\}$ 100 times, each time with different errors, to obtain a statistically meaningful result. Each calculation of the $\{K^b_{3_i}\}$ can be used to reconstruct any image point matches in the two images. Therefore, we projected the house again onto the two image planes, again introducing an error of the same mean deviation. These image points are then reconstructed and compared with the true points. This was done 20 times for each calculation of the $\{K^b_{3_i}\}$. This way we obtained a separation of the calibration and the reconstruction.

The quality measure of a reconstruction is given by the root mean squared error between the locations of the reconstructed points and the true points. That is, we take the root of the mean of the sum of the distances squared between the true and the reconstructed points. We evaluated the RMS error over the 20 reconstructions for each calibration (i.e. calculation of the $\{K^b_{3_i}\}$), and also over all calculations of the $\{K^b_{3_i}\}$ for each mean deviation of the induced error. The former will be called the "RMS/Trial" and the latter the "Total RMS".

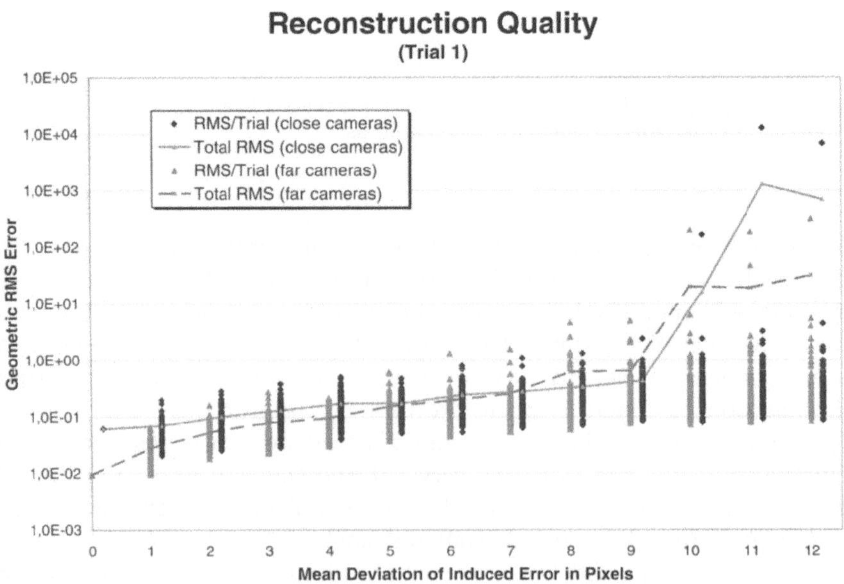

Fig. 15.5. Comparison of reconstruction quality for first trial

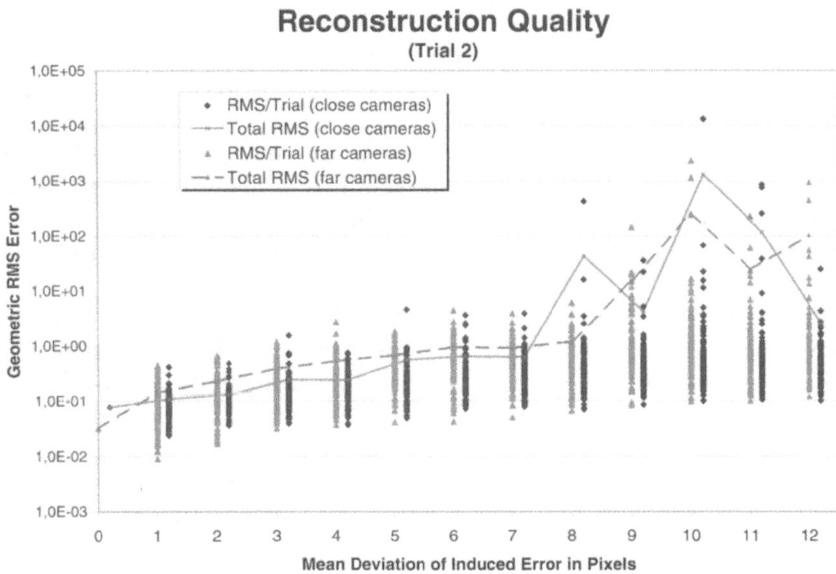

Fig. 15.6. Comparison of reconstruction quality for second trial

Figure 15.5 shows the results when using an orthogonal set of vanishing points and figure 15.6 when using a non-orthogonal set, as described above. Note that the y-axis has a $\log_{10}$ scale. The length of the house is 2 units, its total height 1.5 units and its depth 1 unit. The results for the close camera configuration are slighty displaced to the right, so that they can be distinguished from the far cameras setup.

The first thing we can see from the graphs is that as the induced error increases over 6 pixels we start to get error configurations where the algorithm breaks down. This can be either due to the minimisation getting stuck in local minima or because the absolute minimum is at a wrong position. The latter is possible since the minimisation surface depends on Ψ^{∞} and F.

Furthermore, it can be seen that the far cameras configuration is more immune to induced errors than the close cameras configuration. Also the non-orthogonal set of vanishing points fares worse than the orthogonal one. Curiously, in trial 2 the far cameras configuration is worse than the close cameras configuration.

In general it can be seen, though, that an error with a mean deviation of up to 5 pixels still gives acceptable reconstructions. It might seem odd, though, that if some error is introduced into the image points, the reconstruction can actually be better than with no noise at all. This is because even if no additional error is applied, there is still an error due to the digitisation in the cameras. Particular configurations of induced error can compensate for that by chance. However, the figures also show that the probability of the added error improving the reconstruction is about as high as making the reconstruction worse (relative to the total RMS). Nevertheless, this fact supplies us with an interesting idea: we might be able to improve our reconstructions from real data by *adding noise* to the image points. To be more precise, we could vary the image point coordinates slightly until we obtain an improved reconstruction. Since our calibration algorithm is quite fast it seems feasible to employ maximum entropy methods. We will discuss this in future work.

Note that we have calculated F with a simple method which does not enforce the rank 2 constraint on F. Nevertheless, the reconstruction quality is quite good, which seems to indicate that a highly accurate F is not very important for our algorithm. Therefore, it appears that in certain cases fully constraint evaluations of F are not necessary to obtain good results. Of course, using a fully contraint F might improve the results. Research on calculating F or the trifocal tensor (which is a related problem) optimally can be found in [102, 104, 142, 184, 80, 84, 116].

15.7.2 Real Data

Fig. 15.7. Initial images with parallel lines used for the calculation of the vanishing points and minimisation function indicated

The real test for any reconstruction algorithm is the reconstruction of a real world scene, though. Figure 15.7 shows two views of a chessboard which we used for reconstruction[7]. The original images had a resolution of

[7] These pictures were actually taken by C.Perwass' father, in a different country, with equipment unknown to the authors. They were then sent via email to the authors. That is, the only thing known about the pictures to the authors, are the pictures themselves.

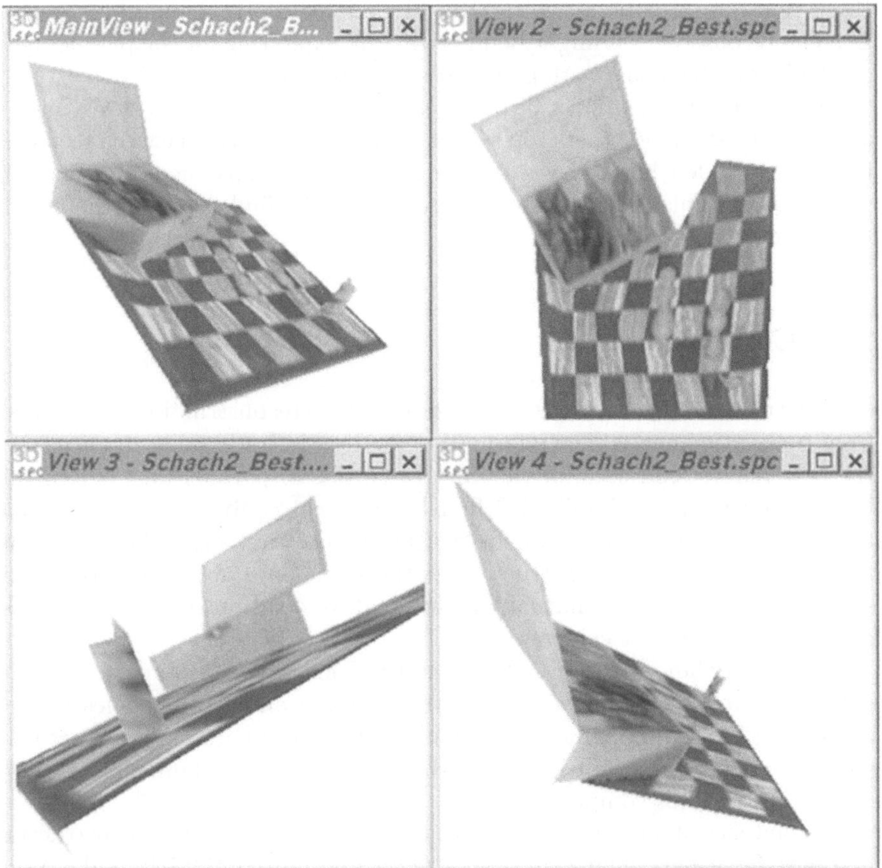

Fig. 15.8. Reconstruction of the chessboard (Schachbrett)

1280 × 960 pixels. The lines indicate the parallel lines used to calculate the vanishing points. The two sets of parallel lines on the front of the chessbox were used in the minimisation routine. The fundamental matrix used was calculated from 13 point matches. The resultant reconstruction[8] can be seen in figure 15.8.

The different views of the reconstruction show that the chessbox was reconstructed quite well. However, the chessboard is not really square. Remember, though, that this is only an affine reconstruction drawn in an orthonormal frame. That is, we assume that the camera frame is orthonormal. Furthermore, we have only used two line pairs and one line triplet to find

[8] This and other reconstructions, as well as some more analysis of the reconstruction algorithm are demonstrated by the program **MVT**, which can be downloaded from C.Perwass' home page.

three vanishing points, of which only two relate to orthogonal directions in $\mathbb{E}^3$. The reconstruction might be improved by exploiting all the parallel lines available, of which there are many on a chessboard.

Also note that the front side of the chessboard is reconstructed very nicely, at a proper right angle to its top side. The chess figure, which can be seen best in the bottom left hand view of figure 15.8, is not reconstructed particularly well, though. This is because it is very difficult to find matching point sets for round objects.

15.8 Conclusions

We have presented here an algorithm for the affine reconstruction of 3D scenes from two static images. The information we need is firstly point matches over the two images, and secondly at least three sets of parallel lines. From this information alone we implicitly[9] find the internal calibration, rotation and translation of the second camera relative to the first one. This allows us to perform an affine reconstruction of the scene. Assuming that the three sets of parallel lines are mutually orthogonal we could also find the internal calibration of the first camera.

Our algorithm is clearly not automatic. This is because apart from the point matches, combinations of vanishing points and parallel lines can be chosen freely. Also the information that certain lines in an image are actually parallel in the world, is a knowledge-based decision that humans are easily capable of, but not computers.

Advantages of our algorithm are that it is fast and that the reconstruction is robust for a particular calibration. On a PentiumII/233MHz under Windows 98 it took on average 160ms for a calibration (10000 trials). This time includes updating of dialog boxes. In an optimised program this time could probably be reduced to less than half. Robustness of the calibration depends mostly on the set of vanishing points used. The more similar the directions the vanishing points describe are, the less robust the calibration is.

We believe that apart from presenting an interesting affine reconstruction algorithm we have also shown that GA is a useful tool which allows us to gain geometric insight into a problem.

[9] Future work will look at how these entities can be found explicitly.

16. Analysis and Computation of the Intrinsic Camera Parameters*

Eduardo Bayro-Corrochano and Bodo Rosenhahn

Institute of Computer Science and Applied Mathematics,
Christian-Albrechts-University of Kiel

16.1 Introduction

The computation of the intrinsic camera parameters is one of the most important issues in computer vision. The traditional way to compute the intrinsic parameters is using a known calibration object. One of the most important methods is based on the absolute conic and it requires as input only information about the point correspondences [158, 103]. As extension a recent approach utilizes the absolute quadric [228]. Other important groups of self-calibration methods either reduce the complexity if the camera motion is known in advance, for example as translation [66], or as rotation about known angles [5, 67], or by using active strategies and e.g. the vanishing point [56].

In this chapter we re-establish the idea of the absolute conic in the context of Pascal's theorem and we get equations different to the Kruppa equations [158, 103]. Although the equations are different, they rely on the same principle of invariance of the mapped absolute conic. The consequence is that we can generate equations so that we require only a couple of images whereas the Kruppa equation method requires at least three views [158]. However, as a prior knowledge the method requires the translational motion direction of the camera and the rotation about at least one fixed axis through a known

* This work has been supported by DFG Grant So-320-2-1.

angle in addition to the point correspondences. The paper will show that although the algorithm requires the extrinsic camera parameters in advance it has the following clear advantages: It is derived from geometric observations, it does not stick in local minima in the computation of the intrinsic parameters and it does not require any initialization at all. We hope that this proposed method derived from geometric thoughts gives a new point of view to the problem of camera calibration.

The chapter is organized as follows. Section two explains the conics and the theorem of Pascal. Section three reformulates the well known Kruppa equations for computer vision in terms of algebra of incidence. Section four presents a new method for computing the intrinsic camera parameters based on Pascal's theorem. Section five is devoted to the experimental analysis and section six to the conclusion part.

16.2 Conics and the Theorem of Pascal

The role of the conics and quadrics is well known in the projective geometry [208] because of their invariant properties with respect to projective transformations. This knowledge lead to the solution of crucial problems in computer vision [172]. The derivation of the Kruppa equations relies on the conic concept. These equations have been used in the last decade to compute the intrinsic camera parameters. In this chapter we will exploit further the conics concept and use Pascal's theorem to establish an equation system with clear geometric transparency. Next, we will explain the role of conics and that of Pascal's theorem in relation with a fundamental projective invariant. This section is mostly based on the interpretation of the linear algebra together with projective geometry in the Clifford algebra framework realized by Hestenes and Ziegler [114].

When we want to use projective geometry in computer vision, we utilize homogeneous coordinate representations. Doing that, we embed the 3–D Euclidean visual space in the 3–D projective space $\mathbb{P}^3$ or $\mathbb{R}^4$ and the 2–D Euclidean space of the image plane in the 2–D projective space $\mathbb{P}^2$ or $\mathbb{R}^3$. In the geometric algebra framework we select for $\mathbb{P}^2$ the 3–D Euclidean geometric algebra $\mathbb{C}_{3,0,0}$ and for $\mathbb{P}^3$ the 4–D geometric algebra $\mathbb{C}_{1,3,0}$. The reader should see chapter 14 for more details about the connection of geometric algebra and projective geometry. Any geometric object of $\mathbb{P}^3$ will be linearly projective mapped to $\mathbb{P}^2$ via a projective transformation, for example the projective mapping of a quadric at infinity in the projective space $\mathbb{P}^3$ results in a conic in the projective plane $\mathbb{P}^2$.

Let us first consider a pencil of lines lying on the plane. Doing that, we will follow the ideas of Hestenes and Ziegler [114]. Any pencil of lines is well defined by a bivector addition of two of its lines: $\boldsymbol{l} = \boldsymbol{l}_a + s\boldsymbol{l}_b$ with $s \in \mathbb{R} \cup \{-\infty, +\infty\}$. If two pencils of lines, $\boldsymbol{l}$ and $\boldsymbol{l}' = \boldsymbol{l}'_a + s'\boldsymbol{l}'_b$, can be related one–to–one so that $\boldsymbol{l} = \boldsymbol{l}'$ for $s = s'$, we can say that they are in projective correspondence. Using this idea, the set of intersecting points of lines in correspondence build a conic. Since the intersecting points $\boldsymbol{x}$ of the line pencils $\boldsymbol{l}$ and $\boldsymbol{l}'$ fulfill for $s = s'$ the following constraints

$$\begin{aligned} \boldsymbol{x}\wedge\boldsymbol{l} &= \boldsymbol{x}\wedge\boldsymbol{l}_a + s\boldsymbol{x}\wedge\boldsymbol{l}_b = 0 \\ \boldsymbol{x}\wedge\boldsymbol{l}' &= \boldsymbol{x}\wedge\boldsymbol{l}'_a + s\boldsymbol{x}\wedge\boldsymbol{l}'_b = 0, \end{aligned} \tag{16.1}$$

the elimination of the scalar s yields a second order geometric product equation in $\boldsymbol{x}$

$$(\boldsymbol{x}\wedge\boldsymbol{l}_a)(\boldsymbol{x}\wedge\boldsymbol{l}'_b) - (\boldsymbol{x}\wedge\boldsymbol{l}_b)(\boldsymbol{x}\wedge\boldsymbol{l}'_a) = 0. \tag{16.2}$$

We can also get the parameterized conic equation simply by computing the intersecting point $\boldsymbol{x}$, taking the meet of the line pencils as follows

$$\boldsymbol{x} = (\boldsymbol{l}_a + s\boldsymbol{l}_b) \vee (\boldsymbol{l}'_a + s\boldsymbol{l}'_b) = \boldsymbol{l}_a \vee \boldsymbol{l}'_a + s(\boldsymbol{l}_a \vee \boldsymbol{l}'_b + \boldsymbol{l}_b \vee \boldsymbol{l}'_a) + s^2(\boldsymbol{l}_b \vee \boldsymbol{l}'_b). \tag{16.3}$$

Let us for now define the involved lines in terms of wedge of points $\boldsymbol{l}_a = \boldsymbol{a}\wedge\boldsymbol{b}$, $\boldsymbol{l}_b = \boldsymbol{a}\wedge\boldsymbol{b}'$, $\boldsymbol{l}'_a = \boldsymbol{a}'\wedge\boldsymbol{b}$ and $\boldsymbol{l}'_b = \boldsymbol{a}'\wedge\boldsymbol{b}'$ such that $\boldsymbol{l}_a \vee \boldsymbol{l}'_a = \boldsymbol{b}$, $\boldsymbol{l}_a \vee \boldsymbol{l}'_b = \boldsymbol{d}$, $\boldsymbol{l}_b \vee \boldsymbol{l}'_a = \boldsymbol{d}'$ and $\boldsymbol{l}_b \vee \boldsymbol{l}'_b = \boldsymbol{b}'$, see Figure 16.1.a. By substituting $\boldsymbol{b}'' = \boldsymbol{l}_a \vee \boldsymbol{l}'_b + \boldsymbol{l}_b \vee \boldsymbol{l}'_a = \boldsymbol{d} + \boldsymbol{d}'$ in the last equation, we get

$$\boldsymbol{x} = \boldsymbol{b} + s\boldsymbol{b}'' + s^2\boldsymbol{b}', \tag{16.4}$$

which represents a nondegenerated conic for $\boldsymbol{b}\wedge\boldsymbol{b}''\wedge\boldsymbol{b}' = \boldsymbol{b}\wedge(\boldsymbol{d}+\boldsymbol{d}')\wedge\boldsymbol{b}' \neq 0$. Now, using this equation let us compute the generating line pencils. Define $\boldsymbol{l}_1 = \boldsymbol{b}''\wedge\boldsymbol{b}'$, $\boldsymbol{l}_2 = \boldsymbol{b}'\wedge\boldsymbol{b}$ and $\boldsymbol{l}_3 = \boldsymbol{b}\wedge\boldsymbol{b}''$. Then using the equation (16.4), its two projective pencils are

$$\begin{aligned} \boldsymbol{b}\wedge\boldsymbol{x} &= s\boldsymbol{b}\wedge\boldsymbol{b}'' + s^2\boldsymbol{b}\wedge\boldsymbol{b}' = s(\boldsymbol{l}_3 - s\boldsymbol{l}_2) \\ \boldsymbol{b}'\wedge\boldsymbol{x} &= \boldsymbol{b}'\wedge\boldsymbol{b} + s\boldsymbol{b}'\wedge\boldsymbol{b}'' = \boldsymbol{l}_2 - s\boldsymbol{l}_1. \end{aligned} \tag{16.5}$$

Considering the points $\boldsymbol{a}$, $\boldsymbol{a}'$, $\boldsymbol{b}$ and $\boldsymbol{b}'$ and some other point $\boldsymbol{c}'$ lying on the conic depicted in Figure 16.1.a, and the equation (16.2) for $s = \rho s'$ slightly different to s', we get the bracket expression

$$\begin{aligned} &[\boldsymbol{c}'\boldsymbol{a}\boldsymbol{b}][\boldsymbol{c}'\boldsymbol{a}'\boldsymbol{b}'] - \rho[\boldsymbol{c}'\boldsymbol{a}\boldsymbol{b}'][\boldsymbol{c}'\boldsymbol{a}'\boldsymbol{b}] = 0 \\ &\qquad \Leftrightarrow \rho = \frac{[\boldsymbol{c}'\boldsymbol{a}\boldsymbol{b}][\boldsymbol{c}'\boldsymbol{a}'\boldsymbol{b}']}{[\boldsymbol{c}'\boldsymbol{a}\boldsymbol{b}'][\boldsymbol{c}'\boldsymbol{a}'\boldsymbol{b}]} \end{aligned} \tag{16.6}$$

for some $\rho \neq 0$. This equation is well known and represents a projective invariant which has been used quite a lot in real applications of computer vision [172]. For a thorough study of the role of this invariant using brackets of points, lines, bilinearities and the trifocal tensor see Bayro and Lasenby [140, 18]. Now evaluating ρ in terms of some other point $\boldsymbol{c}$ we get a conic equation fully represented in terms of brackets

$$\begin{aligned} &[\boldsymbol{c}\boldsymbol{a}\boldsymbol{b}][\boldsymbol{c}\boldsymbol{a}'\boldsymbol{b}'] - \frac{[\boldsymbol{c}'\boldsymbol{a}\boldsymbol{b}][\boldsymbol{c}'\boldsymbol{a}'\boldsymbol{b}']}{[\boldsymbol{c}'\boldsymbol{a}\boldsymbol{b}'][\boldsymbol{c}'\boldsymbol{a}'\boldsymbol{b}]}[\boldsymbol{c}\boldsymbol{a}\boldsymbol{b}'][\boldsymbol{c}\boldsymbol{a}'\boldsymbol{b}] = 0 \\ &\Leftrightarrow [\boldsymbol{c}\boldsymbol{a}\boldsymbol{b}][\boldsymbol{c}\boldsymbol{a}'\boldsymbol{b}'][\boldsymbol{a}\boldsymbol{b}'\boldsymbol{c}'][\boldsymbol{a}'\boldsymbol{b}\boldsymbol{c}'] - [\boldsymbol{c}\boldsymbol{a}\boldsymbol{b}'][\boldsymbol{c}\boldsymbol{a}'\boldsymbol{b}][\boldsymbol{a}\boldsymbol{b}\boldsymbol{c}'][\boldsymbol{a}'\boldsymbol{b}'\boldsymbol{c}'] = 0. \end{aligned} \tag{16.7}$$

Again we get a well known concept, which says that a conic is uniquely determined by the five points in general position $\boldsymbol{a}$, $\boldsymbol{a}'$, $\boldsymbol{b}$, $\boldsymbol{b}'$ and $\boldsymbol{c}$. Now, considering Figure 16.1.b, we assume six points on the conic and we can identify three collinear intersecting points $\boldsymbol{\alpha}_1$, $\boldsymbol{\alpha}_2$ and $\boldsymbol{\alpha}_3$. Using the collinearity constraint and the lines which belong to pencils in projective correspondence we can write down a very useful equation

$$\boldsymbol{\alpha}_1 \wedge \boldsymbol{\alpha}_2 \wedge \boldsymbol{\alpha}_3 = 0$$
$$\Leftrightarrow \Big((\boldsymbol{a}' \wedge \boldsymbol{b}) \vee (\boldsymbol{c}' \wedge \boldsymbol{c})\Big) \wedge \Big((\boldsymbol{a}' \wedge \boldsymbol{a}) \vee (\boldsymbol{b}' \wedge \boldsymbol{c})\Big) \wedge \Big((\boldsymbol{c}' \wedge \boldsymbol{a}) \vee (\boldsymbol{b}' \wedge \boldsymbol{b})\Big) = 0. \tag{16.8}$$

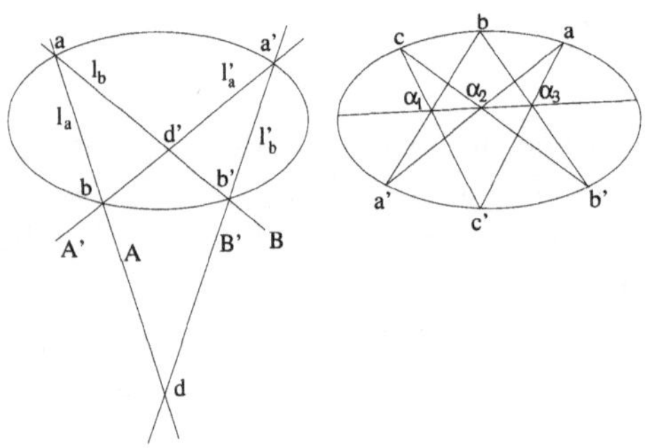

Fig. 16.1. a) Two projective pencils generate a conic b) Pascal's theorem

This expression is a geometric formulation of Pascal's theorem. This theorem proves that the three intersecting points of the lines which connect opposite vertices of a hexagon circumscribed by a conic are collinear ones. The equation (16.8) will be used in later section for computing the intrinsic camera parameters.

16.3 Computing the Kruppa Equations in the Geometric Algebra

In this section we will formulate in two ways the Kruppa equations in the geometric algebra framework. First, we derive the Kruppa equations in its polynomial form using the bracket conic equation (16.7). Secondly, we formulate them in terms of pure brackets. The goal of the section is to compare the bracket representation with the standard one.

16.3.1 The Scenario

Next, we will briefly summarize the scenario for observing a conic at infinity (the absolute conic) in the image planes of multiple views with the aim of self-calibration of the camera. We are applying the standard pinhole camera

model. As described in chapter 14 a pinhole camera can be described by four homogeneous vectors in $\mathbb{P}^3$: One vector gives the optical centre and the other three define the image plane. Let $\{A_\mu\}$ be a reference coordinate system, which consists of four vectors and defines the frame $\mathcal{F}_0$. Let X be a point in a frame $\{Z_\mu\} = \mathcal{F}_1$. The image X_A of the point X on the image plane A of $\{A_\mu\} = \mathcal{F}_0$ can be described by several transformations.
In the first step the frame $\mathcal{F}_1$ can be related to $\mathcal{F}_0$ by a transformation $M_{\mathcal{F}_0}^{\mathcal{F}_1}$. This transformation represents a 3-D rotation R and a 3-D translation $\boldsymbol{t}$ in the 3-D projective space $\mathbb{P}^3$ and depends on six camera parameters. So the frames $\mathcal{F}_0$ and $\mathcal{F}_1$ are first related by a 4×4 matrix

$$M_{\mathcal{F}_0}^{\mathcal{F}_1} = \begin{pmatrix} R & \boldsymbol{t} \\ 0_3^T & 1 \end{pmatrix} . \tag{16.9}$$

The matrix $M_{\mathcal{F}_0}^{\mathcal{F}_1}$ is the matrix of the extrinsic camera parameters.
In the next step changes between the camera planes have to be considered. So the focal length, rotations and translations in the image planes have to be adapted. This affine transformation will be described by the matrix K and has the well known form

$$K = \begin{pmatrix} \alpha_u & \gamma & u_0 \\ 0 & \alpha_v & v_0 \\ 0 & 0 & 1 \end{pmatrix} . \tag{16.10}$$

The parameters u_0, v_0 describe a translation along the image plane and α_u, α_v, γ describe scale changes along the image axes and a rotation in the image plane. So the whole projective transformation can be described by

$$P = K P_0 M_{\mathcal{F}_0}^{\mathcal{F}_1}, \tag{16.11}$$

where $P_0 = [I|0]$ is a 3×4 matrix and I is the 3×3 identity matrix. P_0 describes the projection matrix from the 3-D camera frame $\mathcal{F}_1$ to the normalized camera plane, given in homogeneous coordinates.
The task is to find out the intrinsic camera parameters, which can be found in the matrix K (see equation 16.10) of the affine transformation from the normalized camera coordinate plane to the image coordinate plane. As depicted in Figure 16.2, the images of the points defining the absolute conic are observed from different positions and orientations, and the point correspondences between the images are evaluated. Generally, the relation between points of cameras at different locations depends on both, the extrinsic and the intrinsic parameters. But in case of formulating the Kruppa equations, it will happen that these only depend on intrinsic parameters. An often used notation of equation (16.11), which we want to adopt here for the camera at the i-th frame $\mathcal{F}_i$ with respect to frame $\mathcal{F}_0$, is

$$P_i = K[R|\boldsymbol{t}], \tag{16.12}$$

where $[R|\boldsymbol{t}]$ is a 3×4 matrix constituted by the rotation matrix R and the translation vector $\boldsymbol{t}$, resulting from the fusion of P_0 and $M_{\mathcal{F}_0}^{\mathcal{F}_i}$. For the sake of simplicity, we will set for the first camera $\mathcal{F}_1 \equiv \mathcal{F}_0$, thus, its projective transformation becomes $P_1 = K[I|0]$, where I is the 3×3 identity matrix.

16.3.2 Standard Kruppa Equations

This approach uses the equation (16.7) for the conic in terms of brackets considering five points $\boldsymbol{a}$, $\boldsymbol{b}$, $\boldsymbol{a}'$, $\boldsymbol{b}'$, $\boldsymbol{c}'$ which lie on the conic in the image plane:

$$[\boldsymbol{cab}][\boldsymbol{ca'b'}][\boldsymbol{ab'c'}][\boldsymbol{a'bc'}] - [\boldsymbol{cab'}][\boldsymbol{ca'b}][\boldsymbol{abc'}][\boldsymbol{a'b'c'}] = 0$$

$$[\boldsymbol{abc}][\boldsymbol{a'b'c}] - \frac{[\boldsymbol{a'b'c'}][\boldsymbol{abc'}]}{[\boldsymbol{ab'c'}][\boldsymbol{a'bc'}]}[\boldsymbol{ab'c}][\boldsymbol{a'bc}] = 0. \qquad (16.13)$$

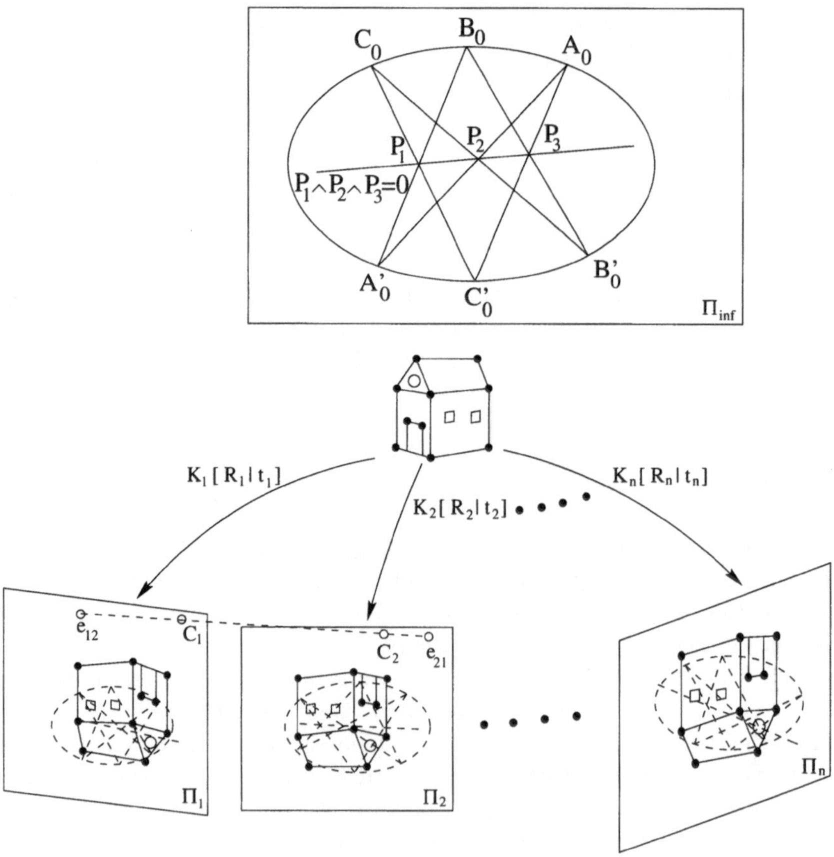

Fig. 16.2. The conics at infinity, the real 3–D visual space and n uncalibrated cameras

These five points are images of points on the absolute conic. A conic at infinity $\Omega_{\inf}$ in $\mathbb{P}^3$ can be defined employing any imaginary five points lying on the conic, e.g.

$$\boldsymbol{A}_0 = \begin{pmatrix} 1 \\ i \\ 0 \\ 0 \end{pmatrix}, \boldsymbol{B}_0 = \begin{pmatrix} i \\ 1 \\ 0 \\ 0 \end{pmatrix}, \boldsymbol{A}_0' = \begin{pmatrix} i \\ 0 \\ 1 \\ 0 \end{pmatrix}, \boldsymbol{B}_0' = \begin{pmatrix} 1 \\ 0 \\ i \\ 0 \end{pmatrix}, \boldsymbol{C}_0' = \begin{pmatrix} 0 \\ i \\ 1 \\ 0 \end{pmatrix}, \quad (16.14)$$

where $i^2 = -1$. Note that we use upper case letters to represent points of the projective space $\mathbb{P}^3$ in $\mathbb{C}_{1,3,0}$. Because these points at infinity fulfill the property $\boldsymbol{A}_0^T\boldsymbol{A}_0 = \boldsymbol{B}_0^T\boldsymbol{B}_0 = \boldsymbol{A}_0'^T\boldsymbol{A}_0' = \boldsymbol{B}_0'^T\boldsymbol{B}_0' = \boldsymbol{C}_0'^T\boldsymbol{C}_0' = 0$ they lie on the absolute conic. In geometric algebra a conic can be described by the points lying on the conic. Furthermore, the image of the absolute conic can be described by the image of the points lying on the absolute conic. In the next step, let us first define the point $\boldsymbol{A}$ as a 3×1-vector which consists of the first three elements of $\boldsymbol{A}_0$. Doing similary with the other points we get the points

$$\boldsymbol{A} = \begin{pmatrix} 1 \\ i \\ 0 \end{pmatrix}, \boldsymbol{B} = \begin{pmatrix} i \\ 1 \\ 0 \end{pmatrix}, \boldsymbol{A}' = \begin{pmatrix} i \\ 0 \\ 1 \end{pmatrix}, \boldsymbol{B}' = \begin{pmatrix} 1 \\ 0 \\ i \end{pmatrix}, \boldsymbol{C}' = \begin{pmatrix} 0 \\ i \\ 1 \end{pmatrix}. \quad (16.15)$$

Since the projection of the points $\boldsymbol{A}_0, \ldots, \boldsymbol{C}_0'$ are translation invariant, their projections $\boldsymbol{x} = P\boldsymbol{X}$ on any image plane are independent of $\boldsymbol{t}$ and thus given by

$$\begin{aligned} \boldsymbol{a} &= K[R|\boldsymbol{t}]\boldsymbol{A}_0 = KR\boldsymbol{A}\ , \ \boldsymbol{b} = K[R|\boldsymbol{t}]\boldsymbol{B}_0 = KR\boldsymbol{B} \\ \boldsymbol{a}' &= K[R|\boldsymbol{t}]\boldsymbol{A}_0' = KR\boldsymbol{A}'\ , \ \boldsymbol{b}' = K[R|\boldsymbol{t}]\boldsymbol{B}_0' = KR\boldsymbol{B}' \\ \boldsymbol{c}' &= K[R|\boldsymbol{t}]\boldsymbol{C}_0' = KR\boldsymbol{C}'\ . \end{aligned} \quad (16.16)$$

In addition the rotated points $R^T\boldsymbol{A}$, $R^T\boldsymbol{B}$, $R^T\boldsymbol{A}'$, $R^T\boldsymbol{B}'$ and $R^T\boldsymbol{C}'$ lie also at the conic, because they fulfill the property

$$\begin{aligned} (R^T\boldsymbol{A})^T(R^T\boldsymbol{A}) = (R^T\boldsymbol{B})^T(R^T\boldsymbol{B}) = (R^T\boldsymbol{A}')^T(R^T\boldsymbol{A}') = \\ (R^T\boldsymbol{B}')^T(R^T\boldsymbol{B}') = (R^T\boldsymbol{C}')^T(R^T\boldsymbol{C}') = 0. \end{aligned} \quad (16.17)$$

Using these rotated points, the rotation R of the camera transformation is canceled and the points on the image of the absolute conic will be described by

$$\boldsymbol{a} = K\boldsymbol{A}, \quad \boldsymbol{b} = K\boldsymbol{B}, \quad \boldsymbol{a}' = K\boldsymbol{A}', \quad \boldsymbol{b}' = K\boldsymbol{B}', \quad \boldsymbol{c}' = K\boldsymbol{C}'. \quad (16.18)$$

To use the points $\boldsymbol{a}, \ldots, \boldsymbol{c}'$ in the bracket notation of conics it is usefull to translate the matrix multiplication $\boldsymbol{x} = K\boldsymbol{X}$ in terms of geometric algebra. Suppose an orthonormal basis $B_1 = \{\mathbf{e}_1, \dots, \mathbf{e}_3\}$ and $\boldsymbol{X}$ as a linear combination of B_1 i.e. $\boldsymbol{X} = \sum_{i=1}^{3} x_i \mathbf{e}_i$. The matrix K describes a linear transformation. As can be seen in chapter 1.3 this linear transformation can be expressed by

$$\underline{K}\mathbf{e}_i = \underline{K}(\mathbf{e}_i) = \sum_{j=1}^{3} \mathbf{e}_j k_{ji} \tag{16.19}$$

with k_{ji} the elements of the matrix K. So the matrix multiplication $K\boldsymbol{X}$ can be substituted by $\underline{K}\boldsymbol{X}$ in terms of geometric algebra. Therefore, the point $\boldsymbol{c}$ lies on the image of the absolute conic iff

$$[(\underline{K}\boldsymbol{A})(\underline{K}\boldsymbol{B})\boldsymbol{c}][(\underline{K}\boldsymbol{A}')(\underline{K}\boldsymbol{B}')\boldsymbol{c}] - \frac{[(\underline{K}\boldsymbol{A}')(\underline{K}\boldsymbol{B}')(\underline{K}\boldsymbol{C}')][(\underline{K}\boldsymbol{A})(\underline{K}\boldsymbol{B})(\underline{K}\boldsymbol{C}')]}{[(\underline{K}\boldsymbol{A})(\underline{K}\boldsymbol{B}')(\underline{K}\boldsymbol{C}')][(\underline{K}\boldsymbol{A}')(\underline{K}\boldsymbol{B})(\underline{K}\boldsymbol{C}')]} \cdot \\ \cdot[(\underline{K}\boldsymbol{A})(\underline{K}\boldsymbol{B}')\boldsymbol{c}][(\underline{K}\boldsymbol{A}')(\underline{K}\boldsymbol{B})\boldsymbol{c}] = 0. \tag{16.20}$$

We can further extract of the brackets the determinant of the intrinsic parameters in the multiplicative ratio of the previous equation. This is explained in chapter 1.3. Now the invariant reduces to a constant

$$\begin{aligned} Inv &= \frac{([\underline{K}\boldsymbol{A}')(\underline{K}\boldsymbol{B}')(\underline{K}\boldsymbol{C}')][(\underline{K}\boldsymbol{A})(\underline{K}\boldsymbol{B})(\underline{K}\boldsymbol{C}')]}{[(\underline{K}\boldsymbol{A})(\underline{K}\boldsymbol{B}')(\underline{K}\boldsymbol{C}'))][(\underline{K}\boldsymbol{A}')(\underline{K}\boldsymbol{B})(K\boldsymbol{C}')]} \\ &= \frac{det(\underline{K})[\boldsymbol{A}'\boldsymbol{B}'\boldsymbol{C}']det(\underline{K})[\boldsymbol{A}\boldsymbol{B}\boldsymbol{C}']}{det(\underline{K})[\boldsymbol{A}\boldsymbol{B}'\boldsymbol{C}']det(\underline{K})[\boldsymbol{A}'\boldsymbol{B}\boldsymbol{C}']} \\ &= \frac{[\boldsymbol{A}'\boldsymbol{B}'\boldsymbol{C}'][\boldsymbol{A}\boldsymbol{B}\boldsymbol{C}']}{[\boldsymbol{A}\boldsymbol{B}'\boldsymbol{C}'][\boldsymbol{A}'\boldsymbol{B}\boldsymbol{C}']}. \end{aligned} \tag{16.21}$$

Substituting the values from equation (16.15) for $\boldsymbol{A}$, $\boldsymbol{B}$, $\boldsymbol{A}'$, $\boldsymbol{B}'$, $\boldsymbol{C}'$ in this equation, we get the value of $Inv = 2$. This value will be used for further computations later on. The equation (16.21) is as expected invariant to the affine transformation $\underline{K}$. Thus, the bracket equation (16.6) of the projective invariant resulting in the image of the absolute conic can be written as

$$[(\underline{K}\boldsymbol{A})(\underline{K}\boldsymbol{B})\boldsymbol{c}][(\underline{K}\boldsymbol{A}')(\underline{K}\boldsymbol{B}')\boldsymbol{c}] - Inv[(\underline{K}\boldsymbol{A})(\underline{K}\boldsymbol{B}')\boldsymbol{c}][(\underline{K}\boldsymbol{A}')(\underline{K}\boldsymbol{B})\boldsymbol{c}] = 0. \tag{16.22}$$

Let be $Q = K^{-T}K^{-1}$ the matrix of the image of the absolute conic, then $\boldsymbol{c}^T Q\boldsymbol{c} = 0$ in matrix notation means that $\boldsymbol{c}$ is a point on the image of the absolute conic. According to the duality principle of points and lines the dual image of the absolute conic, i.e. its matrix $Q^* \sim Q^{-1} = KK^T$ is related to a line $\boldsymbol{l}_c$, tangential to the image of the absolute conic. Because this can be expressed as

$$0 = \boldsymbol{c}^T Q\boldsymbol{c} = \boldsymbol{c}^T Q^T \boldsymbol{c} = (\boldsymbol{c}^T Q^T)Q^{-1}(Q\boldsymbol{c}) = \boldsymbol{l}_c^T Q^* \boldsymbol{l}_c, \tag{16.23}$$

we have $Q\boldsymbol{c} = \boldsymbol{l}_c$ or $\boldsymbol{c} = KK^T\boldsymbol{l}_c$. To use $KK^T\boldsymbol{l}_c$ in the bracket description of conics, it is usefull to translate the matrix multiplications in terms of geometric algebra. The line $\boldsymbol{l}_c$ is tangential to the image of the absolute conic, so it has the form $\sum_{i=1}^{3} l_{c_i}\mathbf{e}_i$. The product $K^T\boldsymbol{l}_c$ can be described using the adjoint $\overline{K}$ of $\underline{K}$ by the expression $\overline{K}\boldsymbol{l}_c$, see chapter 1.3. The expression $\boldsymbol{c} = KK^T\boldsymbol{l}_c$ can thus be formulated as $\boldsymbol{c} = \underline{K}\overline{K}\boldsymbol{l}_c$. We can substitute this line tangent in equation (16.22):

$$\begin{aligned}
&[(\underline{K}\boldsymbol{A})(\underline{K}\boldsymbol{B})\mathrm{c}][(\underline{K}\boldsymbol{A}')(\underline{K}\boldsymbol{B}')\mathrm{c}] - Inv[(\underline{K}\boldsymbol{A})(\underline{K}\boldsymbol{B}')\mathrm{c}][(\underline{K}\boldsymbol{A}')(\underline{K}\boldsymbol{B})\mathrm{c}] = 0\\
\Leftrightarrow\ &[(\underline{K}\boldsymbol{A})(\underline{K}\boldsymbol{B})(\underline{K}\overline{K}\boldsymbol{l}_c)][(\underline{K}\boldsymbol{A}')(\underline{K}\boldsymbol{B}')(\underline{K}\overline{K}\boldsymbol{l}_c)] -\\
&\quad -Inv[(\underline{K}\boldsymbol{A})(\underline{K}\boldsymbol{B}')(\underline{K}\overline{K}\boldsymbol{l}_c)][(\underline{K}\boldsymbol{A}')(\underline{K}\boldsymbol{B})(\underline{K}\overline{K}\boldsymbol{l}_c)] = 0\\
\Leftrightarrow\ &det(\underline{K})[\boldsymbol{AB}(\overline{K}\boldsymbol{l}_c)]det(\underline{K})[\boldsymbol{A}'\boldsymbol{B}'(\overline{K}\boldsymbol{l}_c)] -\\
&\quad -Inv\, det(\underline{K})[\boldsymbol{AB}'(\overline{K}\boldsymbol{l}_c)]det(\underline{K})[\boldsymbol{A}'\boldsymbol{B}(\overline{K}\boldsymbol{l}_c)] = 0\\
\Leftrightarrow\ &[\boldsymbol{AB}(\overline{K}\boldsymbol{l}_c)][\boldsymbol{A}'\boldsymbol{B}'(\overline{K}\boldsymbol{l}_c)] - Inv[\boldsymbol{AB}'(\overline{K}\boldsymbol{l}_c)][\boldsymbol{A}'\boldsymbol{B}(\overline{K}\boldsymbol{l}_c)] = 0. \qquad (16.24)
\end{aligned}$$

To further proceed on the classical way of deriving Kruppa's equations [164, 158, 157], it will be possible to formulate two polynomial constraint equations on the dual of the image of the absolute conic in the frame of epipolar geometry. Let be $\boldsymbol{p} = p_1\mathbf{e}_1 + p_2\mathbf{e}_2 + p_3\mathbf{e}_3$ the epipole of an image and let be $\boldsymbol{q} = \mathbf{e}_1 + \tau\mathbf{e}_2$ a point at infinity. The aim will be to force the line

$$\begin{aligned}
\boldsymbol{l}_c &= (\boldsymbol{p} \wedge \boldsymbol{q})\boldsymbol{I}^{-1}\\
&= \left(\left(\sum_{i=1}^{3} p_i\mathbf{e}_i\right) \wedge (\mathbf{e}_1 + \tau\mathbf{e}_2)\right)(\mathbf{e}_1\mathbf{e}_2\mathbf{e}_3)^{-1}\\
&= (-p_3\tau)\mathbf{e}_1 + (p_3)\mathbf{e}_2 + (p_1\tau - p_2)\mathbf{e}_3, \qquad (16.25)
\end{aligned}$$

to be tangential to the dual of the image of the absolute conic by means of the unknown τ. Then we can substitute the term $\boldsymbol{l}_c$ in equation (16.24). With

$$\begin{aligned}
\overline{K}\boldsymbol{l}_c = (-k_{11}p_3\tau)\mathbf{e}_1 + (-k_{12}p_3\tau + k_{22}p_3)\mathbf{e}_2 +&\\
(-k_{13}p_3\tau + k_{23}p_3 + p_1\tau - p_2)\mathbf{e}_3& \qquad (16.26)
\end{aligned}$$

and the value for $Inv = 2$ the equation (16.24) simplifies to a second order polynomial with respect to τ as follows

$$\begin{aligned}
&[\boldsymbol{AB}(\overline{K}\boldsymbol{l}_c)][\boldsymbol{A}'\boldsymbol{B}'(\overline{K}\boldsymbol{l}_c)] - Inv[\boldsymbol{AB}'(\overline{K}\boldsymbol{l}_c)][\boldsymbol{A}'\boldsymbol{B}(\overline{K}\boldsymbol{l}_c)] =\\
&4p_1\tau p_2 - 2p_1^2\tau^2 - 2k_{22}^2p_3^2 - 4k_{23}p_3p_1\tau + 4k_{23}p_3p_2 - 2k_{13}^2p_3^2\tau^2 -\\
&2k_{12}^2p_3^2\tau^2 - 2k_{23}^2p_3^2 - 2p_2^2 - 2k_{11}^2p_3^2\tau^2 + 4k_{12}p_3^2\tau k_{22} - 4k_{13}p_3\tau p_2 +\\
&4k_{13}p_3^2\tau k_{23} + 4k_{13}p_3\tau^2 p_1. \qquad (16.27)
\end{aligned}$$

Expressing the polynomial in the form $P(\tau) = k_0 + k_1\tau + k_2\tau^2$, we get the following coefficients

$$\begin{aligned}
k_0 &= -2k_{22}^2p_3^2 + 4k_{23}p_3p_2 - 2k_{23}^2p_3^2 - 2p_2^2\\
k_1 &= 4p_1p_2 - 4k_{23}p_3p_1 + 4k_{12}p_3^2k_{22} - 4k_{13}p_3p_2 + 4k_{13}p_3^2k_{23}\\
k_2 &= -2p_1^2 - 2k_{13}^2p_3^2 - 2k_{12}^2p_3^2 - 2k_{11}^2p_3^2 + 4k_{13}p_3p_1. \qquad (16.28)
\end{aligned}$$

Because $\boldsymbol{l}_c$ can be also considered as an epipolar line tangent to the conic in the first camera, according the homography of a point lying at the line at infinity of the second camera, we can use the operator $\underline{F}$ for the describtion of the fundamental matrix F in terms of geometric algebra, and can compute $\boldsymbol{l}_c = \underline{F}(\mathbf{e}_1 + \tau\mathbf{e}_2)$. Using the new expression of $\boldsymbol{l}_c$ we can gain similarly as above new equations for the coefficients of the polynomial $P(\tau)$, now called k_i'. Taking now these equations for the two cameras, we finally can write down the well known Kruppa equations

$$\begin{aligned} k_2k_1' - k_2'k_1 &= 0 \\ k_0k_1' - k_0'k_1 &= 0 \\ k_0k_2' - k_0'k_2 &= 0. \end{aligned} \tag{16.29}$$

We get up to a scalar factor the same Kruppa equations as presented by Luong and Faugeras [157]. The scalar factor is present in all of these equations, thus it can be canceled straightforwardly. The algebraic manipulation of this formulas was checked entirely using a Maple program.

16.3.3 Kruppa's Equations Using Brackets

In this section we will formulate the Kruppa coefficients k_0, k_1, k_2 of the polynomial $P(\tau)$ in terms of brackets. This kind of representation will obviously elucidate the involved geometry. First let us consider again the bracket $[\boldsymbol{AB}(\overline{\boldsymbol{K}}\boldsymbol{l}_c)]$ of equation (16.24). Each bracket can be split in two brackets, one independent of τ and another depending of it

$$[\boldsymbol{AB}(\overline{\boldsymbol{K}}\boldsymbol{l}_c)] = [\boldsymbol{AB}(\overline{\boldsymbol{K}}(p_3\mathbf{e}_2 - p_2\mathbf{e}_3))] + [\boldsymbol{AB}(\overline{\boldsymbol{K}}(-p_3\mathbf{e}_1 + p_1\mathbf{e}_3))]\tau. \tag{16.30}$$

In short, $[\boldsymbol{AB}(\overline{\boldsymbol{K}}\boldsymbol{l}_c)] = a_1 + \tau b_1$. Now using this bracket representation the equation (16.24) can be written as

$$\begin{aligned} &[\boldsymbol{AB}(\overline{\boldsymbol{K}}\boldsymbol{l}_c)][\boldsymbol{A}'\boldsymbol{B}'(\overline{\boldsymbol{K}}\boldsymbol{l}_c)] - Inv[\boldsymbol{AB}'(\overline{\boldsymbol{K}}\boldsymbol{l}_c)][\boldsymbol{A}'\boldsymbol{B}(\overline{\boldsymbol{K}}\boldsymbol{l}_c)] = 0 \\ \Leftrightarrow\ &(a_1 + \tau b_1)(a_2 + \tau b_2) - Inv(a_3 + \tau b_3)(a_4 + \tau b_4) = 0 \\ \Leftrightarrow\ &a_1a_2 + \tau b_1a_2 + a_1\tau b_2 + \tau^2 b_1b_2 - \\ &-Inv(a_3a_4 + a_3a_4\tau + b_3a_4\tau + b_3b_4\tau^2) = 0 \\ \Leftrightarrow\ &\underbrace{a_1a_2 - Inv(a_3a_4)}_{k_0} + \tau\underbrace{(a_1b_2 + b_1a_2 - Inv(a_3b_4 + a_4b_3))}_{k_1} + \\ &+\tau^2\underbrace{(b_1b_2 - Inv(b_3b_4))}_{k_2} = 0. \end{aligned} \tag{16.31}$$

Now let us take a partial vector part of $\overline{\boldsymbol{K}}\boldsymbol{l}_c$ and call it

$$\overline{\boldsymbol{K}}\boldsymbol{l}_{c1} := -k_{11}p_3\mathbf{e}_1 - k_{12}p_3\mathbf{e}_2 + (-k_{13}p_3 + p_1)\mathbf{e}_3$$

and the "rest"-part as

$\overline{K}\boldsymbol{l}_{c2} := (k_{22}p_3)\mathbf{e}_2 + (k_{23}p_3 - p_2)\mathbf{e}_3.$

Using both parts we can write the coefficients of the polynomial in a bracket form as follows:

$$k_0 = [\boldsymbol{AB}(\overline{K}\boldsymbol{l}_{c2})][\boldsymbol{A}'\boldsymbol{B}'(\overline{K}\boldsymbol{l}_{c2})] - Inv[\boldsymbol{AB}'(\overline{K}\boldsymbol{l}_{c2})][\boldsymbol{A}'\boldsymbol{B}(\overline{K}\boldsymbol{l}_{c2})] \tag{16.32}$$

$$\begin{aligned} k_1 = {} & [\boldsymbol{AB}(\overline{K}\boldsymbol{l}_{c1})][\boldsymbol{A}'\boldsymbol{B}'(\overline{K}\boldsymbol{l}_{c2})] + [\boldsymbol{AB}(\overline{K}\boldsymbol{l}_{c2})][\boldsymbol{A}'\boldsymbol{B}'(\overline{K}\boldsymbol{l}_{c1})] \\ & -Inv[\boldsymbol{AB}'(\overline{K}\boldsymbol{l}_{c2})][\boldsymbol{A}'\boldsymbol{B}(\overline{K}\boldsymbol{l}_{c1})] - Inv[\boldsymbol{AB}'(\overline{K}\boldsymbol{l}_{c1})][\boldsymbol{A}'\boldsymbol{B}(\overline{K}\boldsymbol{l}_{c2})] \end{aligned} \tag{16.33}$$

$$k_2 = [\boldsymbol{AB}(\overline{K}\boldsymbol{l}_{c1})][\boldsymbol{A}'\boldsymbol{B}'(\overline{K}\boldsymbol{l}_{c1})] - Inv[\boldsymbol{AB}'(\overline{K}\boldsymbol{l}_{c1})][\boldsymbol{A}'\boldsymbol{B}(\overline{K}\boldsymbol{l}_{c1})]. \tag{16.34}$$

Since $\boldsymbol{A}, \boldsymbol{B}, \boldsymbol{A}', \boldsymbol{B}'$ and Inv are known given an epipole $\boldsymbol{p} = p_1\mathbf{e}_1 + p_2\mathbf{e}_2 + p_3\mathbf{e}_3$, we can finally compute the coefficients k_0, k_1, k_2 straightforwardly. The striking aspect of these equations is twofold. They are expressed in terms of brackets and they depend of the invariant real magnitude Inv. This can certainly help us to explore the involved geometry of the Kruppa equations using brackets.

Let us first analyze the k's. Since the elements of k_1 consists of the elements of k_0 and k_2, it should be sufficient to explore the involved geometry of k_0 and k_2 if these are expressed as follows:

$$\begin{aligned} k_0 &= a_1a_2 - Inv(a_3a_4) \\ &= [\boldsymbol{AB}(\overline{K}\boldsymbol{l}_{c2})][\boldsymbol{A}'\boldsymbol{B}'(\overline{K}\boldsymbol{l}_{c2})] - Inv[\boldsymbol{AB}'(\overline{K}\boldsymbol{l}_{c2})][\boldsymbol{A}'\boldsymbol{B}(\overline{K}\boldsymbol{l}_{c2})] \\ &= ((\mathbf{e}_1 + i\mathbf{e}_2) \wedge (i\mathbf{e}_1 + \mathbf{e}_2) \wedge (k_{22}p_3\mathbf{e}_2 + (k_{23}p_3 - p_2)\mathbf{e}_3)\boldsymbol{I}^{-1}) \\ &\quad ((i\mathbf{e}_1 + \mathbf{e}_3) \wedge (\mathbf{e}_1 + i\mathbf{e}_3) \wedge (k_{22}p_3\mathbf{e}_2 + (k_{23}p_3 - p_2)\mathbf{e}_3)\boldsymbol{I}^{-1}) - \\ &\quad Inv((\mathbf{e}_1 + i\mathbf{e}_2) \wedge (\mathbf{e}_1 + i\mathbf{e}_3) \wedge (k_{22}p_3\mathbf{e}_2 + (k_{23}p_3 - p_2)\mathbf{e}_3)\boldsymbol{I}^{-1}) \\ &\quad ((i\mathbf{e}_1 + \mathbf{e}_3) \wedge (i\mathbf{e}_1 + \mathbf{e}_2) \wedge (k_{22}p_3\mathbf{e}_2 + (k_{23}p_3 - p_2)\mathbf{e}_3)\boldsymbol{I}^{-1}) \end{aligned} \tag{16.35}$$

$$\begin{aligned} k_2 &= b_1b_2 - Inv(b_3b_4) \\ &= [\boldsymbol{AB}(\overline{K}\boldsymbol{l}_{c1})][\boldsymbol{A}'\boldsymbol{B}'(\overline{K}\boldsymbol{l}_{c1})] - Inv[\boldsymbol{AB}'(\overline{K}\boldsymbol{l}_{c1})][\boldsymbol{A}'\boldsymbol{B}(\overline{K}\boldsymbol{l}_{c1})] \\ &\quad ((\mathbf{e}_1 + i\mathbf{e}_2) \wedge (i\mathbf{e}_1 + \mathbf{e}_2) \wedge \\ &\qquad (-k_{11}p_3\mathbf{e}_1 - k_{12}p_3\mathbf{e}_2 + (-k_{13}p_3 + p_1)\mathbf{e}_3)\boldsymbol{I}^{-1}) \\ &\quad ((i\mathbf{e}_1 + \mathbf{e}_3) \wedge (\mathbf{e}_1 + i\mathbf{e}_3) \wedge \\ &\qquad (-k_{11}p_3\mathbf{e}_1 - k_{12}p_3\mathbf{e}_2 + (-k_{13}p_3 + p_1)\mathbf{e}_3)\boldsymbol{I}^{-1}) \\ &\quad -Inv((\mathbf{e}_1 + i\mathbf{e}_2) \wedge (\mathbf{e}_1 + i\mathbf{e}_3) \wedge \\ &\qquad (-k_{11}p_3\mathbf{e}_1 - k_{12}p_3\mathbf{e}_2 + (-k_{13}p_3 + p_1)\mathbf{e}_3)\boldsymbol{I}^{-1}) \\ &\quad ((i\mathbf{e}_1 + \mathbf{e}_3) \wedge (i\mathbf{e}_1 + \mathbf{e}_2) \wedge \\ &\qquad (-k_{11}p_3\mathbf{e}_1 - k_{12}p_3\mathbf{e}_2 + (-k_{13}p_3 + p_1)\mathbf{e}_3)\boldsymbol{I}^{-1}). \end{aligned} \tag{16.36}$$

Let us analyze some effects of camera motions in these two equations. If the camera moves on a straight path parallel to the object, the epipole lies at infinity. Because $p_3 = 0$ in this case, the intrinsic parameters become zero resulting a trivial polynomial, i.e. we can not get the coefficients of the intrinsic camera parameters. On the other hand, for example trying the values $-k_{13}p_3 + p_1 = 0$ or $k_{23}p_3 - p_2 = 0$, the rest of the brackets will have the rank two and their determinant value is also zero. Since the epipole can be normalized with $p_3 = 1$, the equations are equivalent to $k_{13} = p_1$ and $k_{23} = p_2$. This means there is a superposition of the value of the epipole with a parameter of the intrinsic camera parameters. These simple examples show that analyzing the brackets for certain kinds of camera motions can avoid certain camera motions which generate trivial Kruppa equations. It is also interesting to see that for $k_0 = 0$ and $k_2 = 0$ we have also conic equations. So in order to avoid trivial equations we have to consider always $k_0 \neq 0$ and $k_2 \neq 0$. In other words, the splitted parts $\overline{K}\boldsymbol{l}_{c1}$ and $\overline{K}\boldsymbol{l}_{c2}$ of $\overline{K}\boldsymbol{l}_c$ should not lie on the image of the absolute conic.

Now let us consider the invariant real magnitude Inv of the bracket equation (16.24).

$$[\boldsymbol{AB}(\overline{K}\boldsymbol{l}_c)][\boldsymbol{A}'\boldsymbol{B}'(\overline{K}\boldsymbol{l}_c)] - Inv[\boldsymbol{AB}'(\overline{K}\boldsymbol{l}_c)][\boldsymbol{A}'\boldsymbol{B}(\overline{K}\boldsymbol{l}_c)] = 0$$
$$\Leftrightarrow \; Inv = \frac{[\boldsymbol{AB}(\overline{K}\boldsymbol{l}_c)][\boldsymbol{A}'\boldsymbol{B}'(\overline{K}\boldsymbol{l}_c)]}{[\boldsymbol{AB}'(\overline{K}\boldsymbol{l}_c)][\boldsymbol{A}'\boldsymbol{B}(\overline{K}\boldsymbol{l}_c)]}. \tag{16.37}$$

That the invariant value Inv like in the equation (16.6) plays a role in the Kruppa equations is a fact that has been overseen so far. This can be simply explained as the fact that when we formulate the Kruppa equations using the condition $\boldsymbol{c}^T Q\boldsymbol{c} = 0$, we are actually implicitly employing the invariant given by equation (16.37).

16.4 Camera Calibration Using Pascal's Theorem

This section presents a new technique in the geometric algebra framework for computing the intrinsic camera parameters. The previous section used the equation of (16.7) to compute the Kruppa coefficients which in turn can be used to get the intrinsic camera parameters. Along this lines we will proceed here.

In section two it is shown that the equation (16.7) can be reformulated to express the constraint of equation (16.8) known as Pascal's theorem. Since Pascal's theorem fulfills a property of any conic, it should be also possible using this equation to compute the intrinsic camera parameters. Let us consider the three intersecting points which are collinear and fulfill

$$\underbrace{((\boldsymbol{a}'\wedge\boldsymbol{b})\vee(\boldsymbol{c}'\wedge\boldsymbol{c}))}_{\boldsymbol{\alpha}_1}\wedge\underbrace{((\boldsymbol{a}'\wedge\boldsymbol{a})\vee(\boldsymbol{b}'\wedge\boldsymbol{c}))}_{\boldsymbol{\alpha}_2}\wedge\underbrace{((\boldsymbol{c}'\wedge\boldsymbol{a})\vee(\boldsymbol{b}'\wedge\boldsymbol{b}))}_{\boldsymbol{\alpha}_3} = 0. \tag{16.38}$$

Similar to chapter 1.3.2, in Figure 16.3 at the first camera the projected rotated points of the conic at infinity are

$$\boldsymbol{a} = \underline{K}\boldsymbol{A}, \quad \boldsymbol{b} = \underline{K}\boldsymbol{B}, \quad a' = \underline{K}\boldsymbol{A}', \quad \boldsymbol{b}' = \underline{K}\boldsymbol{B}', \quad c' = \underline{K}\boldsymbol{C}'. \qquad (16.39)$$

The point $\boldsymbol{c} = \underline{K}\overline{K}\boldsymbol{l}_c$ depends of the intrinsic parameters and of the line $\boldsymbol{l}_c$ tangent to the conic which is computed in terms of the epipole $\boldsymbol{p} = p_1\mathbf{e}_1 + p_2\mathbf{e}_2 + p_3\mathbf{e}_3$ and a point $\boldsymbol{q} = \mathbf{e}_1 + \tau\mathbf{e}_2$ lying at the line at infinity of the first camera, i.e. $\boldsymbol{l}_c = (\boldsymbol{p} \wedge \boldsymbol{q})\boldsymbol{I}^{-1}$.

Now using this expression for $\boldsymbol{l}_c$ we can simplify equation (16.38) and get the bracket equations of the $\boldsymbol{\alpha}$'s

$$\begin{aligned}
&([\boldsymbol{a}'\boldsymbol{b}\boldsymbol{c}']\boldsymbol{c} - [\boldsymbol{a}'\boldsymbol{b}\boldsymbol{c}]\boldsymbol{c}') \wedge ([\boldsymbol{a}'\boldsymbol{a}\boldsymbol{b}']\boldsymbol{c} - [\boldsymbol{a}'\boldsymbol{a}\boldsymbol{c}]\boldsymbol{b}') \wedge ([\boldsymbol{c}'\boldsymbol{a}\boldsymbol{b}']\boldsymbol{b} - [\boldsymbol{c}'\boldsymbol{a}\boldsymbol{b}]\boldsymbol{b}') = 0\\
\Leftrightarrow\ &([(\underline{K}\boldsymbol{A}')(\underline{K}\boldsymbol{B})(\underline{K}\boldsymbol{C}')](\underline{K}\overline{K}\boldsymbol{l}_c) - [(\underline{K}\boldsymbol{A}')(\underline{K}\boldsymbol{B})(\underline{K}\overline{K}\boldsymbol{l}_c)](\underline{K}\boldsymbol{C}')) \wedge\\
&([(\underline{K}\boldsymbol{A}')(\underline{K}\boldsymbol{A})(\underline{K}\boldsymbol{B}')](\underline{K}\overline{K}\boldsymbol{l}_c) - [(\underline{K}\boldsymbol{A}')(\underline{K}\boldsymbol{A})(\underline{K}\overline{K}\boldsymbol{l}_c)](\underline{K}\boldsymbol{B}')) \wedge\\
&([(\underline{K}\boldsymbol{C}')(\underline{K}\boldsymbol{A})(\underline{K}\boldsymbol{B}')](\underline{K}\boldsymbol{B}) - [(\underline{K}\boldsymbol{C}')(\underline{K}\boldsymbol{A})(\underline{K}\boldsymbol{B})](\underline{K}\boldsymbol{B}')) = 0\\
\Leftrightarrow\ &\Big(det(\underline{K})\underline{K}([\boldsymbol{A}'\boldsymbol{B}\boldsymbol{C}'](\overline{K}\boldsymbol{l}_c) - [\boldsymbol{A}'\boldsymbol{B}(\overline{K}\boldsymbol{l}_c)]\boldsymbol{C}')\Big) \wedge\\
&\Big(det(\underline{K})\underline{K}([\boldsymbol{A}'\boldsymbol{A}\boldsymbol{B}'](\overline{K}\boldsymbol{l}_c) - [\boldsymbol{A}'\boldsymbol{A}(\overline{K}\boldsymbol{l}_c)]\boldsymbol{B}')\Big) \wedge\\
&\Big(det(\underline{K})\underline{K}([\boldsymbol{C}'\boldsymbol{A}\boldsymbol{B}']\boldsymbol{B} - [\boldsymbol{C}'\boldsymbol{A}\boldsymbol{B}]\boldsymbol{B}')\Big) = 0\\
\Leftrightarrow\ &det(\underline{K})^4\Big(([\boldsymbol{A}'\boldsymbol{B}\boldsymbol{C}'](\overline{K}\boldsymbol{l}_c) - [\boldsymbol{A}'\boldsymbol{B}(\overline{K}\boldsymbol{l}_c)]\boldsymbol{C}') \wedge\\
&([\boldsymbol{A}'\boldsymbol{A}\boldsymbol{B}']\overline{K}\boldsymbol{l}_c - [\boldsymbol{A}'\boldsymbol{A}(\overline{K}\boldsymbol{l}_c)]\boldsymbol{B}') \wedge\\
&([\boldsymbol{C}'\boldsymbol{A}\boldsymbol{B}']\boldsymbol{B} - [\boldsymbol{C}'\boldsymbol{A}\boldsymbol{B}]\boldsymbol{B}')\Big) = 0\\
\Leftrightarrow\ &\underbrace{([\boldsymbol{A}'\boldsymbol{B}\boldsymbol{C}'](\overline{K}\boldsymbol{l}_c) - [\boldsymbol{A}'\boldsymbol{B}(\overline{K}\boldsymbol{l}_c)]\boldsymbol{C}') \wedge}_{\boldsymbol{\alpha}_1}\\
&\underbrace{([\boldsymbol{A}'\boldsymbol{A}\boldsymbol{B}'](\overline{K}\boldsymbol{l}_c) - [\boldsymbol{A}'\boldsymbol{A}(\overline{K}\boldsymbol{l}_c)]\boldsymbol{B}') \wedge}_{\boldsymbol{\alpha}_2}\\
&\underbrace{([\boldsymbol{C}'\boldsymbol{A}\boldsymbol{B}']\boldsymbol{B} - [\boldsymbol{C}'\boldsymbol{A}\boldsymbol{B}]\boldsymbol{B}')}_{\boldsymbol{\alpha}_3} = 0.
\end{aligned} \qquad (16.40)$$

Note that the scalar $det(\underline{K})^4$ is cancelled out simplifying the expression for the $\boldsymbol{\alpha}$'s. The computation of the intrinsic parameters will be done first considering that the intrinsic parameters remain stationary under camera motions and second when these parameters change.

16.4.1 Computing Stationary Intrinsic Parameters

Let us assume that the basis $\mathcal{F}_0$ is attached to the optical center of the first camera and consider a second camera which has a motion of $[R_1|\boldsymbol{t}_1]$ with respect to the first one. Accordingly the involved projective transformations are given in matrix notation by

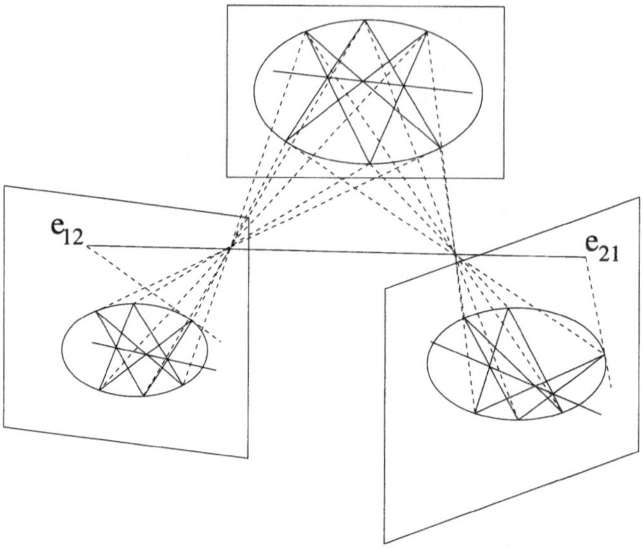

Fig. 16.3. Pascal's theorem at the conic images

$$P_1 = K[I|0] \tag{16.41}$$

$$P_2 = P_1 \begin{pmatrix} R_1 & \boldsymbol{t}_1 \\ 0_3^T & 1 \end{pmatrix}^{-1} = P_1 \left(M_{\mathcal{F}c}^{\mathcal{F}_0} \right)^{-1} \tag{16.42}$$

and their optical centres by $\boldsymbol{C}_1 = (0,0,0,1)^T$ and $\boldsymbol{C}_2 = M_{\mathcal{F}c}^{\mathcal{F}_0} \boldsymbol{C}_1$. In geometric algebra we use the notations $\underline{P_1}$, $\underline{P_2}$, $\boldsymbol{C}_1 = \mathbf{e}_4$ and $\boldsymbol{C}_2 = \underline{M_{\mathcal{F}c}^{\mathcal{F}_0}} \boldsymbol{C}_1$. Thus, we can compute their epipoles as $\boldsymbol{e}_{21} = \underline{P_2} \boldsymbol{C}_1$, $\boldsymbol{e}_{12} = \underline{P_1} \boldsymbol{C}_2$.

Next, we will show by means of an example that the coordinates of the points $\boldsymbol{\alpha}_1$, $\boldsymbol{\alpha}_2$, $\boldsymbol{\alpha}_3$ are entirely independent of the intrinsic parameters. This condition is necessary for solving the problem. Let us choose a camera motion given by

$$[R_1|\boldsymbol{t}_1] = \left(\begin{array}{ccc|c} 0 & -1 & 0 & 2 \\ 1 & 0 & 0 & -1 \\ 0 & 0 & 1 & 3 \end{array} \right). \tag{16.43}$$

For this motion the epipoles are

$$\begin{aligned} \boldsymbol{e}_{12} &= (2k_{11} - k_{12} + 3k_{13})\mathbf{e}_1 + (-k_{22} + 3k_{23})\mathbf{e}_2 + 3\mathbf{e}_3 \quad \text{and} \\ \boldsymbol{e}_{21} &= (k_{11} + 2k_{12} - 3k_{13})\mathbf{e}_1 + (2k_{22} - 3k_{23})\mathbf{e}_2 - 3\mathbf{e}_3. \end{aligned} \tag{16.44}$$

By using the rotated conic points given by the equation (16.15) and replacing $\boldsymbol{e}_{12}$ in the equation (16.40), we can make explicit the $\boldsymbol{\alpha}$'s

$$\begin{aligned}
\boldsymbol{\alpha}_1 &= ((-3+3i)k_{11}\tau)\mathbf{e}_1 + \\
&\quad (3k_{11}\tau - ik_{12}\tau + ik_{22} + 2ik_{11}\tau - 3k_{12}\tau + 3k_{22})\mathbf{e}_2 + \\
&\quad (ik_{11}\tau + 3k_{12}\tau - 3k_{22} + ik_{12}\tau - ik_{22})\mathbf{e}_3 \\
\boldsymbol{\alpha}_2 &= (-3ik_{11}\tau - 2k_{12}\tau + 2k_{22} - 2k_{11}\tau)\mathbf{e}_1 + \\
&\quad (-6i(k_{12}\tau - k_{22}))\mathbf{e}_2 + (-3k_{11}\tau - 4ik_{12}\tau + 4ik_{22} + 2ik_{11}\tau)\mathbf{e}_3 \\
\boldsymbol{\alpha}_3 &= (1-i)\mathbf{e}_1 + (1-i)\mathbf{e}_2 + 2\mathbf{e}_3. \qquad (16.45)
\end{aligned}$$

Note that $\boldsymbol{\alpha}_3$ is fully independent of $\underline{K}$. According to Pascal's theorem these three points lie on the same line, therefore, by replacing these points in the equation (16.38) we get the following second order polynomial in τ

$$\begin{aligned}
&(-40ik_{12}^2 - 52ik_{11}^2 + 16ik_{11}k_{12})\tau^2 + \\
&(-16ik_{11}k_{22} + 80ik_{12}k_{22})\tau - 40ik_{22}^2 = 0. \qquad (16.46)
\end{aligned}$$

Solving this polynomial and choosing one of the solutions which is nothing else than the solution for one of the two lines tangent to the conic we get

$$\tau := \frac{16ik_{11}k_{22} - 80ik_{12}k_{22} + 24\sqrt{14}k_{11}k_{22}}{2(-40ik_{12}^2 - 52ik_{11}^2 + 16ik_{11}k_{12})}. \qquad (16.47)$$

Now considering the homogeneous representation of these intersection points

$$\boldsymbol{\alpha}_i = \alpha_{i1}\mathbf{e}_1 + \alpha_{i2}\mathbf{e}_2 + \alpha_{i3}\mathbf{e}_3 \sim \frac{\alpha_{i1}}{\alpha_{i3}}\mathbf{e}_1 + \frac{\alpha_{i2}}{\alpha_{i3}}\mathbf{e}_2 + \mathbf{e}_3, \qquad (16.48)$$

we can finally express their homogeneous coordinates as follows

$$\alpha_{11} = \frac{-(2k_{11}-10k_{12}+3ik_{11}\sqrt{14}+8ik_{12}+2k_{12}\sqrt{14}-10ik_{11}+2\sqrt{14}k_{11})}{2ik_{11}-10ik_{12}-3\sqrt{14}k_{11}-4k_{12}-4ik_{12}\sqrt{14}-16k_{11}+2ik_{11}\sqrt{14}} \qquad (16.49)$$

$$\alpha_{12} = \frac{2i(-2ik_{12}-3k_{12}\sqrt{14}+13ik_{11})}{2ik_{11}-10ik_{12}-3\sqrt{14}k_{11}-4k_{12}-4ik_{12}\sqrt{14}-16k_{11}+2ik_{11}\sqrt{14}} \qquad (16.50)$$

$$\alpha_{21} = \frac{(1-i)(2ik_{11}-10ik_{12}-3\sqrt{14}k_{11})}{5k_{11}-4k_{12}+ik_{11}\sqrt{14}+2ik_{12}+3k_{12}\sqrt{14}-13ik_{11}+ik_{12}\sqrt{14}} \qquad (16.51)$$

$$\alpha_{22} = \frac{11ik_{11}+8ik_{12}+3\sqrt{14}k_{11}-6k_{12}-ik_{12}\sqrt{14}-3k_{11}+2ik_{11}\sqrt{14}-3k_{12}\sqrt{14}}{5k_{11}-4k_{12}+ik_{11}\sqrt{14}+2ik_{12}+3k_{12}\sqrt{14}-13ik_{11}+ik_{12}\sqrt{14}}. \qquad (16.52)$$

In the case of exactly orthogonal image axis, we can set in previous equation $k_{12} = 0$ and get

$$\alpha_{11} = \frac{2i - 3\sqrt{14} + 10 + 2i\sqrt{14}}{2 + 3i\sqrt{14} + 16i + 2\sqrt{14}} \qquad (16.53)$$

$$\alpha_{12} = 26\frac{i}{2 + 3i\sqrt{14} + 16i + 2\sqrt{14}} \qquad (16.54)$$

$$\alpha_{21} = \frac{(1+i)(-2i + 3\sqrt{14})}{-5i + \sqrt{14} - 13} \qquad (16.55)$$

$$\alpha_{22} = -\frac{-11 + 3i\sqrt{14} - 3i - 2\sqrt{14}}{-5i + \sqrt{14} - 13}. \qquad (16.56)$$

The coordinates of the intersection points are indeed independent of the intrinsic parameters.

After this illustration by an example we will get now the coordinates for any general camera motion. For that it is necessary to separate in the projections the intrinsic parameters from the extrinsic ones. Let us define

$$\boldsymbol{s} = s_1\mathbf{e}_1 + s_2\mathbf{e}_2 + s_3\mathbf{e}_3 = \underline{[I|0]}\ \underline{M_{\mathcal{F}_C}^{\mathcal{F}_0}}\boldsymbol{C}_1. \tag{16.57}$$

Thus, the epipole is

$$\boldsymbol{e}_{12} = \underline{K}\ \underline{[I|0]}\ \underline{M_{\mathcal{F}_C}^{\mathcal{F}_0}}\boldsymbol{C}_1 = \underline{K}\boldsymbol{s}. \tag{16.58}$$

Note that in this expression the intrinsic parameters are separate from the extrinsic ones. Similar as above for the general camera motion with the corresponding epipole value the coordinates for the intersecting points read

$$\alpha_{11} = -\frac{(-s_3s_1s_2+is_3\sqrt{s_3^2(s_1^2+s_2^2+s_3^2)}-is_3^3-is_3s_1^2+s_1\sqrt{s_3^2(s_1^2+s_2^2+s_3^2)}-is_2s_3^2)}{(-is_3s_1s_2-s_3\sqrt{s_3^2(s_1^2+s_2^2+s_3^2)}-s_3^3-s_3s_1^2+is_1\sqrt{s_3^2(s_1^2+s_2^2+s_3^2)}+s_2s_3^2)} \tag{16.59}$$

$$\alpha_{21} = \frac{-2s_3(s_3^2+s_1^2)}{-is_3s_2s_1-s_3\sqrt{s_3^2(s_1^2+s_2^2+s_3^2)}-s_3^3-s_3s_1^2+is_1\sqrt{s_3^2(s_1^2+s_2^2+s_3^2)}+s_2s_3^2} \tag{16.60}$$

$$\alpha_{12} = \frac{(-1-i)(is_1s_2+\sqrt{s_3^2(s_1^2+s_2^2+s_3^2)})s_3}{-is_3s_1s_2-s_3\sqrt{s_3^2(s_1^2+s_2^2+s_3^2)}+s_3s_1^2+s_3^3+s_1\sqrt{s_3^2(s_1^2+s_2^2+s_3^2)}-is_2s_3^2} \tag{16.61}$$

$$\alpha_{22} = \frac{i(is_3s_1s_2+s_3\sqrt{s_3^2(s_1^2+s_2^2+s_3^2)}+is_1\sqrt{s_3^2(s_1^2+s_2^2+s_3^2)}+s_2s_3^2+is_3s_1^2+is_3^3)}{-is_3s_1s_2-s_3\sqrt{s_3^2(s_1^2+s_2^2+s_3^2)}+s_3s_1^2+s_3^3+s_1\sqrt{s_3^2(s_1^2+s_2^2+s_3^2)}-is_2s_3^2}. \tag{16.62}$$

Note that the intrinsic parameters are totally cancelled out. The invariance properties can be used to obtain equations which depend on the four unknown intrinsic camera parameters. The algorithm can be summarized in the following steps.

1. Suppose point correspondences between two cameras and motion between the cameras.
2. Calculate the values of the homogeneous $\boldsymbol{\alpha}_i$ by using the known camera motion and the formulas (16.59–16.62).
3. Calculate $\overline{\boldsymbol{K}\boldsymbol{l}_c}$ with the epipole, evaluated from the point correspondences. To fulfill Pascal's theorem solve the equations system to τ similar to (16.47).
4. Replace τ in (16.45) and calculate the homogeneous representation of these intersection points to get quadratic polynomials which depends on the four unknown intrinsic parameters. Note that the intrinsic parameters are not cancelled out because of the insert of the real values from the epipol. Because of the invariant properties of the $\boldsymbol{\alpha}$'s the polynomials must be equal to the evaluated values of the $\boldsymbol{\alpha}$'s in step 2. This leads to four quadratic equations.

Since we are assuming that the intrinsic parameters remain constant, we can consequently gain a second set of four equations depending again of the four intrinsic parameters from the second epipole.

The interesting aspect here is that we require only one camera motion to find a solvable equation system. Other methods gain for each camera motion only a couple of equations, thus they require at least three camera motions to solve the problem [164, 158]. This particular advantage of our approach relies in the investigation of Pascal's theorem and its formulation in geometric algebra.

16.4.2 Computing Non–stationary Intrinsic Parameters

In this case we will consider that due to the camera motion the intrisic parameters may have been changed. The procedure can be formulated along the same previous ideas with the difference that we compute the line $\boldsymbol{l}_c$ using the operator for the fundamental matrix and a point lying at line at infinite of the second camera as $\boldsymbol{l}_c = \overline{F}(\mathbf{e}_1 + \tau'\mathbf{e}_2)$.

Note that the fundamental matrix can be expressed in terms of the motion between cameras and the K of the camera, i.e. $F = K^{-T}[\boldsymbol{t}]_\times R_{12} K^T$ where $[\boldsymbol{t}]_\times$ is the tensor notation of the antisymmetric matrix representing the translation [158]. The term $E = [\boldsymbol{t}]_\times R_{12}$ is called the essential matrix. The decomposition of F can instantaneous be described by $\underline{F} = \overline{K}^{-1} \underline{[\boldsymbol{t}]_\times}\ \underline{R_{12}} \overline{K}$ in terms of geometric algebra.

Now similar as in previous case we will use an example for facilitating the understanding. We will use the same camera motion given in equation (16.43). The fundamental matrix in terms of the intrinsic parameters of the first camera K and of the second one K', with the assumption of perpendicular pixel grids $k_{12} = k'_{12} = 0$, and the camera motion reads in matrix notation

$$
\begin{aligned}
F &= {K^{-1}}^T [\boldsymbol{t}]_\times R {K'}^{-1} \\
&= \begin{pmatrix} -3\frac{k'_{22}k_{22}}{v_2} & 0 & -\frac{(k'_{11}-3k'_{13})k_{22}k'_{22}}{v_2} \\ 0 & -3\frac{k'_{11}k_{11}}{v_2} & -\frac{k_{11}k'_{11}(2k'_{22}-3k'_{23})}{v_2} \\ \frac{(2k_{11}+3k_{13})k_{22}k'_{22}}{v_2} & -\frac{(k_{22}-3k_{23})k_{11}k'_{11}}{v_2} & 1 \end{pmatrix}
\end{aligned}
\tag{16.63}
$$

where $v_2 = -3k'_{22}k_{22}k_{13}k'_{13} + k_{22}k'_{22}k'_{11}k_{13} + k_{22}k'_{23}k'_{11}k_{11} - 2k_{22}k'_{22}k'_{13}k_{11} + 2k_{23}k'_{22}k'_{11}k_{11} - 3k_{23}k'_{23}k'_{11}k_{11}$.

The value of the line $\boldsymbol{l}_c$ is now computed in terms of the operator of the fundamental matrix, i.e. $\boldsymbol{l}_c = \overline{F}(\mathbf{e}_1 + \tau'\mathbf{e}_2)$ Similar as above we compute the $\boldsymbol{\alpha}$'s and according the Pascal's theorem we gain a polynomial similar as equation (16.46). This reads

$$
10{k'}^2_{11}{\tau'}^2 - 4k'_{22}k'_{11}\tau' + 13{k'}^2_{22} = 0. \tag{16.64}
$$

We select one of both solutions of this second order polynomial

$$
\tau' = \frac{4k'_{22}k'_{11} + 6ik'_{22}k'_{11}\sqrt{14}}{20({k'}^2_{11})} \tag{16.65}
$$

and substitute it in the homogeneous coordinates of the $\boldsymbol{\alpha}$'s

$$\alpha_{11} = -\frac{i(-5i - 4 + i\sqrt{14})}{5i + 2 + 2i\sqrt{14}} \tag{16.66}$$

$$\alpha_{21} = \frac{-2 + 3i\sqrt{14}}{5i + 2 + 2i\sqrt{14}} \tag{16.67}$$

$$\alpha_{12} = \frac{10 - 10i}{-4i - 2 + 3i\sqrt{14} - \sqrt{14}} \tag{16.68}$$

$$\alpha_{22} = -\frac{8 + 6i - \sqrt{14} + 3i\sqrt{14}}{-4i - 2 + 3i\sqrt{14} - \sqrt{14}}, \tag{16.69}$$

where $\boldsymbol{\alpha_3} = (1 - i)\mathbf{e}_1 + (1 - i)\mathbf{e}_2 + 2\mathbf{e}_3$ is again fully independent of the intrinsic parameters.

Finally, we will show the expression when we consider now a general motion

$$[R|\boldsymbol{t}] = \begin{pmatrix} r_{11} & r_{12} & r_{13} & t_1 \\ r_{21} & r_{22} & r_{23} & t_2 \\ r_{31} & r_{32} & r_{33} & t_3 \end{pmatrix}. \tag{16.70}$$

In matrix algebra the fundamental matrix reads

$$F = K^{-T} E K'^{-1} \tag{16.71}$$

$$= \begin{pmatrix} k_{11} & 0 & k_{13} \\ 0 & k_{22} & k_{23} \\ 0 & 0 & 1 \end{pmatrix}^{-T} \begin{pmatrix} E_{11} & E_{12} & E_{13} \\ E_{21} & E_{22} & E_{23} \\ E_{31} & E_{32} & E_{33} \end{pmatrix} \begin{pmatrix} k'_{11} & 0 & k'_{13} \\ 0 & k'_{22} & k'_{23} \\ 0 & 0 & 1 \end{pmatrix}^{-1} \tag{16.72}$$

and in geometric algebra the operator of the fundamental matrix reads $\underline{F} = \overline{\underline{K}}^{-1} \underline{E}\, \underline{K}'^{-1}$.

Using this formulation we compute the homogeneous coordinates of the $\boldsymbol{\alpha}$'s

$$\begin{aligned} \alpha_{11} &= i(iE_{11}E_{22}^2 + iE_{11}E_{32}^2 - iE_{12}E_{21}E_{22} - iE_{12}E_{31}E_{32} \\ &-iE_{12}\sqrt{v_3} + E_{21}E_{12}^2 + E_{21}E_{32}^2 - E_{22}E_{11}E_{12} - E_{22}E_{31}E_{32} \\ &-E_{22}\sqrt{v_3} - E_{31}E_{12}^2 - E_{31}E_{22}^2 + E_{32}E_{11}E_{12} + E_{32}E_{21}E_{22} + E_{32}\sqrt{v_3})/ \\ &(iE_{11}E_{22}^2 + iE_{11}E_{32}^2 - iE_{12}E_{21}E_{22} - iE_{12}E_{31}E_{32} - iE_{12}\sqrt{v_3} - \\ &E_{31}E_{12}^2 - E_{31}E_{22}^2 + E_{32}E_{11}E_{12} + E_{32}E_{21}E_{22} + E_{32}\sqrt{v_3} - \\ &E_{21}E_{12}^2 - E_{21}E_{32}^2 + E_{22}E_{11}E_{12} + E_{22}E_{31}E_{32} + E_{22}\sqrt{v_3}) \end{aligned} \tag{16.73}$$

$$
\begin{aligned}
\alpha_{12} &= 2(-E_{21}E_{12}^2 - E_{21}E_{32}^2 + E_{22}E_{11}E_{12} + E_{22}E_{31}E_{32} + E_{22}\sqrt{v_3})/ \\
&(iE_{11}E_{22}^2 + iE_{11}E_{32}^2 - iE_{12}E_{21}E_{22} - iE_{12}E_{31}E_{32} \\
&-iE_{12}\sqrt{v_3} - E_{31}E_{12}^2 - E_{31}E_{22}^2 + E_{32}E_{11}E_{12} + E_{32}E_{21}E_{22} + \\
&E_{32}\sqrt{v_3} - E_{21}E_{12}^2 - E_{21}E_{32}^2 + E_{22}E_{11}E_{12} + E_{22}E_{31}E_{32} + E_{22}\sqrt{v_3})
\end{aligned} \tag{16.74}
$$

$$
\begin{aligned}
\alpha_{21} &= (1-i)(E_{11}E_{22}^2 + E_{11}E_{32}^2 - E_{12}E_{21}E_{22} - E_{12}E_{31}E_{32} - E_{12}\sqrt{v_3})/ \\
&(-iE_{11}E_{22}^2 - iE_{11}E_{32}^2 + iE_{12}E_{21}E_{22} + iE_{12}E_{31}E_{32} \\
&+iE_{12}\sqrt{v_3} - E_{21}E_{12}^2 - E_{21}E_{32}^2 + E_{22}E_{11}E_{12} + E_{22}E_{31}E_{32} \\
&+E_{22}\sqrt{v_3} - iE_{31}E_{12}^2 - iE_{31}E_{22}^2 + iE_{32}E_{11}E_{12} \\
&+iE_{32}E_{21}E_{22} + iE_{32}\sqrt{v_3})
\end{aligned} \tag{16.75}
$$

$$
\begin{aligned}
\alpha_{22} &= -(E_{11}E_{22}^2 + E_{11}E_{32}^2 - E_{12}E_{21}E_{22} - E_{12}E_{31}E_{32} - E_{12}\sqrt{v_3} \\
&-iE_{31}E_{12}^2 - iE_{31}E_{22}^2 + iE_{32}E_{11}E_{12} + iE_{32}E_{21}E_{22} \\
&+iE_{32}\sqrt{v_3} - E_{21}E_{12}^2 - E_{21}E_{32}^2 + E_{22}E_{11}E_{12} + E_{22}E_{31}E_{32} \\
&+E_{22}\sqrt{v_3})/(-iE_{11}E_{22}^2 - iE_{11}E_{32}^2 + iE_{12}E_{21}E_{22} \\
&+iE_{12}E_{31}E_{32} + iE_{12}\sqrt{v_3} - E_{21}E_{12}^2 - E_{21}E_{32}^2 \\
&+E_{22}E_{11}E_{12} + E_{22}E_{31}E_{32} + E_{22}\sqrt{v_3} - iE_{31}E_{12}^2 \\
&-iE_{31}E_{22}^2 + iE_{32}E_{11}E_{12} + iE_{32}E_{21}E_{22} + iE_{32}\sqrt{v_3})
\end{aligned} \tag{16.76}
$$

where

$$
\begin{aligned}
v_3 &= 2E_{11}E_{12}E_{21}E_{22} + 2E_{11}E_{12}E_{31}E_{32} + 2E_{21}E_{22}E_{31}E_{32} \\
&-E_{12}^2E_{31}^2 - E_{12}^2E_{21}^2 - E_{22}^2E_{31}^2 - E_{22}^2E_{11}^2 - E_{32}^2E_{21}^2 - E_{32}^2E_{11}^2.
\end{aligned} \tag{16.77}
$$

Note that for the general case the α's are fully independent of the intrinsic camera coefficients k_{ij} or k'_{ij}. Together with the equations of the $\boldsymbol{\alpha}$'s obtained using the first epipole the intrinsic parameters can be found solving a quadratic equation system.

16.5 Experimental Analysis

In this section we present tests of the method based on Pascal's theorem using firstly simulated images. We explore the effect of different kinds of camera motion and the effect of increasing noise in the computing of the intrinsic camera parameters. The experiments with real images show that the performance of the method is reliable.

16.5.1 Experiments with Simulated Images

Using a Maple simulation we firstly test the method based on the theorem of Pascal to explore the dependency of the type and the amount of necessary camera motions for solving the problem. The experiments show that at least a rotation about only one axis and a displacement along the three axes are necessary for stabile computations of all intrinsic parameters. Then, we realize a test of our approach by increasing noise.

The camera is rotated about the y–axis with translation along the three camera axes. For the tests we used exact arithmetic of the Maple program instead of floating point arithmetic of the C language. The Table 1 shows the computed intrinsic parameters. The most right column of the table shows the error obtained substituting these parameters in the polynomial (16.64) which gives zero for the case of zero noise. The values in this column show that by increasing noise the computed intrinsic parameters cause a tiny deviation of the ideal value of zero. This indicates that the procedure is relatively stable against noise. We could image that there is a relative flat surface around the global minimum of the polynomial. Note that there are remarkable deviations shown by noise 1.25.

Table 16.1. Intrinsic parameters by rotation about the y–axis and translation along the three axes with increasing noise

Noise(pixels)	k_{11}	k_{13}	k_{22}	k_{23}	Error
0	500	256	500	256	10^{-8}
0.1	505	259	509	261	0.001440
0.5	504	259.5	503.5	258	0.004897
0.75	498	254	503.5	258	0.001668
1	482	242	485	254	0.011517
1.25	473	220	440	238	0.031206
1.5	517	272	518	266	0.015
2	508	262.5	504	258.5	0.006114
2.5	515	268	501.9	257	0.011393
3	510	265	524	276	0.011440

16.5.2 Experiments with Real Images

In this section we present experiments using real images with one general camera motion, see Figure 16.4.

The motion was done about the three coordinate axes. We use a calibration dice and for comparison purposes we compute the intrinsic parameters from the involved projective matrices by splitting the intrinsic parameters from the extrinsic ones. The reference values were: First camera

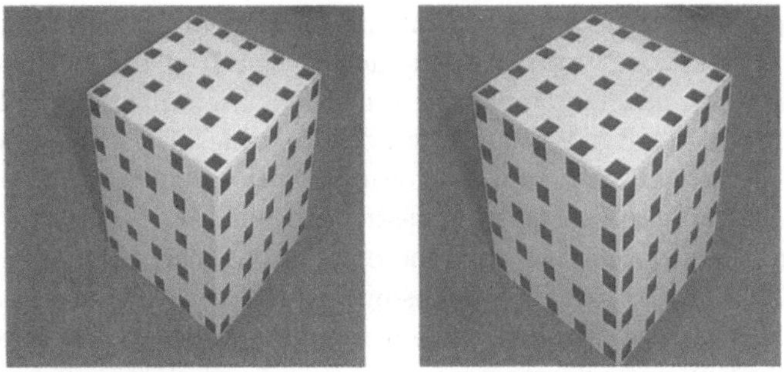

Fig. 16.4. Scenario

$k_{11} = 1200.66$, $k_{22} = 1154.77$, $k_{13} = 424.49$, $k_{23} = 264.389$ and second camera $k_{11} = 1187.82$, $k_{22} = 1141.58$, $k_{13} = 386.797$, $k_{23} = 288.492$ with mean errors of 0.688 and 0.494, respectively.

Thereafter, using the gained extrinsic parameters $[R_1|\boldsymbol{t}_1]$ and $[R_2|\boldsymbol{t}_2]$ we compute the relation $[R|\boldsymbol{t}]$ between cameras which is required for the Pascal's theorem based method. The fundamental matrix is computed using a non-linear method. Using the Pascal's theorem based method with 12 point correspondences unlike 160 point correspondences used by the algorithm with the calibration dice we compute the following intrinsic parameters $k_{11} = 1244$, $k_{22} = 1167$, $k_{13} = 462$ and $k_{23} = 217$. The error is computed using the eight equations gained from the $\boldsymbol{\alpha}$'s of the first and second camera. These values resemble quite well to the reference ones and cause an error of $\sqrt{|eqn_1|^2 + ... + |eqn_8|^2}$: 0.004961 in the error function. The difference with the reference values is attributable to inherent noise in the computation and to the fact that the reference values are not exact, too.

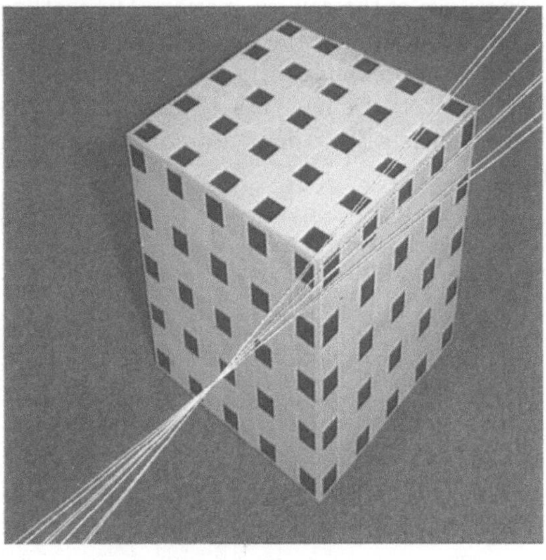

Fig. 16.5. Superimposed epipolar lines using the reference and Pascal's theorem based method

Since this is a system of quadratic equations we resort to an iterative procedure for finding the solution. First we tried the Newton–Raphson method [191] and the continuation method [158]. These methods are not practicable enough due to their complexity. We use instead a variable in size window minima search which through the computation ensure the reduction of the quadratic error. This simply approach work faster and reliable.

In order to visualize how good we gain the epipolar geometry we superimposed the epipolar lines for some points using the reference method and Pascal's theorem based method. In both cases we computed the fundamental matrix in terms of their intrinsic parameters, i.e. $F = K^{-T}[\boldsymbol{t}]_{\times}RK^{-1}$. Figure 16.5 shows this comparison. It is clear that both methods give quite similar epipolar lines and interesting enough it is shown that the intersecting point or epipole coincide almost exactly.

16.6 Conclusions

This paper presents a geometric approach to formulate the Kruppa equations in terms of pure brackets. This can certainly help to explore the geometry of the calibration problem and to find degenerated cases. Furthermore this paper presents an approach to compute the intrinsic camera parameters in the geometric algebra framework using Pascal's theorem. We adopt the projected characteristics of the absolute conic in terms of Pascal's theorem to propose a new camera calibration method based on geometric thoughts. The use of this theorem in the geometric algebra framework allows us the computing of a projective invariant using the conics of only two images. Then, this projective invariant expressed in terms of brackets helps us to set enough equations to solve the calibration problem. Our method requires to know the point correspondences and the values of the camera motion. The method gives a new point of view for the understanding of the problem thanks to the application of Pascal's theorem and it also explains the overseen role of the projective invariant in terms of the brackets. Using synthetic and real images we show that the method performs efficiently without any initialization or getting trapped in local minima.

17. Coordinate-Free Projective Geometry for Computer Vision*

Hongbo Li and Gerald Sommer

Institute of Computer Science and Applied Mathematics,
Christian-Albrechts-University of Kiel

17.1 Introduction

How to represent an image point algebraically? Given a Cartesian coordinate system of the retina plane, an image point can be represented by its coordinates (u, v). If the image is taken by a pinhole camera, then since a pinhole camera can be taken as a system that performs the perspective projection from three-dimensional projective space to two-dimensional one with respect to the optical center [77], it is convenient to describe a space point by its homogeneous coordinates $(x, y, z, 1)$ and to describe an image point by its homogeneous coordinates $(u, v, 1)$. In other words, the space of image points can be represented by the space of 3×1 matrices. This is the coordinate representation of image points.

There are other representations which are coordinate-free. The use of algebras of geometric invariants in the coordinate-free representations can lead to remarkable simplifications in geometric computing. Kanatani [124] uses the three-dimensional affine space for space points, and the space of displacements of the affine space for image points. In other words, he uses vectors fixed at the origin of $\mathbb{R}^3$ to represent space points, and uses free vectors to

* This work has been supported by Alexander von Humboldt Foundation (H.L.) and by DFG Grants So-320-2-1 and So-320-2-2 (G.S.)

represent image points. Then he can use vector algebra to carry out geometric computing. This algebraic representation is convenient for two-dimensional projective geometry, but not for three-dimensional one. The space representing image points depends neither on the retina plane nor on the optical center.

Bayro-Corrochano, Lasenby and Sommer use $\mathbb{R}^4$ for both two-dimensional and three-dimensional projective geometries[19, 17, 216]. They use a coordinate system $\{e_1, e_2, e_3, C\}$ of $\mathbb{R}^4$ to describe a pinhole camera, where the e's are points on the retina plane and C is the optical center. Both space points and image points are represented by vectors fixed at the origin of $\mathbb{R}^4$, the only difference is that an image point is in the space spanned by vectors e_1, e_2, e_3. This algebraic representation is convenient for projective geometric computations using the incidence algebra formulated in Clifford algebra. However, it always needs a coordinate system for the camera. The space representing image points depends only on the retina plane.

We noticed that none of these algebraic representations of image points is related to the optical center. By intuition, it is better to represent image points by vectors fixed at the optical center. The above-mentioned coordinate-free representations do not have this property.

Hestenes [109] proposed a technique called space-time split to realize the Clifford algebra of the Euclidean space in the Clifford algebra of the Minkowskii space. The technique is later generalized to projective split by Hestenes and Ziegler [114] for projective geometry. We find that a version of this technique offers us exactly what we need: three-dimensional linear spaces imbedded in a four-dimensional one, whose origins do not concur with that of the four-dimensional space but whose Clifford algebras are realized in that of the four-dimensional space.

Let C be a vector in $\mathbb{R}^4$. It represents either a space point or a point at infinity of the space. Let M be another vector in $\mathbb{R}^4$. The image of the space point or point at infinity M by a pinhole camera with optical center C can be described by $C \wedge M$. The image points can be represented by the three-dimensional space $C \wedge \mathbb{R}^4 = \{C \wedge X | X \in \mathbb{R}^4\}$. The Clifford algebra of the space $C \wedge \mathbb{R}^4$ can be realized in the Clifford algebra of $\mathbb{R}^4$ by the theorem of projective split proposed later in this chapter. The space representing image points depends only on the optical center. The representation is completely projective and completely coordinate-free.

Using this new representation and the version of Grassmann-Cayley algebra formulated by Hestenes and Ziegler [114] within Clifford algebra, we have reformulated camera modeling and calibration, epipolar and trifocal geometries, relations among epipoles, epipolar tensors and trifocal tensors. Remarkable simplifications and generalizations are obtained through the reformulation, both in conception and in application. In particular, we are to derive and generalize all known constraints on epipolar and trifocal tensors [76, 80, 81, 83] in a systematic way.

This chapter is arranged as follows: in section 17.2 we collect some necessary mathematical techniques, in particular the theorem of projective split in Grassmann-Cayley algebra. In sections 17.3 and 17.4 we reformulate camera modeling and calibration, and epipolar and trifocal geometries. In section 17.5 we derive and generalize the constraints on epipolar and trifocal tensors systematically.

17.2 Preparatory Mathematics

17.2.1 Dual Bases

According to Hestenes and Sobczyk [113], let $\{e_1, \dots, e_n\}$ be a basis of $\mathbb{R}^n$ and $\{e_1^*, \dots, e_n^*\}$ be the corresponding dual (or reciprocal) basis, then

$$\begin{aligned} e_i^* &= (-1)^{i-1}(e_1 \wedge \cdots \wedge \check{e}_i \wedge \cdots \wedge e_n)^\sim, \\ e_i &= (-1)^{i-1}(e_1^* \wedge \cdots \wedge \check{e}_i^* \wedge \cdots \wedge e_n^*)^\sim, \end{aligned} \tag{17.1}$$

for $1 \leq i \leq n$. Here "$\sim$" is the dual operator in $\mathcal{G}_n$ with respect to $e_1 \wedge \cdots \wedge e_n$.

The basis $\{e_1, \dots, e_n\}$ induces a basis $\{e_{j_1} \wedge \cdots \wedge e_{j_s} | 1 \leq j_1 < \ldots < j_s \leq n\}$ for the s-vector subspace $\mathcal{G}_n^s$ of the Clifford algebra $\mathcal{G}_n$ of $\mathbb{R}^n$. We have

$$\begin{aligned} &(e_{j_1} \wedge \cdots \wedge e_{j_s})^* \\ &= e_{j_s}^* \wedge \cdots \wedge e_{j_1}^* \\ &= (-1)^{j_1 + \cdots + j_s + s(s+1)/2}(e_1 \wedge \cdots \wedge \check{e}_{j_1} \wedge \cdots \wedge \check{e}_{j_s} \wedge \cdots \wedge e_n)^\sim. \end{aligned} \tag{17.2}$$

Let $x \in \mathcal{G}_n^s$, then

$$\begin{aligned} x &= \sum_{1 \leq j_1 < \ldots < j_s \leq n} x \cdot (e_{j_1} \wedge \cdots \wedge e_{j_s})^* \; e_{j_1} \wedge \cdots \wedge e_{j_s} \\ &= \sum_{1 \leq j_1 < \ldots < j_s \leq n} (-1)^{j_1 + \cdots + j_s + s(s+1)/2} \; e_{j_1} \wedge \cdots \wedge e_{j_s} \\ &\qquad (e_1 \wedge \cdots \wedge \check{e}_{j_1} \wedge \cdots \wedge \check{e}_{j_s} \wedge \cdots \wedge e_n) \vee x. \end{aligned} \tag{17.3}$$

Let an invertible transformation T of $\mathbb{R}^n$ maps $\{e_1, \dots, e_n\}$ to a basis $\{e'_1, \dots, e'_n\}$. Let $T^* = (T^T)^{-1}$. Then T^* maps the dual basis $\{e_1^*, \dots, e_n^*\}$ to the dual basis $\{e'^*_1, \dots, e'^*_n\}$.

Any linear mapping $T : \mathbb{R}^n \longrightarrow \mathbb{R}^m$ has a tensor representation in $\mathbb{R}^n \otimes \mathbb{R}^m$. Then

$$T = \sum_{i=1}^n e'_i \otimes e_i^*. \tag{17.4}$$

For example, let $\mathit{\Pi}_n$ be the identity transformation of $\mathbb{R}^n$, then in tensor representation, $\mathit{\Pi}_n = \sum_{i=1}^n e_i \otimes e_i^*$ for any basis $\{e_1, \dots, e_n\}$.

17.2.2 Projective and Affine Spaces

An n-dimensional real projective space $\mathbb{P}^n$ can be realized in the space $\mathbb{R}^{n+1}$, where a projective r-space is an $(r+1)$-dimensional linear subspace. In $\mathcal{G}_{n+1}$, a projective r-space is represented by an $(r+1)$-blade, and the representation is unique up to a nonzero scale. Throughout this chapter we use "$x \simeq y$" to denote that if x, y are scalars, they are equal up to a nonzero index-free scale, otherwise they are equal up to a nonzero scale.

An n-dimensional real affine space $\mathcal{A}^n$ can be realized in the space $\mathbb{R}^{n+1}$ as a hyperplane away from the origin. Let e_0 be the vector from the origin to the hyperplane and orthogonal to the hyperplane. When $e_0^2 = 1$, a vector $x \in \mathbb{R}^{n+1}$ is an affine point if and only if $x \cdot e_0 = 1$. An r-dimensional affine plane is the intersection of an $(r+1)$-dimensional linear subspace of $\mathbb{R}^{n+1}$ with $\mathcal{A}^n$, and can be represented by an $(r+1)$-blade of $\mathcal{G}_{n+1}$ representing the subspace.

The space of displacements of $\mathcal{A}^n$ is defined as $\overset{\infty}{\mathcal{A}^n} = \{x - y | x, y \in \mathcal{A}^n\}$. It is an n-dimensional linear subspace of $\mathbb{R}^{n+1}$. Any element of it is called a direction. When $\overset{\infty}{\mathcal{A}^n}$ is taken as an $(n-1)$-dimensional projective space, any element in it is called a point at infinity, and $\overset{\infty}{\mathcal{A}^n}$ is called the space at infinity of $\mathcal{A}^n$.

Let $I_n = e_0 \cdot I_{n+1}$. Then it represents the space $\overset{\infty}{\mathcal{A}^n}$. The mapping

$$\partial_{I_n} : \quad x \mapsto e_0 \cdot x = I_n \vee x, \ \text{ for } x \in \mathcal{G}_{n+1}, \tag{17.5}$$

maps $\mathcal{G}_{n+1}$ to $\mathcal{G}(\overset{\infty}{\mathcal{A}^n})$, called the boundary mapping. When I_n is fixed, ∂_{I_n} is often written as ∂. Geometrically, if I_{r+1} represents an r-dimensional affine space, then ∂I_r represents its space at infinity. For example, when x, y are both affine points, $\partial(x \wedge y) = y - x$ is the point at infinity of line xy.

Let $\{e_1, \ldots, e_{n+1}\}$ be a basis of $\mathbb{R}^{n+1}$. If $e_{n+1} \in \mathcal{A}^n$, $e_1, \ldots, e_n \in \overset{\infty}{\mathcal{A}^n}$, the basis is called a Cartesian coordinate system of $\mathcal{A}^n$, written as $\{e_1, \ldots, e_n; e_{n+1}\}$. The affine point e_{n+1} is called the origin. Let $x \in \mathcal{A}^n$, then $x = e_{n+1} + \sum_{i=1}^{n} \lambda_i e_i$. $(\lambda_1, \ldots, \lambda_n)$ is called the Cartesian coordinates of x with respect to the basis.

Below we list some properties of the three-dimensional projective (or affine) space when described in $\mathcal{G}_4$.

- Two planes N, N' are identical if and only if $N \vee N' = 0$, where N, N' are 3-blades.
- A line L is on a plane N if and only if $L \vee N = 0$, where L is a 2-blade.
- Two lines L, L' are coplanar if and only if $L \vee L' = 0$, or equivalently, if and only if $L \wedge L' = 0$.
- A point A is on a plane N if and only if $A \vee N = 0$, or equivalently, if and only if $A \wedge N = 0$. Here A is a vector.

– A point A is on a line L if and only if $A \wedge L = 0$.
– Three planes N, N', N'' are concurrent if and only if $N \vee N' \vee N'' = 0$.
– For two lines L, L', $L \vee L' = L^{\sim} \vee L'^{\sim}$.
– For point A and plane N, $A \vee N = A^{\sim} \vee N^{\sim}$.

17.2.3 Projective Splits

The following is a modified version of the technique of projective split.

Definition 17.2.1. Let C be a blade in $\mathcal{G}_n$. The projective split P_C of $\mathcal{G}_n$ with respect to C is the following transformation: $x \mapsto C \wedge x$, for $x \in \mathcal{G}_n$.

Theorem 17.2.1. [Theorem of projective split in Grassmann-Cayley algebra [1]] Let C be an r-blade in $\mathcal{G}_n$. Let $C \wedge \mathcal{G}_n = \{C \wedge x | x \in \mathcal{G}_n\}$. Define in it two products "$\wedge_C$" and "$\vee_C$": for $x, y \in \mathcal{G}_n$,

$$\begin{aligned} (C \wedge x) \wedge_C (C \wedge y) &= C \wedge x \wedge y, \\ (C \wedge x) \vee_C (C \wedge y) &= (C \wedge x) \vee (C \wedge y), \end{aligned} \tag{17.6}$$

and define

$$(C \wedge x)^{\sim_C} = C \wedge (C \wedge x)^{\sim}. \tag{17.7}$$

Then vector space $C \wedge \mathcal{G}_n$ equipped with "$\wedge_C$", "$\vee_C$", "$\sim_C$" is a Grassmann-Cayley algebra isomorphic to $\mathcal{G}_{n-r}$, which is taken as a Grassmann-Cayley algebra.

Proof. Let $C \wedge \mathbb{R}^n = \{C \wedge x | x \in \mathbb{R}^n\}$. It is an $(n-r)$-dimensional vector space. By the linear isomorphism of $\{\lambda C | \lambda \in \mathbb{R}\}$ with $\mathbb{R}$, it can be verified that $(C \wedge \mathcal{G}_n, \wedge_C)$ is isomorphic to the Grassmann algebra generated by $C \wedge \mathbb{R}^n$. A direct computation shows that the composition of "$\sim_C$" with itself is the scalar multiplication by $(-1)^{n(n-1)/2} C^2$. That $C \wedge \mathcal{G}_n$ is a Grassmann-Cayley algebra follows from the identity

$$(C \wedge x)^{\sim_C} \vee_C (C \wedge y)^{\sim_C} = ((C \wedge x) \wedge_C (C \wedge y))^{\sim_C}, \tag{17.8}$$

which can be verified by the definitions (17.6) and (17.7).

[1] Theorem 17.2.1 can be generalized to the following one, which is nevertheless not needed in this chapter:
[Theorem of projective split in Clifford algebra] Let C be a blade in $\mathcal{G}_n$. The space $C \wedge \mathcal{G}_n$ equipped with the following outer product "$\wedge_C$" and inner product "$\cdot_C$" is a Clifford algebra isomorphic to $\mathcal{G}(C^{\sim})$:

$$\begin{aligned} (C \wedge x) \wedge_C (C \wedge y) &= C \wedge x \wedge y, \\ (x \wedge C) \cdot_C (C \wedge y) &= C^{-2} C \wedge ((x \wedge C) \cdot (C \wedge y)), \end{aligned}$$

for $x, y \in \mathcal{G}_n$.

Let $\{e_1, \ldots, e_n\}$ be a basis of $\mathbb{R}^n$. The projective split P_C can be written as the composition of the outer product by C and the identity transformation. It has the following tensor representation:

$$P_C = \sum_{s=0}^{n} \sum_{1 \leq j_1 < \ldots < j_s \leq n} (C \wedge e_{j_1} \wedge \cdots \wedge e_{j_s}) \otimes (e_{j_1} \wedge \cdots \wedge e_{j_s})^*. \tag{17.9}$$

For example, when C is a vector and P_C is restricted to $\mathbb{R}^n$, then

$$P_C = \sum_{i=1}^{n} (C \wedge e_i) \otimes e_i^*. \tag{17.10}$$

In particular, when $\{e_1, \ldots, e_{n-1}, C\}$ is a basis of $\mathbb{R}^n$, then

$$P_C = \sum_{i=1}^{n-1} (C \wedge e_i) \otimes e_i^*. \tag{17.11}$$

When P_C is restricted to $\mathcal{G}_n^2$, then

$$P_C = \sum_{1 \leq j_1 < j_2 \leq n} (C \wedge e_{j_1} \wedge e_{j_2}) \otimes (e_{j_1} \wedge e_{j_2})^*. \tag{17.12}$$

In particular, when $\{e_1, \ldots, e_{n-1}, C\}$ is a basis of $\mathbb{R}^n$, then

$$P_C = - \sum_{1 \leq j_1 < j_2 \leq n-1} (C \wedge e_{j_1} \wedge e_{j_2}) \otimes (e_{j_2}^* \wedge e_{j_1}^*). \tag{17.13}$$

When $n = 4$, we use the notation $i \prec i_1 \prec i_2$ to denote that i, i_1, i_2 is an even permutation of $1, 2, 3$. Let

$$\hat{e}_i = e_{i_1} \wedge e_{i_2}, \quad \hat{e}_i^* = e_{i_1}^* \wedge e_{i_2}^*. \tag{17.14}$$

then

$$P_C = - \sum_{i=1}^{3} (C \wedge \hat{e}_i) \otimes \hat{e}_i^*. \tag{17.15}$$

The following theorem establishes a connection between the projective split and the boundary mapping.

Theorem 17.2.2. When C is an affine point, the boundary mapping ∂ realizes an algebraic isomorphism between the Grassmann-Cayley algebras $C \wedge \mathcal{G}_{n+1}$ and $\mathcal{G}(\overset{\infty}{\mathcal{A}^n})$.

17.3 Camera Modeling and Calibration

17.3.1 Pinhole Cameras

According to Faugeras [77], a pinhole camera can be taken as a system that performs the perspective projection from $\mathbb{P}^3$ to $\mathbb{P}^2$ with respect to the optical point $C \in \mathbb{P}^3$. To describe this mapping algebraically, let $\{e_1, e_2, e_3; O\}$ be a fixed Cartesian coordinate system of $\mathcal{A}^3$, called the world coordinate system. Let $\{e_1^C, e_2^C, e_3^C, C\}$ be a basis of $\mathbb{R}^4$ satisfying $(e_1^C \wedge e_2^C \wedge e_3^C \wedge C)^\sim = 1$, called a camera projective coordinate system. When C is an affine point, let e_3^C be the vector from C to the origin O^C of the retina plane (or image plane), and let e_1^C, e_2^C be two vectors in the retina plane. Then $\{e_1^C, e_2^C, e_3^C; C\}$ is a Cartesian coordinate system of $\mathcal{A}^3$, called a camera affine coordinate system.

Let M be a point or point at infinity of $\mathcal{A}^3$, and let m^C be its image. Then M can be represented by its homogeneous coordinates which is a 4×1 matrix, and m^C can be represented by its homogeneous coordinates which is a 3×1 matrix. The perspective projection can then be represented by a 3×4 matrix.

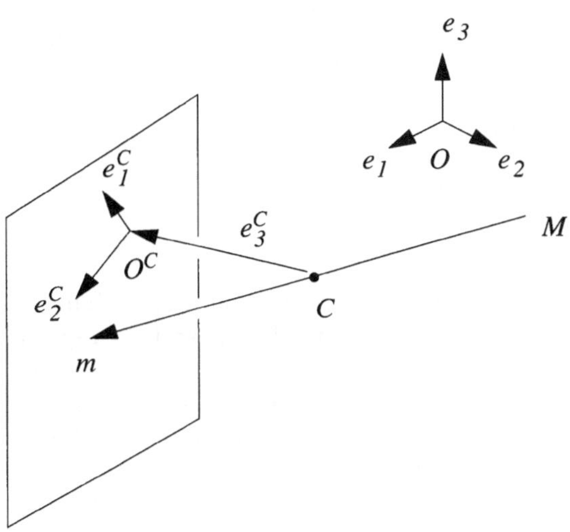

Fig. 17.1. A pinhole camera.

In our approach, we describe a pinhole camera with optical center C, which is either an affine point or a point at infinity of $\mathcal{A}^3$, as a system performing the projective split of $\mathcal{G}_4$ with respect to $C \in \mathbb{R}^4$.

To see how this representation works, we first derive the matrix of the project split P_C restricted to $\mathbb{R}^4$. We consider the case when the camera coordinate system $\{e_1^C, e_2^C, e_3^C, C\}$ is affine. According to (17.11),

$$P_C = \sum_{i=1}^{3} (C \wedge e_i^C) \otimes e_i^{C*}. \tag{17.1}$$

In the camera coordinate system, let the coordinates of e_i, $i = 1, 2, 3$, and O, be $(e_{i1}, e_{i2}, e_{i3}, 0) = (\mathbf{e}_i^T, 0)$, and $(O_1, O_2, O_3, 1) = (-\mathbf{c}^T, 1)$, respectively. Here $\mathbf{e}_i$ and $\mathbf{c}$ represent 3×1 matrices. The following matrix changes $\{e_1^C, e_2^C, e_3^C, C\}$ to $\{e_1, e_2, e_3, O\}$:

$$\begin{pmatrix} \mathbf{e}_1^T & 0 \\ \mathbf{e}_2^T & 0 \\ \mathbf{e}_3^T & 0 \\ -\mathbf{c}^T & 1 \end{pmatrix}. \tag{17.2}$$

Its transpose changes $\{e_1^*, e_2^*, e_3^*, O^*\}$ to $\{e_1^{C*}, e_2^{C*}, e_3^{C*}, C^*\}$. Substituting e_i^{C*}, $i = 1, 2, 3$ expressed by e_1^*, e_2^*, e_3^*, O^* into (17.1), we get the matrix of P_C:

$$\mathbf{P}_C = (\mathbf{e}_1 \;\; \mathbf{e}_2 \;\; \mathbf{e}_3 \;\; -\mathbf{c}). \tag{17.3}$$

When $C = O$, $e_1^C = e_1$, $e_2^C = e_2$ and $e_3^C = -fe_3$, where f is the focal length of the camera,

$$\mathbf{P}_C = \begin{pmatrix} 1 & 0 & 0 & 0 \\ 0 & 1 & 0 & 0 \\ 0 & 0 & -1/f & 0 \end{pmatrix}, \tag{17.4}$$

which is the standard perspective projection matrix. This justifies the representation of the perspective projection by P_C and the representation of image points by vectors in $C \wedge \mathbb{R}^4$.

In the case when the camera coordinate system is projective, let the 4×1 matrices $\mathbf{e}_i^{C*}$, $i = 1, 2, 3$ represent the coordinates of e_i^{C*} with respect to $\{e_1^*, e_2^*, e_3^*, O^*\}$. By (17.1),

$$\mathbf{P}_C = (\mathbf{e}_1^{C*} \;\; \mathbf{e}_2^{C*} \;\; \mathbf{e}_3^{C*})^T. \tag{17.5}$$

Below we derive the matrix of P_C restricted to $\mathcal{G}_4^2$. Let

$$\hat{\mathbf{e}}_i^{C*} = \mathbf{e}_{i_1}^{C*} \times \mathbf{e}_{i_2}^{C*}, \tag{17.6}$$

where $i \prec i_1 \prec i_2$. It represents the coordinates of $\hat{e}_i^{C*}$ with respect to the basis of $\mathcal{G}_4^2$ induced by $\{e_1^*, e_2^*, e_3^*, O^*\}$. According to (17.15), the matrix of P_C is

$$\mathbf{P}_C = -(\hat{\mathbf{e}}_1^{C*} \;\; \hat{\mathbf{e}}_2^{C*} \;\; \hat{\mathbf{e}}_3^{C*})^T. \tag{17.7}$$

17.3.2 Camera Constraints

It is clear that as long as $\det(\mathbf{e}_1\ \mathbf{e}_2\ \mathbf{e}_3) \neq 0$, the matrix $\mathbf{P}_C = (\mathbf{e}_1\ \mathbf{e}_2\ \mathbf{e}_3 - \mathbf{c})$ represents a perspective projection. When there is further information on the pinhole camera, for example vectors e_1^C, e_2^C of the camera affine coordinate system are perpendicular, then $\mathbf{P}_C$ needs to satisfy additional equality constraints in order to represent the perspective projection carried out by such a camera.

Let "$\sim_3$" represent the dual in $\mathcal{G}(\overset{\infty}{\mathcal{A}^3})$. Let the dual bases of $\{e_1, e_2, e_3\}$ and $\{e_1^C, e_2^C, e_3^C\}$ in $\overset{\infty}{\mathcal{A}^3}$ be $\{e_1^{*3}, e_2^{*3}, e_3^{*3}\}$ and $\{e_1^{C*3}, e_2^{C*3}, e_3^{C*3}\}$, respectively. Then

$$\begin{aligned} e_1^C &= (e_2^{C*3} \wedge e_3^{C*3})^{\sim_3} = e_2^{C*3} \times e_3^{C*3}, \\ e_2^C &= (e_3^{C*3} \wedge e_1^{C*3})^{\sim_3} = e_3^{C*3} \times e_1^{C*3}, \end{aligned} \tag{17.8}$$

where "$\times$" is the cross product in vector algebra. The perpendicularity constraint can be represented by

$$e_1^C \cdot e_2^C = (e_2^{C*3} \times e_3^{C*3}) \cdot (e_3^{C*3} \times e_1^{C*3}) = 0. \tag{17.9}$$

Let the 3×1 matrix $\mathbf{e}_i^{C*3}$ represent the coordinates of e_i^{C*3} with respect to $\{e_1^{*3}, e_2^{*3}, e_3^{*3}\}$. Under the assumption that $\{e_1, e_2, e_3\}$ is an orthonormal basis, $e_i^{C*3} \cdot e_j^{C*3} = \mathbf{e}_i^{C*3} \cdot \mathbf{e}_j^{C*3}$ for any $1 \leq i, j \leq 3$. Then (17.9) is changed to

$$(\mathbf{e}_2^{C*3} \times \mathbf{e}_3^{C*3}) \cdot (\mathbf{e}_3^{C*3} \times \mathbf{e}_1^{C*3}) = 0, \tag{17.10}$$

which is a constraint on $\mathbf{P}_C$ because

$$(\mathbf{e}_1^{C*3}\ \ \mathbf{e}_2^{C*3}\ \ \mathbf{e}_3^{C*3}) = (\mathbf{e}_1\ \ \mathbf{e}_2\ \ \mathbf{e}_3)^T. \tag{17.11}$$

17.3.3 Camera Calibration

Let M be a space point or point at infinity, m^C be its image in the retina plane. Assume that m^C is a point, and has homogeneous coordinates $(u, v, 1)$ in the Cartesian coordinate system of the retina plane. Let the 4×1 matrix $\mathbf{M}$ represent the homogeneous coordinates of M in the world coordinate system. Then

$$(u\ \ v\ \ 1)^T \simeq \mathbf{P}_C\mathbf{M} = (\mathbf{e}_1^{C*} \cdot \mathbf{M}\quad \mathbf{e}_2^{C*} \cdot \mathbf{M}\quad \mathbf{e}_3^{C*} \cdot \mathbf{M})^T, \tag{17.12}$$

which can be written as two scalar equations:

$$(\mathbf{e}_1^{C*} - u\mathbf{e}_3^{C*}) \cdot \mathbf{M} = 0, \quad (\mathbf{e}_2^{C*} - v\mathbf{e}_3^{C*}) \cdot \mathbf{M} = 0. \tag{17.13}$$

The matrix $\mathbf{P}_C = (\mathbf{e}_1^{C*}\ \mathbf{e}_2^{C*}\ \mathbf{e}_3^{C*})^T$ can be taken as a vector in the space $\mathbb{R}^4 \times \mathbb{R}^4 \times \mathbb{R}^4$ equipped with the induced inner product from $\mathbb{R}^4$. By this inner product, (17.13) can be written as

$$(\mathbf{M} \quad 0 \quad -u\mathbf{M})^T \cdot \mathbf{P}_C = 0, \quad (0 \quad \mathbf{M} \quad -v\mathbf{M})^T \cdot \mathbf{P}_C = 0. \tag{17.14}$$

Given $\mathbf{M}_i$ and (u_i, v_i) for $i = 1, \ldots, 6$, there are 12 equations of the forms in (17.14). If there is no camera constraint, then since a 3×4 matrix representing a perspective projection has 11 free parameters, $\mathbf{P}_C$ can be solved from the 12 equations if and only if the determinant of the coefficient matrix $\mathbf{A}$ of these equations is zero, i. e.,

$$\Lambda_{i=1}^{6}(\mathbf{M}_i \quad 0 \quad -u_i\mathbf{M}_i) \wedge \Lambda_{i=1}^{6}(0 \quad \mathbf{M}_i \quad -v_i\mathbf{M}_i) = 0, \tag{17.15}$$

where the outer products are in the Clifford algebra generated by $\mathbb{R}^4 \times \mathbb{R}^4 \times \mathbb{R}^4$. Expanding the left-hand side of (17.15), and changing outer products into determinants, we get

$$\sum_{\sigma,\tau} \epsilon(\sigma)\epsilon(\tau) u_{\sigma(1)} u_{\sigma(2)} v_{\tau(1)} v_{\tau(2)} \det(\mathbf{M}_{\sigma(1)} \ \mathbf{M}_{\sigma(2)} \ \mathbf{M}_{\tau(1)} \ \mathbf{M}_{\tau(2)})$$
$$\det(\mathbf{M}_{\sigma(i)})_{i=3..6} \det(\mathbf{M}_{\tau(j)})_{j=3..6} = 0, \tag{17.16}$$

where σ, τ are any permutations of $1, \ldots, 6$ by moving two elements to the front of the sequence, and $\epsilon(\sigma), \epsilon(\tau)$ are the signs of permutation.

For experimental data, (17.16) is not necessarily satisfied because of errors in measurements.

17.4 Epipolar and Trifocal Geometries

17.4.1 Epipolar Geometry

There is no much difference between our algebraic description of the pinhole camera and others if there is only one fixed camera involved, because the underlying Grassmann-Cayley algebras are isomorphic. Let us reformulate the epipolar geometry of two cameras with optical centers C, C' respectively.

The image of C' in camera C is $E^{CC'} = C \wedge C'$, called the epipole of C' in camera C. Similarly, the image of C in camera C' is $E^{C'C} = C' \wedge C$, called the epipole of C in camera C'. An image line passing through the epipole in camera C (or C') is called an epipolar line with respect to C' (or C). Algebraically, an epipolar line is a vector in

$$C \wedge C' \wedge \mathbb{R}^4 = (C \wedge \mathcal{G}_4^2) \cap (C' \wedge \mathcal{G}_4^2). \tag{17.1}$$

An epipolar line $C \wedge C' \wedge M$ corresponds to a unique epipolar line $C' \wedge C \wedge M$, and vice versa.

Let there be two camera projective coordinate systems in the two cameras respectively: $\{e_1^C, e_2^C, e_3^C, C\}$ and $\{e_1^{C'}, e_2^{C'}, e_3^{C'}, C'\}$. Using the relations

$$(C \wedge e_i^C) \vee (C \wedge \hat{e}_i^C) = -C, \text{ for } 1 \leq i \leq 3, \tag{17.2}$$

and

$$(C \wedge \hat{e}_{i_1}^C) \vee (C \wedge \hat{e}_{i_2}^C) = C \wedge \hat{e}_i^C, \text{ for } i \prec i_1 \prec i_2, \tag{17.3}$$

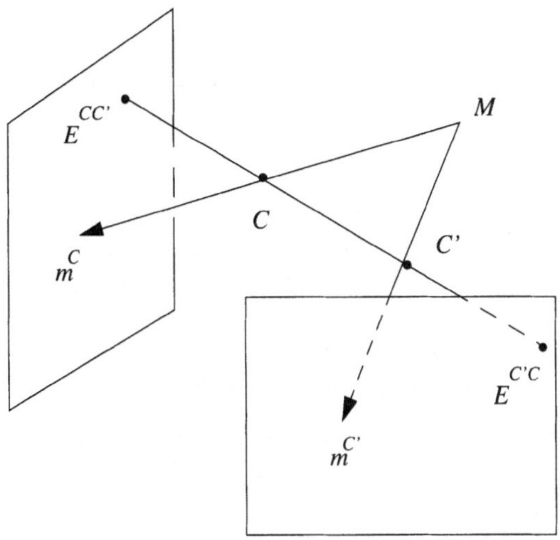

Fig. 17.2. Epipolar geometry.

we get the coordinates of epipole $E^{CC'}$:

$$\begin{aligned}\mathbf{E}^{CC'} &= ((C \wedge \hat{e}_i^C) \vee C')_{i=1..3} \\ &= ((C \wedge \hat{e}_i^C)^\sim \vee C'^\sim)_{i=1..3} \\ &= ((e_1^{C'*} \wedge e_2^{C'*} \wedge e_3^{C'*}) \vee e_i^{C*})_{i=1..3}.\end{aligned} \tag{17.4}$$

The following tensor in $(C \wedge \mathbb{R}^4) \otimes (C' \wedge \mathbb{R}^4)$ is called the epipolar tensor decide by C, C':

$$F^{CC'}(m^C, m^{C'}) = m^C \vee m^{C'}. \tag{17.5}$$

Let $m^C \in C \wedge \mathbb{R}^4$, $m^{C'} \in C' \wedge \mathbb{R}^4$. They are images of the same space point or point at infinity if and only if $F^{CC'}(m^C, m^{C'}) = 0$. This equality is called the epipolar constraint between m^C and $m^{C'}$.

In matrix form, with respect to the bases $\{C \wedge e_1^C, C \wedge e_2^C, C \wedge e_3^C\}$ and $\{C' \wedge e_1^{C'}, C' \wedge e_2^{C'}, C' \wedge e_3^{C'}\}$, $F^{CC'}$ can be represented by

$$\begin{aligned}\mathbf{F}^{CC'} &= ((C \wedge e_i^C) \vee (C' \wedge e_j^{C'}))_{i,j=1..3} \\ &= ((C \wedge e_i^C)^\sim \vee (C' \wedge e_j^{C'})^\sim)_{i,j=1..3} \\ &= (\hat{e}_i^{C*} \vee \hat{e}_j^{C'*})_{i,j=1..3}.\end{aligned} \tag{17.6}$$

(17.6) is called the fundamental matrix.

The epipolar tensor induces a linear mapping $F^{C;C'}$ from $C \wedge \mathbb{R}^4$ to $(C' \wedge \mathbb{R}^4)^* = C' \wedge \mathcal{G}_4^2$, called the epipolar transformation from camera C to camera C':

$$F^{C;C'}(m^C) = C' \wedge m^C. \tag{17.7}$$

Similarly, it induces an epipolar transformation from camera C' to camera C as follows:

$$F^{C';C}(m^{C'}) = C \wedge m^{C'}. \tag{17.8}$$

Both transformations are just projective splits.

The kernel of $F^{C;C'}$ is the one-dimensional subspace of $C \wedge \mathbb{R}^4$ represented by $C \wedge C'$, the range of $F^{C;C'}$ is the two-dimensional space $C' \wedge C \wedge \mathbb{R}^4$. In geometric language, $F^{C;C'}$ maps the epipole of C' to zero, and maps any other point in camera C to an epipolar line with respect to C.

Furthermore, we have the following conclusion:

Proposition 17.4.1. Let L^C be an epipolar line in camera C. If its dual is mapped to epipolar line $L^{C'}$ in camera C' by $F^{C;C'}$, then the dual of $L^{C'}$ is mapped back to L^C by $F^{C';C}$.

The proof follows from the identity that for any vector $M \in \mathbb{R}^4$,

$$C \wedge (C' \wedge (C \wedge C' \wedge M)^{\sim_C})^{\sim_{C'}} \simeq C \wedge C' \wedge M. \tag{17.9}$$

17.4.2 Trifocal Geometry

Let there be three cameras with optical centers C, C', C'' respectively. Let M be a space point or point at infinity. Its images $C \wedge M$, $C' \wedge M$ and $C'' \wedge M$ in the three cameras must satisfy pairwise epipolar constraints. Let us consider the inverse problem: If there are three image points $m^C, m^{C'}, m^{C''}$ in the three cameras respectively, they satisfy the pairwise epipolar constraints, is it true that they are images of the same space point or point at infinity?

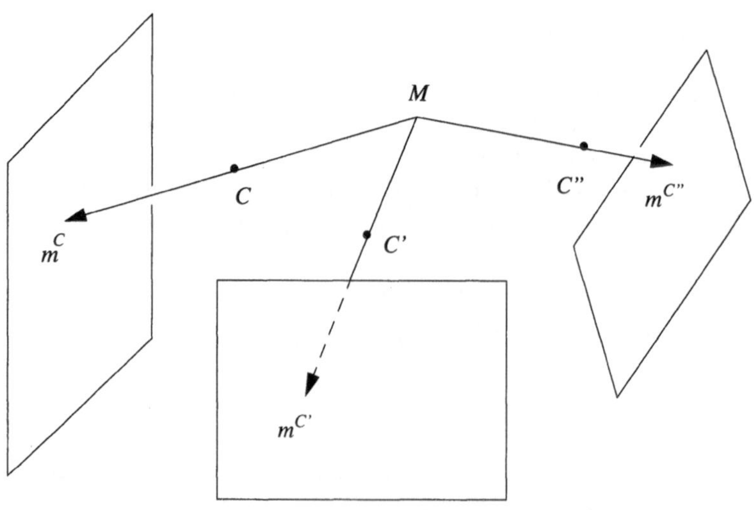

Fig. 17.3. Point correspondence in three cameras.

A simple counter-example shows that the epipolar constraints are not enough. When the 2-blades $m^C, m^{C'}, m^{C''}$ belong to $\mathcal{G}(C \wedge C' \wedge C'')$, the epipolar constraints are always satisfied, but the blades do not necessarily share a common vector.

Assume that the epipolar constraint between $m^{C'}$ and $m^{C''}$ is satisfied. Let M be the intersection of the two lines $m^{C'}$ and $m^{C''}$ in $\mathbb{P}^3$. Then m^C represents the image of M in camera C if and only if $m^C \wedge M = 0$, or equivalently,

$$m^C \vee (M \wedge x) = 0, \text{ for any } x \in \mathbb{R}^4. \tag{17.10}$$

When C', C'', M are not collinear, since

$$M \wedge \mathbb{R}^4 = (C' \wedge M \wedge \mathbb{R}^4) \vee (C'' \wedge M \wedge \mathbb{R}^4), \tag{17.11}$$

(17.10) can be written as

$$m^C \vee (m^{C'} \wedge_{C'} m_0^{C'}) \vee (m^{C''} \wedge_{C''} m_0^{C''}) = 0, \tag{17.12}$$

for any image points $m_0^{C'}, m_0^{C''}$ in cameras C', C'' respectively. When C', C'', M are collinear, since $m^{C'} \simeq m^{C''}$, (17.12) is equivalent to the epipolar constraint between m^C and $m^{C'}$. So the constraint (17.12) must be satisfied for $m^C, m^{C'}, m^{C''}$ to be images of the same space point or point at infinity.

Definition 17.4.1. The following tensor in $(C \wedge \mathbb{R}^4) \otimes (C' \wedge \mathcal{G}_4^2) \otimes (C'' \wedge \mathcal{G}_4^2)$ is called the trifocal tensor [101, 102, 209] of camera C with respect to cameras C', C'':

$$T(m^C, L^{C'}, L^{C''}) = m^C \vee L^{C'} \vee L^{C''}, \tag{17.13}$$

where $m^C \in C \wedge \mathbb{R}^4$, $L^{C'} \in C' \wedge \mathcal{G}_4^2$, $L^{C''} \in C'' \wedge \mathcal{G}_4^2$.

Two other trifocal tensors can be defined by interchanging C with C', C'' respectively:

$$\begin{aligned} T'(m^{C'}, L^C, L^{C''}) &= m^{C'} \vee L^C \vee L^{C''}, \\ T''(m^{C''}, L^C, L^{C'}) &= m^{C''} \vee L^C \vee L^{C'}. \end{aligned} \tag{17.14}$$

In this section we discuss T only. Let $\{e_1^C, e_2^C, e_3^C, C\}$, $\{e_1^{C'}, e_2^{C'}, e_3^{C'}, C'\}$, $\{e_1^{C''}, e_2^{C''}, e_3^{C''}, C''\}$ be camera projective coordinate systems of the three cameras respectively. Then T has the following component representation:

$$\begin{aligned} \mathbf{T} &= ((C \wedge e_i^C) \vee (C' \wedge \hat{e}_j^{C'}) \vee (C'' \wedge \hat{e}_k^{C''}))_{i,j,k=1..3} \\ &= ((C \wedge e_i^C)^\sim \vee ((C' \wedge \hat{e}_j^{C'})^\sim \wedge (C'' \wedge \hat{e}_k^{C''})^\sim))_{i,j,k=1..3} \\ &= (\hat{e}_i^{C*} \vee (e_j^{C'*} \wedge e_k^{C''*}))_{i,j,k=1..3} \\ &= (-(\hat{e}_i^{C*} \wedge e_j^{C'*}) \vee e_k^{C''*})_{i,j,k=1..3}. \end{aligned} \tag{17.15}$$

The trifocal tensor T induces three trifocal transformations:

1. The mapping $T^C : (C' \wedge \mathcal{G}_4^2) \times (C'' \wedge \mathcal{G}_4^2) \longrightarrow (C \wedge \mathbb{R}^4)^* = C \wedge \mathcal{G}_4^2$ is defined as
$$T^C(L^{C'}, L^{C''}) = C \wedge (L^{C'} \vee L^{C''}). \qquad (17.16)$$
When $L^{C'}$ is fixed, T^C induces a linear mapping $T^{CC'}_{L^{C'}} : C'' \wedge \mathcal{G}_4^2 \longrightarrow C \wedge \mathcal{G}_4^2$:
$$T^{CC'}_{L^{C'}}(L^{C''}) = C \wedge (L^{C'} \vee L^{C''}). \qquad (17.17)$$
If $L^{C'}$ is an epipolar line with respect to C, the kernel of $T^{CC'}_{L^{C'}}$ is all epipolar lines with respect to C, the range is the epipolar line represented by $L^{C'}$; else if $L^{C'}$ is an epipolar line with respect to C'', the kernel is the epipolar line represented by $L^{C'}$, the range is all epipolar lines with respect to C''. For other cases, the kernel is zero.
Geometrically, when $T^C(L^{C'}, L^{C''}) \neq 0$, then $L^{C'} \vee L^{C''}$ represents a line or line at infinity L of $\mathcal{A}^3$, both $L^{C'}$ and $L^{C''}$ are images of L. $T^C(L^{C'}, L^{C''})$ is just the image of L in camera C.
2. The mapping $T^{C'} : (C \wedge \mathbb{R}^4) \times (C'' \wedge \mathcal{G}_4^2) \longrightarrow (C' \wedge \mathcal{G}_4^2)^* = C' \wedge \mathbb{R}^4$ is defined as
$$T^{C'}(m^C, L^{C''}) = C' \wedge (m^C \vee L^{C''}). \qquad (17.18)$$
When m^C is fixed, $T^{C'C}$ induces a linear mapping $T^{C'C}_{m^C} : C'' \wedge \mathcal{G}_4^2 \longrightarrow C' \wedge \mathbb{R}^4$:
$$T^{C'C}_{m^C}(L^{C''}) = C' \wedge (m^C \vee L^{C''}). \qquad (17.19)$$
If m^C is the epipole of C'', the kernel of $T^{C'C}_{m^C}$ is all epipolar lines with respect to C, the range is the epipole of C''. For other cases, the kernel is the epipolar line $C'' \wedge m^C$, the range is the two-dimensional subspace of $C' \wedge \mathbb{R}^4$ represented by $C' \wedge m^C$.
Geometrically, when $T^{C'}(m^C, L^{C''}) \neq 0$, then $m^C \vee L^{C''}$ represents a point or point at infinity M of $\mathcal{A}^3$, m^C is its image in camera C, and $L^{C''}$ is the image of a space line or line at infinity passing through M. $T^{C'}(m^C, L^{C''})$ is just the image of M in camera C'.
3. The mapping $T^{C''} : (C \wedge \mathbb{R}^4) \times (C' \wedge \mathcal{G}_4^2) \longrightarrow (C'' \wedge \mathcal{G}_4^2)^* = C'' \wedge \mathbb{R}^4$ is defined as
$$T^{C''}(m^C, L^{C'}) = C'' \wedge (m^C \vee L^{C'}). \qquad (17.20)$$

We prove below two propositions in [81, 83] using the above reformulation of trifocal tensors.

Proposition 17.4.2. Let $L^{C'}$ be an epipolar line in camera C' with respect to C and L^C be the corresponding epipolar line in camera C. Then for any line $L^{C''}$ in camera C'' which is not the epipolar line with respect to C, $T^C(L^{C'}, L^{C''}) \simeq L^C$.

Proof. The hypotheses are $L^{C'} \simeq L^C$, $C \vee L^{C''} \neq 0$. Using the formula that for any $C \in \mathbb{R}^4$, $A_3, B_3 \in \mathcal{G}_4^3$,

$$C \wedge (A_3 \vee B_3) = (C \vee B_3)A_3 - (C \vee A_3)B_3, \tag{17.21}$$

we get

$$T^C(L^{C'}, L^{C''}) \simeq C \wedge (L^C \vee L^{C''}) = (C \vee L^{C''})L^C - (C \vee L^C)L^{C''} \simeq L^C.$$

Proposition 17.4.3. Let $m^{C'}, m^{C''}$ be images of the point or point at infinity M in cameras C', C'' respectively. Let $L^{C'}$ be an image line passing through $m^{C'}$ but not through $E^{C'C}$. Let $L^{C''}$ be an image line passing through $m^{C''}$ but not through $E^{C''C'}$. Then the intersection of $T^C(L^{C'}, L^{C''})$ with the epipolar line $C \wedge m^{C'}$ is the image of M in camera C.

Proof. The hypotheses are $M \vee L^{C'} = M \vee L^{C''} = 0$, $C \vee L^{C'} \neq 0$, $C' \vee L^{C''} \neq 0$. So

$$\begin{aligned}
&T^C(L^{C'}, L^{C''}) \vee (C \wedge m^{C'}) \\
&\quad = (C \wedge (L^{C'} \vee L^{C''})) \vee (C \wedge C' \wedge M) \\
&\quad = ((C \wedge C') \vee L^{C'} \vee L^{C''})(C \wedge M) - ((C \wedge M) \vee L^{C'} \vee L^{C''})(C \wedge C') \\
&\quad = -(C \vee L^{C'})(C' \vee L^{C''})(C \wedge M) \\
&\quad \simeq C \wedge M.
\end{aligned}$$

17.5 Relations among Epipoles, Epipolar Tensors, and Trifocal Tensors of Three Cameras

Consider the following 9 vectors of $\mathbb{R}^4$:

$$ES = \{e_i^{C*}, e_j^{C'*}, e_k^{C''*} | 1 \leq i, j, k \leq 3\}. \tag{17.1}$$

According to (17.4), (17.6) and (17.15), by interchanging among C, C', C'' any of the epipoles, epipolar tensors and trifocal tensors of the three cameras has its components represented as a determinant of 4 vectors in ES. For example,

$$\begin{aligned}
E_i^{CC'} &= (e_i^{C*} \wedge e_1^{C'*} \wedge e_2^{C'*} \wedge e_3^{C'*})^{\sim}; \\
F_{ij}^{CC'} &= (\hat{e}_i^{C*} \wedge \hat{e}_j^{C'*})^{\sim}; \\
T_{ijk} &= (\hat{e}_i^{C*} \wedge e_j^{C'*} \wedge e_k^{C''*})^{\sim}.
\end{aligned} \tag{17.2}$$

Conversely, any determinant of 4 vectors in ES equals a component of one of the epipoles, epipolar tensors and trifocal tensors up to an index-free scale. Since the only constraint on the 9 vectors is that they are all in $\mathbb{R}^4$, theoretically all relations among the epipoles, epipolar tensors and trifocal tensors can be established by manipulating in the algebra of determinants of vectors in ES using the following Cramer's rule [76, 80]:

$$\begin{aligned}(x_2 \wedge x_3 \wedge x_4 \wedge x_5)^{\sim} x_1 = (x_1 \wedge x_3 \wedge x_4 \wedge x_5)^{\sim} x_2 - (x_1 \wedge x_2 \wedge x_4 \wedge x_5)^{\sim} x_3 \\ +(x_1 \wedge x_2 \wedge x_3 \wedge x_5)^{\sim} x_4 - (x_1 \wedge x_2 \wedge x_3 \wedge x_4)^{\sim} x_5,\end{aligned} \tag{17.3}$$

where the x's are vectors in $\mathbb{R}^4$.

In practice, however, we can only select a few expressions from the algebra of determinants and make manipulations, and it is difficult to make the selection. In this section we propose a different approach. Instead of considering the algebra of determinants directly, we consider the set of meets of different blades, each blade being an outer product of vectors in ES. Since the meet operator is associative and anti-commutative in the sense that

$$A_r \vee B_s = (-1)^{rs} B_s \vee A_r, \tag{17.4}$$

for $A_r \in \mathcal{G}_4^r$ and $B_s \in \mathcal{G}_4^s$, for the same expression of meets we can have a variety of expansions. Then we can obtain various equalities on determinants of vectors in ES, which may be changed into equalities, or equalities up to an index-free constant, on components of the epipoles, epipolar tensors and trifocal tensors.

It appears that we need only 7 expressions of meets to derive and further generalize all the known constraints on epipolar and trifocal tensors.

It should be reminded that in this chapter we always use the notation $i \prec i_1 \prec i_2$ to denote that i, i_1, i_2 is an even permutation of $1, 2, 3$.

17.5.1 Relations on Epipolar Tensors

Consider the following expression:

$$Fexp = (e_1^{C'*} \wedge e_2^{C'*} \wedge e_3^{C'*}) \vee (e_1^{C*} \wedge e_2^{C*} \wedge e_3^{C*}) \vee (e_1^{C''*} \wedge e_2^{C''*} \wedge e_3^{C''*}). \tag{17.5}$$

It is the dual of the blade $C' \wedge C \wedge C''$.

Expanding $Fexp$ from left to right, we get

$$\begin{aligned} Fexp &= \sum_{i,k=1}^{3} ((e_1^{C'*} \wedge e_2^{C'*} \wedge e_3^{C'*}) \vee e_i^{C*})(\hat{e}_i^{C*} \vee \hat{e}_k^{C''*}) e_k^{C''*} \\ &= \sum_{i,k=1}^{3} E_i^{CC'} F_{ik}^{CC''} e_k^{C''*}. \end{aligned}$$

Expanding $Fexp$ from right to left, we get

$$\begin{aligned}Fexp &= \sum_{k=1}^{3} \Big(((e_1^{C'*} \wedge e_2^{C'*} \wedge e_3^{C'*}) \vee e_{k_2}^{C''*})((e_1^{C*} \wedge e_2^{C*} \wedge e_3^{C*}) \vee e_{k_1}^{C''*}) \\ &\quad - ((e_1^{C'*} \wedge e_2^{C'*} \wedge e_3^{C'*}) \vee e_{k_1}^{C''*})((e_1^{C*} \wedge e_2^{C*} \wedge e_3^{C*}) \vee e_{k_2}^{C''*}) \Big) \, e_k^{C''*} \\ &= \sum_{k=1}^{3} (E_{k_2}^{C''C'} E_{k_1}^{C''C} - E_{k_1}^{C''C'} E_{k_2}^{C''C}) e_k^{C''*},\end{aligned}$$

where $k \prec k_1 \prec k_2$. So for any $1 \leq i \leq 3$,

$$\sum_{k=1}^{3} E_i^{CC'} F_{ik}^{CC''} \simeq K_k^{C''CC'}, \tag{17.6}$$

where $K_k^{C''CC'} = E_{k_1}^{C''C} E_{k_2}^{C''C'} - E_{k_2}^{C''C} E_{k_1}^{C''C'}$.

(17.6) is a fundamental relation on the epipolar tensor $F^{CC''}$ and the epipoles. In matrix form, it can be written as

$$(\mathbf{F}^{CC''})^T \mathbf{E}^{CC'} \simeq \mathbf{E}^{C''C} \times \mathbf{E}^{C''C'}; \tag{17.7}$$

in Grassmann-Cayley algebra, it can be written as

$$C'' \wedge (C \wedge C') \simeq (C'' \wedge C) \wedge_{C''} (C'' \wedge C'). \tag{17.8}$$

Geometrically, it means that the epipolar line in camera C'' with respect to both C and C' is the image line connecting the two epipoles $E^{C''C}$ and $E^{C''C'}$. One should notice the obvious advantage of Grassmann-Cayley algebraic representation in geometric interpretation.

Since $\mathbf{E}^{C''C} \times \mathbf{E}^{C''C'}$ is orthogonal to $\mathbf{E}^{C''C'}$, an immediate corollary is

$$(\mathbf{E}^{CC'})^T \mathbf{F}^{CC''} \mathbf{E}^{C''C'} = 0, \tag{17.9}$$

which is equivalent to $(C \wedge C') \vee (C'' \wedge C') = 0$. Geometrically, it means that the two epipoles $E^{C''C}$ and $E^{C''C'}$ satisfy the epipolar constraint.

17.5.2 Relations on Trifocal Tensors I

The first idea to derive relations on trifocal tensors is very simple: if the tensor $(T_{ijk})_{i,j,k=1..3}$ is given, then expanding

$$(\hat{e}_i^{C*} \wedge e_j^{C'*}) \vee (e_1^{C''*} \wedge e_2^{C''*} \wedge e_3^{C''*}) \tag{17.10}$$

gives a 2-vector of the $e^{C''*}$'s whose coefficients are known. Similarly, expanding

$$Texp_1 = (\hat{e}_i^{C*} \wedge e_{j_1}^{C'*}) \vee (\hat{e}_i^{C*} \wedge e_{j_2}^{C'*}) \vee (e_1^{C''*} \wedge e_2^{C''*} \wedge e_3^{C''*}). \tag{17.11}$$

from right to left gives a vector of the $e^{C''*}$'s whose coefficients are known. Expanding $Texp_1$ from left to right, we get a vector of the $e^{C''*}$'s whose coefficients depend on epipolar tensors. By comparing the coefficients of the $e^{C''*}$'s, we get a relation on T and epipolar tensors.

Assume that $j \prec j_1 \prec j_2$. Expanding $Texp_1$ from left to right, we get

$$\begin{aligned} Texp_1 &= -(\hat{e}_i^{C*} \wedge e_{j_1}^{C'*} \wedge e_{j_2}^{C'*})^{\sim} \sum_{k=1}^{3} (\hat{e}_i^{C*} \vee \hat{e}_k^{C''*})\, e_k^{C''*} \\ &= -\sum_{k=1}^{3} F_{ij}^{CC'} F_{ik}^{CC''} e_k^{C''*}. \end{aligned}$$

Expanding $Texp_1$ from right to left, we get

$$\begin{aligned} Texp_1 &= \sum_{k=1}^{3} \Big(((\hat{e}_i^{C*} \wedge e_{j_1}^{C'*}) \vee e_{k_2}^{C''*})((\hat{e}_i^{C*} \wedge e_{j_2}^{C'*}) \vee e_{k_1}^{C''*}) \\ &\qquad - ((\hat{e}_i^{C*} \wedge e_{j_1}^{C'*}) \vee e_{k_1}^{C''*})((\hat{e}_i^{C*} \wedge e_{j_2}^{C'*}) \vee e_{k_2}^{C''*}) \Big)\, e_k^{C''*} \\ &= \sum_{k=1}^{3} (T_{ij_1k_2} T_{ij_2k_1} - T_{ij_1k_1} T_{ij_2k_2})\, e_k^{C''*}, \end{aligned}$$

where $k \prec k_1 \prec k_2$. So

$$F_{ij}^{CC'} F_{ik}^{CC''} = t_{ijk}^{C}, \tag{17.12}$$

where

$$t_{ijk}^{C} = T_{ij_1k_1} T_{ij_2k_2} - T_{ij_1k_2} T_{ij_2k_1}. \tag{17.13}$$

Proposition 17.5.1. For any $1 \le i, j, k \le 3$,

$$F_{ij}^{CC'} F_{ik}^{CC''} \simeq t_{ijk}^{C}. \tag{17.14}$$

Corollary 17.5.1. Let $1 \le i, j_1, j_2, k_1, k_2 \le 3$, then

$$\frac{F_{ij_1}^{CC'}}{F_{ij_2}^{CC'}} = \frac{t_{ij_1k}^{C}}{t_{ij_2k}^{C}}, \text{ for any } 1 \le k \le 3; \tag{17.15}$$

$$\frac{F_{ik_1}^{CC''}}{F_{ik_2}^{CC''}} = \frac{t_{ijk_1}^{C}}{t_{ijk_2}^{C}}, \text{ for any } 1 \le j \le 3; \tag{17.16}$$

$$\frac{t_{ij_1k_1}^{C}}{t_{ij_1k_2}^{C}} = \frac{t_{ij_2k_1}^{C}}{t_{ij_2k_2}^{C}}. \tag{17.17}$$

Notice that (17.17) is a constraint of degree 4 on T.

To understand relation (17.14) geometrically, we first express it in terms of Grassmann-Cayley algebra. When $C \wedge e_i^C$ is fixed, T induces a linear mapping $T_i^{C''C} : C' \wedge \mathcal{G}_4^2 \longrightarrow C'' \wedge \mathbb{R}^4$ by

$$T_i^{C''C}(L^{C'}) = C'' \wedge ((C \wedge e_i^C) \vee L^{C'}). \tag{17.18}$$

The matrix of $T_i^{C''C}$ is $(-T_{ijk})_{j,k=1..3}^T$.

Define a linear mapping $t_i^{C''C} : C' \wedge \mathbb{R}^4 \longrightarrow C'' \wedge \mathcal{G}_4^2$ as follows: let $m^{C'} \in C' \wedge \mathbb{R}^4$ and $m^{C'} = L_1^{C'} \vee L_2^{C'}$, where $L_1^{C'}, L_2^{C'} \in C' \wedge \mathcal{G}_4^2$, then

$$t_i^{C''C}(m^{C'}) = T_i^{C''C}(L_1^{C'}) \wedge_{C''} T_i^{C''C}(L_2^{C'}). \tag{17.19}$$

We need to prove that this mapping is well-defined. Using the formula that for any 2-blade $C_2 \in \mathcal{G}_4^2$ and 3-blades $A_3, B_3 \in \mathcal{G}_4^3$,

$$(C_2 \vee A_3) \wedge (C_2 \vee B_3) = -(A_3 \vee B_3 \vee C_2)C_2, \tag{17.20}$$

we get

$$\begin{aligned} t_i^{C''C}(m^{C'}) &= C'' \wedge ((C \wedge e_i^C) \vee L_1^{C'}) \wedge ((C \wedge e_i^C) \vee L_2^{C'}) \\ &= -L_1^{C'} \vee L_2^{C'} \vee (C \wedge e_i^C)\ \ C'' \wedge C \wedge e_i^C \\ &= -m^{C'} \vee (C \wedge e_i^C)\ \ C'' \wedge C \wedge e_i^C \\ &= \sum_{k=1}^{3} m^{C'} \vee (C \wedge e_i^C)\ \ (C'' \wedge C \wedge e_i^C \wedge e_k^{C''})^{\sim}\ C'' \wedge \hat{e}_k^{C''}. \end{aligned} \tag{17.21}$$

So $t_i^{C''C}$ is well-defined. Let $j \prec j_1 \prec j_2$ and $k \prec k_1 \prec k_2$, then since

$$\begin{aligned} t_i^{C''C}(C' \wedge e_j^{C'}) &= T_i^{C''C}(C' \wedge e_j^{C'} \wedge e_{j_1}^{C'}) \wedge_{C''} T_i^{C''C}(C' \wedge e_j^{C'} \wedge e_{j_2}^{C'}) \\ &= -\left(\sum_{k_2=1}^{3} T_{ij_2k_2} C'' \wedge e_{k_2}^{C''}\right) \wedge_{C''} \left(\sum_{k_1=1}^{3} T_{ij_1k_1} C'' \wedge e_{k_1}^{C''}\right) \\ &= \sum_{k=1}^{3} (T_{ij_1k_1}T_{ij_2k_2} - T_{ij_1k_2}T_{ij_2k_1})\, C'' \wedge \hat{e}_k^{C''}, \end{aligned}$$

the matrix of $t_i^{C''C}$ is $(t_{ijk})^T_{j,k=1..3}$.

So (17.14) is equivalent to

$$T_i^{C''C}(L_1^{C'}) \wedge_{C''} T_i^{C''C}(L_2^{C'}) = -m^{C'} \vee (C \wedge e_i^C)\ \ C'' \wedge C \wedge e_i^C. \tag{17.22}$$

Geometrically, $T_i^{C''C}$ maps an image line in camera C' to an image point on the epipolar line $C'' \wedge C \wedge e_i^C$ in camera C''. (17.22) says that the image line connecting the two image points $T_i^{C''C}(L_1^{C'})$ and $T_i^{C''C}(L_2^{C'})$ in camera C'' is just the epipolar line $C'' \wedge C \wedge e_i^C$. This is the geometric interpretation of (17.14).

17.5.3 Relations on Trifocal Tensors II

Now we let the two $\hat{e}^{C*}$'s in $Texp_1$ be different, and let the two $e^{C'*}$ be the same, i. e., we consider the expression

$$Texp_2 = (e_{i_1}^{C*} \wedge e_i^{C*} \wedge e_j^{C'*}) \vee (e_{i_2}^{C*} \wedge e_i^{C*} \wedge e_j^{C'*}) \vee (e_1^{C''*} \wedge e_2^{C''*} \wedge e_3^{C''*}). \tag{17.23}$$

Assume that $i \prec i_1 \prec i_2$. Expanding $Texp_2$ from left to right, we get

$$\begin{aligned} Texp_2 &= -(e_i^{C*} \wedge e_{i_1}^{C*} \wedge e_{i_2}^{C*} \wedge e_j^{C'*})^{\sim} \left(\sum_{k=1}^{3} ((e_i^{C*} \wedge e_j^{C'*}) \vee \hat{e}_k^{C''*}) e_k^{C''*} \right) \\ &= \sum_{k=1}^{3} E_j^{C'C} T''_{kij} e_k^{C''*}. \end{aligned}$$

Expanding $Texp_2$ from right to left, we get

$$\begin{aligned} Texp_2 = &- \sum_{k=1}^{3} \Big(((\hat{e}_{i_2}^{C*} \wedge e_j^{C'*}) \vee e_{k_2}^{C''*})((\hat{e}_{i_1}^{C*} \wedge e_j^{C'*}) \vee e_{k_1}^{C''*}) \\ &\qquad - ((\hat{e}_{i_2}^{C*} \wedge e_j^{C'*}) \vee e_{k_1}^{C''*})((\hat{e}_{i_1}^{C*} \wedge e_j^{C'*}) \vee e_{k_2}^{C''*}) \Big) \; e_k^{C''*} \\ = &- \sum_{k=1}^{3} (T_{i_1 j k_1} T_{i_2 j k_2} - T_{i_1 j k_2} T_{i_2 j k_1}) e_k^{C''*}, \end{aligned}$$

where $k \prec k_1 \prec k_2$. So

$$-E_j^{C'C} T''_{kij} = t_{ijk}^{C'}, \tag{17.24}$$

where

$$t_{ijk}^{C'} = T_{i_1 j k_1} T_{i_2 j k_2} - T_{i_1 j k_2} T_{i_2 j k_1}. \tag{17.25}$$

Proposition 17.5.2. For any $1 \leq i, j, k \leq 3$,

$$E_j^{C'C} T''_{kij} \simeq t_{ijk}^{C'}. \tag{17.26}$$

Corollary 17.5.2. For any $1 \leq i, i_1, i_2, j, k, k_1, k_2 \leq 3$,

$$\frac{T''_{ki_1j}}{T''_{ki_2j}} = \frac{t_{i_1jk}^{C'}}{t_{i_2jk}^{C'}}; \quad \frac{T''_{k_1ij}}{T''_{k_2ij}} = \frac{t_{k_1ij}^{C'}}{t_{k_2ij}^{C'}}. \tag{17.27}$$

Same as before, to understand relation (17.26) geometrically, we first express it in terms of Grassmann-Cayley algebra. When $C' \wedge \hat{e}_j^{C'}$ is fixed, T induces a linear mapping $T_j^{CC'} : C'' \wedge \mathcal{G}_4^2 \longrightarrow C \wedge \mathcal{G}_4^2$ by

$$T_j^{CC'}(L^{C''}) = C \wedge ((C' \wedge \hat{e}_j^{C'}) \vee L^{C''}), \tag{17.28}$$

whose matrix is $(-T_{ijk})_{i,k=1..3}$. T'' also induces a linear mapping ${T''}_j^{CC'} : C'' \wedge \mathbb{R}^4 \longrightarrow C \wedge \mathbb{R}^4$ by

$${T''}_j^{CC'}(m^{C''}) = C \wedge (m^{C''} \vee (C' \wedge \hat{e}_j^{C'})), \tag{17.29}$$

whose matrix is $(-T''_{kij})^T_{k,i=1..3}$. Define a linear mapping $t_j^{CC'} : C'' \wedge \mathbb{R}^4 \longrightarrow C \wedge \mathbb{R}^4$ as follows: let $m^{C''} \in C'' \wedge \mathbb{R}^4$ and $m^{C''} = L_1^{C''} \vee L_2^{C''}$, where $L_1^{C''}, L_2^{C''} \in C'' \wedge \mathcal{G}_4^2$, then

$$t_j^{CC'}(m^{C''}) = T_j^{CC'}(L_1^{C''}) \vee T_j^{CC'}(L_2^{C''}). \tag{17.30}$$

We need to prove that this mapping is well-defined. Using the formula that for any $C \in \mathbb{R}^4$ and $A_3, B_3 \in \mathcal{G}_4^3$,

$$(C \wedge (A_3 \vee B_3)) \vee B_3 = -(B_3 \vee C)(A_3 \vee B_3), \tag{17.31}$$

we get

$$\begin{aligned}
t_j^{CC'}(m^{C''}) &= \left(C \wedge ((C' \wedge \hat{e}_j^{C'}) \vee L_1^{C''})\right) \vee \left(C \wedge ((C' \wedge \hat{e}_j^{C'}) \vee L_2^{C''})\right) \\
&= -C \wedge \left(C \wedge \left((C' \wedge \hat{e}_j^{C'}) \vee L_1^{C''}\right) \vee (C' \wedge \hat{e}_j^{C'}) \vee L_2^{C''}\right) \\
&= (C' \wedge \hat{e}_j^{C'}) \vee C \;\; C \wedge ((C' \wedge \hat{e}_j^{C'}) \vee L_1^{C''} \vee L_2^{C''}) \\
&= (C' \wedge \hat{e}_j^{C'}) \vee C \;\; C \wedge ((C' \wedge \hat{e}_j^{C'}) \vee m^{C''}) \\
&= E_j^{C'C} \;\; T''^{CC'}_j(m^{C''}).
\end{aligned} \tag{17.32}$$

So $t_j^{CC'}$ is well-defined. Using (17.30), it can be verified that the matrix of $t_j^{CC'}$ is $(t_{ijk}^{C'})_{i,k=1..3}$.

Thus, (17.26) is equivalent to

$$T_j^{CC'}(L_1^{C''}) \vee T_j^{CC'}(L_2^{C''}) = (C' \wedge \hat{e}_j^{C'}) \vee C \;\; C \wedge ((C' \wedge \hat{e}_j^{C'}) \vee L_1^{C''} \vee L_2^{C''}). \tag{17.33}$$

Geometrically, $T_j^{CC'}$ maps an image line $L^{C''}$ in camera C'' to the image line in camera C, which is the image of the space line on both planes $C' \wedge \hat{e}_j^{C'}$ and $L^{C''}$. (17.33) says that the intersection of the two image lines $T_j^{CC'}(L_1^{C''})$ and $T_j^{CC'}(L_2^{C''})$ is just the image of the intersection of the plane $C' \wedge \hat{e}_j^{C'}$ with the line $L_1^{C''} \vee L_2^{C''}$ in the space. This is the geometric interpretation of (17.26).

17.5.4 Relations on Trifocal Tensors III

Consider the following expression obtained by changing one of the $e^{C'*}$'s in $Texp_1$ to an $e^{C''*}$:

$$Texp_3 = (\hat{e}_i^{C*} \wedge e_j^{C'*}) \vee (\hat{e}_i^{C*} \wedge e_k^{C''*}) \vee (e_1^{C''*} \wedge e_2^{C''*} \wedge e_3^{C''*}). \tag{17.34}$$

Expanding $Texp_3$ from left to right, we get

$$\begin{aligned} Texp_3 &= (\hat{e}_i^{C*} \wedge e_j^{C'*}) \vee e_k^{C''*} \left(\sum_{l=1}^{3} (\hat{e}_i^{C*} \vee \hat{e}_l^{C''*})\, e_l^{C''*} \right) \\ &= -\sum_{l=1}^{3} T_{ijk} F_{il}^{CC''} e_l^{C''*}. \end{aligned}$$

Expanding $Texp_3$ from right to left, we get

$$\begin{aligned} Texp_3 &= (\hat{e}_i^{C*} \wedge e_j^{C'*}) \vee \Big((\hat{e}_i^{C*} \wedge e_k^{C''*} \wedge e_{k_1}^{C''*})^{\sim} e_{k_2}^{C''*} \wedge e_k^{C''*} \\ &\qquad - (\hat{e}_i^{C*} \wedge e_k^{C''*} \wedge e_{k_2}^{C''*})^{\sim} e_{k_1}^{C''*} \wedge e_k^{C''*} \Big) \\ &= (T_{ijk_1} F_{ik_1}^{CC''} + T_{ijk_2} F_{ik_2}^{CC''}) e_k^{C''*} \\ &\quad - T_{ijk} F_{ik_1}^{CC''} e_{k_1}^{C''*} - T_{ijk} F_{ik_2}^{CC''} e_{k_2}^{C''*}, \end{aligned}$$

where $k \prec k_1 \prec k_2$.

Proposition 17.5.3. For any $1 \le i, j \le 3$,

$$\sum_{k=1}^{3} T_{ijk} F_{ik}^{CC''} = 0. \tag{17.35}$$

By (17.16), $F_{ik}^{CC''} = F_{i1}^{CC''} t_{ijk}^{C} / t_{ij1}^{C}$. So (17.35) is equivalent to

$$\sum_{k=1}^{3} T_{ijk} t_{ijk}^{C} = \det(T_{ijk})_{j,k=1..3} = 0, \tag{17.36}$$

for any $1 \le i, j \le 3$. (17.36) can also be obtained directly by expanding the following expression:

$$(\hat{e}_i^{C*} \wedge e_1^{C'*}) \vee (\hat{e}_i^{C*} \wedge e_2^{C'*}) \vee (\hat{e}_i^{C*} \wedge e_3^{C'*}) \vee (e_1^{C''*} \wedge e_2^{C''*} \wedge e_3^{C''*}). \tag{17.37}$$

Expanding from left to right, (17.37) gives zero; expanding from right to left, it gives $\det(T_{ijk})_{j,k=1..3}$.

To understand (17.36) geometrically, we check the dual form of (17.37), which is

$$\begin{aligned} & C'' \wedge ((C \wedge e_i^{C}) \vee (C' \wedge \hat{e}_1^{C'})) \\ & \wedge ((C \wedge e_i^{C}) \vee (C' \wedge \hat{e}_2^{C'})) \\ & \wedge ((C \wedge e_i^{C}) \vee (C' \wedge \hat{e}_3^{C'})). \end{aligned} \tag{17.38}$$

(17.38) equals zero because the intersections of a line with three planes are always collinear.

Interchanging C with C'', we get $\det(T''_{kij})_{i,j=1..3} = 0$ for any $1 \le k \le 3$. By (17.26), we have

$$\det(t^{C'}_{ijk})_{i,j=1..3} = 0. \tag{17.39}$$

A similar constraint can be obtained by interchanging C and C'.

(17.37) can be generalized to the following one:

$$\begin{aligned}&\left((\sum_{i=1}^{3} \lambda_i \hat{e}_i^{C*}) \wedge e_1^{C'*}\right) \vee \left((\sum_{i=1}^{3} \lambda_i \hat{e}_i^{C*}) \wedge e_2^{C'*}\right)\\ &\quad \vee \left((\sum_{i=1}^{3} \lambda_i \hat{e}_i^{C*}) \wedge e_3^{C'*}\right) \vee (e_1^{C''*} \wedge e_2^{C''*} \wedge e_3^{C''*}).\end{aligned} \tag{17.40}$$

where the λ's are indeterminants. (17.40) equals zero when expanded from the left, and equals a polynomial of the λ's when expanded from the right. The coefficients of the polynomial are expressions of the T_{ijk}'s. Thus, we get 10 constraints of degree 3 on T, called the rank constraints by Faugeras and Papadopoulo [81, 83].

17.5.5 Relations on Trifocal Tensors IV

Now, we let the two $\hat{e}^{C*}$'s in $Texp_3$ be different. Consider

$$Texp_4 = (e_i^{C*} \wedge e_{i_1}^{C*} \wedge e_j^{C'*}) \vee (e_i^{C*} \wedge e_{i_2}^{C*} \wedge e_k^{C''*}) \vee (e_1^{C''*} \wedge e_2^{C''*} \wedge e_3^{C''*}). \tag{17.41}$$

Assume that $i \prec i_1 \prec i_2$. Expanding $Texp_4$ from left to right, we get

$$\begin{aligned} Texp_4 &= (e_i^{C*} \wedge e_{i_1}^{C*} \wedge e_{i_2}^{C*} \wedge e_j^{C'*})^{\sim} (e_1^{C''*} \wedge e_2^{C''*} \wedge e_3^{C''*} \wedge e_i^{C*})^{\sim} e_k^{C''*} \\ &\quad -(e_i^{C*} \wedge e_{i_1}^{C*} \wedge e_j^{C'*} \wedge e_k^{C''*})^{\sim} \\ &\quad \left(\sum_{l=1}^{3} ((e_i^{C*} \wedge e_{i_2}^{C*}) \vee \hat{e}_l^{C''*}) e_l^{C''*}\right) \\ &= (E_j^{C'C} E_i^{CC''} + T_{i_2jk} F_{i_1k}^{CC''}) e_k^{C''*} \\ &\quad + T_{i_2jk} F_{i_1k_1}^{CC''} e_{k_1}^{C''*} + T_{i_2jk} F_{i_1k_2}^{CC''} e_{k_2}^{C''*}. \end{aligned}$$

Expanding $Texp_4$ from right to left, we get

$$\begin{aligned} Texp_4 &= ((e_i^{C*} \wedge e_{i_1}^{C*}) \vee (e_j^{C'*} \wedge e_k^{C''*}))(\hat{e}_{i_1}^{C*} \vee \hat{e}_{k_1}^{C''*}) e_{k_1}^{C''*} \\ &\quad + ((e_i^{C*} \wedge e_{i_1}^{C*}) \vee (e_j^{C'*} \wedge e_k^{C''*}))(\hat{e}_{i_1}^{C*} \vee \hat{e}_{k_2}^{C''*}) e_{k_2}^{C''*} \\ &\quad - \Big(((e_i^{C*} \wedge e_{i_1}^{C*}) \vee (e_j^{C'*} \wedge e_{k_1}^{C''*}))(\hat{e}_{i_1}^{C*} \vee \hat{e}_{k_1}^{C''*}) \\ &\quad + ((e_i^{C*} \wedge e_{i_1}^{C*}) \vee (e_j^{C'*} \wedge e_{k_2}^{C''*}))(\hat{e}_{i_1}^{C*} \vee \hat{e}_{k_2}^{C''*}) \Big) e_k^{C''*} \\ &= T_{i_2jk} F_{i_1k_1}^{CC''} e_{k_1}^{C''*} + T_{i_2jk} F_{i_1k_2}^{CC''} e_{k_2}^{C''*} \\ &\quad - (T_{i_2jk_1} F_{i_1k_1}^{CC''} + T_{i_2jk_2} F_{i_1k_2}^{CC''}) e_k^{C''*}, \end{aligned}$$

where $k \prec k_1 \prec k_2$. So

$$\sum_{k=1}^{3} T_{i_2jk} F_{i_1k}^{CC''} = -E_j^{C'C} E_i^{CC''}. \tag{17.42}$$

Interchanging i_1, i_2 in $Texp_4$, we obtain

$$\sum_{k=1}^{3} T_{i_1jk} F_{i_2k}^{CC''} = E_j^{C'C} E_i^{CC''}. \tag{17.43}$$

Proposition 17.5.4. For any $1 \leq i, j \leq 3$,

$$E_j^{C'C} E_i^{CC''} \simeq W_{ij}, \tag{17.44}$$

where $W_{ij} = \sum_{k=1}^{3} T_{i_1jk} F_{i_2k}^{CC''}$.

From (17.36), (17.42) and (17.43) we get

Proposition 17.5.5. For any $1 \leq i_1, i_2, j \leq 3$,

$$\sum_{k=1}^{3} (T_{i_1jk} F_{i_2k}^{CC''} + T_{i_2jk} F_{i_1k}^{CC''}) = 0. \tag{17.45}$$

In fact, (17.44) can be proved by direct computation:

$$\begin{aligned} W_{ij} &= \sum_{k=1}^{3} (C \wedge e_{i_1}^{C}) \vee (C' \wedge \hat{e}_j^{C'}) \vee (C'' \wedge \hat{e}_k^{C''}) \ \ (C \wedge e_{i_2}^{C}) \vee (C'' \wedge e_k^{C''}) \\ &= -\left(C'' \wedge \left(\sum_{k=1}^{3} (C \wedge e_{i_2}^{C}) \vee (C'' \wedge e_k^{C''})\ \hat{e}_k^{C''}\right)\right) \vee (C \wedge e_{i_1}^{C}) \vee (C' \wedge \hat{e}_j^{C'}) \\ &= (C'' \wedge C \wedge e_{i_2}^{C}) \vee (C \wedge e_{i_1}^{C}) \vee (C' \wedge \hat{e}_j^{C'}) \\ &= (C'' \wedge C \wedge e_{i_1}^{C} \wedge e_{i_2}^{C})^{\sim} (C \wedge C' \wedge \hat{e}_j^{C'})^{\sim} \\ &= E_i^{CC''} E_j^{C'C}. \end{aligned}$$

So (17.44) is equivalent to

$$\begin{aligned} (C'' \wedge C \wedge e_{i_2}^{C}) \vee (C \wedge e_{i_1}^{C}) \vee (C' \wedge \hat{e}_j^{C'}) = \\ (C'' \wedge C \wedge e_{i_1}^{C} \wedge e_{i_2}^{C})^{\sim} (C \wedge C' \wedge \hat{e}_j^{C'})^{\sim}; \end{aligned} \tag{17.46}$$

(17.45) is equivalent to the anti-symmetry of $C'' \wedge C \wedge e_{i_1}^{C} \wedge e_{i_2}^{C}$ with respect to $e_{i_1}^{C}$ and $e_{i_2}^{C}$.

Define

$$u^{C''C}_{i_1 i_2 j_1 j_2} = \sum_{k=1}^{3} t^{C}_{i_1 j_1 k} T_{i_2 j_2 k} \tag{17.47}$$

for $1 \leq i_1, i_2, j_1, j_2 \leq 3$. By (17.12), (17.36), (17.42) and (17.43),

$$\begin{aligned} u^{C''C}_{i_1 i_2 j_1 j_2} &= \sum_{k=1}^{3} F^{CC'}_{i_1 j_1} F^{CC''}_{i_1 k} T_{i_2 j_2 k} \\ &= \begin{cases} 0, & \text{if } i_1 = i_2; \\ -F^{CC'}_{i_1 j_1} E^{C'C}_{j_2} E^{CC''}_{i}, & \text{if } i \prec i_1 \prec i_2; \\ F^{CC'}_{i_1 j_1} E^{C'C}_{j_2} E^{CC''}_{i}, & \text{if } i \prec i_2 \prec i_1. \end{cases} \end{aligned} \tag{17.48}$$

Two corollaries can be drawn immediately:

Corollary 17.5.3. 1. For any $1 \leq i_l, j_l \leq 3$, where $1 \leq k \leq 4$,

$$\frac{u^{C''C}_{i_1 i_2 j_1 j_2}}{u^{C''C}_{i_1 i_2 j_1 j_3}} = \frac{u^{C''C}_{i_3 i_4 j_4 j_2}}{u^{C''C}_{i_3 i_4 j_4 j_3}} = \frac{E^{C'C}_{j_2}}{E^{C'C}_{j_3}}. \tag{17.49}$$

2. Let $i \prec i_1 \prec i_2$. Then for any $1 \leq j_l \leq 3$ where $1 \leq l \leq 4$,

$$\frac{u^{C''C}_{i_1 i_2 j_1 j_2}}{u^{C''C}_{i_1 i j_1 j_2}} = \frac{u^{C''C}_{i_1 i_2 j_3 j_4}}{u^{C''C}_{i_1 i j_3 j_4}} = -\frac{E^{CC''}_{i}}{E^{CC''}_{i_2}}. \tag{17.50}$$

Corollary 17.5.4. 1. For any $1 \leq i_1, i_2, j_1 \leq 3$ where $i_1 \neq i_2$,,

$$\mathbf{E}^{C'C} \simeq (u^{C''C}_{i_1 i_2 j_1 j_2})_{j_2 = 1..3}. \tag{17.51}$$

2. For any $1 \leq j_1, j_2 \leq 3$,

$$\mathbf{E}^{CC''} \simeq (u^{C''C}_{23 j_1 j_2} u^{C''C}_{32 j_1 j_2}, -u^{C''C}_{23 j_1 j_2} u^{C''C}_{31 j_1 j_2}, -u^{C''C}_{21 j_1 j_2} u^{C''C}_{32 j_1 j_2})^T. \tag{17.52}$$

Now we explain (17.48) in terms of Grassmann-Cayley algebra. We have defined two mappings $T^{C''C}_i$ and $t^{C''C}_i$ in (17.18) and (17.19), whose matrices are $(-T_{ijk})_{j,k=1..3}$ and $(t^{C}_{ijk})_{j,k=1..3}$ respectively. By the definition of $u^{C''C}_{i_1 i_2 j_1 j_2}$,

$$u^{C''C}_{i_1 i_2 j_1 j_2} C'' = t^{C''C}_{i_1}(C' \wedge e^{C'}_{j_1}) \vee T^{C''C}_{i_2}(C' \wedge \hat{e}^{C'}_{j_2}). \tag{17.53}$$

Expanding the right-hand side of (17.53), we get

$$u^{C''C}_{i_1 i_2 j_1 j_2} = U^{C''C}(C \wedge e^{C}_{i_1}, C \wedge e^{C}_{i_2}, C' \wedge e^{C'}_{j_1}, C' \wedge \hat{e}^{C'}_{j_2}), \tag{17.54}$$

where $U^{C''C} : (C \wedge \mathbb{R}^4) \times (C \wedge \mathbb{R}^4) \times (C' \wedge \mathbb{R}^4) \times (C' \wedge \mathcal{G}^2_4) \longrightarrow \mathbb{R}$ is defined by

$$U^{C''C}(m^C_1, m^C_2, m^{C'}, L^{C'}) = -m^C_1 \vee m^{C'} \quad C \vee L^{C'} \quad C'' \vee (m^C_1 \wedge_C m^C_2). \tag{17.55}$$

(17.54) is (17.48) in Grassmann-Cayley algebraic form. It means that the $u^{C''C}$'s are components of the mapping $U^{C''C}$.

Notice that (17.49) is a group of degree 6 constraints on T. It is closely related to Faugeras and Mourrain's first group of degree 6 constraints:

$$\begin{aligned}&|\mathbf{T}_{k_1k_2.}\ \mathbf{T}_{k_1l_2.}\ \mathbf{T}_{l_1l_2.}||\mathbf{T}_{k_1k_2.}\ \mathbf{T}_{l_1k_2.}\ \mathbf{T}_{l_1l_2.}|\\ &\quad = |\mathbf{T}_{l_1k_2.}\ \mathbf{T}_{k_1l_2.}\ \mathbf{T}_{l_1l_2.}||\mathbf{T}_{k_1k_2.}\ \mathbf{T}_{l_1k_2.}\ \mathbf{T}_{k_1l_2.}|,\end{aligned} \tag{17.56}$$

where $\mathbf{T}_{k_1k_2.} = (T_{k_1k_2k})_{k=1..3}$.

It is difficult to find the symmetry of the indices in (17.56), so we first express (17.56) in terms of Grassmann-Cayley algebra. Using the fact that $-\mathbf{T}_{k_1k_2.}$ is the coordinates of $C'' \wedge ((C \wedge e^C_{k_1}) \vee (C' \wedge \hat{e}^{C'}_{k_2}))$, we get

$$\begin{aligned}&|\mathbf{T}_{k_1k_2.}\ \mathbf{T}_{k_1l_2.}\ \mathbf{T}_{l_1l_2.}|^{\sim}\\ &\quad = \Big(C'' \wedge \Big((C \wedge e^C_{k_1}) \vee (C' \wedge \hat{e}^{C'}_{k_2})\Big)\Big)\\ &\qquad\qquad \wedge_{C''} \Big(C'' \wedge \Big((C \wedge e^C_{k_1}) \vee (C' \wedge \hat{e}^{C'}_{l_2})\Big)\Big)\\ &\qquad\qquad \wedge_{C''} \Big(C'' \wedge \Big((C \wedge e^C_{l_1}) \vee (C' \wedge \hat{e}^{C'}_{l_s2})\Big)\Big)\\ &\quad = C'' \wedge \Big((C \wedge e^C_{k_1}) \vee (C' \wedge \hat{e}^{C'}_{k_2})\Big) \wedge \Big((C \wedge e^C_{k_1}) \vee (C' \wedge \hat{e}^{C'}_{l_2})\Big)\\ &\qquad\qquad\qquad \wedge \Big((C \wedge e^C_{l_1}) \vee (C' \wedge \hat{e}^{C'}_{l_2})\Big).\end{aligned}$$

By formula (17.20),

$$\begin{aligned}|\mathbf{T}_{k_1k_2.}\ \mathbf{T}_{k_1l_2.}\ \mathbf{T}_{l_1l_2.}| &= -(C \wedge e^C_{k_1}) \vee (C' \wedge \hat{e}^{C'}_{l_2}) \vee (C' \wedge \hat{e}^{C'}_{k_2})\\ &\qquad (C'' \wedge C \wedge e^C_{k_1}) \vee (C \wedge e^C_{l_1}) \vee (C' \wedge \hat{e}^{C'}_{l_2})\\ &= -(C \wedge e^C_{k_1}) \vee (C' \wedge \hat{e}^{C'}_{k_2}) \vee (C' \wedge \hat{e}^{C'}_{l_2})\\ &\qquad C'' \vee (C \wedge e^C_{k_1} \wedge e^C_{l_1})\ \ C \vee (C' \wedge \hat{e}^{C'}_{l_2}).\end{aligned} \tag{17.57}$$

Define a mapping $V^{C''} : (C \wedge \mathbb{R}^4) \times (C \wedge \mathbb{R}^4) \times (C' \wedge \mathcal{G}^2_4) \times (C' \wedge \mathcal{G}^2_4) \longrightarrow \mathbb{R}$ as follows:

$$\begin{aligned}V^{C''}(m^C_1, m^C_2, L^{C'}_1, L^{C'}_2) &= -(C'' \vee (m^C_1 \wedge_C m^C_2))\\ &\qquad (C \vee L^{C'}_2)(m^C_1 \vee L^{C'}_1 \vee L^{C'}_2).\end{aligned} \tag{17.58}$$

Let

$$v^{C''}_{k_1l_1k_2l_2} = V^{C''}(C' \wedge \hat{e}^{C'}_{k_1}, C' \wedge \hat{e}^{C'}_{l_1}, C'' \wedge \hat{e}^{C''}_{k_2}, C'' \wedge \hat{e}^{C''}_{l_2}). \tag{17.59}$$

By (17.57),

$$|\mathbf{T}_{k_1k_2\cdot}\ \mathbf{T}_{k_1l_2\cdot}\ \mathbf{T}_{l_1l_2\cdot}| = v^{C''}_{k_1l_1k_2l_2}. \tag{17.60}$$

Similarly, we can get

$$\begin{aligned} |\mathbf{T}_{k_1k_2\cdot}\ \mathbf{T}_{l_1k_2\cdot}\ \mathbf{T}_{l_1l_2\cdot}| &= v^{C''}_{l_1k_1l_2k_2}, \\ |\mathbf{T}_{l_1k_2\cdot}\ \mathbf{T}_{k_1l_2\cdot}\ \mathbf{T}_{l_1l_2\cdot}| &= v^{C''}_{l_1k_1k_2l_2}, \\ |\mathbf{T}_{k_1k_2\cdot}\ \mathbf{T}_{l_1k_2\cdot}\ \mathbf{T}_{k_1l_2\cdot}| &= v^{C''}_{k_1l_1l_2k_2}. \end{aligned} \tag{17.61}$$

So (17.56) is equivalent to

$$\frac{v^{C''}_{k_1l_1k_2l_2}}{v^{C''}_{k_1l_1l_2k_2}} = \frac{v^{C''}_{l_1k_1k_2l_2}}{v^{C''}_{l_1k_1l_2k_2}}, \tag{17.62}$$

which is simpler than (17.56) in appearance. By (17.58), (17.59), in Grassmann-Cayley algebra, (17.62) is just the following identity:

$$\begin{aligned} &\frac{C'' \vee (m_1^C \wedge_C m_2^C)\ \ C \vee L_2^{C'}\ \ m_1^C \vee L_1^{C'} \vee L_2^{C'}}{C'' \vee (m_1^C \wedge_C m_2^C)\ \ C \vee L_1^{C'}\ \ m_1^C \vee L_2^{C'} \vee L_1^{C'}} \\ &= \frac{C'' \vee (m_2^C \wedge_C m_1^C)\ \ C \vee L_2^{C'}\ \ m_2^C \vee L_1^{C'} \vee L_2^{C'}}{C'' \vee (m_2^C \wedge_C m_1^C)\ \ C \vee L_1^{C'}\ \ m_2^C \vee L_2^{C'} \vee L_1^{C'}}, \end{aligned} \tag{17.63}$$

for any $m_1^C, m_2^C \in C \wedge \mathbb{R}^4$, $L_1^{C'}, L_2^{C'} \in C' \wedge \mathcal{G}_4^2$.

By (17.58), we have

$$v^{C''}_{i_1i_2j_1j_2} = \begin{cases} 0, & \text{if } i_1 = i_2 \text{ or } j_1 = j_2; \\ -F^{CC'}_{i_1j} E^{C'C}_{j_2} E^{CC''}_{i}, & \text{if } i \prec i_1 \prec i_2 \text{ and } j \prec j_1 \prec j_2, \\ & \text{or } i \prec i_2 \prec i_1 \text{ and } j \prec j_2 \prec j_1; \\ F^{CC'}_{i_1j} E^{C'C}_{j_2} E^{CC''}_{i}, & \text{if } i \prec i_1 \prec i_2 \text{ and } j \prec j_2 \prec j_1, \\ & \text{or } i \prec i_2 \prec i_1 \text{ and } j \prec j_1 \prec j_2. \end{cases} \tag{17.64}$$

Corollary 17.5.5. 1. For any $1 \le i_l, j_1, j_2 \le 3$ where $1 \le l \le 4$,

$$\frac{v^{C''}_{i_1i_2j_1j_2}}{v^{C''}_{i_1i_2j_2j_1}} = \frac{v^{C''}_{i_3i_4j_1j_2}}{v^{C''}_{i_3i_4j_2j_1}}. \tag{17.65}$$

2. Let $i \prec i_1 \prec i_2$. Then for any $1 \le j_l \le 3$ where $1 \le l \le 4$,

$$\frac{v^{C''}_{i_1i_2j_1j_2}}{v^{C''}_{i_1ij_1j_2}} = \frac{v^{C''}_{i_1i_2j_3j_4}}{v^{C''}_{i_1ij_3j_4}}. \tag{17.66}$$

(17.56) is a special case of (17.65) where $i_3 = i_2, i_4 = i_1$. Comparing $U^{C''C}$ with $V^{C''}$, we get

$$V^{C''}(m_1^C, m_2^C, L_1^{C'}, L_2^{C'}) = U^{C''C}(m_1^C, m_2^C, L_1^{C'} \vee L_2^{C'}, L_2^{C'}). \qquad (17.67)$$

It appears that we have generalized Faugeras and Mourrain's first group of degree-six constraints furthermore by $U^{C''C}$, because (17.50) is equivalent to (17.66), while (17.65) is a special case of (17.49) where $j_4 = j_1$. An explanation for this phenomenon is that the variables in $V^{C''}$ are less separated than those in $U^{C''C}$, so there are less constraints on T that come from $V^{C''}$ than from $U^{C''C}$.

Interchanging C' and C'' in (17.56), we get Faugeras and Mourrain's second group of degree 6 constraints:

$$\begin{aligned} &|\mathbf{T}_{k_1.k_2}\ \mathbf{T}_{k_1.l_2}\ \mathbf{T}_{l_1.l_2}||\mathbf{T}_{k_1.k_2}\ \mathbf{T}_{l_1.k_2}\ \mathbf{T}_{l_1.l_2}| \\ &\quad = |\mathbf{T}_{l_1.k_2}\ \mathbf{T}_{k_1.l_2}\ \mathbf{T}_{l_1.l_2}||\mathbf{T}_{k_1.k_2}\ \mathbf{T}_{l_1.k_2}\ \mathbf{T}_{k_1.l_2}|, \end{aligned} \qquad (17.68)$$

where $\mathbf{T}_{k_1.k_2} = (T_{k_1 k k_2})_{k=1..3}$.

This group of constraints can be generalized similarly.

17.5.6 Relations on Trifocal Tensors V

Consider the following expression:

$$Texp_5 = (e_i^{C*} \wedge e_{i_1}^{C*} \wedge e_j^{C'*}) \vee (e_{i_2}^{C*} \wedge e_j^{C'*} \wedge e_k^{C''*}) \vee (e_1^{C''*} \wedge e_2^{C''*} \wedge e_3^{C''*}). \qquad (17.69)$$

Assume that $i \prec i_1 \prec i_2$. Expanding $Texp_5$ from left to right, we get

$$Texp_5 = -E_j^{C'C} E_j^{C'C''} e_k^{C''*} - T_{i_2 jk} \sum_{l=1}^{3} T''_{li_2 j} e_l^{C''*}.$$

Expanding $Texp_4$ from right to left, we get

$$\begin{aligned} Texp_5 = &-T''_{k_2 i_2 j} T_{i_2 jk} e_{k_2}^{C''*} - T''_{k_1 i_2 j} T_{i_2 jk} e_{k_1}^{C''*} \\ &+(T''_{k_2 i_2 j} T_{i_2 jk_2} + T''_{k_1 i_2 j} T_{i_2 jk_1}) e_k^{C''*}, \end{aligned}$$

where $k \prec k_1 \prec k_2$. So

$$\sum_{k=1}^{3} T''_{ki_2 j} T_{i_2 jk} = -E_j^{C'C} E_j^{C'C''}. \qquad (17.70)$$

Proposition 17.5.6. For any $1 \le i, j \le 3$,

$$\sum_{k=1}^{3} T''_{kij} T_{ijk} \simeq E_j^{C'C} E_j^{C'C''}. \qquad (17.71)$$

Using the relation (17.26), we get

$$\sum_{k=1}^{3} t_{ijk}^{C'} T_{ijk} = \det(T_{ijk})_{i,k=1..3} = (E_j^{C'C})^2 E_j^{C'C''}. \tag{17.72}$$

Corollary 17.5.6. For any $1 \le i_1, i_2, j_1 \le 3$ where $i_1 \ne i_2$,

$$\mathbf{E}^{C'C''} \simeq \left(\frac{\det(T_{ijk})_{i,k=1..3}}{(u_{i_1 i_2 j_1 j}^{C''C})^2} \right)_{j=1..3}. \tag{17.73}$$

17.5.7 Relations on Trifocal Tensors VI

The second idea of deriving relations on trifocal tensors is as follows: if the tensor $(T_{ijk})_{i,j,k=1..3}$ is given, then expanding

$$(e_j^{C'*} \wedge e_k^{C''*}) \vee (e_1^{C*} \wedge e_2^{C*} \wedge e_3^{C*}) \tag{17.74}$$

gives a vector of the e^{C*}'s whose coefficients are known. Similarly, expanding

$$\begin{aligned} &(e_j^{C'*} \wedge e_j^{C'*}) \vee (e_1^{C*} \wedge e_2^{C*} \wedge e_3^{C*} \wedge e_j^{C'*}), \\ &(e_j^{C'*} \wedge e_k^{C''*}) \vee (e_1^{C*} \wedge e_2^{C*} \wedge e_3^{C*} \wedge e_k^{C''*}) \end{aligned} \tag{17.75}$$

gives two 2-vectors of the $e_j^{C'*} \wedge e_i^{C*}$'s and the $e_k^{C''*} \wedge e_i^{C*}$'s respectively, whose coefficients are known. The meet of two such 2-vectors, i. e.,

$$\begin{aligned} Texp_6 = &\left((e_{j_1}^{C'*} \wedge e_{k_1}^{C''*}) \vee (e_1^{C*} \wedge e_2^{C*} \wedge e_3^{C*} \wedge e_{j_1}^{C'*}) \right) \\ &\vee \left((e_{j_2}^{C'*} \wedge e_{k_2}^{C''*}) \vee (e_1^{C*} \wedge e_2^{C*} \wedge e_3^{C*} \wedge e_{k_2}^{C''*}) \right) \end{aligned} \tag{17.76}$$

is an expression of the T_{ijk}'s. Expanding the meets differently, we get a relation on T, epipoles and epipolar tensors.

Assume that $j \prec j_1 \prec j_2$ and $k \prec k_1 \prec k_2$. Expanding $Texp_6$ according to its parentheses, we get

$$\begin{aligned} Texp_6 &= \left(\sum_{i_1=1}^{3} (e_{j_1}^{C'*} \wedge e_{k_1}^{C''*}) \vee \hat{e}_{i_1}^{C*} \quad e_{i_1}^{C*} \wedge e_{j_1}^{C'*} \right) \\ &\quad \vee \left(\sum_{i_2=1}^{3} (e_{j_2}^{C'*} \wedge e_{k_2}^{C''*}) \vee \hat{e}_{i_2}^{C*} \quad e_{i_2}^{C*} \wedge e_{k_2}^{C''*} \right) \\ &= \sum_{i=1}^{3} (-T_{i_1 j_1 k_1} T_{i_2 j_2 k_2} + T_{i_2 j_1 k_1} T_{i_1 j_2 k_2}) T_{i j_1 k_2}, \end{aligned}$$

where $i \prec i_1 \prec i_2$. Using the fact that the meet of a 4-vector with any multivector in $\mathcal{G}_4$ is a scalar multiplication of the multivector by the dual of the 4-vector, we get

$$\begin{aligned} Texp_6 &= (e_{j_1}^{C'*} \wedge e_{k_1}^{C''*}) \vee (e_{j_2}^{C'*} \wedge e_{k_2}^{C''*}) \vee (e_1^{C*} \wedge e_2^{C*} \wedge e_3^{C*} \wedge e_{j_1}^{C'*}) \\ &\qquad \vee (e_1^{C*} \wedge e_2^{C*} \wedge e_3^{C*} \wedge e_{k_2}^{C''*}) \\ &= -F_{jk}^{C'C''} E_{j_1}^{C'C} E_{k_2}^{C''C}. \end{aligned}$$

So

$$F_{jk}^{C'C''} E_{j_1}^{C'C} E_{k_2}^{C''C} = \sum_{i=1}^{3} (T_{i_1 j_1 k_1} T_{i_2 j_2 k_2} - T_{i_2 j_1 k_1} T_{i_1 j_2 k_2}) T_{i j_1 k_2}. \tag{17.77}$$

Interchanging j_1, j_2 in $Texp_5$, we get

$$-F_{jk}^{C'C''} E_{j_2}^{C'C} E_{k_2}^{C''C} = \sum_{i=1}^{3} (T_{i_1 j_2 k_1} T_{i_2 j_1 k_2} - T_{i_2 j_2 k_1} T_{i_1 j_1 k_2}) T_{i j_2 k_2}. \tag{17.78}$$

Interchanging k_1, k_2 in $Texp_5$, we get

$$-F_{jk}^{C'C''} E_{j_1}^{C'C} E_{k_1}^{C''C} = \sum_{i=1}^{3} (T_{i_1 j_1 k_2} T_{i_2 j_2 k_1} - T_{i_2 j_1 k_2} T_{i_1 j_2 k_1}) T_{i j_1 k_1}. \tag{17.79}$$

Interchanging (j_1, k_1) and (j_2, k_2) in $Texp_5$, we get

$$F_{jk}^{C'C''} E_{j_2}^{C'C} E_{k_1}^{C''C} = \sum_{i=1}^{3} (T_{i_1 j_2 k_2} T_{i_2 j_1 k_1} - T_{i_2 j_2 k_2} T_{i_1 j_1 k_1}) T_{i j_2 k_1}. \tag{17.80}$$

When $j_1 = j_2$ or $k_1 = k_2$, $Texp_5 = 0$ by expanding from left to right. Define

$$v_{j_1 j_2 k_1 k_2}^{C} = -\sum_{i=1}^{3} (T_{i_1 j_1 k_1} T_{i_2 j_2 k_2} - T_{i_2 j_1 k_1} T_{i_1 j_2 k_2}) T_{i j_1 k_2}. \tag{17.81}$$

Then

$$v_{j_1 j_2 k_1 k_2}^{C} = \begin{cases} 0, & \text{if } j_1 = j_2 \text{ or } k_1 = k_2; \\ -F_{jk}^{C'C''} E_{j_1}^{C'C} E_{k_2}^{C''C}, & \text{if } j \prec j_1 \prec j_2 \text{ and } k \prec k_1 \prec k_2, \\ & \text{or } j \prec j_2 \prec j_1 \text{ and } k \prec k_2 \prec k_1; \\ F_{jk}^{C'C''} E_{j_1}^{C'C} E_{k_2}^{C''C}, & \text{if } j \prec j_1 \prec j_2 \text{ and } k \prec k_2 \prec k_1, \\ & \text{or } j \prec j_2 \prec j_1 \text{ and } k \prec k_1 \prec k_2. \end{cases} \tag{17.82}$$

Proposition 17.5.7. For any $1 \le j_1, j_2, k_1, k_2 \le 3$,

$$\frac{v_{j_1 j_2 k_1 k_2}^{C}}{v_{j_2 j_1 k_1 k_2}^{C}} = -\frac{E_{j_1}^{C'C}}{E_{j_2}^{C'C}}; \quad \frac{v_{j_1 j_2 k_1 k_2}^{C}}{v_{j_1 j_2 k_2 k_1}^{C}} = -\frac{E_{k_2}^{C''C}}{E_{k_1}^{C''C}}. \tag{17.83}$$

Corollary 17.5.7. 1. For any $1 \leq j_l, k_l \leq 3$, where $1 \leq l \leq 4$,

$$\frac{v^C_{j_1j_2k_1k_2}}{v^C_{j_2j_1k_1k_2}} = \frac{v^C_{j_1j_2k_3k_4}}{v^C_{j_2j_1k_3k_4}}; \quad \frac{v^C_{j_1j_2k_1k_2}}{v^C_{j_1j_2k_2k_1}} = \frac{v^C_{j_3j_4k_1k_2}}{v^C_{j_3j_4k_2k_1}}. \tag{17.84}$$

2. For any $1 \leq i_l, j, j_l, k_l \leq 3$ where $1 \leq l \leq 2$,

$$\frac{v^C_{j_1j_2k_1k_2}}{v^C_{j_2j_1k_1k_2}} = -\frac{u^{C''C}_{i_1i_2jj_1}}{u^{C''C}_{i_1i_2jj_2}}. \tag{17.85}$$

Notice that (17.84) and (17.85) are groups of degree 6 constraints on T. (17.84) is closely related to Faugeras and Mourrain's third group of degree 6 constraints:

$$\begin{aligned} &|\mathbf{T}_{.k_1k_2}\ \mathbf{T}_{.k_1l_2}\ \mathbf{T}_{.l_1l_2}||\mathbf{T}_{.k_1k_2}\ \mathbf{T}_{.l_1k_2}\ \mathbf{T}_{.l_1l_2}| \\ &\quad = |\mathbf{T}_{.l_1k_2}\ \mathbf{T}_{.k_1l_2}\ \mathbf{T}_{.l_1l_2}||\mathbf{T}_{.k_1k_2}\ \mathbf{T}_{.l_1k_2}\ \mathbf{T}_{.k_1l_2}|, \end{aligned} \tag{17.86}$$

where $\mathbf{T}_{.k_1k_2} = (T_{kk_1k_2})_{k=1..3}$.

Let us express (17.86) in terms of Grassmann-Cayley algebra. Using the fact that $-\mathbf{T}_{.k_1k_2}$ is the coordinates of $C \wedge ((C' \wedge \hat{e}^{C'}_{k_1}) \vee (C'' \wedge \hat{e}^{C''}_{k_2}))$, we get

$$\begin{aligned} &|\mathbf{T}_{.k_1k_2}\ \mathbf{T}_{.k_1l_2}\ \mathbf{T}_{.l_1l_2}|\,C \\ &= \left(C \wedge \left((C' \wedge \hat{e}^{C'}_{k_1}) \vee (C'' \wedge \hat{e}^{C''}_{k_2})\right)\right) \\ &\qquad \vee_C \left(C \wedge \left((C' \wedge \hat{e}^{C'}_{k_1}) \vee (C'' \wedge \hat{e}^{C''}_{l_2})\right)\right) \\ &\qquad \vee_C \left(C \wedge \left((C' \wedge \hat{e}^{C'}_{l_1}) \vee (C'' \wedge \hat{e}^{C''}_{l_2})\right)\right) \\ &= C \left(C \wedge \left((C' \wedge \hat{e}^{C'}_{k_1}) \vee (C'' \wedge \hat{e}^{C''}_{k_2})\right)\right) \\ &\qquad \vee \left(C \wedge \left((C' \wedge \hat{e}^{C'}_{k_1}) \vee (C'' \wedge \hat{e}^{C''}_{l_2})\right)\right) \\ &\qquad \vee (C' \wedge \hat{e}^{C'}_{l_1}) \vee (C'' \wedge \hat{e}^{C''}_{l_2}). \end{aligned}$$

By (17.31), we have

$$\begin{aligned} &|\mathbf{T}_{.k_1k_2}\ \mathbf{T}_{.k_1l_2}\ \mathbf{T}_{.l_1l_2}| \\ &= \left(C \wedge \left((C' \wedge \hat{e}^{C'}_{k_1}) \vee (C'' \wedge \hat{e}^{C''}_{k_2})\right)\right) \\ &\qquad \vee (C' \wedge \hat{e}^{C'}_{k_1}) \vee (C'' \wedge \hat{e}^{C''}_{l_2}) \vee (C' \wedge \hat{e}^{C'}_{l_1})\ \ (C'' \wedge \hat{e}^{C''}_{l_2}) \vee C \\ &= -\,(C' \wedge \hat{e}^{C'}_{k_1}) \vee C\ \ \ (C'' \wedge \hat{e}^{C''}_{l_2}) \vee C \\ &\qquad (C' \wedge \hat{e}^{C'}_{k_1}) \vee (C' \wedge \hat{e}^{C'}_{l_1}) \vee (C'' \wedge \hat{e}^{C''}_{k_2}) \vee (C'' \wedge \hat{e}^{C''}_{l_2}). \end{aligned} \tag{17.87}$$

Define a mapping $V^C : (C' \wedge \mathcal{G}^2_4) \times (C' \wedge \mathcal{G}^2_4) \times (C'' \wedge \mathcal{G}^2_4) \times (C'' \wedge \mathcal{G}^2_4) \longrightarrow \mathbb{R}$ as follows:

$$V^C(L_1^{C'}, L_2^{C'}, L_1^{C''}, L_2^{C''}) = -(L_1^{C'} \vee C)(L_2^{C''} \vee C)(L_1^{C'} \vee L_2^{C'} \vee L_1^{C''} \vee L_2^{C''}). \tag{17.88}$$

Then

$$V^C(C' \wedge \hat{e}_{k_1}^{C'}, C' \wedge \hat{e}_{l_1}^{C'}, C'' \wedge \hat{e}_{k_2}^{C''}, C'' \wedge \hat{e}_{l_2}^{C''}) = v_{k_1 l_1 k_2 l_2}^C, \tag{17.89}$$

According to (17.87),

$$|\mathbf{T}_{.k_1k_2}\ \mathbf{T}_{.k_1l_2}\ \mathbf{T}_{.l_1l_2}| = v_{k_1l_1k_2l_2}^C. \tag{17.90}$$

Similarly, we have

$$\begin{aligned} |\mathbf{T}_{.k_1k_2}\ \mathbf{T}_{.l_1k_2}\ \mathbf{T}_{.l_1l_2}| &= v_{l_1k_1l_2k_2}^C, \\ |\mathbf{T}_{.l_1k_2}\ \mathbf{T}_{.k_1l_2}\ \mathbf{T}_{.l_1l_2}| &= v_{l_1k_1k_2l_2}^C, \\ |\mathbf{T}_{.k_1k_2}\ \mathbf{T}_{.l_1k_2}\ \mathbf{T}_{.k_1l_2}| &= v_{k_1l_1l_2k_2}^C. \end{aligned} \tag{17.91}$$

Now (17.86) is equivalent to

$$\frac{v_{k_1l_1k_2l_2}^C}{v_{k_1l_1l_2k_2}^C} = \frac{v_{l_1k_1k_2l_2}^C}{v_{l_1k_1l_2k_2}^C}, \tag{17.92}$$

or more explicitly, the following identity:

$$\begin{aligned} &\frac{L_1^{C'} \vee C\ \ L_2^{C''} \vee C\ \ L_1^{C'} \vee L_2^{C'} \vee L_1^{C''} \vee L_2^{C''}}{L_1^{C'} \vee C\ \ L_1^{C''} \vee C\ \ L_1^{C'} \vee L_2^{C'} \vee L_2^{C''} \vee L_1^{C''}} \\ &= \frac{L_2^{C'} \vee C\ \ L_2^{C''} \vee C\ \ L_2^{C'} \vee L_1^{C'} \vee L_1^{C''} \vee L_2^{C''}}{L_2^{C'} \vee C\ \ L_1^{C''} \vee C\ \ L_2^{C'} \vee L_1^{C'} \vee L_2^{C''} \vee L_1^{C''}}. \end{aligned} \tag{17.93}$$

(17.84) is a straightforward generalization of it.

17.5.8 A Unified Treatment of Degree-six Constraints

In this section we make a comprehensive investigation of Faugeras and Mourrain's three groups of degree-six constraints. We have defined $u_{i_1i_2j_1j_2}^{C''C}$ in (17.47) to derive and generalize the first group of constraints. We are going to follow the same line to derive and generalize the other two groups of constraints.

The trifocal tensor T induces 6 kinds of linear mappings as shown in table 17.1. We have defined two linear mappings $t_i^{C''C}$ and $t_j^{CC'}$ in (17.19) and (17.30) respectively, which are generated by the T's. There are 6 such linear mappings as shown in table 17.2. Let

$$m^{C'} = L_1^{C'} \vee L_2^{C'},\quad m^{C''} = L_1^{C''} \vee L_2^{C''},\quad L^C = m_1^C \wedge_C m_2^C.$$

Here

$$t_{ijk}^{C''} = T_{i_1j_1k}T_{i_2j_2k} - T_{i_1j_2k}T_{i_2j_1k}, \tag{17.94}$$

where $i \prec i_1 \prec i_2$ and $j \prec j_1 \prec j_2$. t_{ijk}^C and $t_{ijk}^{C'}$ have been defined in (17.13) and (17.25) respectively.

Table 17.1. Linear mappings induced by T

Mapping	Definition	Matrix
$T_i^{C'C}$	$C'' \wedge \mathcal{G}_4^2 \longrightarrow C' \wedge \mathbb{R}^4$ $L^{C''} \mapsto C' \wedge ((C \wedge e_i^C) \vee L^{C''})$	$(-T_{ijk})_{j,k=1..3}$
$T_i^{C''C}$	$C' \wedge \mathcal{G}_4^2 \longrightarrow C'' \wedge \mathbb{R}^4$ $L^{C'} \mapsto C'' \wedge ((C \wedge e_i^C) \vee L^{C'})$	$(-T_{ijk})^T_{j,k=1..3}$
$T_j^{CC'}$	$C'' \wedge \mathcal{G}_4^2 \longrightarrow C \wedge \mathcal{G}_4^2$ $L^{C''} \mapsto C \wedge ((C' \wedge \hat{e}_j^{C'}) \vee L^{C''})$	$(-T_{ijk})_{i,k=1..3}$
$T_j^{C''C'}$	$C \wedge \mathbb{R}^4 \longrightarrow C'' \wedge \mathbb{R}^4$ $m^C \mapsto C'' \wedge (m^C \vee (C' \wedge \hat{e}_j^{C'}))$	$(-T_{ijk})^T_{i,k=1..3}$
$T_k^{CC''}$	$C' \wedge \mathcal{G}_4^2 \longrightarrow C \wedge \mathcal{G}_4^2$ $L^{C'} \mapsto C \wedge (L^{C'} \vee (C'' \wedge \hat{e}_k^{C''}))$	$(-T_{ijk})_{i,j=1..3}$
$T_k^{C'C''}$	$C \wedge \mathbb{R}^4 \longrightarrow C' \wedge \mathbb{R}^4$ $m^C \mapsto C' \wedge (m^C \vee (C'' \wedge \hat{e}_k^{C''}))$	$(-T_{ijk})^T_{i,j=1..3}$

The mappings t's are well-defined because

$$
\begin{aligned}
t_i^{C'C}(m^{C''}) &= -(C \wedge e_i^C) \vee m^{C''} \quad C' \wedge C \wedge e_i^C, \\
t_i^{C''C}(m^{C'}) &= -(C \wedge e_i^C) \vee m^{C'} \quad C'' \wedge C \wedge e_i^C, \\
t_j^{CC'}(m^{C''}) &= (C' \wedge \hat{e}_j^{C'}) \vee C \quad C \wedge ((C' \wedge \hat{e}_j^{C'}) \vee m^{C''}), \\
t_j^{C''C'}(L^C) &= -(C' \wedge \hat{e}_j^{C'}) \vee C \quad C'' \wedge (L^C \vee (C' \wedge \hat{e}_j^{C'})), \\
t_k^{CC''}(m^{C'}) &= (C'' \wedge \hat{e}_k^{C''}) \vee C \quad C \wedge (m^{C'} \vee (C'' \wedge \hat{e}_k^{C''})), \\
t_k^{C'C''}(L^C) &= -(C'' \wedge \hat{e}_k^{C''}) \vee C \quad C' \wedge (L^C \vee (C'' \wedge \hat{e}_k^{C''})).
\end{aligned}
\tag{17.95}
$$

For any $1 \leq i_1, i_2, j_1, j_2, k_1, k_2 \leq 3$, let

Table 17.2. Linear mappings induced by t

Mapping	Definition	Matrix
$t_i^{C'C}$	$C'' \wedge \mathbb{R}^4 \longrightarrow C' \wedge \mathcal{G}_4^2$ $m^{C''} \mapsto T_i^{C'C}(L_1^{C''}) \wedge_{C'} T_i^{C'C}(L_2^{C''})$	$(t_{ijk}^C)_{j,k=1..3}$
$t_i^{C''C}$	$C' \wedge \mathbb{R}^4 \longrightarrow C'' \wedge \mathcal{G}_4^2$ $m^{C'} \mapsto T_i^{C''C}(L_1^{C'}) \wedge_{C''} T_i^{C''C}(L_2^{C'})$	$(t_{ijk}^C)_{j,k=1..3}^T$
$t_j^{CC'}$	$C'' \wedge \mathbb{R}^4 \longrightarrow C \wedge \mathbb{R}^4$ $m^{C''} \mapsto T_j^{CC'}(L_1^{C''}) \vee_C T_j^{CC'}(L_2^{C''})$	$(t_{ijk}^{C'})_{i,k=1..3}$
$t_j^{C''C'}$	$C \wedge \mathcal{G}_4^2 \longrightarrow C'' \wedge \mathcal{G}_4^2$ $L^C \mapsto T_j^{C''C'}(m_1^C) \wedge_{C''} T_j^{C''C'}(m_2^C)$	$(t_{ijk}^{C'})_{i,k=1..3}^T$
$t_k^{CC''}$	$C' \wedge \mathbb{R}^4 \longrightarrow C \wedge \mathbb{R}^4$ $m^{C'} \mapsto T_k^{CC''}(L_1^{C'}) \vee_C T_k^{CC''}(L_2^{C'})$	$(t_{ijk}^{C''})_{i,j=1..3}$
$t_k^{C'C''}$	$C \wedge \mathcal{G}_4^2 \longrightarrow C' \wedge \mathcal{G}_4^2$ $L^C \mapsto T_k^{C'C''}(m_1^C) \wedge_{C'} T_k^{C'C''}(m_2^C)$	$(t_{ijk}^{C''})_{i,j=1..3}^T$

$$
\begin{aligned}
u_{i_1 i_2 j_1 j_2}^{C''C} &= \sum_{k=1}^{3} t_{i_1 j_1 k}^{C} T_{i_2 j_2 k}, \\
u_{i_1 i_2 j_1 j_2}^{C''C'} &= \sum_{k=1}^{3} t_{i_1 j_1 k}^{C'} T_{i_2 j_2 k}, \\
u_{i_1 i_2 k_1 k_2}^{C'C} &= \sum_{j=1}^{3} t_{i_1 j k_1}^{C} T_{i_2 j k_2}, \\
u_{i_1 i_2 k_1 k_2}^{C'C''} &= \sum_{j=1}^{3} t_{i_1 j k_1}^{C''} T_{i_2 j k_2}, \\
u_{j_1 j_2 k_1 k_2}^{CC'} &= \sum_{i=1}^{3} t_{i j_1 k_1}^{C'} T_{i j_2 k_2}, \\
u_{j_1 j_2 k_1 k_2}^{CC''} &= \sum_{i=1}^{3} t_{i j_1 k_1}^{C''} T_{i j_2 k_2}.
\end{aligned}
\tag{17.96}
$$

Then

$$\begin{aligned}
u^{C''C}_{i_1i_2j_1j_2}C'' &= t^{C''C}_{i_1}(C' \wedge e^{C'}_{j_1}) \vee T^{C''C}_{i_2}(C' \wedge \hat{e}^{C'}_{j_2}),\\
u^{C''C'}_{i_1i_2j_1j_2}C'' &= t^{C''C'}_{j_1}(C \wedge \hat{e}^{C}_{i_1}) \vee T^{C''C'}_{j_2}(C \wedge e^{C}_{i_2}),\\
u^{C'C}_{i_1i_2k_1k_2}C' &= t^{C'C}_{i_1}(C'' \wedge e^{C''}_{k_1}) \vee T^{C'C}_{i_2}(C'' \wedge \hat{e}^{C''}_{k_2}),\\
u^{C'C''}_{i_1i_2k_1k_2}C' &= t^{C'C''}_{k_1}(C \wedge \hat{e}^{C}_{i_1}) \vee T^{C'C''}_{k_2}(C \wedge e^{C}_{i_2}),\\
u^{CC'}_{j_1j_2k_1k_2}C &= t^{CC'}_{j_1}(C'' \wedge e^{C''}_{k_1}) \vee T^{CC'}_{j_2}(C'' \wedge \hat{e}^{C''}_{k_2}),\\
u^{CC''}_{j_1j_2k_1k_2}C &= t^{CC''}_{k_1}(C' \wedge e^{C'}_{j_1}) \vee T^{CC''}_{k_2}(C' \wedge \hat{e}^{C'}_{j_2}).
\end{aligned} \tag{17.97}$$

Expanding the right-hand side of the above equalities, we can get a factored form of the u's, from which we get the following constraints.

Constraints from $u^{C''C}_{i_1i_2j_1j_2}$: (see also subsection 17.5.5)

$$u^{C''C}_{i_1i_2j_1j_2} = \begin{cases} 0, \text{ if } i_1 = i_2; \\ -F^{CC'}_{i_1j_1}E^{C'C}_{j_2}E^{CC''}_{i}, \text{ if } i \prec i_1 \prec i_2; \\ F^{CC'}_{i_1j_1}E^{C'C}_{j_2}E^{CC''}_{i}, \text{ if } i \prec i_2 \prec i_1. \end{cases} \tag{17.98}$$

Two constraints can be obtained from $u^{C''C}_{i_1i_2j_1j_2}$:

1. For any $1 \le i_l, j_l \le 3$, where $1 \le l \le 4$,

$$\frac{u^{C''C}_{i_1i_2j_1j_2}}{u^{C''C}_{i_1i_2j_1j_3}} = \frac{u^{C''C}_{i_3i_4j_4j_2}}{u^{C''C}_{i_3i_4j_4j_3}}. \tag{17.99}$$

2. Let $i \prec i_1 \prec i_2$. Then for any $1 \le j_l \le 3$ where $1 \le l \le 4$,

$$\frac{u^{C''C}_{i_1i_2j_1j_2}}{u^{C''C}_{i_1ij_1j_2}} = \frac{u^{C''C}_{i_1i_2j_3j_4}}{u^{C''C}_{i_1ij_3j_4}}. \tag{17.100}$$

Define $U^{C''C} : (C \wedge \mathbb{R}^4) \times (C \wedge \mathbb{R}^4) \times (C' \wedge \mathbb{R}^4) \times (C' \wedge \mathcal{G}^2_4) \longrightarrow \mathbb{R}$ by

$$\begin{aligned} U^{C''C}(m^C_1, m^C_2, m^{C'}, L^{C'}) = \\ -(m^C_1 \vee m^{C'})(C \vee L^{C'})(C'' \vee (m^C_1 \wedge_C m^C_2)). \end{aligned} \tag{17.101}$$

Then

$$U^{C''C}(C \wedge e^C_{i_1}, C \wedge e^C_{i_2}, C' \wedge e^{C'}_{j_1}, C' \wedge \hat{e}^{C'}_{j_2}) = u^{C''C}_{i_1i_2j_1j_2}. \tag{17.102}$$

Constraints from $u^{C''C'}_{i_1i_2j_1j_2}$: If $i_1 \ne i_2$, then

$$u^{C''C'}_{i_1i_2j_1j_2} = \begin{cases} 0, \text{ if } j_1 = j_2; \\ E^{C'C}_{j_1} E^{CC''}_{i_1} F^{CC'}_{i_2j}, \text{ if } j \prec j_1 \prec j_2; \\ -E^{C'C}_{j_1} E^{CC''}_{i_1} F^{CC'}_{i_2j}, \text{ if } j \prec j_2 \prec j_1. \end{cases} \quad (17.103)$$

Two constraints can be obtained from $u^{C''C'}_{i_1i_2j_1j_2}$:

1. Let $i_1 \neq i_2$, $i_3 \neq i_4$. Then for any $1 \leq j_1, j_2 \leq 3$,

$$\frac{u^{C''C'}_{i_1i_2j_1j_2}}{u^{C''C'}_{i_1i_2j_2j_1}} = \frac{u^{C''C'}_{i_3i_4j_1j_2}}{u^{C''C'}_{i_3i_4j_2j_1}}. \quad (17.104)$$

2. Let $i \prec i_1 \prec i_2$. Then for any $1 \leq j_l \leq 3$ where $1 \leq l \leq 4$,

$$\frac{u^{C''C'}_{i_1i_2j_1j_2}}{u^{C''C'}_{ii_2j_1j_2}} = \frac{u^{C''C'}_{i_1i_2j_3j_4}}{u^{C''C'}_{ii_2j_3j_4}}. \quad (17.105)$$

Define $U^{C''C'} : (C \wedge \mathcal{G}_4^2) \times (C \wedge \mathbb{R}^4) \times (C' \wedge \mathcal{G}_4^2) \times (C' \wedge \mathcal{G}_4^2) \longrightarrow \mathbb{R}$ by

$$U^{C''C'}(L^C, m^C, L_1^{C'}, L_2^{C'}) = (L^C \vee C'')(L_1^{C'} \vee C)(m^C \vee L_1^{C'} \vee L_2^{C'}). \quad (17.106)$$

When $i_1 \neq i_2$,

$$U^{C''C'}(C \wedge \hat{e}^C_{i_1}, C \wedge e^C_{i_2}, C' \wedge \hat{e}^{C'}_{j_1}, C' \wedge \hat{e}^{C'}_{j_2}) = u^{C''C'}_{i_1i_2j_1j_2}. \quad (17.107)$$

Constraints from $u^{C'C}_{i_1i_2k_1k_2}$:

$$u^{C'C}_{i_1i_2k_1k_2} = \begin{cases} 0, \text{ if } i_1 = i_2; \\ -E^{C''C}_{k_2} E^{CC'}_{i} F^{CC''}_{i_1k_1}, \text{ if } i \prec i_1 \prec i_2; \\ E^{C''C}_{k_2} E^{CC'}_{i} F^{CC''}_{i_1k_1}, \text{ if } i \prec i_2 \prec i_1. \end{cases} \quad (17.108)$$

Two constraints can be obtained from $u^{C'C}_{i_1i_2k_1k_2}$:

1. For any $1 \leq i_l, k_l \leq 3$ where $1 \leq l \leq 4$,

$$\frac{u^{C'C}_{i_1i_2k_1k_2}}{u^{C'C}_{i_1i_2k_1k_3}} = \frac{u^{C'C}_{i_3i_4k_4k_2}}{u^{C'C}_{i_3i_4k_4k_3}}. \quad (17.109)$$

2. Let $i \prec i_1 \prec i_2$. Then for any $1 \leq k_l \leq 3$ where $1 \leq l \leq 4$,

$$\frac{u^{C'C}_{i_1i_2k_1k_2}}{u^{C'C}_{i_1ik_1k_2}} = \frac{u^{C'C}_{i_1i_2k_3k_4}}{u^{C'C}_{i_1ik_3k_4}}. \quad (17.110)$$

Define $U^{C'C} : (C \wedge \mathbb{R}^4) \times (C \wedge \mathbb{R}^4) \times (C'' \wedge \mathbb{R}^4) \times (C'' \wedge \mathcal{G}_4^2) \longrightarrow \mathbb{R}$ by

$$\begin{aligned} U^{C'C}(m_1^C, m_2^C, m^{C''}, L^{C''}) = \\ -(m_1^C \vee m^{C''})(C \vee L^{C''})(C' \vee (m_1^C \wedge_C m_2^C)). \end{aligned} \tag{17.111}$$

Then

$$U^{C'C}(C \wedge e_{i_1}^C, C \wedge e_{i_2}^C, C'' \wedge e_{k_1}^{C''}, C'' \wedge \hat{e}_{k_2}^{C''}) = u_{i_1 i_2 k_1 k_2}^{C'C}. \tag{17.112}$$

Constraints from $u_{i_1 i_2 k_1 k_2}^{C'C''}$: If $i_1 \neq i_2$, then

$$u_{i_1 i_2 k_1 k_2}^{C'C''} = \begin{cases} 0, \text{ if } k_1 = k_2; \\ E_{k_1}^{C''C} E_{i_1}^{CC'} F_{i_2 k}^{CC''}, \text{ if } k \prec k_1 \prec k_2; \\ -E_{k_1}^{C''C} E_{i_1}^{CC'} F_{i_2 k}^{CC''}, \text{ if } k \prec k_2 \prec k_1. \end{cases} \tag{17.113}$$

Two constraints can be obtained from $u_{i_1 i_2 k_1 k_2}^{C'C''}$:

1. Let $i_1 \neq i_2$ and $i_3 \neq i_4$. Then for any $1 \leq k_1, k_2 \leq 3$,

$$\frac{u_{i_1 i_2 k_1 k_2}^{C'C''}}{u_{i_1 i_2 k_2 k_1}^{C'C''}} = \frac{u_{i_3 i_4 k_1 k_2}^{C'C''}}{u_{i_3 i_4 k_2 k_1}^{C'C''}}. \tag{17.114}$$

2. Let $i \prec i_1 \prec i_2$. Then for any $1 \leq k_l \leq 3$ where $1 \leq l \leq 4$,

$$\frac{u_{i_1 i_2 k_1 k_2}^{C'C''}}{u_{i i_2 k_1 k_2}^{C'C''}} = \frac{u_{i_1 i_2 k_3 k_4}^{C'C''}}{u_{i i_2 k_3 k_4}^{C'C''}}. \tag{17.115}$$

Define $U^{C'C''} : (C \wedge \mathcal{G}_4^2) \times (C \wedge \mathbb{R}^4) \times (C'' \wedge \mathcal{G}_4^2) \times (C'' \wedge \mathcal{G}_4^2) \longrightarrow \mathbb{R}$ by

$$U^{C'C''}(L^C, m^C, L_1^{C''}, L_2^{C''}) = (L^C \vee C'')(L_1^{C''} \vee C)(m^C \vee L_1^{C''} \vee L_2^{C''}). \tag{17.116}$$

When $i_1 \neq i_2$,

$$U^{C'C''}(C \wedge \hat{e}_{i_1}^C, C \wedge e_{i_2}^C, C'' \wedge \hat{e}_{k_1}^{C''}, C'' \wedge \hat{e}_{k_2}^{C''}) = u_{i_1 i_2 k_1 k_2}^{C'C''}. \tag{17.117}$$

Constraints from $u_{j_1 j_2 k_1 k_2}^{CC'}$: If $k_1 \neq k_2$, then

$$u_{j_1 j_2 k_1 k_2}^{CC'} = \begin{cases} 0, \text{ if } j_1 = j_2; \\ -E_{j_1}^{C'C} E_{k_2}^{C''C} F_{j k_1}^{C'C''}, \text{ if } j \prec j_1 \prec j_2; \\ E_{j_1}^{C'C} E_{k_2}^{C''C} F_{j k_1}^{C'C''}, \text{ if } j \prec j_2 \prec j_1. \end{cases} \tag{17.118}$$

Two constraints can be obtained from $u_{j_1 j_2 k_1 k_2}^{CC'}$:

1. Let $k_1 \neq k_2$ and $k_3 \neq k_4$. Then for any $1 \leq j_1, j_2 \leq 3$,

$$\frac{u^{CC'}_{j_1j_2k_1k_2}}{u^{CC'}_{j_2j_1k_1k_2}} = \frac{u^{CC'}_{j_1j_2k_3k_4}}{u^{CC'}_{j_2j_1k_3k_4}}. \tag{17.119}$$

2. Let $k \prec k_1 \prec k_2$, then for any $1 \leq j_l \leq 3$ where $1 \leq l \leq 4$,

$$\frac{u^{CC'}_{j_1j_2k_1k_2}}{u^{CC'}_{j_1j_2k_1k}} = \frac{u^{CC'}_{j_3j_4k_1k_2}}{u^{CC'}_{j_3j_4k_1k}}. \tag{17.120}$$

Define $U^{CC'} : (C' \wedge \mathcal{G}_4^2) \times (C' \wedge \mathcal{G}_4^2) \times (C'' \wedge \mathbb{R}^4) \times (C'' \wedge \mathcal{G}_4^2) \longrightarrow \mathbb{R}$ by

$$\begin{aligned} &U^{CC'}(L_1^{C'}, L_2^{C'}, m^{C''}, L^{C''}) = \\ &\quad -(L_1^{C'} \vee C)(L^{C''} \vee C)(L_1^{C'} \vee L_2^{C'} \vee m^{C''}). \end{aligned} \tag{17.121}$$

When $k_1 \neq k_2$,

$$U^{CC'}(C' \wedge \hat{e}_{j_1}^{C'}, C' \wedge \hat{e}_{j_2}^{C'}, C'' \wedge e_{k_1}^{C''}, C'' \wedge \hat{e}_{k_2}^{C''}) = u^{CC'}_{j_1j_2k_1k_2}. \tag{17.122}$$

Constraints from $u^{CC''}_{j_1j_2k_1k_2}$: If $j_1 \neq j_2$, then

$$u^{CC''}_{j_1j_2k_1k_2} = \begin{cases} 0, \text{ if } k_1 = k_2; \\ E_{j_2}^{C'C} E_{k_1}^{C''C} F_{j_1k}^{C'C''}, \text{ if } k \prec k_1 \prec k_2; \\ -E_{j_2}^{C'C} E_{k_1}^{C''C} F_{j_1k}^{C'C''}, \text{ if } k \prec k_2 \prec k_1. \end{cases} \tag{17.123}$$

Two constraints can be obtained from $u^{CC''}_{j_1j_2k_1k_2}$:

1. Let $j_1 \neq j_2$ and $j_3 \neq j_4$. Then for any $1 \leq k_1, k_2 \leq 3$,

$$\frac{u^{CC''}_{j_1j_2k_1k_2}}{u^{CC''}_{j_1j_2k_2k_1}} = \frac{u^{CC''}_{j_3j_4k_1k_2}}{u^{CC''}_{j_3j_4k_2k_1}}. \tag{17.124}$$

2. Let $j \prec j_1 \prec j_2$, then for any $1 \leq k_l \leq 3$ where $1 \leq l \leq 4$,

$$\frac{u^{CC''}_{j_1j_2k_1k_2}}{u^{CC''}_{j_1jk_1k_2}} = \frac{u^{CC''}_{j_1j_2k_3k_4}}{u^{CC''}_{j_1jk_3k_4}}. \tag{17.125}$$

Define $U^{CC''} : (C' \wedge \mathbb{R}^4) \times (C' \wedge \mathcal{G}_4^2) \times (C'' \wedge \mathcal{G}_4^2) \times (C'' \wedge \mathcal{G}_4^2) \longrightarrow \mathbb{R}$ by

$$\begin{aligned} &U^{CC''}(m^{C'}, L^{C'}, L_1^{C''}, L_2^{C''}) = (L_1^{C''} \vee C)(L^{C'} \vee C) \\ &\quad (m^{C'} \vee L_1^{C''} \vee L_2^{C''}). \end{aligned} \tag{17.126}$$

When $j_1 \neq j_2$,

$$U^{CC''}(C' \wedge e_{j_1}^{C'}, C' \wedge \hat{e}_{j_2}^{C'}, C'' \wedge \hat{e}_{k_1}^{C''}, C'' \wedge \hat{e}_{k_2}^{C''}) = u^{CC''}_{j_1j_2k_1k_2}. \tag{17.127}$$

We have

$$
\begin{aligned}
&V^{C''}(m_1^C, m_2^C, L_1^{C'}, L_2^{C'}) \\
&= \begin{cases} U^{C''C}(m_1^C, m_2^C, L_1^{C'} \vee L_2^{C'}, L_2^{C'}), \text{ if } L_1^{C'} \vee L_2^{C'} \neq 0; \\ -U^{C''C'}(m_1^C \wedge_C m_2^C, m_1^C, L_2^{C'}, L_1^{C'}), \text{ if } m_1^C \vee m_2^C \neq 0. \end{cases}
\end{aligned} \tag{17.128}
$$

Thus

$$
v_{i_1 i_2 j_1 j_2}^{C''} = \begin{cases} u_{i_1 i_2 j j_2}^{C''C}, \text{ if } j \prec j_1 \prec j_2; \\ -u_{i_1 i_2 j j_2}^{C''C}, \text{ if } j \prec j_2 \prec j_1; \\ -u_{i i_1 j_2 j_1}^{C''C'}, \text{ if } i \prec i_1 \prec i_2; \\ u_{i i_1 j_2 j_1}^{C''C'}, \text{ if } i \prec i_2 \prec i_1. \end{cases} \tag{17.129}
$$

Comparing these constraints, we find that the constraints (17.65), (17.66) from $V^{C''}$ are equivalent to the constraints (17.104), (17.105) from $U^{C''C'}$, and are included in the constraints (17.99), (17.100) from $U^{C''C}$. Faugeras and Mourrain's first group of constraints is a special case of any of (17.65), (17.104) and (17.99). Similarly, Faugeras and Mourrain's second group of constraints is a special case of any of (17.109), (17.114).

We also have

$$
\begin{aligned}
&V^{C}(L_1^{C'}, L_2^{C'}, L_1^{C''}, L_2^{C''}) \\
&= \begin{cases} U^{CC'}(L_1^{C'}, L_2^{C'}, L_1^{C''} \vee L_2^{C''}, L_2^{C''}), \text{ if } L_1^{C''} \vee L_2^{C''} \neq 0; \\ -U^{CC''}(L_1^{C'} \wedge_C L_2^{C'}, L_1^{C'}, L_2^{C''}, L_1^{C''}), \text{ if } L_1^{C'} \vee L_2^{C'} \neq 0. \end{cases}
\end{aligned} \tag{17.130}
$$

Thus

$$
v_{j_1 j_2 k_1 k_2}^{C} = \begin{cases} u_{j j_1 k_2 k_1}^{CC''}, \text{ if } j \prec j_1 \prec j_2; \\ -u_{j j_1 k_2 k_1}^{CC''}, \text{ if } j \prec j_2 \prec j_1; \\ -u_{j_1 j_2 k k_2}^{CC'}, \text{ if } k \prec k_1 \prec k_2; \\ u_{j_1 j_2 k k_2}^{CC'}, \text{ if } k \prec k_2 \prec k_1. \end{cases} \tag{17.131}
$$

The constraints (17.84), (17.85) from V^C are equivalent to the constraints (17.124), (17.125) from $U^{C''C'}$, and are also equivalent to the constraints (17.119), (17.120) from $U^{C''C}$. Faugeras and Mourrain's third group of constraints is a special case of any of (17.84), (17.124) and (17.119).

17.6 Conclusion

In this chapter we propose a new algebraic representation for image points obtained from a pinhole camera, based on Hestenes and Ziegler's idea of projective split. We reformulate camera modeling and calibration, epipolar and trifocal geometries with this new representation. We also propose a systematic approach to derive constraints on epipolar and trifocal tensors, by which we have not only derived all known constraints, but also made considerable generalizations.

18. The Geometry and Algebra of Kinematics*

Eduardo Bayro-Corrochano

Institute of Computer Science and Applied Mathematics,
Christian-Albrechts-University of Kiel

18.1 Introduction

This chapter presents the geometric algebra framework for dealing with 3D kinematics. The reader will see the usefulness of this mathematical approach with respect to the modelling of the motion of points, lines and planes. Chapters 19 and 21 illustrate the application of this mathematical system for the direct and inverse kinematics of robot manipulators and for the problem of estimating line motions, respectively.

In the literature we find diverse approaches for representing the kinematics. The foundations of the screw theory can be traced back to the contributions of Chasles and Poinsot in the early 1830s, see e.g. [12]. The dual quaternions were introduced by Clifford in his seminal paper *Preliminary sketch of bi-quaternions* [45]. Later on Study [223] utilized the dual numbers to represent the relative position of two skew lines in space. Selig [207] and McCarthy [167] studied planar manipulators, the former using matrices and the later using coplanar quaternions. Rooney [193] compared matrices and dual numbers approaches for 2D and 3D kinematics. Murray et al. [173] use the twist or infinitesimal generator of the Euclidean group as a Lie algebra matrix approach to describe rigid 3D motions. The work of Chevalier [41]

* This work has been supported by DFG Grant So-320-2-1.

who presented a geometrical formulation of the dual quaternions in the Lie algebra framework is also worth mentioning.

In the area of robotics Gu and Luh [95] used the dual–number transformation for the treatment of the manipulator kinematics and Pennock and Yang [183] presented closed–form solutions for the inverse kinematics problem for various types of robot manipulators employing dual matrices. Funda, and Paul [85] carried out a computational analysis of screw transformations in robotics. They showed that the dual quaternions represent simultaneously the rotation and the translation transformations for dealing with the kinematics of robot chains more efficiently than any other approach. McCarthy [166, 167] analyzed multi–links and similarly to Gu and Luh [95] he computed the dual form of the Jacobian of a manipulator using again dual orthogonal matrices. Kim and Kumar [128] applied the dual quaternion formalism as a line transformation operator and solved the inverse kinematics of a six degree of freedom robot manipulator. White [237] used Grassman–Cayley algebra for analyzing critical configurations of robot manipulators in 3D projective space. Aspragathos and Dimitros [9] confirmed that the homogeneous transformation is the approach commonly used in robotics tasks and that the approaches of dual quaternion and Lie algebra allow the reduction of the number of representation parameters.

Interesting approaches for 3D kinematics have been also developed in the area of visual robotics. For the case of the so called hand–eye calibration problem several authors considered computation descriptions in terms of the rotation axis and angle [211, 231], the use of quaternions [43] and a canonical matrix representation [149]. Using the matrix screw theory Chen [39] found a key invariant of the screw between two 3D axes, namely that the rotation angle and the translation along the screw axis remain constant. Dual quaternions or motors were used for the linearization of the 3D Euclidean transformation for solving the hand–eye calibration problem, see [55]. In other applications authors applied successfully dual quaternions like Walker [236] for estimating 3D location, twists and exponential maps, as well as Bregler and Malik [31] for tracking the kinematic chains of moving persons.

The review shows two key aspects: the use of dual numbers and the representation of screw transformations in terms of matrices or quaternions. In this regard when we use geometric algebra for kinematics we should consider isomorphic dual representations avoiding redundant entries as in the case of the matrix representations.

The chapter is organized as follows: after a brief literature review in the introduction part, section two describes how a 3D geometric algebra can be used to represent rotations in 3D space using rotors. This concept of representing transformations by elements of a geometric algebra is extended to translations employing a 4D geometric algebra, i.e. the algebra of motors, in section three. Section four describes how points, lines and planes in 3D space can be represented in the 3D and 4D geometric algebras introduced in sections two and three, respectively. Finally the modelling of the motion of the geometric entities point, line and plane within these geometric algebras is explained in section five. The chapter ends with the conclusions section six.

18.2 The Euclidean 3D Geometric Algebra

In the case of modelling the Euclidean 3D space we choose the geometric algebra $\mathcal{G}_{3,0,0}$ which has dimension $2^3 = 8$. A basis of $\mathcal{G}_{3,0,0}$ is given by:

$$\underbrace{1}_{scalar},\ \underbrace{\{\sigma_1, \sigma_2, \sigma_3\}}_{vectors},\ \underbrace{\{\sigma_1\sigma_2, \sigma_2\sigma_3, \sigma_3\sigma_1\}}_{bivectors},\ \underbrace{\sigma_1\sigma_2\sigma_3 \equiv I}_{trivector}\ . \tag{18.1}$$

The highest grade element in 3D space is a trivector and is called unit pseudoscalar $I \equiv \sigma_1\sigma_2\sigma_3$ which squares to -1 and commutes with the scalars and bivectors in 3D space. In the space of 3 dimensions we can construct an arbitrary trivector $\boldsymbol{a}\wedge\boldsymbol{b}\wedge\boldsymbol{c} = \lambda I$, where the points are in general position and $\lambda \in \mathbb{R}$.

18.2.1 3D Rotors

Multiplication of the three basis vectors σ_1, σ_2, and σ_3 by I results in the three basis bivectors $\sigma_1\sigma_2 = I\sigma_3$, $\sigma_2\sigma_3 = I\sigma_1$ and $\sigma_3\sigma_1 = I\sigma_2$. These simple bivectors rotate vectors in their own plane by 90^{*3}, e.g. $(\sigma_1\sigma_2)\sigma_2 = \sigma_1$, $(\sigma_2\sigma_3)\sigma_2 = -\sigma_3$ etc. Identifying the $\boldsymbol{i}$, $\boldsymbol{j}$, $\boldsymbol{k}$ of the quaternion algebra with $I\sigma_1, -I\sigma_2, I\sigma_3$ the famous Hamilton relations $\boldsymbol{i}^2 = \boldsymbol{j}^2 = \boldsymbol{k}^2 = \boldsymbol{ijk} = -1$ can be recovered. Since the $\boldsymbol{i}, \boldsymbol{j}, \boldsymbol{k}$ are bivectors, it comes as no surprise that they represent 90^{*3} rotations in orthogonal directions and provide a well-suited system for the representation of general 3D rotations, see Figure 18.1.

In geometric algebra a rotor (short name for rotator), $\boldsymbol{R}$, is an even-grade element of the algebra which satisfies $\boldsymbol{R}\widetilde{\boldsymbol{R}} = 1$, where $\widetilde{\boldsymbol{R}}$ stands for the conjugate of $\boldsymbol{R}$. If $\mathcal{A} = \{a_0, a_1, a_2, a_3\} \in \mathcal{G}_{3,0,0}$ represents a unit quaternion, then the rotor which performs the same rotation is simply given by

$$\boldsymbol{R} = \underbrace{a_0}_{scalar} + \underbrace{a_1(I\sigma_1) - a_2(I\sigma_2) + a_3(I\sigma_3)}_{bivectors}\ . \tag{18.2}$$

The quaternion algebra is therefore seen to be a subset of the geometric algebra of 3-space.

A rotation can be performed by a pair of reflections, see Figure 18.1. It can easily be shown that the result of reflecting a vector $\boldsymbol{a}$ in the plane perpendicular to a unit vector $\boldsymbol{n}$ is $\boldsymbol{a}_\perp - \boldsymbol{a}_\| = -\boldsymbol{nan}$, where $\boldsymbol{a}_\perp$ and $\boldsymbol{a}_\|$ respectively denote projections of $\boldsymbol{a}$ perpendicular and parallel to $\boldsymbol{n}$. Thus, a reflection of $\boldsymbol{a}$ in the plane perpendicular to $\boldsymbol{n}$, followed by a reflection in the plane perpendicular to another unit vector $\boldsymbol{m}$ results in a new vector

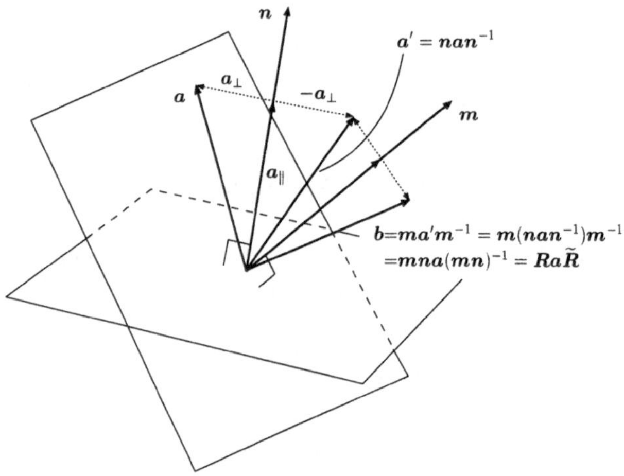

Fig. 18.1. The rotor in the 3D space formed by a pair of reflections

$\boldsymbol{b} = -\boldsymbol{m}(-\boldsymbol{nan})\boldsymbol{m} = (\boldsymbol{mn})\boldsymbol{a}(\boldsymbol{nm}) = \boldsymbol{Ra}\widetilde{\boldsymbol{R}}$. Using the geometric product we can show that the rotor $\boldsymbol{R}$ of equation (18.2) is a multivector consisting of both a scalar part and a bivector part, i.e. $\boldsymbol{R} = \boldsymbol{mn} = \boldsymbol{m}\cdot\boldsymbol{n} + \boldsymbol{m}\wedge\boldsymbol{n}$. These components correspond to the scalar and vector parts of an equivalent unit quaternion in $\mathcal{G}_{3,0,0}$. Considering the scalar and the bivector parts, we can further write the Euler representation of a rotor as follows

$$\boldsymbol{R} = e^{\boldsymbol{n}\frac{\theta}{2}} = cos\frac{\theta}{2} + \boldsymbol{n}sin\frac{\theta}{2}\,, \tag{18.3}$$

where the rotation axis $\boldsymbol{n} = n_1\sigma_2\sigma_3 + n_2\sigma_3\sigma_1 + n_3\sigma_1\sigma_2$ is spanned by the bivector basis.

The transformation in terms of a rotor $\boldsymbol{a} \mapsto \boldsymbol{Ra}\widetilde{\boldsymbol{R}} = \boldsymbol{b}$ is a very general way of handling rotations; it works for multivectors of any grade and in spaces of any dimension in contrast to quaternion calculus. Rotors combine in a straightforward manner, i.e. a rotor $\boldsymbol{R}_1$ followed by a rotor $\boldsymbol{R}_2$ is equivalent to a total rotor $\boldsymbol{R}$ where $\boldsymbol{R} = \boldsymbol{R}_2\boldsymbol{R}_1$.

18.3 The 4D Geometric Algebra for 3D Kinematics

In the case of 3D rigid motion or Euclidean transformation we are confronted with a non–linear mapping. However, if we employ a 4D geometric algebra we will linearize the 3D rigid motion of the 3D Euclidean space. Instead of using homogeneous coordinates we will embed the Euclidean transformation into a degenerate 4D geometric algebra. That is why we choose three basis vectors which square to one and one to zero to provide dual copies of the multivectors of the 3D space. In other words, we extend the Euclidean geometric algebra $\mathcal{G}_{3,0,0}$ to the special or degenerate geometric algebra $\mathcal{G}_{3,0,1}$ which is spanned via following basis

$$\underbrace{1}_{scalar}, \underbrace{\{\gamma_k\}}_{4\ vectors}, \underbrace{\begin{matrix}\{\gamma_2\gamma_3, \gamma_3\gamma_1, \gamma_1\gamma_2, \\ \gamma_4\gamma_1, \gamma_4\gamma_2, \gamma_4\gamma_3\}\end{matrix}}_{6\ bivectors}, \underbrace{\{I\gamma_k\}}_{4\ pseudovectors}, \underbrace{\gamma_1\gamma_2\gamma_3\gamma_4 \equiv I}_{unit\ pseudoscalar}, \quad (18.4)$$

with $k = 1, 2, 3, 4$ and $\gamma_4^2 = 0$ and $\gamma_i^2 = +1$ for $i = 1, 2, 3$. The unit pseudoscalar squares to zero, i.e. $I^2 = 0$.

18.3.1 The Motor Algebra

Clifford introduced the motors with the name bi-quaternions [45]. The word motor is an abbreviation of "moment and vector". Motors are isomorphic to the dual quaternions with the necessary condition of $I^2 = 0$. They can be found in the special 4D even subalgebra of $\mathcal{G}_{3,0,1}$ introduced in the previous section. This even subalgebra will be denominated by $\mathcal{G}^+_{3,0,1}$ and is spanned only via a bivector basis as follows

$$\underbrace{1}_{scalar}, \underbrace{\{\gamma_2\gamma_3, \gamma_3\gamma_1, \gamma_1\gamma_2, \gamma_4\gamma_1, \gamma_4\gamma_2, \gamma_4\gamma_3\}}_{6\ bivectors}, \underbrace{I}_{unit\ pseudoscalar}. \quad (18.5)$$

Note that the bivector basis corresponds to the same basis for spanning 3D lines. Note also that the dual of a scalar is a pseudoscalar and the duals of the first three basis bivectors are the next three ones, that is for example the dual of $\gamma_2\gamma_3$ is $I\gamma_2\gamma_3$ or $\gamma_4\gamma_1$.

According to Clifford [45] a basic geometric interpretation of a motor can be seen as the necessary operation to convert the rotation line axis of one rotor into another one. Each rotor can be geometrically represented as a rotation plane with a rotation axis normal to this plane. Thus, one rotor can be spanned by the bivector basis $\gamma_2\gamma_3$, $\gamma_3\gamma_1$, $\gamma_1\gamma_2$ and the dual one by $\gamma_4\gamma_1$, $\gamma_4\gamma_2$, $\gamma_4\gamma_3$. Figure 18.2 depicts a detailed motor action where the rotor axis is now considered as a rotation line. Let us first turn the orientation of the axis of one rotor, i.e. $\boldsymbol{R}_a$, parallel to the other one, i.e. $\boldsymbol{R}_b$, by applying the rotor $\boldsymbol{R}_s$. Then slide it the distance d along the connecting axis into the position of the axis of the second rotor. These operations can be seen together as forming a screw with the line axis $\boldsymbol{l}$ and with the relation called pitch which equals to $\frac{d}{\theta}$ for $\theta \neq 0$. We said in the last section that a rotor relates two vectors, now in the case of a motor it relates the rotation axes of two rotors. A motor is specified only by its direction and position of the screw axis line, twist angular magnitude and pitch.

18.3.2 Motors, Rotors, and Translators

Since a rigid motion consists of rotation and translation, it should be possible to split a motor multiplicatively in terms of these two transformations which we will call a rotor and a translator.

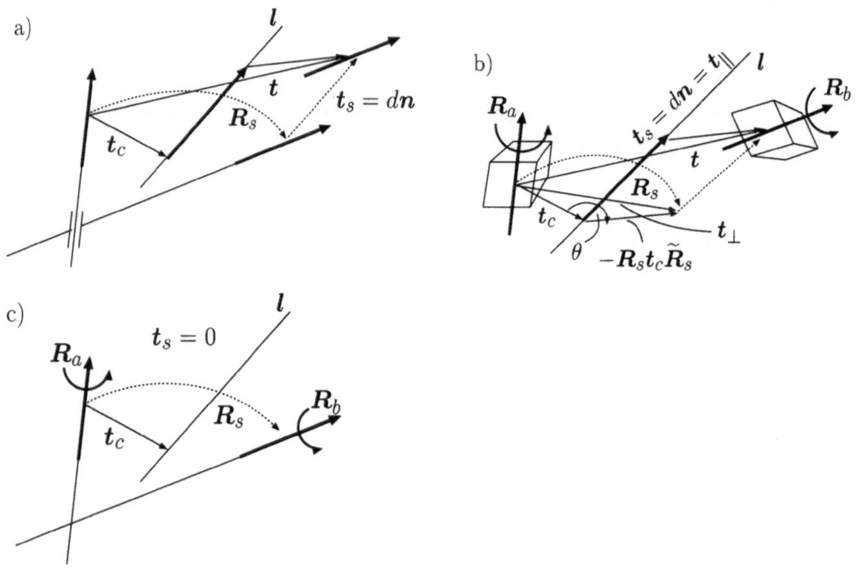

Fig. 18.2. Screw motion about the line axis $\boldsymbol{l}$ ($\boldsymbol{t}_s$: longitudinal displacement in d and rotation in θ) a) the motor relating two axis lines b) motor applied to an object c) degenerate motor relating two coplanar rotors

Let us now express this procedure algebraically. First of all let us consider a simple rotor in its Euler representation for a rotation with angle θ,

$$\begin{aligned} \boldsymbol{R} &= a_0 + a_1\gamma_2\gamma_3 + a_2\gamma_3\gamma_1 + a_3\gamma_1\gamma_2 \\ &= a_0 + \boldsymbol{a} \\ &= cos(\frac{\theta}{2}) + sin(\frac{\theta}{2})\boldsymbol{n} \\ &= a_c + a_s\boldsymbol{n}\ , \end{aligned} \tag{18.6}$$

where $\boldsymbol{n}$ is the unit 3D vector of the rotation–axis spanned by the bivector basis $\gamma_2\gamma_3$, $\gamma_3\gamma_1$, $\gamma_1\gamma_2$ and a_c, $a_s \in \mathbb{R}$. Now dealing with the rotor of a screw motion the rotation axis vector should be represented as screw axis line. For that we have to relate this rotation axis to a reference coordinate system in the distance $\boldsymbol{t}_c$. A translation in 3D in the motor algebra is represented using the dual part of a motor called translator. Applying a translator from the left and its conjugated from the right to the rotor $\boldsymbol{R}$ we get

$$\begin{aligned}
\boldsymbol{R}_s &= \boldsymbol{T}_c\boldsymbol{R}\widetilde{\boldsymbol{T}}_c \\
&= (1+I\frac{\boldsymbol{t}_c}{2})(a_0+\boldsymbol{a})(1-I\frac{\boldsymbol{t}_c}{2}) \\
&= a_0+\boldsymbol{a}+Ia_0\frac{\boldsymbol{t}_c}{2}+I\frac{\boldsymbol{t}_c}{2}\boldsymbol{a}-Ia_0\frac{\boldsymbol{t}_c}{2}-I\boldsymbol{a}\frac{\boldsymbol{t}_c}{2} \\
&= a_0+\boldsymbol{a}+I(\frac{\boldsymbol{t}_c}{2}\boldsymbol{a}-\boldsymbol{a}\frac{\boldsymbol{t}_c}{2}) \\
&= a_0+\boldsymbol{a}+I(\boldsymbol{a}\wedge\boldsymbol{t}_c)\ . \qquad (18.7)
\end{aligned}$$

Here $\boldsymbol{t}_c$ is the 3D vector of the translation spanned by the bivector basis $\gamma_2\gamma_3$, $\gamma_3\gamma_1$, $\gamma_1\gamma_2$. Expressing the last equation in Euler terms we get

$$\begin{aligned}
\boldsymbol{R}_s &= a_0+a_s\boldsymbol{n}+Ia_s\boldsymbol{n}\wedge\boldsymbol{t}_c \\
&= a_c+a_s(\boldsymbol{n}+I\boldsymbol{m}) \\
&= cos(\frac{\theta}{2})+sin(\frac{\theta}{2})(\boldsymbol{n}+I\boldsymbol{m}) \\
&= cos(\frac{\theta}{2})+sin(\frac{\theta}{2})\boldsymbol{l}\ . \qquad (18.8)
\end{aligned}$$

This result is indeed interesting because the new entity called $\boldsymbol{R}_s$ is a rotor to be applied now with respect to an axis line $\boldsymbol{l}$ expressed in dual terms of direction $\boldsymbol{n}$ and moment $\boldsymbol{m}=\boldsymbol{n}\wedge\boldsymbol{t}_c$. Now to finally define the motor let us slide the distance $\boldsymbol{t}_s = d\boldsymbol{n}$ along the rotation axis line $\boldsymbol{l}$. Since a motor is applied from the left and its conjugated from the right we should use the half of $\boldsymbol{t}_s$ when we define the motor

$$\begin{aligned}
\boldsymbol{M} = \boldsymbol{T}_s\boldsymbol{R}_s &= (1+I\frac{\boldsymbol{t}_s}{2})(a_0+\boldsymbol{a}+I\boldsymbol{a}\wedge\boldsymbol{t}_c) \\
&= (1+I\frac{d\boldsymbol{n}}{2})(a_c+a_s\boldsymbol{n}+Ia_s\boldsymbol{n}\wedge\boldsymbol{t}_c) \\
&= a_c+a_s\boldsymbol{n}+Ia_s\boldsymbol{n}\wedge\boldsymbol{t}_c+I\frac{d}{2}a_c\boldsymbol{n}-I\frac{d}{2}a_s\boldsymbol{n}\boldsymbol{n} \\
&= (a_c-I\frac{d}{2}a_s)+(a_s+Ia_c\frac{d}{2})(\boldsymbol{n}+I\boldsymbol{n}\wedge\boldsymbol{t}_c) \\
&= (a_c-Ia_s\frac{d}{2})+(a_s+Ia_c\frac{d}{2})\boldsymbol{l}\ . \qquad (18.9)
\end{aligned}$$

Note that this expression of the motor makes explicit the unit line vector of the screw axis line $\boldsymbol{l}$.

Now let us express a motor as an Euler representation. Substituting the constants $a_c = cos(\frac{\theta}{2})$ and $a_s = sin(\frac{\theta}{2})$ in the motor equation (18.9) and using the Taylor series expansion of a differentiable real function $f:\mathbb{R}\mapsto\mathbb{R}$ with a dual argument $\alpha+I\beta$, where $\alpha,\beta\in\mathbb{R}$ and $0=I^2=I^3=\ldots$, i.e.

$$\begin{aligned}
f(\alpha+I\beta) &= f(\alpha)+If'(\alpha)\beta+I^2f''(\alpha)\frac{\beta^2}{2!}+\ldots \qquad (18.10)\\
&= f(\alpha)+If'(\alpha)\beta\ , \qquad (18.11)
\end{aligned}$$

we get

$$
\begin{aligned}
\boldsymbol{M} = \boldsymbol{T}_s\boldsymbol{R}_s &= \Big(cos(\frac{\theta}{2}) - Isin(\frac{\theta}{2})\frac{d}{2}\Big) + \Big(sin(\frac{\theta}{2}) + Icos(\frac{\theta}{2})\frac{d}{2}\Big)\boldsymbol{l} \\
&= cos(\frac{\theta}{2} + I\frac{d}{2}) + sin(\frac{\theta}{2} + I\frac{d}{2})\boldsymbol{l} \,. \qquad (18.12)
\end{aligned}
$$

Now we will analyze the obtained expressions

$$
\begin{aligned}
\boldsymbol{R} &= cos(\frac{\theta}{2}) + sin(\frac{\theta}{2})\boldsymbol{n} \\
\boldsymbol{R}_s &= cos(\frac{\theta}{2}) + sin(\frac{\theta}{2})\boldsymbol{l} \\
\boldsymbol{M} &= cos(\frac{\theta}{2} + I\frac{d}{2}) + sin(\frac{\theta}{2} + I\frac{d}{2})\boldsymbol{l} \,. \qquad (18.13)
\end{aligned}
$$

We can see how from a simple rotor $\boldsymbol{R}$ expressed in terms of an angle and the rotation axis $\boldsymbol{n}$, we change this axis to a rotation line axis $\boldsymbol{l}$ resulting $\boldsymbol{R}_s$. Finally, the motor information of the sliding distance d is made explicit in terms of dual arguments of the trigonometric functions. It is also nice to see that the expression for the motor simply extends the expression of $\boldsymbol{R}_s$ using dual angles instead.

If we expand the exponential function of the dual bivectors using a Taylor series, the result will follow the general expression $e^{\alpha+I\beta} = e^{\alpha} + Ie^{\alpha}\beta = e^{\alpha}(1 + I\beta)$ as a special case of equation (18.10). We get again the motor expression

$$
e^{\boldsymbol{l}(\frac{\theta}{2}+I\frac{\boldsymbol{t}_s}{2})} = (1 + I\frac{\boldsymbol{t}_s}{2})e^{\boldsymbol{l}\frac{\theta}{2}} = \boldsymbol{T}_s\boldsymbol{R}_s \,, \qquad (18.14)
$$

where $I\frac{\boldsymbol{t}_s}{2} = \frac{1}{2}I(t_{s_1}\sigma_2\sigma_3 + t_{s_2}\sigma_3\sigma_1 + t_{s_3}\sigma_1\sigma_2) = \frac{1}{2}(t_{s_1}\sigma_4\sigma_1 + t_{s_2}\sigma_4\sigma_2 + t_{s_3}\sigma_4\sigma_3)$.

If we want to express the motor using only rotors, we proceed as follows

$$
\begin{aligned}
\boldsymbol{M} = \boldsymbol{T}_s\boldsymbol{R}_s &= (1 + I\frac{\boldsymbol{t}_s}{2})\boldsymbol{R}_s \\
&= \boldsymbol{R}_s + I\frac{\boldsymbol{t}_s}{2}\boldsymbol{R}_s \,. \qquad (18.15)
\end{aligned}
$$

Let us consider in detail the dual part of the motor. This is the geometric product of the bivector $\boldsymbol{t}_s$ and the rotor $\boldsymbol{R}_s$. Since both are expressed in terms of the same bivector basis, their geometric product will be also expressed in this basis and this can be seen as a new rotor $\boldsymbol{R}'_s$. Thus, we can further write

$$
\boldsymbol{M} = \boldsymbol{R}_s + I\frac{\boldsymbol{t}_s}{2}\boldsymbol{R}_s = \boldsymbol{R}_s + I\boldsymbol{R}'_s \,. \qquad (18.16)
$$

In this equation the line axes of the rotors are skew ones, see Figure 18.2.a. That means that they represent the general case of non-coplanar rotors. If the sliding distance $\boldsymbol{t}_s$ is zero, then the motor will degenerate to a rotor

$$
\boldsymbol{M} = \boldsymbol{T}_s\boldsymbol{R}_s = (1 + I\frac{0}{2})\boldsymbol{R}_s = \boldsymbol{R}_s \,. \qquad (18.17)
$$

In this case the two axes lines of the rotors are coplanar, thus the motor is called a degenerate one, see Figure 18.2.c.

The bivector $\boldsymbol{t}_s$ can be expressed in terms of the rotors using previous results

$$\boldsymbol{R}'_s\widetilde{\boldsymbol{R}}_s = (\frac{\boldsymbol{t}_s}{2}\boldsymbol{R}_s)\widetilde{\boldsymbol{R}}_s \;, \tag{18.18}$$

therefore,

$$\boldsymbol{t}_s = 2\boldsymbol{R}'_s\widetilde{\boldsymbol{R}}_s \;. \tag{18.19}$$

Figure 18.2 shows that $\boldsymbol{t}$ is the 3D vector, expressed in the bivector basis, referred to the rotation axis of a rotor, and $\boldsymbol{t}_s$ is a bivector along the motor axis line. Thus, $\boldsymbol{t}$ considered here as a bivector can be computed in terms of the bivectors $\boldsymbol{t}_c$ and $\boldsymbol{t}_s$ as follows

$$\begin{aligned}
\boldsymbol{t} &= \boldsymbol{t}_\perp + \boldsymbol{t}_\parallel \\
\boldsymbol{t} &= (\boldsymbol{t}_c - \boldsymbol{R}_s\boldsymbol{t}_c\widetilde{\boldsymbol{R}}_s) + (\boldsymbol{t}\cdot\boldsymbol{n})\boldsymbol{n} = (\boldsymbol{t}_c - \boldsymbol{R}_s\boldsymbol{t}_c\widetilde{\boldsymbol{R}}_s) + d\boldsymbol{n} \\
&= \boldsymbol{t}_c - \boldsymbol{R}_s\boldsymbol{t}_c\widetilde{\boldsymbol{R}}_s + \boldsymbol{t}_s \\
&= \boldsymbol{t}_c - \boldsymbol{R}_s\boldsymbol{t}_c\widetilde{\boldsymbol{R}}_s + 2\boldsymbol{R}'_s\widetilde{\boldsymbol{R}}_s \;.
\end{aligned} \tag{18.20}$$

18.3.3 Properties of Motors

A general motor can be expressed as

$$\boldsymbol{M}_\alpha = \alpha\boldsymbol{M} \;, \tag{18.21}$$

where $\alpha \in \mathbb{R}$ and $\boldsymbol{M}$ is a unit motor as in previous sections. In this section we deal further with unit motors. The norm of a motor $\boldsymbol{M}$ is defined as follows

$$\begin{aligned}
|\boldsymbol{M}| &= \boldsymbol{M}\widetilde{\boldsymbol{M}} = \boldsymbol{T}_s\boldsymbol{R}_s\widetilde{\boldsymbol{R}}_s\widetilde{\boldsymbol{T}}_s = (1 + I\frac{\boldsymbol{t}_s}{2})\boldsymbol{R}_s\widetilde{\boldsymbol{R}}_s(1 - I\frac{\boldsymbol{t}_s}{2}) \\
&= 1 + I\frac{\boldsymbol{t}_s}{2} - I\frac{\boldsymbol{t}_s}{2} = 1 \;,
\end{aligned} \tag{18.22}$$

where $\widetilde{\boldsymbol{M}}$ is the conjugate motor and 1 is the identity of the motor multiplication. Now using the equation (18.16) and considering the unit motor magnitude we find two useful properties

$$\begin{aligned}
|\boldsymbol{M}| &= \boldsymbol{M}\widetilde{\boldsymbol{M}} = (\boldsymbol{R}_s + I\boldsymbol{R}'_s)(\widetilde{\boldsymbol{R}}_s + I\widetilde{\boldsymbol{R}'_s}) \\
&= \boldsymbol{R}_s\widetilde{\boldsymbol{R}}_s + I(\widetilde{\boldsymbol{R}}_s\boldsymbol{R}'_s + \widetilde{\boldsymbol{R}}_s\boldsymbol{R}_s) = 1 \;.
\end{aligned} \tag{18.23}$$

This requires the following two constraints equations

$$\begin{aligned}
\boldsymbol{R}_s\widetilde{\boldsymbol{R}}_s &= 1 \\
\widetilde{\boldsymbol{R}}_s\boldsymbol{R}'_s + \widetilde{\boldsymbol{R}}_s\boldsymbol{R}_s &= 0 \;.
\end{aligned} \tag{18.24}$$

The combination of two rigid motions can be expressed using two concatenated motors. The resultant motor describes the overall displacement, namely

$$\begin{aligned}\boldsymbol{M}_c = \boldsymbol{M}_a\boldsymbol{M}_b &= (\boldsymbol{R}_{s_a} + I\boldsymbol{R}'_{s_a})(\boldsymbol{R}_{s_b} + I\boldsymbol{R}'_{s_b})\\ &= \boldsymbol{R}_{s_a}\boldsymbol{R}_{s_b} + I(\boldsymbol{R}_{s_a}\boldsymbol{R}'_{s_b} + \boldsymbol{R}'_{s_a}\boldsymbol{R}_{s_b})\\ &= \boldsymbol{R}_{s_c} + I\boldsymbol{R}'_{s_c}\ .\end{aligned} \tag{18.25}$$

Note that pure rotations combine multiplicatively and dual parts containing the translation combine components additively.

Using the equation (18.16) let us express a motor in terms of a scalar, bivector, dual scalar and dual bivector

$$\begin{aligned}\boldsymbol{M} &= \boldsymbol{T}_s\boldsymbol{R}_s = \boldsymbol{R}_s + I\boldsymbol{R}'_s\\ &= (a_0 + a_1\gamma_2\gamma_3 + a_2\gamma_3\gamma_2 + a_3\gamma_2\gamma_1) +\\ &\quad I(b_0 + b_1\gamma_2\gamma_3 + b_2\gamma_3\gamma_2 + b_3\gamma_2\gamma_1)\\ &= (a_0 + \boldsymbol{a}) + I(b_0 + \boldsymbol{b})\ .\end{aligned} \tag{18.26}$$

We can use another notation to enhance the components of the real and dual parts of the motor as follows

$$\boldsymbol{M} = (a_0, \boldsymbol{a}) + I(b_0, \boldsymbol{b})\ , \tag{18.27}$$

where each term within the brackets consists of a scalar part and a 3D bivector.

A motor expressed in terms of a translator and a rotor is applied similarly as in the case of a rotor from the left and its conjugate from the right (motor reflections) to build an automorphism equivalent to the screw. Yet conjugating only the rotor or only the translator for the second reflection we can derive different types of automorphisms.

Changing the sign of the scalar and bivector in the real and the dual parts of the motor, we get the following variants of a motor

$$\begin{aligned}\boldsymbol{M} &= (a_0 + \boldsymbol{a}) + I(b_0 + \boldsymbol{b}) = \boldsymbol{T}_s\boldsymbol{R}_s\\ \widetilde{\boldsymbol{M}} &= (a_0 - \boldsymbol{a}) + I(b_0 - \boldsymbol{b}) = \widetilde{\boldsymbol{R}}_s\widetilde{\boldsymbol{T}}_s\\ \overline{\boldsymbol{M}} &= (a_0 + \boldsymbol{a}) - I(b_0 + \boldsymbol{b}) = \boldsymbol{R}_s\widetilde{\boldsymbol{T}}_s\\ \widetilde{\overline{\boldsymbol{M}}} &= (a_0 - \boldsymbol{a}) - I(b_0 - \boldsymbol{b}) = \widetilde{\boldsymbol{R}}_s\boldsymbol{T}_s\ .\end{aligned} \tag{18.28}$$

The first, the second and the third versions will be used for the modelling of the motion of points, lines and planes.

Using the relations from above it is straightforward to compute the expressions for the individual components

$$\begin{aligned}a_0 &= \frac{1}{4}(\boldsymbol{M} + \widetilde{\boldsymbol{M}} + \overline{\boldsymbol{M}} + \widetilde{\overline{\boldsymbol{M}}})\\ Ib_0 &= \frac{1}{2}(\boldsymbol{M} - \widetilde{\boldsymbol{M}}) = \frac{1}{4}(\boldsymbol{M} + \widetilde{\boldsymbol{M}} - \overline{\boldsymbol{M}} - \widetilde{\overline{\boldsymbol{M}}})\\ \boldsymbol{a} &= \frac{1}{4}(\boldsymbol{M} - \widetilde{\boldsymbol{M}} + \overline{\boldsymbol{M}} - \widetilde{\overline{\boldsymbol{M}}})\\ I\boldsymbol{b} &= \frac{1}{4}(\boldsymbol{M} - \widetilde{\boldsymbol{M}} - \overline{\boldsymbol{M}} + \widetilde{\overline{\boldsymbol{M}}})\ .\end{aligned} \tag{18.29}$$

This expressions are useful for the straightforward computation of the individual components.

18.4 Representation of Points, Lines, and Planes Using 3D and 4D Geometric Algebras

This section introduces the representation of lines, points and planes in 3D and 4D geometric algebra for applications in computer vision and kinematics. Let us start with the representations in the 3D space by reformulating the classical expressions of the vector calculus using the multivector concept of the geometric algebra. Thereafter we will extend these representations in the 4D space in a natural manner.

18.4.1 Representation of Points, Lines, and Planes in the 3D GA

The modelling of points, lines and planes in the 3D Euclidean space will be done using the Euclidean geometric algebra $\mathcal{G}_{3,0,0}$ where the pseudoscalar $I^2 = -1$. A point in the 3D space represents a position, thus it can be simply spanned using the vector basis of $\mathcal{G}_{3,0,0}$

$$\boldsymbol{x} = x\sigma_1 + y\sigma_2 + z\sigma_3 \,, \tag{18.30}$$

where $x, y, z \in \mathbb{R}$.

In the classical vector calculus a line is described by a position vector $\boldsymbol{x}$ touching any point of the line and a vector $\boldsymbol{n}$ for the line direction, i.e. $\boldsymbol{l} = \boldsymbol{x} + \alpha\boldsymbol{n}$, where $\alpha \in \mathbb{R}$. In geometric algebra we have the multivector concept, thus we can represent a line compactly using a vector $\boldsymbol{n}$ for its direction and a bivector $\boldsymbol{m}$ for the moment, namely

$$\boldsymbol{l} = \boldsymbol{n} + \boldsymbol{x}\wedge\boldsymbol{n} = \boldsymbol{n} + \boldsymbol{m} \,, \tag{18.31}$$

note that the moment $\boldsymbol{m}$ is a bivector computed as the outer product of the position vector $\boldsymbol{x}$ and the vector $\boldsymbol{n}$ for the line direction.

The representation of the plane is even more striking. The plane is a geometric entity one grade higher than the line, so we should expect that the multivector representation of the plane should be a natural multivector grade extension from that of a line. In the classical vector calculus a plane is described in terms of the Hesse distance from the origin to the plane and a vector indicating the plane orientation, i.e. $\{d\,,\,\boldsymbol{n}\}$. Again in the geometric algebra we can resort to a compact expression with clear geometric sense. The extension of the line expression to a plane should be done in terms of a bivector and a trivector as follows

$$\boldsymbol{h} = \boldsymbol{n} + \boldsymbol{x}\wedge\boldsymbol{n} = \boldsymbol{n} + Id \,, \tag{18.32}$$

where $\boldsymbol{n}$ is now a bivector indicating the plane orientation, and the outer product of the position vector $\boldsymbol{x}$ and the bivector $\boldsymbol{n}$ builds a trivector

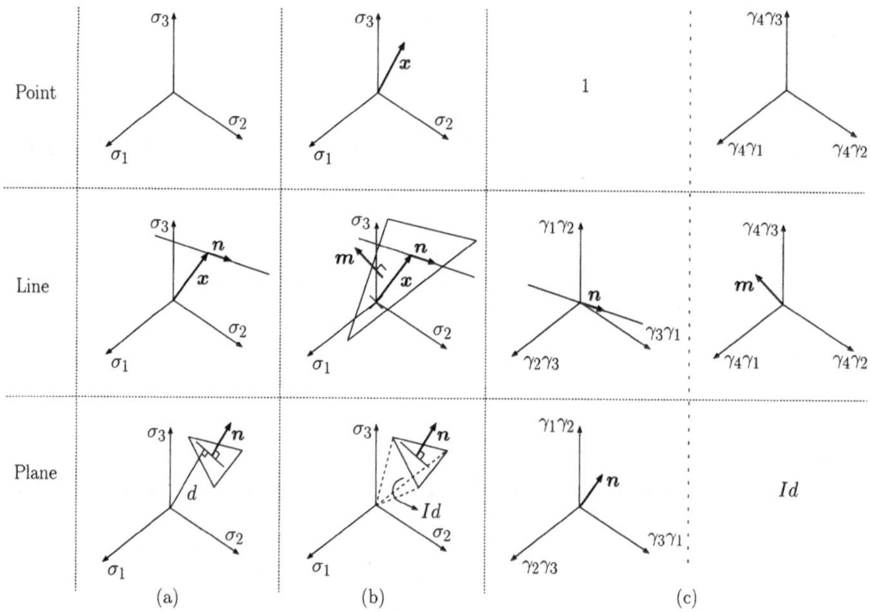

Fig. 18.3. Comparison of representations of points, lines and planes using (a) vector calculus, (b) $\mathcal{G}_{3,0,0}$ Euclidean 3D geometric algebra and (c) $\mathcal{G}^+_{3,0,1}$ motor algebra

which can be expressed using the Hesse distance d, a scalar value, and the pseudoscalar I. Figure 18.3 presents a comparison of the representations using classical vector calculus, the Euclidean geometric algebra $\mathcal{G}_{3,0,0}$ and the motor algebra $\mathcal{G}^+_{3,0,1}$.

18.4.2 Representation of Points, Lines, and Planes in the 4D GA

Now we will model points, lines and planes in the 4D space. For that we choose the special algebra of motors $\mathcal{G}^+_{3,0,1}$ which spans in 4D the line space using bivector basis.

For the case of the point representation, we proceed embedding a 3D point on the hyperplane $X_4 = 1$, the equation of the point $\boldsymbol{X}$ in $\mathcal{G}^+_{3,0,1}$ reads

$$\begin{aligned}
\boldsymbol{X} &= 1 + x_1\gamma_4\gamma_1 + x_2\gamma_4\gamma_2 + x_3\gamma_4\gamma_3 \\
&= 1 + I(x_1\gamma_2\gamma_3 + x_2\gamma_3\gamma_1 + x_3\gamma_1\gamma_2) \\
&= 1 + I\boldsymbol{x} \qquad (18.33)
\end{aligned}$$

or $\boldsymbol{X} = (1,0) + I(0,\boldsymbol{x})$. We can see that in this expression the real part consists of the scalar 1 and the dual part of only of 3D bivector.

A line will be expressed in $\mathcal{G}^+_{3,0,1}$ using the bivector basis $\{\gamma_2\gamma_3, \gamma_3\gamma_1, \gamma_1\gamma_2\}$ and the dual bivector basis $\{\gamma_4\gamma_1, \gamma_4\gamma_2, \gamma_4\gamma_3\}$. In the degenerate geometric algebra $\mathcal{G}^+_{3,0,1}$ the line is represented by

$$\boldsymbol{L} = \boldsymbol{n} + I\boldsymbol{m} \;, \tag{18.34}$$

where the coefficients of the bivectors for the line direction and moment are computed using two bivector points $\boldsymbol{x}_1$ and $\boldsymbol{x}_2$ lying on the line as follows

$$\begin{aligned}
\boldsymbol{n} &= (\boldsymbol{x}_2 - \boldsymbol{x}_1) \\
&= (x_{21} - x_{11})\gamma_2\gamma_3 + (x_{22} - x_{12})\gamma_3\gamma_1 + (x_{23} - x_{13})\gamma_1\gamma_2 \\
&= L_{n_1}\gamma_2\gamma_3 + L_{n_2}\gamma_3\gamma_1 + L_{n_3}\gamma_1\gamma_2 \\
\boldsymbol{m} &= (x_{12}x_{23} - x_{13}x_{22})\gamma_2\gamma_3 + (x_{13}x_{21} - x_{11}x_{23})\gamma_3\gamma_1 + \\
&= (x_{11}x_{22} - x_{12}x_{21})\gamma_1\gamma_2 \\
&= L_{m_1}\gamma_2\gamma_3 + L_{m_2}\gamma_3\gamma_1 + L_{m_3}\gamma_1\gamma_2 \;.
\end{aligned} \tag{18.35}$$

This line representation using dual numbers is easier to understand and to manipulate algebraically and it is fully equivalent to the one in terms of Plücker coordinates. Using the notation with brackets the line equation reads $\boldsymbol{L_d} = (0, \boldsymbol{n}) + I(0, \boldsymbol{m})$, where the $\boldsymbol{n}$ and $\boldsymbol{m}$ are spanned with a 3D bivector basis.

For the equation of the plane we can proceed similarly as for the equation (18.32). We represent the orientation of the plane via the bivector $\boldsymbol{n}$ and the the outer product between a bivector $\boldsymbol{x}$ touching the plane and its orientation $\boldsymbol{n}$. Since this outer product results in a 4-vector, we can express it as the Hesse distance $d = (\boldsymbol{x} \cdot \boldsymbol{n})$ multiplied by the unit pseudoscalar

$$\boldsymbol{H} = \boldsymbol{n} + \boldsymbol{x}\wedge\boldsymbol{n} = \boldsymbol{n} + I(\boldsymbol{x} \cdot \boldsymbol{n}) = \boldsymbol{n} + Id = \boldsymbol{n} + Id$$

or $\boldsymbol{H} = (0, \boldsymbol{n}) + I(d, 0)$. Note that the plane equation is the dual of the point equation

$$\boldsymbol{H} = (d + I\boldsymbol{n})^* = (I\boldsymbol{n})^* + (d)^* = \boldsymbol{n} + Id \;. \tag{18.36}$$

where instead of the plane orientation we consider the unit bivector $\boldsymbol{n}$ and for the scalar 1 the Hesse distance d.

18.5 Modeling the Motion of Points, Lines, and Planes Using 3D and 4D Geometric Algebras

This section concerns the modelling of the motion of basic geometric entities in the 3D and 4D space, respectively. The comparison of these motion models will show the power of the geometric algebra representation and the linearization of the translation transformation achieved in the 4D geometric algebra.

18.5.1 Motion of Points, Lines, and Planes in the 3D GA

The 3D motion of a point $\boldsymbol{x}$ in $\mathcal{G}_{3,0,0}$ has the following equation

$$x' = Rx\widetilde{R} + t\,. \tag{18.37}$$

Using the equation (18.31), the motion of the line reads

$$\begin{aligned} l' &= n' + m' = n' + x' \wedge n' \\ &= Rn\widetilde{R} + (Rx\widetilde{R} + t) \wedge (Rn\widetilde{R}) \\ &= Rn\widetilde{R} + Rx\widetilde{R} \wedge Rn\widetilde{R} + t \wedge Rn\widetilde{R} \\ &= Rn\widetilde{R} + Rx\widetilde{R} \wedge Rn\widetilde{R} + \frac{t}{2} Rn\widetilde{R} - Rn\widetilde{R}\frac{t}{2} \\ &= Rn\widetilde{R} + Rn\frac{t}{2}\widetilde{R} + \frac{t}{2} Rn\widetilde{R} + Rm\widetilde{R}\,, \end{aligned} \tag{18.38}$$

where x' stands for the rotated and shifted position vector, n' stands for the rotated orientation vector and m' for the new line moment.

The model of the motion of the plane in $\mathcal{G}_{3,0,0}$ can be expressed in terms of the multivector Hesse equation (18.32) as follows

$$\begin{aligned} h' &= n' + Id' = n' + x' \wedge n' \\ &= Rn\widetilde{R} + (Rx\widetilde{R} + t) \wedge (Rn\widetilde{R}) \\ &= Rn\widetilde{R} + t \wedge Rn\widetilde{R} + Rx \wedge n\widetilde{R} \\ &= Rn\widetilde{R} + t \wedge Rn\widetilde{R} + R(Id)\widetilde{R} \\ &= Rn\widetilde{R} + t^* \cdot Rn\widetilde{R} + Id \\ &= Rn\widetilde{R} + I(t \cdot Rn\widetilde{R} + d)\,, \end{aligned} \tag{18.39}$$

where n' stands for the rotated bivector plane orientation, x' stands for the rotated and shifting position vector and d' is the new Hesse distance. Here we use the concept of duality to claim that $t \wedge Rn\widetilde{R} = t^* \cdot Rn\widetilde{R} = (It) \cdot Rn\widetilde{R}$.

18.5.2 Motion of Points, Lines, and Planes in the 4D GA

The modelling of the 3D motion of the geometric primitives using the motor algebra $\mathcal{G}^+_{3,0,1}$ takes place in a 4D space where the rotation and translation are applied as multiplicative operators; as a result the 3D general motion becomes linear. Having a linear method we can then compute for example the unknown rotation and translation simultaneously using the motor extended Kalman filter or in cases like the hand-eye problem estimate motion [55]. In these kind of problems if we would use instead the 3D geometric algebra $\mathcal{G}_{3,0,0}$ we were unfortunately compelled to compute the translation decoupled of rotation increasing therefore the inaccuracy.

Using the representation of points given in (18.33), we can model the transformation of a point X under a rigid motion represented by $M = T_s R_s$ in the following way

$$\begin{aligned}
\boldsymbol{X}' &= 1 + I\boldsymbol{x}' = \boldsymbol{M}\boldsymbol{X}\widetilde{\overline{\boldsymbol{M}}} = \boldsymbol{M}(1 + I\boldsymbol{x})\widetilde{\overline{\boldsymbol{M}}} \\
&= \boldsymbol{T}_s\boldsymbol{R}_s(1 + I\boldsymbol{x})\widetilde{\boldsymbol{R}}_s\boldsymbol{T}_s \\
&= (1 + I\frac{\boldsymbol{t}_s}{2})\boldsymbol{R}_s(1 + I\boldsymbol{x})\widetilde{\boldsymbol{R}}_s(1 + I\frac{\boldsymbol{t}_s}{2}) \\
&= (1 + I\frac{\boldsymbol{t}_s}{2})(1 + I\boldsymbol{R}_s\boldsymbol{x}\widetilde{\boldsymbol{R}}_s)(1 + I\frac{\boldsymbol{t}_s}{2}) \\
&= 1 + I\frac{\boldsymbol{t}_s}{2} + I\boldsymbol{R}_s\boldsymbol{x}\widetilde{\boldsymbol{R}}_s + I\frac{\boldsymbol{t}_s}{2} \\
&= 1 + I(\boldsymbol{R}_s\boldsymbol{x}\widetilde{\boldsymbol{R}}_s + \boldsymbol{t}_s)\,. \qquad (18.40)
\end{aligned}$$

Note that the dual part of this equation in the 4D space is in the 3D space fully equivalent to the equation (18.37).

Using the line equation (18.34), we can express the transformation of a line $\boldsymbol{L}$ under a rigid motion as follows

$$\begin{aligned}
\boldsymbol{L}' &= \boldsymbol{n}' + I\boldsymbol{m}' = \boldsymbol{M}\,\boldsymbol{L}\,\widetilde{\boldsymbol{M}} = \boldsymbol{M}(\boldsymbol{n} + I\boldsymbol{m})\widetilde{\boldsymbol{M}} \\
&= \boldsymbol{T}_s\boldsymbol{R}_s(\boldsymbol{n} + I\boldsymbol{m})\widetilde{\boldsymbol{R}}_s\widetilde{\boldsymbol{T}}_s \\
&= (1 + I\frac{\boldsymbol{t}_s}{\boldsymbol{2}})\boldsymbol{R}_s(\boldsymbol{n} + I\boldsymbol{m})\widetilde{\boldsymbol{R}}_s(1 - I\frac{\boldsymbol{t}_s}{2}) \\
&= (1 + I\frac{\boldsymbol{t}_s}{\boldsymbol{2}})(\boldsymbol{R}_s\boldsymbol{n}\widetilde{\boldsymbol{R}}_s + I\boldsymbol{R}_s\boldsymbol{m}\widetilde{\boldsymbol{R}}_s - I\boldsymbol{R}_s\boldsymbol{n}\widetilde{\boldsymbol{R}}_s\frac{\boldsymbol{t}_s}{2}) \\
&= \boldsymbol{R}_s\boldsymbol{n}\widetilde{\boldsymbol{R}}_s + I(-\boldsymbol{R}_s\boldsymbol{n}\widetilde{\boldsymbol{R}}_s\frac{\boldsymbol{t}_s}{2} + \frac{\boldsymbol{t}_s}{2}\boldsymbol{R}_s\boldsymbol{n}\widetilde{\boldsymbol{R}}_s + \boldsymbol{R}_s\boldsymbol{m}\widetilde{\boldsymbol{R}}_s) \\
&= \boldsymbol{R}_s\boldsymbol{n}\widetilde{\boldsymbol{R}}_s + I(\boldsymbol{R}_s\boldsymbol{n}\widetilde{\boldsymbol{R}}_s{}' + \boldsymbol{R}_s'\boldsymbol{n}\widetilde{\boldsymbol{R}}_s + \boldsymbol{R}_s\boldsymbol{m}\widetilde{\boldsymbol{R}}_s)\,. \qquad (18.41)
\end{aligned}$$

Note that in equation (18.41) before we merge the bivector $\frac{\boldsymbol{t}_s}{2}$ with the rotors $\boldsymbol{R}_s$ or $\widetilde{\boldsymbol{R}}_s$ the real and the dual parts are fully equivalent with the elements of the line equation (18.38) of $\mathcal{G}_{3,0,0}$.

The transformation of a plane under a rigid motion in $\mathcal{G}^+_{3,0,1}$ can be seen as the motion of the dual of the point, thus using the expression of equation (18.36), the motion equation of the plane is

$$\begin{aligned}
\boldsymbol{H}' &= \boldsymbol{n}' + Id' = \boldsymbol{M}\,\boldsymbol{H}\,\widetilde{\overline{\boldsymbol{M}}} = \boldsymbol{M}(\boldsymbol{n} + Id)\widetilde{\overline{\boldsymbol{M}}} \\
&= \boldsymbol{T}_s\boldsymbol{R}_s(\boldsymbol{n} + Id)\widetilde{\boldsymbol{R}}_s\boldsymbol{T}_s \\
&= (1 + I\frac{\boldsymbol{t}_s}{\boldsymbol{2}})(\boldsymbol{R}_s\boldsymbol{n}\widetilde{\boldsymbol{R}}_s + Id)(1 + I\frac{\boldsymbol{t}_s}{\boldsymbol{2}}) \\
&= \boldsymbol{R}_s\boldsymbol{n}\widetilde{\boldsymbol{R}}_s + I(\boldsymbol{R}_s\boldsymbol{n}\widetilde{\boldsymbol{R}}_s\frac{\boldsymbol{t}_s}{2} + \frac{\boldsymbol{t}_s}{2}\boldsymbol{R}_s\boldsymbol{n}\widetilde{\boldsymbol{R}}_s + d) \\
&= \boldsymbol{R}_s\boldsymbol{n}\widetilde{\boldsymbol{R}}_s + I((\boldsymbol{R}_s\boldsymbol{n}\widetilde{\boldsymbol{R}}_s)\cdot\boldsymbol{t}_s + d)\,. \qquad (18.42)
\end{aligned}$$

Note that the real part and the dual part of this expression are fully equivalent to the bivector and trivector parts of the equation (18.39) in $\mathcal{G}_{3,0,0}$.

18.6 Conclusion

This chapter has presented the motor algebra as a suitable geometric algebra for kinematics. The modelling of motion of points, lines and planes in that degenerate geometric algebra results in linearization of the Euclidean transformation of lines. This can be comfortably used if the motion of lines is of interest as in the following chapters. While the effect of an Euclidean transformation on points is a linear one in Euclidean geometric algebra, this nice property is lost in motor algebra. This is the price to be payed by gaining linear expressions for the motion of a higher order entity. The motor algebra does also not linearize rigid motion of planes. Chapter 19 illustrates the application of the 4D motor algebra for the computation of the direct and inverse kinematics of robot manipulators. Another application with respect to estimation of line motion by Kalman filtering can be found in Chapter 21. In Chapter 20 the dual quaternion algebra which is isomorphic to the motor algebra is presented and studied with respect to motion alignment. The author believes that this framework is a modern geometric approach with computational advantages for the solution of problems of visually guided robotics.

19. Kinematics of Robot Manipulators in the Motor Algebra*

Eduardo Bayro-Corrochano and Detlef Kähler

Institute of Computer Science and Applied Mathematics,
Christian-Albrechts-University of Kiel

19.1 Introduction

In the literature we find a variety of mathematical approaches for solving problems in robotics which we will review now briefly. Denavit and Hartenberg [60] introduced the mostly used kinematic notation for lower pair mechanisms based on matrix algebra, Walker [236] used the epsilon algebra for the treatment of the manipulator kinematics, Gu and Luh [95] utilized dual–matrices for computing the Jacobians useful for kinematics and robot dynamics and Pennock and Yang [183] derived closed–form solutions for the inverse kinematics problem for various types of robot manipulators employing dual–matrices. McCarthy [166] used the dual form of the Jacobian for the analysis of multi–links similarly. Funda and Paul [85] gave a detailed computational analysis of the use of screw transformations in robotics. These authors explained that since the dual quaternion can represent the rotation and translation transformations simultaneously it is more effective than the unit quaternion formalism for dealing with the kinematics of robot chains. Kim and Kumar [128] computed a closed–form solution of the inverse kinematics of a 6 degree of freedom robot manipulator in terms of line transformations using dual quaternions. Aspragathos and Dimitros [9] confirmed once again

* This work has been supported by DFG Grant So-320-2-1.

that the use of dual quaternion and Lie algebra in robotics were overseen so far and that their use helps to reduce the number of representation parameters.

We can see in all these mathematical approaches that the authors take into account basically two key aspects: the obvious use of dual numbers and the representation of the screw transformations in terms of matrices or dual quaternions. In this regard in this chapter we are concerned with the extension of the representation capabilities of the dual numbers, particularly using the motor algebra beside the point and line representation we are able to model the motion of planes. This widens up the possibilities for the modelling of the motion of the basic geometric objects referred to frames attached to the robot manipulator which according to the circumstances simplify the complexity of the problem preserving the underlying geometry. After giving the modelling of prismatic and revolute transformations of a robot manipulator using points, lines and planes we solve the direct and inverse kinematics of robot manipulators. Using the motion of points, lines and planes in terms of motors we present constraints for a simple grasping task. The chapter shows clearly the advantages of the use of representations in motor algebra for solving problems related to robot manipulators.

The organization of the chapter is as follows: section two describes the prismatic and revolute transformations of robot manipulators in the motor algebra framework. The third section deals with the computation of the direct kinematics of robot manipulators. The fourth section is dedicated to the solution of the inverse kinematics of one standard robot manipulator. Finally, section five presents the conclusions.

19.2 Motor Algebra for the Kinematics of Robot Manipulators

The study of the rigid motion of objects in 3D space plays an important role in robotics. In order to linearize the rigid motion of the Euclidean space homogeneous coordinates are normally utilized. That is why in the geometric algebra framework we choose the special or degenerated geometric algebra to extend the algebraic system from 3D Euclidean space to the 4D space. In this system we can nicely model the motion of points, lines and planes with computational advantages and geometric insight, see chapter 18 for more details. Let us start with a description of the basic elements of robot manipulators in terms of the special or degenerated geometric algebra $\mathcal{G}_{3,0,1}^{+}$ or motor algebra. The most basic parts of a robot manipulator are revolute joints, prismatic joints, connecting links and the end–effectors. In the next subsections we will treat the kinematics of the prismatic and revolute manipulator parts using the 4D geometric algebra $\mathcal{G}_{3,0,1}^{+}$ and will illustrate an end–effector grasping task.

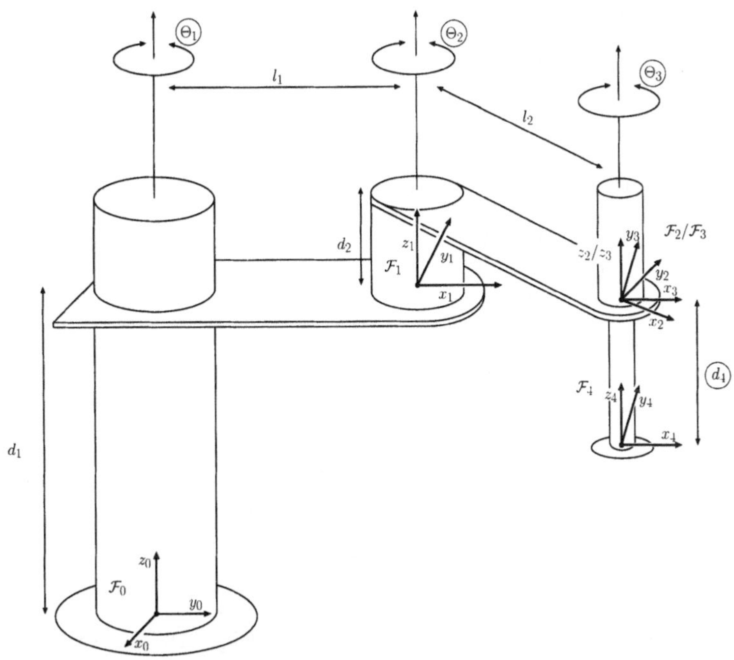

Fig. 19.1a. SCARA type manipulator according to the DH parameters in table 19.1. Variable parameters are encircled

19.2.1 The Denavit–Hartenberg Parameterization

The computation of the direct or inverse kinematics requires both the exact description of the robot manipulators structure and its configuration. The mostly used description approach is known as Denavit–Hartenberg procedure [60]. This is based on the uniform description of the position of the reference coordinate system of a joint relative to the next one in consideration. Figure 19.2a shows how coordinate frames are attached to a joint of a robot manipulator. Table 19.1 presents the specifications of two robot manipulators: the SCARA and the Stanford manipulator as shown in figures 19.1a and 19.1b, respectively.

In table 19.1 a variable parameter is indicated by the letter v and a constant one by c. This tells us whether the joint is for rotation (revolute) or for translation (prismatic). The transformation of the reference coordinate system between two joints will be called joint–transition. Figure 19.2b shows the involved screws in a joint–transition according to the Denavit–Hartenberg parameters . The frame or reference coordinate system related to the i-th joint is attached at the end of this link and it is called $\mathcal{F}_i$. The position and orientation of the end–effector in relation to the reference coordinate system of the robot basis can be computed by linking all joint–transitions. In this way we get straightforwardly the direct kinematics.

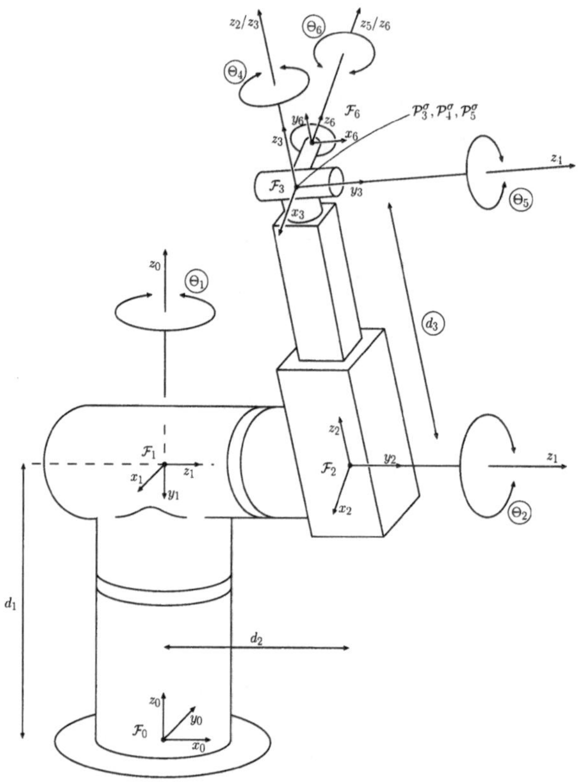

Fig. 19.1b. Stanford type manipulator according to the DH parameters in table 19.1. Variable parameters are encircled

Conversely for the inverse kinematics given the position and orientation of the end–effector we have to find values of the variable parameters of the joint–transitions which satisfy this requirement. In the next sections we will go more into details about the computation of direct and inverse kinematics of robot manipulators.

19.2.2 Representations of Prismatic and Revolute Transformations

The transformation of any point, line or plane between coordinate systems $\mathcal{F}_{i-1}$ and $\mathcal{F}_i$ is a revolute one when the degree of freedom is only a variable angle θ_i and a prismatic one when the degree of freedom is only a variable length d_i. The transformation motor ${}^{i-1}\boldsymbol{M}_i$ between $\mathcal{F}_i$ and $\mathcal{F}_{i-1}$ consists of a sequence of two screw transformations, one fixed, i.e. $\boldsymbol{M}^x_{\hat{\alpha}_i}$, and another variable , i.e. $\boldsymbol{M}^z_{\hat{\theta}_i}$, see figure 19.2b. Note that we use dual angles $\hat{\theta}_i = \theta_i + Id_i$ and $\hat{\alpha}_i = \alpha_i + Il_i$, see chapter 18. In the revolute case the latter has as a variable parameter the angle θ_i and in the prismatic case the displacement d_i. The transformation reads

Table 19.1. Kinematic configuration of two robot manipulators

Robot type	link	revolute θ_i	v/c	prismatic d_i	v/c	twist angle α_i	link length l_i
SCARA	1	θ_1	v	d_1	c	0	l_1
	2	θ_2	v	d_2	c	0	l_2
	3	θ_3	v	0		0	0
	4	0		d_4	v	0	0
Stanford	1	θ_1	v	d_1	c	-90 deg	0
	2	θ_2	v	d_2	c	90 deg	0
	3	0		d_3	v	0	0
	4	θ_4	v	0		-90 deg	0
	5	θ_5	v	0		90 deg	0
	6	θ_6	v	d_6	c	0	0

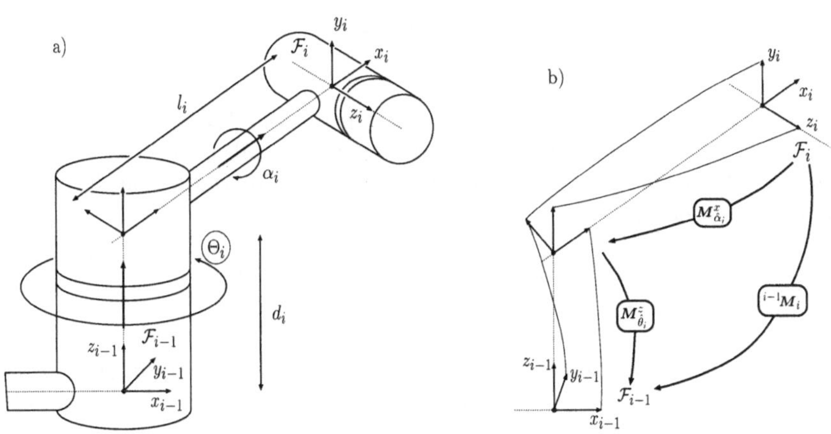

Fig. 19.2. a) The i-th joint of a robot manipulator and the attached coordinate frames according to the Denavit–Hartenberg procedure. Here the encircled θ_i is the variable parameter, b) the transformation from frame $\mathcal{F}_i$ to $\mathcal{F}_{i-1}$ is represented by ${}^{i-1}\boldsymbol{M}_i$. The motor ${}^{i-1}\boldsymbol{M}_i$ consists of two screw transformations $\boldsymbol{M}^x_{\hat{\alpha}_i}$ and $\boldsymbol{M}^z_{\hat{\theta}_i}$

$$
\begin{aligned}
{}^{i-1}\boldsymbol{M}_i &= \boldsymbol{M}^z_{\hat{\theta}_i}\boldsymbol{M}^x_{\hat{\alpha}_i} = \boldsymbol{T}^z_{d_i}\boldsymbol{R}^z_{\theta_i}\boldsymbol{T}^x_{l_i}\boldsymbol{R}^x_{\alpha_i} \\
&= (1+\frac{I}{2}\begin{pmatrix}0\\0\\d_i\end{pmatrix})\boldsymbol{R}^z_{\theta_i}(1+\frac{I}{2}\begin{pmatrix}l_i\\0\\0\end{pmatrix})\boldsymbol{R}^x_{\alpha_i}\,. \qquad (19.1)
\end{aligned}
$$

For the sake of clearness the dual bivectors of translators are given as a column vector simply to make the variable parameters explicit.

Since ${}^{i-1}\boldsymbol{M}_i\,{}^{i-1}\widetilde{\boldsymbol{M}}_i = 1$, we obtain

$$
{}^{i}\boldsymbol{M}_{i-1} = \widetilde{\boldsymbol{M}}^x_{\hat{\alpha}_i}\widetilde{\boldsymbol{M}}^z_{\hat{\theta}_i} = \widetilde{\boldsymbol{T}}^x_{l_i}\widetilde{\boldsymbol{R}}^x_{\alpha_i}\widetilde{\boldsymbol{T}}^z_{d_i}\widetilde{\boldsymbol{R}}^z_{\theta_i}\,. \qquad (19.2)
$$

Be aware for the rest of the chapter that ${}^j\boldsymbol{M}_i$ denotes a motor transformation from $\mathcal{F}_i$ to $\mathcal{F}_j$.

We will now give general expressions for the transformation of points, lines and planes with one of the parameters θ_i and d_i, respectively, as a variable and with two fixed parameters α_i and l_i. In the joint depicted in figure 19.2b a revolute transformation will take place only when θ_i varies and a prismatic transformation only when d_i varies. Now taking a point $\boldsymbol{X}$ represented in the frame $\mathcal{F}_{i-1}$, we can describe its transformation from $\mathcal{F}_{i-1}$ to $\mathcal{F}_i$ in the motor algebra according to chapter 18 with either θ_i or d_i as variable parameter. We will call this transformation a *forward transformation* .

The multivector representation of point $\boldsymbol{X}$ related to the frame $\mathcal{F}_i$ will be expressed as ${}^i\boldsymbol{X}$ with

$$\begin{aligned}
{}^i\boldsymbol{X} &= {}^i\boldsymbol{M}_{i-1}\ {}^{i-1}\boldsymbol{X}\ {}^i\overline{\widetilde{\boldsymbol{M}}}_{i-1} = \widetilde{\boldsymbol{M}}^x_{\hat{\alpha}_i}\widetilde{\boldsymbol{M}}^z_{\hat{\theta}_i}\ {}^{i-1}\boldsymbol{X}\ \overline{\boldsymbol{M}}^z_{\hat{\theta}_i}\overline{\boldsymbol{M}}^x_{\hat{\alpha}_i} \\
&= \widetilde{\boldsymbol{T}}^x_{l_i}\widetilde{\boldsymbol{R}}^x_{\alpha_i}\widetilde{\boldsymbol{T}}^z_{d_i}\widetilde{\boldsymbol{R}}^z_{\theta_i}\ {}^{i-1}\boldsymbol{X}\ \boldsymbol{R}^z_{\theta_i}\widetilde{\boldsymbol{T}}^z_{d_i}\boldsymbol{R}^x_{\alpha_i}\widetilde{\boldsymbol{T}}^x_{l_i} \\
&= 1 + I\,{}^i\boldsymbol{x}\,,
\end{aligned} \tag{19.3}$$

where ${}^i\boldsymbol{x}$ is a bivector representing the 3D position of $\boldsymbol{X}$ referred to $\mathcal{F}_i$. Thinking in a transformation in the reverse sense we call it a *backward transformation* which transforms a point $\boldsymbol{X}$ represented in the frame $\mathcal{F}_i$ to the frame $\mathcal{F}_{i-1}$ as follows

$$\begin{aligned}
{}^{i-1}\boldsymbol{X} &= {}^{i-1}\boldsymbol{M}_i\ {}^i\boldsymbol{X}\ {}^{i-1}\overline{\widetilde{\boldsymbol{M}}}_i = \boldsymbol{M}^z_{\hat{\theta}_i}\boldsymbol{M}^x_{\hat{\alpha}_i}\ {}^i\boldsymbol{X}\ \overline{\widetilde{\boldsymbol{M}}}^x_{\hat{\alpha}_i}\overline{\widetilde{\boldsymbol{M}}}^z_{\hat{\theta}_i} \\
&= 1 + I\,{}^{i-1}\boldsymbol{x}\,.
\end{aligned} \tag{19.4}$$

Note that the motor applied from the right side is not purely conjugated as in the line case. This will be also the case for a plane, see chapter 18 for details of the point and plane transformations.

Consider a line $\boldsymbol{L}$ represented in the frame $\mathcal{F}_{i-1}$ by ${}^{i-1}\boldsymbol{L} = {}^{i-1}\boldsymbol{n} + I^{i-1}\boldsymbol{m}$, where $\boldsymbol{n}$ and $\boldsymbol{m}$ are bivectors indicating the orientation and moment of the line, respectively. We can write its forward transformation related to the frame $\mathcal{F}_i$ according to chapter 18 as follows

$$\begin{aligned}
{}^i\boldsymbol{L} &= {}^i\boldsymbol{M}_{i-1}\ {}^{i-1}\boldsymbol{L}\ {}^i\widetilde{\boldsymbol{M}}_{i-1} = \widetilde{\boldsymbol{M}}^x_{\hat{\alpha}_i}\widetilde{\boldsymbol{M}}^z_{\hat{\theta}_i}\ {}^{i-1}\boldsymbol{L}\ \boldsymbol{M}^z_{\hat{\theta}_i}\boldsymbol{M}^x_{\hat{\alpha}_i} \\
&= {}^i\boldsymbol{n} + I\,{}^i\boldsymbol{m}\,.
\end{aligned} \tag{19.5}$$

Its backward transformation reads

$$\begin{aligned}
{}^{i-1}\boldsymbol{L} &= {}^{i-1}\boldsymbol{M}_i\ {}^i\boldsymbol{L}\ {}^{i-1}\widetilde{\boldsymbol{M}}_i = \boldsymbol{M}^z_{\hat{\theta}_i}\boldsymbol{M}^x_{\hat{\alpha}_i}\ {}^i\boldsymbol{L}\ \widetilde{\boldsymbol{M}}^x_{\hat{\alpha}_i}\widetilde{\boldsymbol{M}}^z_{\hat{\theta}_i} \\
&= {}^{i-1}\boldsymbol{n} + I\,{}^{i-1}\boldsymbol{m}\,.
\end{aligned} \tag{19.6}$$

Finally, the forward transformation of a plane $\boldsymbol{H}$ represented in $\mathcal{F}_{i-1}$ reads

$$\begin{aligned}
{}^i\boldsymbol{H} &= {}^i\boldsymbol{M}_{i-1}\ {}^{i-1}\boldsymbol{H}\ {}^i\overline{\widetilde{\boldsymbol{M}}}_{i-1} = \widetilde{\boldsymbol{M}}^x_{\hat{\alpha}_i}\widetilde{\boldsymbol{M}}^z_{\hat{\theta}_i}\ {}^{i-1}\boldsymbol{H}\ \overline{\boldsymbol{M}}^z_{\hat{\theta}_i}\overline{\boldsymbol{M}}^x_{\hat{\alpha}_i} \\
&= {}^i\boldsymbol{n} + I\,{}^i d_H\,.
\end{aligned} \tag{19.7}$$

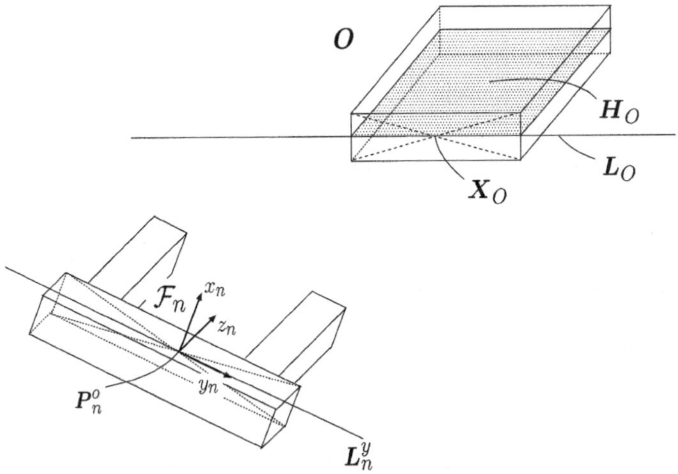

Fig. 19.3. Two finger grasper approaching to an object

and similarly as above, its backward transformation equation is

$$
\begin{aligned}
{}^{i-1}\boldsymbol{H} &= {}^{i-1}\boldsymbol{M}_i \, {}^{i}\boldsymbol{H} \, {}^{i-1}\widetilde{\overline{\boldsymbol{M}}}_i = \boldsymbol{M}^z_{\hat{\theta}_i} \boldsymbol{M}^x_{\hat{\alpha}_i} \, {}^{i}\boldsymbol{H} \, \widetilde{\overline{\boldsymbol{M}}}^x_{\hat{\alpha}_i} \widetilde{\overline{\boldsymbol{M}}}^z_{\hat{\theta}_i} \\
&= {}^{i-1}\boldsymbol{n} + I \, {}^{i-1}d_H \, . \qquad (19.8)
\end{aligned}
$$

19.2.3 Grasping by Using Constraint Equations

In this subsection we will illustrate grasping as a manipulation related task. grasping operation. This task involves the positioning of a two finger grasper in front of a static object. Figure 19.3 shows the grasper and the considered object $\boldsymbol{O}$. The manipulator moves the grasper near to the object and together they should fulfill some conditions to grasp the object firmly. In order to determine the overall transformation ${}^{0}\boldsymbol{M}_n$, which moves the grasper to an appropriate grasping position, we claim that ${}^{0}\boldsymbol{M}_n$ has to fulfill three constraints. For the formulation of these constraints we can take advantages of the point, line and plane representations of the motor algebra. In the following we assume that the representations of geometric entities attached to the object $\boldsymbol{O}$ in frame $\mathcal{F}_0$ are known.

Attitude condition: The grasping movement of the two fingers should be in the reference plane $\boldsymbol{H}_O$ of $\boldsymbol{O}$. That is, the yz-plane of the end–effector frame $\mathcal{F}_n$ should be equal to the reference plane $\boldsymbol{H}_O$. The attitude condition can be simply formulated in terms of a plane equation as follows

$$
{}^{0}\boldsymbol{M}_n \, {}^{n}\boldsymbol{H}^{yz}_n \, {}^{0}\widetilde{\overline{\boldsymbol{M}}}_n - {}^{0}\boldsymbol{H}_O \approx 0 \, , \qquad (19.9)
$$

where ${}^{n}\boldsymbol{H}^{yz}_n = (1,0,0)^T + I\,0 = (1,0,0)^T$, see figure 19.3.

Alignment condition: The grasper and object should be aligned parallel after the application of the motor ${}^{0}\boldsymbol{M}_n$. That is, the direction of the y-axis

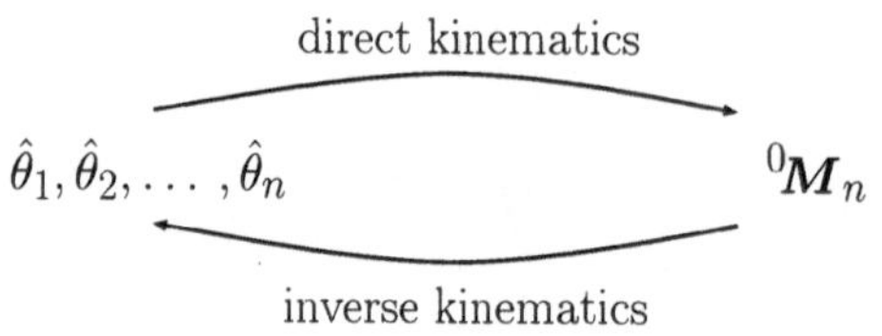

Fig. 19.4. Direct and inverse kinematics

and the line $\boldsymbol{L}_O$ should be the same. This condition can be simply expressed in terms of a line equation

$$\langle {}^0\boldsymbol{M}_n\, {}^n\boldsymbol{L}_n^y\, {}^0\widetilde{\boldsymbol{M}}_n \rangle_d - \langle {}^0\boldsymbol{L}_O \rangle_d \approx 0\,, \tag{19.10}$$

where ${}^n\boldsymbol{L}_n^y = (0,1,0)^T + I(0,0,0)^T = (0,1,0)^T$ and $\langle \boldsymbol{L} \rangle_d$ denotes the components of direction of line $\boldsymbol{L}$.

Touching condition: The motion ${}^0\boldsymbol{M}_n$ should also guarantee that the grasper is in the right grasping position. That is, the origin $\boldsymbol{P}_n^o$ of the end–effector frame $\mathcal{F}_n$ should touch the reference point $\boldsymbol{X}_O$ of $\boldsymbol{O}$. A formulation of this constraint in our framework is

$${}^0\boldsymbol{M}_n\, {}^n\boldsymbol{P}_n^o\, {}^0\widetilde{\overline{\boldsymbol{M}}}_n - {}^0\boldsymbol{X}_O \approx 0\,. \tag{19.11}$$

By these three conditions we get constraints for the components of ${}^0\boldsymbol{M}_n$, and we can determine ${}^0\boldsymbol{M}_n$ numerically. The next step is to determine the variable joint parameters of the robot manipulator which leads to the position and orientation of the end–effector frame $\mathcal{F}_n$ described by ${}^0\boldsymbol{M}_n$. This problem is called the inverse kinematics problem of robot manipulators and will be treated in section 19.4.

19.3 Direct Kinematics of Robot Manipulators

The direct kinematics involves the computation of the position and orientation of the end–effector or frame $\mathcal{F}_n$ given the parameters of the joint–transitions, see figure 19.4. In this section we will show how the direct kinematics can be computed when we use as geometric object a point, line or plane. The notation for points, lines and planes we will use in the next sections is illustrated in figure 19.5. The direct kinematics for the general case of a manipulator with n joints can be written as follows

$${}^0\boldsymbol{M}_n = {}^0\boldsymbol{M}_1{}^1\boldsymbol{M}_2{}^2\boldsymbol{M}_3 \cdots {}^{n-1}\boldsymbol{M}_n = \prod_{i=1}^{n} {}^{i-1}\boldsymbol{M}_i\,. \tag{19.12}$$

Now we can formulate straightforwardly the direct kinematics in terms of

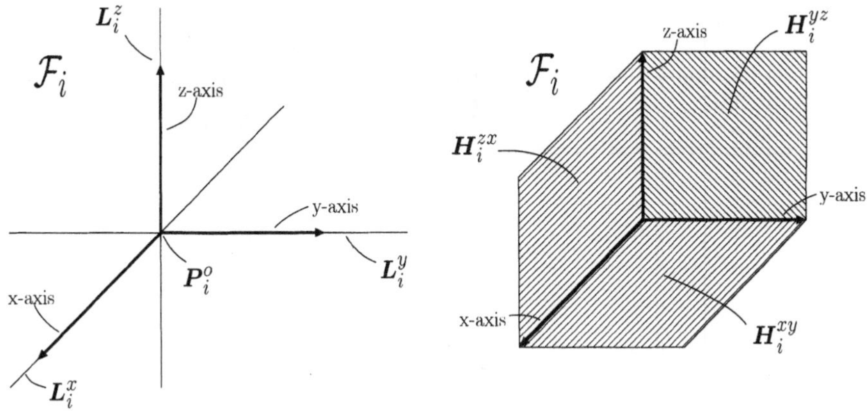

Fig. 19.5. Notations for frame specific entities as the origin, the coordinate axis and coordinate planes

point, line or plane representations as follows

$$
\begin{aligned}
{}^0\boldsymbol{X} &= {}^0\boldsymbol{M}_n\ {}^n\boldsymbol{X}\ {}^0\widetilde{\overline{\boldsymbol{M}}}_n = \prod_{i=1}^{n} {}^{i-1}\boldsymbol{M}_i\ {}^n\boldsymbol{X}\ \prod_{i=1}^{n} {}^{n-i}\widetilde{\overline{\boldsymbol{M}}}_{n+1-i}\ ,\\
{}^0\boldsymbol{L} &= \prod_{i=1}^{n} {}^{i-1}\boldsymbol{M}_i\ {}^n\boldsymbol{L}\ \prod_{i=1}^{n} {}^{n-i}\widetilde{\boldsymbol{M}}_{n+1-i}\ ,\\
{}^0\boldsymbol{H} &= \prod_{i=1}^{n} {}^{i-1}\boldsymbol{M}_i\ {}^n\boldsymbol{H}\ \prod_{i=1}^{n} {}^{n-i}\widetilde{\overline{\boldsymbol{M}}}_{n+1-i}\ .
\end{aligned}
\tag{19.13}
$$

Let us now write the motor ${}^0\boldsymbol{M}_4$ for the direct kinematics for points, lines and planes like equation (19.13) for the SCARA manipulator specified by the Denavit–Hartenberg parameters of table 19.1. Firstly, using equation (19.12) with n=4, we can write down straightforwardly the required motor ${}^0\boldsymbol{M}_4$ as follows

$$
\begin{aligned}
{}^0\boldsymbol{M}_4 &= {}^0\boldsymbol{M}_1{}^1\boldsymbol{M}_2{}^2\boldsymbol{M}_3{}^3\boldsymbol{M}_4 = (\boldsymbol{M}^z_{\hat{\theta}_1}\boldsymbol{M}^x_{\hat{\alpha}_1})\cdots(\boldsymbol{M}^z_{\hat{\theta}_4}\boldsymbol{M}^x_{\hat{\alpha}_4})\\
&= (\boldsymbol{T}^z_{d_1}\boldsymbol{R}^z_{\theta_1}\boldsymbol{T}^x_{l_1}\boldsymbol{R}^x_{\alpha_1})\cdots(\boldsymbol{T}^z_{d_4}\boldsymbol{R}^z_{\theta_4}\boldsymbol{T}^x_{l_4}\boldsymbol{R}^x_{\alpha_4})\\
&= (1+\frac{I}{2}\begin{pmatrix}0\\0\\d_1\end{pmatrix})\boldsymbol{R}^z_{\theta_1}(1+\frac{I}{2}\begin{pmatrix}l_1\\0\\0\end{pmatrix})(1+\frac{I}{2}\begin{pmatrix}0\\0\\d_2\end{pmatrix})\\
&\quad \boldsymbol{R}^z_{\theta_2}(1+\frac{I}{2}\begin{pmatrix}l_2\\0\\0\end{pmatrix})\boldsymbol{R}^z_{\theta_3}(1+\frac{I}{2}\begin{pmatrix}0\\0\\d_4\end{pmatrix})\ .
\end{aligned}
\tag{19.14}
$$

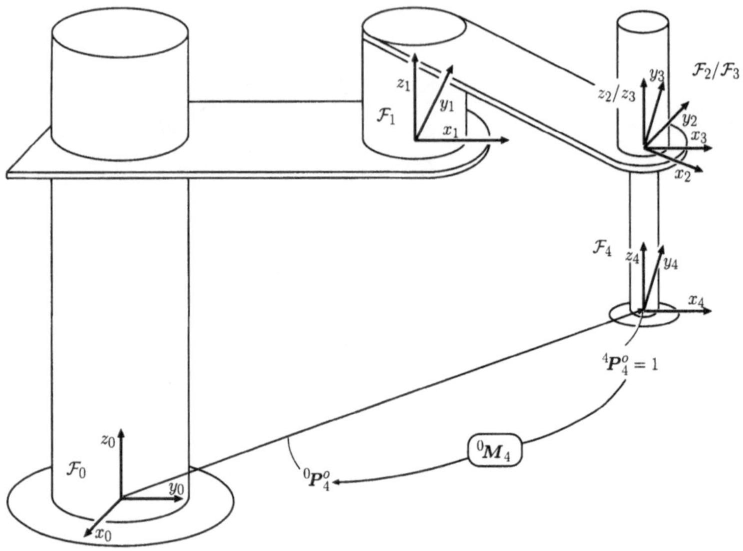

Fig. 19.6. The representation ${}^0\boldsymbol{P}_4^o$ of $\boldsymbol{P}_4^o$ in frame $\mathcal{F}_0$ is computed using ${}^0\boldsymbol{M}_4$

Note that translators with zero translation and rotors with zero angle become 1.

Applying the motor ${}^0\boldsymbol{M}_4$ from the left and ${}^0\widetilde{\overline{\boldsymbol{M}}}_4$ from the right for point and plane equations and the motor ${}^0\boldsymbol{M}_4$ from the left and ${}^0\widetilde{\boldsymbol{M}}_4$ from the right for line equations as indicated by equations (19.13), we get the direct kinematics equations of points, lines and planes for the SCARA robot manipulator.

19.3.1 Maple Program for Motor Algebra Computations

Since the nature of our approach requires symbolic computation we chose Maple to implement a program suitable for computations in the motor algebra framework $\mathcal{G}^+_{3,0,1}$. We have developed a comfortable program for computations in the frame of different geometric algebras. When dealing with the motor algebra we have simply to specify its vector basis. The program has a variety of useful algebraic operators to carry out computations involving reversion, Clifford conjugations, inner and wedge operations, rotations, translations, motors, extraction of the i–blade of a multivector etc.

As a first illustration using our Maple program, we computed the direct kinematic equation of the origin $\boldsymbol{P}_4^o$ of $\mathcal{F}_4$ for the SCARA manipulator specified by the Denavit–Hartenberg parameters of table 19.1. The figure 19.6 shows the frames and the point $\boldsymbol{P}_4^o$ refered to $\mathcal{F}_0$. The final result is

$$
\begin{aligned}
{}^{0}\boldsymbol{P}^{o}_{4} = {}^{0}\boldsymbol{M}_4\, {}^{4}\boldsymbol{P}^{o}_{4}\, {}^{0}\widetilde{\overline{\boldsymbol{M}}}_4 &= {}^{0}\boldsymbol{M}_4 \left(1 + I \begin{pmatrix} 0 \\ 0 \\ 0 \end{pmatrix}\right) {}^{0}\widetilde{\overline{\boldsymbol{M}}}_4 \\
&= 1 + I \begin{pmatrix} l_2 \cos(\theta_1 + \theta_2) + l_1 \cos(\theta_1) \\ l_2 \sin(\theta_1 + \theta_2) + l_1 \sin(\theta_1) \\ d_1 + d_2 + d_4 \end{pmatrix} .
\end{aligned}
\tag{19.15}
$$

19.4 Inverse Kinematics of Robot Manipulators

Since the inverse kinematics is more complex than the direct kinematics our aim should be to find a systematic way to solve it exploiting the point, line and plane motor algebra representations. Unfortunately the procedure is not amenable for a general formulation as in the case of the direct kinematics equation (19.12). That is why we better choose a real robot manipulator and compute its inverse kinematics in order to show all the characteristics of the computational assumptions.

The Stanford robot manipulator is well known among researchers concerned with the design of strategies for the symbolic computation of the inverse kinematics. According to table 19.1 the variable parameters to be computed are θ_1, θ_2, θ_4, θ_5, θ_6 and d_3. By means of this example we will show that in the motor algebra approach we have the freedom to switch between the point, line or plane representation according to the geometrical circumstances. This is one of the most important advantages of our motor algebra approach.

According to the mechanical characteristics of the Stanford manipulator we can divide it into two basic parts: one dedicated for the positioning involving the joints 1,2 and 3 and one dedicated for the orientation of the end–effector like a wrist comprising the joints 4 to 6. Since the philosophy of our approach relies on the application of point, line or plane representation where it is needed, we should firstly recognize whether a point or a line or a plane representation is the suitable representation for the joint–transitions. As a result on the one hand a better geometric insight is guaranteed and on the other hand the solution method is easier to be developed. The first three joints of the Stanford manipulator are used to position the origin of the coordinate frame $\mathcal{F}_3$. Therefore we apply a point representation to describe this part of the problem. The last three joints are used to achieve to desired orientation of the end–effector frame. For the formulation of this subproblem we use a line and a plane representation because with these entities we can model orientations.

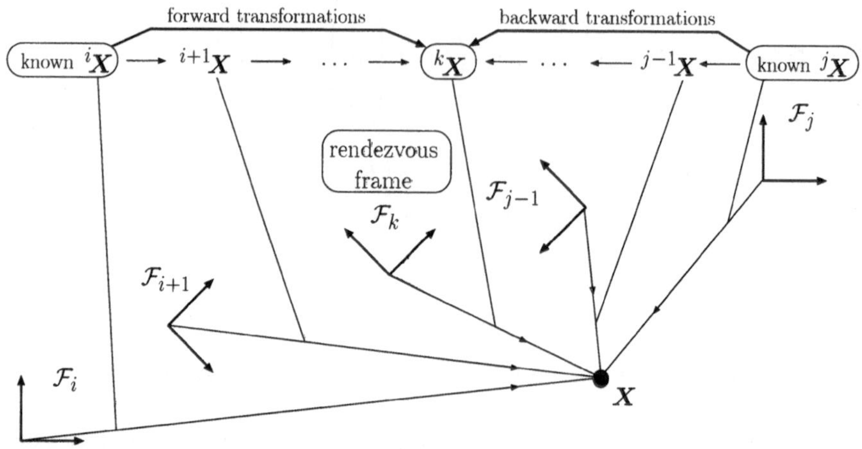

Fig. 19.7. Rendezvous method: If ${}^{i}\boldsymbol{X}$ and ${}^{j}\boldsymbol{X}$ are known, we can compute ${}^{k}\boldsymbol{X}$ for each $i \leq k \leq j$ in two different ways: by successive forward transformations of ${}^{i}\boldsymbol{X}$ and by successive backward transformation of ${}^{j}\boldsymbol{X}$

19.4.1 The Rendezvous Method

The next important step is to represent the motor transformations from the beginning of a chain of joint–transitions to the end and vice versa as it is depicted in figure 19.7. As a result we gain a set of equations for each meeting point. In each of these points the forward equation is equal with the backward equation. Using these equalities we have a guideline to compute the unknowns. We will call this procedure the *rendezvous method* . This simple idea has proved to be very useful as a strategy for the solution of the inverse kinematics.

19.4.2 Computing θ_1, θ_2 and d_3 Using a Point Representation

In the case of the Stanford manipulator the orientation and position of frame $\mathcal{F}_6$ uniquely determines the position of frame $\mathcal{F}_3$. This will be explained in the following.

The position of frame $\mathcal{F}_3$ with respect to $\mathcal{F}_0$ is described by the multivector representation ${}^{0}\boldsymbol{P}_3^o$ of $\boldsymbol{P}_3^o$ in $\mathcal{F}_0$. By successive forward transformation applied on ${}^{3}\boldsymbol{P}_3^o = 1$ we get the representation ${}^{6}\boldsymbol{P}_3^o$ of $\boldsymbol{P}_3^o$ in $\mathcal{F}_6$ by

$$ {}^{6}\boldsymbol{P}_3^o = {}^{6}\boldsymbol{M}_3 \, {}^{3}\boldsymbol{P}_3^o \, {}^{6}\widetilde{\overline{\boldsymbol{M}}}_3 = 1 - I \begin{pmatrix} 0 \\ 0 \\ d_6 \end{pmatrix} . \tag{19.16} $$

Now we can compute ${}^{0}\boldsymbol{P}_3^o$ by

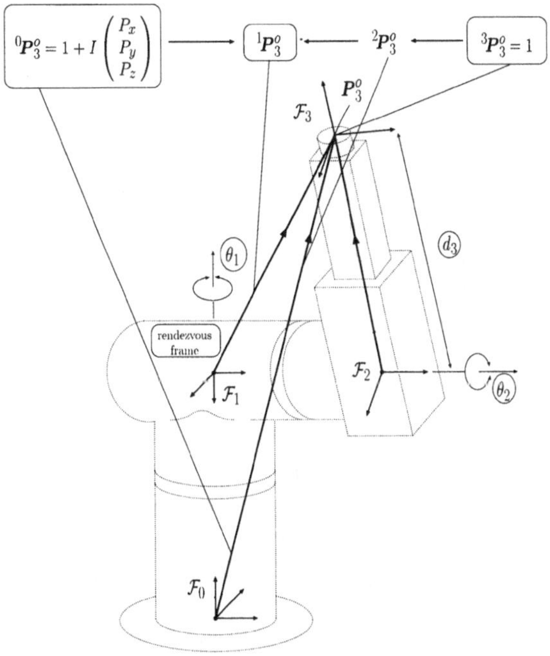

Fig. 19.8. The rendezvous method applied to P_3^o in order to determine the equations shown in table 19.2. The equations of rendezvous frame $\mathcal{F}_1$ are choosen to compute the variable parameters θ_1, θ_2 and d_3

$$
\begin{aligned}
{}^0P_3^o &= {}^0M_6\ {}^6P_3^o\ {}^0\widetilde{\overline{M}}_6 = {}^0M_6\ (1 - I \begin{pmatrix} 0 \\ 0 \\ d_6 \end{pmatrix})\ {}^0\widetilde{\overline{M}}_6 \\
&= 1 + I \begin{pmatrix} P_x \\ P_y \\ P_z \end{pmatrix}, \qquad (19.17)
\end{aligned}
$$

note that ${}^0M_6$ is given. The vector $(P_x, P_y, P_z)^T$ describes the position of the origin P_3^o of frame $\mathcal{F}_3$ in frame $\mathcal{F}_0$ for a given overall transformation ${}^0M_6$. Now we can apply the rendezvous method since we know the representation of P_3^o in the two different frames $\mathcal{F}_0$ and $\mathcal{F}_3$, see figure 19.8.

Applying successive forward transformations we obtain

$$
\begin{aligned}
{}^1P_3^o &= {}^1M_0\ {}^0P_3^o\ {}^1\widetilde{\overline{M}}_0 , \\
{}^2P_3^o &= {}^2M_1\ {}^1P_3^o\ {}^2\widetilde{\overline{M}}_1 , \\
{}^3P_3^o &= {}^3M_2\ {}^2P_3^o\ {}^3\widetilde{\overline{M}}_2 . \qquad (19.18)
\end{aligned}
$$

These computations were carried out with our Maple program getting the left hand sides of the four groups of equations of the table 19.2.

On the other hand, applying successive backward transformations to the origin of $\mathcal{F}_3$ given by

$$ {}^{3}\boldsymbol{P}_3^o = 1 + I \begin{pmatrix} 0 \\ 0 \\ 0 \end{pmatrix} = 1 \,, \tag{19.19} $$

we get

$$ {}^{2}\boldsymbol{P}_3^o = {}^{2}\boldsymbol{M}_3 \; {}^{3}\boldsymbol{P}_3^o \; {}^{2}\widetilde{\overline{\boldsymbol{M}}}_3 = 1 + I \begin{pmatrix} 0 \\ 0 \\ d_3 \end{pmatrix} , $$

$$ {}^{1}\boldsymbol{P}_3^o = {}^{1}\boldsymbol{M}_2 \; {}^{2}\boldsymbol{P}_3^o \; {}^{1}\widetilde{\overline{\boldsymbol{M}}}_2 = 1 + I \begin{pmatrix} d_3 \sin(\theta_2) \\ -d_3 \cos(\theta_2) \\ d_2 \end{pmatrix} , $$

$$ {}^{0}\boldsymbol{P}_3^o = {}^{0}\boldsymbol{M}_1 \; {}^{1}\boldsymbol{P}_3^o \; {}^{0}\widetilde{\overline{\boldsymbol{M}}}_1 = 1 + I \begin{pmatrix} d_3 \sin(\theta_2)\cos(\theta_1) - d_2 \sin(\theta_1) \\ d_3 \sin(\theta_2)\sin(\theta_1) + d_2 \cos(\theta_1) \\ d_3 \cos(\theta_2) + d_1 \end{pmatrix} \tag{19.20} $$

These equations correspond to the right hand sides of the four groups of equations of table 19.2. For simplicity we use the abbreviations s_i for $\sin(\theta_i)$ and c_i for $\cos(\theta_i)$. Using the third equation of the rendezvous frame $\mathcal{F}_1$, we

Table 19.2. Rendezvous equations obtained for $\boldsymbol{P}_3^o$ regarding frames $\mathcal{F}_0, \mathcal{F}_1, \mathcal{F}_2$ and $\mathcal{F}_3$

Frame	Eq.	forward		backward
	1	P_x	$=$	$d_3 s_2 c_1 - d_2 s_1$
$\mathcal{F}_0$	2	P_y	$=$	$d_3 s_2 c_1 + d_2 c_1$
	3	P_z	$=$	$d_3 c_2 + d_1$
	1	$P_y s_1 + P_x c_1$	$=$	$d_3 s_2$
$\mathcal{F}_1$	2	$d_1 - P_z$	$=$	$-d_3 c_2$
	3	$P_y c_1 - P_x s_1$	$=$	d_2
	1	$-P_z s_2 + d_1 s_2 + P_x c_1 c_2 + P_y s_1 c_2$	$=$	0
$\mathcal{F}_2$	2	$d_2 - P_y c_1 + P_x s_1$	$=$	0
	3	$P_z c_2 - d_1 c_2 + P_x c_1 s_2 + P_y s_1 s_2$	$=$	d_3
	1	$-P_z s_2 + d_1 s_2 + P_x c_1 c_2 + P_y s_1 c_2$	$=$	0
$\mathcal{F}_3$	2	$d_2 - P_y c_1 + P_x s_1$	$=$	0
	3	$P_z c_2 - d_1 c_2 + P_x c_1 s_2 + P_y s_1 s_2 - d_3$	$=$	0

compute

$$\theta_1 = \arctan_2(x_{1/2}, y_{1/2}) \,, \tag{19.21}$$

where

$$x_{1/2} = \frac{d_2 - P_y y_{1/2}}{-P_x} \,, \qquad y_{1/2} = \frac{P_y d_2 \pm P_x \sqrt{P_x^2 + P_y^2 - d_2^2}}{P_x^2 + P_y^2} \tag{19.22}$$

and

$$\arctan_2(x,y) = \begin{cases} \arctan(\frac{x}{y}) & : \; y > 0 \\ \frac{\pi}{2} & : \; y = 0 \text{ and } x > 0 \\ \text{undefined} & : \; y = 0 \text{ and } x = 0 \\ -\frac{\pi}{2} & : \; y = 0 \text{ and } x < 0 \\ \arctan(\frac{x}{y}) + \pi & : \; y < 0 \end{cases} \,. \tag{19.23}$$

This gives two values for θ_1. Now let us look for d_3 and θ_2. For that we consider the first and second equation of the rendezvous frame $\mathcal{F}_1$. With $a_{1/2} = P_y x_{1/2} + P_x y_{1/2}$ and $b = P_z - d_1$ we get two values for d_3. Since for the Stanford manipulator d_3 must be positive, we choose

$$d_{3_{1/2}} = \sqrt{a_{1/2}^2 + b^2} \,. \tag{19.24}$$

Using this value in equations 1 and 2, we compute straightforwardly

$$\theta_2 = \arctan_2\left(\frac{a_{1/2}}{d_{3_{1/2}}}, \frac{b}{d_{3_{1/2}}}\right) . \tag{19.25}$$

19.4.3 Computing θ_4 and θ_5 Using a Line Representation

These variables will be computed using the joint–transition from $\mathcal{F}_3$ to $\mathcal{F}_6$. According to the geometric characteristics of the manipulator it appears appealing that we should use the line representation to set up an appropriate equation system. The representation ${}^0\boldsymbol{L}_6^z$ of the line $\boldsymbol{L}_6^z$ in frame $\mathcal{F}_0$ can be computed using ${}^0\boldsymbol{M}_6$

$${}^0\boldsymbol{L}_6^z = {}^0\boldsymbol{M}_6 \, {}^6\boldsymbol{L}_6^z \, {}^0\widetilde{\boldsymbol{M}}_6 = {}^0\boldsymbol{M}_6 \left(\begin{pmatrix} 0 \\ 0 \\ 1 \end{pmatrix} + I \begin{pmatrix} 0 \\ 0 \\ 0 \end{pmatrix} \right) {}^0\widetilde{\boldsymbol{M}}_6 \,. \tag{19.26}$$

Since the z-axis of $\mathcal{F}_6$ frame crosses the origin of $\mathcal{F}_3$, we can see that the z-axis line related to this frame has zero moment. Thus we can claim that L_6^z in $\mathcal{F}_3$ frame is

$$ {}^{3}\boldsymbol{L}_6^z = {}^{3}\boldsymbol{M}_0\ {}^{0}\boldsymbol{L}_6^z\ {}^{3}\widetilde{\boldsymbol{M}}_0 = \begin{pmatrix} A_x \\ A_y \\ A_z \end{pmatrix} + I \begin{pmatrix} 0 \\ 0 \\ 0 \end{pmatrix} . \tag{19.27} $$

Note that ${}^{3}\boldsymbol{M}_0$ is known since we have already computed θ_1, θ_2 and d_3.

Now applying successively forward transformations as follows

$$ \begin{aligned} {}^{4}\boldsymbol{L}_6^z &= {}^{4}\boldsymbol{M}_3\ {}^{3}\boldsymbol{L}_6^z\ {}^{4}\widetilde{\boldsymbol{M}}_3 , \\ {}^{5}\boldsymbol{L}_6^z &= {}^{5}\boldsymbol{M}_4\ {}^{4}\boldsymbol{L}_6^z\ {}^{5}\widetilde{\boldsymbol{M}}_4 , \\ {}^{6}\boldsymbol{L}_6^z &= {}^{6}\boldsymbol{M}_5\ {}^{5}\boldsymbol{L}_6^z\ {}^{6}\widetilde{\boldsymbol{M}}_5 , \end{aligned} \tag{19.28} $$

we get the left hand sides of the four groups of equations of table 19.3. The z-axis line $\boldsymbol{L}_6^z$ of $\mathcal{F}_6$ represented in $\mathcal{F}_6$ has zero moment, thus it can be expressed as

$$ {}^{6}L_6^z = \begin{pmatrix} 0 \\ 0 \\ 1 \end{pmatrix} + I \begin{pmatrix} 0 \\ 0 \\ 0 \end{pmatrix} . \tag{19.29} $$

Now applying successive backward transformations, we have

$$ \begin{aligned} {}^{5}\boldsymbol{L}_6^z &= {}^{5}\boldsymbol{M}_6\ {}^{6}\boldsymbol{L}_6^z\ {}^{5}\widetilde{\boldsymbol{M}}_6 , \\ {}^{4}\boldsymbol{L}_6^z &= {}^{4}\boldsymbol{M}_5\ {}^{5}\boldsymbol{L}_6^z\ {}^{4}\widetilde{\boldsymbol{M}}_5 , \\ {}^{3}\boldsymbol{L}_6^z &= {}^{3}\boldsymbol{M}_4\ {}^{4}\boldsymbol{L}_6^z\ {}^{3}\widetilde{\boldsymbol{M}}_4 . \end{aligned} \tag{19.30} $$

Using our Maple program, we compute the right hand sides of the four groups of equations of table 19.3. We will consider the equations of rendezvous frame

Table 19.3. Rendezvous equations obtained for $\boldsymbol{L}_6^z$ regarding frames $\mathcal{F}_3, \mathcal{F}_4, \mathcal{F}_5$ and $\mathcal{F}_6$

Frame	Eq.	forward		backward
	1	A_x	$=$	$-c_4 s_5$
$\mathcal{F}_3$	2	A_y	$=$	$-s_4 s_5$
	3	A_z	$=$	$-c_5$
	1	$A_y s_4 + A_x c_4$	$=$	$-s_5$
$\mathcal{F}_4$	2	A_z	$=$	$-c_5$
	3	$A_y c_4 - A_x s_4$	$=$	0
	1	$-A_z s_5 + A_x c_4 c_5 + A_y s_4 c_5 c_6$	$=$	0
$\mathcal{F}_5$	2	$A_y c_4 - A_x s_4$	$=$	0
	3	$-A_z c_5 - A_x c_4 s_5 - A_y s_4 s_5$	$=$	1
	1	$A_x s_4 s_6 - A_y c_4 s_6 + A_y s_4 c_5 c_6 + A_x c_4 c_5 c_6 - A_z s_5 c_6$	$=$	0
$\mathcal{F}_6$	2	$-A_x s_4 c_6 + A_y c_4 c_6 + A_y s_4 c_5 c_6 + A_x c_4 c_5 s_6 - A_z s_5 s_6$	$=$	0
	3	$-A_z c_5 - A_x c_4 s_5 - A_y s_4 s_5$	$=$	1

$\mathcal{F}_4$. Using the third equation, we compute

$$\theta_4 = \arctan_2(x_{1/2}, y_{1/2}) \,, \tag{19.31}$$

where

$$x_{1/2} = -\frac{A_y y_{1/2}}{-Ax} = \pm\frac{A_y}{\sqrt{A_x^2 + A_y^2}} \,, \qquad y_{1/2} = \pm\frac{A_x}{\sqrt{A_x^2 + A_y^2}} \,. \tag{19.32}$$

This results in two values for θ_4 which substituted in the first and second equation helps us to find two solutions for θ_5

$$\theta_5 = \arctan_2(s_5, c_5) = \arctan_2\left((-A_y s_4 - A_x c_4), -A_z\right) . \tag{19.33}$$

19.4.4 Computing θ_6 Using a Plane Representation

Since θ_1, θ_2, d_3, θ_4 and θ_5 are now known, we can compute the motor ${}^5\boldsymbol{M}_0$. The yz–plane $\boldsymbol{H}_6^{yz}$ represented in $\mathcal{F}_6$ has the Hesse distance 0, thus

$${}^6\boldsymbol{H}_6^{yz} = \begin{pmatrix} 1 \\ 0 \\ 0 \end{pmatrix} + I0 = \begin{pmatrix} 1 \\ 0 \\ 0 \end{pmatrix} . \tag{19.34}$$

Its transformation to $\mathcal{F}_0$ reads

$${}^0\boldsymbol{H}_6^{yz} = {}^0\boldsymbol{M}_6 \, {}^6\boldsymbol{H}_6^{yz} \, {}^0\widetilde{\overline{\boldsymbol{M}}}_6 = {}^0\boldsymbol{M}_6 \begin{pmatrix} 1 \\ 0 \\ 0 \end{pmatrix} {}^0\widetilde{\overline{\boldsymbol{M}}}_6 \,. \tag{19.35}$$

Now we compute ${}^5\boldsymbol{H}_6^{yz}$ by

$${}^5\boldsymbol{H}_6^{yz} = {}^5\boldsymbol{M}_0 \, {}^0\boldsymbol{H}_6^{yz} \, {}^5\widetilde{\overline{\boldsymbol{M}}}_0 = \begin{pmatrix} N_x \\ N_y \\ N_z \end{pmatrix} + I \, {}^5d_{H_6^{yz}} \,. \tag{19.36}$$

The orientation bivector $(N_x, N_y, N_z)^T$ describes the orientation of the yz-plane of frame $\mathcal{F}_6$ in frame $\mathcal{F}_5$ given the values of the joint variables $\theta_1, \theta_2, \theta_4, \theta_5$ and d_3. Now applying forward transformation from $\mathcal{F}_5$ to $\mathcal{F}_6$, we obtain

$${}^6\boldsymbol{H}_6^{yz} = {}^6\boldsymbol{M}_5 \, {}^5\boldsymbol{H}_6^{yz} \, {}^6\widetilde{\overline{\boldsymbol{M}}}_5 \,. \tag{19.37}$$

Using our Maple program, we get the left hand sides of the two groups of equations of the table 19.4. Since the values for θ_1, θ_2, d_3, θ_4 and θ_5 are not

Table 19.4. Rendezvous equations obtained for $\boldsymbol{H}_6^{yz}$ regarding frames $\mathcal{F}_5$ and $\mathcal{F}_6$

Frame	Eq.	forward		backward
	1	N_x	=	c_6
$\mathcal{F}_5$	2	N_y	=	s_6
	3	N_z	=	0
	1	$N_y s_6 + N_x c_6$	=	1
$\mathcal{F}_6$	2	$N_x s_6 - N_y c_6$	=	0
	3	N_z	=	0

unique we, will get different values for the equations. Applying ${}^5\boldsymbol{M}_6$ to ${}^6\boldsymbol{H}_6^{yz}$ we get, the right hand sides of the two groups of equations of table 19.4 by

$$
{}^5\boldsymbol{H}_6^{yz} = {}^5\boldsymbol{M}_6\, {}^6\boldsymbol{H}_6^{yz}\, {}^5\widetilde{\boldsymbol{M}}_6 = {}^5\boldsymbol{M}_6 \begin{pmatrix} 1 \\ 0 \\ 0 \end{pmatrix} {}^5\widetilde{\boldsymbol{M}}_6 = \begin{pmatrix} \sin(\theta_6) \\ \cos(\theta_6) \\ 0 \end{pmatrix} . \tag{19.38}
$$

We will consider the equations of the rendezvous frame $\mathcal{F}_5$. Using the first and second equation, we can compute θ_6 by

$$
\theta_6 = \arctan_2(s_6, c_6) = \arctan_2(N_x, N_y) . \tag{19.39}
$$

Note that since we had two values for θ_4 and two values for θ_5, there is more than one solution for θ_6.

19.5 Conclusion

This chapter presented the application of the algebra of motors for the treatment of the direct and inverse kinematics of robot manipulators. When dealing with 3D rigid motion it is usual to use homogeneous coordinates in the 4D space to linearize this non–linear 3D transformation. With the same effect we model the prismatic and revolute motion of points, lines and planes using motors which are equivalent to screws. The fact that in our approach we can also use the representation of planes widens up the geometric language for the treatment of robotic problems.

The chapter has shown the flexibility of the motor algebra approach for the solution of the direct and inverse kinematics of robot manipulators. Using a standard robot manipulator, we show that according to the need we can resort for solving its inverse kinematics either to a point, a line or a plane representation. Thus, the main contribution of this chapter is to show that while preserving the geometric insight during the computation our approach gains more flexibility. The authors of this chapter believe that the increasing complexity of future multi–links mechanisms will profit from the versatility of the motor algebra framework.

20. Using the Algebra of Dual Quaternions for Motion Alignment

Kostas Daniilidis

GRASP Laboratory,
University of Pennsylvania, Philadelphia

20.1 Introduction

Whenever measurements have to be taken with respect to two different coordinate frames the problem arises how to relate these measurements to each other. When these measurements are rigid 3D-displacements we obtain descriptions of motions with respect to two different coordinate systems. These systems might for example be the motor coordinate system of a vehicle and the coordinate system of a sensor mounted on the vehicle. If we use conventional homogeneous coordinates notation we usually obtain the well known equation $\boldsymbol{AX} = \boldsymbol{XB}$ where all the variables are 4×4 matrices representing rigid motions. On the other hand, we might have line measurements with respect to two coordinate frames in which case we usually have a problem of the form $\boldsymbol{P} = \boldsymbol{QX}$ where $\boldsymbol{P}, \boldsymbol{Q}$ are matrices containing the Plücker coordinates of the lines and $\boldsymbol{X}$ a matrix encoding the rigid motion which we will describe later. We will first make a short break in our motivation in order to introduce the Clifford algebra we will use. Our geometric algebra treatment is inspired by [154], [167], and [22]. Then, we will describe 3D-lines and 3D-motions in this framework and we will present two examples of the algorithmic superiority of this representation.

20.2 Even Subalgebras of Non-degenerate $\mathcal{R}^{p,q,r}$

We briefly repeat some facts in order to facilitate a smooth transition from the other chapters to our notation. Assume n basis vectors $e_1, e_2, \dots, e_n$ of an n-dimensional vector space on the reals and define a vector product as follows

$$\begin{aligned} e_i e_j &= -e_j e_i && \text{for} \quad i \neq j \\ e_i^2 &= 1 && \text{for} \quad i = 1, \dots, p \\ e_i^2 &= -1 && \text{for} \quad i = p+1, \dots, p+q \\ e_i^2 &= 0 && \text{for} \quad i = p+q+1, \dots, p+q+r = n. \end{aligned} \tag{20.1}$$

Choose then m basis vectors and consider all linear combinations of the $\binom{n}{m}$ products of them. The result is defined as a multivector of rank p. For example, for $n = 3$ and $m = 2$ we obtain the bivector $a_1 e_2 e_3 + a_2 e_1 e_3 + a_3 e_1 e_2$. The multivector of rank n contains only one component $e_1 \dots e_n$ called *pseudoscalar.*

If we consider the sum of multivectors of all ranks we obtain a vector space of dimension 2^n which with the above product defines an 2^n-dimensional associative algebra called the *Clifford algebra* $\mathcal{R}^{p,q,r}$. The exponents p, q and r denote the cardinalities of the three different kinds of basis vectors in (20.1).

Algebras $\mathcal{R}^{p,q,r}$ with $r \neq 0$ – which means with no basis vectors squaring to zero – are called *non-degenerate algebras.*

Consider now the subset of $\mathcal{R}^{p,q,r}$ containing multivectors of only *even* rank. This is a subalgebra of $\mathcal{R}^{p,q,r}$ since the product of two products with even numbers of basis vectors contains also an even number of basis vectors.

Our first example is the even subalgebra of $\mathcal{R}^{2,0,0}$ consisting of all numbers

$$a + b e_1 e_2.$$

The square of the pseudoscalar $(e_1 e_2)(e_1 e_2) = -e_1 e_2 e_2 e_1 = -1$. Hence, this even subalgebra is isomorphic to the complex numbers $a + bi$ if we identify i with the pseudoscalar $e_1 e_2$. We have the freedom to write a point or vector (x, y) in the plane either as $x + y e_1 e_2$ or as $x e_1 + y e_2$. The rotation of a vector by angle ϕ can then be written either as

$$(x + y e_1 e_2)(\cos\phi + \sin\phi e_1 e_2)$$

or as

$$(\cos\frac{\phi}{2} - \sin\frac{\phi}{2} e_1 e_2)(x e_1 + y e_2)(\cos\frac{\phi}{2} + \sin\frac{\phi}{2} e_1 e_2).$$

We increase the dimension by one and consider the even subalgebra of $\mathcal{R}^{3,0,0}$ with elements of the form

$$q_0 + q_1 \boldsymbol{e}_2\boldsymbol{e}_3 + q_2 \boldsymbol{e}_1\boldsymbol{e}_3 + q_3 \boldsymbol{e}_1\boldsymbol{e}_2.$$

If we set $i = \boldsymbol{e}_2\boldsymbol{e}_3$, $j = \boldsymbol{e}_1\boldsymbol{e}_3$, and $k = \boldsymbol{e}_1\boldsymbol{e}_2$ we obtain $i^2 = j^2 = k^2 = -1$ and $ij = k = -ji, jk = i = -kj, ki = j = -ik$. The above described subalgebra is the associative algebra of quaternions.

If we summarize the three real coefficients of the *bivector* part of a quaternion $\boldsymbol{q}$ into a vector $\boldsymbol{q}$ and write the quaternion as a pair $(q_0, \boldsymbol{q})$ then the quaternion product can be written in terms of inner and cross products, $\boldsymbol{p}^T\boldsymbol{q}$ and $\boldsymbol{p} \times \boldsymbol{q}$, respectively:

$$\boldsymbol{p}\boldsymbol{q} = (p_0 q_0 - \boldsymbol{p}^T\boldsymbol{q}, p_0\boldsymbol{q} + q_0\boldsymbol{p} + \boldsymbol{p} \times \boldsymbol{q}). \tag{20.2}$$

The norm of a quaternion is defined via the conjugate quaternion $\bar{\boldsymbol{q}}$ as $|\boldsymbol{q}|^2 = \boldsymbol{q}\bar{\boldsymbol{q}}$. The unit quaternions ($\boldsymbol{q}\bar{\boldsymbol{q}} = 1$) act isomorphically to the group of rotations in three-dimensions. If the rotation axis is the unit vector (n_x, n_y, n_z) and the angle of rotation is θ then the unit quaternion representing this rotation reads

$$\boldsymbol{q} = \cos\frac{\theta}{2} + \sin\frac{\theta}{2}(n_x i + n_y j + n_z k). \tag{20.3}$$

Using the unit quaternions we have two ways to describe a rotation in 3D-space dependent on whether we stay inside the even subalgebra or not. We can describe points of $\mathbb{R}^3$ with bivectors $xi + yj + kz$ also called vector-quaternions.

On the other hand we can describe vectors as $x\boldsymbol{e}_1 + y\boldsymbol{e}_2 + z\boldsymbol{e}_3$ which is the natural representation for vectors in geometric algebra. In the former case the bivector $\boldsymbol{x}$ is rotated into the bivector $\boldsymbol{q}\boldsymbol{x}\bar{\boldsymbol{q}}$. In the latter case, the vector $\boldsymbol{x}$ is rotated into the vector $\boldsymbol{r}\boldsymbol{x}\bar{\boldsymbol{r}}$ where $\boldsymbol{r}$ is called a rotor and equals

$$\begin{aligned} \boldsymbol{r} &= q_0 + q_1\boldsymbol{e}_2\boldsymbol{e}_3 - q_2\boldsymbol{e}_1\boldsymbol{e}_3 + q_3\boldsymbol{e}_1\boldsymbol{e}_2 \\ &= q_0 + \omega(q_1\boldsymbol{e}_1 + q_2\boldsymbol{e}_2 + q_3\boldsymbol{e}_3) \\ &= q_0 + \omega\boldsymbol{q}, \end{aligned} \tag{20.4}$$

where ω is the pseudoscalar $\boldsymbol{e}_1\boldsymbol{e}_2\boldsymbol{e}_3$. It can be easily proved that the pseudoscalar commutes with vectors and that $\omega^2 = -1$.

20.3 Even Subalgebras of Degenerate $\mathcal{R}^{p,q,r}$

We already mentioned that if any of the basis vectors $\boldsymbol{e}_i$ squares to zero the algebra is called degenerate. Let us consider the even subalgebra of $\mathcal{R}^{1,0,1}$ with $\boldsymbol{e}_1^2 = 1$ and $\boldsymbol{e}_2^2 = 0$. The pseudoscalar squares then also to zero:

$$\boldsymbol{e}_1\boldsymbol{e}_2\boldsymbol{e}_1\boldsymbol{e}_2 = -\boldsymbol{e}_1\boldsymbol{e}_2\boldsymbol{e}_2\boldsymbol{e}_1 = 0.$$

We call it ϵ and the elements of the even subalgebra are the *dual* numbers invented by Clifford [45] and further developed by Study [222]:

$$a + b\boldsymbol{e}_1\boldsymbol{e}_2 = a + b\epsilon \quad \text{where} \quad \epsilon^2 = 0.$$

Considering the sum and the product, the dual numbers are an abelian ring but not a field because only dual numbers with not vanishing real part possess an inverse element. An important property is associated with the derivatives of functions with dual arguments. Since all powers greater equal two of ϵ vanish a Taylor expansion always yields

$$f(a + \epsilon b) = f(a) + \epsilon b f'(a). \tag{20.5}$$

We jump to the four-dimensional Clifford algebra $\mathcal{R}^{3,0,1}$ with

$$\boldsymbol{e}_1^2 = \boldsymbol{e}_2^2 = \boldsymbol{e}_3^2 = 1 \quad \text{and} \quad \boldsymbol{e}_4^2 = 0.$$

Its even subalgebra consists of elements $q_0 + q_1\boldsymbol{e}_2\boldsymbol{e}_3 + q_2\boldsymbol{e}_1\boldsymbol{e}_3 + q_3\boldsymbol{e}_1\boldsymbol{e}_2 + q_1'\boldsymbol{e}_4\boldsymbol{e}_1 + q_2'\boldsymbol{e}_2\boldsymbol{e}_4 + q_3'\boldsymbol{e}_4\boldsymbol{e}_3 + q_0'\boldsymbol{e}_1\boldsymbol{e}_2\boldsymbol{e}_2\boldsymbol{e}_3$. Let us denote the pseudoscalar with ϵ. Then we can easily prove the following facts:

1. $\epsilon^2 = 0$;
2. ϵ commutes with bivectors: $\epsilon\boldsymbol{e}_i\boldsymbol{e}_j = \boldsymbol{e}_i\boldsymbol{e}_j\epsilon$;
3. $$q_1'\boldsymbol{e}_4\boldsymbol{e}_1 + q_2'\boldsymbol{e}_2\boldsymbol{e}_4 + q_3'\boldsymbol{e}_4\boldsymbol{e}_3 = \epsilon(q_1'\boldsymbol{e}_2\boldsymbol{e}_3 + q_2'\boldsymbol{e}_1\boldsymbol{e}_3 + q_3'\boldsymbol{e}_1\boldsymbol{e}_2).$$

Because of (2) and (3) we can write every element of the even subalgebra of $\mathcal{R}^{3,0,1}$ as

$$(q_0 + \epsilon q_0') + (q_1 + \epsilon q_1')\boldsymbol{e}_2\boldsymbol{e}_3 + (q_2 + \epsilon q_2')\boldsymbol{e}_1\boldsymbol{e}_3 + (q_3 + \epsilon q_3')\boldsymbol{e}_1\boldsymbol{e}_2$$

It is now obvious that this is a quaternion with the real coefficients having been replaced by dual numbers. It is also known as *dual quaternion* and has been discovered almost simultaneously with the dual numbers [45, 222]. With respect to addition and multiplication the even subalgebra of $\mathcal{R}^{3,0,1}$ is a non-abelian ring with unit element $(1, \boldsymbol{0})$ and an associative algebra over the dual numbers. We will denote the dual quaternions with $\check{\boldsymbol{q}}$ in order to differentiate them from the quaternions. Thus, a dual quaternion can be written as the sum of a non-dual and a dual part $\check{\boldsymbol{q}} = \boldsymbol{q} + \epsilon\boldsymbol{q}'$, but also as a pair of a dual scalar and a dual bivector $\check{\boldsymbol{q}} = (\check{q}_0, \check{\boldsymbol{q}})$. Dual bivectors $\check{\boldsymbol{q}}$ can be written as dual bivectors $(0, \check{\boldsymbol{q}})$ and their product property

$$(0, \check{\boldsymbol{q}}_1)(0, \check{\boldsymbol{q}}_2) = (-\check{\boldsymbol{q}}_1^T\check{\boldsymbol{q}}_2, \check{\boldsymbol{q}}_1 \times \check{\boldsymbol{q}}_2). \tag{20.6}$$

The norm of a dual quaternion is defined as $\|\check{\boldsymbol{q}}\|^2 = \check{\boldsymbol{q}}\bar{\check{\boldsymbol{q}}}$ and is a dual number with positive real part. If the norm has a non vanishing real part then the dual quaternion has an inverse $\check{\boldsymbol{q}}^{-1} = \|\check{\boldsymbol{q}}\|^{-1}\bar{\check{\boldsymbol{q}}}$. If the norm is equal one then an inverse element exists and is equal to the conjugate quaternion. If $\check{\boldsymbol{q}} = \boldsymbol{q} + \epsilon \boldsymbol{q}'$ then the unity condition $\check{\boldsymbol{q}}\bar{\check{\boldsymbol{q}}} = 1$ can be written

$$\boldsymbol{q}\bar{\boldsymbol{q}} = 1 \quad \text{and} \quad \bar{\boldsymbol{q}}\boldsymbol{q}' + \bar{\boldsymbol{q}}'\boldsymbol{q} = 0. \tag{20.7}$$

As we shall describe in the following, unit dual quaternions represent general motions of lines and the expression $\check{\boldsymbol{q}}\check{\boldsymbol{x}}\bar{\check{\boldsymbol{q}}}$ valid for rotation of points in case of real quaternions is also true for general motion of lines in case of dual quaternions.

A line in space with direction $\boldsymbol{l}$ through a point $\boldsymbol{p}$ can be represented with the 6-tuple of the Plücker coordinates $(\boldsymbol{l}, \boldsymbol{m})$ where $\boldsymbol{m}$ is called the line moment and is equal to $\boldsymbol{p} \times \boldsymbol{l}$. The line moment is normal to the plane through the line and the origin with magnitude equal to the distance from the line to the origin. The constraints $\|\boldsymbol{l}\| = 1$ and $\boldsymbol{l}^T \boldsymbol{m} = 0$ guarantee that the degrees of freedom of an arbitrary line in space are four. On the other hand a line can be written as the outer product of two points (vectors) and is thus a bivector which can be written as

$$l_1 \boldsymbol{e}_2 \boldsymbol{e}_3 + l_2 \boldsymbol{e}_1 \boldsymbol{e}_3 + l_3 \boldsymbol{e}_1 \boldsymbol{e}_2 + m_1 \boldsymbol{e}_4 \boldsymbol{e}_1 + m_2 \boldsymbol{e}_2 \boldsymbol{e}_4 + m_3 \boldsymbol{e}_4 \boldsymbol{e}_3$$

where $\boldsymbol{l} = (l_1, l_2, l_3)$ and $\boldsymbol{m} = (m_1, m_2, m_3)$ the Plücker coordinates defined above.

20.4 Line Transformation

We will next prove that the law that rotates points $\boldsymbol{q}\boldsymbol{x}\bar{\boldsymbol{q}}$ is exactly the same as the law that translates *and* rotates lines in space if we replace the vector with a line and the quaternion with a dual quaternion.

Proposition 20.4.1. *If a line given by the bivector $\check{\boldsymbol{l}}_a = \boldsymbol{l}_a + \epsilon \boldsymbol{m}_a$ is transformed with a rotation $\boldsymbol{R}$ and a translation $\boldsymbol{t}$ into a line $\check{\boldsymbol{l}}_b$ then a unit dual quaternion $\check{\boldsymbol{q}}$ exists such that $\check{\boldsymbol{l}}_a = \check{\boldsymbol{q}}\check{\boldsymbol{l}}_b\bar{\check{\boldsymbol{q}}}$.*

Proof. Applying a rotation $\boldsymbol{R}$ and a translation $\boldsymbol{t}$ to a line $(\boldsymbol{l}_b, \boldsymbol{m}_b)$ we obtain the transformed line $(\boldsymbol{l}_a, \boldsymbol{m}_a)$

$$\boldsymbol{l}_a = \boldsymbol{R}\boldsymbol{l}_b \tag{20.8}$$

$$\begin{aligned} \boldsymbol{m}_a &= \boldsymbol{p}_a \times \boldsymbol{l}_a = (\boldsymbol{R}\boldsymbol{p}_b + \boldsymbol{t}) \times \boldsymbol{R}\boldsymbol{l}_b \\ &= \boldsymbol{R}(\boldsymbol{p}_b \times \boldsymbol{l}_b) + \boldsymbol{t} \times \boldsymbol{R}\boldsymbol{l}_b \\ &= \boldsymbol{R}\boldsymbol{m}_b + \boldsymbol{t} \times \boldsymbol{R}\boldsymbol{l}_b. \end{aligned} \tag{20.9}$$

We change from vector to quaternion notation which means that the vector $\boldsymbol{l}$ is represented by a quaternion with zero scalar part $\boldsymbol{l} = (0, \boldsymbol{l})$. The terms containing rotation can be easily written with quaternions. The difficulty with the cross-product is tackled with the identity

$$(0, \boldsymbol{t} \times \boldsymbol{q}) = \frac{1}{2}(\boldsymbol{q}\bar{\boldsymbol{t}} + \boldsymbol{t}\boldsymbol{q}) \qquad (20.10)$$

where $\boldsymbol{t}$ is the translation quaternion $(0, \boldsymbol{t})$ and $\boldsymbol{q}$ the rotation quaternion $(0, \boldsymbol{q})$. Using the identity (20.10) we obtain

$$\begin{aligned} \boldsymbol{l}_a &= \boldsymbol{q}\boldsymbol{l}_b\bar{\boldsymbol{q}} \\ \boldsymbol{m}_a &= \boldsymbol{q}\boldsymbol{m}_b\bar{\boldsymbol{q}} + \frac{1}{2}(\boldsymbol{q}\boldsymbol{l}_b\bar{\boldsymbol{q}}\bar{\boldsymbol{t}} + \boldsymbol{t}\boldsymbol{q}\boldsymbol{l}_b\bar{\boldsymbol{q}}). \end{aligned} \qquad (20.11)$$

We define a new quaternion $\boldsymbol{q}' = \frac{1}{2}\boldsymbol{t}\boldsymbol{q}$ and a dual quaternion $\check{\boldsymbol{q}} = \boldsymbol{q} + \epsilon\boldsymbol{q}'$. It can be easily shown that (20.11) is equivalent to

$$\boldsymbol{l}_a + \epsilon\boldsymbol{m}_a = (\boldsymbol{q} + \epsilon\boldsymbol{q}')(\boldsymbol{l}_b + \epsilon\boldsymbol{m}_b)(\bar{\boldsymbol{q}} + \epsilon\bar{\boldsymbol{q}'}). \qquad (20.12)$$

Denoting also the lines by dual quaternions $\check{\boldsymbol{l}}_a$ and $\check{\boldsymbol{l}}_b$ we obtain

$$\check{\boldsymbol{l}}_a = \check{\boldsymbol{q}}\check{\boldsymbol{l}}_b\bar{\check{\boldsymbol{q}}}$$

□

Lines can thus be rigidly transformed using a single operation (multiplying left and right with the conjugate) in the non-abelian ring of dual quaternions. The norm

$$|\check{\boldsymbol{q}}|^2 = \check{\boldsymbol{q}}\bar{\check{\boldsymbol{q}}} = \boldsymbol{q}\bar{\boldsymbol{q}} + \epsilon(\boldsymbol{q}\bar{\boldsymbol{q}'} + \boldsymbol{q}'\bar{\boldsymbol{q}}) = \boldsymbol{q}\bar{\boldsymbol{q}} + \epsilon/2(\boldsymbol{q}\bar{\boldsymbol{q}}\bar{\boldsymbol{t}} + \boldsymbol{t}\boldsymbol{q}\bar{\boldsymbol{q}}) = 1$$

hence $\check{\boldsymbol{q}}$ is a unit dual quaternion. The above relations give also explicitly the transformation from $(\boldsymbol{R}, \boldsymbol{t})$ to $\boldsymbol{q} + \epsilon\boldsymbol{q}'$. The dual part $\boldsymbol{q}' = \frac{1}{2}\boldsymbol{t}\boldsymbol{q}$ and the quaternion $\boldsymbol{q}$ can be obtained from the rotation matrix by finding the axis and the angle of rotation. If $\check{\boldsymbol{q}}$ is a solution then $-\check{\boldsymbol{q}}$ is also a solution. It is sufficient to enforce like in non-dual quaternions that the scalar non-dual part is positive in order to eliminate this ambiguity.

Reversely, the translation $\boldsymbol{t}$ can be recovered from the dual quaternion as

$$\boldsymbol{t} = 2\boldsymbol{q}'\bar{\boldsymbol{q}}. \qquad (20.13)$$

The unit dual quaternion $\check{\boldsymbol{q}}$ can be written as the concatenation of a pure translational unit dual quaternion and a pure rotational quaternion with dual part equal zero i.e.

$$\check{q} = (1, \epsilon\frac{\boldsymbol{t}}{2})\boldsymbol{q}.$$

20.5 Motion Estimation from 3D-Line Matches

If $((\boldsymbol{l}_a, \boldsymbol{m}_a), (\boldsymbol{l}_b, \boldsymbol{m}_b))$ are the Plücker coordinates before and after the motion, respectively, then given at least two non-parallel lines we can estimate the rotation from eq. (20.8) and then insert it into (20.9) in order to solve for the translation. We next propose a novel algorithm which simultaneously solves for rotation and translation without requiring a non-linear minimization as in existing algorithms [196].

We split eq. (20.12) in its non-dual part

$$\boldsymbol{l}_a = \boldsymbol{q}\boldsymbol{l}_b\bar{\boldsymbol{q}} \tag{20.14}$$

and its dual part

$$\boldsymbol{m}_a = \boldsymbol{q}\boldsymbol{l}_b\bar{\boldsymbol{q}}' + \boldsymbol{q}\boldsymbol{m}_b\bar{\boldsymbol{q}} + \boldsymbol{q}'\boldsymbol{l}_b\bar{\boldsymbol{q}}. \tag{20.15}$$

Multiplying both equations on the right with $\boldsymbol{q}$ and applying the identity $\bar{\boldsymbol{q}}\boldsymbol{q}' + \bar{\boldsymbol{q}}'\boldsymbol{q} = 0$ in the first term of the right hand side of the second equation we obtain

$$\begin{aligned}\boldsymbol{l}_a\boldsymbol{q} &= \boldsymbol{q}\boldsymbol{l}_b\\ \boldsymbol{m}_a\boldsymbol{q} &= -\boldsymbol{l}_a\boldsymbol{q}' + \boldsymbol{q}\boldsymbol{m}_b + \boldsymbol{q}'\boldsymbol{l}_b.\end{aligned}$$

The scalar parts of all line quaternions are zero, hence each of the above equations consists actually of three scalar equations. We introduce again the direction and moment vectors of the lines and we rewrite the above equations into a homogeneous linear system

$$\begin{pmatrix} \boldsymbol{l}_a - \boldsymbol{l}_b & [\boldsymbol{l}_a + \boldsymbol{l}_b]_\times & \mathbf{0}_{3\times1} & \mathbf{0}_{3\times3} \\ \boldsymbol{m}_a - \boldsymbol{m}_b & [\boldsymbol{m}_a + \boldsymbol{m}_b]_\times & \boldsymbol{l}_a - \boldsymbol{l}_b & [\boldsymbol{l}_a + \boldsymbol{l}_b]_\times \end{pmatrix} \begin{pmatrix} \boldsymbol{q} \\ \boldsymbol{q}' \end{pmatrix} = 0 \tag{20.16}$$

where the matrix - we will call $\boldsymbol{S}$ - is a 6×8 matrix and the vector of unknowns $(\boldsymbol{q}^T, \boldsymbol{q}'^T)$ is 8-dimensional.

Recall that we have two constraints on the unknowns so that the result is a unit dual quaternion

$$\boldsymbol{q}^T\boldsymbol{q} = 1 \qquad \text{and} \qquad \boldsymbol{q}^T\boldsymbol{q}' = 0. \tag{20.17}$$

Unfortunately, the six equations are dependent because the vectors $\boldsymbol{l}_a$ and $\boldsymbol{l}_b$ are unit vectors and the vectors $\boldsymbol{m}_a$ and $\boldsymbol{m}_b$ are perpendicular to $\boldsymbol{l}_a$ and $\boldsymbol{l}_b$, respectively, so that two equations are redundant. As already known [196] we need two non-parallel lines correspondences to solve the absolute orientation problem.

Suppose now that $\mathrm{N} \geq 2$ correspondences are given. We construct the $6n \times 8$ matrix

$$\boldsymbol{T} = \left(\boldsymbol{S}_1^T \; \boldsymbol{S}_2^T \; \dots \; \boldsymbol{S}_n^T \right)^T \tag{20.18}$$

which in the noise-free case has rank 6. Since in the noise-free case the equations arise from natural constraints the null-space contains at least the actual solution $(\boldsymbol{q}, \boldsymbol{q}')$. It is trivial to see that an additional orthogonal solution is $(\mathbf{0}_{4\times1}, \boldsymbol{q})$. Hence, the matrix is maximally of rank 6. The solution is the element of the two-parametric kernel that satisfies the two conditions for the dual quaternion to be a unit dual quaternion (20.17).

20.6 The Principle of Transference

We have shown in the last section that the same law is valid for vector rotation as well as for general line displacement. In this section we will show that the dual quaternion representing the rigid displacement has exactly the same form as the rotation quaternion if we replace the rotation angle with a special dual angle and the rotation axis with the Plücker coordinates of the screw axis. According to Chasles' theorem [39] a rigid transformation can be modeled as a rotation *with the same angle* about an axis not through the origin and a translation along this axis. As the screw axis is a line in space it depends on four parameters which together with the rotation angle θ and the translation along the axis d (pitch) constitute the six degrees of freedom of a rigid transformation.

In the following we will compute the pitch d as well as the screw axis given by its direction and moment pair $(\boldsymbol{l}, \boldsymbol{m})$ as a function of the rotation $\boldsymbol{R}$ about an axis through the origin and a translation $\boldsymbol{t}$.

The direction $\boldsymbol{l}$ is parallel to the rotation axis. The pitch d is the projection of translation on the rotation axis, therefore equal $\boldsymbol{t}^T\boldsymbol{l}$. The not mentioned angle θ is the same in both the $(\boldsymbol{R}, \boldsymbol{t})$ and the screw representation. In order to recover the moment $\boldsymbol{m}$ we introduce a point $\boldsymbol{c}$ on the screw axis being the projection of the origin on the axis (Fig. 20.1).

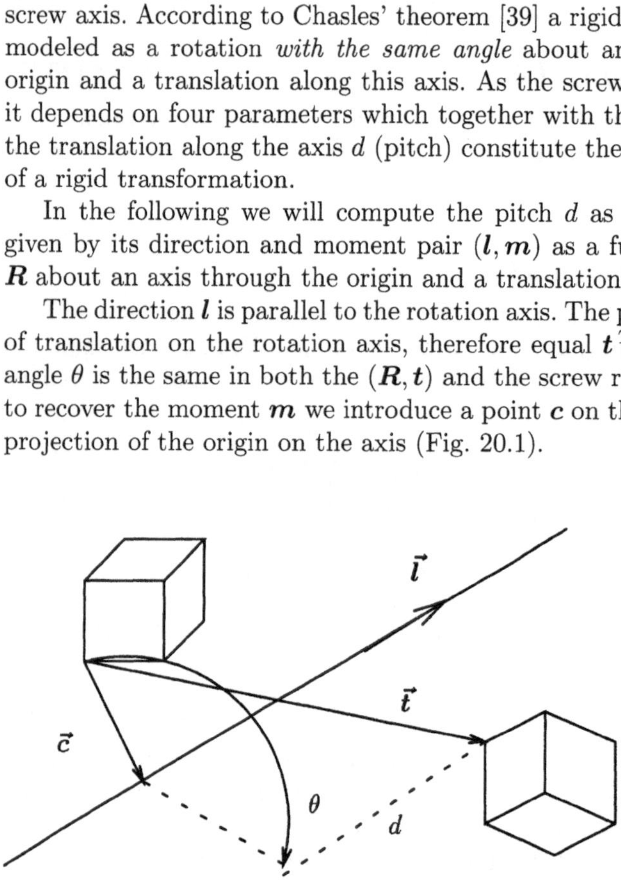

Fig. 20.1. The geometry of a screw: Every motion can be modeled as a rotation with angle θ about an axis at $\boldsymbol{c}$ with direction $\boldsymbol{l}$ and a subsequent translation d along the axis.

The coordinate system is shifted to this point and then transformed. The resulting translation is then $d\boldsymbol{l}+(I-\boldsymbol{R})\boldsymbol{c}$. The so called pitch d reads $d = \boldsymbol{l}^T\boldsymbol{t}$. Using the Rodrigues formula

$$\boldsymbol{R}\boldsymbol{c} = \boldsymbol{c} + \sin(\theta)\boldsymbol{l} \times \boldsymbol{c} + (1 - \cos\theta)\boldsymbol{l} \times (\boldsymbol{l} \times \boldsymbol{c})$$

and $\boldsymbol{c}^T\boldsymbol{l} = 0$ it follows that

$$\boldsymbol{c} = \frac{1}{2}(\boldsymbol{t} - (\boldsymbol{t}^T\boldsymbol{l})\boldsymbol{l} + \cot\frac{\theta}{2}\boldsymbol{l} \times \boldsymbol{t}). \tag{20.19}$$

This point $\boldsymbol{c}$ and hence the screw axis is not defined if the angle θ is either 0 or 180 degrees. Otherwise the moment vector reads then

$$\boldsymbol{m} = \boldsymbol{c} \times \boldsymbol{l} = \frac{1}{2}(\boldsymbol{t} \times \boldsymbol{l} + \boldsymbol{l} \times (\boldsymbol{t} \times \boldsymbol{l}) \cot \frac{\theta}{2}). \tag{20.20}$$

We proceed then with the computation of the corresponding dual quaternion: given the screw parameters $(\theta, d, \boldsymbol{l}, \boldsymbol{m})$.

The quaternion derived from the rotation matrix $\boldsymbol{R}$ reads

$$(q_0, \boldsymbol{q}) = (\cos \frac{\theta}{2}, \sin \frac{\theta}{2} \boldsymbol{l}) \tag{20.21}$$

hence the moment equation (20.20) can be written

$$\sin \frac{\theta}{2} \boldsymbol{m} = \frac{1}{2}(\boldsymbol{t} \times \boldsymbol{q} + q_0 \boldsymbol{t} - \cos \frac{\theta}{2} (\boldsymbol{l}^T \boldsymbol{t}) \boldsymbol{l}).$$

Using $\boldsymbol{l}^T \boldsymbol{t} = d$ and rewriting

$$\sin \frac{\theta}{2} \boldsymbol{m} + \frac{d}{2} \cos \frac{\theta}{2} \boldsymbol{l} = \frac{1}{2}(\boldsymbol{t} \times \boldsymbol{q} + q_0 \boldsymbol{t})$$

which is the vector part of the dual part $\boldsymbol{q}'$ of the dual quaternion $\check{\boldsymbol{q}}$. Applying (20.21) and $\boldsymbol{q}' = \frac{1}{2} \boldsymbol{t} \boldsymbol{q}$ we obtain

$$\check{\boldsymbol{q}} = \begin{pmatrix} q_0 \\ \boldsymbol{q} \end{pmatrix} + \epsilon \begin{pmatrix} -\frac{1}{2} \boldsymbol{q}^T \boldsymbol{t} \\ \frac{1}{2}(q_0 \boldsymbol{t} + \boldsymbol{t} \times \boldsymbol{q}) \end{pmatrix} = \begin{pmatrix} \cos \frac{\theta}{2} \\ \sin \frac{\theta}{2} \boldsymbol{l} \end{pmatrix} + \epsilon \begin{pmatrix} -\frac{d}{2} \sin \frac{\theta}{2} \\ \sin \frac{\theta}{2} \boldsymbol{m} + \frac{d}{2} \cos \frac{\theta}{2} \boldsymbol{l} \end{pmatrix}. \tag{20.22}$$

Every function f of dual numbers obeys the rule

$$f(a + \epsilon b) = f(a) + \epsilon b f'(a)$$

hence

$$\cos(\frac{\theta + \epsilon d}{2}) = \cos \frac{\theta}{2} - \epsilon \frac{d}{2} \sin \frac{\theta}{2} \quad \text{and} \quad \sin(\frac{\theta + \epsilon d}{2}) = \sin \frac{\theta}{2} + \epsilon \frac{d}{2} \cos \frac{\theta}{2}.$$

It is now straightforward to see that a dual quaternion can also be written as

$$\check{\boldsymbol{q}} = \begin{pmatrix} \cos(\frac{\theta + \epsilon d}{2}) \\ \sin(\frac{\theta + \epsilon d}{2})(\boldsymbol{l} + \epsilon \boldsymbol{m}) \end{pmatrix}. \tag{20.23}$$

This representation is very powerful since it algebraically separates the angle and pitch information from the line information characterizing the pose of the screw axis. Moreover, writing the dual angle $\check{\theta} = \theta + \epsilon d$ and the dual vector $\check{\boldsymbol{l}} = \boldsymbol{l} + \epsilon \boldsymbol{m}$ (20.23) becomes equivalent to the pure rotation non-dual equation (20.21). We can easily verify that

$$\check{\boldsymbol{q}} = (\cos\check{\theta}/2, \check{\boldsymbol{l}}\sin\check{\theta}/2)$$

is a unit quaternion $\check{\boldsymbol{q}}\bar{\check{\boldsymbol{q}}} = 1$.

This completes the last part of the proof of one aspect of the well-known *principle of transference* which we quote from Rooney as cited in [162]:

> All valid laws and formulae relating to a system of intersecting line vectors (and hence involving real variables) are equally valid to an equivalent system of skew unit line vectors, if each real variable a in the formulae is replaced by the corresponding dual variable $a + \epsilon a'$.

In the sense of the even subalgebra of $\mathcal{R}^{3,0,1}$ described here we must emphasize that the vectors in the rotation formulae must be written as bivectors before being replaced by the dual quantities. Chevallier [42] shows counterexamples for which the substitution with dual numbers of a theorem for vectors does not lead to a theorem true for lines.

20.7 Relating Coordinate Systems to Each Other

Suppose that we describe a rigid motion with respect to two different coordinate systems, for example, with respect to a camera sensor $\boldsymbol{A}$ as well as with respect to an infrared position measurement device $\boldsymbol{B}$. The two motions are related by the transformation $\boldsymbol{X}$ between these two coordinate systems. This is well known – mainly from results in hand-eye calibration – that it yields the equation $\boldsymbol{AX} = \boldsymbol{XB}$ which can be decomposed in one matrix equation

$$\boldsymbol{R}_A\boldsymbol{R}_X = \boldsymbol{R}_X\boldsymbol{R}_B \tag{20.24}$$

and one vector equation

$$(\boldsymbol{R}_A - \boldsymbol{I})\boldsymbol{t}_X = \boldsymbol{R}_X\boldsymbol{t}_B - \boldsymbol{t}_A. \tag{20.25}$$

The majority of the approaches regards the rotation estimation in (20.24) decoupled from translation estimation, the latter following the former. At least two rotations containing motions with not parallel rotation axes are required to solve the problem [231, 211, 43].

Let $\check{\boldsymbol{a}}$ denote the screw described in coordinate system $\boldsymbol{A}$ and $\check{\boldsymbol{b}}$ denote the screw as described in coordinate system $\boldsymbol{B}$. The rigid transformation between them is unknown and it will be denoted by the unit dual quaternion $\check{\boldsymbol{q}}$. The screw concatenation yields then

$$\check{\boldsymbol{a}} = \check{\boldsymbol{q}}\check{\boldsymbol{b}}\bar{\check{\boldsymbol{q}}}. \tag{20.26}$$

The scalar part of a dual quaternion $\check{\boldsymbol{a}}$ is $(\check{\boldsymbol{a}} + \bar{\check{\boldsymbol{a}}})/2$, hence

$$\begin{aligned} Sc(\breve{\boldsymbol{a}}) &= \tfrac{1}{2}(\breve{\boldsymbol{a}} + \bar{\breve{\boldsymbol{a}}}) = \tfrac{1}{2}(\breve{\boldsymbol{q}}\breve{\boldsymbol{b}}\bar{\breve{\boldsymbol{q}}} + \breve{\boldsymbol{q}}\bar{\breve{\boldsymbol{b}}}\bar{\breve{\boldsymbol{q}}}) \\ &= \tfrac{1}{2}\breve{\boldsymbol{q}}(\breve{\boldsymbol{b}} + \bar{\breve{\boldsymbol{b}}})\bar{\breve{\boldsymbol{q}}} = \breve{\boldsymbol{q}}Sc(\breve{\boldsymbol{b}})\bar{\breve{\boldsymbol{q}}} = Sc(\breve{\boldsymbol{b}})\,\breve{\boldsymbol{q}}\bar{\breve{\boldsymbol{q}}} = Sc(\breve{\boldsymbol{b}}). \end{aligned} \tag{20.27}$$

According to (20.23) the scalar parts are equal to the cosine of the respective dual angles:

$$\cos\frac{(\theta_a + \epsilon d_a)}{2} = \cos\frac{(\theta_b + \epsilon d_b)}{2}.$$

which is equivalent to

$$\cos\frac{\theta_a}{2} = \cos\frac{\theta_b}{2} \quad \text{and} \quad d_a \sin\frac{\theta_a}{2} = d_b \sin\frac{\theta_b}{2}.$$

Hence, the angle and the pitch remain invariant under coordinate transformations. This is also known as the Screw Congruence Theorem [39], its proof without dual unit quaternions is, however, considerably longer than the one line proof in (20.27).

The fundamental equation $\breve{\boldsymbol{a}} = \breve{\boldsymbol{q}}\breve{\boldsymbol{b}}\bar{\breve{\boldsymbol{q}}}$ consists of four dual equations. Since the scalar parts are equal, only the vector components contribute to the computation of the unknown $\breve{\boldsymbol{q}}$:

$$\sin\frac{\breve{\theta}_a}{2}(0, \breve{\boldsymbol{a}}) = \breve{\boldsymbol{q}}(0, \sin\frac{\breve{\theta}_b}{2}\breve{\boldsymbol{b}})\bar{\breve{\boldsymbol{q}}} = \sin\frac{\breve{\theta}_b}{2}\breve{\boldsymbol{q}}(0, \breve{\boldsymbol{b}})\bar{\breve{\boldsymbol{q}}}.$$

If the angles $\theta_{a,b}$ are not 0 or 360 degrees the sines can be simplified yielding

$$(0, \breve{\boldsymbol{a}}) = \breve{\boldsymbol{q}}(0, \breve{\boldsymbol{b}})\bar{\breve{\boldsymbol{q}}} \tag{20.28}$$

which is nothing else than the motion of the lines of the screw axes.

Thus:

1. The angle and the pitch of a rigid motion are independent of the coordinate frame where the motion is described.
2. Relating two coordinate systems to each other is equivalent to the 3D motion estimation problem from 3D-line correspondences where the lines are the screw axes of the motions. The solution requires the computation of the kernel of a matrix (20.18) using the Singular Value Decomposition.

In [54] we show the details of the computational algorithm and extensive results on real experiments on the hand-eye calibration problem.

20.8 Conclusion

In this chapter we studied the properties of the even subalgebra of $\mathcal{R}^{3,0,1}$ known also as dual quaternions. Although we agree with Hestenes's argument that quaternions fail to distinguish among vectors and bivectors we believe that this argument does not apply here because 3D-lines are bivectors per se. On the contrary, we showed how the use of dual quaternions enables a better insight into the problem and facilitates a novel, elegant, and computationally simple approach to relating motions to each other that has not been possible with any other representation.

21. The Motor Extended Kalman Filter for Dynamic Rigid Motion Estimation from Line Observations*

Yiwen Zhang, Gerald Sommer, and Eduardo Bayro-Corrochano

Institute of Computer Science and Applied Mathematics, Christian-Albrechts-University of Kiel

21.1 Introduction

The motion estimation of a moving object in front of an observer is fundamental for various tasks in visual robotics like tracking, object collision avoidance, surveillance and visual navigation.

The issue we are here interested in is the estimation of the rigid motion of an object in observer frame or equivalently, the motion of an observer in a world frame. Fig. 21.1 gives a more detailed illustration. The 3-D coordinate frame A is supposed as observer frame, the coordinate frame B is fixed on a moving rigid body. The position and orientation of the rigid body in frame A are sampled by the observer at discrete time t_i, $i = 0, 1, \cdots$. At time t_0, frame A and frame B are duplicate. At time t_i, frame B goes to B_i, and an observed feature $\boldsymbol{L}$ on the surface of the rigid body goes to $\boldsymbol{L}_i$ with respect to frame A. We use state vector $\boldsymbol{X_i}$ to describe the position and orientation of the coordinate frame B_i relative to the frame A. $\boldsymbol{X_i}$ satisfies the dynamic model (which is also known as the plant model)

* This work has been supported by DFG Grants So-320-2-1, So-320-2-2, and Graduiertenkolleg No. 357.

$$\boldsymbol{X}_i = \boldsymbol{\Phi}_{i/i-1}(\boldsymbol{X}_{i-1}, \boldsymbol{W}_i), \tag{21.1}$$

where $\boldsymbol{W}_i$ is independent normally distributed noise with zero mean and known statistics. We will assume that the measurement of the feature $\boldsymbol{L_i}$ is also corrupted by independent normally distributed noise $\boldsymbol{V}_i$, which is also zero mean and with known statistics, and it is uncorrelated with $\boldsymbol{W}_i$. The real observed measurement $\boldsymbol{L_i}$ is expressed as

$$\boldsymbol{L}_i = \boldsymbol{L}_i' + \boldsymbol{V}_i. \tag{21.2}$$

where $\boldsymbol{L}_i'$ are the accurate data. The relationship between the measurements and the state is given by the measurement model as

$$\boldsymbol{f}(\boldsymbol{L}_0, \boldsymbol{L}_i, \boldsymbol{X}_i, \boldsymbol{V}_0, \boldsymbol{V}_i) = \boldsymbol{0}. \tag{21.3}$$

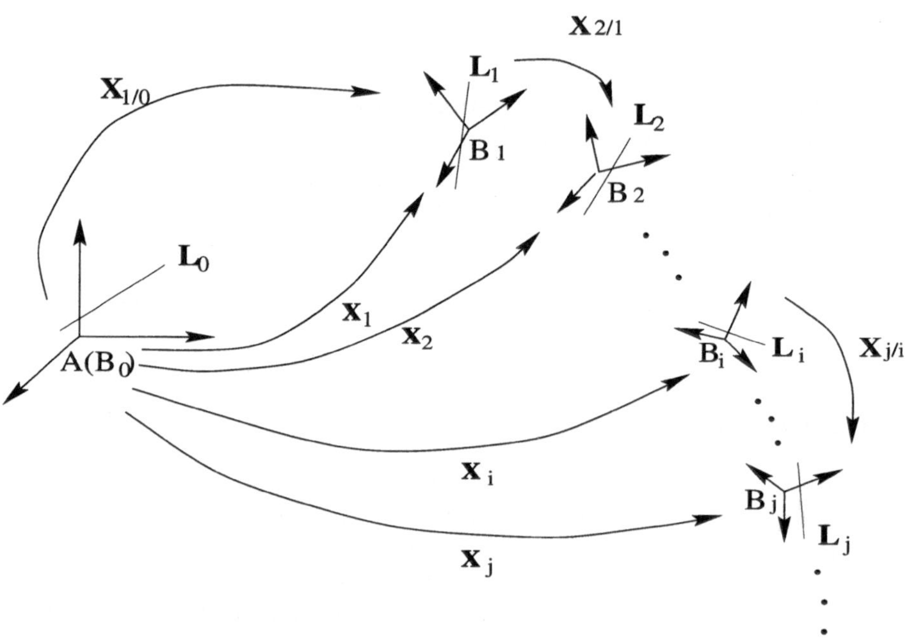

Fig. 21.1. Coordinate frames for observation of rigid motions

In such a noisy scenario we urgently require a method able to estimate a "best" state variable vector $\hat{\boldsymbol{X}}_i$.

The basic 3-D geometric primitives of the visual space for the motion registration are points (corners) or lines (edges). These local features are sensitive to noise and quantization errors which jeopardize to some extent the motion estimation. Alternatively the use of global features such as planes or surfaces makes the motion estimation process more robust, however with higher computational complexity.

In the literature we distinguish basically two main groups of estimation methods: batch and sequential processing.

The batch approaches include SVD and the analytical solutions by minimization techniques in terms of least square error. They use all the features' measurements observed at time t_i and t_j to estimate the optimal motion parameters $\boldsymbol{X}_{j/i}$ (so called two-view motion parameters) [119] [8] [196]. These batch methods do not use *a priori* information given by (21.1).

The sequential processing scheme is also called Kalman filtering [122] [165] [218]. The state $\boldsymbol{X}_i$ is estimated from the current predicted *a priori* state (using (21.1)) and the current measurements. The Kalman filter is a recursive algorithm: the new solution is based on the new measurements and the old solution. If the model equations (21.1) and (21.3) are nonlinear, the extended Kalman filter (EKF) can be used. In computer vision the measurement models are usually nonlinear. For applying the extended Kalman filter, such nonlinear models must be linearized about the current observations and current predicted state.

Former research shows that when we use both the batch and EKF algorithms to estimate the motion parameters with the same given measurements, the later gains better estimates [239].This results from the use of additional *a priori* information of the dynamic model (21.1) in case of Kalman filter processing. In other words, Kalman filtering is the best solution to our problem stated above.

The application of the Kalman filter as a recursive minimum variance estimator has become popular since the sixties. In order to estimate dynamic motion parameters, authors used the Kalman filter together with different types of state variable representations. For instance, Bar-Itzhack et al. used point sets for the quaternion EKF to estimate dynamic rotation [13] and Zhang and Faugeras used line segments with their midpoints to estimate all dynamic motion parameters with a standard EKF [239]. Recently Azarbayejani and Pentland [11] applied the EKF for estimation of motion and structure using relative orientation constraints in terms of quaternions. These methods are all based on point measurements (a line segment is defined by its midpoint and direction). We have not yet seen a method using straight line measurements.

With recently developed Hough transformation techniques [144], [181] one can extract a 2-D straight line from the image of the object boundary and then reconstruct a 3-D straight line by calibrated images. The coordinates of 3-D reconstructed straight lines are more reliable than 3-D reconstructed points. This motivated us to develop a Kalman filter from straight lines observations.

In this paper, we present the development of a novel EKF in the geometric algebra framework. The key for the filter design is that the measurement model of straight lines is established in the geometric algebra $\mathcal{G}^+_{3,0,1}$ called motor algebra, which is of the homogeneously extended Euclidean space E_3.

This aims at the useful property that the nonlinear motion model of a straight line in space E_3 can be written linearly in motor algebra. The modeling of the problem at hand in algebra $\mathcal{G}^+_{3,0,1}$ corresponds to the implicit assumption of a line geometry. That means that lines are the basic primitive entities (instead of points in E_3) and the known approaches of Kalman filter can be used in this algebraic framework. The real experiments show that the motor extended Kalman filter (MEKF) is indeed an attractive estimation approach. Compared with a batch method, the MEKF gives more accurate results in the dynamic motion estimation problem.

This paper is organized as follows. Section 21.2 reviews the basic knowledge of Kalman filter techniques. Section 21.3 represents the 3-D line motion model in geometric algebra $\mathcal{G}^+_{3,0,1}$ and gives an outline of the geometric algebra of rotors and motors. In section 21.4 we present the motor extended Kalman filter algorithm. Section 21.5 provides the experimental results of our MEKF, and finally, the conclusions are presented in section 21.6.

21.2 Kalman Filter Techniques

We will review in this section the principal equations for both the Kalman filter and the extended Kalman filter [122], [165] in order to introduce the necessary notations for the following sections.

21.2.1 The Kalman Filter

Consider a dynamical system whose state is described by a linear, vector difference equation. The system dynamic model is given by

$$\boldsymbol{X}_i = \boldsymbol{\Phi}_{i/i-1}\boldsymbol{X}_i + \boldsymbol{W}_i. \tag{21.4}$$

The state of the system at t_i is given by the n-dimensional vector $\boldsymbol{X}_i$. $\boldsymbol{\Phi}_{i/i-1}$ is an $n \times n$ matrix and $\boldsymbol{W}_i$ is a vector random sequence with known statistics

$$E[\boldsymbol{W}_i] = \boldsymbol{0}, \qquad i = 0, 1, ... \tag{21.5}$$

$$E[\boldsymbol{W}_i \boldsymbol{W}_j^T] = \boldsymbol{\mathcal{Q}}_i \delta_{ij} \tag{21.6}$$

where δ_{ij} is the Kronecker delta function. The matrix $\boldsymbol{\mathcal{Q}}_i$ is assumed to be nonnegative-definite.

Suppose that at each time t_i there is available an m-dimensional vector of measurement $\boldsymbol{Z}_i$ that is linearly related to the state and which is corrupted by additive noise $\boldsymbol{V}_i$.

$$\boldsymbol{Z}_i = \boldsymbol{\mathcal{H}}_i \boldsymbol{X}_i + \boldsymbol{V}_i \tag{21.7}$$

$\boldsymbol{\mathcal{H}}_i$ is a known $m \times n$ observation matrix. The vector $\boldsymbol{V}_i$ is an additive, random sequence with known statistics

$$E[\boldsymbol{V}_i] = \boldsymbol{0}, \qquad i = 0, 1, ... \tag{21.8}$$

$$E[\boldsymbol{V}_i \boldsymbol{V}_j^T] = \mathcal{C}_i \delta_{ij}. \tag{21.9}$$

The matrix $\mathcal{C}_i$ is assumed to be nonnegative-definite.

Further, assume that the random processes $\boldsymbol{W}_i$ and $\boldsymbol{V}_i$ are mutually uncorrelated. These processes will also be called white noise sequence. That means

$$E[\boldsymbol{W}_i \boldsymbol{V}_j^T] = \mathcal{O} \qquad i = 0, 1, ... \tag{21.10}$$

matrix $\mathcal{O}$ is null matrix.

Given the preceding models ((21.4) and (21.7)), we shall determine an estimate $\hat{\boldsymbol{X}}_i$ of the state at t_i that is a linear combination of an estimate $\hat{\boldsymbol{X}}_{i-1}$ at t_{i-1} and the measurement data $\boldsymbol{Z}_i$ at t_i. By defining an unknown gain matrix $\mathcal{K}_i$ $(n \times m)$, the estimate $\hat{\boldsymbol{X}}_i$ is given by

$$\hat{\boldsymbol{X}}_i = \boldsymbol{\Phi}_{i/i-1} \hat{\boldsymbol{X}}_{i-1} + \mathcal{K}_i [\boldsymbol{Z}_i - \mathcal{H}_i \boldsymbol{\Phi}_{i/i-1} \hat{\boldsymbol{X}}_{i-1}] \tag{21.11}$$

The matrix $\mathcal{K}_i$ shall be determined so that the estimate must be "best" in the sense that the expected value of the sum of the squares of the error in the estimate is a minimum. That is, the $\hat{\boldsymbol{X}}_i$ is to be chosen so that

$$E_{min} = Min\{E[(\hat{\boldsymbol{X}}_i - \boldsymbol{X}_i)^T (\hat{\boldsymbol{X}}_i - \boldsymbol{X}_i)]\}. \tag{21.12}$$

Equation (21.12) is equivalent to minimization of the trace of state error covariance matrix $\mathcal{P}_i$

$$E_{min} = Min\{trace \mathcal{P}_i\} = Min\{trace E[(\hat{\boldsymbol{X}}_i - \boldsymbol{X}_i)(\hat{\boldsymbol{X}}_i - \boldsymbol{X}_i)^T]\}. \tag{21.13}$$

By substituting (21.7) into (21.11), and then substituting (21.11) and (21.4) into (21.13), we can see that the trace of matrix $\mathcal{P}_i$ will be minimized by choosing the optimal gain matrix $\mathcal{K}_i$ such as

$$\mathcal{K}_i = \mathcal{P}_{i/i-1} \mathcal{H}_i^T (\mathcal{H}_i \mathcal{P}_{i/i-1} \mathcal{H}_i^T + \mathcal{C}_i)^{-1}, \tag{21.14}$$

where $\mathcal{P}_{i/i-1}$ is called predicted state error covariance matrix

$$\mathcal{P}_{i/i-1} = \boldsymbol{\Phi}_{i/i-1} \mathcal{P}_i \boldsymbol{\Phi}_{i/i-1}^T + \mathcal{Q}_i, \tag{21.15}$$

which is the error covariance matrix of the predicted state $\hat{\boldsymbol{X}}_{i/i-1}$

$$\hat{\boldsymbol{X}}_{i/i-1} = \boldsymbol{\Phi}_{i/i-1} \hat{\boldsymbol{X}}_i. \tag{21.16}$$

With this optimal gain matrix $\mathcal{K}_i$, the matrix $\mathcal{P}_i$ reduces to

$$\mathcal{P}_i = \mathcal{P}_{i/i-1} - \mathcal{K}_i \mathcal{H}_i \mathcal{P}_{i/i-1}. \tag{21.17}$$

Equations (21.11), (21.15), (21.14) and (21.17) constitute the Kalman filter for the model of the system (21.4) and that of the measurement (21.7), respectively.

Looking at (21.14), we see that as the measurement error covariance matrix $\mathcal{C}_i$ approaches zero, the gain matrix $\mathcal{K}_i$ weights the residual more heavily. Specifically,

$$\lim_{\mathcal{C}_i \to \mathcal{O}} \mathcal{K}_i = \mathcal{H}_i^{-1}. \tag{21.18}$$

On the other hand, as the estimated state error covariance $\mathcal{P}_i$ approaches zero, the gain $\mathcal{K}_i$ weights the residual less heavily. Specifically,

$$\lim_{\mathcal{P}_i \to \mathcal{O}} \mathcal{K}_i = \mathcal{O}. \tag{21.19}$$

Another way of thinking about the weighting by $\mathcal{K}_i$ is that as the measurement error covariance matrix $\mathcal{C}_i$ approaches zero, the actual measurement $\boldsymbol{Z}_i$ is "trusted" more and more, while the predicted state $\boldsymbol{\Phi}_{i/i-1}\hat{\boldsymbol{X}}_i$ is trusted less and less. On the other hand, as the estimated state error covariance $\mathcal{P}_i$ approaches zero the actual measurement $\boldsymbol{Z}_i$ is trusted less and less, while the predicted state $\boldsymbol{\Phi}_{i/i-1}\hat{\boldsymbol{X}}_i$ (the dynamic model) is trusted more and more.

21.2.2 The Extended Kalman Filter

As described in section 21.2.1, the Kalman filter addresses the general problem of trying to estimate the state $\boldsymbol{X}_i$ of a discrete-time controlled process that is governed by a linear stochastic difference equation. But what happens if the process and (or) the relation between the measurement and the state is non-linear? Some of the most interesting and successful applications of Kalman filtering are concerned with such situations. A Kalman filter that linearizes about the current predicated state $\hat{\boldsymbol{X}}_{i/i-1}$ and measurement $\boldsymbol{Z}_i$ is referred to as an extended Kalman filter or EKF.

In computer vision the measurement model is usually found to be described by a nonlinear observation equation $\boldsymbol{f}_i(\boldsymbol{Z}_{0,i}, \boldsymbol{X}_i) = \mathbf{0}$. The parameter $\boldsymbol{Z}_{0,i}$ is the accurate measurement. In practice, such measurement is affected by random errors. We assume that the measurement system is disturbed by additive white noise, i.e., the real observed measurement $\boldsymbol{Z}_i$ is expressed as

$$\boldsymbol{Z}_i = \boldsymbol{Z}_{0,i} + \boldsymbol{V}_i, \tag{21.20}$$

the statistics of noise $\boldsymbol{V}_i$ are given by (21.8) and (21.9).

For applying the Kalman filter technique, we must expand the nonlinear observation equation into a first order Taylor series about $(\boldsymbol{Z}_i, \hat{\boldsymbol{X}}_{i/i-1})$

$$\begin{aligned} \boldsymbol{f}_i(\boldsymbol{Z}_{0,i}, \boldsymbol{X}_i) &= \boldsymbol{f}_i(\boldsymbol{Z}_i, \hat{\boldsymbol{X}}_{i/i-1}) + \\ &+ \frac{\partial \boldsymbol{f}_i(\boldsymbol{Z}_i, \hat{\boldsymbol{X}}_{i/i-1})}{\partial \boldsymbol{Z}_{0,i}} (\boldsymbol{Z}_{0,i} - \boldsymbol{Z}_i) + \\ &+ \frac{\partial \boldsymbol{f}_i(\boldsymbol{Z}_i, \hat{\boldsymbol{X}}_{i/i-1})}{\partial \boldsymbol{X}_i} (\boldsymbol{X}_i - \hat{\boldsymbol{X}}_{i/i-1}) + \boldsymbol{R}_2 = \mathbf{0}. \end{aligned} \tag{21.21}$$

By ignoring the second order term $\boldsymbol{R}_2$, the linearized measurement equation (21.21) becomes

$$\boldsymbol{Y}_i = \mathcal{H}_i \boldsymbol{X}_i + \boldsymbol{N}_i, \tag{21.22}$$

where $\boldsymbol{Y}_i$ is the new measurement vector, $\boldsymbol{N}_i$ is the noise vector of the new measurement, and $\mathcal{H}_i$ is the linearized transformation matrix. The components of the equation (21.22) are given by

$$\begin{aligned}
\boldsymbol{Y}_i &= -\boldsymbol{f}_i(\boldsymbol{Z}_i, \hat{\boldsymbol{X}}_{i/i-1}) + \frac{\partial \boldsymbol{f}_i(\boldsymbol{Z}_i, \hat{\boldsymbol{X}}_{i/i-1})}{\partial \boldsymbol{X}_i} \hat{\boldsymbol{X}}_{i/i-1}, \\
\mathcal{H}_i &= \frac{\partial \boldsymbol{f}_i(\boldsymbol{Z}_i, \hat{\boldsymbol{X}}_{i/i-1})}{\partial \boldsymbol{X}_i}, \\
\boldsymbol{N}_i &= \frac{\partial \boldsymbol{f}_i(\boldsymbol{Z}_i, \hat{\boldsymbol{X}}_{i/i-1})}{\partial \boldsymbol{Z}_{0,i}} (\boldsymbol{Z}_{0,i} - \boldsymbol{Z}_i),
\end{aligned}$$

$$\begin{aligned}
E[\boldsymbol{N}_i] &= \boldsymbol{0}, \\
E[\boldsymbol{N}_i \boldsymbol{N}_i^T] &= \mathcal{C}_{i/i-1} \\
&= \frac{\partial \boldsymbol{f}_i(\boldsymbol{Z}_i, \hat{\boldsymbol{X}}_{i/i-1})}{\partial \boldsymbol{Z}_{0,i}} \mathcal{C}_i \frac{\partial \boldsymbol{f}_i(\boldsymbol{Z}_i, \hat{\boldsymbol{X}}_{i/i-1})}{\partial \boldsymbol{Z}_{0,i}}^T,
\end{aligned}$$

where $\mathcal{C}_i$ is given by the statistics of measurement (21.9). This linearized equation (21.22) is a general form for a nonlinear model. We will use this form for our particular nonlinear measurement model later in section 21.4.

21.3 3-D Line Motion Model

A line is one of the basic rigid geometric entities. In Euclidean space E_3, the operation of the line rigid motion is nonlinear. Whereas using the 4-D geometric algebra $\mathcal{G}^+_{3,0,1}$, also called motor algebra, the transformation becomes linear. In this section we first introduce the structure of the geometric algebra $\mathcal{G}^+_{3,0,1}$, and then give the Plücker line model and its motion model in $\mathcal{G}^+_{3,0,1}$.

21.3.1 Geometric Algebra $\mathcal{G}^+_{3,0,1}$ and Plücker Line Model

Given a homogeneous extension of the Euclidean space E_3 by an orthonormal set of vectors $\gamma_1, \gamma_2, \gamma_3, \gamma_4$, which in geometric algebra $\mathcal{G}^+_{3,0,1}$ satisfy:

$$\gamma_i^2 = 1 \qquad for \quad i = 1,2,3, \tag{21.23}$$

$$\gamma_4^2 = 0, \tag{21.24}$$

$$\gamma_i\gamma_j = -\gamma_j\gamma_i \quad for \quad i \neq j. \tag{21.25}$$

The basis of the linear space spanned by $\mathcal{G}^+_{3,0,1}$ is composed by one scalar, six bivectors, and one pseudoscalar, that means the basis $\mathcal{B}_{\mathcal{G}^+_{3,0,1}}$ is

$$\mathcal{B}_{\mathcal{G}^+_{3,0,1}} = \{1, \gamma_2\gamma_3, \gamma_3\gamma_1, \gamma_1\gamma_2, \gamma_4\gamma_1, \gamma_4\gamma_2, \gamma_4\gamma_3, I = \gamma_1\gamma_2\gamma_3\gamma_4\}, \tag{21.26}$$

with $I^2 = 0$.

A multivector $\boldsymbol{A} \in \mathcal{G}^+_{3,0,1}$,

$$\begin{aligned} \boldsymbol{A} &= a_0 + a_1\gamma_2\gamma_3 + a_2\gamma_3\gamma_1 + a_3\gamma_1\gamma_2 + \\ &\quad + I(a'_0 + a'_1\gamma_2\gamma_3 + a'_2\gamma_3\gamma_1 + a'_3\gamma_1\gamma_2), \end{aligned} \tag{21.27}$$

can be also expressed in a condensed dual form

$$\boldsymbol{A} = \boldsymbol{B} + I\boldsymbol{B}', \tag{21.28}$$

where $\boldsymbol{B}$ and $\boldsymbol{B}'$ are equivalent to quaternions.

A line $\boldsymbol{L}$ with Plücker coordinates in $\mathcal{G}^+_{3,0,1}$ can be represented as

$$\boldsymbol{L} = \boldsymbol{n} + I\boldsymbol{m}, \tag{21.29}$$

where $\boldsymbol{n}$ and $\boldsymbol{m}$ are bivectors,

$$\boldsymbol{n} = n_1\gamma_2\gamma_3 + n_2\gamma_3\gamma_1 + n_3\gamma_1\gamma_2 \tag{21.30}$$

$$\boldsymbol{m} = m_1\gamma_2\gamma_3 + m_2\gamma_3\gamma_1 + m_3\gamma_1\gamma_2. \tag{21.31}$$

Here $\boldsymbol{n}$ is the direction of the line and $\boldsymbol{m}$ is its moment. Any point $\boldsymbol{p}$ on the line,

$$\boldsymbol{p} = p_1\gamma_2\gamma_3 + p_2\gamma_3\gamma_1 + p_3\gamma_1\gamma_2, \tag{21.32}$$

satisfies

$$\boldsymbol{m} = \boldsymbol{p} \wedge \boldsymbol{n}. \tag{21.33}$$

If $\boldsymbol{n}$ is the normal direction of the line, then the norm of the moment calculated by (21.33) is the distance from the origin to the line (see Fig. 21.2).

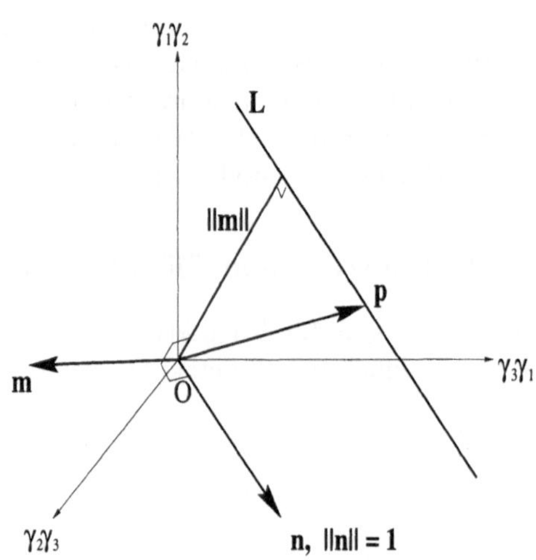

Fig. 21.2. Plücker coordinates of a line $\boldsymbol{L}$

21.3.2 Plücker Line Motion Model in $\mathcal{G}_{3,0,1}^{+}$

In general, rigid motion consists of rotation and translation. The rotation is defined by both its rotation axis and rotation angle. A certain rigid motion has a unique rotation angle and a unique rotation axis direction, but the rotation axis can be placed anywhere in a 3-D coordinate system, the corresponding translation is then dependent on the position of the rotation axis. There are two positions of rotation axis having particular meaning. One is the axis passing through the origin of a reference coordinate system, the translation is applied after rotation. The other is so called screw motion, the rotation axis is in such a place that a rigid motion consists of rotation about this axis in space through an angle of θ, followed by translation along the same axis by an amount d. The screw motion plays a very important role in rigid motion study [173]. In this section, we will discuss the features of motion of lines in Plücker coordinates.

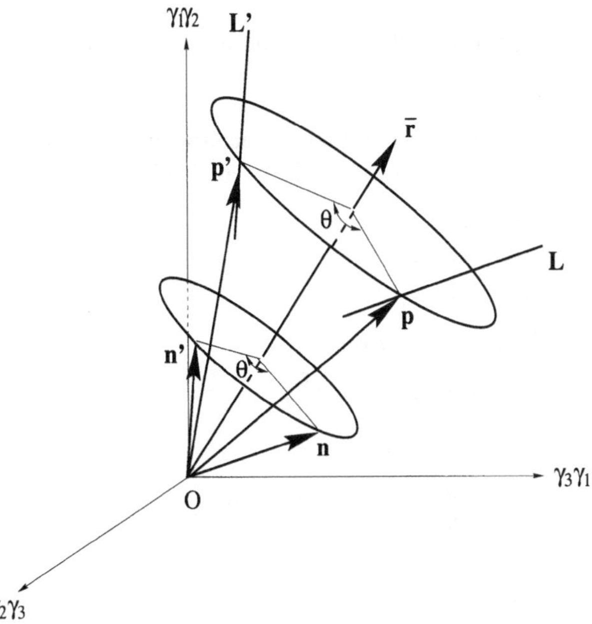

Fig. 21.3. The representation of pure rotation of a line

We first discuss the case of pure rotation as depicted in Fig. 21.3. The line is rotated by an angle θ about an axis $\bar{\boldsymbol{r}}$ going through the origin O, $\bar{\boldsymbol{r}}$ is a unit bivector. We can use a unit rotor $\boldsymbol{R}$ to represent this rotation,

$$\begin{aligned} \boldsymbol{R} &= r_0 + r_1\gamma_2\gamma_3 + r_2\gamma_3\gamma_1 + r_3\gamma_1\gamma_2 \\ &= r_0 + \boldsymbol{r} \\ &= cos(\theta/2) + sin(\theta/2)\bar{\boldsymbol{r}} \end{aligned} \tag{21.34}$$

$$\tilde{\boldsymbol{R}} = r_0 - r_1\gamma_2\gamma_3 - r_2\gamma_3\gamma_1 - r_3\gamma_1\gamma_2 = r_0 - \boldsymbol{r}, \tag{21.35}$$

where $\tilde{\boldsymbol{R}}$ is the inversion of $\boldsymbol{R}$ with the constraint

$$\boldsymbol{R}\tilde{\boldsymbol{R}} = r_0^2 - \boldsymbol{r}\cdot\boldsymbol{r} = 1. \tag{21.36}$$

$\boldsymbol{L} = \boldsymbol{n} + I\boldsymbol{m}$ and $\boldsymbol{L}' = \boldsymbol{n}' + I\boldsymbol{m}'$ are line coordinates before and after motion. $\boldsymbol{p}$ is a point on the line $\boldsymbol{L}$, after motion it goes to $\boldsymbol{p}'$. Then

$$\begin{aligned} \boldsymbol{L}' &= \boldsymbol{n}' + I\boldsymbol{m}' \\ &= \boldsymbol{R}\boldsymbol{n}\tilde{\boldsymbol{R}} + I(\boldsymbol{R}\boldsymbol{p}\tilde{\boldsymbol{R}})\wedge(\boldsymbol{R}\boldsymbol{n}\tilde{\boldsymbol{R}}) \\ &= \boldsymbol{R}\boldsymbol{n}\tilde{\boldsymbol{R}} + I\boldsymbol{R}(\boldsymbol{p}\wedge\boldsymbol{n})\tilde{\boldsymbol{R}} \\ &= \boldsymbol{R}(\boldsymbol{n} + I\boldsymbol{m})\tilde{\boldsymbol{R}} \\ &= \boldsymbol{R}\boldsymbol{L}\tilde{\boldsymbol{R}}, \end{aligned} \tag{21.37}$$

In the case of pure translation $\boldsymbol{t}$, where $\boldsymbol{t}$ is the bivector,

$$\boldsymbol{t} = t_1\gamma_2\gamma_3 + t_2\gamma_3\gamma_1 + t_3\gamma_1\gamma_2,$$

the direction $\boldsymbol{n}$ of the line $\boldsymbol{L}$ remains unchanged. A point $\boldsymbol{p}$ on the line is moving to $\boldsymbol{p}' = \boldsymbol{p} + \boldsymbol{t}$. The translated line $\boldsymbol{L}'$ is given by

$$\begin{aligned} \boldsymbol{L}' &= \boldsymbol{n}' + I\boldsymbol{m}' \\ &= \boldsymbol{n} + I(\boldsymbol{p} + \boldsymbol{t})\wedge\boldsymbol{n} \\ &= \boldsymbol{n} + I(\boldsymbol{m} + \boldsymbol{t}\wedge\boldsymbol{n}) \\ &= \boldsymbol{n} + I(\boldsymbol{m} + (\boldsymbol{t}\boldsymbol{n} - \boldsymbol{n}\boldsymbol{t})/2) \\ &= (1 + I\frac{\boldsymbol{t}}{2})(\boldsymbol{n} + I\boldsymbol{m})(1 - I\frac{\boldsymbol{t}}{2}) \\ &= \boldsymbol{T}\boldsymbol{L}\tilde{\boldsymbol{T}} \end{aligned} \tag{21.38}$$

With line rotation model (21.37) and translation model (21.38), the transformation of a line (see Fig. 21.4) can be modeled by, e.g., applying a rotation $\boldsymbol{R}$ followed by a translation $\boldsymbol{T}$

$$\begin{aligned} \boldsymbol{L}' &= \boldsymbol{T}\boldsymbol{R}\boldsymbol{L}\tilde{\boldsymbol{R}}\tilde{\boldsymbol{T}} \\ &= \boldsymbol{M}\boldsymbol{L}\tilde{\boldsymbol{M}}, \end{aligned} \tag{21.39}$$

where $\boldsymbol{M}$ is a motor ,

$$\boldsymbol{M} = \boldsymbol{T}\boldsymbol{R} = (1 + I\frac{\boldsymbol{t}}{2})\boldsymbol{R} = \boldsymbol{R} + I\boldsymbol{R}' = r_0 + \boldsymbol{r} + I(r_0' + \boldsymbol{r}') \tag{21.40}$$

$$\tilde{\boldsymbol{M}} = \tilde{\boldsymbol{R}}\tilde{\boldsymbol{T}} = \tilde{\boldsymbol{R}}(1 - I\frac{\boldsymbol{t}}{2}) = \tilde{\boldsymbol{R}} + I\tilde{\boldsymbol{R}}' \tag{21.41}$$

and

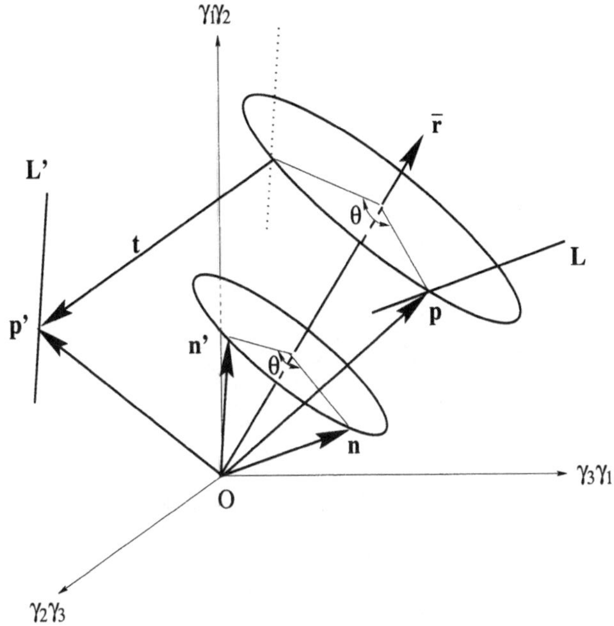

Fig. 21.4. The representation of rigid motion of a line

$$
\begin{aligned}
\boldsymbol{M}\tilde{\boldsymbol{M}} &= (1 + I\frac{\boldsymbol{t}}{2})\boldsymbol{R}\tilde{\boldsymbol{R}}(1 - I\frac{\boldsymbol{t}}{2}) = 1 \\
&= (\boldsymbol{R} + I\boldsymbol{R}')(\tilde{\boldsymbol{R}} + I\tilde{\boldsymbol{R}'}) = 1 + I(\boldsymbol{R}\tilde{\boldsymbol{R}'} + \boldsymbol{R}'\tilde{\boldsymbol{R}}).
\end{aligned} \tag{21.42}
$$

Deduced from the dual part of (21.42) we then get the following constraint

$$
\boldsymbol{R}\tilde{\boldsymbol{R}'} + \boldsymbol{R}'\tilde{\boldsymbol{R}} = 2(r_0 r_0' - \boldsymbol{r} \cdot \boldsymbol{r}') = 0. \tag{21.43}
$$

As we mentioned above, the motion can also be seen as a screw motion. First let us consider a line $\boldsymbol{L}$ rotating about another straight line $\boldsymbol{L}_s = \bar{\boldsymbol{r}} + I\boldsymbol{t}_c \wedge \bar{\boldsymbol{r}}$ by an angle θ, where the rotation axis $\boldsymbol{L}_s$ is in some general position of a 3-D coordinate system and $\boldsymbol{t}_c$ is pointing to an arbitrary point on $\boldsymbol{L}_s$. We call such a rotation as general rotation $\boldsymbol{R}_s$. $\boldsymbol{R}_s$ can be seen as a combined motion, represented by a translation $-\boldsymbol{t}_c$ first, then a rotation by rotor $\boldsymbol{R}$, finally followed by translation $\boldsymbol{t}_c$. That means, we first translate the rotation axis $\boldsymbol{L}_s$ to pass the origin of the 3-D coordinate system, after that we perform a rotation and finally we translate this axis $\boldsymbol{L}_s$ back to its original position:

$$
\begin{aligned}
\boldsymbol{R}_s &= (1 + I\boldsymbol{t}_c/2)(cos(\theta/2) + sin(\theta/2)\bar{\boldsymbol{r}})(1 - I\boldsymbol{t}_c/2) \\
&= cos(\theta/2) + sin(\theta/2)(\bar{\boldsymbol{r}} + I\boldsymbol{t}_c \wedge \bar{\boldsymbol{r}}) \\
&= cos(\theta/2) + sin(\theta/2)\boldsymbol{L}_s.
\end{aligned} \tag{21.44}
$$

A line rotated in this way can be easily given by

$$\begin{aligned} \boldsymbol{L}' &= \boldsymbol{n}' + \boldsymbol{I}\boldsymbol{m}' \\ &= \boldsymbol{R}_s \boldsymbol{L} \tilde{\boldsymbol{R}}_s. \end{aligned} \tag{21.45}$$

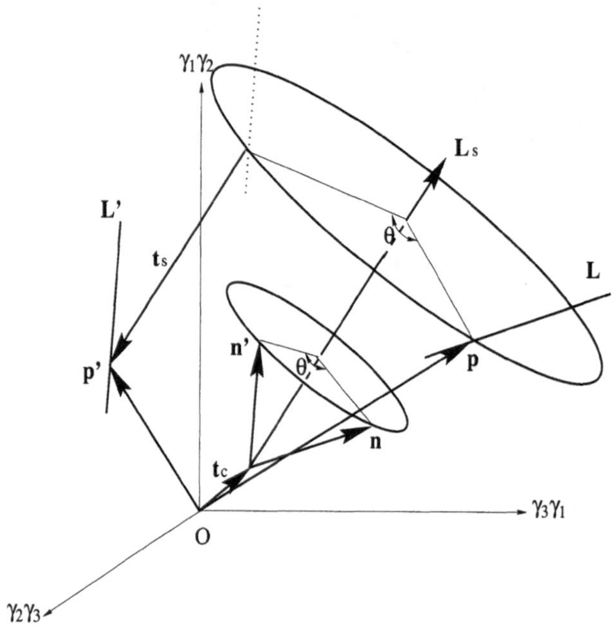

Fig. 21.5. The representation of screw motion of a line

Now, we can describe a screw motion easily. A screw motion is the combination of a general rotation, represented by $\boldsymbol{R}_s$ and a translation $\boldsymbol{t}_s$ which is parallel to the line $\boldsymbol{L}_s$, see Fig. 21.5. The screw motion equation is

$$\begin{aligned} \boldsymbol{L}' &= \boldsymbol{T}_s \boldsymbol{R}_s \boldsymbol{L} \tilde{\boldsymbol{R}}_s \tilde{\boldsymbol{T}}_s \\ &= \boldsymbol{M} \boldsymbol{L} \tilde{\boldsymbol{M}}. \end{aligned} \tag{21.46}$$

From the above discussion we see that using motor algebra, we can deal with rigid motion easily and efficiently. For example, if we use matrix algebra to describe a rigid motion, we must deal with 12 parameters, 9 for rotation and 3 for translation, using 6 constraints. In motor algebra, on the other hand, we only deal with 8 parameters and 2 constraints given by equations (21.36) and (21.43), respectively. Furthermore, motor algebra is algebraically isomorphic to dual-quaternion algebra. In reference [85], J. Funda et al. compared several methods of line-oriented representations of general spatial displacements of rigid bodies and drew the conclusion that dual-quaternion algebra is the best for the line transformations. As pointed out by D. Hestenes et al. in chapter 1 , "the drawback of quaternions is that they are limited to 3-D applications, and even there they fail to make the important distinction between vectors and bivectors". "It should be clear that geometric algebra retains all of the advantages and none of the drawbacks of quaternions, while extending the range of applications enormously". Another important advantage is that motors and rotors are both spinors. In spinor representation of Euclidean transformation, the group of several transformations corresponds to the geometric product of the spinors representing these transformations. We will use (21.46) for the motion estimation which will be discussed in the following section.

21.3.3 Interpretation of the Plücker Line Motion Model in Linear Algebra

The Plücker line motion model presented in the last section is considered in geometric algebra $\mathcal{G}_{3,0,1}^{+}$. Because the EKF algorithm is computed in linear algebra, we should interpret the line motion model $\boldsymbol{L}' = \boldsymbol{M}\boldsymbol{L}\tilde{\boldsymbol{M}}$ in the frame of linear algebra. This can be done by remembering that $\mathcal{G}_{3,0,1}^{+}$ spans the 8-dimensional linear space represented by (21.26), which is the union of a real and a dual 4-dimensional subspace, respectively. In that space the lines are the basic geometric entities and their mutual relations correspond to linear transformations by rotors or motors. This is just as the rotation of points in E_3 can be linearly transformed using a rotation matrix $\boldsymbol{\mathcal{R}}$.

First let us see some basic conversions.

The multiplication of two rotors $\boldsymbol{U}$ and $\boldsymbol{V}$ in geometric algebra $\mathcal{G}_{3,0,1}^{+}$ reads

$$\begin{aligned} \boldsymbol{W} = \boldsymbol{U}\boldsymbol{V} &= (u_0 + \boldsymbol{u})(v_0 + \boldsymbol{v}) \\ &= u_0 v_0 + \boldsymbol{u} \cdot \boldsymbol{v} + u_0 \boldsymbol{v} + v_0 \boldsymbol{u} + \boldsymbol{u} \wedge \boldsymbol{v}. \end{aligned} \tag{21.47}$$

Multiplication of these two rotors in linear algebra is

$$\boldsymbol{W} = \boldsymbol{\mathcal{U}}_{Rl} \boldsymbol{V} = \boldsymbol{\mathcal{V}}_{Rr} \boldsymbol{U}, \tag{21.48}$$

where $\boldsymbol{U} = (u_0 \quad u_1 \quad u_2 \quad u_3)^T$, $\boldsymbol{V} = (v_0 \quad v_1 \quad v_2 \quad v_3)^T$ and

$$\boldsymbol{\mathcal{U}}_{Rl} = \begin{pmatrix} u_0 & -u_1 & -u_2 & -u_3 \\ u_1 & u_0 & u_3 & -u_2 \\ u_2 & -u_3 & u_0 & u_1 \\ u_3 & u_2 & -u_1 & u_0 \end{pmatrix},$$

$$\boldsymbol{\mathcal{V}}_{Rr} = \begin{pmatrix} v_0 & -v_1 & -v_2 & -v_3 \\ v_1 & v_0 & -v_3 & v_2 \\ v_2 & v_3 & v_0 & -v_1 \\ v_3 & -v_2 & v_1 & v_0 \end{pmatrix}.$$

We call $\mathcal{U}_{Rl}$ "left-multiplication matrix of motor $\boldsymbol{U}$" and $\mathcal{V}_{Rr}$ "right-multiplication matrix of motor $\boldsymbol{V}$".

Multiplication of two motors $\boldsymbol{S} = \boldsymbol{U} + I\boldsymbol{U}'$ and $\boldsymbol{T} = \boldsymbol{V} + I\boldsymbol{V}'$ in geometric algebra $\mathcal{G}^+_{3,0,1}$ results

$$\begin{aligned} \boldsymbol{Q} &= \boldsymbol{ST} = (\boldsymbol{U} + I\boldsymbol{U}')(\boldsymbol{V} + I\boldsymbol{V}') \\ &= \boldsymbol{UV} + I(\boldsymbol{UV}' + \boldsymbol{U}'\boldsymbol{V}). \end{aligned} \tag{21.49}$$

Here $\boldsymbol{U}$, $\boldsymbol{U}'$, $\boldsymbol{V}$ and $\boldsymbol{V}'$ are all in the form of rotors. Multiplication of these two motors in linear algebra is given by

$$\boldsymbol{Q} = \mathcal{S}_{Ml}\boldsymbol{T} = \mathcal{T}_{Mr}\boldsymbol{S}, \tag{21.50}$$

where

$$\begin{aligned} \boldsymbol{S} &= (u_0 \quad u_1 \quad u_2 \quad u_3 \quad u_0' \quad u_1' \quad u_2' \quad u_3')^T, \\ \boldsymbol{T} &= (v_0 \quad v_1 \quad v_2 \quad v_3 \quad v_0' \quad v_1' \quad v_2' \quad v_3')^T, \\ \mathcal{S}_{Ml} &= \begin{pmatrix} \mathcal{U}_{Rl} & \mathbf{0}_{4\times4} \\ \mathcal{U}'_{Rl} & \mathcal{U}_{Rl} \end{pmatrix}, \\ \mathcal{T}_{Mr} &= \begin{pmatrix} \mathcal{V}_{Rr} & \mathbf{0}_{4\times4} \\ \mathcal{V}'_{Rr} & \mathcal{V}_{Rr} \end{pmatrix}, \end{aligned}$$

We call $\mathcal{S}_{Ml}$ "left-multiplication matrix of motor $\boldsymbol{S}$" and $\mathcal{T}_{Mr}$ "right-multiplication matrix of motor $\boldsymbol{T}$".

To convert the Plücker line motion model (21.46) to linear algebra we can handle the real and dual components $\boldsymbol{n}$, $\boldsymbol{m}$, $\boldsymbol{n}'$ and $\boldsymbol{m}'$ of the lines $\boldsymbol{L}$ and $\boldsymbol{L}'$ as rotors with zero scalar. By right multiplication of both sides of equation (21.46) by $\boldsymbol{M}$ we get

$$\boldsymbol{L}'\boldsymbol{M} - \boldsymbol{ML} = 0. \tag{21.51}$$

This results in the following linear motion equation

$$(\mathcal{L}'_{Ml} - \mathcal{L}_{Mr})\boldsymbol{M} = \mathcal{A}_M\boldsymbol{M} = \mathbf{0}. \tag{21.52}$$

The constraints of equations (21.36) and (21.43), respectively, now are

$$\boldsymbol{R}^T\boldsymbol{R} = 1, \tag{21.53}$$

$$\boldsymbol{R}^T\boldsymbol{R}' = 0, \tag{21.54}$$

with $\boldsymbol{R} = (r_0\ r_1\ r_2\ r_3)^T$, $\boldsymbol{R}' = (r_0'\ r_1'\ r_2'\ r_3')^T$ and $\boldsymbol{M} = \boldsymbol{R} + I\boldsymbol{R}'$.

These properties will be used for the implementation of the MEKF algorithm in the next section.

21.4 The Motor Extended Kalman Filter

In this section, we will formulate the motor extended Kalman filter (MEKF) algorithm . For applying Kalman filter techniques which were introduced in section 21.2, we know that we must be given both a dynamic model and a measurement model. We will first present the dynamic model using motor as state, then linearize the measurement equation (21.52) to get a linearized measurement equation, use (21.53) and (21.54) to modify the estimation to construct a proper motor estimation and finally, we present the MEKF algorithm.

21.4.1 Discrete Dynamic Model Using Motor State

Let us assume that we have a rigid object moving in 3-D space with approximately known trajectory. The object includes a number of lines ($\boldsymbol{L}^1$, $\boldsymbol{L}^2, ..., \boldsymbol{L}^n, n \geq 2$), we use the notation $\boldsymbol{L}$ to represent any one of these lines. The 3-D coordinates of these lines are sampled at a number of time instants $t_0, t_1, ..., t_N$. Suppose at time t_i, the rigid motion parameters with respect to time t_0 are described by the motor $\boldsymbol{M}_i$, the relationship of the Plücker coordinates of a line at time t_0 (denoted as $\boldsymbol{L}_0$) and at time t_i ($\boldsymbol{L}_i$) in $\mathcal{G}^+_{3,0,1}$ is

$$\boldsymbol{L}_i = \boldsymbol{M}_i \boldsymbol{L}_0 \tilde{\boldsymbol{M}}_i. \tag{21.55}$$

The change of motion parameters from time t_{i-1} to t_i is described by the motor $\boldsymbol{V}_{i/i-1}$, that is

$$\boldsymbol{L}_i = \boldsymbol{V}_{i/i-1} \boldsymbol{L}_{i-1} \tilde{\boldsymbol{V}}_{i/i-1}. \tag{21.56}$$

By substituting (21.55) into (21.56), we get

$$\begin{aligned} \boldsymbol{L}_i &= \boldsymbol{V}_{i/i-1} \boldsymbol{M}_{i-1} \boldsymbol{L}_0 \tilde{\boldsymbol{M}}_{i-1} \tilde{\boldsymbol{V}}_{i/i-1} \\ &= \boldsymbol{M}_i \boldsymbol{L}_0 \tilde{\boldsymbol{M}}_i. \end{aligned} \tag{21.57}$$

Then we get the ideal dynamic motion model

$$\boldsymbol{M}_i = \boldsymbol{V}_{i/i-1} \boldsymbol{M}_{i-1}. \tag{21.58}$$

The motor $\boldsymbol{V}_{i/i-1}$ encodes the velocity information. For example, suppose the motion is a screw motion with rotation of constant angular velocity ω about an axis of known line ($\boldsymbol{L}_s = \bar{\boldsymbol{r}} + \boldsymbol{I} t_c \wedge \bar{\boldsymbol{r}}$) and with constant translation velocity $\boldsymbol{v}_s$ which is parallel to the axis. The data are sampled by a constant time interval and such a time interval is normalized to 1, then

$$\boldsymbol{V}_{i/i-1} = \boldsymbol{V} = (1 + \boldsymbol{I}\boldsymbol{v}_s/2)(cos(\omega/2) + sin(\omega/2)\boldsymbol{L}_s). \tag{21.59}$$

In real applications we can only know the relation between $\boldsymbol{M}_{i-1}$ and $\boldsymbol{M}_i$ approximately. That means that such a dynamic motion model has to contain a process noise $\boldsymbol{W}_i$. Thus, the real dynamic model is given by

$$\boldsymbol{M}_i = \boldsymbol{V}_{i/i-1}\boldsymbol{M}_{i-1} + \boldsymbol{W}_i, \tag{21.60}$$

where the statistics of $\boldsymbol{W}_i$ is given by (21.5) and (21.6). In linear algebra, (21.60) is expressed as

$$\boldsymbol{M}_i = \boldsymbol{\mathcal{V}}_{i/i-1,Ml}\boldsymbol{M}_{i-1} + \boldsymbol{W}_i. \tag{21.61}$$

It must be noted that the motion parameters $\boldsymbol{M}$ and $\boldsymbol{V}$ should be described in the same coordinate system of the line $\boldsymbol{L}$, which is spanned by the algebra $\mathcal{G}^+_{3,0,1}$.

21.4.2 Linearization of the Measurement Model

It is obvious that in (21.52) the relation between the measurement $\boldsymbol{\mathcal{A}}_M$ and the state $\boldsymbol{M}$ is nonlinear, we must therefore first linearize it.

Suppose the measurement $\boldsymbol{\mathcal{A}}_{\boldsymbol{M}i}$ is the true data $\boldsymbol{\mathcal{A}}_{\boldsymbol{M}0,i}$ contaminated by measurement noise $\boldsymbol{\mathcal{N}}_{\mathcal{A}_M,i}$

$$\boldsymbol{\mathcal{A}}_{\boldsymbol{M}i} = \boldsymbol{\mathcal{A}}_{\boldsymbol{M}0,i} + \boldsymbol{\mathcal{N}}_{\mathcal{A}_M,i}. \tag{21.62}$$

The noise matrix $\boldsymbol{\mathcal{N}}_{\mathcal{A}_M,i}$ is zero mean and we know the covariance of every component of the noise matrix. We define a function $\boldsymbol{f}_{M,i}$ depending on the variables $(\boldsymbol{\mathcal{A}}_{\boldsymbol{M}0,i}, \boldsymbol{M}_i)$ as follows

$$\boldsymbol{f}_{M,i}(\boldsymbol{\mathcal{A}}_{\boldsymbol{M}0,i}, \boldsymbol{M}_i) = \boldsymbol{\mathcal{A}}_{\boldsymbol{M}0,i}\boldsymbol{M}_i = \boldsymbol{0}. \tag{21.63}$$

Expanding (21.63) into a first order Taylor series about the measurement and the predicted state $(\boldsymbol{\mathcal{A}}_{\boldsymbol{M}i}, \hat{\boldsymbol{M}}_{i/i-1})$, we get

$$\begin{aligned}
&\boldsymbol{f}_{M,i}(\boldsymbol{\mathcal{A}}_{\boldsymbol{M}0,i}, \boldsymbol{M}_i) \\
&= \boldsymbol{f}_{M,i}(\boldsymbol{\mathcal{A}}_{\boldsymbol{M}i}, \hat{\boldsymbol{M}}_{i/i-1}) + \\
&\quad + \frac{\partial \boldsymbol{f}_{M,i}(\boldsymbol{\mathcal{A}}_{\boldsymbol{M}i}, \hat{\boldsymbol{M}}_{i/i-1})}{\partial \boldsymbol{M}_i}(\boldsymbol{M}_i - \hat{\boldsymbol{M}}_{i/i-1}) + \\
&\quad + (\boldsymbol{\mathcal{A}}_{\boldsymbol{M}0,i} - \boldsymbol{\mathcal{A}}_{\boldsymbol{M}i})\frac{\partial \boldsymbol{f}_{M,i}(\boldsymbol{\mathcal{A}}_{\boldsymbol{M}i}, \hat{\boldsymbol{M}}_{i/i-1})}{\partial \boldsymbol{\mathcal{A}}_{\boldsymbol{M}0,i}} + \boldsymbol{R}_2 \\
&= \boldsymbol{0},
\end{aligned} \tag{21.64}$$

where

$$\frac{\partial \boldsymbol{f}_{M,i}(\boldsymbol{\mathcal{A}}_{\boldsymbol{M}i}, \hat{\boldsymbol{M}}_{i/i-1})}{\partial \boldsymbol{M}_i} = \boldsymbol{\mathcal{A}}_{\boldsymbol{M}i}, \tag{21.65}$$

$$\frac{\partial \boldsymbol{f}_{M,i}(\boldsymbol{\mathcal{A}}_{\boldsymbol{M}i}, \hat{\boldsymbol{M}}_{i/i-1})}{\partial \boldsymbol{\mathcal{A}}_{\boldsymbol{M}0,i}} = \hat{\boldsymbol{M}}_{i/i-1}. \tag{21.66}$$

Substituting (21.65) and (21.66) into (21.64), omitting the second order terms $\boldsymbol{R}_2$, and using (21.62), (21.64) can be written as follows

$$\begin{aligned}
&\boldsymbol{\mathcal{A}}_{\boldsymbol{M}i}\hat{\boldsymbol{M}}_{i/i-1} + \boldsymbol{\mathcal{A}}_{\boldsymbol{M}i}(\boldsymbol{M}_i - \hat{\boldsymbol{M}}_{i/i-1}) + \\
&+(\boldsymbol{\mathcal{A}}_{\boldsymbol{M}0,i} - \boldsymbol{\mathcal{A}}_{\boldsymbol{M}i})\hat{\boldsymbol{M}}_{i/i-1} \\
&= \boldsymbol{\mathcal{A}}_{\boldsymbol{M}i}\hat{\boldsymbol{M}}_{i/i-1} + \boldsymbol{\mathcal{A}}_{\boldsymbol{M}i}(\boldsymbol{M}_i - \hat{\boldsymbol{M}}_{i/i-1}) - \boldsymbol{\mathcal{N}}_{\mathcal{A}_M,i}\hat{\boldsymbol{M}}_{i/i-1} \\
&= \mathbf{0}.
\end{aligned} \tag{21.67}$$

Then the linearized measurement equation for MEKF at step i is

$$\begin{aligned}
\boldsymbol{Z}_i &= -\boldsymbol{\mathcal{A}}_{\boldsymbol{M}i}\boldsymbol{M}_i + \boldsymbol{\mathcal{N}}_{\mathcal{A}_M,i}\hat{\boldsymbol{M}}_{i/i-1} \\
&= \boldsymbol{\mathcal{H}}_i\boldsymbol{M}_i + \boldsymbol{N}_{Z,i} \\
&= \mathbf{0},
\end{aligned} \tag{21.68}$$

where $\boldsymbol{\mathcal{H}}_i = -\boldsymbol{\mathcal{A}}_{\boldsymbol{M}i}$ and $\boldsymbol{N}_{Z,i} = \boldsymbol{\mathcal{N}}_{\boldsymbol{\mathcal{A}}_M,i}\hat{\boldsymbol{M}}_{i/i-1}$. The covariance matrix of $\boldsymbol{N}_{Z,i}$ is $\boldsymbol{\mathcal{C}}_i$.

21.4.3 Constraints Problem

According to the Kalman filter algorithm ((21.11), (21.15), (21.14) and (21.17)), we can compute the estimation $\boldsymbol{M}_i^*$ as

$$\begin{aligned}
\boldsymbol{M}_i^* &= \boldsymbol{\Phi}_{i/i-1}\hat{\boldsymbol{M}}_{i-1} + \boldsymbol{\mathcal{K}}_i(\boldsymbol{Z}_i - \boldsymbol{\mathcal{H}}_i\boldsymbol{\Phi}_{i/i-1}\hat{\boldsymbol{M}}_{i-1}) \\
&= \boldsymbol{\mathcal{V}}_{i/i-1,Ml}\hat{\boldsymbol{M}}_{i-1} + \boldsymbol{\mathcal{K}}_i(-\boldsymbol{\mathcal{H}}_i\boldsymbol{\mathcal{V}}_{i/i-1,Ml}\hat{\boldsymbol{M}}_{i-1}) \\
&= (\boldsymbol{R}^{*T}_i \quad \boldsymbol{R'}^{*T}_i)^T
\end{aligned} \tag{21.69}$$

The 4-dimensional vectors $\boldsymbol{R}_i^*$ and $\boldsymbol{R'}_i^*$ are the first 4 components and the last 4 components of M_i^*, respectively. They must be modified to satisfy the constraints (21.53) and (21.54). For the constraint (21.53), this can be done simply by

$$\hat{\boldsymbol{R}}_i = \frac{\boldsymbol{R}_i^*}{\parallel \boldsymbol{R}_i^* \parallel}. \tag{21.70}$$

But to satisfy the constraint (21.54) is not so simple. Now, we rewrite (21.54) as $\boldsymbol{R'}^T\boldsymbol{R} = 0$, this equation means that the rotor $\boldsymbol{R}$ and the dual rotor $\boldsymbol{R}'$, in their vector form, must be orthogonal to each other. Unfortunately, the estimated rotor $\boldsymbol{R}_i^*$ is usually not orthogonal to the estimated dual rotor $\boldsymbol{R'}_i^*$, see Figure 21.6. Suppose the angle between estimates $\boldsymbol{R}_i^*$ and $\boldsymbol{R'}_i^*$ is φ, then

$$cos(\varphi) = \frac{\boldsymbol{R'}^{*T}_i \boldsymbol{R}_i^*}{\parallel \boldsymbol{R'}_i^* \parallel \cdot \parallel \boldsymbol{R}_i^* \parallel} \tag{21.71}$$

Using (21.70), (21.71) can be simplified by introducing the unit rotor $\hat{\boldsymbol{R}}_i$ as

$$cos(\varphi) = \frac{\boldsymbol{R'}^{*T}_i \hat{\boldsymbol{R}}_i}{\parallel \boldsymbol{R'}_i^* \parallel}. \tag{21.72}$$

It can be easily understood that the best modified dual rotor $\hat{\boldsymbol{R}}'_i$ should be closest to the estimated dual rotor $\boldsymbol{R'}_i^*$. That means that the difference of

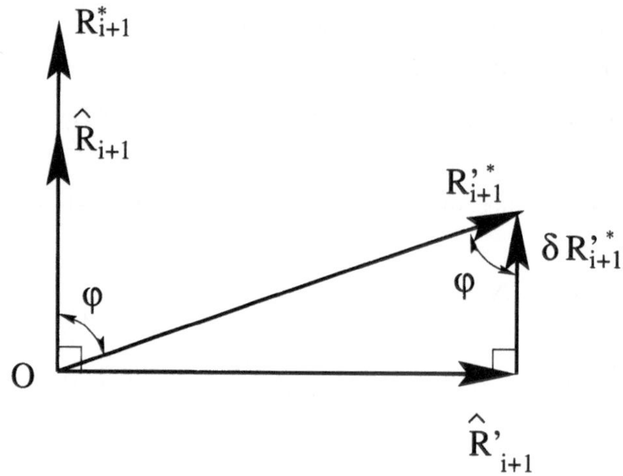

Fig. 21.6. Constraint of $\hat{\boldsymbol{R}}^{*T}\hat{\boldsymbol{R}}'^{*} = 0$

these two vectors, $\delta\boldsymbol{R}'^{*}_{i}$, should be orthogonal to the modified dual rotor $\hat{\boldsymbol{R}}'_i$ and should be parallel to the rotor $\hat{\boldsymbol{R}}_i$. In other words, the length of $\delta\boldsymbol{R}'^{*}_{i}$ is $\| \boldsymbol{R}'^{*}_{i} \| cos(\varphi)$ and the direction of it is equal to that of the rotor $\hat{\boldsymbol{R}}_i$. Then,

$$\delta\boldsymbol{R}'^{*}_{i} = \| \boldsymbol{R}'^{*}_{i} \| cos(\varphi)\hat{\boldsymbol{R}}_i = (\boldsymbol{R}'^{*T}_{i}\hat{\boldsymbol{R}}_i)\hat{\boldsymbol{R}}_i, \tag{21.73}$$

so that

$$\hat{\boldsymbol{R}}'_i = \boldsymbol{R}'^{*}_{i} - (\boldsymbol{R}'^{*T}_{i}\hat{\boldsymbol{R}}_i)\hat{\boldsymbol{R}}_i. \tag{21.74}$$

$\hat{\boldsymbol{R}}_i$ and $\hat{\boldsymbol{R}}'_i$ are the modified estimations at i and satisfy the constraints (21.53) and (21.54).

21.4.4 The MEKF Algorithm

The MEKF algorithm is summarized in Fig. 21.7. At time 0, it begins with a given initial predicted state $\hat{M}_{1/0}$ and the initial predicted state error covariance matrix $\mathcal{P}_{1/0}$ as a prediction of time 1. If we do not know the initial predicted state, we can simply set

$$\hat{\boldsymbol{M}}_{1/0} = [1 \quad 0 \quad 0 \quad 0 \quad 0 \quad 0 \quad 0 \quad 0]^T \tag{21.75}$$

$$\mathcal{P}_{1/0} = \boldsymbol{I}_{8\times 8} \tag{21.76}$$

At time 1, we first compute the matrix $\mathcal{H}_1$ of the linearized measurement equation and the Kalman gain matrix $\mathcal{K}_1$, then we can calculate the estimation $\boldsymbol{M}^*_1$. This estimation must be modified to be $\hat{\boldsymbol{M}}_1$ which satisfies the motor constraints. $\hat{\boldsymbol{M}}_1$ serves as the result of the estimation and then we can get the prediction $\hat{\boldsymbol{M}}_{2/1}$ of time 2 by dynamic model. The MEKF will run recursively till time N. The MEKF algorithm is listed in Fig. 21.7.

It must be noted that the numerical instability of Kalman filter implementation is well known. Several techniques are developed to overcome those

- Initialization of the prediction

$$\hat{\boldsymbol{M}}_{1/0} = E[\boldsymbol{M}_{1/0}]$$
$$\boldsymbol{\mathcal{P}}_{1/0} = E[(\boldsymbol{M}_{1/0} - \hat{\boldsymbol{M}}_{1/0})(\boldsymbol{M}_{1/0} - \hat{\boldsymbol{M}}_{1/0})^T]$$

- Linearization

$$\boldsymbol{\mathcal{H}}_i = -\boldsymbol{\mathcal{A}}_{\boldsymbol{M}i}$$

- Kalman gain matrix

$$\boldsymbol{\mathcal{K}}_i = \boldsymbol{\mathcal{P}}_{i/i-1}\boldsymbol{\mathcal{H}}_i^T(\boldsymbol{\mathcal{H}}_i\boldsymbol{\mathcal{P}}_{i/i-1}\boldsymbol{\mathcal{H}}_i^T + \boldsymbol{\mathcal{C}}_i)^{-1}$$

- Update

$$\boldsymbol{M}_i^* = \hat{\boldsymbol{M}}_{i/i-1} + \boldsymbol{\mathcal{K}}_i(-\boldsymbol{\mathcal{H}}_i\hat{\boldsymbol{M}}_{i/i-1})$$
$$\boldsymbol{\mathcal{P}}_i = (\boldsymbol{I} - \boldsymbol{\mathcal{K}}_i\boldsymbol{\mathcal{H}}_i)\boldsymbol{\mathcal{P}}_{i/i-1}$$

- Modification of the state estimation

$$\hat{\boldsymbol{R}}_i = \boldsymbol{R}_i^* \| \boldsymbol{R}_i^* \|^{-1}$$
$$\hat{\boldsymbol{R}}'_i = \boldsymbol{R}'^*_i - (\boldsymbol{R}'^{*T}_i\hat{\boldsymbol{R}}_i)\hat{\boldsymbol{R}}_i$$
$$\hat{\boldsymbol{M}}_i = (\hat{\boldsymbol{R}}_i^T \quad \hat{\boldsymbol{R}}'^T_i)^T$$

- Prediction of the state for next time

$$\hat{\boldsymbol{M}}_{i+1/i} = \boldsymbol{\Phi}_{i+1/i}\hat{\boldsymbol{M}}_i$$
$$\boldsymbol{\mathcal{P}}_{i+1/i} = \boldsymbol{\Phi}_{i+1/i}\boldsymbol{\mathcal{P}}_i\boldsymbol{\Phi}_{i+1/i}^T + \boldsymbol{\mathcal{C}}_{\boldsymbol{W}i}$$

Fig. 21.7. MEKF algorithm

problems, such as square-root filtering and U-D factorization. See [165] for a thorough discussion.

21.4.5 A Batch Method of Analytical Solution

In [239], Zhang and Faugeras have presented an analytical method to recover the motion parameters from Plücker line measurements. We will introduce it here for the purpose of comparing it with the method of the MEKF. This analytical solution can also be used for estimating the initial prediction in our MEKF algorithm.

Assume that there are n lines of the rigid object, which are measured before and after motion $\boldsymbol{M}_i$. The coordinates of these lines are $\boldsymbol{L}_0^k = \boldsymbol{n}_0^k + \boldsymbol{I}\boldsymbol{m}_0^k$ and $\boldsymbol{L}_i^k = \boldsymbol{n}_i^k + \boldsymbol{I}\boldsymbol{m}_i^k$, $k = 1, 2, ..., n$, where the subscript numbers 0 and i correspond to the case before and after motion, respectively. The motor $\boldsymbol{M}_i$ can also be seen as a combined motion of a rotation $\boldsymbol{R}_i$ and a translation $\boldsymbol{t}_i$, which in $\mathcal{G}_{3,0,1}^+$ satisfies

$$\boldsymbol{M}_i = (1 + \boldsymbol{I}\boldsymbol{t}_i/2)\boldsymbol{R}_i. \qquad (21.77)$$

Then, the relation between a line $\boldsymbol{L}_0 = \boldsymbol{L}_0^k$ and the transformed line $\boldsymbol{L}_i = \boldsymbol{L}_i^k$ is given by

$$\begin{aligned}
\boldsymbol{L}_i &= \boldsymbol{n}_i + I\boldsymbol{m}_i \\
&= \boldsymbol{M}_i\boldsymbol{L}_0\tilde{\boldsymbol{M}}_i \\
&= (1 + I\boldsymbol{t}_i/2)\boldsymbol{R}_i(\boldsymbol{n}_0 + I\boldsymbol{m}_0)\tilde{\boldsymbol{R}}_i(1 - I\boldsymbol{t}_i/2) \\
&= \boldsymbol{R}_i\boldsymbol{n}_0\tilde{\boldsymbol{R}}_i + I(\boldsymbol{R}_i\boldsymbol{m}_0\tilde{\boldsymbol{R}}_i + (\boldsymbol{t}_i(\boldsymbol{R}_i\boldsymbol{n}_0\tilde{\boldsymbol{R}}_i) - (\boldsymbol{R}_i\boldsymbol{n}_0\tilde{\boldsymbol{R}}_i)\boldsymbol{t}_i)/2) \\
&= \boldsymbol{R}_i\boldsymbol{n}_0\tilde{\boldsymbol{R}}_i + I(\boldsymbol{R}_i\boldsymbol{m}_0\tilde{\boldsymbol{R}}_i + \boldsymbol{t}_i\wedge(\boldsymbol{R}_i\boldsymbol{n}_0\tilde{\boldsymbol{R}}_i)).
\end{aligned} \tag{21.78}$$

By separating the real and dual part of above equation, we get

$$\boldsymbol{n}_i = \boldsymbol{R}_i\boldsymbol{n}_0\tilde{\boldsymbol{R}}_i \tag{21.79}$$
$$\boldsymbol{m}_i = \boldsymbol{R}_i\boldsymbol{m}_0\tilde{\boldsymbol{R}}_i + \boldsymbol{t}_i\wedge\boldsymbol{n}_i. \tag{21.80}$$

Because $\boldsymbol{L}_0^k$ and $\boldsymbol{L}_i^k$ are the noisy measurements, we use a least square method to estimate a best solution of rotation and translation. We determine first the rotation using (21.79) by minimizing the following criterion

$$E_{min} = Min\{\sum_{k=1}^{n} \parallel \boldsymbol{n}_i^k - \boldsymbol{R}_i\boldsymbol{n}_0^k\tilde{\boldsymbol{R}}_i \parallel^2\}. \tag{21.81}$$

After right-multiplying both sides of (21.79) with the rotor $\boldsymbol{R}_i$, we get

$$\boldsymbol{n}_i\boldsymbol{R}_i - \boldsymbol{R}_i\boldsymbol{n}_0 = \mathbf{0}. \tag{21.82}$$

In linear algebra, (21.82) is expressed as

$$(\boldsymbol{n}_i)_{Rl}\boldsymbol{R}_i - (\boldsymbol{n}_0)_{Rr}\boldsymbol{R}_i = \boldsymbol{\mathcal{A}}_R\boldsymbol{R}_i = \mathbf{0}. \tag{21.83}$$

Then, (21.81) can be further restated as

$$E'_{min} = Min\{\sum_{k=1}^{n} \boldsymbol{R}_i^T\boldsymbol{\mathcal{A}}_{\boldsymbol{R}}^{\boldsymbol{k}\,T}\boldsymbol{\mathcal{A}}_R^k\boldsymbol{R}_i\} = Min\{\boldsymbol{R}_i^T\boldsymbol{\mathcal{A}}\boldsymbol{R}_i\}, \tag{21.84}$$

where

$$\boldsymbol{\mathcal{A}}_R^k = \sum_{k=1}^{n}((\boldsymbol{n}_i^k)_{Rl} - (\boldsymbol{n}_0^k)_{Rr}), \tag{21.85}$$

$$\boldsymbol{\mathcal{A}} = \sum_{k=1}^{n}\boldsymbol{\mathcal{A}}_{\boldsymbol{R}}^{\boldsymbol{k}\,T}\boldsymbol{\mathcal{A}}_R^k. \tag{21.86}$$

Since $\boldsymbol{\mathcal{A}}$ is a symmetric matrix and $\parallel \boldsymbol{R}_i \parallel = 1$, the solution to this problem is the 4-dimensional vector $\hat{\boldsymbol{R}}_i$ corresponding to the smallest eigenvalue of $\boldsymbol{\mathcal{A}}$.

With the recovered rotation $\hat{\boldsymbol{R}}_i$ we can then determine the translation using (21.80). In linear algebra, (21.80) is expressed as

$$\boldsymbol{m}_i = \boldsymbol{\mathcal{R}}_i\boldsymbol{m}_0 - (\boldsymbol{n}_i)_\times\boldsymbol{t}_i, \tag{21.87}$$

where

$$\boldsymbol{\mathcal{R}}_i = (\boldsymbol{R}_i)_{Rl}(\tilde{\boldsymbol{R}}_i)_{Rr}, \tag{21.88}$$

and the matrix $(\boldsymbol{n}_i)_\times$ is the skew-symmetric matrix of $\boldsymbol{n}_i$, which performs the outer product of the bivector $\boldsymbol{n}_i$ with another bivector. If $n_{1,i}$, $n_{2,i}$ and $n_{3,i}$ are three components of the bivector $\boldsymbol{n}_i$, then

$$(\boldsymbol{n}_i)_\times = \begin{pmatrix} 0 & n_{3,i} & -n_{2,i} \\ -n_{3,i} & 0 & n_{1,i} \\ n_{2,i} & -n_{1,i} & 0 \end{pmatrix}. \tag{21.89}$$

We estimate the translation $\hat{\boldsymbol{t}}_i$ by minimizing the following criterion

$$E''_{min} = Min\{\sum_{k=1}^{n} \| \boldsymbol{m}_i^k - \hat{\boldsymbol{\mathcal{R}}}_i \boldsymbol{m}_0^k + (\boldsymbol{n}_i^k)_\times \hat{\boldsymbol{t}}_i \|^2\}. \tag{21.90}$$

By differentiating the criterion (21.90) with respect to $\boldsymbol{t}_0$ and setting the result equal to zero, we obtain

$$\sum_{k=1}^{n} 2\left(\boldsymbol{m}_i^k - \hat{\boldsymbol{\mathcal{R}}}_i \boldsymbol{m}_0^k + (\boldsymbol{n}_i^k)_\times \hat{\boldsymbol{t}}_i\right)^T (\boldsymbol{n}_i^k)_\times = \boldsymbol{0}. \tag{21.91}$$

Then, $\hat{\boldsymbol{t}}_i$ can be solved by the equation:

$$\left(\sum_{k=1}^{n} (\boldsymbol{n}_i^k)_\times^T (\boldsymbol{n}_i^k)_\times\right) \hat{\boldsymbol{t}}_i = \sum_{k=1}^{n} (\boldsymbol{n}_i^k)_\times^T (\hat{\boldsymbol{\mathcal{R}}}_i \boldsymbol{m}_0^k - \boldsymbol{m}_i^k). \tag{21.92}$$

It can be shown that the matrices $\boldsymbol{\mathcal{A}}$ and $\boldsymbol{\mathcal{B}} = \sum_{k=1}^{n} (\boldsymbol{n}_i^k)_\times (\boldsymbol{n}_i^k)_\times^T$ are always of full rank if two of the lines $\boldsymbol{L}_i^k (k = 1...n)$ are non-parallel. In other words, to determine a unique motion displacement there must be at least two non-parallel lines.

21.5 Experimental Analysis of the MEKF

To further verify the analyses presented above and to demonstrate the performance of the MEKF algorithm, experiments using both simulated data and real 3-D reconstructed lines have been performed.

21.5.1 Simulation

The routine of the MEKF is programmed in MATLAB. The goal of the simulated experiments is to test the routine of MEKF, and by filter tuning to improve the accuracy and the converge rate of the estimate.

Let us suppose a rigid object is moving along a screw in 3-D with constant angular velocity $\omega/2 = -\pi/15$ about an axis of known line ($\boldsymbol{L}_s = \bar{\boldsymbol{r}} + I t_c \wedge \bar{\boldsymbol{r}}$) and constant translation velocity $\boldsymbol{v_s} = 0.3\bar{\boldsymbol{r}}$ which is parallel to the axis. A

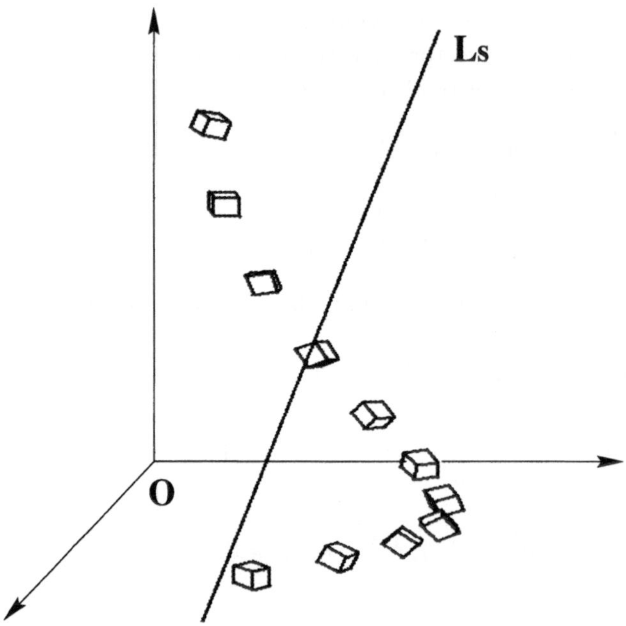

Fig. 21.8. The object moving in 3-D with screw trajectory

box moving in this way is shown in Fig. 21.8. The line $\boldsymbol{L}_s$ is given by

$$\begin{aligned}\boldsymbol{L}_s &= \bar{\boldsymbol{r}} + \boldsymbol{I}\boldsymbol{t}_c \wedge \bar{\boldsymbol{r}} \\ &= 0.7071\gamma_2\gamma_3 + 0.3536\gamma_3\gamma_1 + 0.6124\gamma_1\gamma_2 + \\ &\quad + I(-0.7418\gamma_2\gamma_3 + 0.3813\gamma_3\gamma_1 + 0.6364\gamma_1\gamma_2).\end{aligned}$$

Assume the measurements are sampled by equal time intervals which are normalized to 1. Then the motion between times $i-1$ and i can be described by the motor $\boldsymbol{V}$, which can be calculated by (21.59).

$$\begin{aligned}\boldsymbol{V} &= (1 + \boldsymbol{I}\boldsymbol{v}_{s0}/2)(cos(\omega/2) + sin(\omega/2)\boldsymbol{L}_s) \\ &= (1 + I0.3\bar{\boldsymbol{r}}/2)(cos(-\pi/15) + sin(-\pi/15)\boldsymbol{L}_s \\ &= 0.9832 - 0.1289\gamma_2\gamma_3 - 0.0645\gamma_3\gamma_1 - 0.1117\gamma_1\gamma_2 + \\ &\quad + I(-0.0266 + 0.2367\gamma_2\gamma_3 - 0.0188\gamma_3\gamma_1 - 0.0283\gamma_1\gamma_2). \qquad (21.93)\end{aligned}$$

Then we get the dynamic motion equation in linear algebra

$$\boldsymbol{M}_i = \boldsymbol{\mathcal{V}}_{Ml}\boldsymbol{M}_{i-1}. \qquad (21.94)$$

In simulation, the real applied motion parameters $\boldsymbol{V}_{i/i-1}$ between times $i-1$ and i are contaminated by noise:

$$\begin{aligned}\omega_i &= \omega + n_{\omega_i}, \\ v_{si} &= v_s + n_{v_{si}}, \\ \boldsymbol{V}_{i/i-1} &= (1 + \boldsymbol{I}\boldsymbol{v}_{si}/2)(cos(\omega_i/2) + sin(\omega_i/2)\boldsymbol{L}_s),\end{aligned}$$

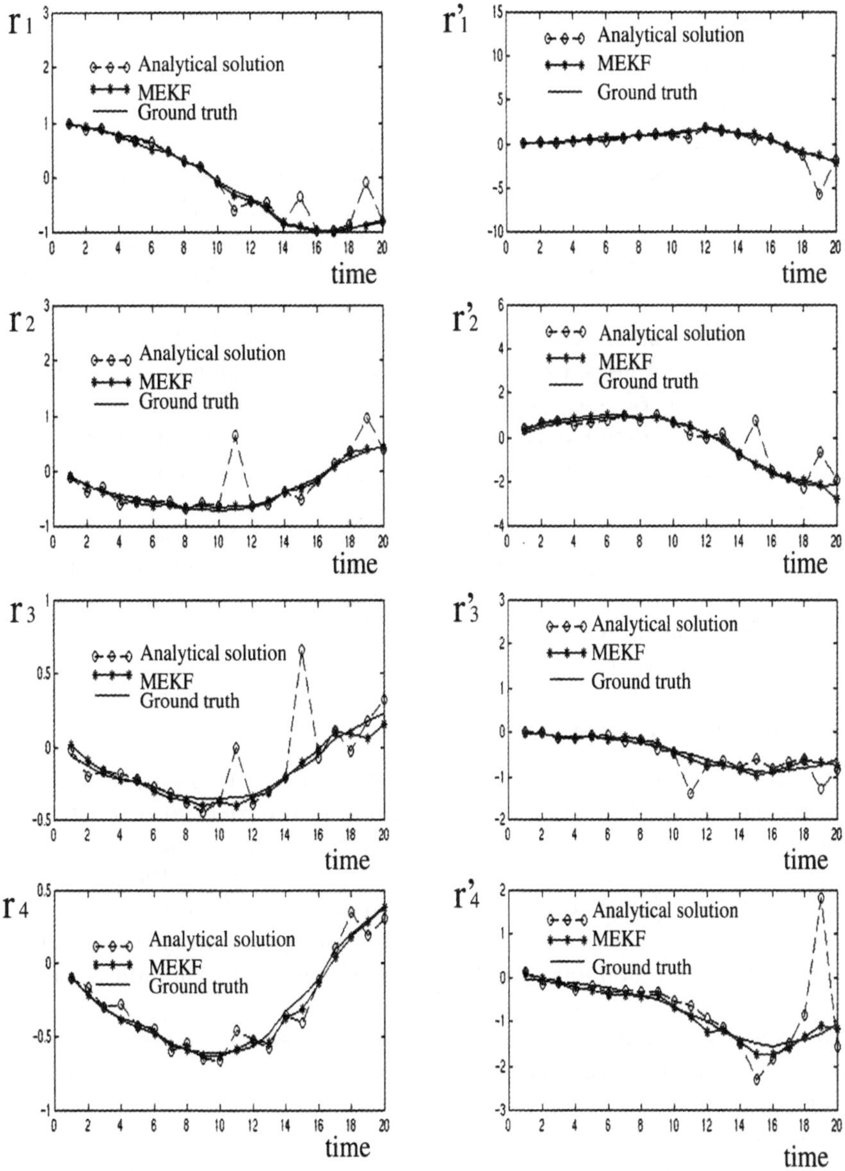

Fig. 21.9. The estimation results of the motor parameters by simulation

where the noises n_{ω_i} and $n_{v_{si}}$ are independent normally distributed with zero mean and known deviation σ_ω and σ_{v_s}. Then the ground truth of motor trajectories $\boldsymbol{M}_{0,i}$ can be computed by

$$\begin{aligned}\boldsymbol{M}_{0,i} &= \left(r_{0i}\ r_{1i}\ r_{2i}\ r_{3i}\ r'_{0i}\ r'_{1i}\ r'_{2i}\ r'_{3i}\right)^T \\ &= \boldsymbol{\mathcal{V}}_{i/i-1_{Ml}}\boldsymbol{M}_{0,i-1},\end{aligned} \tag{21.95}$$

with initial state $\boldsymbol{M}_{0,0} = (1\ 0\ 0\ 0\ 0\ 0\ 0\ 0)$.

At i-th ($i = 1, 2, ..., N$) time step of MEKF algorithm we first generate two 3-D points $\boldsymbol{x}'_0$ and $\boldsymbol{y}'_0$ to define a line $\boldsymbol{L}'_0$ in an observer coordinate frame A. The points are then moved to $\boldsymbol{x}'_i$ and $\boldsymbol{y}'_i$ by the motor $\boldsymbol{M}_i = (\boldsymbol{V}_{i/i-1})_{Ml}\boldsymbol{M}_{i-1}$ which can be decomposed to rotation $\boldsymbol{R}_i$ and translation $\boldsymbol{t}_i$, where $\boldsymbol{R}_i$ is the real part of $\boldsymbol{M}_i$. If $\boldsymbol{R}'_i$ is the dual part of $\boldsymbol{M}_i$, from (21.41) we can get $\boldsymbol{t}_i = 2\boldsymbol{R}'_i\tilde{\boldsymbol{R}}_i$.

The coordinate of this line after motion is $\boldsymbol{L}'_i$ relative to the frame A. We obtain thus a pair of noise-free coordinates of the same line in two positions (the initial position $\boldsymbol{L}'_0$ and the position $\boldsymbol{L}'_i$ at time i). To simulate the noisy observation, independent Gaussian noise with zero mean and known standard deviation σ is added to both lines $\boldsymbol{L}'_0$ and $\boldsymbol{L}'_{i+1}$ and we obtain thus the noisy observation $\boldsymbol{L}_0$ and $\boldsymbol{L}_i$.

In Fig. 21.9, we show the eight components of motor trajectories estimated by MEKF algorithm and by batch method of the analytical solution. In MEKF algorithm we use the analytical solution to estimate the initial prediction. Comparing with ground truth, we can see that the MEKF gives more accurate and more stable estimates.

21.5.2 Real Experiment

Fig. 21.10 shows the physical setup of our experiment. Two grey-scale-CCD 640 × 480 cameras are fastened to the last joint of the robot arm RX90. The RX90 has six rotation joints which can be controlled by six parameters (x, y, z,roll, pitch, and yaw). The coordinates (x, y, z) that describe the position of the end joint are referred to the base coordinate system W which is fixed on the base of the arm. The rotation parameters (roll, pitch, yaw) that describe the orientation of the end joint are Z-Y-Z Euler angles [50]. The sample object is placed below the cameras.

We want to estimate the relative motion between the end joint and the sample object based on the cameras' images while the arm is moving with a given trajectory.

In practice, we use 3 cameras to reconstruct a 3-D line. In the experimental setup, the third camera was realized by applying a certain motion to one of the cameras.

We have no ground truth of the relative motion of the sample object. But we can compare the estimation with the given motion trajectory of the robot arm. A coordinate system T which is fixed on the end joint is called a tool coordinate system.We control the robot arm by controlling the relative

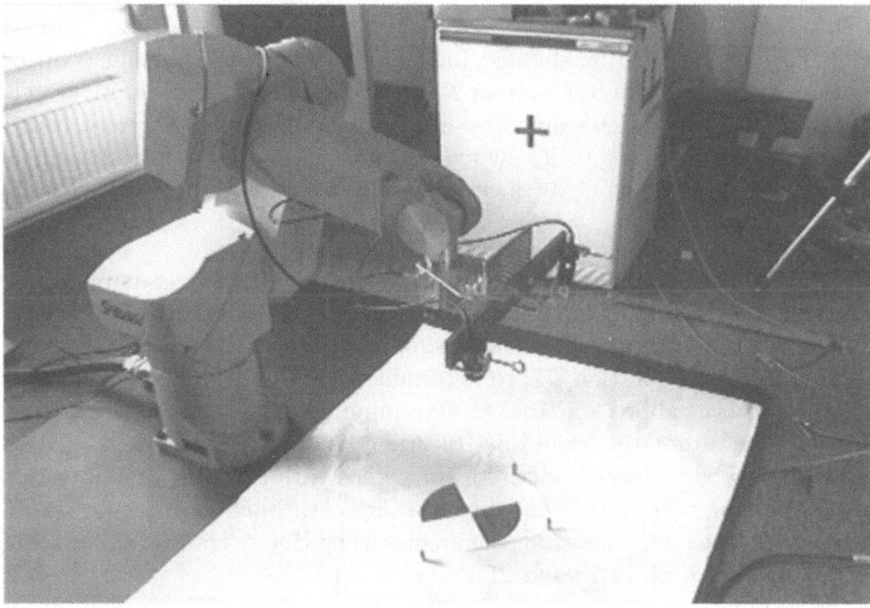

Fig. 21.10. The physical setup of the experiment

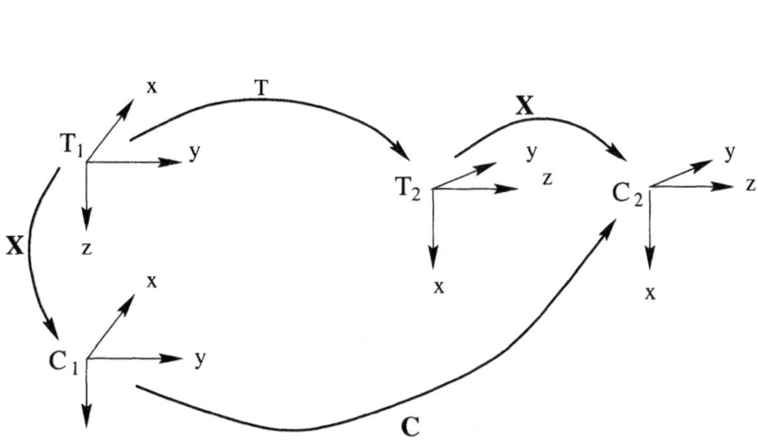

Fig. 21.11. The relationship between the tool system T and the camera system C

position and orientation between the tool system T and the base system W. After camera calibration, we get a matrix $\boldsymbol{\mathcal{P}}$, which describes the relationship between the point coordinates in the 2-D image and the correspondent 3-D world point coordinates with respect to a system C (up to a scalar λ). The system C is fixed on the end joint and there exists a certain transformation $\boldsymbol{X}$ between the tool system T and the system C, see Fig. 21.11. The transformation $\boldsymbol{X}$ is determined by hand-eye calibration. If the tool system is transformed from T_1 to T_2 by transformation $\boldsymbol{T}$, the system C will be transformed from C_1 to C_2 by a certain transformation $\boldsymbol{C}$, which is given by

$$\boldsymbol{C} = \boldsymbol{X}\boldsymbol{T}\boldsymbol{X}^{-1}. \tag{21.96}$$

Using (21.96) we can compare the relative motion $\boldsymbol{C}$ of the sample object referring to system C with the given motion $\boldsymbol{T}$ of the robot arm.

The steps of the experiment are as follows:

1) Cameras calibration [77] to determine the $\boldsymbol{\mathcal{P}}$ matrix;
2) Hand-eye calibration [134] to determine the unknown transformation $\boldsymbol{X}$; (an alternative would be [16] using motors)
3) Taking the images in discrete time steps with constant time intervals while the robot arm is moving, see Figs. 21.12 and 21.14;
4) Extracting 2-D lines from the images using Hough transformation [144] [181], see Figs. 21.13 and 21.15;
5) 3-D line reconstruction by 3 matched image lines [77], see Tab. 21.1 ;
6) Estimation the motion based on 3-D line observations using MEKF, see Fig. 21.16 .

The algorithm of motion estimation will run online recursively from step 3) to 6).

In our experiment, the given relative motion of the sample object with respect to system C is a screw motion with constant angular velocity $\omega = -\pi/90$ and constant translation velocity $\boldsymbol{v}_s = 0.2$ which is parallel to the rotation axis. The rotation axis $\boldsymbol{L}_s$ is parallel to the z axis of the system C, and one point on $\boldsymbol{L}_s$ is (1.5, 0, 0). In $\mathcal{G}^+_{3,0,1}$ the screw axis $\boldsymbol{L}_s$ is given by

$$\begin{aligned}\boldsymbol{L}_s &= \gamma_1\gamma_2 + I(1.5\gamma_2\gamma_3)\wedge(\gamma_1\gamma_2)\\ &= \gamma_1\gamma_2 + I1.5\gamma_3\gamma_1.\end{aligned} \tag{21.97}$$

Just like (21.93), the motor $\boldsymbol{V}$ can be calculated as

$$\begin{aligned}\boldsymbol{V} &= (1 + I\boldsymbol{v}_{s0}/2)(cos(\omega/2) + sin(\omega/2)\boldsymbol{L}_s)\\ &= 0.9994 - 0.0349\gamma_1\gamma_2 + I(0.0035 - 0.0523\gamma_3\gamma_1 + 0.0999\gamma_1\gamma_2).\end{aligned} \tag{21.98}$$

The motor $\boldsymbol{M}_{i+1}$ is in linear algebra given by

$$\boldsymbol{M}_i = \boldsymbol{\mathcal{V}}_{Ml}\boldsymbol{M}_{i-1}, \tag{21.99}$$

with initial data $\boldsymbol{M}_0 = (1 \quad 0 \quad 0 \quad 0 \quad 0 \quad 0 \quad 0 \quad 0)^T$.

We use the reconstructed 3-D lines listed in Tab. 21.1 to estimate the relative motion of the sample object. The results are shown in Fig. 21.16, which

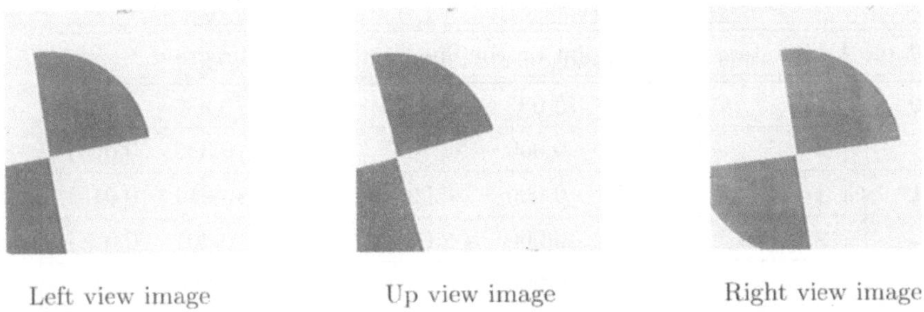

Fig. 21.12. A stereo triplet of a sample object at time $i = 0$

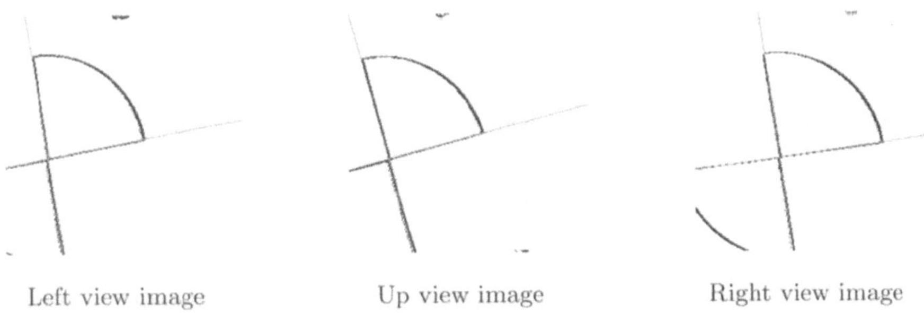

Fig. 21.13. Edge images of Fig. 21.12 overlapped by extracted 2-D lines

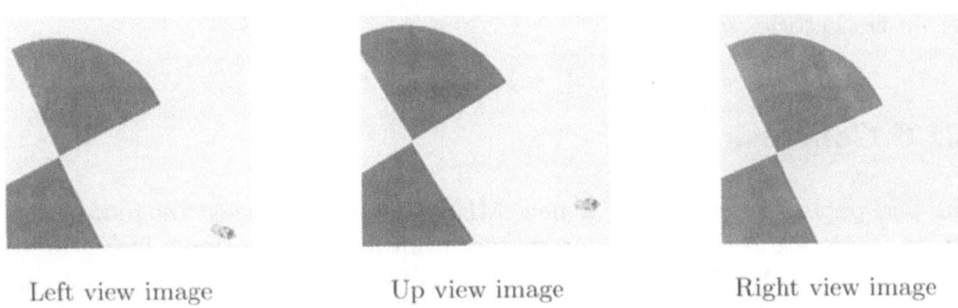

Fig. 21.14. A stereo triplet of a sample object at time $i = 4$

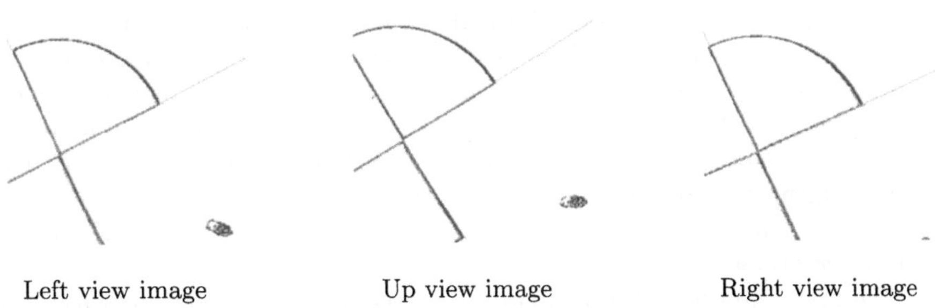

Fig. 21.15. Edge images of Fig. 21.14 overlapped by extracted 2-D lines

Table 21.1. Reconstructed 3-D lines

Time	Line item	A point on the line			direction		
0	1	(0.000	3.087	-2.327)	(-0.345	0.937	-0.027)
	2	(0.556	0.000	-2.250)	(0.941	0.336	0.023)
1	1	(1.125	0.000	-2.027)	(-0.404	0.914	0.013)
	2	(0.701	0.000	-2.049)	(0.915	0.401	0.029)
2	1	(1.111	0.000	-1.82)	(-0.462	0.886	0.017)
	2	(0.794	0.000	-1.83)	(0.880	0.471	0.055)
⋮							
14	1	(0.018	0.000	0.648)	(-0.971	0.236	-0.036)
	2	(1.103	0.000	0.538)	(0.241	0.965	0.103)
15	1	(-0.680	0.000	0.753)	(-0.986	0.159	-0.025)
	2	(0.000	-6.341	0.783)	(0.171	0.985	-0.003)

shows the trajectories of eight components of the estimated motor trajectories $\hat{\boldsymbol{M}}_i$ (star-solid lines) and the given motor trajectories $\boldsymbol{M}_i$ (solid lines). Although we use an inaccurate initial predicted motor for the algorithm, after three or four time steps the estimations approach the truth and follow the given trajectories very well.

21.6 Conclusion

In this paper, we presented a new MEKF algorithm based on motor algebra to estimate 3-D motion parameters from line observations. Using motor algebra, we modeled in the 4-D space the motion of lines and the dynamic motion system. This kind of modeling linearizes the 3-D Euclidean rigid motion transformation and describes the discrete dynamic system straightforwardly.

The MEKF has the virtue that it can estimate the motion parameters from Plücker line observations. Since all recursive algorithms of the literature estimate motion parameters from observations of points or line segments with its middle point, we can claim that the use of Plücker lines is one of the most important advantages of the MEKF. Additionally, using the modeling of the lines in the motor algebra, we could linearize the nonlinear measurement model which dose not face singularities, this was also a big problem of many researchers who tried in some way to apply the Kalman filter using Plücker line observations.

We first introduced the Kalman filter techniques and then presented the measurement model based on motor algebra and its constraints. This mea-

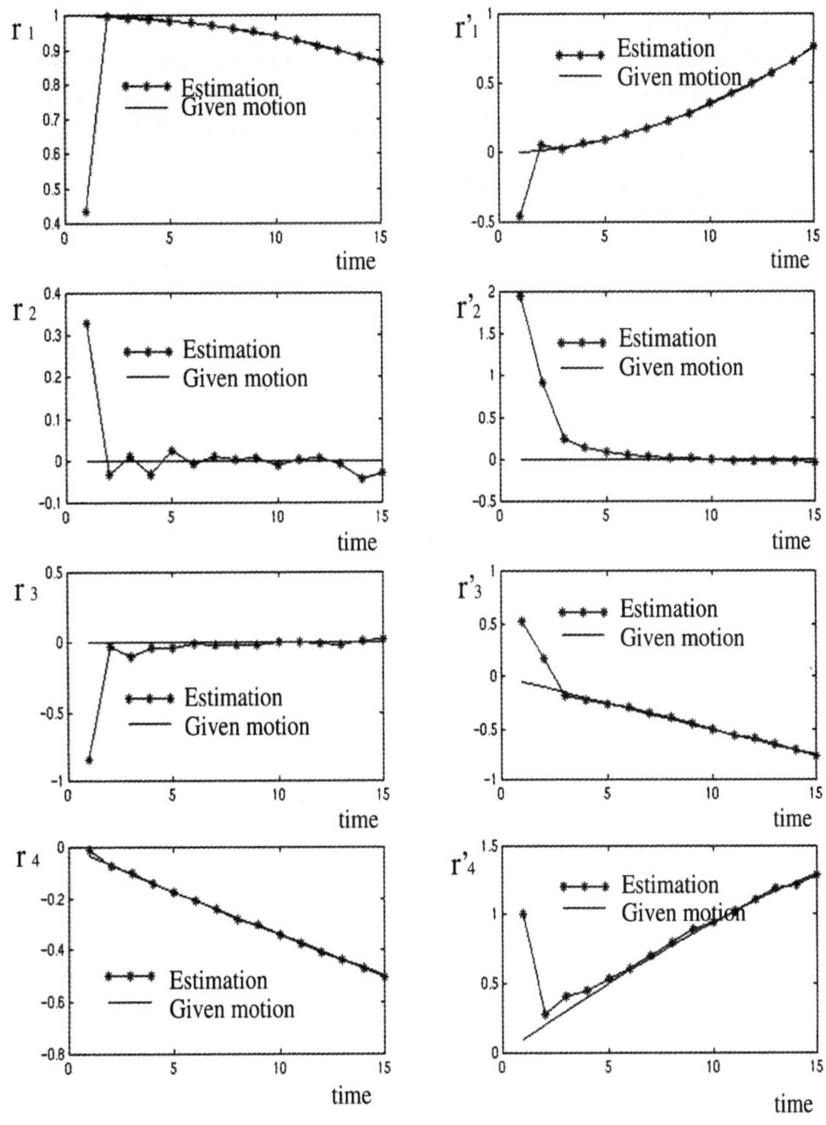

Fig. 21.16. The estimation results of the motor parameters by MEKF in real experiment

surement model was then linearized for Kalman filtering. We also described the dynamic motion model using motors as states from which we observe that the motor algebra is useful to effectively formulate and to compute the screw motion of a line as minimal rigid entity. In the algorithm of MEKF, we modified the estimation to satisfy the constraints, which made the estimation converge to a proper motor state.

Tests with both simulated data and real experimental data showed that the MEKF algorithm is effective to dynamically estimate the motion parameters from Plücker line observations. We also compared the MEKF with an analytical solution using least squares and the results show that the MEKF gives more accurate and more stable estimations.

References

1. L.V. Ahlfors. Möbius transformations and Clifford numbers. In I. Chavel and H.M. Farkas, editors, *Differential Geometry and Complex Analysis*. Springer-Verlag, Berlin, Heidelberg, 1985.
2. L.V. Ahlfors. Möbius transformations in ∇^n expressed through 2×2 matrices of Clifford numbers. *Complex Variables Theory*, 5:215–224, 1986.
3. Y. Akishige. Perceptual space and the law of conservation of perceptual information. In *XVIII Int. Congr. Psycho. Symp. 19: Perception of Space and Time*, 1966.
4. Y. Akishige. Studies on constancy problem in Japan. *Psychologia*, 11(1/2):43–55, 1968.
5. B. Anup. Active calibration: alternative strategy and analysis. *IEEE Conf. on Computer Vision and Pattern Recognition*, pages 495–500, 1993.
6. P. Arena, L. Fortuna, R. Re, and M. G. Xibilia. Multilayer perceptrons to approximate complex valued functions. *Neural Systems*, 6:435–446, 1995.
7. P. Arena et al. Multilayer perceptrons to approximate quaternion valued functions. *Neural Networks*, 9:1–8, 1996.
8. K.S. Arun, T.S. Huang, and S. D. Blostein. Least-squares fitting of two 3d point sets. *IEEE Trans. Pattern Anal. Machine Intell. PAMI*, 9(5):698–700, 1987.
9. N.A. Aspragathos and J.K. Dimitros. A comparative study of three methods for robot kinematics. *IEEE Transactions an Systems, Man and Cybernetics - Part B: Cybernetics*, 28(2):135–145, April 1998.
10. M. F. Atiyah and I. G. MacDonald. *Introduction to Commutative Algebra.* Addison-Wesley Publishing Co., London, 1969.
11. A.J. Azarbayejani, H. Bradley, and A. Pentland. Recursive estimation of structure and motion using relative orientation constraints. In *Proc. IEEE Conf. Computer Vision and Pattern Recognition, Los Alamitos, CA, June, 1993*, 1993.
12. R.S. Ball. *A Treatise on the Theory of Screws.* Cambridge University Press, 1900.
13. LY. Bar-Itzhack and Y. Oshman. Attitude determination from vector observations: quaternion estimation. *IEEE Trans. on Aerospace and Electronic Systems*, 21(1):128–135, 1985.
14. M. Barnabei, A. Brini, and G-C. Rota. On the exterior calculus of invariant theory. *J. of Algebra*, 96:120–160, 1985.
15. E. Bayro-Corrochano and S. Buchholz. Geometric neural networks. In G. Sommer and J. J. Koenderink, editors, *Algebraic Frames for the Perception–Action Cycle*, Lecture Notes in Computer Science, vol. 1315, pages 379–394. Springer-Verlag, Berlin, Heidelberg, 1997.
16. E. Bayro-Corrochano, K. Daniilidis, and G. Sommer. Hand-eye calibration in terms of motions of lines using geometric algebra. In *10th Scandinavian Conference an Image Analysis, Lappeenranta*, volume 1, pages 397–404, 1997.

17. E. Bayro-Corrochano and J. Lasenby. A unified language for computer vision and robotics. In G. Sommer and J.J. Koenderink, editors, *Algebraic Frames for the Perception–Action Cycle*, volume 1315 of *Lecture Notes in Computer Science*, pages 219–234. Springer–Verlag, Berlin, Heidelberg, 1997.
18. E. Bayro-Corrochano and J. Lasenby. Geometric techniques for the computation of projective invariants using n uncalibrated cameras. In *Proceedings of the Indian Conference on Computer Vision and Image Processing*, pages 95–100, New Delhi, India, 1998.
19. E. Bayro-Corrochano, J. Lasenby, and G. Sommer. Geometric algebra: A framework for computing point and line correspondences and projective structure using n uncalibrated cameras. In *IEEE Proceedings of ICPR'96*, volume I, pages 334–338, Viena, Austria, August 1996.
20. R. Bellman. *Introduction to Matrix Analysis.* McGraw-Hill Book Company, Inc., New York, Toronto, London, 1960.
21. E. Beltrami. Saggio di interpetrazione della geometria non-euclidea. *Giorn. Mat.*, 6:248–312, 1868.
22. W. Blaschke. *Kinematik und Quaternionen.* VEB Deutscher Verlag der Wissenschaften, Berlin, 1960.
23. L. M. Blumenthal. *Theory and Applications of Distance Geometry.* Cambridge University Press, Cambridge, 1953. reprinted by Chelsea, London, 1970.
24. L. M. Blumenthal. *A Modern View of Geometry.* Dover, New York, 1961.
25. J. Bolyai. Appendix, scientiam spatii absolute veram exhibens. In *tentamen Juventutem studiosam in elementa Matheseos purae.* W. Bolyai, Maros Vasarhelyini, 1832.
26. A.C. Bovik, M. Clark, and W. Geisler. Multichannel texture analysis using localized spatial filters. *IEEE Trans. Pattern Analysis and Machine Intelligence*, 12(1):55–73, 1990.
27. C. B. Boyer. *The History of the Calculus and its Conceptual Development.* Dover Publications Inc., New York, 1959 (1949).
28. R. Bracewell. *The Fourier Transform and its Applications.* McGraw Hill, 1986.
29. R.N. Bracewell. Affine theorem for the Hartley transform of an image. *Proceedings of the IEEE*, 82:388–390, 1994.
30. F. Brackx, R. Delanghe, and F. Sommen. *Clifford Analysis.* Pitman, Boston, 1982.
31. A. Bregler and J. Malik. Tracking people with twists and exponential maps. In *IEEE Computer Society Conference on Computer Vision and Pattern Recognition, Brisbane*, pages 8–15, 1998.
32. P. Brodatz. *Textures: A Photographic Album for Artists and Designers.* New York: Dover, 1966.
33. S. Buchholz. Algebraische Einbettungen Neuronaler Netze. Master's thesis, Cognitive Systems Group, Inst. of Comp. Sci., Univ. of Kiel, Germany, 1997.
34. T. Bülow. *Global and Local Hypercomplex Spectral Signal Representations for Image Processing and Analysis.* PhD thesis, Christian-Albrechts-University of Kiel, 1999.
35. T. Bülow and G. Sommer. Algebraically Extended Representation of Multi-Dimensional Signals. In *Proceedings of the 10th Scandinavian Conference on Image Analysis, Lappeenranta, vol. 2*, pages 559–566, 1997.
36. T. Bülow and G. Sommer. Multi-dimensional signal processing using an algebraically extended signal representation. In G. Sommer and J.J. Koenderink, editors, *Int. Workshop on Algebraic Frames for the Perception-Action Cycle*, volume 1315 of *Lecture Notes in Computer Science*, pages 148–163. Springer-Verlag, Berlin, Heidelberg, 1997.

37. J. W. Cannon, W. J. Floyd, R. Kenyon, and W. R. Parry. Hyperbolic geometry. In S. Levy, editor, *Flavors of Geometry*. Cambridge University Press, Cambridge, 1997.
38. B. Caprile and V. Torre. Using vanishing points for camera calibration. *International Journal of Computer Vision*, 4:127–140, 1990.
39. H. Chen. A screw motion approach to uniqueness analysis of head-eye geometry. In *IEEE Conf. Computer Vision and Pattern Recognition*, pages 145–151, Maui, Hawaii, June 3-6, 1991.
40. V. M. Chernov. Discrete orthogonal transforms with data representation in composition algebras. In *In Proceedings of the 9th Scandinavian Conference on Image Analysis, Uppsala, Sweden, 1995*, pages 357–364, 1995.
41. D.P. Chevallier. Lie algebras, modules, dual quaternions, and algebraic methods in kinematics. *Mechanics and Machine Theory*, 26:613–627, 1991.
42. D.P. Chevallier. On the transference principle in kinematics: its various forms and limitations. *Mechanics and Machine Theory*, 31:57–76, 1996.
43. J.C.K. Chou and M. Kamel. Finding the position and orientation of a sensor on a robot manipulator using quaternions. *Intern. Journal of Robotics Research*, 10(3):240–254, 1991.
44. Roberto Cipolla and Edmond Boyer. 3d model acquisition from uncalibrated images. In *Proc. IAPR Workshop on Machine Vision Applications, Chiba, Japan*, pages 559–568, 1998.
45. W. K. Clifford. Preliminary sketch of bi-quaternions. *Proceedings of the London Mathematical Society*, 4:381–395, 1873.
46. W. K. Clifford. Applications of Grassman's extensive algebra. *American Journal of Mathematics*, I:350–358, 1878.
47. W. K. Clifford. *Mathematical Papers.* Edited by R. Tucker. Macmillan, 1882.
48. S. Cornbleet. Geometrical optics reviewed: A new light on an old subject. *Proceedings of the IEEE*, 71(4), April 1983.
49. Digital Equipment Corporation. Digital Semiconductor Alpha 21164PC Microprocessor Data Sheet[1], 1997.
50. J.J. Craig. *Introduction to Robotics: Mechanics and Control.* Addison Wesley Publishing Company, 1989.
51. T. W. Cronin and N. J. Marschal. A retina with at least ten spectral types of photoreceptors in a mantis shrimp. *Nature*, 339:137–140, 1989.
52. M. J. Crowe. *A History of Vector Analysis.* Dover Publications Inc., New York, 1994 (1967).
53. G. Cybenko. Approximation by superposition of a sigmoidal function. *Mathematics of Control, Signals and Systems*, 2:303–314, 1989.
54. K. Daniilidis. Hand-eye calibration using dual quaternions. *International Journal of Robotics Research*, 18:286–298, 1999.
55. K. Daniilidis and E. Bayro-Corrochano. The dual–quaternion approach to hand–eye calibration. In *Proc. 13th Int. Conf. on Pattern Recognition*, volume A, pages 318–322. IEEE Computer Soc. Press, Vienna, 1996.
56. K. Daniilidis and J. Ernst. Active intrinsic calibration using vanishing points. *Pattern Recognition Letters*, 17:1179–1189, 1996.
57. J. G. Daugman. Two-dimensional spectral analysis of cortical receptive field profiles. *Vision Res.*, 20:847–856, 1980.
58. C.M. Davenport. A Commutative hypercomplex algebra with associated function theory. In R. Ablamowicz, editor, *Clifford Algebras with Numeric and Symbolic Computations*, pages 213–227. Birkhäuser, Boston, 1996.

[1] `http://ftp.digital.com/pub/DECinfo/semiconductor/literature/164pcds.pdf`

59. M.A. Delsuc. Spectral representation of 2D NMR spectra by hypercomplex numbers. *Journal of Magnetic Resonance*, 77:119–124, 1988.
60. J. Denavit and R. S. Hartenberg. A kinematic notation for lower-pair mechanisms based on matrices. *ASME Journal of Applied Mechanics*, 22:215–221, 1955.
61. R. Descartes. *La Geometrie*. Dover Publ. Inc., New York, 1954 (1637,1925).
62. F. Devernay and O. D. Faugeras. From Projective to Euclidean Reconstruction. Technical Report 2725, INRIA, Sophia Antipolis, 1995.
63. C. Doran, D. Hestenes, F. Sommen, and N. Van Acker. Lie groups as spin groups[2]. *J. Math. Phys.*, 34:3642–3669, 1993.
64. C. Doran, A. Lasenby, and S. Gull. Chapter 6: Linear algebra. In W.E. Baylis, editor, *Clifford (Geometric) Algebras with Applications in Physics, Mathematics and Engineering*. Birkhäuser, Boston, 1996.
65. A. Dress and T. Havel. Distance geometry and geometric algebra. *Foundations of Physics*, 3642–3669(23), 1993.
66. L. Dron. Dynamic camera self–calibration from controlled motion sequences. In *IEEE Conf. on Computer Vision and Pattern Recognition*, pages 501–506, 1993.
67. F. Du and M. Brady. Self-calibration of the intrinsic parameters of cameras for active vision systems. *In IEEE Conf. on Computer Vision and Pattern Recognition*, pages 477–482, 1993.
68. L. Eisenhart. *A Treatise on the Differential Geometry of Curves and Surfaces*. Ginn & Co, The Athaenum Press, Boston, Massachussetts, 1909.
69. T. A. Ell. *Hypercomplex Spectral Transformations*. PhD thesis, University of Minnesota, 1992.
70. T.A. Ell. Quaternion Fourier transforms for analysis of 2-dimensional linear time-invariant partial-differential systems. In *Proc. 32nd IEEE Conf. on Decision and Control, San Antonio, TX, USA, 15-17 Dec.*, pages 1830–1841, 1993.
71. R. Enciso and T. Vieville. Self-calibration from four views with possibly varying intrinsic parameters. *Image and Vision Computing*, 15:293–305, 1997.
72. R.R. Ernst, W.P. Aue, P. Bachmann, J. Karhan, A. Kumar, and L. Müller. Two-dimensional NMR spectroscopy. In *Proc. 4th Ampère Int. Summer School, Pula, Yugoslavia*, 1976.
73. R.R. Ernst, G. Bodenhausen, and A. Wokaun. *Principles of Nuclear Magnetic Resonance in One and Two Dimensions*. Oxford Science Publications, 1985.
74. C. Daul et al. A fast image processing algorithm for quality control of woven textiles. In P. Levi et al., editor, *Levi, P. and et al. (Eds.), 20. DAGM Symposium Mustererkennung, Stuttgart*, pages 471–479. Springer-Verlag, Berlin, Heidelberg, 1998.
75. L. Euler. Trigonometria sphaerica universa ex primis principiis breviter et dilucide derivata. *Acta Acad. Sci. Petrop.*, 3:72–86, 1782.
76. O. Faugeras and B. Mourrain. On the geometry and algebra of the point and line correspondences between *n* images. In *Proceedings of ICCV'95*, pages 951–956, Boston, 1995. IEEE Computer Society Press.
77. O. D. Faugeras. *Three-Dimensional Computer Vision: A Geometrie Viewpoint*. MIT Press, Cambridge, MA, 1993.
78. O. D. Faugeras. Stratification of three dimensional vision: projective, affine and metric representations. *Journal of the Optical Society of America - A*, 12(3), 1995.

[2] Available at the Geometric Calculus Web Site:
`http://ModelingNTS.la.asu.edu/GC_R&D.html`

79. O. D. Faugeras and S. Maybank. Motion from point matches: Multiplicity of solutions. *International Journal of Computer Vision*, 4:225–246, 1990.
80. O. D. Faugeras and B. Mourrain. On the Geometry and Algebra of the Point and Line Correspondences between N Images. Technical Report No. 2665, INRIA, Sophia Antipolis, 1995.
81. O. D. Faugeras and T. Papadopoulo. Grassmann-Cayley Algebra for Modelling Systems of Cameras and the Algebraic Equations of the Manifold of Trifocal Tensors. Technical Report No. 3225, INRIA, 1997.
82. O. D. Faugeras and T. Papadopoulo. Grassmann-Cayley Algebra for Modelling Systems of Cameras and the Algebraic Equations of the Manifold of Trifocal Tensors. *Phil. Trans. R. Soc. Lond. A*, 356(1740):1123–1152, 1998.
83. O. D. Faugeras and T. Papadopoulo. A nonlinear method for estimating the projective geometry of three views. In *Proceedings of ICCV'98*, pages 477–484, 1998.
84. O. D. Faugeras and Theodore Papadopoulo. A Nonlinear Method for Estimating the Projective Geometry of Three Views. Technical Report 3221, INRIA, Sophia Antipolis, 1997.
85. J. Funda and R.P. Paul. A computational analysis of screw transformations in robotics. *IEEE Trans. Robotics and Automation*, 6:348–356, 1990.
86. D. Gabor. Theory of communication. *Journal of the IEE*, 93:429–457, 1946.
87. G. Georgiou and C. Koutsougeras. Complex domain backpropagation. *IEEE Trans. Circ. and Syst. II*, 39:330–334, 1992.
88. G.H.Granlund and H. Knutsson. *Signal Processing for Computer Vision.* Kluwer Academic Publishers, 1995.
89. J. W. Gibbs. Quaternions and the algebra of vectors. *Nature*, 47:463–464, March 16 1893.
90. J. J. Gibson. Optical motions and transformations as stimuli for visual perception. *Psych. Rev.*, 64(5):288–295, 1957.
91. J. J. Gibson. *The Ecological Approach to Visual Perception.* Houghton Mifflin Company, 1979.
92. G.H. Golub and C.F. van Loan. *Matrix computations.* Johns Hopkins University Press, 1983.
93. H. Grassmann. Der Ort der Hamilton'schen Quaternionen in der Ausdehnungslehre. *Mathematische Annalen*, 12:375, 1877.
94. H. Grassmann. "Linear Extension Theory" (Die Lineare Ausdehnungslehre) translated by L. C. Kannenberg. In *The Ausdehnungslehre of 1844 and Other Works*, Chicago, 1995. La Salle: Open Court Publ.
95. Y.-L. Gu and J.Y.S. Luh. Dual-number transformation and its application to robotics. *IEEE Journal of Robotics and Automation*, 3:615–623, 1987.
96. S. L. Hahn. Multidimensional complex signals with single-orthant spectra. *Proc. IEEE*, 80(8):1287–1300, 1992.
97. S. L. Hahn. *Hilbert Transforms in Signal Processing.* Artech House, Boston, London, 1996.
98. W. R. Hamilton. *Elements of Quaternions*, volume I-II. Chelsea Publ. Co., New York, 1969 (1899).
99. R.M. Haralick. Statistical and structural approaches to texture. *Proceedings of the IEEE*, 67:786–804, 1979.
100. J. Harris. *Algebraic Geometry.* Springer-Verlag, New York, 1992.
101. R. Hartley. Lines and points in three views – a unified approach. In *ARPA Image Understanding Workshop*, Monterey, California, 1994.
102. R. Hartley. Lines and points in three views and the trifocal tensor. *The International Journal of Computer Vision*, 22(2):125–140, 1997.

103. R. I. Hartley. An algorithm for self–calibration from several views. In *In Proc. Conference on Computer Vision and Pattern Recognition*, pages 908–912, 1994.
104. R. I. Hartley. In defence of the 8-point algorithm. *In IEEE Int. Conf. Computer Vision*, pages 1064–1070, 1995.
105. R. Hartshorne. *Algebraic Geometry*. Springer-Verlag, New York, 1977.
106. T. Havel. Distance geometry: Theory, algorithms and chemical applications. In *Encyclopedia of Computational Chemistry*. J. Wiley & Sons, 1998.
107. O. Heaviside. Vectors versus quaternions. *Nature*, 47:533–534, April 6 1893.
108. E. Hecht and A. Zajac. *Optics*. Addison-Wesley Publishing Company, 1974.
109. D. Hestenes. *Space-Time Algebra*. Gordon and Breach, New York, 1966.
110. D. Hestenes. The design of linear algebra and geometry. *Acta Applicandae Mathematicae*, 23:65–93, 1991.
111. D. Hestenes. Grassmann's vision[2]. In Gert Schubring, editor, *Hermann Gunther Grassmann (1809-1877): Visionary Mathematician, Scientist and Neohumanist Scholar*, Dordrecht, 1996. Kluwer Academic Publishers.
112. D. Hestenes. *New Foundations for Classical Mechanics*. D. Reidel, Dordrecht/Boston, 2nd edition, 1998.
113. D. Hestenes and G. Sobczyk. *Clifford Algebra to Geometric Calculus: A Unified Language for Mathematics and Physics*. D. Reidel Publishing Co., Dordrecht, 1984 (1992).
114. D. Hestenes and R. Ziegler. Projective geometry with Clifford algebra[2]. *Acta Applicandae Mathematicae*, 23:25–63, 1991.
115. A. Heyden and G. Sparr. Reconstruction from calibrated cameras - a new proof of the Kruppa-demazure theorem. *Journal of Mathematical Imaging and Vision*, 10:123–142, 1999.
116. Anders Heyden. A common framework for multiple view tensors. In H. Burkhardt and B. Neumann, editors, *Proceedings ECCV 98*, number 1406 in LNCS, pages 3–19. Springer-Verlag, Berlin, Heidelberg, 1998.
117. K. Hornik, M. Stinchcombe, and H. White. Multilayer feedforward networks are universal approximators. *Neural Networks*, 2:359–366, 1989.
118. M. K. Hu. Visual pattern recognition by moment invariants. *IEEE Trans. on Information Theory.*, 8:179–187, 1962.
119. T.S. Huang, S.D. Blostein, and E.A. Margerum. Least-squares estimation of motion parameters from 3d point correspondences. In *Proc. IEEE Conf. Computer Vision and Pattern Recognition. Miami Beach,FL*, 1986.
120. B. Iversen. *Hyperbolic Geometry*. Cambridge University Press, Cambridge, 1992.
121. B. Jähne. *Digitale Bildverarbeitung*. Springer-Verlag, Berlin, 1997.
122. A. Jazwinsky. *Stochastic Processes and Filtering Theory*. Academic Press, New York, 1970.
123. B. Julesz. Textons, the elements of texture perception, and their interpretations. *Nature*, 290(12):91–97, 1981.
124. K. Kanatani. *Geometric Computation for Machine Vision*. Oxford University Press, Oxford, 1993.
125. I. L. Kantor and A. S. Solodovnikov. *Hypercomplex Numbers*. Springer-Verlag, New York, 1989.
126. G. Kienle. Experiments concerning the non–Euclidian structure of the visual space. *Bioastronautics.*, 4:386–400, 1964.
127. W. Killing. Über zwei Raumformen mit constanter positiver Krümmung. *J. Reine Angew. Math.*, 86:72–83, 1878.
128. J. Kim and V.R. Kumar. Kinematics of robot manipulators via line transformations. *Journal of Robotic Systems*, 7:649–674, 1990.

129. F. Klein. Ueber Liniengeometrie und metrische Geometrie. *Math. Ann.*, 5:257–277, 1872.
130. F. Klein. Ueber die sogenannte Nicht-Euklidische Geometrie (Zweiter Aufsatz.). *Math. Ann.*, 6:112–145, 1873.
131. F. Klein. *Elementary Mathematics from an Advanced Standpoint, Geometry.* Reprinted Dover (1939), New York, 1908.
132. G. Kowalewski. *Vorlesungen über Allgemeine natürliche Geometrie und Liesche Transformationsgruppen.* Walter de Gruyter, Berlin, 1931.
133. E. Kruppa. Zur Ermittlung eines Objektes aus zwei Perspektiven mit innerer Orientierung. *Sitz.-Ber. Akad. Wiss., Wien, math. naturw. Kl. Abt. IIa.*, 122:1939–1948, 1913.
134. S. Kunze. Ein Hand-Auge-System zur visuell basierten Lokalisierung und Identifikation von Objekten. Master's thesis, Christian-Albrechts-Universität zu Kiel, Institut für Informatik und Praktische Mathematik, Kiel, 1999.
135. E. V. Labunets. Group-Theoretical Methods in Image Recognition. Technical report, LiTH–ISY–R–1855. Linkoping University, 1996.
136. E. V. Labunets, V. G. Labunets, and M.V. Assonov. Modular invariants used in pattern invariant recognition. *In New Informations Methods in Research of Discrete Structures.*, Ekatarinburg. IMM UD RAS:52–58, 1996. in Russian.
137. E. V. Labunets, V. G. Labunets, and R. Creutzburg. Fast fractional trigonometrical transform. In *Second Workshop on Transforms and Filterbanks*, Brandenburg, 1999.
138. E. V. Labunets, V. G. Labunets, K. Egiazarian, and J. Astola. Hypercomplex moments application in invariant image recognition. *In Int. Conf. on Image Processing 98*, pages 256–261, 1998.
139. J. H. Lambert. Observations trigonometriques. *Mem. Acad. Sci. Berlin*, 24:327–354, 1770.
140. J. Lasenby and E. Bayro-Corrochano. Computing 3–d projective invariants from points and lines. In G. Sommer, K. Daniilidis, and J. Pauli, editors, *Computer Analysis of Images and Patterns*, volume 1296 of *LNCS*, pages 82–89. Springer-Verlag, Berlin, Heidelberg, 1997.
141. J. Lasenby and E. Bayro-Corrochano. Computing Invariants in Computer Vision using Geometric Algebra. Technical Report CUED/F - INFENG/TR. 224, Cambridge University, Engineering Department, 1997.
142. J. Lasenby and A. N. Lasenby. Estimating tensors for matching over multiple views. *Phil. Trans. R. Soc. Lond. A*, 356(1740):1267–1282, 1998.
143. J. Lasenby, A. N. Lasenby, C. J. L. Doran, and W. J. Fitzgerald. New geometric methods for computer vision – an application to structure and motion estimation. *International Journal of Computer Vision*, 26(3):191–213, 1998.
144. V. Leavers. Survey: Which Hough transform? *CVGIP: Image Understanding*, 58:250–264, 1993.
145. G. W. Leibniz. *Nova methodus pro maximis et minimis, itemque tangentibus, quae nec fractas nec irrationales quantitates moratur, et singulare pro illis calculi genus (A new method for maxima and minima, als well as tangents, which ist not obstructed by fractional and irrational quantities, and a curious type of calculus for it).* Acta Eruditorium, 1684.
146. G. W. Leibniz. *The Early Mathematical Manuscripts.* Trans. from the latin texts of C. I. Gerhardt, Chicago, 1920.
147. H. Leung and S. Haykin. The complex backpropagation algorithm. *IEEE Transactions on Signal Processing*, 39(9):2101–2104, 1992.
148. H. Li. Hyperbolic geometry with Clifford algebra. *Acta Appl. Math.*, 48(3):317–358, 1997.

149. M. Li and D. Betsis. Hand–eye calibration. In *Proc. Int. Conf. on Computer Vision*, pages 40–46, 1995.
150. S. Lie. *Geometrie der Berührungstransformationen.* Chelsea Publishing Co., New York, (1892) 1977.
151. J. Liouville. Extension au cas trois dimensions de la question du tracé géographique. *In Applications de l'analyse à géometrie*, pages 609–616, G. Monge, Paris (1850).
152. H. C. Longuet-Higgins. A computer algorithm for reconstructing a scene from two projections. *Nature*, 293:133–135, 1981.
153. P. Lounesto. Marcel Riesz's work on Clifford algebras. In E.F. Bolinder and P. Lounesto, editors, *Clifford Numbers and Spinors*, pages 119–241. Kluwer Academic Publishers, 1993.
154. P. Lounesto. *Clifford Algebras and Spinors.* Cambridge University Press, Cambridge, 1997.
155. R. K. Luneburg. Metric methods in binocular visual perception. *Studies and Essays. Courant Anniv.*, 1:215–239, 1948.
156. R. K. Luneburg. The metric methods in binocular visual space. *J. Opt. Soc. Amer.*, 40(10):627–642, 1950.
157. Q. T. Luong and O. D. Faugeras. An optimization framework for efficient self–calibration and motion determination. In *Proc. Conference on Computer Vision and Pattern Recognition, Jerusalen, Israel*, volume A, pages 248–252, 1994.
158. Q. T. Luong and O. D. Faugeras. Self-Calibration of a moving camera from point correspondences and fundamental matrices. *International Journal of Computer Vision*, 22(3):261–289, 1997.
159. Q.T. Luong and O. D. Faugeras. Self-Calibration of a Stereo Rig from Unknown Camera Motions and Point Correspondences. Technical Report 2014, INRIA, Sophia Antipolis, 1993.
160. D. MacKay. MacOpt - a nippy wee optimizer. http://wol.ra.phy.cam.ac.uk/mackay/c/macopt.html.
161. F. J. MacWilleams and N. J. A. Sloane. *The Theory of Error–Correcting Codes.* N.J., 1976.
162. J. M. Rico Martinez and J. Duffy. The principle of transference: History, statement, and proof. *Mech. Machine Theory*, 28:165–177, 1993.
163. J. C. Maxwell. *A Treatise on Electricity and Magnetism*, volume I-II. Dover Publications Inc., N.Y., 1954 (1891).
164. S. J. Maybank and O. D.Faugeras. A theory of self–calibration of a moving camera. *International Journal of Computer Vision*, 8(2):123–151, 1992.
165. P. Maybeck. *Stochastic Models, Estimation and Control.* Academic Press, New York, 1979.
166. J.M. McCarthy. Dual orthogonal matrices in manipulator kinematics. *Intern. Journal of Robotics Research*, 5(2):45–51, 1986.
167. J.M. McCarthy. *Introduction to Theoretical Kinematics.* MIT Press, Cambridge, MA, 1990.
168. K. Menger. New foundation of Euclidean geometry. *Am. J. Math.*, 53(721–745), 1931.
169. M. Michaelis. *Low Level Image Processing Using Steerable Filters.* PhD thesis, Christian-Albrechts University, Kiel, 1995.
170. C.W. Misner, K.S. Thorne, and J.A. Wheeler. *Gravitation.* W.H. Freeman, New York, 1973.
171. A. F. Möbius. Die Theorie der Kreisverwandtschaft in rein geometrischer Darstellung. *Abh. Königl. Sächs. Ges. Wiss. Math.-Phys*, Kl. 2:529–595, 1855.

172. J.L. Mundy and A. Zisserman. *Geometric Invariance in Computer Vision.* MIT Press, Cambridge, Massachusetts, USA., 1992.
173. R. M. Murray, Z. Li, and S. S. Sastry. *A Mathematical Introduction to Robotic Manipulation.* CRC Press, Boca Raton, 1994.
174. A. Naeve. Projective Line Geometry of the Visual Operator. Technical Report TRITA-NA-8606, Computational Vision and Active Perception Laboratory (CVAP-29), KTH, 1986.
175. A. Naeve. *Focal Shape Geometry on Surfaces in Euclidean Space.* PhD thesis, Computational Vision and Active Perception Laboratory (CVAP-130), TRITA-NA-P9319, KTH, 1993.
176. A. Naeve and J. O. Eklundh. On projective geometry and the recovery of 3D-structure. In *Proceedings of the first International Conference on Computer Vision (ICCV)*, pages 128–135, London, 1987.
177. I. Newton. *Philosophiae Naturalis Principia Mathematica.* Cambridge, 1687.
178. I. Newton. De analysi per aequationes numero terminorum infinitas. *Opera omnia*, I:257–282, (written 1669, published 1711). Opulusca, I, pp. 3–28.
179. T. Papadopoulo and O.D. Faugeras. A New Characterization of the Trifocal Tensor. On INRIA Sophia Antipolis Web-Site.
180. R.C. Pappas. Oriented projective geometry with Clifford algebra. In R. Abłamowicz, P. Lounesto, and J.M. Parra, editors, *Clifford Algebras with Numeric and Symbolic Computations*, pages 233–250. Birkhäuser, Boston, 1996.
181. J. Pauli. Geometric/photometric consensus and regular shape (quasiinvariants for object localization and boundary extraction). Technical Report 9805, Christian-Albrechts-Universität zu Kiel, Institut fü r Informatik und Praktische Mathematik, Kiel, 1998.
182. J. K. Pearson. *Clifford Networks.* PhD thesis, Univ. of Kent, 1994.
183. G.R. Pennoc and A.T. Yang. Application of dual-number matrices to the inverse kinematics problem of robot manipulators. *Journal of Mechanisms, Transmissions, and Automation in Design*, 107:201–208, 1985.
184. C.B.U. Perwass and J. Lasenby. A geometric analysis of the trifocal tensor. In R. Klette, G. Gimel'farb, and R. Kakarala, editors, *Image and Vision Computing New Zealand, IVCNZ'98, Proceedings*, pages 157–162. The University of Auckland, 1998.
185. C.B.U. Perwass and J. Lasenby. A Geometric Derivation of the Trifocal Tensor and its Constraints. Technical Report CUED/F - INFENG/TR. 331, Cambridge University, Engineering Department, 1998.
186. C.B.U. Perwass and J. Lasenby. A Geometric Algebra Approach to 3D-Reconstruction from Vanishing Points. Technical Report CUED/F - INFENG/TR. 364, Cambridge University, Engineering Department, 1999.
187. P. Plücker. On a new geometry of space. *Phil. Trans. R. Soc.Lond.*, 155, 1865.
188. H. Poincaré. *Science et Méthode.* E. Flammarion, Paris, 1908.
189. I. R. Porteous. *Clifford Algebras and the Classical Groups.* Cambridge University Press, Cambridge, 1995.
190. W. H. Press, B. P. Flannery, S. A. Teukolsky, and W. T. Vetterling. *Numerical Recipes in Pascal.* Cambridge University Press, 1990.
191. W. Press et al. *Numerical Recipes in C.* Cambridge University Press, 1994.
192. M. Riesz. *Clifford numbers and spinors.* Edited by E.F. Bolinder and P. Lounesto. Kluwer Academic Publishers, 1993.
193. J. Rooney. On the three types of complex numbers and planar transformations. *Environment and Planning B*, 5:89–99, 1978.
194. C. Rothwell, G. Csurka, and O. D. Faugeras. A Comparison of Projective Reconstruction Methods for Pairs of Views. Technical Report 2538, INRIA, Sophia Antipolis, 1995.

195. C. Rothwell, O. D. Faugeras, and G. Csurka. A comparison of projective reconstruction methods for pairs of views. *Computer Vision and Image Understanding*, 68-1:37–58, 1997.
196. B. Sabata and J.K. Aggarwal. Estimation of motion from a pair of range images: a review. *CVGIP: Image Understanding*, 54:309–324, 1991.
197. P. Samuel. *Projective Geometry*. Springer-Verlag, New York, 1988.
198. S. J. Sangwine. Fourier-transforms of color images using quaternion of hypercomplex numbers. *Electronic Letters*, 32(21):1979–1980, October 1996.
199. S.J. Sangwine. Colour image edge detector based on quaternionic convolution. *Electronics Letters*, 34(10):969–971, 1998.
200. S.J. Sangwine and R.E.N. Horne, editors. *The Colour Image Processing Handbook*. Chapman & Hall, 1998.
201. R. Sauer. *Projektive Liniengeometrie*. Göschens Lehrbücherei, Gruppe Reine und Angewandte Mathematik, Band 23, Walter de Gruyer & Co, Berlin, 1937.
202. L. Schläfli. *Theorie der vielfachen Kontinuität*, volume 38 of *Denkschriften der Schweizerischen naturforschenden Gesellschaft*. Zürcher & Furrer, Zürich, 1901.
203. H. Seidel. Quaternionen in der graphischen Datenverarbeitung. In *Geometrische Verfahren der Graphischen Datenverarbeitung*. Springer-Verlag, Berlin, Heidelberg, 1990.
204. J. J. Seidel. Angles and distance in n-dimensional Euclidean and non-Euclidean geometry, I–III. *Indag. Math.*, 17(329–340):65–76, 1952.
205. J. J. Seidel. Distance-geometric development of two-dimensional Euclidean, hyperbolic and spherical geometry I, II. *Simon Stevin*, 5355–541(29):32–50, 1955. reprinted by *Proc. Ned. Akad. Wetensch.*
206. J. M. Selig. *Geometrical Methods in Robotics*. Springer-Verlag, New York, 1996.
207. J.M. Selig. *Introductory Robotics*. Prentice Hall, 1992.
208. J.G. Semple and G.T.Kneebone. *Algebraic Projective Geometry*. Oxford University Press, 1952.
209. A. Shashua. Trilinearity in visual recognition by alignment. In J.O. Eklundh, editor, *Computer Vision - ECCV'94*, volume 800 of *LNCS*, pages 479–484. Springer-Verlag, Berlin, Heidelberg, 1994.
210. A. Shashua. Trilinear tensor: the fundamental construct of multiple-view geometry and its applications. In G. Sommer and J.J. Koenderink, editors, *Algebraic Frames for the Perception-Action Cycle*, number 1315 in LNCS, pages 190–206. Springer-Verlag, Berlin, Heidelberg, 1997.
211. Y.C. Shiu and S. Ahmad. Calibration of wrist-mounted robotic sensors by solving homogeneous transform equations of the form $ax = xb$. *IEEE Trans. Robotics and Automation*, 5:16–27, 1989.
212. Silicon Graphics, Inc. MIPS RISC Technology R10000 Microprocessor Technical Brief[3], 1998.
213. E. Snapper and R. J. Troyer. *Metric Affine Geometry*. Dover Publications Inc., New York, 1989. First publ. by Academic Press, Inc., New York, 1971.
214. G. Sobczyk. Simplicial calculus with geometric algebra[2]. In A. Micali et al., editors, *Clifford Algebras and their Applications in Mathematical Physics*. Kluwer Academic Publishers, Dordrecht/Boston, 1992.
215. G. Sommer. Algebraic aspects of designing behavior based systems. In G. Sommer and J.J. Koenderink, editors, *Algebraic Frames for the Perception–Action Cycle*, volume 1315 of *Lecture Notes in Computer Science*, pages 1–28. Springer–Verlag, Berlin, Heidelberg, 1997.

[3] `http://www.sgi.com/processors/r10k/tech_info/Tech_Brief.html`

216. G. Sommer, E. Bayro-Corrochano, and T. Bülow. Geometric algebra as a framework for the perception–action cycle. In F. Solina, W.G. Kropatsch, R. Klette, and R. Bajcsy, editors, *Advances in Computer Vision*, pages 251–260. Springer-Verlag, Wien, New York, 1997.
217. G. Sommer and J.J. Koenderink. *Algebraic Frames for the Perception-Action Cycle*, volume 1315 of *LNCS*. Springer-Verlag, Berlin, Heidelberg, 1997.
218. H.W. Sorenson. *Kalman filtering techniques. Advances in Control Systems Theory and Applications. 3, Edited by Leondes, C.T., 219-292.* Academic Press, New York, 1966.
219. H. Stark. An extension of the Hilbert transform product theorem. *Proc. IEEE*, 59:1359–1360, 1971.
220. J. Stolfi. *Oriented Projective Geometry.* Academic Press, 1991.
221. D. J. Struik. *Lectures on Classical Differential Geometry.* Addison-Wesley, Reading, Massachusetts, 1950.
222. E. Study. Von den Bewegungen und Umlegungen. *Mathematische Annalen*, 39:441–566, 1891.
223. E. Study. *Geometrie der Dynamen.* Leipzig, 1903.
224. Sun Microsystems, Inc. UltraSPARC-II Data Sheet[4], 1998.
225. L. Svensson. On the Use of the Double Algebra in Computer Vision. Printed as a Computational Vision and Active Perception Laboratory report (CVAP-122), ISRN KTH/NA/P-93/10, KTH, 1993, Nice, France, June 1992. Talk at the INRIA Workshop on Invariants in Computer Vision.
226. M. Swain and D. Ballard. Color indexing. *International Journal of Computer Vision.*, 7:11–32, 1992.
227. M. Teague. Image analyses via the general theory of moments. *J. Opt. Soc. Am.*, 70:920–930, 1980.
228. B. Triggs. Autocalibration and the absolute quadric. *In IEEE Conf. on Computer Vision and Pattern Recognition*, pages 609–614, 1997.
229. B. Triggs. Linear projective reconstruction from matching tensors. *Image and Vision Computing*, 15:617–625, 1997.
230. B. Triggs. Autocalibration from planar scenes. In H. Burkhardt and B. Neumann, editors, *Computer Vision - ECCV'98*, volume 1406 of *LNCS*, pages 89–105. Springer-Verlag, Berlin, Heidelberg, 1998.
231. R.Y. Tsai and R.K. Lenz. A new technique for fully autonomous and efficient 3d robotics hand/eye calibration. *IEEE Trans. Robotics and Automation*, 5:345–358, 1989.
232. K. Vahlen. Über Bewegungen und komplexe Zahlen. *Math. Ann.*, 55:585–593, 1902.
233. R. L. De Valois and K. K. De Valois. *Spatial Vision.* Oxford University Press, New York, 1988.
234. S. Venkatesh, J. Cooper, and B. White. Local energy and pre-envelope. *Pattern Recognition*, 28(8):1127–1134, 1995.
235. S. Venkatesh and R. Owens. On the classification of image features. *Pattern Recognition Letters*, 11:339–349, 1990.
236. M.W. Walker. Manipulator kinematics and the epsilon algebra. *IEEE Journal of Robotics and Automation*, 4:186–192, 1988.
237. N. White. Geometric applications of the Grassmann-Cayley algebra. In J.E. Goodman and J. O'Rourke, editors, *Handbook of Discrete and Computational Geometry.* CRC Press, Florida, 1997.

[4] `http://www.sun.com/microelectronics/datasheets/stp1031/index2.html`

238. C. Zeller and O. D. Faugeras. Camera Self-Calibration from Video Sequences: the Kruppa Equations Revisited. Technical Report 2793, INRIA, Sophia Antipolis, 1996.
239. Z. Zhang and O. D. Faugeras. *3-D Dynamic Scene Analysis.* Springer-Verlag, Berlin, Heidelberg, 1992.

Author Index

Subject Index